Methods in Enzymology

Volume 343

G PROTEIN PATHWAYS

Part A

Receptors

METHODS IN ENZYMOLOGY

Methods in Enzymology

Volume 343

G Protein Pathways

Part A
Receptors

EDITED BY

Ravi Iyengar

MOUNT SINAI SCHOOL OF MEDICINE
NEW YORK, NEW YORK

John D. Hildebrandt

MEDICAL UNIVERSITY OF SOUTH CAROLINA
CHARLESTON, SOUTH CAROLINA

ACADEMIC PRESS

San Diego London Boston New York Sydney Tokyo Toronto

This book is printed on acid-free paper. ∞

Academic Press
A Harcourt Science and Technology Company
525 B Street, Suite 1900, San Diego, California 92101-4495, USA
http://www.academicpress.com

Academic Press
Harcourt Place, 32 Jamestown Road, London NW1 7BY, UK
http://www.academicpress.com

International Standard Book Number: 0-12-182244-3

PRINTED IN THE UNITED STATES OF AMERICA
01 02 03 04 05 06 07 SB 9 8 7 6 5 4 3 2 1

Table of Contents

Section I. G Protein-Coupled Receptors

A. Theoretical Evaluation of Receptor Function

B. Design and Use of Receptor Ligands

C. Structural Characterization of Receptor Proteins

D. Design and Use of Engineered Receptor Proteins

H. The Study of Receptor Trafficking

Section II. Regulators of GPCR Function

A. G Protein-Coupled Receptor Kinases (GRKs)

B. Arrestins and Novel Proteins

Contributors to Volume 343

Article numbers are in parentheses following the names of contributors.
Affiliations listed are current.

ZSOLT ABLONCZY (9), *Department of Ophthalmology, Medical University of South Carolina, Charleston, South Carolina 29425*

RICHARD S. AGNES (5), *Department of Chemistry, University of Arizona, Tucson, Arizona 85721*

JUNG-MO AHN (4), *Department of Chemistry, The Scripps Research Institute, La Jolla, California 92037*

ARLENE D. ALBERT (15), *Department of Molecular and Cell Biology, University of Connecticut, Storrs, Connecticut 06269*

ASTRID E. ALEWIJNSE (25), *Leiden/Amsterdam Center for Drug Research, Division of Medicinal Chemistry, Vrije Universiteit, 1081HV Amsterdam, The Netherlands*

REMKO A. BAKKER (27), *Leiden/Amsterdam Center for Drug Research, Division of Medicinal Chemistry, Vrije Universiteit, 1081HV Amsterdam, The Netherlands*

LAUREN E. BALL (9), *Department of Pharmacology, Medical University of South Carolina, Charleston, South Carolina 29425*

JUAN A. BALLESTEROS (22), *Novasite Pharmaceuticals, Inc., San Diego, California 92121*

PREETI M. BALSE-SRINIVASAN (4), *American Peptide Company, Inc., Sunnyvale, California 94086*

JEFFREY L. BENOVIC (31, 34, 37), *Department of Microbiology and Immunology, Kimmel Cancer Institute, Thomas Jefferson University, Philadelphia, Pennsylvania 19107*

LUTZ BIRNBAUMER (24), *Department of Anesthesiology, University of California Los Angeles School of Medicine, Los Angeles, California 90095*

JANINA BUCZYŁKO (36), *Department of Ophthalmology, University of Washington, Seattle, Washington 98195*

ETHAN S. BURSTEIN (27), *ACADIA Pharmaceuticals Inc., San Diego, California 92121*

CHAOZHONG CAI (5), *Department of Chemistry, Duke University, Durham, North Carolina 27708*

BELINDA S. W. CHANG (19), *Laboratory of Molecular Biology and Biochemistry, Rockefeller University, New York, New York 10021*

PETER CHIDIAC (1), *Department of Pharmacology and Toxicology, Faculty of Medicine and Dentistry, University of Western Ontario, London, Ontario N6A 5C1, Canada*

RICHARD B. CLARK (32), *Department of Integrative Biology and Pharmacology, University of Texas Medical School, Houston, Texas 77025*

BRUCE R. CONKLIN (16), *Gladstone Institute of Cardiovascular Disease, Departments of Medicine and Pharmacology, University of California, San Francisco, California 94141*

M. CARTER CORNWALL (3), *Department of Physiology, Boston University School of Medicine, Boston, Massachusetts 02118*

PETER COWARD (16), *Tularik Inc., San Francisco, California 94080*

SCOTT M. COWELL (4), *Department of Chemistry, University of Arizona, Tucson, Arizona 85721*

ROSALIE K. CROUCH (3, 9, 10), *Department of Ophthalmology, Medical University of South Carolina, Charleston, South Carolina 29425*

DAVID P. DAVIS (13), *Department of Pathology, Committee on Immunology, University of Chicago, Chicago, Illinois 60637*

BARBARA J. EBERSOLE (7), *Department of Neurology, Mount Sinai School of Medicine, New York, New York 10029*

MARKUS EILERS (14), *Department of Biochemistry and Cell Biology, State University of New York, Stony Brook, New York 11794*

WOLFGANG GÄRTNER (3), *Max-Planck-Institut für Strahlenchemie, D-45470 Mülheim an der Ruhr, Germany*

JAMES C. GARRISON (23), *Department of Pharmacology, University of Virginia Health System, Charlottesville, Virginia 22908*

ANDREW K. GELASCO (9, 10), *Division of Nephrology, Department of Medicine, Medical University of South Carolina, Charleston, South Carolina 29425*

ULRIK GETHER (11), *Department of Cellular Physiology, Institute of Medical Physiology, The Panum Institute, University of Copenhagen, DK-2200 Copenhagen, Denmark*

KAZUKO HAGA (35), *Institute for Biomolecular Science, Gakushuin University, Toshima-ku, Tokyo 171-8588, Japan*

TATSUYA HAGA (35), *Institute for Biomolecular Science, Gakushuin University, Toshima-ku, Tokyo 171-8588, Japan*

RANDY A. HALL (38), *Department of Pharmacology, Emory University School of Medicine, Atlanta, Georgia 30322*

DENNIS P. HEALY (28), *Department of Pharmacology, Mount Sinai School of Medicine, New York, New York 10029*

MARCEL HOFFMANN (25), *Leiden/Amsterdam Center for Drug Research, Division of Medicinal Chemistry, Vrije Universiteit, 1081HV Amsterdam, The Netherlands*

KLAUS-PETER HOFMANN (36), *Institut für Medizinische Physik und Biophysik, Charité, Medizinische Fakultät der Humboldt-Universität zu Berlin, 10098 Berlin, Germany*

VICTOR J. HRUBY (4, 5, 6), *Department of Chemistry, University of Arizona, Tucson, Arizona 85721*

MARY HUNZICKER-DUNN (24), *Department of Cell and Molecular Biology, Northwestern University Medical School, Chicago, Illinois 60611*

JONATHAN A. JAVITCH (8), *Center for Molecular Recognition and Departments of Psychiatry and Pharmacology, College of Physicians and Surgeons, Columbia University, New York, New York 10032*

SULIN JIANG (21), *Department of Cell and Molecular Pharmacology, Medical University of South Carolina, Charleston, South Carolina 29425*

LORENA KALLAL (31), *GlaxoSmithKline Pharmaceuticals, King of Prussia, Pennsylvania 19406*

KIMIHIKO KAMEYAMA (35), *Molecular Neurophysiology Group, Neuroscience Research Institute, National Institute of Advanced Industrial Science and Technology, Tsukuba, Ibaraki 305-8565, Japan*

SADASHIVA S. KARNIK (17), *Department of Molecular Cardiology, Lerner Research Institute, The Cleveland Clinic Foundation, Cleveland, Ohio 44195*

MANIJA A. KAZMI (19), *Howard Hughes Medical Institute, Laboratory of Molecular Biology and Biochemistry, Rockefeller University, New York, New York 10021*

JEFFREY R. KEEFER (33), *Department of Pediatrics, Johns Hopkins School of Medicine, Baltimore, Maryland 21205*

VLADIMIR KEFALOV (3), *Department of Neuroscience, Johns Hopkins School of Medicine, Baltimore, Maryland 21205*

H. GOBIND KHORANA (14), *Department of Biology, Massachusetts Institute of Technology, Cambridge, Massachusetts 02039*

DANIEL R. KNAPP (9, 10), *Department of Pharmacology, Medical University of South Carolina, Charleston, South Carolina 29425*

BRIAN J. KNOLL (32), *Department of Pharmacological and Pharmaceutical Sciences, University of Houston College of Pharmacy, Houston, Texas 77204*

BRIAN K. KOBILKA (11), *Howard Hughes Medical Institute, Department of Molecular and Cellular Physiology, Stanford University Medical School, Stanford, California 94305*

EHUD M. LANDAU (12), *Department of Physiology and Biophysics, The Membrane Protein Laboratory, University of Texas Medical Branch, Galveston, Texas 77555*

PAUL LEFF (2), *AstraZeneca, EST-Bio, Mereside, Alderley Park, Cheshire SK10 4TG, United Kingdom*

ROB LEURS (25), *Leiden/Amsterdam Center for Drug Research, Division of Medicinal Chemistry, Vrije Universiteit, 1081HV Amsterdam, The Netherlands*

GEORGE LIAPAKIS (8), *Department of Pharmacology, School of Medicine, University of Crete, 71110 Heraklion, Greece*

STEPHEN B. LIGGETT (29), *Departments of Medicine, Molecular Genetics, and Pharmacology, University of Cincinnati, Cincinnati, Ohio 45267*

LEE E. LIMBIRD (33), *Department of Pharmacology, Vanderbilt University School of Medicine, Nashville, Tennessee 37232*

ROBERT P. LOUDON (34), *Department of Biology, Haverford College, Haverford, Pennsylvania 19041*

GAVIN MACCLEERY (23), *Department of Pharmacology, University of Virginia Health System, Charlottesville, Virginia 22908*

DAVID R. MANNING (26), *Department of Pharmacology, University of Pennsylvania School of Medicine, Philadelphia, Pennsylvania 19104*

WILLIAM E. MCINTIRE (23), *Department of Pharmacology, University of Virginia Health System, Charlottesville, Virginia 22908*

GRAEME MILLIGAN (18), *Molecular Pharmacology Group, Division of Biochemistry and Molecular Biology, Institute of Biomedical and Life Sciences, University of Glasgow, Glasgow G12 8QQ, Scotland, United Kingdom*

STUART J. MUNDELL (37), *Department of Pharmacology, School of Medical Sciences, University of Bristol, Bristol BS8 1TD, United Kingdom*

CHANG-SEON MYUNG (23), *Department of Pharmacology, University of Virginia Health System, Charlottesville, Virginia 22908*

JAVIER NAVARRO (12), *Department of Physiology and Biophysics, The Membrane Protein Laboratory, University of Texas Medical Branch, Galveston, Texas 77555*

COLLEEN M. NISWENDER (30), *Department of Pharmacology, University of Washington, Seattle, Washington 98195*

PETER NOLLERT (12), *Department of Biochemistry and Biophysics, University of California, San Francisco, California 94134*

TORU OKAYAMA (6), *Fuji Chemical Industries, Ltd. Toyama 933-8511, Japan*

MICHAEL J. ORSINI (37), *Johnson and Johnson, Raritan, New Jersey 08869*

KRZYSZTOF PALCZEWSKI (36), *Departments of Ophthalmology, Chemistry and Pharmacology, University of Washington School, Seattle, Washington 98195*

RICHARD T. PREMONT (38), *Liver Center, Department of Medicine, Duke University Medical Center, Durham, North Carolina 27710*

ALEXEY N. PRONIN (34), *Senomyx, Inc., La Jolla, California 92037*

ALEXANDER PULVERMÜLLER (36), *Institut für Medizinische Physik und Biophysik, Charité, Medizinische Fakultät der Humboldt-Universität zu Berlin, 10098 Berlin, Germany*

WEI QIU (6), *Tularik, Inc., San Francisco, California 94080*

DEBORAH A. RATHZ (29), *Department of Pharmacology, University of Cincinnati, Cincinnati, Ohio 45267*

CHARLES H. REDFERN (16), *Sidney Kimmel Cancer Center, San Diego, California 92123*

PHILIP J. REEVES (14), *Department of Biology, Massachusetts Institute of Technology, Cambridge, Massachusetts 02039*

THOMAS P. SAKMAR (19), *Howard Hughes Medical Institute, Laboratory of Molecular Biology and Biochemistry, Rockefeller University, New York, New York 10021*

CHRISTINE SAUNDERS (33), *Department of Biochemistry, University of Texas Health Science Center, San Antonio, Texas 78229*

CLARE SCARAMELLINI (2), *Novartis Horsham Research Centre, Horsham, West Sussex RH12 5AB, United Kingdom*

KIMBERLY SCEARCE-LEVIE (16), *Gladstone Institute of Cardiovascular Disease, Departments of Medicine and Pharmacology, University of California, San Francisco, California 94141*

NICOLE SCHRAMM (33), *Department of Molecular Physiology and Biophysics, Vanderbilt University School of Medicine, Nashville, Tennessee 37232*

STUART C. SEALFON (7), *Department of Neurology, Mount Sinai School of Medicine, New York, New York 10029*

DEBORAH L. SEGALOFF (13), *Department of Physiology and Biophysics, The University of Iowa, Iowa City, Iowa 52242*

LEI SHI (8), *Center for Molecular Recognition and Department of Pharmacology, College of Physicians and Surgeons, Columbia University, New York, New York 10032*

KERSTEN M. SMALL (29), *Department of Medicine, University of Cincinnati, Cincinnati, Ohio 45267*

MARTINE J. SMIT (25, 27), *Leiden/Amsterdam Center for Drug Research, Division of Medicinal Chemistry, Vrije Universiteit, 1081HV Amsterdam, The Netherlands*

STEVEN O. SMITH (14), *Department of Biochemistry and Cell Biology, State University of New York, Stony Brook, New York 11794*

IZABELA SOKAL (36), *Department of Ophthalmology, University of Washington, Seattle, Washington 98195*

VADIM A. SOLOSHONOK (6), *Department of Chemistry, University of Arizona, Tucson, Arizona 85721*

HENK TIMMERMAN (25), *Leiden/Amsterdam Center for Drug Research, Division of Medicinal Chemistry, Vrije Universiteit, 1081HV Amsterdam, The Netherlands*

HIROFUMI TSUGA (35), *Pharmaceutical Research Laboratories, Ajinomoto Co., Inc., Kawasaki 210-8681, Japan*

ILYA A. VAKSER (21), *Department of Cell and Molecular Pharmacology, Medical University of South Carolina, Charleston, South Carolina 29425*

IRACHE VISIERS (22), *Department of Physiology and Biophysics, Mount Sinai School of Medicine, New York, New York 10029*

QI WANG (23), *Department of Pharmacology, University of Virginia Health System, Charlottesville, Virginia 22908*

HAREL WEINSTEIN (22), *Department of Physiology and Biophysics and Institute for Computational Biomedicine, Mount Sinai School of Medicine, New York, New York 10029*

JÜRGEN WESS (20), *Laboratory of Bioorganic Chemistry, National Institute of Diabetes and Digestive and Kidney Diseases, Bethesda, Maryland 20892*

ROLF T. WINDH (26), *Adolor Corporation, Malvern, Pennsylvania 19104*

MAGDALENA WOZNIAK (33), *Renal Division, Washington University School of Medicine, St. Louis, Missouri 63110*

PHILIP L. YEAGLE (15), *Department of Molecular and Cell Biology, University of Connecticut, Storrs, Connecticut 06269*

WEIWEN YING (14), *Department of Biochemistry and Cell Biology, State University of New York, Stony Brook, New York 11794*

NORIHIRO YOSHIDA (35), *Department of Neurochemistry, University of Tokyo, Tokyo 113-0033, Japan*

Preface

The heterotrimeric G proteins are the central component of one of the primary mechanisms used by eukaryotic cells to receive, interpret, and respond to extracellular signals. Many of the basic concepts associated with the entire range of fields collectively referred to in terms such as "signal transduction" originate from the pioneering work of Sutherland, Krebs and Fischer, Rodbell and Gilman, Greengard, and others on systems that are primary examples of G protein signaling pathways. The study of these proteins has been and remains at the forefront of research on cell signaling mechanisms. The G Protein Pathways volumes (343, 344, and 345) of *Methods in Enzymology* have come about as part of a continuing attempt to use the methods developed in studying the G protein signaling pathways as a resource both within this field and throughout the signal transduction field.

Several volumes of this series have been devoted in whole or in part to approaches for studying the heterotrimeric G proteins. Volumes 109 and 195 were the earliest to devote substantial parts to G protein-mediated signaling systems. In 1994 Volumes 237 and 238 comprehensively covered this field. The continued growth, the ever increasing impact of this field, and the continued evolution of approaches and questions generated by research related to G proteins have led inevitably to the need for a new and comprehensive treatment of the approaches used to study these proteins. Each volume of G Protein Pathways brings together varied topics and approaches to this central theme.

Very early in the development of the concepts of G protein signaling mechanisms, Rodbell and Birnbaumer recognized at least three components of these signaling systems. They compared them to a receiver, a transducer, and an amplifier. The receiver was the receptor for an extracellular signal. The transducer referred to those mechanisms and components required for converting an extracellular signal into an intracellular response. The amplifier was synonymous with the effector enzymes that generate the beginning of the intracellular signal. Over the years these ideas have evolved in many ways. We now know an immense amount about the receptors and their great range of diversity. Through the work of Gilman, along with his associates and contemporaries, the transducer component turned out to be nearly synonymous with the heterotrimeric G proteins themselves. Nevertheless, the complexity of this component of the system continues to become more and more apparent with the recognition of the diversity of these proteins, the many ways they interact, and the increasing number of regulatory influences on their function. The key concepts associated with the effector enzymes, such as adenylyl cyclase that produces the intracellular "second messenger" cAMP, have been broadened substantially to include other enzymes, ion channels, and the components of other signaling systems that form an interacting network of systems inside cells.

The organization of Volumes 343, 344, and 345 is still conveniently centered on these three components of the G proteins signaling pathway: receptor, G protein, and effector. The evolution of the field, however, inevitably left a mark on the form of these volumes. So, for example, receptors in Volume 343 and G proteins in Volume 344 no longer stand alone as individual components in these volumes, but share space with other directly interacting proteins that influence their function. In addition, Volume 345 addresses, more generally, effector mechanisms and forms a bridge between the many different cell regulatory mechanisms that cooperate to control cellular function. Thus, there are chapters that include methods related to small GTP binding proteins, ion channels, gene regulation, and novel signaling compounds.

As we learn more about G protein signaling systems, we acquire an ever increasing appreciation of their complexity. In the previous volumes on G proteins it was already evident that there were many different isoforms of each of the three heterotrimeric G protein subunits. Initial analysis of the Human Genome Project suggests some constraints on the number of members of these proteins with 27 α subunit isoforms, 5 β subunits, and 13 γ subunits. It is interesting that in recent years we had nearly accounted for all or most of the β and γ subunit isoforms, but that there are nearly twice as many potential α subunits as the ones we currently understand. These recent discoveries, along with the recognition of the existence of between 600 and 700 G protein-coupled receptors in the human genome, may place limits on the complexity of the G protein signaling system itself. These potential limits, however, are balanced by the immense number of possible combinations of interactions that can be generated from all these components, by the possible variation of all of these proteins at levels after their genomic structure, and by our continual discovery of additional interacting components of the system.

Perhaps one of the really substantial gains in our understanding of this system since the earlier volumes is the increasing recognition of the role of accessory proteins in G protein signaling pathways. Those proteins recognized nearly 20 years ago that work at the level of the receptor continue to grow and have a prominent place in Volume 343. One of the real breakthroughs though has been the rapid development of our knowledge of accessory proteins that interact with the G proteins themselves. Prominent among these are the RGS (regulators of G protein signaling) proteins that act as GTPase activating proteins for selective G protein α subunits. A fairly substantial section of these volumes is devoted to these proteins. One apparent aside related to the RGS proteins though, is that as much as we have rapidly learned about them, there is much more yet to be learned, because there is wide variation in the structure of these proteins outside their G protein interaction sites. These proteins likely have many different stories yet to be developed based on their interactions with the G proteins, perhaps mediating functions that we do not yet know about. The RGS proteins are not the end of the interacting proteins either, however, with the description of additional G protein-interacting proteins such as the AGS (activators of G protein signaling) proteins. These are likely a heterogeneous group of proteins with several different mechanisms of interaction and roles in G protein signaling mechanisms.

The range of topics covered in these volumes turned out to be quite large. This is a result of the wide range of approaches that creative scientists can develop to gain an understanding of a complex and rapidly evolving field. There are several chapters that provide a theoretical basis for the analysis and interpretation of data, several chapters on the application of modeling techniques at several different levels, and chapters on structural biology approaches, classical biochemical techniques juxtaposed to protein engineering, molecular biology, gene targeting strategies addressing physiological questions, and DNA array approaches to evaluating the effects of pathway activation. In all likelihood, the G protein signaling field will continue to be one that moves at the forefront of scientific approaches to studying events at the interface between biochemistry and molecular biology, on the one hand, and physiology and cell biology, on the other. Thus, it is our hope that these volumes will serve a scientific readership beyond those that study G proteins per se, or even those that study cell signaling mechanisms. We would hope that the approaches and techniques described here would hold relevance for those large number of scientists involved in many different kinds of projects that address the interface between the molecular/biochemical world and the cell/tissue/organism world.

We owe a tremendous debt of gratitude to our colleagues who so readily contributed chapters to these volumes. Truly, without their so willing participation this work would not have evolved into as substantial and comprehensive a work as it ultimately became. We would also like to thank Ms. Shirley Light for her support, encouragement, and patience throughout this long process.

JOHN D. HILDEBRANDT
RAVI IYENGAR

METHODS IN ENZYMOLOGY

VOLUME I. Preparation and Assay of Enzymes
Edited by SIDNEY P. COLOWICK AND NATHAN O. KAPLAN

VOLUME II. Preparation and Assay of Enzymes
Edited by SIDNEY P. COLOWICK AND NATHAN O. KAPLAN

VOLUME III. Preparation and Assay of Substrates
Edited by SIDNEY P. COLOWICK AND NATHAN O. KAPLAN

VOLUME IV. Special Techniques for the Enzymologist
Edited by SIDNEY P. COLOWICK AND NATHAN O. KAPLAN

VOLUME V. Preparation and Assay of Enzymes
Edited by SIDNEY P. COLOWICK AND NATHAN O. KAPLAN

VOLUME VI. Preparation and Assay of Enzymes (*Continued*)
Preparation and Assay of Substrates
Special Techniques
Edited by SIDNEY P. COLOWICK AND NATHAN O. KAPLAN

VOLUME VII. Cumulative Subject Index
Edited by SIDNEY P. COLOWICK AND NATHAN O. KAPLAN

VOLUME VIII. Complex Carbohydrates
Edited by ELIZABETH F. NEUFELD AND VICTOR GINSBURG

VOLUME IX. Carbohydrate Metabolism
Edited by WILLIS A. WOOD

VOLUME X. Oxidation and Phosphorylation
Edited by RONALD W. ESTABROOK AND MAYNARD E. PULLMAN

VOLUME XI. Enzyme Structure
Edited by C. H. W. HIRS

VOLUME XII. Nucleic Acids (Parts A and B)
Edited by LAWRENCE GROSSMAN AND KIVIE MOLDAVE

VOLUME XIII. Citric Acid Cycle
Edited by J. M. LOWENSTEIN

VOLUME XIV. Lipids
Edited by J. M. LOWENSTEIN

VOLUME XV. Steroids and Terpenoids
Edited by RAYMOND B. CLAYTON

VOLUME XVI. Fast Reactions
Edited by KENNETH KUSTIN

VOLUME XVII. Metabolism of Amino Acids and Amines (Parts A and B)
Edited by HERBERT TABOR AND CELIA WHITE TABOR

VOLUME XVIII. Vitamins and Coenzymes (Parts A, B, and C)
Edited by DONALD B. MCCORMICK AND LEMUEL D. WRIGHT

VOLUME XIX. Proteolytic Enzymes
Edited by GERTRUDE E. PERLMANN AND LASZLO LORAND

VOLUME XX. Nucleic Acids and Protein Synthesis (Part C)
Edited by KIVIE MOLDAVE AND LAWRENCE GROSSMAN

VOLUME XXI. Nucleic Acids (Part D)
Edited by LAWRENCE GROSSMAN AND KIVIE MOLDAVE

VOLUME XXII. Enzyme Purification and Related Techniques
Edited by WILLIAM B. JAKOBY

VOLUME XXIII. Photosynthesis (Part A)
Edited by ANTHONY SAN PIETRO

VOLUME XXIV. Photosynthesis and Nitrogen Fixation (Part B)
Edited by ANTHONY SAN PIETRO

VOLUME XXV. Enzyme Structure (Part B)
Edited by C. H. W. HIRS AND SERGE N. TIMASHEFF

VOLUME XXVI. Enzyme Structure (Part C)
Edited by C. H. W. HIRS AND SERGE N. TIMASHEFF

VOLUME XXVII. Enzyme Structure (Part D)
Edited by C. H. W. HIRS AND SERGE N. TIMASHEFF

VOLUME XXVIII. Complex Carbohydrates (Part B)
Edited by VICTOR GINSBURG

VOLUME XXIX. Nucleic Acids and Protein Synthesis (Part E)
Edited by LAWRENCE GROSSMAN AND KIVIE MOLDAVE

VOLUME XXX. Nucleic Acids and Protein Synthesis (Part F)
Edited by KIVIE MOLDAVE AND LAWRENCE GROSSMAN

VOLUME XXXI. Biomembranes (Part A)
Edited by SIDNEY FLEISCHER AND LESTER PACKER

VOLUME XXXII. Biomembranes (Part B)
Edited by SIDNEY FLEISCHER AND LESTER PACKER

VOLUME XXXIII. Cumulative Subject Index Volumes I-XXX
Edited by MARTHA G. DENNIS AND EDWARD A. DENNIS

VOLUME XXXIV. Affinity Techniques (Enzyme Purification: Part B)
Edited by WILLIAM B. JAKOBY AND MEIR WILCHEK

VOLUME XXXV. Lipids (Part B)
Edited by JOHN M. LOWENSTEIN

VOLUME XXXVI. Hormone Action (Part A: Steroid Hormones)
Edited by BERT W. O'MALLEY AND JOEL G. HARDMAN

VOLUME XXXVII. Hormone Action (Part B: Peptide Hormones)
Edited by BERT W. O'MALLEY AND JOEL G. HARDMAN

VOLUME XXXVIII. Hormone Action (Part C: Cyclic Nucleotides)
Edited by JOEL G. HARDMAN AND BERT W. O'MALLEY

VOLUME XXXIX. Hormone Action (Part D: Isolated Cells, Tissues, and Organ Systems)
Edited by JOEL G. HARDMAN AND BERT W. O'MALLEY

VOLUME XL. Hormone Action (Part E: Nuclear Structure and Function)
Edited by BERT W. O'MALLEY AND JOEL G. HARDMAN

VOLUME XLI. Carbohydrate Metabolism (Part B)
Edited by W. A. WOOD

VOLUME XLII. Carbohydrate Metabolism (Part C)
Edited by W. A. WOOD

VOLUME XLIII. Antibiotics
Edited by JOHN H. HASH

VOLUME XLIV. Immobilized Enzymes
Edited by KLAUS MOSBACH

VOLUME XLV. Proteolytic Enzymes (Part B)
Edited by LASZLO LORAND

VOLUME XLVI. Affinity Labeling
Edited by WILLIAM B. JAKOBY AND MEIR WILCHEK

VOLUME XLVII. Enzyme Structure (Part E)
Edited by C. H. W. HIRS AND SERGE N. TIMASHEFF

VOLUME XLVIII. Enzyme Structure (Part F)
Edited by C. H. W. HIRS AND SERGE N. TIMASHEFF

VOLUME XLIX. Enzyme Structure (Part G)
Edited by C. H. W. HIRS AND SERGE N. TIMASHEFF

VOLUME L. Complex Carbohydrates (Part C)
Edited by VICTOR GINSBURG

VOLUME LI. Purine and Pyrimidine Nucleotide Metabolism
Edited by PATRICIA A. HOFFEE AND MARY ELLEN JONES

VOLUME LII. Biomembranes (Part C: Biological Oxidations)
Edited by SIDNEY FLEISCHER AND LESTER PACKER

VOLUME LIII. Biomembranes (Part D: Biological Oxidations)
Edited by SIDNEY FLEISCHER AND LESTER PACKER

VOLUME LIV. Biomembranes (Part E: Biological Oxidations)
Edited by SIDNEY FLEISCHER AND LESTER PACKER

VOLUME LV. Biomembranes (Part F: Bioenergetics)
Edited by SIDNEY FLEISCHER AND LESTER PACKER

VOLUME LVI. Biomembranes (Part G: Bioenergetics)
Edited by SIDNEY FLEISCHER AND LESTER PACKER

VOLUME LVII. Bioluminescence and Chemiluminescence
Edited by MARLENE A. DELUCA

VOLUME LVIII. Cell Culture
Edited by WILLIAM B. JAKOBY AND IRA PASTAN

VOLUME LIX. Nucleic Acids and Protein Synthesis (Part G)
Edited by KIVIE MOLDAVE AND LAWRENCE GROSSMAN

VOLUME LX. Nucleic Acids and Protein Synthesis (Part H)
Edited by KIVIE MOLDAVE AND LAWRENCE GROSSMAN

VOLUME 61. Enzyme Structure (Part H)
Edited by C. H. W. HIRS AND SERGE N. TIMASHEFF

VOLUME 62. Vitamins and Coenzymes (Part D)
Edited by DONALD B. MCCORMICK AND LEMUEL D. WRIGHT

VOLUME 63. Enzyme Kinetics and Mechanism (Part A: Initial Rate and Inhibitor Methods)
Edited by DANIEL L. PURICH

VOLUME 64. Enzyme Kinetics and Mechanism (Part B: Isotopic Probes and Complex Enzyme Systems)
Edited by DANIEL L. PURICH

VOLUME 65. Nucleic Acids (Part I)
Edited by LAWRENCE GROSSMAN AND KIVIE MOLDAVE

VOLUME 66. Vitamins and Coenzymes (Part E)
Edited by DONALD B. MCCORMICK AND LEMUEL D. WRIGHT

VOLUME 67. Vitamins and Coenzymes (Part F)
Edited by DONALD B. MCCORMICK AND LEMUEL D. WRIGHT

VOLUME 68. Recombinant DNA
Edited by RAY WU

VOLUME 69. Photosynthesis and Nitrogen Fixation (Part C)
Edited by ANTHONY SAN PIETRO

VOLUME 70. Immunochemical Techniques (Part A)
Edited by HELEN VAN VUNAKIS AND JOHN J. LANGONE

VOLUME 71. Lipids (Part C)
Edited by JOHN M. LOWENSTEIN

VOLUME 72. Lipids (Part D)
Edited by JOHN M. LOWENSTEIN

VOLUME 73. Immunochemical Techniques (Part B)
Edited by JOHN J. LANGONE AND HELEN VAN VUNAKIS

VOLUME 74. Immunochemical Techniques (Part C)
Edited by JOHN J. LANGONE AND HELEN VAN VUNAKIS

VOLUME 75. Cumulative Subject Index Volumes XXXI, XXXII, XXXIV–LX
Edited by EDWARD A. DENNIS AND MARTHA G. DENNIS

VOLUME 76. Hemoglobins
Edited by ERALDO ANTONINI, LUIGI ROSSI-BERNARDI, AND EMILIA CHIANCONE

VOLUME 77. Detoxication and Drug Metabolism
Edited by WILLIAM B. JAKOBY

VOLUME 78. Interferons (Part A)
Edited by SIDNEY PESTKA

VOLUME 79. Interferons (Part B)
Edited by SIDNEY PESTKA

VOLUME 80. Proteolytic Enzymes (Part C)
Edited by LASZLO LORAND

VOLUME 81. Biomembranes (Part H: Visual Pigments and Purple Membranes, I)
Edited by LESTER PACKER

VOLUME 82. Structural and Contractile Proteins (Part A: Extracellular Matrix)
Edited by LEON W. CUNNINGHAM AND DIXIE W. FREDERIKSEN

VOLUME 83. Complex Carbohydrates (Part D)
Edited by VICTOR GINSBURG

VOLUME 84. Immunochemical Techniques (Part D: Selected Immunoassays)
Edited by JOHN J. LANGONE AND HELEN VAN VUNAKIS

VOLUME 85. Structural and Contractile Proteins (Part B: The Contractile Apparatus and the Cytoskeleton)
Edited by DIXIE W. FREDERIKSEN AND LEON W. CUNNINGHAM

VOLUME 86. Prostaglandins and Arachidonate Metabolites
Edited by WILLIAM E. M. LANDS AND WILLIAM L. SMITH

VOLUME 87. Enzyme Kinetics and Mechanism (Part C: Intermediates, Stereochemistry, and Rate Studies)
Edited by DANIEL L. PURICH

VOLUME 88. Biomembranes (Part I: Visual Pigments and Purple Membranes, II)
Edited by LESTER PACKER

VOLUME 89. Carbohydrate Metabolism (Part D)
Edited by WILLIS A. WOOD

VOLUME 90. Carbohydrate Metabolism (Part E)
Edited by WILLIS A. WOOD

VOLUME 91. Enzyme Structure (Part I)
Edited by C. H. W. HIRS AND SERGE N. TIMASHEFF

VOLUME 92. Immunochemical Techniques (Part E: Monoclonal Antibodies and General Immunoassay Methods)
Edited by JOHN J. LANGONE AND HELEN VAN VUNAKIS

VOLUME 93. Immunochemical Techniques (Part F: Conventional Antibodies, Fc Receptors, and Cytotoxicity)
Edited by JOHN J. LANGONE AND HELEN VAN VUNAKIS

VOLUME 94. Polyamines
Edited by HERBERT TABOR AND CELIA WHITE TABOR

VOLUME 95. Cumulative Subject Index Volumes 61–74, 76–80
Edited by EDWARD A. DENNIS AND MARTHA G. DENNIS

VOLUME 96. Biomembranes [Part J: Membrane Biogenesis: Assembly and Targeting (General Methods; Eukaryotes)]
Edited by SIDNEY FLEISCHER AND BECCA FLEISCHER

VOLUME 97. Biomembranes [Part K: Membrane Biogenesis: Assembly and Targeting (Prokaryotes, Mitochondria, and Chloroplasts)]
Edited by SIDNEY FLEISCHER AND BECCA FLEISCHER

VOLUME 98. Biomembranes (Part L: Membrane Biogenesis: Processing and Recycling)
Edited by SIDNEY FLEISCHER AND BECCA FLEISCHER

VOLUME 99. Hormone Action (Part F: Protein Kinases)
Edited by JACKIE D. CORBIN AND JOEL G. HARDMAN

VOLUME 100. Recombinant DNA (Part B)
Edited by RAY WU, LAWRENCE GROSSMAN, AND KIVIE MOLDAVE

VOLUME 101. Recombinant DNA (Part C)
Edited by RAY WU, LAWRENCE GROSSMAN, AND KIVIE MOLDAVE

VOLUME 102. Hormone Action (Part G: Calmodulin and Calcium-Binding Proteins)
Edited by ANTHONY R. MEANS AND BERT W. O'MALLEY

VOLUME 103. Hormone Action (Part H: Neuroendocrine Peptides)
Edited by P. MICHAEL CONN

VOLUME 104. Enzyme Purification and Related Techniques (Part C)
Edited by WILLIAM B. JAKOBY

VOLUME 105. Oxygen Radicals in Biological Systems
Edited by LESTER PACKER

VOLUME 106. Posttranslational Modifications (Part A)
Edited by FINN WOLD AND KIVIE MOLDAVE

VOLUME 107. Posttranslational Modifications (Part B)
Edited by FINN WOLD AND KIVIE MOLDAVE

VOLUME 108. Immunochemical Techniques (Part G: Separation and Characterization of Lymphoid Cells)
Edited by GIOVANNI DI SABATO, JOHN J. LANGONE, AND HELEN VAN VUNAKIS

VOLUME 109. Hormone Action (Part I: Peptide Hormones)
Edited by LUTZ BIRNBAUMER AND BERT W. O'MALLEY

VOLUME 110. Steroids and Isoprenoids (Part A)
Edited by JOHN H. LAW AND HANS C. RILLING

VOLUME 111. Steroids and Isoprenoids (Part B)
Edited by JOHN H. LAW AND HANS C. RILLING

VOLUME 112. Drug and Enzyme Targeting (Part A)
Edited by KENNETH J. WIDDER AND RALPH GREEN

VOLUME 113. Glutamate, Glutamine, Glutathione, and Related Compounds
Edited by ALTON MEISTER

VOLUME 114. Diffraction Methods for Biological Macromolecules (Part A)
Edited by HAROLD W. WYCKOFF, C. H. W. HIRS, AND SERGE N. TIMASHEFF

VOLUME 115. Diffraction Methods for Biological Macromolecules (Part B)
Edited by HAROLD W. WYCKOFF, C. H. W. HIRS, AND SERGE N. TIMASHEFF

VOLUME 116. Immunochemical Techniques (Part H: Effectors and Mediators of Lymphoid Cell Functions)
Edited by GIOVANNI DI SABATO, JOHN J. LANGONE, AND HELEN VAN VUNAKIS

VOLUME 117. Enzyme Structure (Part J)
Edited by C. H. W. HIRS AND SERGE N. TIMASHEFF

VOLUME 118. Plant Molecular Biology
Edited by ARTHUR WEISSBACH AND HERBERT WEISSBACH

VOLUME 119. Interferons (Part C)
Edited by SIDNEY PESTKA

VOLUME 120. Cumulative Subject Index Volumes 81–94, 96–101

VOLUME 121. Immunochemical Techniques (Part I: Hybridoma Technology and Monoclonal Antibodies)
Edited by JOHN J. LANGONE AND HELEN VAN VUNAKIS

VOLUME 122. Vitamins and Coenzymes (Part G)
Edited by FRANK CHYTIL AND DONALD B. MCCORMICK

VOLUME 123. Vitamins and Coenzymes (Part H)
Edited by FRANK CHYTIL AND DONALD B. MCCORMICK

VOLUME 124. Hormone Action (Part J: Neuroendocrine Peptides)
Edited by P. MICHAEL CONN

VOLUME 125. Biomembranes (Part M: Transport in Bacteria, Mitochondria, and Chloroplasts: General Approaches and Transport Systems)
Edited by SIDNEY FLEISCHER AND BECCA FLEISCHER

VOLUME 126. Biomembranes (Part N: Transport in Bacteria, Mitochondria, and Chloroplasts: Protonmotive Force)
Edited by SIDNEY FLEISCHER AND BECCA FLEISCHER

VOLUME 127. Biomembranes (Part O: Protons and Water: Structure and Translocation)
Edited by LESTER PACKER

VOLUME 128. Plasma Lipoproteins (Part A: Preparation, Structure, and Molecular Biology)
Edited by JERE P. SEGREST AND JOHN J. ALBERS

VOLUME 129. Plasma Lipoproteins (Part B: Characterization, Cell Biology, and Metabolism)
Edited by JOHN J. ALBERS AND JERE P. SEGREST

VOLUME 130. Enzyme Structure (Part K)
Edited by C. H. W. HIRS AND SERGE N. TIMASHEFF

VOLUME 131. Enzyme Structure (Part L)
Edited by C. H. W. HIRS AND SERGE N. TIMASHEFF

VOLUME 132. Immunochemical Techniques (Part J: Phagocytosis and Cell-Mediated Cytotoxicity)
Edited by GIOVANNI DI SABATO AND JOHANNES EVERSE

VOLUME 133. Bioluminescence and Chemiluminescence (Part B)
Edited by MARLENE DELUCA AND WILLIAM D. MCELROY

VOLUME 134. Structural and Contractile Proteins (Part C: The Contractile Apparatus and the Cytoskeleton)
Edited by RICHARD B. VALLEE

VOLUME 135. Immobilized Enzymes and Cells (Part B)
Edited by KLAUS MOSBACH

VOLUME 136. Immobilized Enzymes and Cells (Part C)
Edited by KLAUS MOSBACH

VOLUME 137. Immobilized Enzymes and Cells (Part D)
Edited by KLAUS MOSBACH

VOLUME 138. Complex Carbohydrates (Part E)
Edited by VICTOR GINSBURG

VOLUME 139. Cellular Regulators (Part A: Calcium- and Calmodulin-Binding Proteins)
Edited by ANTHONY R. MEANS AND P. MICHAEL CONN

VOLUME 140. Cumulative Subject Index Volumes 102–119, 121–134

VOLUME 141. Cellular Regulators (Part B: Calcium and Lipids)
Edited by P. MICHAEL CONN AND ANTHONY R. MEANS

VOLUME 142. Metabolism of Aromatic Amino Acids and Amines
Edited by SEYMOUR KAUFMAN

VOLUME 143. Sulfur and Sulfur Amino Acids
Edited by WILLIAM B. JAKOBY AND OWEN GRIFFITH

VOLUME 144. Structural and Contractile Proteins (Part D: Extracellular Matrix)
Edited by LEON W. CUNNINGHAM

VOLUME 145. Structural and Contractile Proteins (Part E: Extracellular Matrix)
Edited by LEON W. CUNNINGHAM

VOLUME 146. Peptide Growth Factors (Part A)
Edited by DAVID BARNES AND DAVID A. SIRBASKU

VOLUME 147. Peptide Growth Factors (Part B)
Edited by DAVID BARNES AND DAVID A. SIRBASKU

VOLUME 148. Plant Cell Membranes
Edited by LESTER PACKER AND ROLAND DOUCE

VOLUME 149. Drug and Enzyme Targeting (Part B)
Edited by RALPH GREEN AND KENNETH J. WIDDER

VOLUME 150. Immunochemical Techniques (Part K: *In Vitro* Models of B and T Cell Functions and Lymphoid Cell Receptors)
Edited by GIOVANNI DI SABATO

VOLUME 151. Molecular Genetics of Mammalian Cells
Edited by MICHAEL M. GOTTESMAN

VOLUME 152. Guide to Molecular Cloning Techniques
Edited by SHELBY L. BERGER AND ALAN R. KIMMEL

VOLUME 153. Recombinant DNA (Part D)
Edited by RAY WU AND LAWRENCE GROSSMAN

VOLUME 154. Recombinant DNA (Part E)
Edited by RAY WU AND LAWRENCE GROSSMAN

VOLUME 155. Recombinant DNA (Part F)
Edited by RAY WU

VOLUME 156. Biomembranes (Part P: ATP-Driven Pumps and Related Transport: The Na, K-Pump)
Edited by SIDNEY FLEISCHER AND BECCA FLEISCHER

VOLUME 157. Biomembranes (Part Q: ATP-Driven Pumps and Related Transport: Calcium, Proton, and Potassium Pumps)
Edited by SIDNEY FLEISCHER AND BECCA FLEISCHER

VOLUME 158. Metalloproteins (Part A)
Edited by JAMES F. RIORDAN AND BERT L. VALLEE

VOLUME 159. Initiation and Termination of Cyclic Nucleotide Action
Edited by JACKIE D. CORBIN AND ROGER A. JOHNSON

VOLUME 160. Biomass (Part A: Cellulose and Hemicellulose)
Edited by WILLIS A. WOOD AND SCOTT T. KELLOGG

VOLUME 161. Biomass (Part B: Lignin, Pectin, and Chitin)
Edited by WILLIS A. WOOD AND SCOTT T. KELLOGG

VOLUME 162. Immunochemical Techniques (Part L: Chemotaxis and Inflammation)
Edited by GIOVANNI DI SABATO

VOLUME 163. Immunochemical Techniques (Part M: Chemotaxis and Inflammation)
Edited by GIOVANNI DI SABATO

VOLUME 164. Ribosomes
Edited by HARRY F. NOLLER, JR., AND KIVIE MOLDAVE

VOLUME 165. Microbial Toxins: Tools for Enzymology
Edited by SIDNEY HARSHMAN

VOLUME 166. Branched-Chain Amino Acids
Edited by ROBERT HARRIS AND JOHN R. SOKATCH

VOLUME 167. Cyanobacteria
Edited by LESTER PACKER AND ALEXANDER N. GLAZER

VOLUME 168. Hormone Action (Part K: Neuroendocrine Peptides)
Edited by P. MICHAEL CONN

VOLUME 169. Platelets: Receptors, Adhesion, Secretion (Part A)
Edited by JACEK HAWIGER

VOLUME 170. Nucleosomes
Edited by PAUL M. WASSARMAN AND ROGER D. KORNBERG

VOLUME 171. Biomembranes (Part R: Transport Theory: Cells and Model Membranes)
Edited by SIDNEY FLEISCHER AND BECCA FLEISCHER

VOLUME 172. Biomembranes (Part S: Transport: Membrane Isolation and Characterization)
Edited by SIDNEY FLEISCHER AND BECCA FLEISCHER

VOLUME 173. Biomembranes [Part T: Cellular and Subcellular Transport: Eukaryotic (Nonepithelial) Cells]
Edited by SIDNEY FLEISCHER AND BECCA FLEISCHER

VOLUME 174. Biomembranes [Part U: Cellular and Subcellular Transport: Eukaryotic (Nonepithelial) Cells]
Edited by SIDNEY FLEISCHER AND BECCA FLEISCHER

VOLUME 175. Cumulative Subject Index Volumes 135–139, 141–167

VOLUME 176. Nuclear Magnetic Resonance (Part A: Spectral Techniques and Dynamics)
Edited by NORMAN J. OPPENHEIMER AND THOMAS L. JAMES

VOLUME 177. Nuclear Magnetic Resonance (Part B: Structure and Mechanism)
Edited by NORMAN J. OPPENHEIMER AND THOMAS L. JAMES

Volume 213. Carotenoids (Part A: Chemistry, Separation, Quantitation, and Antioxidation)
Edited by Lester Packer

Volume 214. Carotenoids (Part B: Metabolism, Genetics, and Biosynthesis)
Edited by Lester Packer

Volume 215. Platelets: Receptors, Adhesion, Secretion (Part B)
Edited by Jacek J. Hawiger

Volume 216. Recombinant DNA (Part G)
Edited by Ray Wu

Volume 217. Recombinant DNA (Part H)
Edited by Ray Wu

Volume 218. Recombinant DNA (Part I)
Edited by Ray Wu

Volume 219. Reconstitution of Intracellular Transport
Edited by James E. Rothman

Volume 220. Membrane Fusion Techniques (Part A)
Edited by Nejat Düzguünes

Volume 221. Membrane Fusion Techniques (Part B)
Edited by Nejat Düzgünes

Volume 222. Proteolytic Enzymes in Coagulation, Fibrinolysis, and Complement Activation (Part A: Mammalian Blood Coagulation Factors and Inhibitors)
Edited by Laszlo Lorand and Kenneth G. Mann

Volume 223. Proteolytic Enzymes in Coagulation, Fibrinolysis, and Complement Activation (Part B: Complement Activation, Fibrinolysis, and Nonmammalian Blood Coagulation Factors)
Edited by Laszlo Lorand and Kenneth G. Mann

Volume 224. Molecular Evolution: Producing the Biochemical Data
Edited by Elizabeth Anne Zimmer, Thomas J. White, Rebecca L. Cann, and Allan C. Wilson

Volume 225. Guide to Techniques in Mouse Development
Edited by Paul M. Wassarman and Melvin L. DePamphilis

Volume 226. Metallobiochemistry (Part C: Spectroscopic and Physical Methods for Probing Metal Ion Environments in Metalloenzymes and Metalloproteins)
Edited by James F. Riordan and Bert L. Vallee

Volume 227. Metallobiochemistry (Part D: Physical and Spectroscopic Methods for Probing Metal Ion Environments in Metalloproteins)
Edited by James F. Riordan and Bert L. Vallee

Volume 228. Aqueous Two-Phase Systems
Edited by Harry Walter and Göte Johansson

VOLUME 338. Nuclear Magnetic Resonance of Biological Macromolecules (Part A)
Edited by THOMAS L. JAMES, VOLKER DÖTSCH, AND ULI SCHMITZ

VOLUME 339. Nuclear Magnetic Resonance of Biological Macromolecules (Part B)
Edited by THOMAS L. JAMES, VOLKER DÖTSCH, AND ULI SCHMITZ

VOLUME 340. Drug–Nucleic Acid Interactions
Edited by JONATHAN B. CHAIRES AND MICHAEL J. WARING

VOLUME 341. Ribonucleases (Part A)
Edited by ALLEN W. NICHOLSON

VOLUME 342. Ribonucleases (Part B)
Edited by ALLEN W. NICHOLSON

VOLUME 343. G Protein Pathways (Part A: Receptors)
Edited by RAVI IYENGAR AND JOHN D. HILDEBRANDT

VOLUME 344. G Protein Pathways (Part B: G Proteins and Their Regulators) (in preparation)
Edited by RAVI IYENGAR AND JOHN D. HILDEBRANDT

VOLUME 345. G Protein Pathways (Part C: Effector Mechanisms)
Edited by RAVI IYENGAR AND JOHN D. HILDEBRANDT

VOLUME 346. Gene Therapy Methods (in preparation)
Edited by M. IAN PHILLIPS

VOLUME 347. Protein Sensors and Reactive Oxygen Species (Part A: Selenoproteins and Thioredoxin) (in preparation)
Edited by HELMUT SIES AND LESTER PACKER

VOLUME 348. Protein Sensors and Reactive Oxygen Species (Part B: Thiol Enzymes and Proteins) (in preparation)
Edited by HELMUT SIES AND LESTER PACKER

VOLUME 349. Superoxide Dismutase (in preparation)
Edited by LESTER PACKER

Section I

G Protein-Coupled Receptors

[1] Considerations in the Evaluation of Inverse Agonism and Protean Agonism at G Protein-Coupled Receptors

By PETER CHIDIAC

Introduction

Over the past decade it has become accepted that G protein-coupled receptors (GPCRs) can activate G proteins and thus initiate signaling in the absence of agonist; this *spontaneous receptor activity* can be inhibited to varying degrees by antagonists, a phenomenon termed *inverse agonism*. The modulation of spontaneous GPCR activity by the binding of specific ligands implies that receptors can interconvert between active (R*) and inactive forms (R), with agonists promoting the former and inverse agonists the latter. Thus, inverse agonists produce biochemical effects opposite to those of agonists. This stands in contrast to earlier theories that antagonists lack intrinsic activity and merely preclude the binding of agonists to receptors.

Inverse agonism has been observed in a wide variety of systems, with both endogenously and heterologously expressed GPCRs.[1] Notwithstanding these repeated observations *in vitro,* however, the contribution of inverse agonism to the overall therapeutic effects of antagonists is difficult to determine due to the continual presence of endogenous agonists under normal physiological conditions. Proof of the therapeutic relevance of inverse agonism awaits evidence that spontaneous receptor signaling is significant *in vivo* and/or that changes in receptor density,[2] distribution,[3,4] or posttranslational modification,[5] attributable to inverse agonist effects at the cellular and subcellular levels, also occur in intact organisms.

An emerging concept related to inverse agonism is that of *protean agonism,*[6] wherein some ligands display both agonist and inverse agonist properties at a single GPCR. Although this phenomenon has only been observed at a handful of GPCRs and is poorly understood at present, it may ultimately provide a key to understanding how all ligands modulate GPCR behavior.

[1] R. A. de Ligt, A. P. Kourounakis, and A. P. IJzerman, *Br. J. Pharmacol.* **130,** 1 (2000).
[2] M. J. Smit, R. Leurs, A. E. Alewijnse, J. Blauw, G. P. Nieuw Amerongen, D. Van, V. E. Roovers, and H. Timmerman, *Proc. Natl. Acad. Sci. U.S.A.* **93,** 6802 (1996).
[3] D. F. McCune, S. E. Edelmann, J. R. Olges, G. R. Post, B. A. Waldrop, D. J. Waugh, D. M. Perez, and M. T. Piascik, *Mol. Pharmacol.* **57,** 659 (2000).
[4] M. Rinaldi-Carmona, A. Le Duigou, D. Oustric, F. Barth, M. Bouaboula, P. Carayon, P. Casellas, and G. le Fur, *J. Pharmacol. Exp. Ther.* **287,** 1038 (1998).
[5] M. Bouaboula, D. Dussossoy, and P. Casellas, *J. Biol. Chem.* **274,** 20397 (1999).
[6] T. Kenakin, *Pharmacol. Rev.* **48,** 413 (1996).

0076-6879/02 $35.00

Experimental Detection of Inverse Agonist Activity

In theory, any preparation used to measure GPCR stimulation by agonists *in vitro* can also be used to study the inhibitory effects of inverse agonists. Because both agonists and inverse agonists appear to produce their effects by modulating the balance between active and inactive receptors, some level of spontaneous receptor activity presumably is needed to measure the effects of either class of ligand. The detection of inverse agonism additionally requires that spontaneous receptor activity be clearly distinguishable from background noise, which in practice frequently is not the case. Measuring inverse agonism thus is intrinsically more difficult, and the best conditions for doing so will not necessarily be the same as for measuring receptor activation by agonists. Optimizing protein expression levels, buffer components, and so on may help reveal spontaneous receptor activity. If a ligand is found to have a negative effect on agonist-independent GPCR signaling, the experimenter must verify that it is genuine. Thus, the possible presence of endogenous activating ligands needs to be ruled out; if that cannot be done, one should at least demonstrate that the effects of strong inverse agonists are competitively inhibited by neutral antagonists or weak inverse agonists.[7] Also, the possible confounding effects of related receptor subtypes, if any are present, should be considered. Inverse agonism can be assessed at the level of the receptor, G protein, effector, or events further downstream. Which technique works best may depend on the system under investigation. All end points will not necessarily yield equivalent results, and therefore it is worthwhile to assay ligand activity at multiple levels.

At the receptor, the binding of inverse agonists tends to be increased by guanine nucleotides, whereas that of agonists is decreased.[8] Thus, in competition experiments with radiolabeled antagonists, inverse agonist binding profiles are left-shifted by the inclusion of GTP analogues, implying a nucleotide-associated increase in affinity, whereas agonist binding profiles are rightshifted, implying a decrease.[9] Using fluorescently labeled β_2-adrenergic receptors, Gether and coworkers[10] have shown that agonists and inverse agonists appear to favor different receptor conformations, with the change in fluorescence induced by a given ligand corresponding to its ability to modulate receptor activity. GPCR conformation is also important for recognition by GPCR-selective kinases (GRKs), which preferentially phosphorylate agonist-bound receptors and thus appear to favor the active state. Accordingly, basal phosphorylation of the CB2 cannabinoid receptor is decreased by inverse agonists.[5] Finally, receptor localization within cells is

[7] P. Chidiac, T. E. Hebert, M. Valiquette, M. Dennis, and M. Bouvier, *Mol. Pharmacol.* **45,** 490 (1994).

[8] E. L. Barker, R. S. Westphal, D. Schmidt, and E. Sanders-Bush, *J. Biol. Chem.* **269,** 11687 (1994).

[9] A. Newman-Tancredi, L. Verriele, C. Chaput, and M. J. Millan, *Naunyn Schmiedebergs Arch. Pharmacol.* **357,** 205 (1998).

[10] U. Gether, S. Lin, and B. K. Kobilka, *J. Biol. Chem.* **270,** 28268 (1995).

sensitive to agonists, which typically promote internalization upon prolonged cell exposure. Emerging evidence indicates that inverse agonists can have the opposite effect, promoting GPCR movement from intracellular compartments to the plasma membrane.[3,4]

At the level of the G protein, activated GPCRs promote the dissociation of GDP; inverse agonists thus are expected to decrease GDP off rates, although this approach is not used commonly. In the presence of micromolar concentrations of unlabeled GDP, the binding of $[^{35}S]GTP\gamma S$ to G proteins is decreased by inverse agonists, opposite to the effect of agonists.[11,12] (It should be noted that receptor-related changes in $[^{35}S]GTP\gamma S$ binding are typically GDP dependent and thus reflect changes in the affinity of GDP but not necessarily that of $[^{35}S]GTP\gamma S$, as is frequently assumed.) GPCR effects on G proteins are also manifested as changes in GTP turnover, with agonists increasing and inverse agonists decreasing the rate of steady-state GTP hydrolysis.[13] While measurements of G protein activity usually are made using cell membranes, inverse agonism has also been observed in GTPase assays with purified receptors and G proteins coreconstituted into phospholipid vesicles,[14] although that approach is technically challenging. In addition to their effects on nucleotide binding and hydrolysis, inverse agonists in whole cells may also influence G protein synthesis and/or degradation, as exposure of cells expressing the CB2 cannabinoid receptor to an inverse agonist upregulates G_i, the target G protein.[15]

At the level of effector proteins, inverse agonism can be observed via changes in second messenger production, both in intact cells and in membrane-based assays. The production of cyclic AMP by adenylyl cyclase is decreased by inverse agonists acting on G_s-coupled receptors[7] and is increased by inverse agonists acting on G_i-coupled receptors. Particularly in the latter case, the ability to detect inverse agonism may be enhanced in the presence of the adenylyl cyclase-stimulating diterpene forskolin.[11,16] Also, IP_3 production by phospholipase $C\beta$ is inhibited by inverse agonists to receptors coupled to that effector system, via G_q.[18,19] Inhibition

[11] J. C. Shryock, M. J. Ozeck, and L. Belardinelli, *Mol. Pharmacol.* **53,** 886 (1998).

[12] R. Brys, K. Josson, M. P. Castelli, M. Jurzak, P. Lijnen, W. Gommeren, and J. E. Leysen, *Mol. Pharmacol.* **57,** 1132 (2000).

[13] T. Costa and A. Herz, *Proc. Natl. Acad. Sci. U.S.A.* **86,** 7321 (1989).

[14] R. A. Cerione, J. Codina, J. L. Benovic, R. J. Lefkowitz, L. Birnbaumer, and M. G. Caron, *Biochemistry* **23,** 4519 (1984).

[15] M. Bouaboula, N. Desnoyer, P. Carayon, T. Combes, and P. Casellas, *Mol. Pharmacol.* **55,** 473 (1999).

[16] A. E. Alewijnse, M. J. Smit, M. S. Rodriguez Pena, D. Verzijl, H. Timmerman, and R. Leurs, *FEBS Lett.* **419,** 171 (1997).

[17] Deleted in proof.

[18] J. Labrecque, A. Fargin, M. Bouvier, P. Chidiac, and M. Dennis, *Mol. Pharmacol.* **48,** 150 (1995).

[19] R. A. Bakker, K. Wieland, H. Timmerman, and R. Leurs, *Eur. J. Pharmacol.* **387,** R5 (2000).

of phospholipase Cβ activity by inverse agonists has also been implied by ligand-induced decreases in intracellular calcium.[20] Other downstream events subject to inhibition by inverse agonists include GPCR-dependent MAP kinase activity[21] and agonist-independent, β-adrenergic receptor-mediated inotropic effects in isolated myocardium.[22,23]

Factors Affecting the Measurement of Inverse Agonism

Multiple factors influence the measurement of inverse agonism, most importantly the levels of receptor and G protein present, the G protein-effector pathways available, and the end point being measured. In addition, ligand effects can be influenced by other variables, such as assay buffer components, receptor mutations, receptor heterogeneity, and the presence of auxiliary proteins. Note that the same combination of receptor and ligand may yield differing results depending on the experimental context. Indeed, protean agonists were first identified by their ability to stimulate receptor activity in one assay and inhibit in another.[7]

Typically, inverse agonism is detected in systems in which GPCRs are expressed at high levels (>100 fmol/mg of membrane protein). Agonist-independent signal tends to increase proportionally as receptor expression level increases.[7] Therefore, maximizing receptor expression increases the likelihood of being able to detect inverse agonism, although other signaling components, such as G protein or effector, might become limiting as the receptor is increased. With overexpressed receptors it may be difficult to detect guanine nucleotide-induced changes in agonist and inverse agonist binding, but this can be remedied by the coexpression of an appropriate G protein. Increasing G protein levels may also make inverse agonism easier to detect[12,24]; however the increased spontaneous receptor signal associated with higher G protein expression sometimes is accompanied by increased receptor-independent background noise (e.g., second messenger production or [^{35}S]GTPγS binding). Many receptors are capable of activating multiple G protein subtypes and sometimes even different families of G proteins, and many G proteins in turn can activate multiple effector pathways.[25] The types and subtypes of G poteins and effectors available, either endogenously expressed or cotransfected

[20] J. A. Garcia-Sainz and M. E. Torres-Padilla, *FEBS Lett.* **443,** 277 (1999).

[21] M. Bouaboula, S. Perrachon, L. Milligan, X. Canat, M. Rinaldi-Carmona, M. Portier, F. Barth, B. Calandra, F. Pecceu, J. Lupker, J. P. Maffrand, G. le Fur, and P. Casellas, *J. Biol. Chem.* **272,** 22330 (1997).

[22] D. R. Varma, *Can. J. Physiol Pharmacol.* **77,** 943 (1999).

[23] D. R. Varma, H. Shen, X. F. Deng, K. G. Peri, S. Chemtob, and S. Mulay, *Br. J. Pharmacol.* **127,** 895 (1999).

[24] E. S. Burstein, T. A. Spalding, and M. R. Brann, *Mol. Pharmacol.* **51,** 312 (1997).

[25] L. Birnbaumer, J. Abramowitz, and A. M. Brown, *Biochim. Biophys. Acta* **1031,** 163 (1990).

with a receptor, can affect receptor–G protein coupling efficiency,[26,27] as well as events downstream. Note also that the same receptor, when coupled to different G protein-effector pathways, may exhibit different rank orders of ligand potency or efficacy.[28]

In addition to the densities and subtypes of receptors, G proteins, and effectors contained in an experimental preparation, the presence of auxiliary proteins, such as RGS proteins, scaffolding proteins, phosducin, and arrestins, all can influence the final readout of receptor–ligand interactions. The organization and availability of these auxiliary proteins may be altered by manipulations such as cell lysis and membrane preparation, which in turn can modulate GPCR responsiveness. Disruption of the cytoskeleton, for example, can increase G_s-dependent cAMP production in S49 cells via several different pathways.[29] Finally, spontaneous receptor activity can be highly sensitive to the ingredients of the experimental buffer. For example, NaCl decreases the agonist-independent GTPase activity of G_i coupled to 5HT1-serotonergic[30] and δ-opioid receptors in membrane-based assays,[13] whereas KCl has the opposite effect at the δ-opioid receptor.[13] NaCl thus decreases one's ability to detect spontaneous receptor activity in these systems.

Whereas ionic conditions and receptor expression levels influence primarily the magnitude of GPCR stimulatory and inhibitory responses, other factors presumably exist that can switch the activity of protean agonists between stimulatory and inhibitory modes. Labetalol increases intracellular cAMP in Sf9 cells expressing the β_2-adrenergic receptor, but conversely inhibits adenylyl cyclase activity in membranes derived from those same cells.[7] Although the determinants of protean activity are poorly understood, some chemical modifications of the receptor itself, such as the phosphorylation or substitution of key amino acid residues, appear to be important. Prolonged agonist stimulation of GPCRs in intact cells leads to desensitization, characterized by decreased agonist responsiveness and increased GPCR phosphorylation.[31] In contrast to the loss of stimulation with agonists, the ability of labetalol to inhibit spontaneous β_2-adrenergic receptor activity in Sf9 membranes is *increased* by GPCR desensitization.[32] Moreover, dichloroisoproterenol, which is strongly stimulatory in whole cells (~50% of isoproterenol activity),[7] variably acts as a weak agonist or weak inverse agonist in nondesensitized membranes (±20% of ligand-independent adenylyl cyclase activity) from Sf9 cells

[26] D. J. Carty, E. Padrell, J. Codina, L. Birnbaumer, J. D. Hildebrandt, and R. Iyengar, *J. Biol. Chem.* **265,** 6268 (1990).

[27] Q. Yang and S. M. Lanier, *Mol. Pharmacol.* **56,** 651 (1999).

[28] P. Leff, C. Scaramellini, C. Law, and K. McKechnie, *Trends Pharmacol. Sci.* **18,** 355 (1997).

[29] D. Leiber, J. R. Jasper, A. A. Alousi, J. Martin, D. Bernstein, and P. A. Insel, *J. Biol. Chem.* **268,** 3833 (1993).

[30] D. W. Gray, H. Giles, V. Barrett, and G. R. Martin, *Ann. N.Y. Acad. Sci.* **812,** 236 (1997).

[31] R. J. Lefkowitz, W. P. Hausdorff, and M. G. Caron, *Trends Pharmacol. Sci.* **11,** 190 (1990).

[32] P. Chidiac, S. Nouet, and M. Bouvier, *Mol. Pharmacol.* **50,** 662 (1996).

expressing β_2-adrenergic receptors, but consistently causes a 40% decrease in activity in membranes from desensitized cells.[32] Receptor phosphorylation thus appears to act as a switch for these protean ligands.

Similar to the effects of posttranslational modifications, mutations in GPCR amino acid sequences can significantly alter GPCR responsiveness to ligands. The most striking examples of this are perhaps the single amino acid substitutions that result in highly elevated spontaneous receptor activity.[33] These mutants allow a larger window through which to view the effects of inverse agonists. Mirroring the effect of desensitization on weak β_2AR partial agonists described in the preceding paragraph, activating receptor mutations at the B_2 bradykinin receptor cause drugs that are weak inverse agonists at wild-type receptors to behave as partial agonists.[34,35] In contrast, a spontaneously active form of the secretin receptor containing two point mutations was found to be inhibited by secretin, the natural activating ligand.[36]

While they occasionally occur in humans, it should be kept in mind that with constitutively activated GPCRs one is no longer dealing with the normal physiological target. Any alteration in GPCR structure can potentially influence both isomerization and affinity. Thus, the spontaneous activity of a receptor, the selectivity of ligands among receptor states, or both may be affected; note that it may be possible to misinterpret a change in isomerization as a change in affinity, as Colquhoun[37] has pointed out. Changes in GPCR responsiveness to ligand binding therefore need to be interpreted with care. Furthermore, compared to their wild-type counterparts, constitutively activated mutants are likely to be more highly phosphorylated because they may be better substrates for G protein receptor kinases (GRKs),[33] cause increased activation of second messenger-dependent kinases, or both.

Quantifying Inverse Agonist Activity

The fraction of spontaneous receptor activity that can be inhibited by an inverse agonist can be taken as an empirical measure of its activity in an experiment. A full inverse agonist would have an activity of 1.0, corresponding to 100% inhibition, whereas a partial inverse agonist would have an activity between 0 and 1.0, and a neutral antagonist lacking intrinsic activity would fail to inhibit and thus yield a value of 0. This is roughly analogous to assigning values of 1.0 or less for intrinsic

[33] P. Samama, S. Cotecchia, T. Costa, and R. J. Lefkowitz, *J. Biol. Chem.* **268,** 4625 (1993).

[34] J. Marie, C. Koch, D. Pruneau, J. L. Paquet, T. Groblewski, R. Larguier, C. Lombard, B. Deslauriers, B. Maigret, and J. C. Bonnafous, *Mol. Pharmacol.* **55,** 92 (1999).

[35] D. B. Fathy, T. Leeb, S. A. Mathis, and L. M. Leeb-Lundberg, *J. Biol. Chem.* **274,** 29603 (1999).

[36] S. C. Ganguli, C. G. Park, M. H. Holtmann, E. M. Hadac, T. P. Kenakin, and L. J. Miller, *J. Pharmacol. Exp. Ther.* **286,** 593 (1998).

[37] D. Colquhoun, *Br. J. Pharmacol.* **125,** 924 (1998).

agonist activity, as proposed originally by Ariens. Note that in both cases the parameter is system dependent.

To calculate inverse agonist activity, one must know the level of spontaneous receptor activity, which can be determined by subtracting the ligand-independent signal measured in the absence of receptor from that measured in the presence of receptor. While straightforward to do with heterologously expressed receptors, where cells transfected (or infected) with empty vector can be used to get an approximation of receptor-free signal,[7] this may not be feasible with endogenously expressed receptors (although specific alkylating ligands are available for some).

Figure 1 shows a hypothetical experiment with increasing concentrations of an inverse agonist: *A* represents the observed signal in the presence of spontaneously

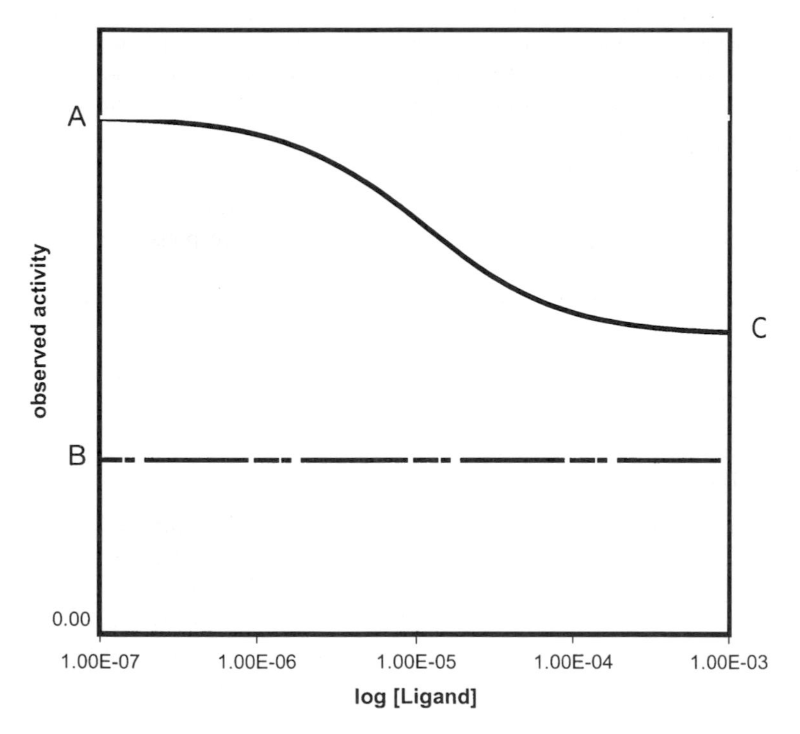

FIG. 1. Quantification of inverse agonist activity. Two hypothetical concentration dependence profiles with the same ligand in two equivalent preparations are shown: one in the absence and the other in the presence of a spontaneously active receptor. *A* represents observed activity in the presence of the receptor but in the absence of ligand, *B* represents the basal activity of the system in the absence of the receptor, and *C* represents the maximally inhibited signal at a saturating concentration of ligand. Maximal inverse agonist activity is defined as $(A-C)/(A-B)$, which equals the fraction of spontaneous receptor activity that can be inhibited by the ligand.

active receptor but in the absence of ligand, B represents the signal in the absence of receptor, and C represents the signal in the presence of receptor and a saturating concentration of inverse agonist. Spontaneous receptor activity equals the difference between A and B. Maximal inverse agonist activity equals $(A-C)/(A-B)$, corresponding to the fraction of spontaneous receptor activity that can be inhibited. Findings from various laboratories show that the same ligand may vary in maximal activity depending on the experimental conditions used to measure receptor function. For example, at the β_2-adrenergic receptor, propranolol has an inverse agonist activity of up to 0.8 in some systems[7] but is essentially inactive (i.e., zero) in others.[38]

The apparent equilibrium binding affinity (K_d) of an inverse agonist can be measured using standard agonist equilibrium-binding techniques, i.e., either by the direct binding to a receptor of increasing concentrations of a radiolabeled form of the inverse agonist or via competition between the unlabeled inverse agonist and a radioligand that binds to the same receptor (sometimes referred to as K_i). K_d is taken as equivalent to the half-saturating concentration of inverse agonist in a direct binding experiment (assuming mass-action behavior), or calculated from competition binding results, accounting for the rightward shift in the inverse agonist-binding profile caused by the presence of the radioligand (e.g., Cheng–Prusoff). Note that for both agonists and inverse agonists, K_d actually represents an amalgam of the individual affinities of the various receptor states.[37,39]

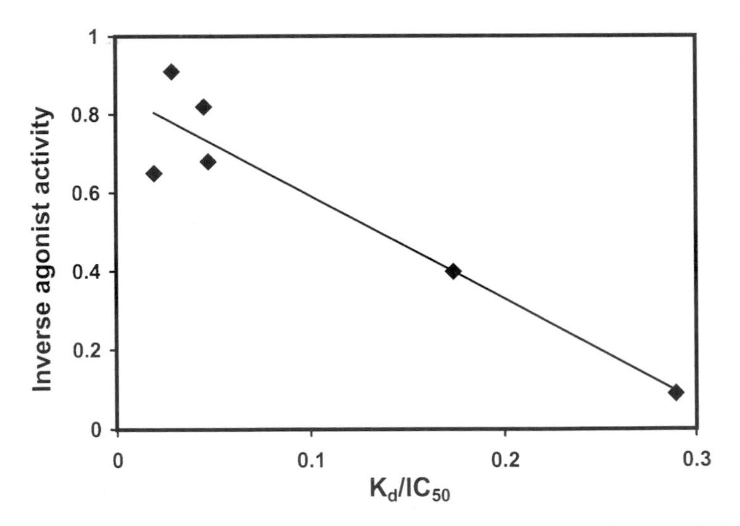

FIG. 2. Negative relationship between inverse agonist activity and K_d/IC_{50}. Apparent binding affinity, potency, and inverse agonist activity were measured with a variety of ligands in membranes prepared from Sf9 cells expressing the β_2-adrenergic receptor. Values of IC_{50} were higher than the corresponding K_d values, and this discrepancy between affinity and potency decreased with maximal inverse agonist activity. Plotted from data in Ref. 7.

The IC_{50} of an inverse agonist, i.e., the concentration required to produce 50% of its maximal inhibitory effect on spontaneous receptor activity, can be taken as an indication of its potency. Interestingly, IC_{50} values for inverse agonists sometimes tend to be *higher* than their K_d values. As shown in Fig. 2, this discrepancy is greater for strong inverse agonists than for weak inverse agonists at the β_2-adrenergic receptor. Although reports of such observations are limited, this trend is the reverse of what is found with GPCR agonists. Disproportionately high agonist activities at partial receptor occupancies and the analogous discrepancy found with inverse agonists may represent different manifestations of the same underlying phenomenon, which in the case of agonists is believed to reflect the existence of spare receptors.

Models of GPCR Activity

The simplest scheme for describing spontaneous receptor activity and inverse agonism is the two-state model:

$$\begin{array}{ccc} & K_I & \\ L + R & \leftrightarrows & L + R^* \\ K_L \downarrow\uparrow & & \downarrow\uparrow \alpha K_L \\ LR & \leftrightarrows & LR^* \\ & \alpha K_I & \end{array}$$

In this model, the receptor spontaneously isomerizes between inactive (R) and active (R^*) conformations or states, and the ratio of $[R]/[R^*]$ is described by the unimolecular equilibrium constant K_I. A ligand (L) binds to the inactive state with an affinity $K_L = [L][R]/[LR]$ (equilibrium dissociation constant) and to the active state with an affinity $\alpha K_L = [L][R^*]/[LR^*]$. The selectivity factor α describes the mutual effect of the ligand on isomerization to R^* and of that isomerization on ligand affinity. If $\alpha < 1$, the ligand will bind with higher affinity to R^* and also promote the isomerization of R to R^*, and therefore will act as an agonist; analogously, if $\alpha > 1$, the ligand will favor R and act as an inverse agonist. Ligands with only a weak binding preference will fail to drive the receptor completely into either state, and thus will act as partial agonists or partial inverse agonists. Ligands for which $\alpha = 1$ (sometimes referred to as neutral or true antagonists) will have no effect on receptor activity, but will competitively inhibit the effects of both agonists and inverse agonists.

[38] P. Samama, G. Pei, T. Costa, S. Cotecchia, and R. J. Lefkowitz, *Mol. Pharmacol.* **45,** 390 (1994).

[39] J. W. Wells, *in* "Receptor-Ligand Interactions: A Practical Approach" (E. C. Hulme, ed.). Oxford Univ. Press, Oxford, 1992.

The two-state model is formally analogous to the *basic ternary complex model*[40] provided that G protein is not limiting. In that model, the receptor is considered to alternate between a free form (R) and a G protein-bound (RG) form, rather than between two conformations with different activities (i.e., R* and R). Samama and co-workers[33] combined the two-state model and the ternary complex model together to create the *extended ternary complex model* based on the idea that a constitutively active mutant form of the β_2AR could exhibit high-affinity agonist binding when free from G protein. Krumins and Barber[41] have argued that the extended ternary complex model reduces to a form of the two-state model wherein the receptor has two binding sites, one for ligands and one for G proteins, but still isomerizes between only two states, R and R*.

Some observations suggest that the two-state model may be too simplistic. For example, the activity of LR* can differ from one ligand to the next,[41] suggesting that ligands may distinguish between multiple active and inactive receptor conformations and/or induce distinct conformational changes on binding. Analogously, the observation of dissimilar agonist rank orders for the coupling of a single receptor to two different G proteins suggests that the latter also may recognize different active receptor states.[6] Still, it has been proposed that "R" and "R*" can be taken as representing two clusters of microstates drawn within the entire conformational space of a GPCR,[42] in which case differences among ligands and G proteins may be viewed as essentially consistent with the two-state model, although there might be preferences for different microstates within those clusters.

While the individual effects of agonists and inverse agonists are arguably consistent with the simple two-state model, some data remain difficult to explain in that context. Specifically, the phenomenon of protean agonism, wherein a single ligand can act as an agonist or inverse agonist at a single receptor, is problematic. It is difficult, indeed, to understand how one ligand can both increase and decrease GPCR activity. The *cubic ternary complex model* (CTCM)[6] is capable of accounting for protean agonism; for instance, if a ligand can promote the isomerization of R to R* but simultaneously disfavor the binding of receptor to G protein, then its net activity will depend on how these two effects balance out under a given set of experimental conditions. Protean agonism can also be accounted for in terms of the three-state model,[28] wherein a single receptor can couple to two different G proteins (termed G1 and G2), both of which are freely available to interact with the receptor, but which recognize different conformations (e.g., R* and R**). In the context of the three-state model, if an activating ligand promotes the formation of R*G1 more than R**G2, then assays based on an end point dependent on R**G2 will show agonism if no G1 is present, but increasing the availability of G1 will

[40] A. De Lean, J. M. Stadel, and R. J. Lefkowitz, *J. Biol. Chem.* **255,** 7108 (1980).

[41] A. M. Krumins and R. Barber, *Mol. Pharmacol.* **52,** 144 (1997).

[42] H. O. Onaran and T. Costa, *Ann. N.Y. Acad. Sci.* **812,** 98 (1997).

create a sink for the receptor, leading to a diminution of the R**G2 response and ultimately manifesting as inverse agonism (provided that sufficient spontaneous receptor activity is present). Hence, the same ligand can appear either stimulatory or inhibitory, depending on the available G protein complement.

The models discussed up to this point are based on the premise that a single receptor can assume two or more conformations that differ in affinity for ligands and/or G proteins, with the underlying assumption that the population of receptors under consideration is homogeneous. While it is clear that cloned receptors can be expressed as single gene products in cultured cells, a uniform amino acid sequence does not guarantee a homogeneous population of proteins. Thus, a GPCR may normally exist as a mixture of posttranslationtially modified forms of a common primary structure. This would be significant if more than one structurally modified form were to contribute to the experimental readout of receptor function. As noted earlier, relatively minor changes in the structure of a GPCR can profoundly affect its basal activity and regulation by ligands. In particular, it appears that the phosphorylation that accompanies receptor desensitization does not really turn off receptors but rather serves to alter their responsiveness to agonists and inverse agonists.[32]

The idea that desensitized receptors can still interact productively with their G proteins runs counter to the prevailing notion that phosphorylation causes the functional uncoupling of receptor and G protein.[43] Receptor phosphorylation precedes or accompanies internalization from the plasma membrane; however, blocking that sequestration does not prevent desensitization *per se*.[31] The notion that receptor phosphorylation prevents coupling to G proteins presumably stems from the observed similarity of the changes in the concentration dependence of agonist activity that occur after either receptor alkylation (i.e., loss of spare receptors) or receptor desensitization. In both cases, agonist potency and subsequently maximal activity decrease in a characteristic pattern[44]; interestingly, the same pattern can be produced using the two-state model shown earlier (Fig. 3), indicating that desensitization is superficially consistent with at least two possible mechanisms, namely a decrease in receptor number and a loss of agonist selectivity for R*.

Equating desensitization to a decrease in the number of functional receptors dictates that there should be a proportional decrease in spontaneous receptor activity, but such a relationship has seldom, if ever, been described. Alternatively, desensitization can be modeled as a decrease in the ability of agonists to distinguish between R and R*. The simulations in Fig. 3 show that decreasing the selectivity of an agonist for R* (i.e., increasing α) mimics observed patterns of agonist desensitization, without having to assume a decrease in receptor density.

[43] J. G. Krupnick and J. L. Benovic, *Annu. Rev. Pharmacol. Toxicol.* **38,** 289 (1998).

[44] M. Bouvier, S. Collins, B. F. O'Dowd, P. T. Campbell, A. De Blasi, B. K. Kobilka, C. MacGregor, G. P. Irons, M. G. Caron, and R. J. Lefkowitz, *J. Biol. Chem.* **264,** 16786 (1989).

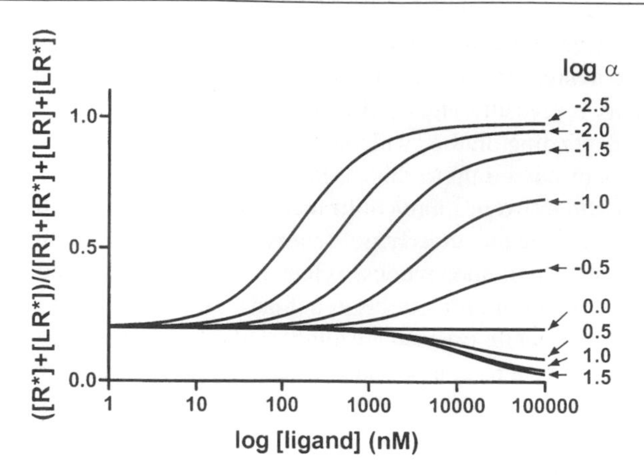

FIG. 3. Effect of varying α on ligand behavior. In the simulations shown, which were carried out according to the two-state model, one in five free receptors is assumed to be in the active conformation (i.e., $K_I = 4$), and K_L, the affinity for R, is set to 10 μM throughout. Agonist activity is modeled by setting the selectivity factor α to values of less than one so that the ligand binds with higher affinity to R* than to R. When the preference for R* is 300-fold or greater (i.e., $\alpha < 0.003$), the ligand is able to fully activate the receptor, with smaller values of α shifting the curve increasingly to the left. As α approaches unity, there is a rightward shift in the ligand concentration dependence, and eventually a complete loss of agonist activity. Inverse agonist activity analogously is modeled by setting α to values of greater than one. Maximal inhibition increases with α; however, once inhibition is complete (i.e., $\alpha \sim 30$), there is no further effect of increasing α.

This interpretation is preferable to the prevailing "functional uncoupling" hypothesis because it allows spontaneous receptor activity to be maintained. Another advantage of the two-state model of course is that it accounts for inverse agonism; moreover, desensitization-induced changes in inverse agonism are consistent with a decrease in ligand affinity for R* relative to R. The simulations in Fig. 3 show that an increase in α increases the maximal effect of a partial inverse agonist but has no effect on a full inverse agonist, again recalling the results seen with desensitization of the β_2AR.[32] For both agonists and inverse agonists, a 10-fold or lower decrease in ligand affinity for R* with no change in affinity for R appears consistent with the effects of receptor desensitization. In addition, a similar change in α can account for protean agonism; for example, in Fig. 3, increasing α from 0.32 to 3.2 (i.e., changing log α from -0.5 to 0.5) shifts the ligand from being a partial agonist to being a partial inverse agonist, recalling experiments with dichloroisoproterenol before and after desensitization at the β_2AR.[32]

The effects of phosphorylation (and presumably other posttranslational modifications) on GPCR behavior argue that each posttranslationally modified form of a GPCR should be considered as a distinct protein capable of isomerizing between active and inactive conformations. It follows that any measurement of receptor activity would represent the summation of the activities of all of these related structures. The relative amount of each differentially modified form of the receptor would be in constant flux in whole cells, where phosphorylation, dephosphorylation, palmitoylation, and so on occur continuously; subcellular preparations and purified proteins would tend to have fixed receptor subpopulations, as the factors governing their interconversion would largely be lost or inactivated. From this perspective, the two-state model is clearly too simplistic; following the same basic principle, the minimum number of receptor states would be twice the number of structural forms, in essence begetting a multi two-state model. Briefly, the observed response to a ligand would be equal to the sum of its effects on each structural variant of the target receptor. For protean agonists, where the changes in receptor activity would occur simultaneously in two opposite directions, the predominating effect would dictate whether the ligand appeared to be stimulatory or inhibitory. Thus, such ligands could appear to function as agonists or inverse agonists, depending on the relative amount of each posttranslationally modified form available in a given experiment. For an illustration of this concept, the reader is referred to previously published simulations carried out assuming a mixture of two subpopulations of receptor with "normal" and "desensitized" ligand-binding properties.[32]

$$\begin{array}{ccc} K_{\mathrm{I}} & & K'_{\mathrm{I}} & & K''_{\mathrm{I}} & \\ \mathrm{L}+\mathrm{R} \leftrightarrows \mathrm{L}+\mathrm{R}^{*} & & \mathrm{L}+\mathrm{R}' \leftrightarrows \mathrm{L}+\mathrm{R}'^{*} & & \mathrm{L}+\mathrm{R}'' \leftrightarrows \mathrm{L}+\mathrm{R}''^{*} & \\ K_{\mathrm{L}} \downarrow\uparrow \quad \downarrow\uparrow \alpha K_{\mathrm{L}} & + & K'_{\mathrm{L}} \downarrow\uparrow \quad \downarrow\uparrow \alpha' K'_{\mathrm{L}} & + & K''_{\mathrm{L}} \downarrow\uparrow \quad \downarrow\uparrow \alpha'' K''_{\mathrm{L}} & + \cdots \\ \mathrm{LR} \leftrightarrows \mathrm{LR}^{*} & & \mathrm{LR}' \leftrightarrows \mathrm{LR}'^{*} & & \mathrm{LR}'' \leftrightarrows \mathrm{LR}''^{*} & \\ \alpha K_{\mathrm{I}} & & \alpha' K'_{\mathrm{I}} & & \alpha'' K''_{\mathrm{I}} & \end{array}$$

multi two-state model

The multi two-state model offers an alternative explanation for the phenomenon of protean agonism that has some advantages over the three-state and cubic ternary complex models. For example, the three-state model requires that the receptor can interact freely with multiple pools of G protein, a situation that has been difficult to verify experimentally.[45] Also, the CTCM may be unable to account for protean agonism in cases where receptors and G proteins are precoupled or form stable complexes. The basic premise of the multi two-state model is that altering the structure of a receptor changes its function; the same idea could be incorporated into modified three-state or cubic ternary models, as these are essentially extensions of the two-state model. The implied number of parameters might become unwieldy, however, especially in the case of a multi-CTCM.

[45] R. R. Neubig, *FASEB J.* **8,** 939 (1994).

The implications of the multi two-state model go beyond its ability to account for protean agonism and the effects of receptor desensitization on agonist and inverse agonist activities. Indeed, several studies have found that for receptors that couple to multiple G proteins, the phosphorylation state of the receptor influences which G protein pathway will be activated.[46,47] A logical extension of this is that the inconsistencies in ligand rank orders of potency and activity found when one receptor couples to two different G proteins may reflect the existence of receptor populations with different posttranslational modifications.

Conclusions

The discovery of spontaneous receptor activity and inverse agonism has brought about significant advances in our understanding of how GPCR activity can be modulated by the binding of stimulatory and inhibitory drugs. Because agonists and inverse agonists appear to bind to different receptor conformations, it may be possible to design improved therapeutics by targeting them to specific receptor states. Moreover, the sensitivity of inverse agonists and protean agonists to receptor desensitization implies an additional level of complexity in structure–activity relationships. While inverse agonism and protean agonism reveal unforseen complexities in GPCR regulation, the elucidation of the molecular events underlying these phenomena ultimately should increase our understanding of GPCR-related diseases as well as our ability to design more selective and more efficacious therapeutics.

Acknowledgments

Much useful information was gleaned from the ongoing conferences on inverse agonism and related topics posted on-line at the ASPET Molecular Pharmacology Bulletin Board (http://webboard.med.umich.edu/~molpharm). Thanks to Rick Neubig for maintaining this unique resource. The author is a Heart and Stroke Foundation of Canada Research Scholar.

[46] Y. Daaka, L. M. Luttrell, and R. J. Lefkowitz, *Nature* **390,** 88 (1997).

[47] L. M. Luttrell, S. S. Ferguson, Y. Daaka, W. E. Miller, S. Maudsley, G. J. Della Rocca, F. Lin, H. Kawakatsu, K. Owada, D. K. Luttrell, M. G. Caron, and R. J. Lefkowitz, *Science* **283,** 655 (1999).

[2] Theoretical Implications of Receptor Coupling to Multiple G Proteins Based on Analysis of a Three-State Model

By CLARE SCARAMELLINI and PAUL LEFF

Introduction

In the last few years, we, and many other researchers, have become interested in the possibility that receptors exist in multiple conformational states. Our interest derives primarily from observations that a single receptor type can exhibit different pharmacological behavior, depending on the experimental system in which the receptor is placed. In particular, evidence has accumulated for G protein-coupled receptors for so-called receptor promiscuity, i.e., the ability of the same receptor type to couple to different effector pathways. In turn, this promiscuous coupling results in different agonist potency orders[1–6] or efficacy patterns[7] depending on which effector pathway is used to record receptor activity. In traditional pharmacological receptor theory, such differences would be attributed to differences in receptor types, and it was this apparent contradiction that led us to consider how existing theoretical models of receptor activation needed to be extended.

Our starting point was the simple two-state model of receptor activation,[8,9] in which receptors are hypothesized to exist in an inactive conformation, R, and in an active conformation, R*. Both this model and the extended ternary complex model[10] assume a single active receptor state. We reasoned that such models could be extended to account for receptor promiscuity by assuming that the active receptor state is capable of interacting with multiple G proteins, but that

[1] D. Spengler, C. Waeber, C. Pantaloni, F. Holsboer, J. Bockaert, P. H. Seeburg, and L. Journot, *Nature* **365,** 170 (1993).
[2] S. Robb, T. R. Cheek, F. L. Hannan, L. M. Hall, J. M. Midgley, and P. D. Evans, *EMBO J.* **13,** 1325 (1994).
[3] D. M. Perez, J. Hwa, R. Gaivin, M. Mathur, F. Brown, and R. M. Graham, *Mol. Pharmacol.* **49,** 112 (1996).
[4] M. G. Eason, M. T. Jacinto, and S. B. Liggett, *Mol. Pharmacol.* **45,** 696 (1994).
[5] M. Negishi, A. Irie, Y. Sugimoto, T. Namba, and A. Ichikawa, *J. Biol. Chem.* **270,** 16122 (1995).
[6] M. H. Richards and P. L. M. van Giersbergen, *Life Sci.* **57,** 397 (1995).
[7] K. A. Berg, S. Maayani, J. Goldfarb, C. Scaramellini, P. Leff, and W. P. Clarke, *Mol. Pharmacol.* **54,** 94 (1998).
[8] J. Monod, J. Wyman, and J.-P. Changeux, *J. Mol. Biol.* **12,** 88 (1965).
[9] P. Leff, *Trends Pharmacol. Sci.* **16,** 89 (1995).
[10] P. Samama, S. Cotecchia, T. Costa, and R. J. Lefkowitz, *J. Biol. Chem.* **268,** 4625 (1993).

0076-6879/02 $35.00

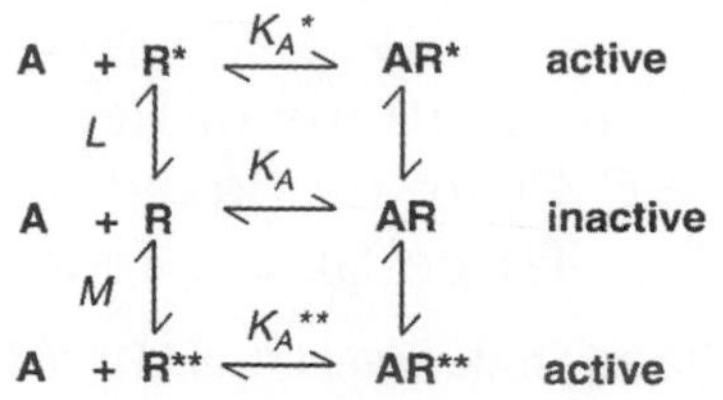

$$\begin{array}{ccccl} A + R^* & \underset{}{\overset{K_A^*}{\rightleftharpoons}} & AR^* & & \text{active} \\ L \updownarrow & & \updownarrow & & \\ A + R & \overset{K_A}{\rightleftharpoons} & AR & & \text{inactive} \\ M \updownarrow & & \updownarrow & & \\ A + R^{**} & \overset{K_A^{**}}{\rightleftharpoons} & AR^{**} & & \text{active} \end{array}$$

FIG. 1. The simple three-state model: intact mode.

this would not allow for altered pharmacology through the different pathways. This is because agonist activity in such systems would be determined by the extent to which individual ligands displaced receptors from the R to the R* state, and that these events at the receptor level would imprint themselves on downstream events regardless of the effector pathway measured. In order for promiscuity to produce altered agonist pharmacology, we proposed the existence of a second active receptor state. The result was a three-state model of receptor activation.[11,12]

Initially, we explored the theoretical predictions of the three-state model considering events purely at the receptor level. Thus, it was assumed that the two active states of the receptor couple to distinct pathways, but the interactions with G proteins were not modeled explicitly. Subsequently, we extended the model to incorporate those interactions, in particular to explore the influence of receptor:G protein stoichiometry. This article briefly reviews the predictions of the simple three-state model and then describes in greater detail the results of the later, mechanistically more complete, analysis.

The Simple Three-State Model

In this model (Fig. 1), receptors are assumed to exist in three receptor conformations: an inactive state, R, and two active states, R^* and R^{**}. The basal ratio among [R], [R^*], and [R^{**}] is governed by receptor distribution constants L and M, and agonist activity is determined by the three ligand-specific dissociation equilibrium constants K_A, K_A^*, and K_A^{**}.

In order to account for the different patterns of agonist pharmacology that promiscuous receptor systems have been observed to exhibit in practice, it was necessary to propose that the model could operate in two different modes. These

[11] P. Leff, C. Scaramellini, C. Law, and K. McKechnie, *Trends Pharmacol. Sci.* **18,** 355 (1997).

[12] P. Leff and C. Scaramellini, *in* "Receptor Mechanisms: Principles of Agonism" (D. Hoyer, P. Leff, P. P. A. Humphrey, N. P. Shankley, and D. G. Trist, eds.), IUPHAR Media (1998).

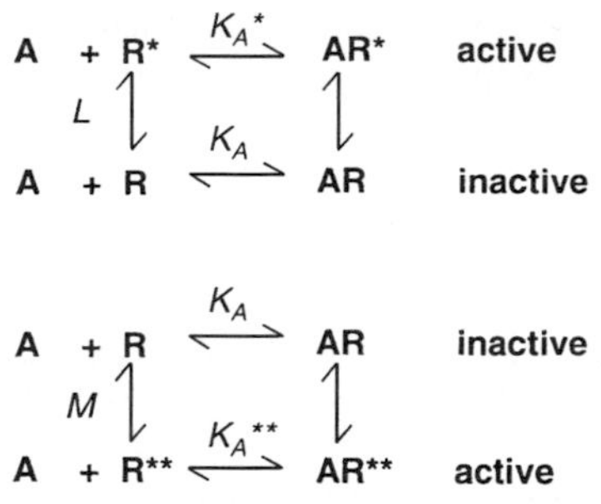

$$\begin{array}{ccccl} A + R^* & \xrightleftharpoons{K_A^*} & AR^* & & \text{active} \\ L \updownarrow & & \updownarrow & & \\ A + R & \xrightleftharpoons{K_A} & AR & & \text{inactive} \end{array}$$

$$\begin{array}{ccccl} A + R & \xrightleftharpoons{K_A} & AR & & \text{inactive} \\ M \updownarrow & & \updownarrow & & \\ A + R^{**} & \xrightleftharpoons{K_A^{**}} & AR^{**} & & \text{active} \end{array}$$

FIG. 2. The simple three-state model: isolated mode.

were termed *intact* and *isolated.* Figure 1 illustrates the model operating in intact mode and Fig. 2 in isolated mode.

The essential differences between the two modes of operation are explained and summarized as follows.

Intact Mode

In this situation, equilibria governing the distribution of all the receptor states are linked. Thus, the distribution of receptors toward R^* is at the expense of distribution toward R^{**} and vice versa. So, for example, an agonist that has a higher affinity for R^* than for R^{**} (i.e., $K_A^* < K_A^{**} < K_A$) is predicted to have higher efficacy in the R^*-linked pathway than in the R^{**}-linked pathway. The linkage between equilibria also means that, for a particular ligand, the displacement of receptors away from R and toward R^* and R^{**} is governed by the same affinity constants. Thus, the measured affinity for an agonist will depend on K_A, K_A^*, and K_A^{**} regardless of the effector pathway. Therefore, agonist potency orders are predicted to be identical through the two pathways.

The constitutive activation of receptors is determined by the values of L and M. When the system is constitutively activated a ligand with, for example, affinity constants, $K_A^{**} < K_A < K_A^*$ is predicted to act as an inverse agonist through the R^*-linked pathway, but as an agonist through the R^{**}-linked pathway. Moreover, due to the linkage between equilibria in the scheme, inverse agonism is predicted to be a system-dependent variable. For example, a ligand characterized by the affinity constants $K_A^{**} < K_A^* < K_A$ intrinsically has higher affinity for R^* (and R^{**}) than for R and would therefore be expected to act as an agonist through the R^*-linked pathway. However, when, under basal conditions, there is a high level of distribution of receptors toward R^{**}, the influence of K_A^{**} is more marked, resulting in a net distribution of receptors *away* from R^*, meaning that the ligand is predicted to behave as an *inverse* agonist.

In summary, the following predictions are made for the model operating in intact mode: (i) same agonist potency orders in the R*- and R**-linked pathways; (ii) converse agonist efficacy orders between R*- and R**-linked pathways; and (iii) inverse agonism is pathway dependent and system dependent.

Isolated Mode

In this case, the model operates as two independent two-state systems. The two pathways are unlinked and so there are no mutual effects on receptor distribution between them. Therefore, in accordance with the predictions of the two-state model,[9] agonist potency and efficacy depend on K_A and K_A^* in the R*-linked pathway and on K_A and K_A^{**} in the R**-linked pathway.

Because K_A^* and K_A^{**} can have different values for the same agonist and vary among different agonists, potency orders, as well as efficacy values, are predicted to differ between the two pathways.

Similarly, the influences of L and M are independent of one another and they determine, separately, the level of constitutive activation in the two pathways. Agonism and inverse agonism depend solely on the relative values of K_A and K_A^* in the R*-linked pathway and K_A and K_A^{**} in the R**-linked pathway.

Therefore, the following predictions are made by the model operating in isolated mode: (i) different agonist potency orders in the R*- and R**-linked pathways; (ii) different agonist efficacy orders between R*- and R**-linked pathways; and (iii) inverse agonism is purely ligand dependent.

Limitations of the Simple Three-State Model

As explained earlier, the simple three-state model deals with events at the receptor level. Thus, in order to construct agonist–concentration effect curves, it is assumed that pharmacological effects are proportional to the concentrations of receptors in the active states. This assumption is obviously simplistic in that it excludes the influence of the signal transduction pathways coupled to the two receptor states. Having said this, the model does appear to provide a basis to account for the observed pharmacological behavior of agonists in promiscuous receptor systems. Perhaps more importantly, the model does not provide a *mechanistic* interpretation for the predicted behavior. The two modes of operation of the model, intact and isolated, describe experimental data but they do not explain it.

A logical extension to the model was to introduce, explicitly, interactions between active receptor states and G proteins. In addition to giving the model increased mechanistic realism, it allowed us to explore the influence of receptor:G protein stoichiometry as a potentially important experimental factor.

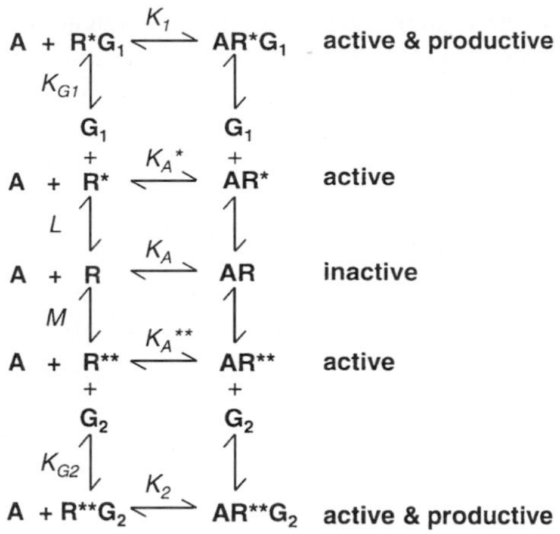

$$\begin{array}{ccccl}
A + R^{*}G_1 & \overset{K_1}{\rightleftharpoons} & AR^{*}G_1 & & \text{active \& productive} \\
K_{G1} \updownarrow & & \updownarrow & & \\
G_1 + A + R^{*} & \overset{K_A^{*}}{\rightleftharpoons} & G_1 + AR^{*} & & \text{active} \\
L \updownarrow & & \updownarrow & & \\
A + R & \overset{K_A}{\rightleftharpoons} & AR & & \text{inactive} \\
M \updownarrow & & \updownarrow & & \\
A + R^{**} + G_2 & \overset{K_A^{**}}{\rightleftharpoons} & AR^{**} + G_2 & & \text{active} \\
K_{G2} \updownarrow & & \updownarrow & & \\
A + R^{**}G_2 & \overset{K_2}{\rightleftharpoons} & AR^{**}G_2 & & \text{active \& productive}
\end{array}$$

FIG. 3. The extended three-state model.

The Extended Three-State Model

In this model (illustrated in Fig. 3), it was assumed that the two active receptor conformations, R^* and R^{**}, interact, respectively and exclusively with different G proteins, G_1 and G_2. The productive receptor complexes are R^*G_1 and AR^*G_1 for the G_1-linked pathway and $R^{**}G_2$ and $AR^{**}G_2$ for the G_2-linked pathway. Pharmacological effects through each pathway are taken to be proportional to the fraction of receptors in the productive receptor complexes.

From inspection of the model it can be appreciated that the basal distribution of receptors is determined by L, M, K_{G_1} and K_{G_2}. Thus, constitutive activity in the G_1-linked pathway is dependent on L and K_{G_1} and that in the G_2-linked pathway is dependent on M and K_{G_2}.

It can also be seen that the activity of agonists depends on five affinity constants: K_A, K_A^*, K_A^{**}, K_1, and K_2. In fact, in the simulations of this model that follow, it is assumed that $K_1 = K_A^*$ and that $K_2 = K_A^{**}$. This implicitly makes the assumption that the binding of G proteins has no effect on receptor conformation as regarding agonist binding.

Another theoretical assumption made was that R^* and R^{**} are inherently unstable in the absence of G_1 and G_2. This was done by choosing values of L and M that favor the inactive receptor state, R. The interaction with G proteins then has the effect of stabilizing and enriching the active receptor conformations. This condition is in accordance with thermodynamic evidence,[13] but in fact it is necessary

[13] U. Gethert, J. A. Ballesteros, R. Seifert, E. Sanders-Bush, H. Weinstein, and B. K. Kobilka, *J. Biol. Chem.* **272,** 2587 (1997).

in the model for G protein interactions to have a significant effect on receptor distribution. When this assumption was not made in simulations of the model, those interactions had virtually no impact.

Theoretical Derivation

The different receptor species in the extended three-state model are related by the following equilibria:

$$L = \frac{[R]}{[R^*]} \quad M = \frac{[R]}{[R^{**}]} \quad K_A = \frac{[A][R]}{[AR]} \quad K_A^* = \frac{[A][R^*]}{[AR^*]} \quad K_A^{**} = \frac{[A][R^{**}]}{[AR^{**}]}$$

$$K_1 = \frac{[A][R^*G_1]}{[AR^*G_1]} \quad K_2 = \frac{[A][R^{**}G_2]}{[AR^{**}G_2]} \quad K_{G_1} = \frac{[R^*][G_1]}{[R^*G_1]} \quad K_{G_2} = \frac{[R^{**}][G_2]}{[R^{**}G_2]}$$

The total receptor concentration is

$$\begin{aligned}[R_{tot}] = {} & [R] + [R^*] + [R^{**}] + [R^*G_1] + [R^{**}G_2] + [AR] + [AR^*] \\ & + [AR^{**}] + [AR^*G_1] + [AR^{**}G_2]\end{aligned}$$

The total concentration of G_1 is $[G_{1tot}] = [G_1] + [R^*G_1] + [AR^*G_1]$; and the total concentration of G_2 is $[G_{2tot}] = [G_2] + [R^{**}G_2] + [AR^{**}G_2]$.

In order to solve these equations and calculate the concentration of each species it is necessary first to determine the free concentration of receptors, [R]. This requires solving a cubic equation in [R], of the form

$$0 = [R]^3 + a_2[R]^2 + a_1[R] + a_0$$

with coefficients

$$\begin{aligned} a_0 &= \frac{-[R_{tot}]}{xyz} \\ a_1 &= \frac{z + x\left([G_{1tot}] - [R_{tot}]\right) + y\left([G_{2tot}] - [R_{tot}]\right)}{xyz} \\ a_2 &= \frac{z(x+y) + xy\left([G_{1tot}] + [G_{2tot}] - [R_{tot}]\right)}{xyz} \end{aligned}$$

where

$$\begin{aligned} x &= \frac{1}{K_{G_1}L}\left(1 + \frac{[A]}{K_1}\right) \\ y &= \frac{1}{K_{G_2}M}\left(1 + \frac{[A]}{K_2}\right) \\ z &= 1 + \frac{1}{L} + \frac{1}{M} + [A]\left(\frac{1}{K_A} + \frac{1}{LK_A^*} + \frac{1}{MK_A^{**}}\right) \end{aligned}$$

The roots of the cubic equation can be calculated as follows[14]:

$$Q = \frac{3a_1 - (a_2)^2}{9} \qquad R = \frac{9a_1a_2 - 27a_0 - 2(a_2)^3}{54}$$

$$D = Q^3 + R^2 \qquad S = \sqrt[3]{R + \sqrt{D}} \qquad T = \sqrt[3]{R - \sqrt{D}}$$

$$A = S - T \qquad B = S + T$$

The roots are then given by

$$\text{root1} = -\frac{1}{3}a_2 + B$$

$$\text{root2} = -\frac{1}{3}a_2 - \frac{1}{2}B + \frac{1}{2}i\sqrt{3}A$$

$$\text{root3} = -\frac{1}{3}a_2 - \frac{1}{2}B - \frac{1}{2}i\sqrt{3}A$$

Root 1 is found to be the physiologically relevant root.

The response through the G_1-linked pathway is taken to be the fraction of receptors in the R^*G_1 and AR^*G_1 states ($f_{R^*G_1}$), and the response through the G_2-linked pathway is taken to be the fraction of receptors in the $R^{**}G_2$ and $AR^{**}G_2$ states ($f_{R^{**}G_2}$):

$$f_{R^*G_1} = \frac{\frac{[G_1]}{K_{G_1}L}\left(1 + \frac{[A]}{K_1}\right)}{\frac{[R_{tot}]}{[R]}} \qquad f_{R^{**}G_2} = \frac{\frac{[G_2]}{K_{G_2}M}\left(1 + \frac{[A]}{K_2}\right)}{\frac{[R_{tot}]}{[R]}}$$

Using Excel 7.0 on a Compaq Deskpro personal computer, the cubic equation in [R] was solved with various parameter values in order to calculate $f_{R^*G_1}$ and $f_{R^{**}G_2}$, and thus generate concentration effect curves for a range of different ligands under different conditions.

Predictions of the Model

Receptor:G Protein Stoichiometry. The model was simulated using a variety of parameter values in order to examine the dependence of agonist activity on receptor:G protein stoichiometry. In particular, we were interested to determine whether the pharmacological predictions of the simple three-state model, in terms of intact and isolated behavior, were associated with particular stoichiometric conditions.

[14] E. W. Weisstein, Cubic equation. www.astro.virginia.edu/~eww6n/math/CubicEquation.html, 1996.

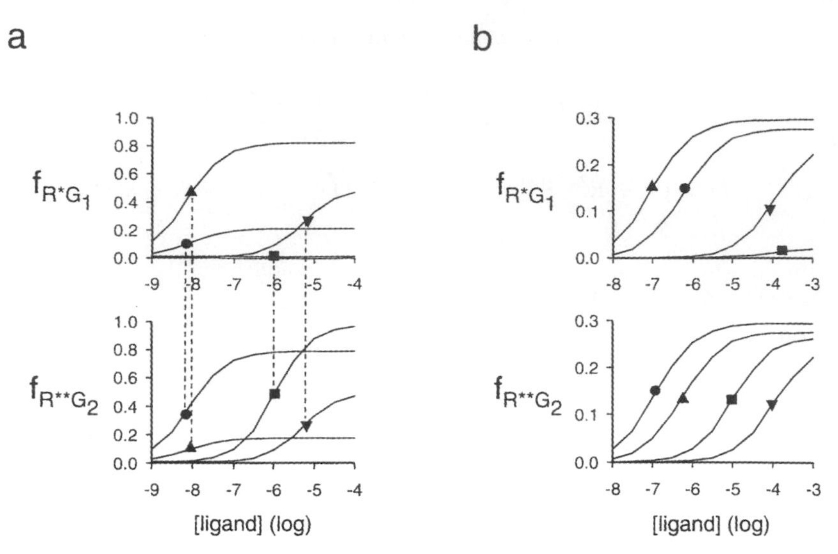

FIG. 4. Simulations using the extended three-state model. (a) **[R]<[G]**; $[R_{tot}] = 1 \times 10^{-4}$; $[G_{1tot}] = 1 \times 10^{-3}$; $[G_{2tot}] = 1 \times 10^{-3}$; $L = M = 1 \times 10^{+7}$; $K_{G_1} = K_{G_2} = 1 \times 10^{-8}$. (b) **[R] > [G]**; $[R_{tot}] = 1 \times 10^{-4}$; $[G_{1tot}] = 3 \times 10^{-5}$; $[G_{2tot}] = 3 \times 10^{-5}$; $L = M = 1 \times 10^{+7}$; $K_{G_1} = K_{G_2} = 1 \times 10^{-8}$. In each case the top half shows the G_1-linked pathway, and the bottom half shows the G_2-linked pathway. Four agonists are shown with the following parameter values: agonist 1, $K_A = 1 \times 10^{-5}$, $K_A^* = 4 \times 10^{-10}$, $K_A^{**} = 1 \times 10^{-10}$, $K_1 = 4 \times 10^{-10}$, $K_2 = 1 \times 10^{-10}$; agonist 2, $K_A = 1 \times 10^{-4}$, $K_A^* = 1 \times 10^{-6}$, $K_A^{**} = 1 \times 10^{-8}$, $K_1 = 1 \times 10^{-6}$, $K_2 = 1 \times 10^{-8}$; agonist 3, $K_A = 1 \times 10^{-3}$, $K_A^* = 1 \times 10^{-7}$, $K_A^{**} = 1 \times 10^{-7}$, $K_1 = 1 \times 10^{-7}$, $K_2 = 1 \times 10^{-7}$; and agonist 4, $K_A = 1 \times 10^{-5}$, $K_A^* = 8 \times 10^{-11}$, $K_A^{**} = 4 \times 10^{-10}$, $K_1 = 8 \times 10^{-11}$, $K_2 = 4 \times 10^{-10}$. Circles, agonist 1; squares, agonist 2; triangles, agonist 3; and inverted triangles, agonist 4.

[R] < [G]

Under this condition, the receptor concentration is limiting, meaning that G proteins essentially compete for the same receptor pool. Therefore, it is intuitively predictable that the generation of active receptor:G protein complexes in each pathway is mutually depleting so that formation of R^*G_1 and AR^*G_1 is at the expense of $R^{**}G_2$ and $AR^{**}G_2$ and vice versa. Thus, the receptor distribution equilibria in the two pathways are effectively linked. This situation is characteristic of intact behavior.

These predictions are borne out by the model simulations. Figure 4a illustrates the concentration–effect curves for the same four agonists acting through G_1- and G_2-linked pathways. Evidently, the agonists have identical potency orders in the two pathways. Also, agonists that exhibit high efficacy in one pathway have low efficacy in the other.

[R] > [G]

Under this condition, the receptor concentration is in excess so that G proteins have access to an unlimited receptor pool. Therefore, it is predicted that receptors can be distributed freely into G_1- and G_2-linked pathways with no associated mutual depletion. Formation of R^*G_1 and AR^*G_1 occurs independently of $R^{**}G_2$ and $AR^{**}G_2$ and vice versa. Thus, the two pathways operate as if they are separated from one another, a situation that characterizes isolated behavior.

Again, these predictions are borne out by the model simulations. Figure 4b shows concentration–effect curves for the same set of agonists as those illustrated in Fig. 4a. Now, the agonists demonstrate clearly different potency orders in G_1- and G_2-linked pathways.

Constitutive Activation and Inverse Agonism. The simple three-state model predicts that inverse agonism may be a signaling pathway-dependent and a system-dependent phenomenon. This section uses the extended model to demonstrate the conditions under which such behavior may be observed.

Figure 5 illustrates the predicted concentration curves for four ligands acting through the two pathways in a constitutively activated system: [R] < [G] and [R] > [G].

Pathway-Dependent Inverse Agonism. Under the condition [R] < [G], as already explained, the model operates in intact mode, meaning that events in one pathway influence those in the other. Ligands that act to distribute receptors into one pathway do so at the expense of the other. In a constitutively activated system, depletion of receptors in a pathway is expected to manifest as inverse agonism. Therefore, it is predicted that a ligand that expresses high efficacy in one pathway will act as an inverse agonist through the other. Figure 5a illustrates this prediction, showing that a ligand with high efficacy in the G_1-linked pathway operates as an inverse agonist in the G_2-linked pathway (inverted triangles), and the converse case (squares). Ligands with weaker efficacy in either pathway are able to operate in the same "direction" through both pathways, i.e., as agonists in both or as inverse agonists in both.

Under the condition [R] > [G], again as already explained, the model operates in isolated mode. In this case, there is no mutual influence between the pathways, the two pathways individually operate as two-state systems, and agonism and inverse agonism are purely ligand-dependent properties. Figure 5b illustrates this. Ligands act as agonists or inverse agonists simply according to their affinity values in each pathway.

System-Dependent Inverse Agonism. Figure 5 illustrates how different receptor:G protein stoichiometries may influence agonist activity. Thus, a ligand (squares) that acts as an inverse agonist in the G_1-linked pathway when [R] < [G] acts as an agonist in the same pathway when [R] > [G]. A similar example is

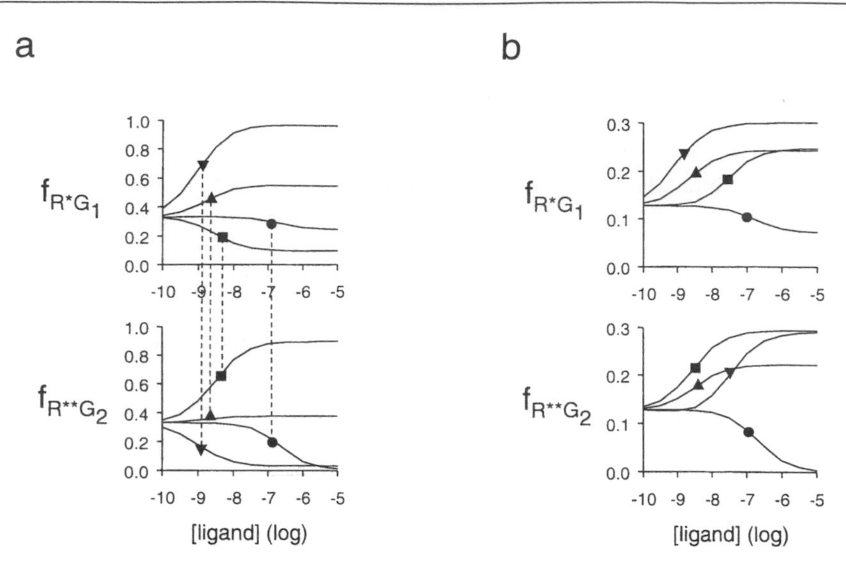

FIG. 5. Simulations using the extended three-state model in the presence of constitutive activity. (a) **[R] < [G]**; $[R_{tot}] = 1 \times 10^{-4}$; $[G_{1tot}] = 1 \times 10^{-3}$; $[G_{2tot}] = 1 \times 10^{-3}$; $L = M = 1 \times 10^{+5}$; $K_{G_1} = K_{G_2} = 1 \times 10^{-8}$. (b) **[R] > [G]**; $[R_{tot}] = 1 \times 10^{-4}$; $[G_{1tot}] = 3 \times 10^{-5}$; $[G_{2tot}] = 3 \times 10^{-5}$; $L = M = 1 \times 10^{+4}$; $K_{G_1} = K_{G_2} = 1 \times 10^{-8}$. In each case the top half shows the G_1-linked pathway, and the bottom half shows the G_2-linked pathway. Four agonists are shown with the following parameter values: agonist 1, $K_A = 1 \times 10^{-7}$, $K_A^* = 3 \times 10^{-7}$, $K_A^{**} = 3 \times 10^{-5}$, $K_1 = 3 \times 10^{-7}$, $K_2 = 3 \times 10^{-5}$; agonist 2, $K_A = 1 \times 10^{-7}$, $K_A^* = 1 \times 10^{-8}$, $K_A^{**} = 1 \times 10^{-9}$, $K_1 = 1 \times 10^{-8}$, $K_2 = 1 \times 10^{-9}$; agonist 3, $K_A = 8 \times 10^{-9}$, $K_A^* = 1 \times 10^{-9}$, $K_A^{**} = 1.5 \times 10^{-7}$, $K_1 = 1 \times 10^{-9}$, $K_2 = 1.5 \times 10^{-9}$; and agonist 4, $K_A = 1 \times 10^{-6}$, $K_A^* = 3 \times 10^{-10}$, $K_A^{**} = 1 \times 10^{-8}$, $K_1 = 3 \times 10^{-10}$, $K_2 = 1 \times 10^{-8}$. Circles, agonist 1; squares, agonist 2; triangles, agonist 3; and inverted triangles, agonist 4.

shown in the G_2-linked pathway (inverted triangles). Receptor:G protein stoichiometry may be considered to be a system-dependent quantity so inverse agonism is predicted by the model to be a system-dependent phenomenon.

This prediction of the model is further demonstrated in Fig. 6, which shows the dependence of ligand activity on G protein concentration. Concentration–effect curves are plotted for the effects in the G_1-linked pathway of a single ligand characterized by affinity constants $K_A^{**} < K_A^* < K_A$. All constants are fixed in the model except $[G_2]$. When $[G_2]$ is relatively low, there is little influence of the G_2-linked pathway on events in the G_1-linked pathway and so the ligand exhibits its agonist effects according to its higher affinity for R* than for R. As $[G_2]$ is increased, the influence of the G_2-linked pathway increases and the intrinsically higher affinity of the ligand for R** over R* now takes effect. Ultimately, this results in a net distribution of receptors away from the R* and toward R** forms of the receptor, manifested as inverse agonism in the G_1-linked pathway.

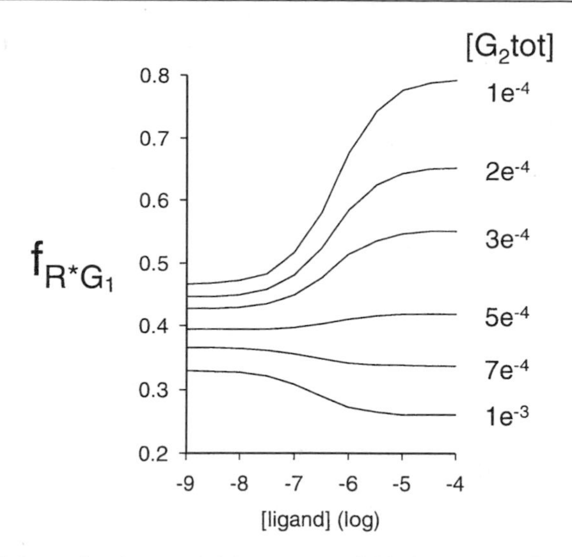

FIG. 6. Simulations using the extended three-state model in the presence of constitutive activity–system-dependent inverse agonism. $[R_{tot}] = 1 \times 10^{-4}$; $[G_{1tot}] = 1 \times 10^{-3}$; $[G_{2tot}] = 1 \times 10^{-3}$, 7×10^{-4}, 5×10^{-4}, 3×10^{-4}, 2×10^{-4}, 1×10^{-4}; $L = M = 1 \times 10^{+7}$; $K_{G_1} = K_{G_2} = 1 \times 10^{-10}$; $K_A = 1 \times 10^{-2}$, $K_A^* = 3 \times 10^{-7}$, $K_A^{**} = 1 \times 10^{-7}$, $K_1 = 3 \times 10^{-7}$, $K_2 = 1 \times 10^{-7}$.

Summary: Implications of the Three-State Model

The foregoing analysis provides a mechanistic basis for explaining variable agonist pharmacology in so-called promiscuous receptor systems. Thus, it explains how a single receptor type, when coupled to more than one signaling pathway, can exhibit different potency orders, different efficacy orders, and variable propensity to inverse agonism. It predicts that such variable pharmacological behavior can arise from the use of different response measurements and under conditions of variable receptor:G protein stoichiometry. It is important to emphasize that such pharmacological observations would, according to earlier theoretical models, be attributed to differences in receptor types.

In practice, it will be obvious when different response measurements are the source of altered pharmacology, but in the case of variable receptor:G protein stoichiometry, the situation may not be so clear. Such differences may arise in a number of ways: the use of different recombinant receptor expression systems; the use of different primary cell lines; the use of G protein toxins; and receptor/G protein compartmentalization. Thus, receptor:G protein stoichiometry may not be under the researcher's control and so an awareness of its theoretical impact is clearly important in the pharmacological analysis of agonist activity in such experimental systems.

As explained at the outset, our theoretical studies were stimulated primarily by experimental data, and it is appropriate to close this article by interpreting

TABLE I
PROMISCUOUSLY COUPLED RECEPTOR SYSTEMS

Receptor	Responses measured	Experimental protocol	Results	Theoretical interpretation
PACAP-1 [1]	Stimulation of adenylyl cyclase (G_s) and phospholipase C	Two different responses measured in the same receptor expression system	Different agonist potency order	Isolated behavior [R] > [G] in one or both pathways
Octopamine [2]	Stimulation of adenylyl cyclase (G_s) and $[Ca^{2+}]_i$			
α_{1B} [3]	Stimulation of phospholipase C and phospholipase A_2			
α_2C4 [4]	Stimulation (G_s) and inhibition (G_i) of adenylyl cyclase	Toxin treatment used to isolate the two responses, measured in the same receptor expression system	Different agonist potency order	Isolated behavior [G] = 0 in converse pathway
EP3D [5]	Stimulation (G_s) and inhibition (G_i) of adenylyl cyclase			
m1 [6]	Stimulation of phospholipase C	Same response measured in two different receptor expression systems	Different agonist potency order	Isolated behavior [R]:[G] ratio differs between the two expression systems
5-HT_{2C} [7]	Stimulation of phospholipase C and phospholipase A_2	Two different responses measured simultaneously in the same receptor expression system	Same agonist potency order different agonist efficacy order	Intact behavior [R] < [G] in both pathways

those observations in the context of the model predictions. Table I summarizes relevant literature data on promiscuously coupled receptor systems, and in each case, based on considerations of pathway and stoichiometry, we offer an explanation for the results.

[3] Use of Retinal Analogues for the Study of Visual Pigment Function

By ROSALIE K. CROUCH, VLADIMIR KEFALOV, WOLFGANG GÄRTNER, and M. CARTER CORNWALL

Introduction

All the G-protein receptor proteins are integral membrane proteins, crossing the membrane seven times and having the majority of the protein composed of hydrophobic amino acids within the membrane. This has made the study of these proteins difficult by many standard biophysical techniques. The visual pigments are unique as they contain a chromophore, the 11-*cis* form of retinal or one of a few closely related retinals, which is attached to the protein via a protonated Schiff base linkage. The site of attachment is at a lysine in the center of the seventh helix. A tuning of the wavelength of maximal absorbance is accomplished by electrostatic interactions of the ligand with the protein. Isomerization of this chromophore from its 11-*cis* configuration to the all-*trans* configuration upon absorption of a photon of light activates this G-protein receptor. The chromophore is therefore acting as an inverse agonist, locking the protein in its inactive conformation. In the all-*trans* form, the chromophore becomes an agonist, moving the protein into its active conformation.

The virtue of this system for the study of G-protein receptor function is that the native chromophore can be removed and replaced by retinal derivatives that have been modified structurally. *In vitro,* exposure of these proteins to light results in the complete detachment of the retinal; subsequent washing the membranes with bovine serum albumin (1%) removes any residual retinal. This procedure generates the opsin apoprotein, from which pigment can be reformed on the addition of 11-*cis* retinal in the dark. The apoproteins have been found to accommodate various retinal isomers and analogues, forming pigments or complexes that can then be tested for functionality and physical properties. This procedure can be carried out on *in vitro* preparations,[1] with intact photoreceptors[2] or *in vivo* if the animal is

[1] K. Nakanishi and R. K. Crouch, *Israel J. Chem.* **35,** 253 (1996).

[2] D. W. Corson and R. K. Crouch, *Photochem. Photobiol.* **63,** 595 (1996).

0076-6879/02 $35.00

vitamin A deprived.[3] This article discusses results obtained on intact rods and cones with the aim of demonstrating that this is a powerful tool for the study of the function of these two G protein-coupled receptors.

A large number of retinals have been synthesized and tested with rod opsins from various species.[1] Little work to date has been conducted on cone opsins due to problems in obtaining reasonable quantities of the proteins. However, with the development of both improved expression systems and purification techniques, *in vitro* studies are now being initiated on the cone opsins as well. Constraints of the chromophore-binding site of rod opsin have been determined to be as follows: (1) all isomers (including di- and tri-*cis* isomers) except for 13-*cis* and all-*trans* can be accommodated; (2) the methyl at C-9 on the polyene chain is critical to the absorption properties and activity of the protein; and (3) the binding cavity has little tolerance for additional bulk in the region of the cyclohexyl ring; however, the ring itself is not essential for pigment formation if one methyl group is present, corresponding to the C-1 or C-6 methyl of the native chromophore. The experiments discussed in this article demonstrate that the chromophore of these G-protein receptors can be used for studies of both the structure and the function of the protein (Fig. 1).

The application of retinal derivatives has been critical to the understanding that retinal has a most precise role in the control of the activity of this G-protein receptor. The retinal is linked to the protein by a protonated Schiff base and interacts with it so as to absorb light at a wavelength specific to the particular opsin. A photon of light isomerizes the critical 11-*cis* double bond to the all-*trans* conformation, bringing the receptor from its inactive state to the active state. 11-*cis* Retinal therefore acts as an inverse agonist, locking the receptor in an inactive conformation, which of course is critical in keeping the receptor "quiet" in the dark. The only action of the light is to initiate this specific isomerization in the retinal.[4] Movement from the 11-*cis* to the all-*trans* conformation is proposed to break the glutamate-113/lysine-296 Schiff base salt bridge and to change the conformation of the receptor at the cytoplasmic surface to allow maximal interaction with the α subunit of transducin, the rod G protein.[5] In the all-*trans* conformation, the retinal is therefore acting as an agonist, activating the receptor. At least two forms of the rod rhodopsin containing the all-*trans* conformation are proposed to have activity: (1) the metarhodopsin II (or R*) form, which is most effective in activating the transduction process (for review see Ref. 6), and (2) a second less potent form of

[3] R. Crouch, B. R. Nodes, J. I. Perlman, D. R. Pepperberg, H. Akita, and K. Nakanishi, *Invest. Ophthalmol. Vis. Sci.* **25,** 419 (1984).

[4] G. Wald, *Nature* **219,** 800 (1956).

[5] T. P. Sakmar, *Progr. Nucl. Acid Res. Mol. Biol.* **59,** 1 (1998).

[6] P. A. Hargrave, H. E. Hamm, and K. P. Hofmann, *Bioessays* **15,** 43 (1993).

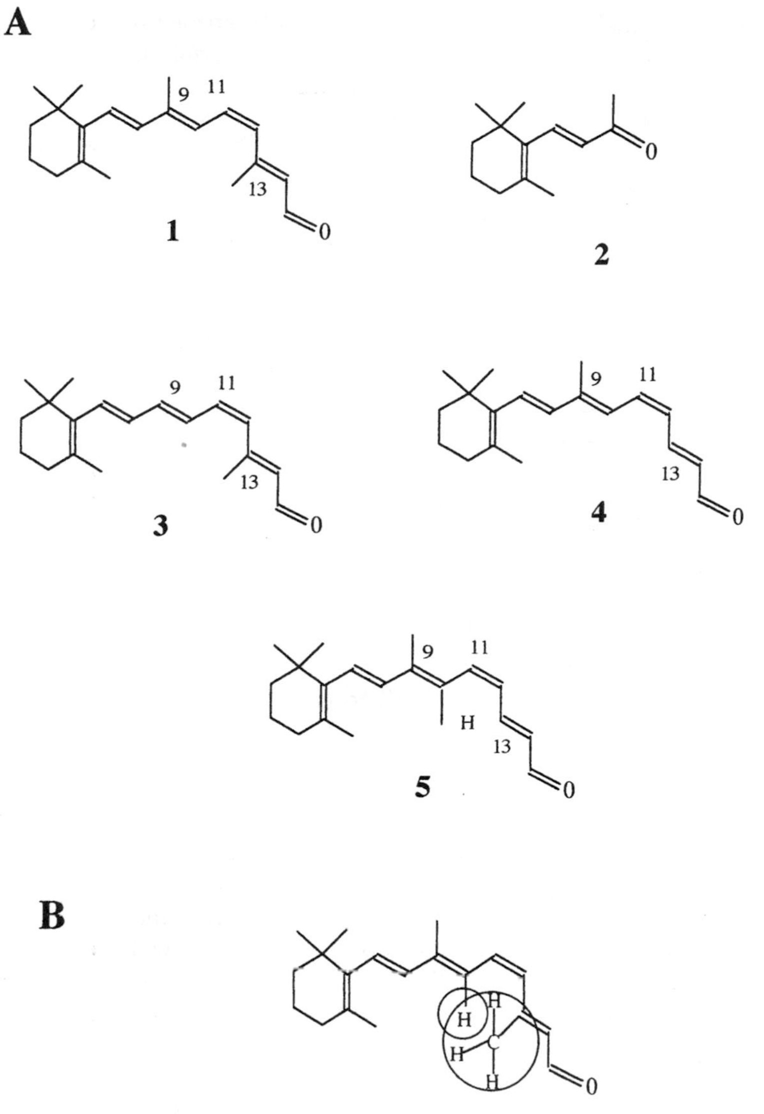

A

1

2

3

4

5

B

FIG. 1. Structures of 11-*cis* retinal and related analogues. (A) 1, 11-*cis* retinal; 2, β-ionone; 3, 11-*cis* 9 desmethylretinal; 4, 11-*cis* 13-desmethylretinal; and 5, 11-*cis* 10-methyl-13-desmethylretinal. (B). Schematic demonstrating the proposed interaction of the 10-hydrogen with the 13-methyl group of 11-*cis* retinal.

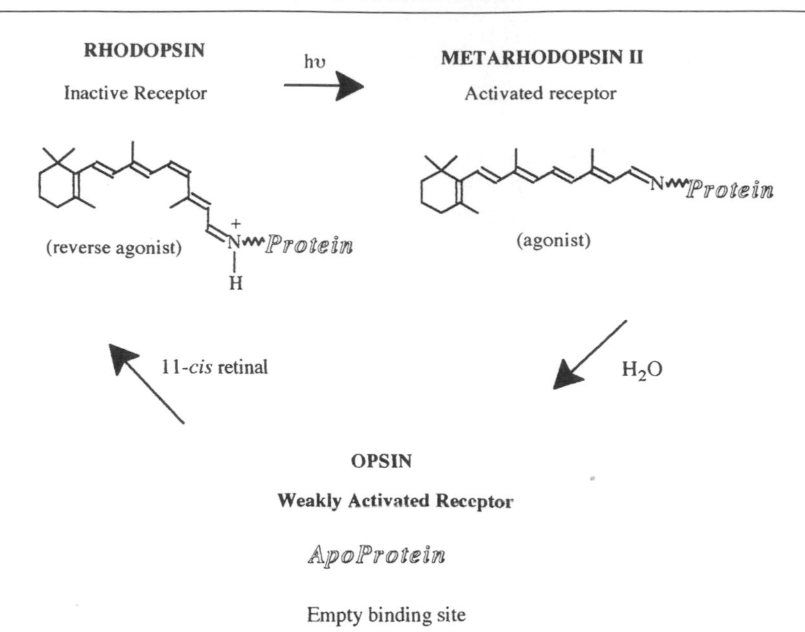

FIG. 2. Schematic of the role of retinal in the activation/deactivation of opsins.

metarhodopsin II, proposed to be phosphorylated[7] and bound to arrestin, having an activity that is about 10^{-5} that of R*.[8] The apoprotein opsin with no ligand in the binding site has also some basal activity, estimated at 10^{-7} that of R*.[9,10] While this is decidedly a low level, it is quite meaningful on the physiological scale, as has been demonstrated using both isolated rods and cones. A number of laboratories have also determined that combination of the apoprotein opsin with all-*trans* retinal forms a complex that has activity resembling that of the photolyzed rhodopsin as measured by phosphodiesterase,[11] phosphorylation by rhodopsin kinase,[12] and arrestin binding.[13] The role, if any, of this complex in the physiological process is as yet unclear (Fig. 2).

4-Hydroxyretinal was the first analogue to be incorporated into isolated photoreceptors in the manner discussed here and demonstrated the applicability of this

[7] S. Jäger, K. Palczewski, and K. P. Hofmann, *Biochemistry* **35,** 2901 (1996).
[8] C. S. Leibrock, T. Reuter, and T. D. Lamb, *Eye* **12,** 511 (1998).
[9] M. C. Cornwall and G. L. Fain, *J. Physiol.* **480,** 261 (1994).
[10] T. J. Melia, Jr., C. W. Cowan, J. K. Angleson, and T. G. Wensel, *Biophys. J.* **73,** 3182 (1997).
[11] Y. Fukada and T. Yoshizawa, *Biochim. Biophys. Acta* **675,** 195 (1981).
[12] J. Buczylko, J. C. Saari, R. K. Crouch, and K. Palczewski, *J. Biol. Chem.* **271,** 20621 (1996).
[13] K. P. Hofmann, A. Pulvermüller, J. Buczylko, P. Van Hooser, and K. Palczewski, *J. Biol. Chem.* **267,** 15701 (1992).

technique.[14] Bleached rods treated with this analogue regenerated a visual pigment, promoted recovery of sensitivity, and induced photon-like noise in recordings in darkness. This latter finding has yet to be explained. Also, various retinals incorporating cyclic rings in the side chains, which lock the polyene chain into a fixed *cis*-like conformation,[15,16] have been incorporated into bleached photoreceptors. These compounds appear to combine with bleached opsin to form a visual pigment, but because the 11–12 bond is constrained in a ring structure, isomerization is not possible. Consequently, recovery of sensitivity is only partial because the newly formed pigment cannot activate the transduction cascade. As the ability of these retinoids to occupy the binding site without perturbing the protein is as yet unclear, these compounds will not be discussed here. Finally, Makino *et al.*[17] have used a series of dihydro retinals to examine the spectral properties of the red rod and blue and red cones. They have concluded that the chromophore ring is important for governing the spectra of the two cones but not the red rod.

This article concentrates on analogues in which methyl groups have been deleted from the polyene side chain and on analogues in which the polyene side chain has been truncated (see Fig. 1 for structures). These analogues have proved to be particularly useful for the control of various aspects of rhodopsin function, as will be discussed.

Animal Model

We have selected the salamander as our experimental model for several reasons. There is an extensive literature describing the physiology and pharmacology of salamander retina, and it has been the animal of choice for many (and perhaps even the majority) of the physiological studies of isolated photoreceptors (for a recent reviews, see Refs. 2, 18, 19). Both rod and cone cell types are large and abundant in the retina, and the photoreceptor cells can be dissociated easily and maintained in primary culture for up to 2 weeks.[20,21] In addition, stable extracellular recordings of membrane current can be made over this entire period to assess changes in sensitivity under different conditions of adaptation. Finally as will

[14] D. W. Corson, M. C. Cornwall, E. F. MacNichol, V. Mani, and R. K. Crouch, *Biophys. J.* **57,** 109 (1990).

[15] D. W. Corson, M. C. Cornwall, E. F. MacNichol, J. Jin, R. Johnson, F. Derguini, R. K. Crouch, and K. Nakanishi, *Proc. Natl. Acad. Sci. U.S.A.* **87,** 6823 (1990).

[16] C. L. Makino, T. W. Kraft, R. A. Mathies, J. Lugtenburg, M. E. Miley, R. van der Steen, and D. A. Baylor, *J. Physiol.* **424,** 545 (1990).

[17] C. L. Makino, M. Groesbeek, J. Lugtenburg, and D. A. Baylor, *Biophys. J.* **77,** 1024 (1999).

[18] G. L. Fain, H. R. Matthews, and M. C. Cornwall, *Trends Neurosci.* **19,** 502 (1996).

[19] E. N. Pugh, Jr., S. Nikonov, and T. D. Lamb, *Cur. Opin. Neurobiol.* **9,** 410 (1999).

[20] D. W. Corson, S. Gunasinghe, and T. W. Fleury, *Invest. Ophthal. Vis. Sci.* **37,** 5239 (1996).

[21] C. R. Bader, P. R. MacLeish, and E. A. Schwartz, *Proc. Natl. Acad. Sci. U.S.A.* **75,** 3507 (1978).

be described here, methods have been worked out that allow replacement of the native chromophore with various retinoid analogues that have been designed to probe specific molecular interactions between opsin and the chromophore.

The salamander retina contains five different photoreceptors: red rod (61.1%), green rod (1.3%), red cone (31.7%), blue cone (3.2%), and UV cone (2.6%).[22] Each of these can be identified by microspectroscopy and physiological recordings.[23,24] In addition, there have been reports of the cloning and sequencing of the salamander red rod[25] and two of the cones.[26] α Subunits of the rod and cone transducins have also been cloned.[27] The correlation of the physiology with the biochemistry makes the salamander a particularly interesting species to examine. In the experiments discussed here, only the red rod and the red cone have been used.

Electrophysiological Methods

Membrane current responses are recorded with suction microelectrodes from rod and cone photoreceptors in darkness before and following bleaching light exposure, and following treatment with solutions containing 11-*cis* retinal or other retinoid analogues. These methods allow one to examine the visual cycle under controlled conditions and to compare the physiological properties of dark-adapted photoreceptors to those where the chromophore has been removed (bleached or replaced with an analogue). A recent review details the methods we have used in these studies.[28] These techniques are outlined next.

Salamanders are dark adapted for 12 hr before use. Following decapitation, the head and body are pithed, and the eyes are removed from the animal and hemisected in physiological saline solution. The retina is isolated from the retinal pigment epithelium and torn into small pieces with fine forceps. The resulting suspension containing fragments of the retina, as well as intact cells, is transferred to a recording chamber that has been placed on the stage of an inverted microscope. The optical system of the microscope is fitted with an infrared television camera and video monitor to allow visualization of the cells using infrared light. An individual intact rod or cone photoreceptor is drawn,

[22] D. M. Sherry, D. D. Bui, and W. J. Degrip, *Vis. Neurosci.* **15,** 1175 (1998).

[23] R. J. Perry and P. A. McNaughton, *J. Physiol.* **433,** 561 (1991).

[24] F. Harosi, *J. Gen. Physiol.* **66,** 357 (1975).

[25] N. Chen, J. X. Ma, D. W. Corson, E. S. Hazard, and R. K. Crouch, *Invest. Ophthalmol. Vis. Sci.* **37,** 1907 (1996).

[26] L. Xu, E. S. Hazard III, D. K. Lockman, R. K. Crouch, and J. Ma, *Mol. Vis.* **4,** 10 (1998).

[27] J. C. Ryan, S. Znoiko, L. Xu, R. K. Crouch, and J.-X. Ma, *Vis. Neurosci.* **17,** 847 (2000).

[28] M. C. Cornwall, G. J. Jones, V. J. Kefalov, G. L. Fain, and H. R. Matthews, *Methods Enzymol.* **316,** 224 (2000).

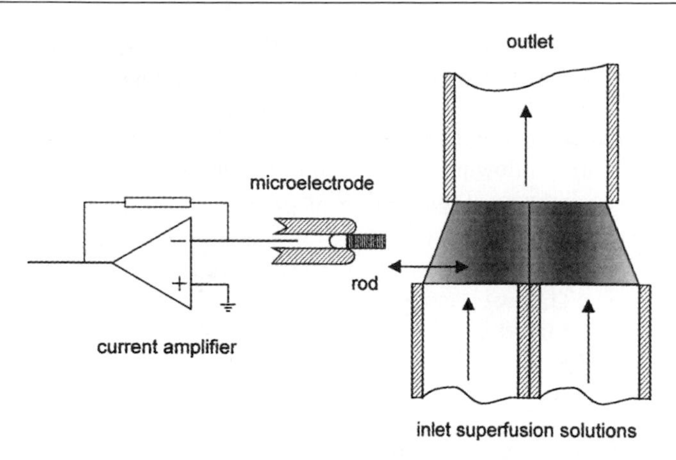

FIG. 3. Diagrammatic representation of the apparatus for physiological measurements. A glass micropipette filled with physiological medium (center) is shown with the inner segment of a rod located in the orifice. A silver/silver chloride wire inserted into the lumen of the pipette is connected to the current amplifier (left). A rapid superfusion system (right) consists of two inlet pipettes and a single outlet pipette through which flow is regulated by gravity. A computer-generated pulse actuates a stepping motor that produces lateral movement of the pipettes so as to bring the exposed portion of the cell into contact with solution flowing between the opposed orifices. See text for further details.

inner segment first, into a tight-fitting glass micropipette filled with physiological solution to prepare for recording. In this condition, the outer segment of the cell is free for optical stimulation, as well as treatment with drugs and retinal analogues, while at the same time, physiological measurements are made by sampling extracellular membrane current from the inner segment. The fluid inside the pipette is connected to the head stage of a patch clamp amplifier. The current recorded from the cell is converted to voltage, amplified, low pass filtered, and finally stored on a computer for further analysis. Figure 3 shows a rod cell held in the glass recording pipette and connected to the current amplifier. Also shown in Fig. 3 are superfusion pipettes arranged in such a way as to allow rapid exposure of the cell to solutions containing either drugs or retinoid analogues. Two inlet pipettes (bottom), about 100 μm in diameter, are fused together at their tips and held so that their openings are directly opposite an outlet pipette (top). As illustrated, the laminar gravity-driven solution flow occurs from inlet to outlet within the bulk medium in the recording chamber. Both inlet and outlet pipettes are held on a micrometer-driven stage that is connected to a stepping motor. Computer-generated pulses trigger the stepping motor, causing lateral movement of the superfusion pipettes so as to expose the cell to one or the other of the solutions between the pipettes for a specified time.

Test flashes (20 msec) assess response kinetics as well as sensitivity in darkness before and following bleaching background light, and then following retinoid incorporation. An example of such an experiment is illustrated in Fig. 4. Families of rod responses elicited by test flashes of different intensities in darkness prior to (Fig. 4A), 35 min following a bright light that bleached 20% of the rhodopsin (Fig. 4B), and in darkness 30 min following treatment with a solution containing

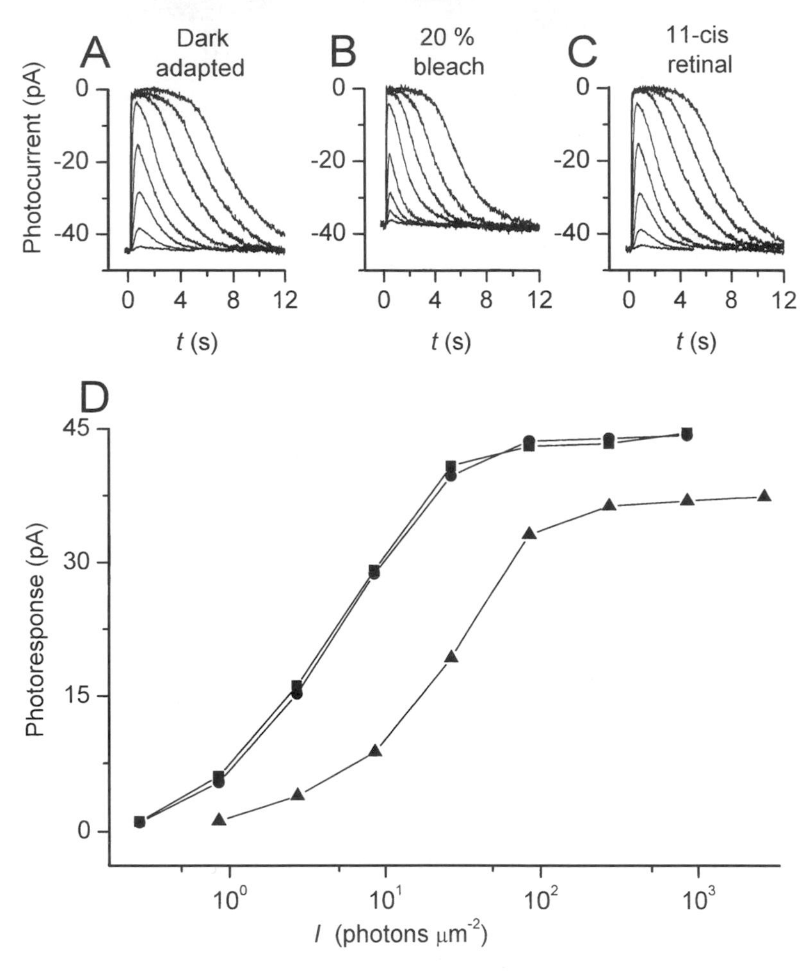

FIG. 4. Effect of bleaching and regeneration with 11-*cis* retinal on the dark current and sensitivity of a rod. Shown at the top are superimposed responses of the cell to 20-msec flashes in its dark-adapted state (A), 35 min following a 20% bleach (B), and 30 min after exposure to a vesicle solution containing 100 μM 11-*cis* retinal (C). The corresponding intensity–response relations are shown (D): dark adapted (■), bleach adapted (▲), and following treatment with 11-*cis* retinal (●). See text for additional details.

lipid vesicles loaded with 11-*cis* retinal (Fig. 4C) are shown. The graphs in Fig. 4D show intensity–response relations plotted from these data. It is apparent from the graphs that bleaching produces a significant reduction of response amplitude and sensitivity that can be fully reversed following treatment with 11-*cis* retinal.

The fractional pigment bleaching is calculated according to the relation (1):

$$F = 1 - \exp(-IPt), \tag{1}$$

where F is the fraction of bleached pigment, I is the light intensity in photons μm^{-2} sec^{-1}, and t is the duration of the light exposure in seconds. The value of the photosensitivity, P, used is 6.2×10^{-9} μm^2 for rods and 6.0×10^{-9} μm^2 for cones.[29]

β-Ionone

A fundamental difference has been observed between rods and cones in the interaction of opsin with retinal analogues such as β-ionone that contain a truncated polyene chain (see Fig. 1A, 2). Biochemical measurements of the rates of pigment regeneration in rod outer segment preparation show that β-ionone acts as a competitive inhibitor with 11-*cis* retinal for the binding site.[30,31] Physiological studies on isolated bleached salamander cones demonstrate that this analogue acts as an inverse agonist, resensitizing the receptor and reversing the bleach-induced increase in guanylyl cyclase activity.[32] Examples of this are shown in Figs. 5 and 6 (modified from Ref. 32). Figure 5 shows sets of photoresponses elicited from a cone in its dark-adapted state (Fig. 5A), following a 90% bleach (Fig. 5B), and then in the presence of β-ionone (Fig. 5C). Clearly, β-ionone reversed partially the effect of the bleach on the response amplitude. The corresponding intensity–response relations plotted in Fig. 5D demonstrate that when it was treated with β-ionone, the cone regained some of the sensitivity lost as a result of the bleach. The residual 10-fold difference in the sensitivity of the dark-adapted and the bleach-adapted β-ionone-treated cell is due to the reduction of the pigment by 90% produced by the bleach.

The effect of the noncovalent binding of β-ionone to cone opsin can also be seen from the change in the kinetics of the dimflash response, which is a function of the kinetics of the phototransduction cascade. Activation of transduction by bleach or by background light accelerates the dim flash response, which is reversed by dark adaptation.[33,34] Figure 6 illustrates that acceleration of the cone dim flash response caused by bleach can be completely reversed by β-ionone.

[29] G. J. Jones, *J. Physiol.* **487,** 441 (1995).

[30] H. Matsumoto and T. Yoshizawa, *Nature* **258,** 523 (1975).

[31] R. K. Crouch, C. D. Veronee, and M. E. Lacy, *Vis. Res.* **22,** 1451 (1982).

[32] J. Jin, R. K. Crouch, D. W. Corson, B. M. Katz, E. F. MacNichol, and M. C. Cornwall, *Neuron* **11,** 513 (1993).

[33] M. C. Cornwall, A. Fein, and E. F. MacNichol, Jr., *J. Gen. Physiol.* **96,** 345 (1990).

[34] H. R. Matthews, G. L. Fain, R. L. Murphy, and T. D. Lamb, *J. Physiol.* **420,** 447 (1990).

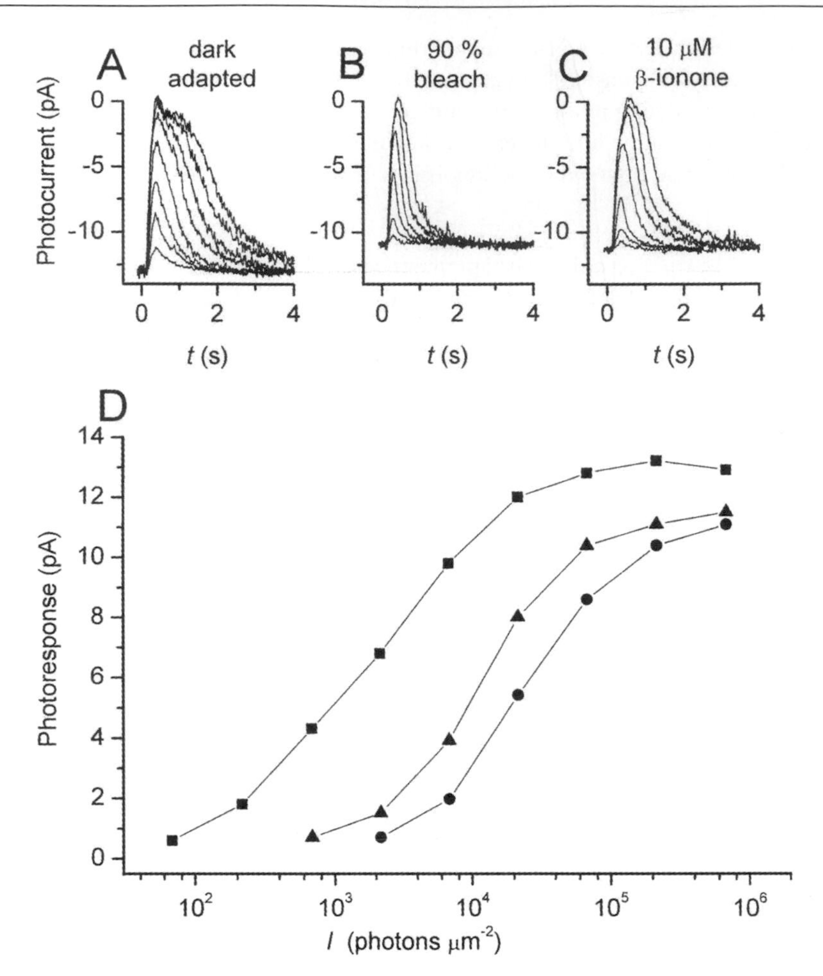

FIG. 5. Effect of β-ionone on the dark current and on the sensitivity of a cone. A series of superimposed flash responses to 20-msec flashes from one cell in its dark-adapted state (A), after a 90% bleach (B), and subsequently in the presence of 10 μM β-ionone (C) are shown. The corresponding intensity–response curves are shown (D): dark adapted (■), bleach adapted (●), and in β-ionone (▲). β-Ionone caused a reversible increase in the dark current and sensitivity of the bleach-adapted cone.

Interestingly, these results on cones are at variance with results from biochemical experiments performed on outer segment preparations that were composed mostly of rod cells. There it has been shown both by transducin activation[35] and

[35] F. Jäger, S. Jäger, O. Kräutle, N. Friedman, M. Sheves, K. P. Hofmann, and F. Siebert, *Biochemistry* **33,** 7389 (1994).

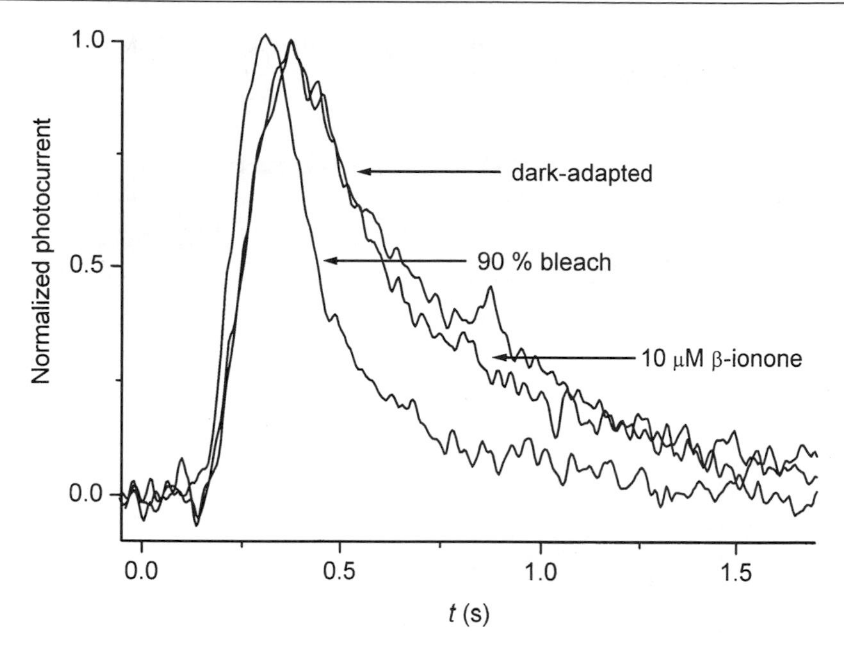

FIG. 6. Effect of β-ionone on the kinetics of dim flash responses in a cone. Normalized dim flash responses from one cone in a dark-adapted state, after a 90% bleach, and subsequently in the presence of 10 μM β-ionone. Treatment of the bleach-adapted cone with β-ionone reversibly slowed the time course of the dim flash responses. Reprinted with permission from Kefalov *et al.*[36]

by phosphorylation by rhodopsin kinase[12] that this same analogue acts as an agonist, activating the receptor. Physiological experiments on isolated bleached rods have resolved this apparent conflict by demonstrating that β-ionone does indeed act as an agonist in rods and that activation of the receptor is obtained.[36] Figure 7, which is modified from Fig. 2 of Kefalov and co-workers,[36] shows the effect on dark current and sensitivity in a salamander rod. Superimposed responses to test flashes are shown in darkness (Fig. 7A) following a light that had bleached 20% of the pigment (Fig. 7B), in the presence of 10 μm β-ionone (Fig. 7C), and following removal of the analogue from the superfusion bath (Fig. 7D). From comparison of data in Figs. 7B and 7C, it is apparent that treatment with β-ionone resulted in a reversible loss of dark current and sensitivity. β-Ionone also causes a reversible acceleration of the dim flash response in bleach-adapted rods (Fig. 8), which suggests that its binding to rod opsin leads to activation of the phototransduction cascade.

[36] V. J. Kefalov, M. C. Cornwall, and R. K. Crouch, *J. Gen. Physiol.* **113,** 491 (1999).

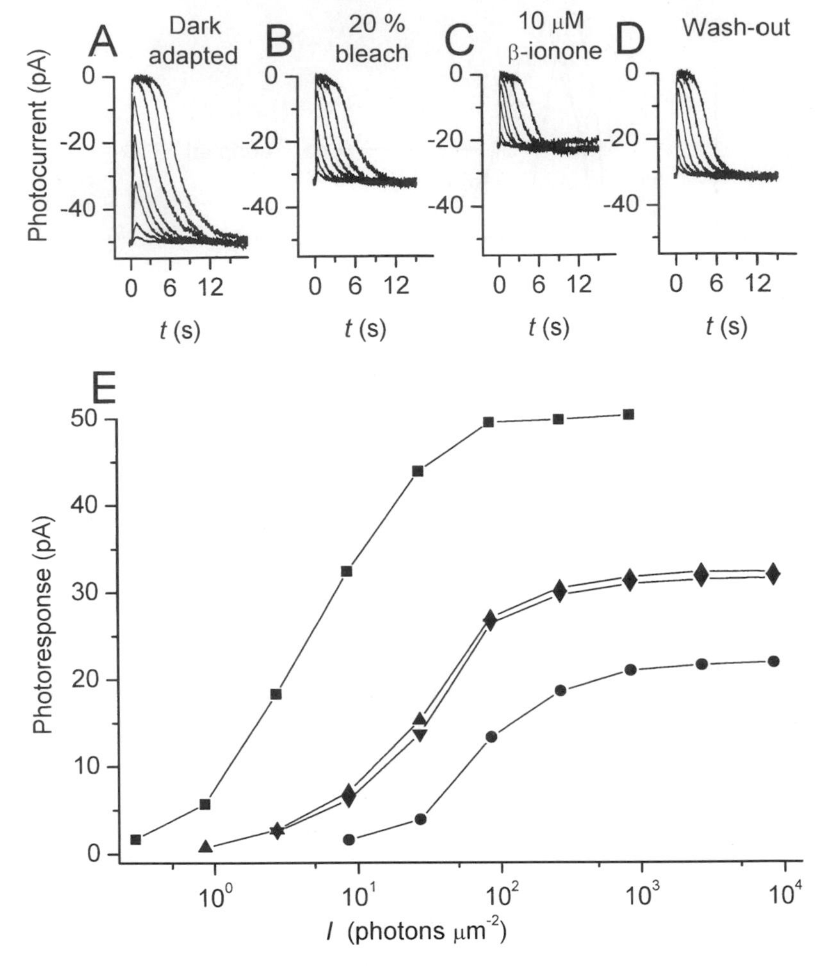

FIG. 7. Effect of β-ionone on the dark current and on the sensitivity of a rod. A series of flash responses from one cell in its dark-adapted state (A), after a 20% bleach (B), in 10 μM β-ionone (C), and after washing out the β-ionone (D) are shown. The corresponding intensity–response curves are shown (E): dark adapted (■), bleach adapted (▲), in β-ionone (●), and after washing out the β-ionone (▼). β-Ionone caused a reversible decrease in the dark current and in the sensitivity of the cell. Reprinted with permission from Kefalov *et al.*[36]

Kefalov *et al.*[36] have made physiological measurements of the rates of guanylyl cyclase and cGMP phosphodiesterase in both rods and cones under these experimental conditions. These studies demonstrate that treatment with β-ionone activates these enzymes in bleached rods whereas these same enzymes are inactivated in bleached cones. These results clearly demonstrate that the interactions

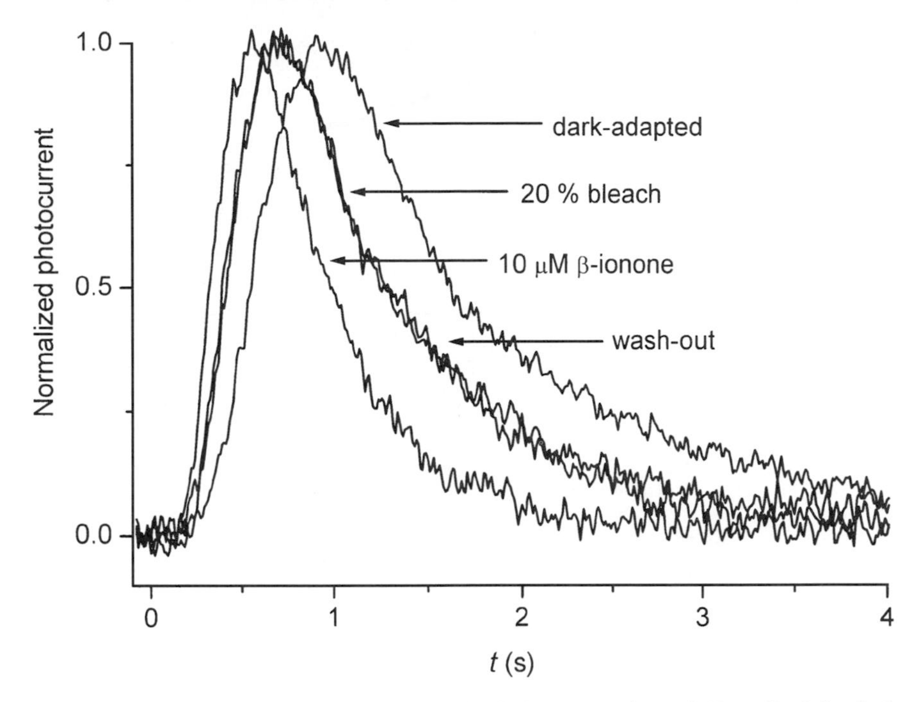

FIG. 8. Effect of β-ionone on the kinetics of dim flash responses in a rod. Normalized dim flash responses from one rod in the dark-adapted state, after a 20% bleach, in the presence of 10 μM β-ionone, and after washing out the β-ionone. β-Ionone caused a reversible acceleration of the dim flash response of the bleach-adapted rod. Reprinted with permission from Kefalov *et al.*[36]

of the rod and cone opsins with at least the β-ionone portion of the chromophore 11-*cis* retinal differ. As the expressed cone pigments become more available for biochemical studies, it will be critical to study these mechanisms in detail.

13-Desmethylretinal

The removal of the 13-methyl group from the polyene side chain has a remarkable effect on the properties of rhodopsin. In the 11-*cis* isomer of retinal, there is steric hindrance between the C-10 hydrogen and the C-13 methyl (see Fig. 1B), and the polyene side chain twists to accommodate this interaction. Removal of 13-methyl relieves this strain. The pigment formed between 11-*cis* 13-desmethylretinal (Fig. 1A, 4) *in vitro* was reported to activate phosphodiesterase in the absence of light.[37] In more recent studies, rhodopsin kinase was also shown to be activated in the absence of light with the activity decreasing over time.

[37] T. G. Ebrey, M. Tsuda, G. Sassenrath, J. L. West, and W. H. Waddell, *FEBS Lett.* **116,** 217 (1980).

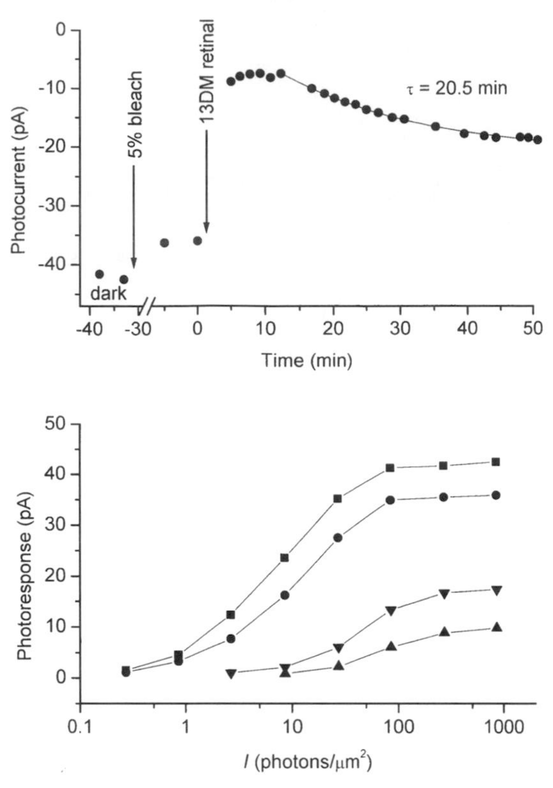

FIG. 9. Effect of 11-*cis* 13-desmethylretinal on dark current and sensitivity in a rod. (Bottom) Intensity–response curves: dark adapted (■), following a 5% bleach (●), immediately following exposure to 11-*cis* 13 desmethylretinal (▲), and 30 min following exposure to 11-*cis* 13-desmethylretinal (▼). See text for additional details. (Top) Time course of changes in a light-suppressible dark current in a cell that was first dark adapted, subsequently exposed to a bleaching light that bleached 5% of the visual pigment, and finally exposed to a solution containing 100 μM 11-*cis* 13-desmethylretinal.

Pigment formation between 11-*cis* 13-desmethylretinal and opsin is quite slow (one-ninth the rate of rhodopsin). This slow rate of analogue pigment formation has been used in physiological experiments to separate the effects of non-covalent and covalent interactions between 11-*cis* 13-desmethylretinal and opsin in intact photoreceptors.[38] In bleach-adapted rods, 11-*cis* 13-desmethylretinal caused an initial activation of the phototransduction cascade, similar to the effect of β-ionone. Figure 9 (bottom) shows that a significant decrease in the response amplitude and sensitivity occurred when the bleached rod was treated with 11-*cis* 13-desmethylretinal (compare circles and triangles). However, this effect was transient and was

[38] D. W. Corson, V. J. Kefalov, M. C. Cornwall, and R. K. Crouch, *J. Gen. Physiol.* **116,** 283 (2000).

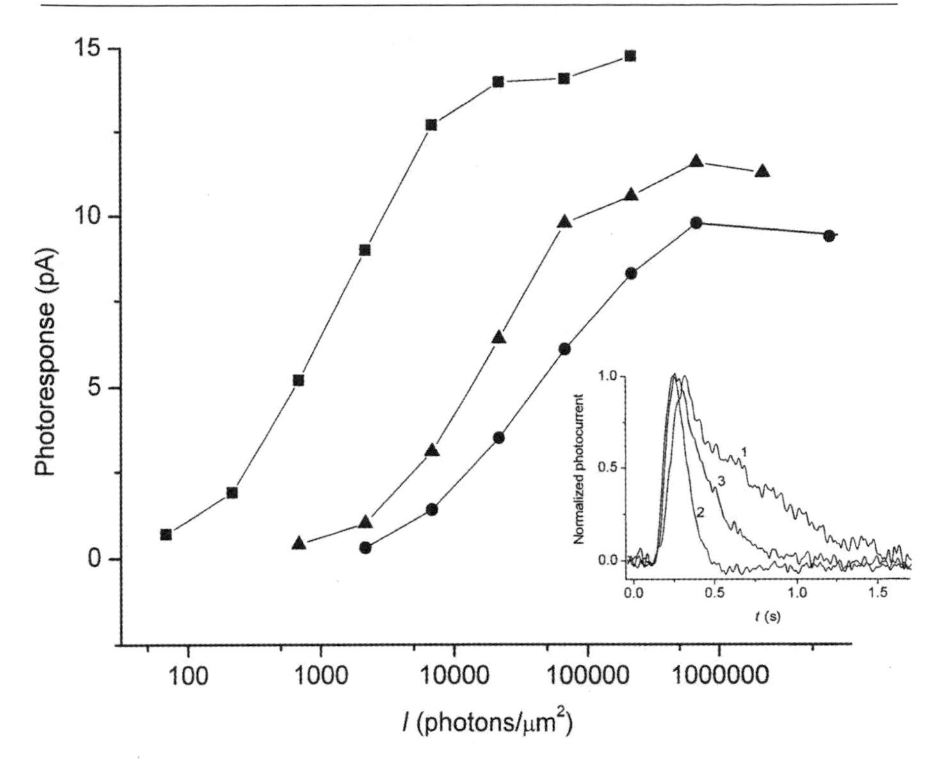

FIG. 10. Effect of 11-*cis* 13-desmethylretinal on sensitivity and dim flash response kinetics in a cone. Intensity–response curves from a cone recorded in its dark-adapted state (■), after bleaching 90% of the pigment (●), and subsequently in the presence of 100 μM 11-*cis* 13-desmethylretinal (▲). (Inset) The corresponding normalized dim flash responses from the cone: 1, dark-adapted state; 2, after a 90% bleach; and 3, in the presence of 11-*cis* 13-desmethylretinal. 11-*cis* 13-Desmethylretinal caused an increase in dark current and sensitivity as well as deceleration of the dim flash response.

partially reversed (compare upside down triangles and triangles). The time course of the effect can also be seen from the examination of changes in the dark current (Fig. 9B, top). As expected, bleaching 5% of the pigment produced a small steady decrease in the dark current. Treatment of the bleach-adapted rod with 11-*cis* 13-desmethylretinal caused an additional immediate large decline in the current, followed by its slow recovery along an exponential time course ($\tau = 20.5$ min).

As was the case with β-ionone, the effect of 11-*cis* 13-desmethylretinal in bleach-adapted cones is opposite to that in rods. Figure 10 shows that the large decrease in sensitivity and dark current, as well as the acceleration of flash responses (Fig. 10, inset) produced by the bleaching light, was partially reversed following treatment with 11-*cis* 13-desmethylretinal.

One interpretation of these results is that the initial complex formed between the 11-*cis* 13-desmethylretinal and the opsin places the protein into a conformation that is recognized by transducin and by rhodopsin kinase as a metarhodopsin II conformation. This results in activation of the transduction cascade and also in phosphorylation of the pigment. In biochemical experiments, if the pigment is allowed to fully regenerate (24 hr) before exposure to transducin and rhodopsin kinase, no activation is obtained until the pigment is photolyzed.[12] Therefore, either the 13-methyl group has a critical interaction with the protein that facilitates the Schiff base formation or the change in the conformation of the polyene chain as a result of the relief of the strain due to the interaction between the 10-hydrogen and the 10-methyl moves the polyene chain into a position that is no longer optimum for interaction with the critical lysine.

10-Methyl 13-desmethylretinal

To further investigate the effect of torsion in the polyene chain, we have undertaken studies on 10-methyl-13-desmethylretinal. In this compound, the 10-hydrogen is replaced by a methyl group and the 13-methyl group is replaced by a hydrogen (Fig. 1A, 5). It is then reasonable to expect that the polyene side chain has a conformation similar to native 11-*cis* retinal, as the interaction of the 10-hydrogen with the 13-methyl (Fig. 1B) has been replaced with an interaction of the 10-methyl with the 13-hydrogen. There has been some dispute in the literature on this subject.[39,40] Figure 11 illustrates that treatment of a bleached rod in darkness with 11-*cis* 10 methyl-13-desmethylretinal gave results similar to those obtained with 11-*cis* 13-desmethylretinal. These effects are a reduction in dark current and sensitivity of flash responses (Fig. 11), as well as an accelerated time course of dim flash responses (Fig. 11, inset) and an activation of guanylyl cyclase (data not shown). The results of these experiments have led us to the interesting and somewhat surprising conclusion that it is the interaction of the 13-methyl group with the protein that controls formation of the stable rhodopsin pigment rather than the twist in the side chain that results from the interaction of the methyl and hydrogen at the 10 and 13 positions.

11-*cis* Retinal

Experiments with β-ionone and 11-*cis* 13-desmethylretinal demonstrate that modulation of the activity of the transduction cascade in intact bleached photoreceptors occurs as a retinoid ligand enters the opsin-binding site. In rods, this is

[39] G. G. Kochendoerfer, P. J. Verdegem, I. van der Hoef, J. Lugtenburg, and R. A. Mathies, *Biochemistry* **35,** 16230 (1996).

[40] D. Koch and W. Gärtner, *Photochem. Photobiol.* **65,** 181 (1997).

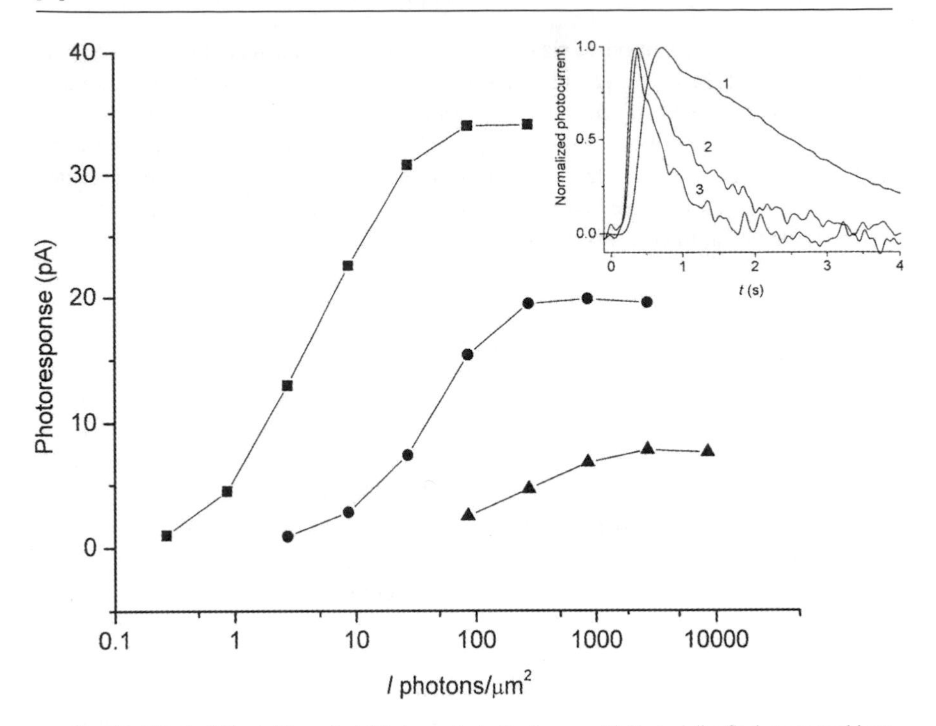

FIG. 11. Effect of 11-*cis* 10-methyl-13-desmethylretinal on sensitivity and dim flash response kinetics in a rod. Intensity–response curves from a rod recorded in its dark-adapted state (■), after bleaching 80% of the pigment (●), and subsequently in the presence of 100 μM 10-methyl-13-desmethylretinal (▲). (Inset) The corresponding normalized dim flash responses from the rod: 1, dark-adapted state; 2, after a 80% bleach; and 3, in the presence of 11-*cis* 10-methyl-13-desmethylretinal. 10-Methyl-13-desmethylretinal caused a substantial decrease in dark current and sensitivity as well as acceleration of the dim flash response.

seen as a activation of the transduction cascade; in cones, deactivation occurs. Our experiments allow us to conclude that these effects are exclusive of pigment formation. We then asked if these same effects might be observed when bleached rods and cones are treated with the native chromophore 11-*cis* retinal. This has been observed to be the case.[40a] In bleached rods, treatment with 11-*cis* retinal caused a transient decrease in the dark current and sensitivity, together with transient activation of guanylyl cyclase and phosphodiesterase. Following this initial activation, cell sensitivity increased, and all parameters returned to their prebleach values with a time constant of about 4 min, several times faster than the corresponding recovery with 11-*cis* 13-desmethylretinal. However, similar treatment of bleached cones caused an immediate increase in dark current and sensitivity

[40a] V. J. Kefalov, R. K. Crouch, and M. C. Cornwall, *Neuron* **29,** 749 (2001).

and a decrease in guanylyl cyclase activity without any transient increase. Our interpretation of these results is that the initial effects on the transduction cascade in both rods and cones are caused by the noncovalent binding of the retinoid in the chromophore pocket of the opsin. We propose that the initial deceleration of the transduction cascade that occurs in cones but not in rods, together with the faster pigment formation in these cells, contributes substantially to the much faster rate of dark adaptation in cones compared to rods.

9-Desmethylretinal

Interestingly, removal of the methyl group from position 9 on the polyene chain of the retinal (Figs. 1A, 3) has a striking effect on the function of salamander rods. This compound has not yet been tested on cones. The 9-methyl has an interaction with the protein, which is critical to the absorption properties, as demonstrated by the large shift (33 nm) to shorter wavelengths observed in 9-desmethyl pigments.[41] The pigment also shows a limited activation of transducin (about 8%) of the normal value[42] and a much reduced level (25%) of phosphorylation.[12,43] These data suggest that the receptor is not completely in the activated state and that the interaction of the 9-methyl group with the protein is quite critical. Somewhat surprisingly, early photointermediates, known as the batho and lumi states, appear normal.[42] However, the active metarhodopsin state is, by biochemical measurements, clearly not fully active and does not phosphorylate completely, suggesting that the deactivation of the pigment cannot proceed in a normal fashion. Groesbeek and colleagues[44] have explored the interaction of this position with the protein by using mutants of rhodopsin and have found that Gly-121 is implicated as the amino acid that is in contact with the 9-methyl group.

When 9-desmethylretinal is incorporated into isolated salamander rods, the quantal response is about 30 times smaller and the decay is about 5 times slower than for the native pigment.[45,46] The abnormally long lifetime of the photoactivated pigment is in agreement with the findings on phosphorylation.

An example of these findings is illustrated in Fig. 12 (reproduced from Corson *et al.*[46] with permission). Physiological responses to test flashes at three different wavelengths under three experimental conditions are shown. The left column

[41] A. Kropf, B. P. Whittenberger, S. P. Goff, and A. S. Waggoner, *Exp. Eye Res.* **17,** 591 (1973).

[42] U. M. Ganter, E. D. Schmid, D. Perez-Sala, R. R. Rando, and F. Siebert, *Biochemistry* **28,** 5954 (1989).

[43] D. F. Morrison, T. D. Ting, V. Vallury, Y. K. Ho, R. K. Crouch, D. W. Corson, N. J. Mangini, and D. R. Pepperberg, *J. Biol. Chem.* **270,** 6718 (1995).

[44] M. Han, M. Groesbeek, S. O. Smith, and T. P. Sakmar, *Biochemistry* **37,** 538 (1998).

[45] D. W. Corson, M. C. Cornwall, and D. R. Pepperberg, *Vis. Neurosci.* **11,** 91 (1994).

[46] D. W. Corson, M. C. Cornwall, E. F. MacNichol, S. Tsang, F. Derguini, R. K. Crouch, and K. Nakanishi, *Proc. Natl. Acad. Sci. U.S.A.* **91,** 6958 (1994).

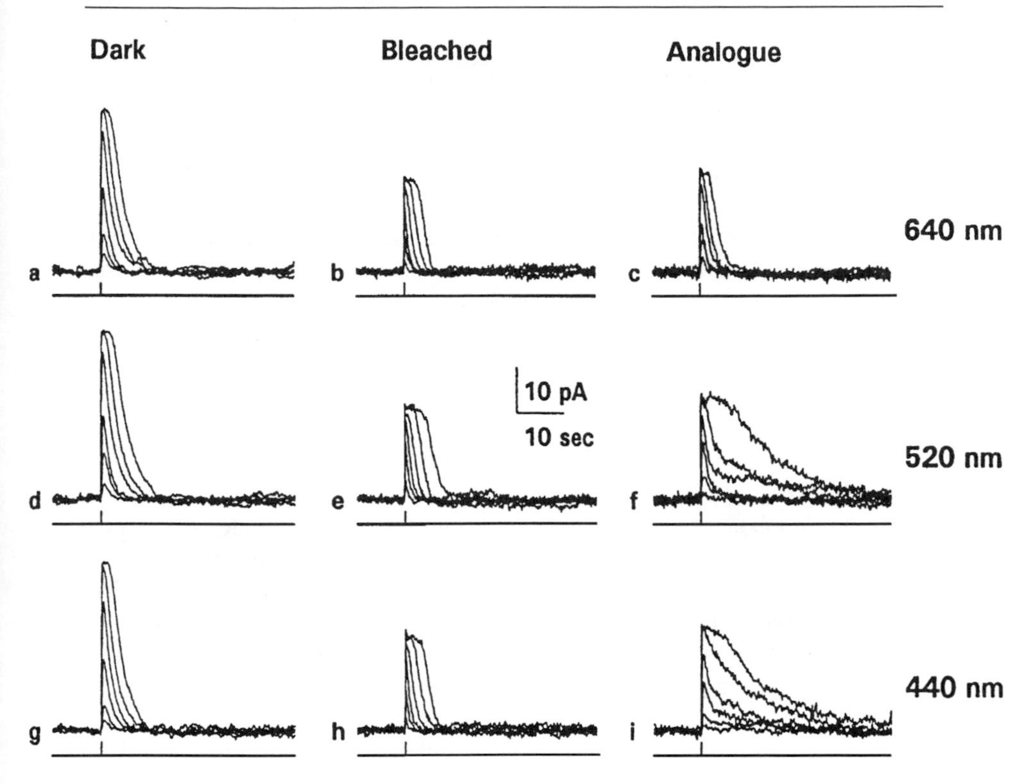

FIG. 12. Changes in response waveform induced by bleaching and by application of 11-*cis* 9-desmethylretinal. Series of test flashes (20 msec) ascending in brightness in 0.5 log unit steps were applied to a rod until the response reached saturation for three wavelengths (rows as indicated) in three experimental conditions: dark adapted (column 1), bleached (column 2), and resensitized with 9-desmethylretinal (column 3). Bleaching results in a reduction of maximum response amplitude and an acceleration of the response kinetics at all wavelengths. Treatment with 9-desmethylretinal resulted in partial resensitization of the rod and wavelength dependency of the response time course. Long wavelength stimulation elicited responses that were essentially normal, whereas short wavelength stimulation resulted in responses that were sluggish to recover. See text for additional details. Reprinted with permission from Corson *et al.*[46]

shows responses when the cell was in a dark-adapted condition. As expected, long or short wavelength stimulation elicits responses that have the same time course. The middle column of responses illustrates the reduction of amplitude and acceleration of the time course that occurs following bleaching more than 95% of the pigment. The column on the right shows responses elicited following treatment with 9-desmethylretinal and the subsequent formation of 9-desmethyl rhodopsin. In addition to a partial recovery of sensitivity, the response time course after treatment depended on the wavelength of stimulation. Short wavelength light (440 nm), expected to be absorbed by the regenerated analogue visual pigment

(λ_{max} = 470 nm), elicited responses that were abnormally long to decay. In contrast, long wavelength flashes (640 nm), expected to stimulate primarily the residual native pigment (λ_{max} = 520 nm), elicited responses that decayed normally. Experiments using this analogue have demonstrated that a calcium-dependent step involved in adaption in rods is associated with the photopigment quenching by phosphorylation.[47]

Taken together, biochemical and physiological data suggest that the 9-methyl group is critical to controlling the conformation of the cytoplasmic loops of the protein so that the required interaction with the various enzymes involved in the transduction pathway can occur. In addition, it is clear that this analogue is a potentially important tool for the study of the mechanisms that terminate light responses in normal photoreceptors.

Conclusions

The chromophore of the G protein-coupled receptor rhodopsin demonstrates a remarkable ability to control the activity of that receptor. When the pigment is in its regenerated state as rhodopsin, the chromophore is acting as an inverse agonist, locking the receptor in its inactive state. If the chromophore-binding pocket is vacant, the apoprotein shows a low level of activity, which is eliminated on the formation of pigment. The action of a single photon of light is to isomerize a double bond in the chromophore, converting the compound into an agonist. The agonist results in the receptor moving to its active conformation by inducing protein conformational changes and forming a binding site for the G protein. We have shown here that in the process of the chromophore entering the binding pocket, a series of conformations are produced that transiently activate rod receptors before the ligand locks the receptor into its inactive state. Cones, however, are inactivated immediately by noncovalent ligand binding. We have further demonstrated that removal of one methyl group from the chromophore reduces the efficiency with which the deactivation of the receptor can occur, providing another tool with which to examine the receptor cascade. This ability to control the activity of the receptor by the simple use of retinal analogues provides a useful tool for the study of the activation and deactivation of this G protein-coupled receptor.

[47] H. R. Matthews, M. C. Cornwall, and R. K. Crouch, *Invest. Ophthalmol. Vis. Res.* **41S,** 1692 (2000).

[4] Design and Synthesis of Peptide Antagonists and Inverse Agonists for G Protein-Coupled Receptors

By SCOTT M. COWELL, PREETI M. BALSE-SRINIVASAN, JUNG-MO AHN, and VICTOR J. HRUBY

Introduction

Knowledge of G protein-coupled receptors (GPCR), signal transduction, and information transfer depends on the symbiotic development of agonists and antagonists. Ligand design, of course, requires a receptor/acceptor to evaluate the agonist and antagonist being developed, but it must be emphasized that receptor antagonists have been developed long before specific receptors were identified. Moreover, there is only a limited understanding of either signal transduction or *ab initio* ligand design. Awareness of the relationships between the structure of a ligand and the structural requirements of a particular receptor must be exploited in the design of ligands. Thus, the design and synthesis of a ligand for a G protein-coupled receptor offer a unique opportunity to exercise pure logic in an ironic situation: only by design can we discover what to design next.[1] Finally, ligand design uses different areas of scientific endeavors, such as peptide and peptidomimetic chemistry, computational chemistry, biophysics, and pharmacology, and thus one must be fully aware of the powers and limitations of each of these areas of science.

Many ligands interact with the body through signal transduction. Signal transduction occurs across a cellular membrane via specialized proteins called G protein-coupled receptors. GPCRs are involved in a myriad of regulatory processes in the body, including pain response, heart rate, blood pressure, anxiety, and feeding. Whereas only specific ligands interact with a given receptor, all receptors share some common features in that they are composed of a long chain of amino acids spanning the membrane seven times. The transmembrane sections are packed together so that the entire structure resembles an oval looking down its axis perpendicular to the membrane.

When ligands interact with the extracellular portion of the receptor, a binding process occurs and there is a conformational change in the receptor, which, in the case of the agonist, causes the intracellular portion to interact with G proteins. G protein components then trigger various second messengers leading to signal transduction. Generally, there is a basal activity of the second messenger function that is mediated spontaneously by G protein interactions. A ligand whose binding

[1] V. J. Hruby, *in* "Progress in Brain Research" (J. Joosse, R. M. Buijs, and F. J. H. Tilders, eds.), p. 215. Elsevier, New York, 1992.

METHODS IN ENZYMOLOGY, VOL. 343
0076-6879/02 $35.00

causes a measurable increase in basal activity of the G proteins or second messengers is defined as an agonist. A ligand that interacts with the receptor in the same binding pocket as the agonist and causes no marked change in the activity of G proteins and its second messenger functions is defined as a competitive antagonist. If a ligand interacts with the receptor and causes a marked decrease in basal activity within the cell, it is known as an inverse agonist.

Discussion of inverse agonism is a fairly recent phenomenon, and there are relatively few well-documented examples in the literature, although undoubtedly many examples exist that were overlooked or defined simply as antagonists. However, they represent an exciting opportunity in the field as a new possibility of therapeutics that can directly control the basal tone of a system. This article describes approaches to the design and synthesis of antagonists in detail and how the same principles can be used to synthesize an inverse agonist.

Uses of Antagonists

Antagonists are used in pharmacology to evaluate agonists. The strength of an agonist is often determined by its ability to displace or be displaced by an antagonist. Therefore, the specificity of the antagonist gives refined characterization of an agonist. Antagonists are also used as therapeutics to adjust the basal tone of receptor systems to overcome the presence of excess agonist *in vivo* or as an antidote to an overdose of a particular agonist.

Evaluation of an Antagonist

The qualitative and quantitative interaction of the ligand with the receptor can be evaluated by a dose–response curve as seen in Fig. 1. Since, by definition, an antagonist should have no response, one evaluates the interaction of the antagonist by measuring how it affects the interaction of the receptor with an agonist. Thermodynamically, measurement of the ability of an antagonist to bind to a binding site in a GPCR is called its dissociation constant (K_d), which is best determined directly using association and dissociation rate constants with a radiolabeled or fluorescent-labeled ligand.

To determine the potency of an antagonist, several dose–response experiments with different fixed concentrations of the antagonist vs agonists are done. Data are then used to construct a Schild's plot,[2] which gives the pA_2 value of the antagonist. A linear shift to the right of the dose–response curve as a function of increasing concentration of the antagonist with the same slope is a sign that the antagonist is competitive. Alternatively, a Hill plot with $n = 1$ is normally interpreted as indicating a competitive antagonist.

[2] H. Schild, *Br. J. Pharmacol.* **2,** 189 (1947).

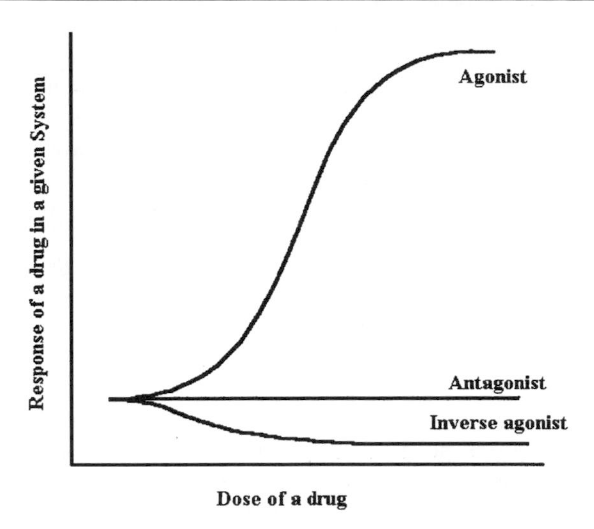

FIG. 1. Stylized representation of a dose–response curve. Differences among an agonist, antagonist, and inverse agonist as a function of increasing concentrations of drug are shown. The response can be anything one measures from GTPγS to frog skin assay.

General Strategies for Developing an Antagonist

The development of an antagonist ligand often starts with the characterization of the native agonist until one obtains a lead. This gives insight into the mechanism of interaction of the agonist with the receptor. Many peptide agonists can be roughly divided into a "message" region and an "address" region. The "address" region is responsible for the binding of the peptide into the receptor, whereas the "message" region is responsible for changing the conformation of the receptor, initiating G protein interactions in cytosol. Analysis of the native peptide to define the regions of binding and signal transduction is accomplished by a study of structure–activity relationships.[3] Several structure–activity experiments are utilized in the characterization of a native agonist. Among them are truncation, alanine scan, D-amino acid scans, changing net charge, and changing hydrophobic character.

Truncation

Simple N-terminal and C-terminal truncation of the native ligand can provide a wealth of information on both the binding and the message regions using binding and second messenger assays.

[3] V. J. Hruby, H. I. Yamamura, and F. Porreca, *Ann. N.Y. Acad. Sci.* **757,** 7 (1995).

Alanine Scan

An alanine scan is the systematic replacement of each residue with an alanine to determine the relative importance of the side chain groups of each residue in the interactions of the peptide with the receptor. Glycine scans have also been used.

D-Amino Acids Scans

While most native peptides have L-enantiomer residues, the replacement of each residue with the D-enantiomeric form of the residue can render the ligand in a unique conformation, which may give insight into bioactive conformation.

Changing Net Charge

With the inclusion of polar residues, native ligands have a net charge. Often charges are involved in the electrostatic interaction with the receptor. Changing the net charge of the residues can be used to evaluate whether electrostatic interactions of the ligand with the receptor are important.

Changing Hydrophobic Character

Native ligands with hydrophobic regions can also interact with hydrophobic counterparts in the receptor-binding pocket. Placing polar residues in these portions of the ligand can test the importance of the hydrophobic regions in the ligand, e.g., by changing aliphatic side chains with aromatic residues and vice versa.

Generating an Antagonist

Deletion of the message region of the native agonist while leaving only the address region sometimes can generate an antagonist if the peptide fragment will bind to the receptor. However, usually residues of the message region are critical for binding and may also influence the conformation of the address region. Therefore, critical parts of the message often must be retained. If the message sequence is not sequentially concurrent, this is especially critical.

Biophysical Studies

Both binding and message functions of the native ligand are conformationally driven and usually intertwine. Therefore, it is paramount to consider conformation when designing novel antagonists using either amino acids or peptidomimetics. Biophysical studies on lead antagonists usually are critical to define the conformational properties of the ligand.[1,4–7] Because the conformation of the ligand at the binding site in GPCRs is difficult to obtain, close approximations can be made using a variety of techniques, including conformational constraint, computational studies, nuclear magnetic resonance (NMR) spectroscopy, X-ray crystallography, and other biophysical methods.[7–12]

Refining a Lead Antagonist

In theory, suitable restrictions of the conformation of the lead antagonist can improve the binding affinity and selectivity of the antagonist. These conformational restrictions include use of unnatural amino acids and cyclization.

Cyclization of the antagonist restricts the ϕ and ψ angles in the amino acid residues. This provides insight into the optimum backbone conformation of the ligand. Restricted ϕ and ψ angles can also cause diminished degrees of freedom in side chain functional groups. Use of novel amino acids further restricts the orientation of the side chain functionalities by restricting χ angles of residues.[13] Unnatural amino acids can also be used to enhance certain properties in the lead antagonist, such as hydrophobicity, charge density, and polarity, provide routes for cyclization, and prevent enzymatic degradation.

Peptidomimetics

With further understanding of the conformation, topography, and dynamics of the antagonist, in principle one can design an antagonist that is not a peptide, although it possesses several difficult problems.[14,15] More recently, when the exact geometry is known or can be reasonably postulated, one may use combinatorial chemistry to generate varied functional groups on the same scaffold molecule while maintaining the required geometry of the system to explore this possibility.

The above discussion provides the basic outline as to how an antagonist is designed and refined. The next section gives three examples of how this has been implemented. The first entails the design of an antagonist for the glucagon receptor.

---

[4] J.-P. Meraldi, V. J. Hruby, and A. I. R. Brewster, *Proc. Natl. Acad. Sci. U.S.A.* **74,** 1373 (1977).

[5] V. J. Hruby, *in* "Perspectives in Peptide Chemistry" (A. Eberle, R. Geiger, and T. Wieland, eds.), p. 207. Karger, Basel, Switzerland, 1981.

[6] V. J. Hruby and H. I. Mosberg, *in* "Hormone Antagonists" (M. K. Agarwal, ed.), p. 433. Walter de Gruyter & Co., Berlin, 1982.

[7] V. J. Hruby and H. I. Mosberg, *Life Sci.* **31,** 189 (1982).

[8] V. J. Hruby, *in* "The Peptides: Analysis, Synthesis, Biology" (V. J. Hruby, ed.), Vol. 7, p. 1. Academic Press, New York, 1985.

[9] V. J. Hruby, *Trends Pharmacol. Sci.* **8,** 336 (1987).

[10] G. R. Marshall, *Tetrahedron* **49,** 3547 (1993).

[11] V. J. Hruby, F. Al-Obeidi, and W. M. Kazmierski, *Biochem. J.* **268,** 249 (1990).

[12] V. J. Hruby, *Biopolymers* **33,** 1073 (1993).

[13] V. J. Hruby, G. Li, C. Haskell-Luevano, and M. D. Shenderkovich, *Biopolymers* **43,** 219 (1997).

[14] V. J. Hruby, *Drug Discov. Today* **2,** 165 (1997).

[15] V. J. Hruby and C. Slate, *in* "Advances in Amino Acid Mimetics and Peptidomimetics" (A. Abell, ed.), p. 199. JAI Press, Greenwich, 1999.

The second part outlines advances in melanocortin antagonists. The last part gives the evolution of an antagonist for the human δ opioid receptor to a dipeptide inverse agonist, TMT-L-Tic.

Glucagon Antagonists

Glucagon is a 29 amino acid peptide hormone involved in the release of glucose from the glycogen stored in the liver and in fat metabolism (gluconeogenesis). The development of glucagon antagonists has relied on structure–activity relationship studies, but the systematic development of antagonists has been difficult due to the number of residues involved in binding and signal transduction. Still, the development of a glucagon antagonist may have a therapeutic role in lowering glucose levels in the blood.

Attempts to replace residues of the glucagon sequence generally resulted in significant loss of binding affinity.[16] This indicated that most of residues in the glucagon sequence are important to maintain binding affinity. In addition, truncation of glucagon from N-terminal or C-terminal failed to produce fragments with substantial binding affinity. However, SAR has shown that the C-terminal region is important for receptor recognition, serving as the "address" region, whereas the N-terminal region serves as the "message" region[16] for signal transduction.

Extensive structure–activity relationship studies have identified a number of key residues involved in signal transduction and binding affinity. The partial agonist/antagonist activity of [*des*His^1]-glucagon indicates that an important residue for signal transduction is the His at position 1. This led to the development of the first antagonist, [N^a-trinitrophenyl-His^1, homoArg^{12}]glucagon.[17] Later, discovery of the importance of the Asp at position 9 in glucagon activity led to the development of the potent antagonist [*des*His^1, Glu^9]glucagon amide.[18] X-ray crystallography[19] and two-dimensional NMR spectroscopy[20] indicated that the C-terminal "address" region of glucagon adopted an amphiphilic α-helical conformation. Importance of the α-helical conformation at the C-terminal region for increased binding affinity was demonstrated by the agonist [Lys^{17,18}, Glu^{21}]glucagon.[21] This agonist incorporated a salt bridge between Lys^{18} and Glu^{21}, which was observed by X-ray crystallography[21] and was presumed to enhance receptor

[16] V. J. Hruby, J. Krstenansky, B. Gysin, J. T. Penton, D. Trivedi, and R. L. McKee, *Biopolymers* **S135,** (1986).

[17] M. D. Bregman, D. Trivedi, and V. J. Hruby, *J. Biol. Chem.* **255,** 11725 (1980).

[18] C. G. Unson, D. Andreu, E. M. Gurzenda, and R. B. Merrifield, *Proc. Natl. Acad. Sci. U.S.A.* **84,** 4083 (1987).

[19] K. Sasaki, S. Dockerill, D. A. Adamiak, I. J. Tickle, and T. Blundell, *Proc. Natl. Acad. Sci. U.S.A.* **257,** 751 (1975).

[20] W. Braun, G. Wider, K. H. Lee, and K. Wuthrich, *Mol. Biol.* **169,** 921 (1983).

[21] N. S. Sturm, Y. Liin, S. K. Burley, J. Krstenansky, J.-M. Ahn, B. Y. Azizeh, D. Trivedi, and V. J. Hruby, *J. Med. Chem.* **41,** 2693 (1998).

TABLE I
POTENT GLUCAGON ANTAGONISTS

Compound	Receptor binding		Adenylate cyclase activity		
	IC_{50} (n*M*)	Rel. binding potency (%)	EC_{50} (n*M*)	Max. stimulation (%)	pA_2
Glucagon	1.5	100.0	5.9	100	—
[N^a-Trinitrophenyl-His[1], homoArg[12]]-Glucagon	15.0	10.0	i.a.[a]	0	8.16
[*des*His[1],Glu[9]]-Glucagon-NH_2	3.75	40.0	i.a.[b]	0	7.25
[*des*His[1], *des*Phe[6], Glu[9]]-Glucagon-NH_2	48.0	3.1	i.a.[a]	0	8.20

[a] Inactive up to 10 μM.
[b] Inactive up to 100 μM.

recognition of the agonist. In order to improve the binding affinity of glucagon antagonists, [Lys[17,18], Glu[21]] residues were incorporated in the sequences of several antagonists, but this resulted in a loss of binding affinity.

The X-ray crystal structure of glucagon revealed a hydrophobic area in the N-terminal region consisting of the residues His[1], Phe[6], Tyr[10,13], and Leu[14]. An attempt at reducing the strength of the hydrophobic area was made by deletion of the Phe at position 6. The reduced hydrophobic area was confirmed by computational study.[22] This resulted in the potent glucagon antagonist [*des*His[1], *des*Phe[6], Glu[9]]glucagon amide, which was a 10 times more potent antagonist than [*des*His[1], Glu[9]]glucagon amide.

Glucagon antagonists have been shown to lower blood glucose concentration in diabetic rats[23,24] by blocking the effect of endogenous glucagon. The three most widely used antagonists (Table I) are [N^a-trinitrophenyl-His[1], homoArg[12]]glucagon,[17] [*des*His[1], Glu[9]]glucagon amide,[18] and [*des*His[1], *des*Phe[6], Glu[9]]glucagon amide,[24] although only the latter analogue is a pure antagonist.[24] This was demonstrated using cAMP accumulation, a second messenger bioassay that has high sensitivity for determining any partial agonist activity.[24] cAMP accumulation showed that [N^a-trinitrophenyl-His[1], homoArg[12]]glucagon, and [*des*His[1], Glu[9]] glucagon amide possessed partial agonist activity. However, [*des*His[1], *des*Phe[6], Glu[9]]glucagon amide did not accumulate any cAMP, demonstrating that it is a pure antagonist. This study suggests that as well as the development of antagonists,

[22] B. Y. Azizeh, J.-M. Ahn, R. Caspari, M. D. Shenderovich, D. Trivedi, and V. J. Hruby, *J. Med. Chem.* **40,** 2555 (1997).

[23] D. G. Johnson, C. A. Goebel, V. J. Hruby, M. D. Bregman, and D. Trivedi, *Science* **215,** 1115 (1982).

[24] B. A. Van Tine, B. Y. Azizeh, D. Trivedi, J. R. Phelps, M. D. Houslay, D. G. Johnson, and V. J. Hruby, *Endocrinology* **137,** 3316 (1996).

development of assays for carefully differentiating second messenger activity is crucial for obtaining clean antagonists.

Synthesis and Purification of [des His^{1}, des Phe^{6}, Glu^{9}]Glucagon Amide.[22] Synthesis is carried out on an Applied Biosystems (ABI) 431A peptide synthesizer on a 0.25-mmol scale. Using an Fmoc strategy, the glucagon is synthesized by solid-phase methodology using a 4-(2′,4′-dimethoxyphenyl-Fmoc-aminomethyl) phenoxy resin. The following protecting groups are used: Arg (2,2,5,7,8-pentamethylchroman-6-sulfonyl, Pmc); Asn (trityl); Asp (*t*-butyl); Gln (trityl); Glu (*t*-butyl); Lys (Boc); Ser (*t*-butyl); Thr (*t*-butyl), and Tyr (*t*-butyl). The analogue is cleaved from the resin using standard techniques and a cleavage mixture of 50% trifluoroacetic acid (TFA), 5% anisole, 2.5% methyl sulfide, and 2.5% 1,2-ethanedithiol. Purification is carried out on a Perkin-Elmer Model 410-BIO instrument by preparative high-performance liquid chromatography (HPLC) with a Vydac 218TBP-16 column (16 × 250 mm). The sample is eluted by applying a gradient of 10–90% acetonitrile in 0.1% aqueous TFA over a 30-min period using a flow rate of 5 ml/min. The eluent containing the product is collected and lyophilized. Homogeneity is checked by HPLC on an analytical Vydac 218TBP-16 column (4.6 × 250 mm). The analogue is characterized by amino acid analysis, TLC in three solvent systems, and by electrospray mass spectrometry (ES-MS) where the calculated mass is 3211.5 and the experimental value is 3212.0.

Preparation of Plasma Membranes from Rat Liver. Partially purified plasma membranes from rat liver are prepared by a modification of the procedure of Neville[25] as described by Pohl and co-workers[26] and stored in liquid nitrogen.

Assay for Adenylate Cyclase Activity. The assay medium consists of [α-^{32}P]ATP (about 40 cpm/pmol) at 2 m*M*; $MgCl_2$, 5 m*M*; cyclic AMP, 0.4 m*M*; dithiothreitol (DTT), 1 m*M*; bovine serum albumin (BSA), 2 mg/ml; creatine phosphate, 5 m*M*; creatine phosphokinase, 0.3 mg (50 units)/ml; and Tris–Cl buffer, 30 m*M* at pH 7.5 in a final volume of 100 μl. Liver membranes, suspended in Tris–Cl buffer containing 1 m*M* DTT, are added to a final concentration of 0.2–0.5 mg/ml to initiate the reaction. After 5 min at 30°, the reaction is terminated by adding 100 ml of stopping solution. The cyclic AMP (cAMP) formed is determined with the use of Dowex-50 and alumina columns.[27]

Receptor Binding Assay for Glucagon Antagonists. The following binding assay is performed according to a method similar to that reported by Lin *et al.*[28] where the amount of radioiodinated glucagon displaced from receptor sites by increasing concentrations of antagonist is measured. Triplicate determinations are made in all binding experiments, and each experiment is carried out at least twice.

[25] D. M. Neville, *Biochim. Biophys. Acta* **154,** 540 (1968).

[26] S. L. Pohl, L. Birnbaumer, and M. Rodbell, *J. Biol. Chem.* **246,** 1849 (1971).

[27] Y. Salomon, C. Londos, and M. Rodbell, *Anal. Biochem.* **58,** 541 (1974).

[28] M. C. Lin, D. E. Wright, V. J. Hruby, and M. Rodbell, *Biochemistry* **14,** 1559 (1975).

[^{125}I]Glucagon is prepared as described previously.[29] The concentration of biologically active glucagon in the preparations of labeled hormone is estimated from assays of adenylate cyclase activity (see earlier discussion); activities are compared with those generated by native glucagon over a wide range of concentrations. The incubation medium used for the binding assay contains [^{125}I]glucagon at $1–5 \times 10^{-9}$ *M* (specific activity 10^6 cpm/pmol); BSA, 3 mg/ml; liver membrane, 50–100 μg/ml; and Tris–Cl buffer, 20 m*M* at pH 7.5 or as indicated in a final volume of 1 ml. The assay is initiated by the addition of liver membranes, and the whole mixture is poured on an oxoid membrane filter after 10 min at 30°. The filter is washed twice quickly with 1 ml of cold Tris–Cl buffer containing 2 mg/ml of BSA. Suction is applied during the washing so that each wash takes less than 5 sec to complete. Then the filter is counted in well-type Packard γ counter.

Isolation of Rat Hepatocytes and Challenge with Peptides. Hepatocytes are isolated by the method of Berry and Friend,[30] with modifications by Heyworth and Houslay,[31] from 200- to 250-g male Sprague–Dawley rats. Cells are preincubated at 37° for 45 min with gassing of 95% O_2/5% CO_2. Cell viability is determined using trypan blue exclusion, and preparations with greater than 95% viability are used. The final volume (4.5 mg cell dry weight) is 1.0 ml. Hepatocytes are then incubated with hormone in the concentration range of 0–1 μM. Inhibitors or the controls [dimethyl sulfoxide (DMSO)] are added to hepatocytes with the addition of hormone. After 3 min, the reaction is stopped by adding 4% perchloric acid at a 50% volume to give a final concentration of 2%. The reaction is neutralized with 0.5 *M* triethylamine in 2 *M* KOH and stored at 4°.

cAMP Accumulation Assay. Determination of cAMP accumulation in cells is performed using a methodology similar to the competition method of Brown *et al.*[32] in which the cAMP-binding protein is isolated from bovine adrenal glands. Briefly, 50 μl of sample is added to 100 μl [^{3}H]cAMP (13,000 dpm/tube), 50 μl assay buffer (50 m*M* Tris and 4 m*M* EDTA, pH 7.4, at 4°), and 100 μl binding protein. A binding protein dilution is determined for each new binding protein preparation so that optimum sensitivity could be achieved. The samples are incubated for 2 hr at 4°, and the reaction is stopped by adding 250 μl 2% Norit GSX charcoal–1% BSA assay buffer solution, followed by centrifugation at 14,000*g* in a tabletop centrifuge for 2 min. Three hundred-microliter aliquots are counted using liquid scintillation cocktail and an LS 5000TD Beckman counter (Beckman, Palo Alto, CA). Samples are then computed with the help of a standard curve of cAMP concentration ranging from 0.0625 to 16 p*M*. Data are represented as either

[29] M. Rodbell, H. M. J. Krans, S. L. Pohl, and L. Birnbaumer, *J. Biol. Chem.* **246,** 1861 (1971).

[30] M. N. Berry and D. S. Friend, *J. Biol. Chem.* **43,** 541 (1969).

[31] C. M. Heyworth and M. D. Houslay, *Biochem. J.* **214,** 93 (1983).

[32] B. L. Brown, R. P. Ekins, and J. M. P. Albano, *Adv. Cyclic Nucleotide Res.* **2,** 25 (1972).

fold over basal, as in the case of glucagon, or a percentage of the maximal glucagon stimulation over basal. All determinations are made using three experiments performed in triplicate compared to basal levels.

Melanotropic Peptide Antagonists

α-Melanotropin(α-MSH, Ac-Ser-Tyr-Ser-Met-Glu-His-Phe-Arg-Trp-Gly-Lys-Pro-Val- NH_2) is an important peptide hormone and neurotransmitter that has been studied extensively for its important role in skin pigmentation.[32,33] It is derived from a large precursor protein, proopiomelanocortin, which also contains within its structure the sequences of other melanotropic peptides: β-MSH, γ-MSH, and adrenocorticotropin (ACTH), as well as β-endorphin and β-lipotropin. α-MSH performs its classic activity by stimulating pigmentation on interaction with the melanocortin 1 receptor (MC1R). Considerable additional interest has developed with the cloning of the MC1R, the ACTH receptor (MC2R), the MC3R and the MC4Rs (both found primarily in the brain), and the MC5R (found in the brain and in peripheral tissues).[35–42] These new receptors appear to be involved in several critical biological functions, including cardiovascular function, erectile function, and feeding behavior.

MSH Message Sequence. Frog (*Rana pipiens*) and lizard (*Anolis carolinensis*) skin bioassays have been used traditionally to study the bioactivities of MSH and related analogues.[43,44] In these assays, light reflectance from the surface of the skin is monitored as a measure of darkening of the skin due to melanosome dispersion in response to an agonist melanotropin; in the presence of an antagonist, the darkening produced by an agonist, such as α-MSH, is blocked. EC_{50} and IC_{50} values are then obtained from dose–response curves. In both these assays,

[33] M. E. Hadley, "Source, Synthesis, Chemistry, Secretion, and Metabolism." CRC Press, Boca Raton, FL, 1988.

[34] A. N. Eberle, "The Melanotropins: Chemistry, Physiology, and Mechanisms of Action." Karger, Basel, Switzerland 1988.

[35] K. G. Mountjoy, L. S. Robbins, M. T. Mortrud, and R. D. Cone, *Science* **257,** 1248 (1992).

[36] V. Chhajlani, R. Muceniece, and J. E. S. Wikberg, *Biochem. Biophys. Res. Commun.* **195,** 866 (1993).

[37] F. Desarnaud, O. Labbé, D. Eggerickx, G. Vassart, and M. Parmentier, *Biochem. J.* **299,** 366 (1994).

[38] Z. Fathi, L. G. Iben, and E. M. Parker, *Neurochem. Res.* **20,** 107 (1995).

[39] I. Gantz, Y. Konda, T. Tashiro, Y. Shimoto, H. Miwa, G. Munzert, S. J. Watson, J. Del Valle, and T. Yamada, *J. Biol. Chem.* **268,** 8246 (1993).

[40] I. Gantz, H. Miwa, Y. Konda, Y. Shimoto, T. Tashiro, S. J. Watson, V. Del Valle, and T. Yamada, *J. Biol. Chem.* **268,** 15174 (1993).

[41] N. Griffon, V. Mignon, P. Facchinetti, J. Diaz, J. C. Schwartz, and P. Sokoloff, *Biochem. Biophys. Res. Commun.* **200,** 1007 (1994).

[42] O. Labbé, F. Desarnaud, D. Eggerickx, G. Vassart, and M. Parmentier, *Biochemistry* **33,** 4543 (1994).

[43] A. M. L. Castrucci, M. E. Hadley, and V. J. Hruby, *Gen. Comp. Endocrinol.* **55,** 104 (1984).

[44] M. E. Hadley and A. M. L. Castrucci, *in* "The Melanotropic Peptides" (M. E. Hadley, ed.), Vol. III, p. 1. CRC Press, Boca Raton, FL, 1988.

the central 6–9 tetrapeptide, His-Phe-Arg-Trp, was found to be the melanotropic message sequence of α-MSH.[43–45] Potency is enhanced as the sequence is extended toward either the N or the C termini. Development of antagonists to MSH receptors, as revealed later in this section, involves structural changes within this central 6–9 sequence.

MSH Structure–Activity. Several modifications made within the primary structure of α-MSH have yielded peptides with superpotent activity.[46,47] For example, substitution of Nle for Met^4 with α-MSH and its fragments generally results in more potent derivatives. In addition, substitution of the D-enantiomer of Phe^7 into such $[Nle^4]$-modified-α-MSH analogues gave analogues that are 10–100 times more potent than L-amino acid-containing analogues, as well as exhibiting prolonged biological activity.[48] Hruby *et al.* postulated that the increased potency of these $[Nle^4,D\text{-}Phe^7]$-substitued analogues might be due to a reverse turn within the α-MSH(4–10) sequence, which was stabilized conformationally by the D-Phe^7 substitution. Based on this hypothesis, a series of α-MSH analogues were synthesized wherein Cys was substituted for Met^4 and Gly^{10} and these were cyclized by disulfide bridges.[49,50] These conformationally restricted analogues proved to be superpotent in the frog skin bioassay.

Finally, α-MSH has a very short half-life *in vivo,* about 3 min or less in humans and rodents. Thus, an important property of the just-mentioned MSH analogues was their improved metabolic stability *in vivo* or under *in vitro* cell culture conditions. In this respect, $[Nle^4,D\text{-}Phe^7]$-substituted analogues and cyclic $[Cys^{4,10}]\alpha$-MSH analogues were resistant to biological inactivation by proteolysis under various conditions.[51,52] These studies provide insights toward the design of melanotropin antagonists that can function as biologically stable probes for biological and biomedical studies.

[45] V. J. Hruby, B. C. Wilkes, M. E. Hadley, F. A. Al-Obeidi, T. K. Sawyer, D. J. Staples, A. E. de Vaux, O. Dym, A. M. L. Castrucci, M. F. Hintz, J. R. Riehm, and R. R. Rao, *J. Med. Chem.* **30,** 2126 (1987).

[46] V. J. Hruby, B. C. Wilkes, W. L. Cody, T. K. Sawyer, and M. E. Hadley, *in* "Peptide Protein Reviews" (M. T. W. Hearn, ed.), Vol. 3, p. 1. Dekker, New York, 1984.

[47] V. J. Hruby, S. D. Sharma, K. Toth, J. Y. Jaw, F. A. Al-Obeidi, T. K. Sawyer, A. M. L. Castrucci, and M. E. Hadley, *Ann. N.Y. Acad. Sci.* **680,** 51 (1993).

[48] T. K. Sawyer, P. J. Sanfilippo, V. J. Hruby, M. H. Engel, C. B. Heward, J. B. Burnett, and M. E. Hadley, *Proc. Natl. Acad. Sci. U.S.A.* **77,** 5754 (1980).

[49] T. K. Sawyer, V. J. Hruby, P. S. Darman, and M. E. Hadley, *Proc. Natl. Acad. Sci. U.S.A.* **79,** 1751 (1982).

[50] W. L. Cody, M. Mahoney, J. J. Knittel, V. J. Hruby, A. M. L. Castrucci, and M. E. Hadley, *J. Med. Chem.* **29,** 583 (1985).

[51] A. M. L. Castrucci, M. E. Hadley, and V. J. Hruby, *in* "The Melanotropic Peptides" (M. E. Hadley, ed.), Vol. II, p. 171. CRC Press, Boca Raton, FL, 1988.

[52] K. Akiyama, H. I. Yamamura, B. C. Wilkes, W. L. Cody, V. J. Hruby, A. M. L. Castrucci, and M. E. Hadley, *Peptides* **5,** 1191 (1984).

MSH Receptor Interactions. MSH initially interacts with MSH receptors and activates adenylate cyclase, which leads to elevated intracellular cAMP levels.[53–55] Further, the actions of MSH are Ca^{2+} dependent.[56] Melatonin, fluphenazine, melittin, a synthetic octapeptide analogue of the calmodulin-binding domain of a myosin light chain kinase M5, Ca^{2+}-dependent chelating agents, and so on have been reported to inhibit α-MSH-stimulated melanocytes and melanoma cells.[57] However, these antagonists exhibit noncompetitive inhibitory properties and exemplify the diversity of potential sites of intervention in the α-MSH receptor-binding and/or receptor-mediated signaling mechanism.

MSH Receptor Antagonists

MSH inhibitory sequence. Ac-α-MSH(7–10)-NH_2 (Ac-Phe-Arg-Trp-Gly-NH_2) is devoid of any agonist activity in both frog and lizard skin assays.[58] However, this compound is a very weak competitive antagonist ($pA_2 = 4.3$) of α-MSH in the lizard skin assay. Other inactive α-MSH fragments [e.g., Ac-α-MSH (9–13)-NH_2, Ac-α-MSH(10–13)-NH_2, Ac-α-MSH(11–13)-NH_2, Ac-α-MSH(6–8)-NH_2], however, showed no such inhibitory activity even at high concentrations.

D-Amino acid-substituted peptide antagonists. The following synthetic derivatives of α-MSH have specific D-amino acid substitutions that have resulted in the identification of leads for the development of metabolically stable competitive antagonists of the native sequence.

H-[Xaa^7,Yaa^8,D-Phe^{10}]α-MSH(6–11)-NH_2 template.[55] Several analogues (**I–VI**) in Table II[59] of the generic formula H-[Xaa^7,Yaa^8,D-Phe^{10}]α-MSH(6–11)-NH_2 have been investigated for their melanotropic activities in both the frog and the lizard skin assays as shown in Table III. In the frog skin assay, peptide **III** was inactive and **II** was a full but weak agonist; derivatives **I** and **IV–VI** were inhibitors of MSH activity. In the lizard skin assay, three analogues (**I, III, VI**) were inactive, one was a full but weak agonist (**II**), and two were weak antagonists (**IV, V**). **IV** and **V** were inhibitors in both assays, but the other two inhibitors (**I, VI**) in the frog skin assay were inactive in the lizard skin assay.

Ac-[Xaa^7, Yaa^{10}]-α-MSH(7–10)-NH_2 template. Using α-MSH(7–10) as a template, modifications of the generic formula Ac-[Xaa^7, Yaa^{10}]-α-MSH(7–10)-NH_2

[53] M. E. Hadley and A. M. L. Castrucci, *in* "The Melanotropic Peptides" (M. E. Hadley, ed.), Vol. III, p. 15. CRC Press, Boca Raton, FL, 1988.

[54] T. K. Sawyer, M. E. Hadley, B. C. Wilkes, and V. J. Hruby, *in* "The Melanotropic Peptides" (M. E. Hadley, ed.), Vol. III, p. 59. CRC Press, Boca Raton, FL, 1988.

[55] M. D. Bregman, T. K. Sawyer, M. E. Hadley, and V. J. Hruby, *Arch. Biochem. Biophys.* **200,** 1 (1989).

[56] M. E. Hadley, B. Anderson, C. B. Heward, T. K. Sawyer, and V. J. Hruby, *Science* **213,** (1981).

[57] J. E. Gerst and Y. Salomon, *Endocrinology* **121,** 1766 (1987).

[58] T. K. Sawyer, D. J. Staples, A. M. L. Castrucci, and M. E. Hadley, *Peptide Res.* **2,** (1989).

[59] A. M. L. Castrucci, T. K. Sawyer, F. A. Al-Obeidi, V. J. Hruby, and M. E. Hadley, "Melanotropic Peptide Antagonists: Recent Discoveries and Biomedical Implications." Barcelona, 1990.

TABLE II
PRIMARY STRUCTURES OF α-MSH ANALOGUES[a]

I	H-His-D-Trp-Ala-Trp-D-Phe-Lys-NH_2
II	H-His-D-Phe-Ala-Trp-D-Phe-Lys-NH_2
III	H-His-D-Ala-Ala-Trp-D-Phe-Lys-NH_2
IV	H-His-D-Arg-Ala-Trp-D-Phe-Lys-NH_2
V	H-His-D-Trp-Arg-Trp-D-Phe-Lys-NH_2
VI	H-His-Phe-Arg-Trp-D-Phe-Lys-NH_2
VII	Ac-Phe-Arg-Trp-Gly-NH_2
VIII	Ac-Phe-Arg-Trp-D-Phe-NH_2
IX	Ac-D-Trp-Arg-Trp-D-Phe-NH_2
X	Ac-D-Phe-Arg-Trp-D-Phe-NH_2
XI	Ac-D-Phe-Arg-Trp-Gly-NH_2
XII	Ac-Nle-Asp-His-D-Phe-Arg-Trp-Lys-NH_2
XIII	Ac-Nle-Asp-Trp-D-Phe-Arg-Trp-Lys-NH_2
XIV	Ac-Nle-Asp-His-D-Phe-Nle-Trp-Lys-NH_2
XV	Ac-Nle-Asp-Trp-D-Phe-Nle-Trp-Lys-NH_2
XVI	Ac-Nle-c[Asp-Trp-D-Phe-Nle-Trp-Lys]-NH_2
XVII	Ac-Nle-c[Asp-His-D-Phe-Nle-Trp-Lys]-NH_2
XVIII	Ac-Nle-Asp-Trp-D-Phe-Ala-Trp-Lys-NH_2
XIX	Ac-Nle-Asp-Trp-D-Phe-Pro-Trp-Lys-NH_2
XX	Ac-Nle-Asp-Trp-Phe-Nle-Trp-Lys-NH_2
XXI	Ac-Nle-Asp-D-Trp-D-Phe-Nle-Trp-Lys-NH_2
XXII	Ac-Nle-Asp-Trp-D-Phe-Nle-D-Trp-Lys-NH_2
XXIII	Ac-Nle-Asp-D-Trp-D-Phe-Nle-D-Trp-Lys-NH_2

[a] See Ref. 59 for further discussion.

(**VII–XI;** Table II) were studied (shown in Table IV).[60] A significant increase (7-fold) in inhibitory activity was demonstrated for Ac-Phe-Arg-Trp-D-Phe-NH_2 (**VIII,** $pA_2 = 5$) in the lizard skin assay, while it was a weak, full agonist in the frog skin assay. Substitution of Phe^7 by D-Trp in **VIII** yielded Ac-D-Trp-Arg-Trp-D-Phe-NH_2 (**IX**), a weak antagonist in both assays ($pA_2 = 4.8$, frog; $pA_2 = 5.7$, lizard). The side chain structural specificity of the D-Trp^7 substitution in **IX** was further examined by testing Ac-[D-$Phe^{7,10}$]α-MSH(7–10)-NH_2 (**X**), which affected inhibitory activity ($pA_2 = 4.8$) in the lizard skin assay but was a weak agonist in the frog skin assay. Finally, D-Phe^7 substituted into the α-MSH(7–10) template yielded **XI,** which was a weak agonist on both frog and lizard skins.

Ac-[Nle^4, Waa^5, Xaa^6, D-Phe^7, Yaa^8, Zaa^{10}]α-MSH(4–10)-NH_2 template. Considerations from structure–activity studies in conjunction with conformational analysis using NMR, molecular mechanics, and molecular dynamics calculations for a series of analogues based on the type Ac-[Nle^4, Waa^5, Xaa^6, D-Phe^7,Yaa^8,

[60] T. K. Sawyer, D. J. Staples, A. M. L. Castrucci, M. E. Hadley, F. A. Al-Obeidi, W. L. Cody, and V. J. Hruby, *Peptides* **11,** 351 (1990).

TABLE III
MELANOTROPIC PROPERTIES OF H-[Xaa^7, Yaa^8, D-Phe^{10}]α-MSH(6–11)-NH_2 ANALOGUES

	H-His-Xaa-Yaa-Trp-D-Phe-Lys-NH_2		Frog skin		Lizard skin	
Compound	Xaa	Yaa	EC_{50}, M[a]	pA_2[b]	EC_{50}, M	pA_2
I	D-Trp	Ala	Inactive[c]	4.7	Inactive	NI[d]
II	D-Phe	Ala	10^{-6}	—	10^{-6}	—
III	D-Ala	Ala	Inactive	NI	Inactive	NI
IV	D-Arg	Ala	Inactive	5.0	Inactive	6.0
V	D-Trp	Arg	Inactive	5.5	Inactive[c]	5.6
VI	Phe	Arg	Inactive	5.8	Inactive	NI

[a] The concentration of agonist effecting one-half maximal stimulation. The EC_{50} of α-MSH in the frog and lizard skin assays is 1.0×10^{-10} *M* and 1.0×10^{-9} *M*, respectively.
[b] The -log of the inhibitor concentration, which requires twice the concentration of an agonist (α-MSH) to affect its original response.
[c] The lack of measurable melanotropic activity (agonism) at concentrations $<10^{-5}$ *M*.
[d] No inhibition at concentrations $>10^{-5}$ *M*.
[e] Peptide V was found to be a partial agonist in the lizard skin assay. At $\sim10^{-5}$ *M*, peptide V induced a maximum response, which is about one-half the maximum response of α-MSH.

Zaa^{10}]α-MSH(4–10) led to the discovery of an α-MSH antagonist.[43,44,46] Compounds **XII–XXIII** (Table II) were investigated, and Ac-[Nle^4,Asp^5,Trp^6,D-Phe^7, Nle^8,Lys^{10}]α-MSH(4–10)-NH_2 (**XV,** Table II) had high antagonist potency (10^{-7} *M*, $pA_2 = 8.4$) with specificity for the α-MSH receptor in frog skin, but all other analogues in Table II were agonists. Interestingly, cyclization via a lactam bridge

TABLE IV
MELANOTROPIC PROPERTIES OF Ac-[Xaa^7, Yaa^{10}]α-MSH(7–10)-NH_2 ANALOGUES

	Ac-Xaa-Arg-Trp-Yaa-NH_2		Frog skin		Lizard skin	
Compound	Xaa	Yaa	EC_{50}, M[a]	pA_2[b]	EC_{50}, M	pA_2
VII	Phe	Gly	Inactive[c]	NI[d]	Inactive	4.3
VIII	Phe	D-Phe	2.5×10^{-6}	—	Inactive	5.0
IX	D-Trp	D-Phe	Inactive	4.8	Inactive	5.7
X	D-Phe	D-Phe	1.5×10^{-5}	—	Inactive	4.8
XI	D-Phe	Gly	1.4×10^{-5}	—	2.5×10^{-6}	—

[a] The concentration of agonist effecting one-half maximal stimulation. The EC_{50} of α-MSH in the frog and lizard skin assays is 1.0×10^{-10} *M* and 1.0×10^{-9} *M*, respectively.
[b] The -log of the inhibitor concentration, which requires twice the concentration of an agonist (α-MSH) to affect its original response.
[c] The lack of measurable melanotropic activity (agonism) at concentrations $<10^{-5}$ *M*.
[d] No inhibition at concentrations $>10^{-5}$ *M*.

TABLE V

BIOACTIVITIES OF D-AMINO ACID 7-SUBSTITUTED ANALOGUES OF SUPERPOTENT CYCLIC LACTAM α-MELANTROPIC ANALOGUES

Compound[a] (amino acid in position 7)	Biological activities (EC_{50} nM)					
	Frog skin[b]	Lizard skin[c]	hMC1R[d]	hMC3R[d]	hMC4R[d]	hMC5R[d]
D-Phe(pI) (**1**)	Antagonist ($pA_2 = 10.3$)	Agonist (0.16)	Agonist (0.055)	Partial agonist (1.1) ($pA_2 = 8.0$)	Partial agonist (0.57) ($pA_2 = 9.5$)	Partial agonist (0.68)
D-Nal(2′) (**2**)	Antagonist ($pA_2 > 10.5$)	Agonist	Agonist (0.036)	Partial agonist (2.8) ($pA_2 = 8.3$)	Antagonist ($pA_2 = 9.3$)	Agonist (0.44)

[a] Analogues of Ac-Nle[4]-c[Asp-His-D-Phe[7]-Arg-Trp-Lys[10]]-NH_2.
[b] *Rana pipiens* skins.
[c] *Anolis carolinensis* skins.
[d] cAMP-dependent β-galactosidase activity in cloned receptors.

between Asp[5] β-COOH and Lys[10] ε-NH_2 led to an analogue with full agonist activity in the frog skin assay. Structure–function analysis indicated that the key residues responsible for antagonist activity are positions 6 and 7. Conformational analysis suggested that when the D-Phe[7] χ_1 angle is constrained into a *gauche*(+) conformation, transduction cannot occur and an antagonist results.

A highly potent cyclic α-MSH antagonist containing naphthylalanine. Based on the results of the studies just described, a series of p-substituted unusual aromatic D-amino acid residues were substituted into position 7 of the superagonist cyclic lactam α-MSH analogue Ac-Nle[4]-c[Asp[5],D-Phe[7],Lys[10]]α-MSH(4–10)-NH_2[61] and were examined in frog and lizard skins and at the cloned MC1, MC3, MC4, and MC5 receptors.[62] The important results are summarized (Table V). The D-Phe(pI)[7] (**1**) and D-Nal(2′)[7] (**2**) analogues are potent antagonists in the classic frog skin (pA_2 10.3 and 10.5, respectively), but agonists in the lizard skin assay. Furthermore, **2** is a potent antagonist at the human MC4R (hMC4R), but a weak partial agonist/potent antagonist at the hMC3R and a full agonist at the hMC5R. **1**, however, is a weak partial agonist at all three human MC3, MC4, and MC5 receptors, but has potent antagonist activities at MC3 and MC4 receptors. All the other substitutions [e.g., D-Phe(pF)[7], D-Phe(pCl)[7], and D-Nal(1′)[7]] were potent agonists at all receptors. Even more unexpectedly, all the analogues in this series are agonists at both mouse and human MC1Rs.

A conformationally constrained somatostatin template with antagonistic activity at μ opioid receptors. Previously, a β-turn scaffold related to somatostatin

[61] F. A. Al-Obeidi, M. E. Hadley, B. M. Pettitt, and V. J. Hruby, *J. Am. Chem. Soc.* **111,** 3413 (1989).
[62] V. J. Hruby, D. Lu, S. D. Sharma, A. M. L. Castrucci, R. A. Kesterson, F. A. Al-Obeidi, M. E. Hadley, and R. D. Cone, *J. Med. Chem.* **38,** 3454 (1995).

TABLE VI
ANTAGONIST MELANOTROPIC ACTIVITY FOR MODIFIED ANALOGUES OF THE μ-OPIOID PEPTIDE CTAP USING THE FROG SKIN ASSAY

Compound		Antagonist activity[a] IC_{50} (*M*)
H-D-Phe-c[Cys-Tyr-D-Trp-Arg-Thr-Pen]-Thr-NH_2 (CTAP)		$\sim 10^{-5}$
H-D-Phe-c[Cys-Tyr-D-Trp-Arg-Trp-Pen]-Thr-NH_2	(**1**)	10^{-6}
H-D-Phe-c[Cys-Phe-D-Trp-Arg-Trp-Pen]-Thr-NH_2	(**2**)	$\sim 5 \times 10^{-7}$
H-D-Phe-c[Cys-His-D-Phe(*p*I)-Arg-Trp-Cys]-Thr-NH_2	(**3**)	10^{-7}

[a] Frog skin (*Rana pipiens*) assay.

was used to convert the somatostatin–receptor ligand to peptide analogues such as CTOP (D-Phe-c[Cys-Tyr-D-Trp-Orn-Thr-Pen]-Thr-NH_2) and CTAP (D-Phe-c[Cys-Tyr-D-Trp-Arg-Thr-Pen]-Thr-NH_2) that are potent and selective μ opioid receptor antagonists.[63,64] Because somatostatin, opioid, and melanocortin receptors are all structurally related and belong to the seven transmembrane GPCRs, their ligands may also share some similarities, more specifically a related backbone scaffold. Extensive NMR and molecular dynamics studies on CTAP demonstrated that a β-turn structure about the D-Trp-Arg sequence was a preferred stable conformation.[65,66] Putting these separate observations together, a series of chimeric peptides related to somatostatin and to the μ-opioid ligands were designed that were melanocortin receptor antagonis (Table VI). CTAP itself shows weak antagonist potency in the frog skin assay. This potency is enhanced by replacement of the ring Thr residue in CTAP by a Trp residue (**1**) and is enhanced further by replacement of the Tyr residue in **1** by a Phe as in analogue **2.** Finally, replacement of the Phe by a His and the D-Trp by a D-Phe(*p*I) led to antagonist **3** with an IC_{50} of 10^{-7} *M*.

In conclusion, these studies illustrate the power of *de novo* design once one understands the significance of template conformation to biological potency for a particular classes of receptors. Potent and specific agonists and antagonists are extremely valuable tools for determining the physiological roles of these receptors. Finally, in terms of their use in therapeutic intervention, MSH antagonists may be

[63] J. T. Pelton, K. Gulya, V. J. Hruby, S. P. Duckles, and H. I. Yamamura, *Proc. Natl. Acad. Sci. U.S.A.* **82,** 236 (1985).
[64] W. M. Kazmierski, W. S. Wire, G. K. Lui, R. J. Knapp, R. E. Shook, T. F. Burks, H. I. Yamamura, and V. J. Hruby, *J. Med. Chem.* **31,** 2170 (1988).
[65] W. M. Kazmierski, R. D. Ferguson, A. Lipkowski, and V. J. Hruby, *Int. J. Peptide Protein Res.* **46,** 265 (1995).
[66] W. M. Kazmierski, H. I. Yamamura, and V. J. Hruby, *J. Am. Chem. Soc.* **113,** 2275 (1991).

used to block the proposed autocrine and/or paracrine actions of α-MSH in the proliferation of melanoma tumors. Furthermore, the structural features of these peptides suggest that they may have facile passage across the blood–brain barrier. This allows their potential use in the intervention of various physiological processes mediated in the brain and periphery by α-MSH, such as learning and memory processes, feeding behavior, sexual behavior, regulation of body temperature, cardiovascular function, and immune response.

*Synthesis of Ac-Nle^4-c[Asp^5,D-Phe(pI)^7,Lys^{10}]α-MSH-(4–10)-NH$_2$(**1**).* The title compound is prepared using methods similar to those reported previously for cyclic lactam analogues of α-MSH-(4–10)-NH_2 with cyclization on solid support.[61,67] The protected peptide resin to the title compound is prepared from 1.5 g of *p*-methylbenzhydrylamine (pMBHA) resin (0.35 mmol of NH_2/g of resin) by first coupling N^α-Boc-Lys(N^ε-Fmoc) to the resin. The following amino acids are then added to the growing peptide chain by the stepwise addition of N^α-Boc-Trp(N^i-For), N^α-Boc-Arg(N^G-Tosyl), N^α-Boc-D-Phe(*p*I), N^α-Boc-His(N^π-Bom), N^α-Boc-Asp(β-OFm), and N^α-Boc-Nle using standard solid-phase methods. Each coupling reaction is achieved using a 3-fold excess of diisopropylcarbodiimide (DIC) and a 2.4-fold excess of *N*-hydroxybenzotriazole (HOBt). After coupling the last amino acid, the N^ε-Fmoc and β-OFm protecting groups are removed by treating the N^α-Boc-protected peptide resin with 20% piperidine in 1-methyl-2-pyrrolidinone (NMP) for 30 min. The peptide resin is washed with DMF (3×40 ml), dichloromethane (DCM) (3×40 ml), and 10% diisopropylethylamine (DIEA) (3×40 ml) and is then suspended in 15 ml of NMP and mixed with a 6-fold excess of benzotriazol-1-yl-oxytris(dimethylamino)phosphonium hexafluorophosphate (BOP reagent) in the presence of an 8-fold excess of DIEA for 2 hr. The coupling is repeated twice if needed until the resin gives a negative ninhydrin test.[68] Then the N^α-Boc protecting group is removed in the usual manner with 50% TFA in DCM. The amino group is neutralized with 10% DIEA in DCM and is acetylated with 25% acetic anhydride in DCM for 20 min. The peptide resin is cleaved with 15 ml of anhydrous HF in the presence of 10% anisole at 0° for 45 min. The HF–anisole is removed rapidly by vacuum distillation at 4°. The residue is washed with ethyl ether (2×25 ml), and the cleaved peptide is dissolved in acetic acid (2×25 ml) and 30% aqueous acetic acid (2×30 ml). The pooled acetic acid and aqueous acid phase fractions are lyophilized to give a white powder that is purified by preparative HPLC on a C18-bonded silica gel column (Vydac 218TP 1010, 1.0×25 cm) eluted with a linear gradient of acetonitrile (20–40%) in aqueous 0.1% TFA (v/v). Purification is monitored at 280 nm, and fractions corresponding to the major peak are collected, combined, and lyophilized to give

[67] F. Al-Obeidi, A. M. L. Castrucci, M. E. Hadley, and V. J. Hruby, *J. Med. Chem.* **32,** 2555 (1989).
[68] F. Kaiser, R. L. Colescott, C. D. Bossinger, and P. I. Cook, *Anal. Biochem.* **34,** 595 (1970).

TABLE VII
ANALYTICAL PROPERTIES OF CYCLIC LACTAM α-MELANOTROPIN ANALOGUES

Compound[a] (amino acid in position 7)	$[\alpha]^{22}_{589}$ in 10% HOAc (°)	TLC R_f[a] values			HPLC[b] K'	FAB-MS,[e] M + H found (calcd)
		A	B	C		
D-Phe(*p*I) (**1**)	−45.4 (*c* 0.032)	0.84	0.02	0.68	6.83[c] 8.14[d]	1151 (1150.1)
D-Nal(2′) (**2**)	−65.2 (*c* 0.018)	0.79	0.01	0.65	7.28[c] 9.59[d]	1075 (1075.3)

[a] Solvent systems: (A) 1-butanol/HOAc/pyridine/H_2O (5 : 1 : 5 : 4), (B) EtOAc/pyridine/HOAc/H_2O (5 : 5 : 1 : 3), and (C) 1-butanol/HOAc/H_2O (4 : 1 : 5).
[b] Analytical HPLC performed on a C18 column (Vydac 218TP 104) using a gradient of acetonitrile in 0.1% aqueous TFA for 30 min at 1.5 ml/min.
[c] 20–30% acetonitrile in 30 min.
[d] 10–40% acetonitrile in 30 min.
[e] Fast atom bombardment mass spectrometry.

the title compound as a pure (>98%) white powder. Analytical data are given in Table VII.

Synthesis of Ac-Nle^4-c[Asp^5,D-Nal(2)^7,Lys^{10}]α-MSH-(4–10)-NH_2. The title compound is prepared by methods very similar to those reported earlier for analogue **1** except that N^α-Boc-D-Nal(2′) is added to the growing peptide chain in the appropriate sequence. The peptide product is purified as before for compound **1** to give **2** as a white powder. Analytical data are given in Table VII.

β-Galactosidase Activity Assay.[69] Clonal 293 cell lines expressing the human MSH receptor, human MC3 receptor, human MC4 receptor, and mouse MC5 receptor are transfected with a pCRE/β-galactosidase construct using a $CaPO_4$ method.[70] Four micrograms of pCRE/β-gal DNA is used for transfection of a 10-cm dish of cells. After 15–24 hr posttransfection, cells are split into 96-well plates with 20,000–30,000 cells/well and incubated at 37° in a 5% CO_2 incubator until 48 hr posttransfection. Cells are then stimulated with different α-MSH analogues diluted in stimulation medium (Dulbecco's modified Eagle's medium containing 0.1 mg/ml BSA and 0.1 m*M* isobutylmethylxanthine) for 6 hr at 37° in a 5% CO_2 incubator. Agonist activity is measured by stimulating cells with various concentrations of α-MSH, Ac-[Nle^4,D-Phe^7]α-MSH, and cyclic analogues **1** and **2** (Table V). Antagonist activity is measured by stimulating MC3, MC4, and MC5 receptor cell lines with various concentrations of α-MSH or α-MSH plus various concentrations of the compounds. After stimulation, cells are lysed in 50 μl of

[69] W. Chen, T. S. Shields, P. J. S. Stork, and R. D. Cone, *Anal. Biochem.* **226,** 349 (1995).
[70] C. Chen and H. Okayama, *Mol. Cell Biol.* **7,** 2745 (1987).

lysis buffer (250 mM Tris–HCl, pH 8.0, 0.1% Triton X-100), frozen, thawed, and then assayed for β-galactosidase activity as described.[69] Data represent means and standard deviations from triplicate data points, and curves are fitted by linear regression using GraphPad Prism software. Antagonist pA_2 values are determined using the method of Schild.[2]

Synthesis of Melanotropin Analogues Based on the Somatostatin Template. All peptides are synthesized manually in a stepwise fashion via the solid-phase method using 1 or 2% cross-linked pMBHA resin (0.30–0.45 mmol/g). For N^{α}-Boc-protected amino acids, the following side chain protecting groups are used: Arg(N^{G}-Tos), Cys(S-pMb), D-Trp(N^{i}-For), His(N^{π}-Bom), Pen(S-pMb), Thr(O-Bn), Trp (N^{i}-For), and Tyr(O-2, 6-dichlorobenzyl). Coupling of the amino acids to the resin is carried out using a 3-fold excess of the HOBt, and couplings are monitored by the Kaiser test. Each natural amino acid is coupled in this fashion, whereas novel amino acids are coupled using 1.2- to 2-fold excess HBTU with HOBt (1 equivalent) and DIEA. Deprotection of the N^{α}-Boc protecting group is carried out with 50% TFA/DCM containing 0.2% methionine. After coupling the N^{α}-terminal residue, the final N^{α}-Boc-peptide resin is cleaved directly by HF. All protecting groups of the side chains are cleaved by HF in the presence of scavengers [commonly used: thioanisole and p-cresol (1 : 1 mixture)]. The crude peptide, scavengers, and resin mixture are washed three times with anhydrous ether that was discarded. The filter cake, crude peptide, and resin mixture are washed successively with glacial acetic acid three times followed by deionized water once. The solution mixture is then lyophilized.

The lyophilized linear crude peptide mixture is cyclized directly using potassium ferric(III) hexacyanide [$K_3Fe(CN)_6$] as follows: In a 250-ml round-bottom flask equipped with a magnetic stirring bar, 25–30 ml of 0.1 N $K_3Fe(CN)_6$ in water is added together with an equivalent volume of acetonitrile or methanol. Then, the pH of the solution is adjusted to 8.5–10 by the addition of 10% ammonia solution. The crude linear peptide (0.1 mmol) is dissolved in the 15–30 ml mixture of water and methanol. With vigorous stirring of the flask, the peptide solution is injected into the flask at a steady speed of around 2–3 ml/hr using an automated syringe pump. After the addition is complete, the reaction mixture is stirred an extra hour. The final cyclized peptide solution is acidified with dilute acetic acid in water to pH 4–5. A tablespoon of activated Amberlite IR-120 (ion-exchange resin) is added to the reaction flask with vigorous stirring. After 2 hr, the stirring is stopped, and after the resin has settled at the bottom of the flask, if the color of the solution is still green, more resin is added. This process is repeated until the solution is colorless. Then the resin is filtered and washed twice with acetic acid. The filtrates are combined and dried under reduced vacuum and lyophilized.

The cyclized crude peptide is purified by HPLC using a preparative C18 column (reverse phase) eluted with a linear gradient of acetonitrile (10–90%) in aqueous 0.1% TFA (v/v) for over 40 min. The absorbance of the eluent is detected by a Rainin UVD detector, monitored at 230 and 280 nm. The fractions collected are checked for purity by an analytical reverse HPLC column using a Hewlett Packard HPLC 1090II with a diode array detector monitored at various wavelengths, typically at 230, 254, and 280 nm. The pure fractions are then pooled and lyophilized. The identity of the synthesized peptides is confirmed by positive ion fast atom bombardment mass spectroscopy (FAB-MS) and amino acid analysis (AAA). AAA is performed using an Applied Biosystems 420A amino acid analyzer with automatic hydrolysis (vapor phase hydrolysis at 160° for 1 hr using 6 *N* HCl) or with prior hydrolysis (110° for 24 hr using 6 *N* HCl), with precolumn phenylthiocarbamoylamino acid (PTC-AA) analysis. However, some amino acids are not reliably analyzed quantitatively under these conditions. High-resolution mass spectroscopy is also used. Purity of the final peptide is also determined by TLC in three different solvent systems [solvent A: *n*-BuOH/AcOH/pyridine/H_2O (5 : 5 : 1 : 4); solvent B: EtOAc/pyridine/AcOH/H_2O (5 : 5 : 1 : 3); and solvent C: *n*-BuOH/AcOH/H_2O = 4 : 1 : 1] (Table VII).

In Vitro Frog and Lizard Skin Bioassays.[71,72] Skin from the thigh and legs of *R. pipiens* frogs (30–50 g) is removed and cleaned of large vessels. The skin samples are placed in a frame consisting of two rings. The outer ring is made of Bakelite, and the inner of aluminum. Notches and holes in the Bakelite frame are made in order to prevent the accumulation of air bubbles and to allow free circulation of the test solution. Excess skin is cut off, and the frame with the surface of the skin downward is placed in a 50-ml beaker containing 20 ml of frog Ringer's solution (NaCl 6.5 g/liter, KCl 0.14 g/liter, $CaCl_2$ 0.12 g/liter, $NaHCO_3$ 0.2 g/liter). A photoelectric reflectometer is used to measure light reflection from the skin. The search unit is placed on the table with its opening pointing upward, and the beaker containing the frame with the skin is placed on top of it so that the skin in the frame fits over the search unit opening. Although the frame is made to fit the search unit opening of the reflection meter, the light from this source is not of equal intensity throughout the field. For this reason, a mark is put both on the frame and on the search unit, and the beaker is placed to conform to these markings so that the position of the skin with respect to the search unit opening is constant throughout every reading. For the reading of frog skin, no filter is used. Room light must be kept constant during the assay.

After the skin is put into the 20 ml of Ringer's solution, an increase in reflection occurs gradually because of the spontaneous lightening in color of the melanocytes. After about 1 hr, lightening of the melanocytes reaches a maximum, during which time the Ringer's solution is changed once. The reading at the end of 1 hr is taken as the baseline value. If the reflectance is more than 100 or less than 70, the skin is discarded in order to approximate constancy of color among the skin samples.

[71] K. Shizume, A. B. Lerner, and T. B. Fitzpatrick, *Endocrinology* **54,** 553 (1954).
[72] J. M. Goldman and M. E. Hadley, *J. Pharmacol. Exp. Ther.* **166,** 1 (1969).

In response to melanotropins, melanosomes within integumental melanophores migrate from a perinuclear position into dendritic processes of pigment cells. This centrifugal organellar movement results in a darkening of the skin and hence a rapid decrease of reflection results. Subsequent removal of the melanotropins (Ringer rinse) usually results in a rapid perinuclear (centripetal) reaggregation of melanosomes, leading to lightening of the skins back to their original base reflectance value.

For the lizard skin bioassay, both male and female lizards (*A. carolinensis*) are used. Animals are sacrificed by decapitation, the back skin from each animal is removed and mounted on the frame as described earlier, and the experiment is then performed in a similar manner as described for the frog skin bioassay.

δ-Opioid Receptor Antagonists and Inverse Agonists

δ-Opioid Antagonists as Therapeutic Agents. While antagonists are used for the biochemical and pharmacological characterization of both agonists and receptors, they are also useful as therapeutic agents. Thus μ-opioid receptor antagonists, such as naloxone, are used commonly as an antidote for patients overdosing on morphine or heroin. A recent hypothesis suggests that antagonists could possibly correct aberrations in the basal tone of the opioid system. Overactivation of the opioid system, facilitating dopaminergic release, has been theorized as a contributory factor in cocaine abuse. Blocking this hyperactivation with an opioid antagonist causes cocaine cravings to diminish.[73]

Again, overactivation of the opioid systems in the brain stem area of the brain is considered a contributory factor in autism.[74] Improvement was shown when naltrexone, a broad spectrum opioid antagonist, was given to an autistic patient. Improvements may have been more dramatic if an opioid receptor-specific antagonist had been given.

Other human diseases have shown improvement with the use of a therapeutic δ-opioid receptor antagonist, naltrindole: suppression of immune function[75] and a decrease in the craving for alcohol in patients.[76]

Evolution of Peptide-Based Antagonists. *N,N*-Diallyl-Leu-enkephalin was the first δ-opioid antagonist and was called 139462.[77] Unfortunately, this molecule degraded easily *in vivo.* To slow the enzymatic degradation, the amide bond in the molecule was converted to a (CH_2S) bond, which stabilized the peptide to

[73] K. Menkens, L. D. Reid, P. S. Portoghese, K. D. Wild, E. J. Bilsky, and F. Porreca, *Eur. J. Pharmacol.* **219,** 345 (1992).

[74] J. Panksepp, *Trends Neurosci.* **174** (1979).

[75] R. V. House, J. T. Kozak, P. T. Thomas, and H. Bhargava, *Neurosci. Lett.* **198,** 119 (1995).

[76] J. C. Froehlich, D. E. McCullough, R. W. Zink, N. E. Badia-Elder, and P. S. Portoghese, *J. Pharmacol. Exp. Ther.* **287,** 284 (1998).

[77] J. S. Shaw, J. J. Gormley, J. J. Turnbull, L. Miller, and J. S. Morley, *Life Sci.* **31,** 1259 (1982).

enzymatic degradation. The resultant peptide was (*N,N*-bisallyl-Tyr-Gly-Gly-Ψ-(CH_2S)-Phe-Leu-OH (ICI 154129).[78] ICI 154129 had a dissociation constant (K_d value) of 254 *nM* and 7.4 μM against [^{3}H]Leu-enkephalins and [^{3}H]normorphine, respectively. The low-affinity ICI 154129 required that higher affinity δ-opioid receptor antagonists were needed.

Modifications of ICI 154129 resulted in better δ-opioid receptor antagonists such as *N,N*-diallyl-Tyr-Aib-Phe-Leu-OH (ICI 174864).[79] The difference between ICI 174,864 and ICI 154,129 lies in the replacement of the Gly–Gly moiety with the unusual amino acid residue α-aminoisobutyric acid (Aib) and the absence of the CH_2S bond. Aib is known to conformationally constrain a peptide. Conformationally constraining the peptide with Aib increased the selectivity of the antagonist at the δ-opioid receptor. In fact, ICI 174,864 demonstrated a K_e value of 30.1 and 30.6 *nM* against the δ selective agonists [^{3}H]Leu-enkephalin and [^{3}H][D-Thr^2,Leu^5,Thr^6]enkephalin (DSLET), respectively.

Further constraints of the peptide backbone were investigated with the use of *N*-methylphenylalanine and tetrahydro-3-isoquinoline carboxylic acid (Tic).[80] Although these amino acids are similar structurally, Tic modification provides more constraint through a covalent attachment of the *N*-methyl group to the aromatic ring.

This unnatural constrained amino acid was also explored in the TIP(P) series of peptides; Tyr-Tic-Phe (TIP) and Tyr-Tic-Phe-Phe (TIPP). Similar to ICI 174864, the constraint of the molecule was made at the second amino acid position. D- and L-Tic provided a μ/δ selectivity of 1410 for the antagonists.[80] The Tic residue differentiated the antagonistic character of the peptide by enantiomer alone. Incorporation of a D-Tic produced a μ-opioid agonist, whereas the incorporation of an L-Tic produced a δ-opioid antagonist. Both TIPP and TIP were potent and selective at the δ-opioid receptor.

While the TIPP peptide was very selective and potent, it was found that the incorporation of Tic^2 led to slow spontaneous diketopiperazine formation between the Tyr and the Tic with subsequent cleavage of the Tic–Phe bond. In order to overcome this problem, the TIPP peptide was modified further to produce the peptide TIPP[Ψ].[81] This peptide contains a reduced peptide bond between the Tic^2 and the Phe^3 residue, which would not form diketopiperazines. TIPP[Ψ] was found to be highly potent and selective for the δ-opioid receptor and proved to be highly stable against enzymatic degradation.[82]

[78] T. Priestley, J. J. Turnbull, and E. Wei, *Neuropharmacology* **24,** 107 (1985).

[79] R. Cotton, J. S. Shaw, L. Miller, M. G. Giles, and D. Timms, *Eur. J. Pharmacol.* **97,** 331 (1984).

[80] N. N. Chung, C. Lemieux, B. J. Marsden, B. C. Wilkes, G. Weltrowska, T. M.-D. Nguyen, and P. W. Schiller, *Proc. Natl. Acad. Sci. U.S.A.* **89,** 11871 (1992).

[81] G. W. Pasternak, P. W. Schiller, K. M. Standifier, and L. M. Visconti, *Neurosci. Lett.* **181,** 47 (1994).

[82] C. Lemieux, N. N. Chung, B. C. Wilkes, T. M.-D. Nguyen, G. Weltrowska, and P. W. Schiller, *J. Med. Chem.* **36,** 3182 (1993).

Dipeptide Antagonists. The formation of δ-antagonists has been based primarily on changes in the message domain of enkephalins. Because the message domain for both dermorphins (μ selective) and deltorphins (δ selective) is the same (Tyr-D-Ala-Phe), it has been suggested that the orientation in aqueous solution contributed to whether the agonists were μ or δ selective.[83] Spatial orientation of the TIPP series of peptides, as well as enkephalins, dermorphins, and deltorphins, was established to be important. A hypothesis was then formulated to suggest that the spatial orientation of the antagonist naltrindole (NTI) was similar to the dipeptide analogue Tyr-Tic.

Analogues of Tyr-Tic were then synthesized that demonstrated good selectivity. Tyr-Tic was also modeled and compared with NTI and was found to be in good agreement with the aromatic portions of NTI. Tyr-Tic adopted a low energy conformation similar to that of *N*-methyl naltrindole.[84] This represents a convergence of peptide and nonpeptide constraints in antagonists. This constraint convergence between peptide and nonpeptide antagonists offers insight in both the structural basis of peptides and the binding site of the receptors.

A second generation of analogues was synthesized to explore the impact of increased lipophilicity of the Tyr moiety of the molecule. The tyrosine was replaced with dimethyl tyrosine (Dmt).[85] The result was a δ-opioid antagonist that surpassed all other antagonists in its δ selectivity and its binding affinity. The affinity of Dmt-Tic for the rat brain δ receptors was reported to be 0.022 n*M* with a selectivity of 150,000 over the μ-opioid receptor.[83] Interestingly, however, the dissociation constant of Dmt-Tic at the mouse vas deferens δ-opioid receptor was only 5 n*M*. Its potencies at the human DOR or MOR have not been investigated.

Inverse Agonism in hDOR

In order to achieve an even better antagonist, we hypothesized that further constrain of the χ space in Dmt-L-Tic would lead to a better antagonist. Incorporation of the χ^1 and χ^2 constrained novel amino acids, β-methyl-2′,6′-dimethyltyrosine (TMT) (all four isomers), led to the analogue (2S,3R)TMT-Tic, which has very high binding affinity, with inverse agonist activity[86,87] at the δ-opioid receptor. This compound now provides a potent and selective δ ligand to explore the role of inverse agonists in opioid pharmacology and physiology.

[83] L. H. Lazarus, P. S. Cooper, S. Salvadori, and S. D. Bryant, *Trends Pharmacol. Sci.* **19,** 42 (1998).

[84] T. Tancredi, D. Picone, L. H. Lazarus, R. Tomatis, R. Guerrini, C. Bianchi, P. Amodeo, S. Salvadori, and P. A. Temussi, *Biochem. Biophys. Res. Commun.* **198,** 933 (1994).

[85] S. Salvadori, M. Attila, G. Balboni, C. Bianchi, S. D. Bryant, O. G. R. Crescenzi, D. Picone, T. Tandredi, and P. A. Temussi, *Mol. Med.* **1,** 678 (1995).

[86] S. Liao, J. Lin, M. D. Shenderovich, Y. Han, K. Hosohata, P. Davis, W. Qiu, F. Porreca, H. I. Yamamura, and V. J. Hruby, *Bioorg. Med. Chem. Lett.* **7,** 3049 (1997).

[87] K. Hosohata, T. H. Burkey, J. Alfaro-Lopez, V. J. Hruby, W. R. Roeske, and H. I. Yamamura, *Eur. J. Pharmacol.* **380,** R9 (1999).

Synthesis of TMT-L-Tic. The four optically pure diastereoisomers of N^{α}-Boc-β-methyl-2′,6′-dimethyltyrosine (N^{α}-Boc-TMT) were synthesized by literature methods.[88] The synthesis of TMT-Tic dipeptide started from commercially available N^{α}-Boc-Tic. The compound was first converted to its benzyl ester via reaction with benzyl bromide in the presence of phase transfer catalyst triethylbenzylammonium chloride. Then the N^{α}-Boc protecting group was removed with 1 *N* hydrochloric acid in acetic acid. The coupling of the intermediate with N^{α}-Boc-TMT was the most difficult step in the synthesis. It did not work using DIC/HOBt coupling reagents. However, it could be accomplished in 60–75% yields using HATU in DCC in the presence of diisopropylethylamine (DIEA) to give the protected dipeptide. The protected dipeptide Boc-TMT-Tic-OBn was hydrogenolyzed in the presence of 10% Pd/C to cleave the benzyl ester and was then treated with 1 *N* hydrochloric acid to remove the N^{α}-Boc protection. The final product was precipitated with ether, filtered, and purified with RP-HPLC with 0.1% TFA/acetonitrile as the eluent.

Conclusion

Many disciplines are needed in order to design a drug, and it is quite clear that by appropriate design we can improve on Nature. As a drug develops toward perfection, and as techniques become better, it is theoretically possible that one may be able to invent the "perfect" drug with the "perfect" binding and bioactivities. While it is theoretically possible to design a drug that binds perfectly to the receptor, it may be difficult to, at same time, design a molecule with "perfect" biodistribution, diffusion, pharmacokinetics, etc. In any case, GPCR antagonist ligands will continue to play a major role in the evaluation of these receptors and as drugs. As science begins to understand the role of second messengers in the body, one can foresee that antagonists will be used to help balance the effects of the second messengers on some receptors in order to correct dysfunctions of mind and body.

Acknowledgments

Supported by grants from the U.S. Public Health Service (USPHS) and the National Institute of Drug Abuse. The opinions expressed are those of the authors and do not necessarily reflect those of the USPHS.

[88] X. Qian, K. C. Russell, L. W. Boteju, and V. J. Hruby, *Tetrahedron* **51,** 1033 (1995).

[5] Design of Peptide Agonists

By VICTOR J. HRUBY, RICHARD S. AGNES, and CHAOZHONG CAI

Introduction

Peptides and protein hormones that interact with receptors, in particular G protein-coupled receptors (GPCRs), and neurotransmitter ligands, which interact with GPCRs, have been targets for drug design for many years, constituting over 50% of all current drugs. An important goal of research in this area has been to determine the important functional groups and structural, conformational, topographical, and dynamic properties of the peptide ligands responsible for a specific biological activity. These peptide analogue modifications are designed with several goals in mind: to have agonist bioactivity that retains or improves the affinity and selectivity for a specific receptor or receptor subtype; to have agonist bioactivity, either for signal transduction in a second messenger system or in a physiological response; to have stability against proteolytic degradation; and to have improved bioavailability.

The design of peptides resulting in potent and selective agonist activities is difficult because of several factors, but success can be normally obtained with a systematic approach.[1–4] This approach requires highly interdisciplinary considerations that combine knowledge gathered from structure–biological activity relationships, conformational analysis, computer-assisted calculations and molecular design, biophysical analysis of structures (spectroscopy, crystallography, etc.), and *in vivo* and *in vitro* assays. In addition, new asymmetric syntheses and other synthetic methodologies are often important in preparing the designed compounds. The following sections briefly describe a systematic approach to designing potent and selective agonists for GPCRs. Sufficient and detailed experimentals are included as a model to synthesize the mentioned peptide agonists successfully.

Minimal Requirements

Once a lead compound (whether from Nature or from a peptide library prepared by combinatorial chemistry,[5,6]) has been defined based on the desired agonist activity, it is often desirable to determine the minimal structural requirements for bioactivity. Truncation at N- and/or C-terminal residues is the first step in reducing

[1] V. J. Hruby, F. Al-Obeidi, and W. M. Kazmierski, *Biochem. J.* **268,** 249 (1990).
[2] G. R. Marshall, *Tetrahedron* **49,** 3547 (1993).
[3] V. J. Hruby, *Biopolymers* **33,** 1073 (1993).

METHODS IN ENZYMOLOGY, VOL. 343
0076-6879/02 $35.00

the complexity of a peptide. By identifying the important amino acid residues responsible for bioactivity, for retaining binding affinity and agonist potency compared to the parent peptide through a shortened sequence of the peptide, a more simple peptide pharmacophore can be identified. This is done by the sequential removal of residues from the C- and/or the N-terminal sides of the parent peptide. A major advantage of this initial approach is that it will subsequently allow an easier and faster syntheses of analogues. Also, because the sequence is shortened, the complexity of biophysical analysis is reduced.

Next, one must consider the contributions of specific functional groups in the primary sequence. In an alanine scan, a specific side chain group of a specific residue on bioactivity is examined to determine whether it plays an important role in binding affinity and/or agonist bioactivity by systematically replacing each amino acid side chain moiety (except glycine) with the methyl group of alanine. Residues found not to be very important are considered and utilized as sites for further substitutions for explorations of potency, selectivity, stability, and so on and as points for introducing further constraints. Critical residues can also be modified to increase potency, selectivity, stability, and other properties, but this requires a more comprehensive analysis.

To demonstrate the principles discussed earlier, structure–activity relationship studies of a linear peptide ligand, neurotensin, will be discussed, as shown in Table I. Neurotensin is a 13 amino acid residue peptide hormone that has roles in hypothermia and analgesia, among others.[7] Truncation studies revealed that only the C-terminal hexapeptide sequence (Arg^8-Arg^9-Pro^{10}-Tyr^{11}-Ile^{12}-Leu^{13}-OH) was required for maintaining the biological effects of neurotensin. Specifically, the sequence Arg^9-Pro^{10}-Tyr^{11} appeared to contain the most critical chemical groups responsible for the intrinsic activity at neurotensin receptors.[8]

To evaluate this hypothesis, the role of each amino acid side chain of neurotensin in the interaction between the peptide and its receptor was investigated using an alanine scan strategy. The alanine scan study, which sequentially replaced each amino acid of NT(8–13) with alanine, suggested the relative importance of the individual amino acid side chains for bioactivity to be as follows: $Leu^{13} > Tyr^{11} \gg Ile^{12} > Arg^9 > Pro^{10} > Arg^8$; this showed that the isopropyl group of leucine and the aromatic ring of tyrosine were particularly important for binding

[4] M. E. Hadley, V. J. Hruby, J. Blanchard, R. T. Dorr, N. Levine, B. V. Dawson, F. Al-Obeidi, and T. K. Sawyer, "Integration of Pharmaceutical Discovery and Development: Case Studies" (R. Borchard, ed.), p. 575. Plenum Press, New York, 1998.

[5] R. A. Houghten, C. Pinella, S. E. Blondelle, J. R. Appel, C. T. Dooley, and J. H. Cuervo, *Nature* **354,** 84 (1991).

[6] K. S. Lam, S. E. Salmon, E. M. Hersh, V. J. Hruby, W. M. Kazmierski, and R. J. Knapp, *Nature* **354,** 82 (1991).

[7] J.-P. Vincent, J. Mazell, and P. Kitabgi, *Trends Pharmacol. Sci.* **20,** 302 (1993).

[8] C. Grainer, J. Van Rietschoten, P. Kitabgi, and P. Freychet, *Eur. J. Biochem.* **124,** 117 (1982).

TABLE I
BINDING AFFINITIES OF NEUROTENSIN AGAINST RADIOLABELED NEUROTENSIN AT VARIOUS CELL PREPARATIONS[a]

Compound	Binding (IC_{50}, n*M*)
NT(1–13)	1.2[b]
Arg^8-Arg^9-Pro^{10}-Tyr^{11}-Ile^{12}-Leu^{13}	1[b]
NT(9–13)	90[b]
NT(10–13)	>4000[b]
Lys^8-NT(8–13)	1[b]
[Ala^8]-NT(8–13)	0.25[c]
[Ala^9]-NT(8–13)	2.1[c]
[Ala^{10}]-NT(8–13)	0.90[c]
[Ala^{11}]-NT(8–13)	1100[c]
[Ala^{12}]-NT(8–13)	4.5[c]
[Ala^{13}]-NT(8–13)	2000[c]
[Trp^{11}]-NT(8–13)	10[d]
L-[1'-Nal^{11}]-NT(8–13)	6[d]

[a] These sets of compounds show the minimum required elements in neurotensin binding.
[b] From C. Granier, J. Van Rietschoten, P. Kitabgi, and P. Freychet, *Eur. J. Biochem.* **124,** 117 (1982).
[c] From J. A. Henry, D. C. Horwell, K. G. Meecham, and D. C. Rees, *Bioorg. Med. Chem. Lett.* **3,** 949 (1993).
[d] From R. Quiron, F. Regoli, F. Rioux, and S. St. Pierre, *Br. J. Pharmacol.* **69,** 689 (1980).

affinity *in vitro.*[9] In fact, substitution of the Tyr residue with Trp and naphthalanine further supported the importance of an aromatic ring.[10,11] Truncation studies of NT(8–13) also suggested a requirement of at least one basic residue, as NT(8–13) and NT(9–13) have comparable binding affinities, whereas NT(10–13) has a significantly lower binding affinity. This basic residue, however, is not required to be a guanidium ion, as the analogue in which residue 9 was substituted with Lys retained binding affinity.[12] Furthermore, a free C-terminal carboxylate also appears to be essential for high receptor affinity, as the NT(8–13) C-terminal amide was two orders of magnitude less potent in binding affinity than the carboxylate terminal peptide.

[9] J. A. Henry, D. C. Horwell, K. G. Meecham, and D. C. Rees, *Bioorg. Med. Chem. Lett.* **3,** 949 (1993).
[10] B. Cusack, K. Ghorshan, D. J. McCormick, Y.-P. Pang, C.-T. Phung, T. Souder, and E. Richelson, *J. Biol. Chem.* **271,** 15054 (1986).
[11] R. Quiron, F. Regoli, F. Rioux, and S. St. Pierre, *Br. J. Pharmacol.* **69,** 689 (1980).
[12] C. Granier, J. Van Rietschoten, P. Kitabgi, C. Poustis, and P. Freychet, *Eur. J. Biochem.* **124,** 117 (1982).

Conformational Local Constraints

Once the minimum active sequence and key residues required for agonist bioactivity have been determined from the primary sequence of the parent peptide, the secondary structure of the peptide has to be explored. Because most small linear peptides are inherently flexible, it is difficult to determine the bioactive conformation of the peptide when it interacts with the receptor (for GPCRs and their ligands, this could change if X-ray crystal structures or other biophysical studies, which provide three-dimensional information of the ligand–receptor complex, could be obtained). One thus seeks to constrain the local backbone conformation (ϕ, Ψ, and ω angles) of the peptide to a preferred secondary structure. One general approach that can provide suggestive insight into possible secondary structures important to ligand bioactivity is to do a D-residue scan. In this approach, a systematic replacement of each L-amino acid residue by the corresponding D-amino acid is examined. This provides insight into the stereochemical requirements, as well as the possible spatial orientations of the side chain residues for peptide–receptor interactions. Because most enzyme proteases are stereoselective for natural L-amino acid residues, peptides with D-amino acid residues generally are more stable against degradation in a biological setting. For example, the incorporation of D-Phe in α-MSH analogues led to [Nle^4, D-Phe^7]α-MSH (MT I), which is super potent, with prolonged biological activity in pigment cell receptors and with greatly enhanced stability.[13] It was proposed that the increased biological activity of MT I was due to stabilization of a β turn by the D-Phe residue, which was tested by the preparation of cyclic analogues with excellent results.[13–15]

Another way to determine whether a turn in the peptide backbone is important for bioactivity is to scan for residues that induce secondary structure more easily. For example, proline is known to induce a certain class of reverse turns (β turns), although this is not always the case. Also, X-proline amide bonds are known to assume both *trans* and *cis* conformations. In addition, other amino acid residues are at times N^{α}-alkylated or arylated for this purpose. The N^{α}-alkylation of the backbone in peptides reduces the freedom of the amide bond two- to fourfold, eliminates the hydrogen bond donation ability, and affects the backbone torsional angle. This constraint has been applied to many peptide systems. For example, N^{α}-methylnorleucine was incorporated in cholecystokinin (CCK) at position 31 and resulted in an increased selectivity for the CCK-B receptor.[16]

[13] T. K. Sawyer, P. J. Sanfilippo, V. J. Hruby, M. H. Engel, C. B. Heward, J. B. Burnett, and M. E. Hadley, *Proc. Natl. Acad. Sci. U.S.A.* **77,** 5754 (1980).

[14] T. K. Sawyer, V. J. Hruby, P. S. Darman, and M. E. Hadley, *Proc. Natl. Acad. Sci. U.S.A.* **79,** 1751 (1982).

[15] V. J. Hruby, B. C. Wilkes, W. L. Cody, T. K. Sawyer, and M. E. Hadley, *Peptide Protein Rev.* **3,** 1 (1984).

[16] V. J. Hruby, S. Fang, R. Knapp, W. Kazmierski, G. K. Lui, and H. I. Yamamura, "Peptide: Chemistry, Structure, and Biology. Proceedings of the 12th American Peptide Symposium" (J. E. Rivier and G. R. Marshall, eds.), p. 53, ESCOM, Lieden, 1990.

TABLE II
AGONIST MELANOTROPIC ACTIVITY FOR MODIFIED ANALOGUES OF THE μ OPIOID PEPTIDE CTOP USING THE FROG SKIN ASSAY[a]

Compound	Agonist activity, EC_{50} (M)
α-MSH	1.5×10^{-10}
Ac-Nle-c[Asp-His-D-Phe-Arg-Trp Lys]-Thr-NH_2	10^{-11}
H-D-Phe-c[Cys-Tyr-Orn-Arg-Trp-Pen]-Thr-NH_2 (CTOP)	10^{-7}
H-D-Phe-c[Cys-His-D-Phe-Arg-Trg-Pen]-Thr-NH_2	10^{-7}
H-D-Phe-c[Cys-His-D-Phe-Arg-Trp-Cys]-Thr-NH_2	10^{-8}
H-D-Phe-c[HCys-His-D-Phe-Arg-Trp-Cys]-Thr-NH_2	3×10^{-10}
H-D-Phe-c[Asp-His-D-Phe-Arg-Trp-Lys]-Thr-NH_2	10^{-10}

[a] From V. J. Hruby, G. Han, and M. E. Hadley, *Lett. Peptide Sci.* **5,** 117 (1998).

Another approach used to obtain a stabilized local secondary structure is to use a stabilized cyclic template that has served previously as a secondary structure scaffold for a peptide ligand for a particular GPCR, but that can be substituted with specific amino acid residues or fragments in a peptide ligand that will retain its secondary structure but became a potent and selective ligand for a different GPCR. This approach was first used in our laboratory to design a μ opioid receptor antagonist based on a β-turn fragment of somatostatin.[17,18] The same approach was shown to be effective for designing agonists and antagonists for the melanocortin-1 receptor (MC1R), which is also a GPCR.[19] Insights that led to this approach came from our observation that both α-melanotropin and CTOP (H-D-Phe-c[Cys-Tyr-D-Trp-Orn-Thr-Pen]-Thr-NH_2), a μ opioid receptor antagonist, have a β turn about an aromatic residue in its bioactive conformation. Thus, substituting the turn scaffold from CTOP with appropriate α-MSH residues resulted in α-MSH-related analogues with comparable bioactivity to α-MSH at the MC1R as the parent peptide hormone, as shown in Table II.[19]

Stabilizing Bioactive Conformations by Global Constraints

Computational modeling and biophysical studies (circular dichroism, nuclear magnetic resonance spectroscopy, FT-IR, Raman spectroscopy, fluoresence spectroscopy, electron transfer reactions, etc.) can provide important information regarding secondary structures (α helix, β turns, and β sheets, for example) of peptides. The global structure of a peptide can be constrained by cyclization,

[17] J. T. Pelton, K. Gulya, V. J. Hruby, S. P. Duckles, and H. I. Yamamura, *Proc. Natl. Acad. Sci. U.S.A.* **82,** 236 (1985).

[18] J. T. Pelton, W. M. Kazmierski, K. Gulya, H. I. Yamamura, and V. J. Hruby, *J. Med Chem.* **29,** 2370 (1986).

[19] V. J. Hruby, G. Han, and M. E. Hadley, *Lett. Peptide. Sci.* **5,** 117 (1998).

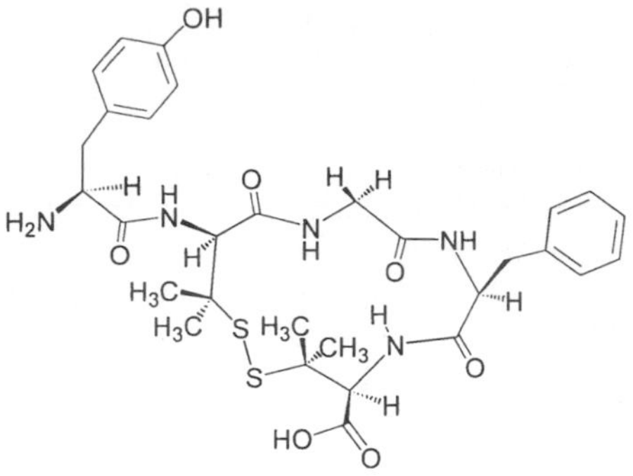

FIG. 1. The structure of c[D-Pen^2, D-Pen^5]enkephalin (DPDPE).

which induces or stabilizes a secondary structure favorable for agonist activity.[20–22] From an alanine scan and other related methods, residues that are not considered important for binding and potency can be uncovered and these residues can then be considered as potential points of cyclization. Several factors can be explored here such as macrocyclic ring size and ring type (end to end, end to side chain, side chain to side chain, backbone to side chain, etc.). Depending on the type of side chains involved, one can make cyclic disulfide bonds, cyclic amide bonds, and many others.[1,20,21] Because of the variety of ways in introducing a cyclic structure, the synthetic methods that are used are many and varied.

Enkephalins, the endogenous opioid pentapeptides, have been shown to interact with several classes of opioid receptors. However, like most small and linear peptides, native enkephalins possess considerable conformational flexibility and are not selective or stable enough to be considered as potential nonaddictive pain relievers. Considerable effort has been made to develop highly potent and selective ligands for δ opioid receptors. One method for affecting conformational restriction in enkephalins is through internal disulfide cyclization of the peptide, resulting in analogues that can assume a compact topography. One successful approach used in our laboratory led to the development on one of the most selective analogues for δ opioid receptors, c[D-Pen^2, D-Pen^5]enkephalin (DPDPE, Fig. 1),[23] a peptide with a highly constrained 14-membered ring, which was also highly resistant to enzymatic degradation.

In another example based on computer-assisted molecular dynamic simulations, a class of superpotent α-melanotropin agonists with prolonged activity at

[20] H. Kessler, *Angew. Chem. Int. Ed. Engl.* **21,** 512 (1982).

[21] V. J. Hruby, *Life Sci.* **31,** 189 (1982).

[22] J. Rizo and L. M. Gierasch, *Annu, Rev. Biochem.* **61,** 387 (1992).

[23] H. I. Mosberg, R. Hurst, V. J. Hruby, K. Gee, H. I. Yamamura, J. J. Galligan, and T. F. Burks, *Proc. Natl. Acad. Sci. U.S.A.* **80,** 5871 (1983).

melanocortin receptors were developed by designing cyclic analogues of the linear analogue Ac-[Nle^4, D-Phe^7]α-MSH (MT-I).[24,25] A side chain to side chain cyclic lactam bridge was introduced flanking the His-D-Phe-Arg-Trp sequence of MT-I, producing Ac-Nle^4-c[Asp^5, D-Phe^7, Lys^{10}]α-MSH(4–10)-NH_2 (MT-II).[25] MT-II has superpotent and prolonged activity at pigment cell receptors and cloned human receptors; interestingly, it is also equipotent to α-MSH in the frog skin assay, suggesting that cyclization is an effective tool in determining the bioactive conformation of α-MSH at a variety of receptors, and subsequent NMR studies and computation studies gave evidence of a β-turn structure.[26]

As a formal example, conformational studies suggested that the message sequence of dynorphin A (Dyn A) may have an α-helical conformation and the address sequence of an induced helical structure.[27] Also, an α-helical structure from Gly^3 to Arg^9 was found in Dyn A bound to a lipid micelle in a ^{1}H-NMR study.[28] A side chain to side chain lactam cyclization between the side chain groups of aspartic acid and lysine at positions i and $i + 4$ has been found to stabilize helical peptides.[29] Thus a series of Dyn A analogues with cyclic lactam rings were designed and synthesized with these substitutions at the i and $i + 4$ positions along the side chain. c[D-Asp^3, Lys^7]DynA(1–11)-NH_2 was found to have higher affinity, selectivity, and activity at the k-opioid receptors in guinea pig ileum assays than the linear [D-Asp^3, Lys^7]DynA(1–11)-NH_2 counterpart in support of this hypothesis.[30]

Topographical Constraints

Once critical insights into the key pharmacophore elements in terms of amino acid side chain groups and the proper backbone template for these elements have been obtained, it is still necessary to determine the three-dimensional topographical relationships of the pharmacophore elements.[1,31] Substitution at diastereotopic β positions of amino acid residues can confer appropriate conformational constraints in χ space to explore topography for many amino acids. β-Substituted isomers of the same amino acid have similar physicochemical properties, such as electronegativity and hydrophobicity. The only difference is their side chain

[24] F. Al-Obeidi, V. J. Hruby, B. M. Pettit, and M. E. Hadley, *J. Am. Chem. Soc.* **111,** 3413 (1989).

[25] F. Al-Obeidi, V. J. Hruby, A. M. L. Castrucci, and M. E. Hadley, *J. Med. Chem.* **32,** 2555 (1989).

[26] F. Al.-Obeidi, S. D. O'Connor, C. Job, V. J. Hruby, and B. M. Pettit, *J. Peptide Res.* **51,** 420 (1998).

[27] N. Collins and V. J. Hruby, *Biopolymers* **24,** 1231 (1994).

[28] D. A. Kallick, *J. Am. Chem. Soc.* **115,** 9317 (1993).

[29] A. M. Felix, C.-T. Want, R. M. Campbell, V. Toome, D. C. Fry, and V. S. Madison, "Peptide: Chemistry and Biology. Proceedings of the 12th American Peptide Symposium" (J. E. Rivier and G. R. Marshall, eds.), p. 77, ESCOM, Leiden, 1992.

[30] T. F. Lung, N. Collins, D. Stropova, P. Davis, H. I. Yamamura, F. Porreca, and V. J. Hruby, *J. Med. Chem.* **39,** 1136 (1996).

[31] V. J. Hruby, G. Li, C. Haskell-Luevano, and M. D. Shenderovich, *Biopolymers (Peptide Sci.)* **43,** 219 (1997).

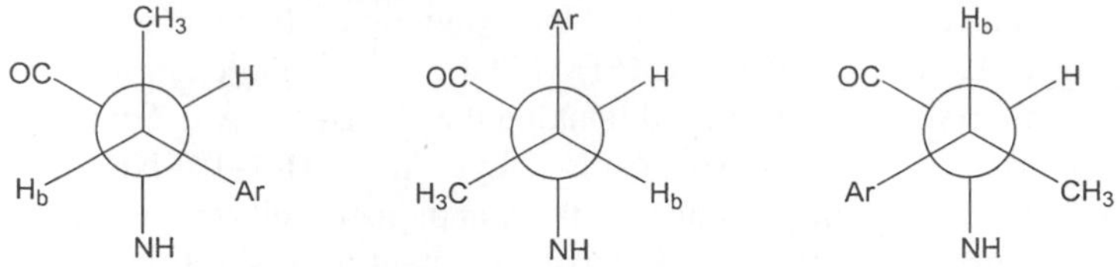

FIG. 2. Low energy side chain conformations for (2*S*,3*S*)-β-methyl-2′,6′-dimethyltyrosine (TMT) about the bond between the α and the β carbons (Ar = 2′,6′-dimethyl-4′-hydroxyphenyl) [*gauche* (−) = −60°, *trans* = ± 180°, *gauche* (+) = +60°].

conformations, which are differentially biased by nonbonding or other interactions between vicinal substituents. In our group, we have systematically developed efficient asymmetric synthetic methods to produce all four optically pure isomers of several β-substituted phenylalanines, tryptophanes, tyrosines, glutamic acids, and their derivatives, etc.[31,32] One representative example is the highly topographically constrained β-methyl-2′,6′-dimethyltyrosines (TMT, all four isomers). With methyl substituents at the 2′ and 6′ positions of the aromatic ring and β position, not only is the χ^1 torsional angle of this amino acid constrained and specific torsional angles energetically favored, but the χ^2 torsional angle also becomes highly restricted.[31–33] As shown in Fig. 2, (2*S*,3*S*) TMT has three staggered rotamers about the χ^1 torsional angle: *gauche*(−), *trans,* and *gauche*(+). For the (2*S*,3*S*) isomer, the *gauche*(−) side chain conformation is the energetically favored χ^1 conformation because it has the lowest amount of unfavorable steric interactions. Indeed, each of the other three diastereoisomers has only one energetically highly preferred rotamer of the side chain χ^1 torsional angle. Incorporation of all four isomers of β-substituted amino acids into specific peptide templates performs a systematic topographical scan and, with the help of molecular modeling, nuclear magnetic resonance (NMR) spectroscopy, X-ray crystallography, and careful binding and biological activity studies, provides critical insight into the topographical requirements for the specificity of peptide ligand–GPCR binding affinity, selectivity, and agonist/antagonist/inverse agonist biological activity recognition.

DPDPE (Tyr-c[D-Pen^2-Gly-Phe-D-Pen^5]-OH)[23] and deltorphin I (DELT I, Tyr-D-Ala-Phe-Asp-Val-Val-Gly-NH_2)[34] are two highly selective δ opioid agonists. Incorporation of all four TMT isomers in position 1 resulted in different profiles of biological activities for DPDPE and DELT I analogues.[35–37] In the case of [TMT^1]DPDPE, only the [(2*S*,3*R*)-TMT^1]DPDPE isomer showed both highly binding affinity and exceptional selectivity for the δ opioid receptor agonist

[32] V. A. Soloshonok, C. Cai, and V. J. Hruby, *Tetrahedron Lett.* **41,** 135 (2000).

[33] D. Jiao, K. C. Russell, and V. J. Hruby, *Tetrahedron* **49,** 3511 (1993).

[34] V. Erspamer, P. Melchiorri, G. Falconieri Erspamer, L. Negri, R. Corsi, C. Severini, D. Barra, M. Simmaco, and G. Kreil, *Proc. Natl. Acad. Sci. U.S.A.* **86,** 5188 (1989).

TABLE III
BIOLOGICAL POTENCIES OF [TMT[1]]DPDPE[a]

Peptide	EC_{50}, nM GPI(μ)	EC_{50}, nM MVD(δ)	Selectivity (μ/δ)
DPDPE	7300	4.1	1780
[(2*S*,3*S*)-TMT[1]]DPDPE	290	170	2
[(2*S*,3*R*)-TMT[1]]DPDPE	0% at 60 μM antagonist (IC_{50} 5μM)	1.8	>33,000
[(2*R*,3*R*)-TMT[1]]DPDPE	49,900	2200	23
[(2*R*,3*S*)-TMT[1]]DPDPE	75% at 82 μM	28% at 10 μM	N/A

[a] From X. Qian, M. D. Shendorovich, K. E. Kövér, P. Davis, R. Horvath, T. Zalewska, H. I. Yamamura., F. Porreca, and V. J. Hruby, *J. Am. Chem. Soc.* **118,** 7280 (1996).

activity with potent activity (Table III). In the series of DELT I analogues, two [TMT[1]]DELT I isomers possessed considerable δ receptor-binding affinity. The [(2*S*,3*R*)-TMT[1]]DELT I isomer provided a superpotent, but moderately δ selective agonist, whereas the [(2*S*,3*S*)TMT[1]]DELT I isomer showed the highest selectivity for the δ receptors.

These and many other examples[31] demonstrate the significant advantages that understanding topographical requirements for molecular recognition and signal transduction can provide for peptide ligand design for seven transmembrane G protein-coupled receptors.

Experimental Methods

Synthesis of N^{α}-Boc-β-methyl 2′, 6′-dimethyltyrosines (N^{α}-Boc-TMTs)[35] (Scheme I)

4-Methoxy-2,6-dimethylbenzaldehyde (**2**). 3, 5-Dimethylanisole **1** (87.5 ml, 0.619 mol) is dissolved in CCl_4 (1.0 liter). The mixture is cooled to −25°, and a solution of Br_2 (35.0 ml, 0.679 mol) in CCl_4 (200 ml) is added over 8 hr until the bromine color persists. The solution is allowed to warm to room temperature, poured into H_2O (1.0 liter), and stirred for 1.5 hr. The reaction mixture is separated, and the aqueous phase is extracted with Et_2O (3 × 200 ml). The combined organic phases are dried over $MgSO_4$, filtered, and solvents removed by rotary evaporation. Residue oil is distilled under a reduced pressure to yield 113.5 g (85%) of 4-bromo-3, 5-dimethylanisole as a clear colorless liquid (bp 64°, 2.1 torr).

[35] X. Qian, K. C. Russell, L. W. Boteju, and V. J. Hruby, *Tetrahedron* **51,** 1033 (1995).

[36] X. Qian, K. E. Kövér, M. D. Shendorovich, B. S. Lou, A. Misicka, T. Zalewska, R. Horvath, P. Davis, E. J. Bilsky, F. Porreca, H. I. Yamamura, and V. J. Hruby, *J. Med. Chem.* **37,** 1746 (1994).

[37] X. Qian, M. D. Shendorovich, K. E. Kövér, P. Davis, R. Horvath, T. Zalewska, H. I. Yamamura, F. Porreca, and V. J. Hruby, *J. Am. Chem. Soc.* **118,** 7280 (1996).

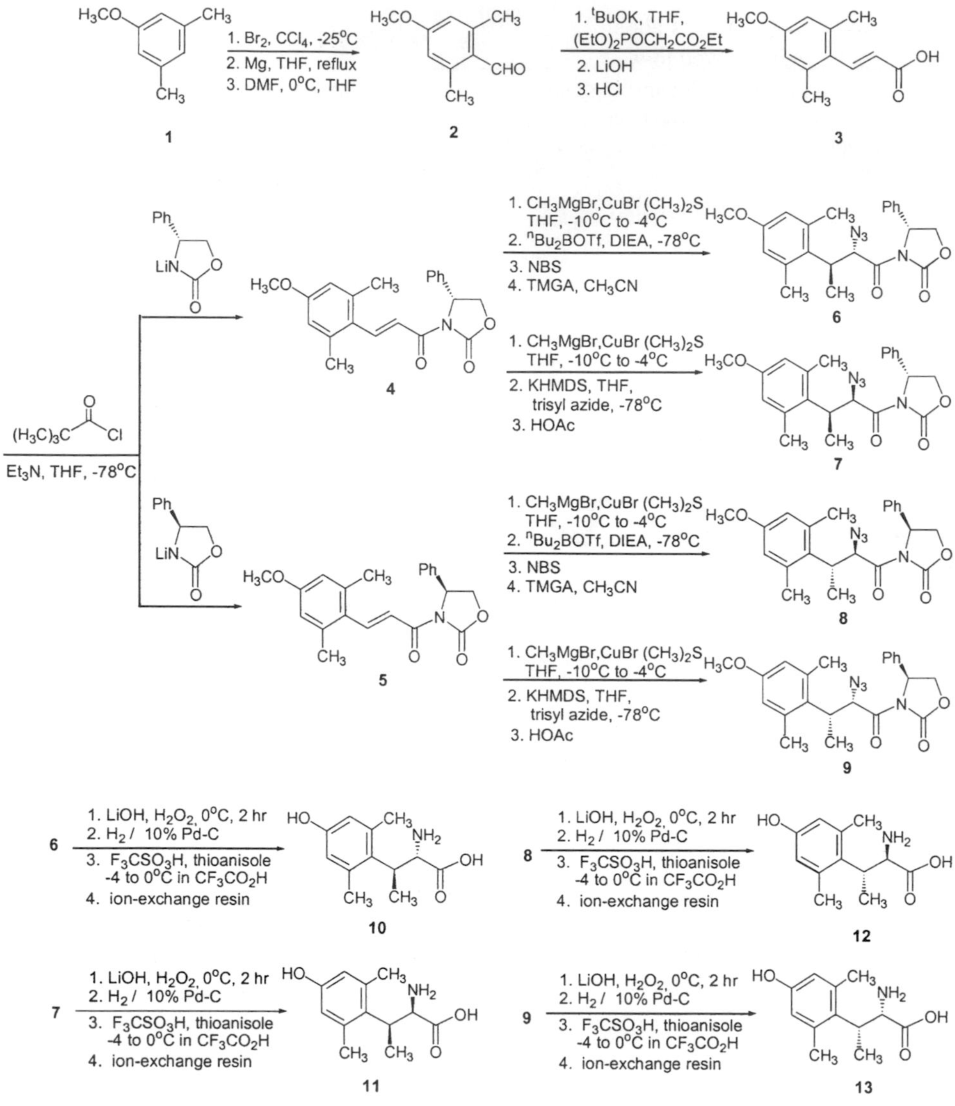

SCHEME 1. Synthesis of β-methyl-2′,6′-dimethyltyrosines (TMTs).

A mixture of the just-described 4-bromo-3,5-dimethylanisole (68.0 g, 0.316 mol), polished Mg strips (9.0 g, 0.158 mol), and dry THF (1.0 liter) is heated gently. The heating source is removed until the reaction ceases to reflux. The black colloidal suspension is reheated to reflux for 4 hr and then recooled to 0°. A solution of anhydrous *N,N*-dimethylformamide (DMF) (28.7 mL, 0.371 mol) in dry THF (30 ml) is added to the suspension dropwise over 20 min. The gray suspension is allowed to warm to room temperature and stirred for 1 hr. The reaction is quenched by decanting the reaction solution into an aqueous NH_4Cl solution (1.0 liter). The reaction flask is washed with additional THF (2 × 30 ml), which is decanted into the NH_4Cl solution. The reaction mixture is separated, and the aqueous phase is extracted with Et_2O (3 × 200 ml). The combined organic phases are dried over $MgSO_4$, filtered, and the solvents removed by rotary evaporation. The oily residue is dried further *in vacuo* to give a yellow solid (96%). The crude product is recrystallized from hexanes at −20° to give 44.9 g (86%) of a pure product (mp 44.0–45.5°).

*(2E)3-(4′-Methoxy-2′,6′-dimethylphenyl)propenoic acid (**3**).* To a solution of potassium *tert*-butoxide (53.1 g, 0.450 mol) and triethyl phosphonoacetate (93 ml, 0.450 mol) in dry THF (1.2 liter) is added 4-methoxy-2,6-dimethylbenzaldehyde **2** (41.4 g, 0.252 mol) in one portion at room temperature. The reaction is stirred at room temperature for 2 hr, and MeOH (400 ml), H_2O (400 ml), and $LiOH \cdot H_2O$ (75 g, 1.8 mol) are added. After the reaction mixture is stirred overnight, volatiles are removed by rotary evaporation. The aqueous layer, now containing a white precipitate, is diluted to 1.5 liter with H_2O and heated gently to dissolve solids. The warm solution is washed with $CHCl_3$ (3 × 200 ml) and acidified to pH 1 with HCl (6 *M*). The white solid is filtered, washed with H_2O (1.0 liter), and dried *in vacuo* to give 48.6 g (93%) of off-white solid (mp 174.5–175.5°).

General Methods for the Synthesis of (N-Acetyl)oxazolidinones. To a stirred −78° precooled solution of (2*E*)3-(4′-methoxy-2′,6′-dimethylphenyl)propenoic acid **3** (6.50 g, 0.0315 mol) in dry THF (650 ml) is added triethylamine (TEA) (5.3 ml, 0.0378 mol) via syringe, followed by pivaloyl chloride (4.3 ml, 0.0349 mol). The suspension is stirred for 15 min at −78°, for 45 min at 0°, and then recooled to −78°. The suspension is then transferred via cannula to a stirring slurry of the lithiated (4*R*)-, or (*S*)-4-phenyl-2-oxazolidinone at −78° [prepared 20 min in advance at −78° by the addition of *n*-butyllithium (1.6 *M* in hexanes, 18.0 ml, 0.0288 mol) to a solution of the (4*R*) or (*S*)-4-phenyl-2-oxazolidinone (4.70 g, 0.0288 mol) in dry THF(120 ml)]. The resulting suspension is stirred at −78° for 20 min and at room temperature for 2 hr. The reaction is quenched by the addition of saturated aqueous NH_4Cl solution (70 ml). Volatiles are removed by rotary evaporation, and the residual aqueous slurry is extracted with $CHCl_3$ (3 × 10 ml). The combined extracts are washed with dilute aqueous $NaHCO_3$ (3 × 70 ml) and brine, dried over $MgSO_4$, filtered, and rotary evaporated to dryness. The crude product is chromatographed on silica gel [EtOAc/hexane, (v/v): 2/8].

3(2*E*),(4*R*)-3-[3′-(4″-Methoxy-2″,6″-dimethylphenyl)propenoyl]-4-phenyl-2-oxazolidinone **4** [75%, mp 158.5–160.0°, $[\alpha]_D^{22}$ −26.1 (c 1.05, $CHCl_3$)]

3(2*E*), (4*S*)-3-[3′-(4″-Methoxy-2″,6″-dimethylphenyl)propenoyl]-4-phenyl-2-oxozolidinone **5** [77%, mp 157.5–158.0°, $[\alpha]_D^{25}$ +32.0 (c 0.29, $CHCl_3$)]

General Procedure for Conjugate Additions. A MeMgBr solution (3.0 M in Et_2O, 7.5 ml, 0.0225 mol) is added by a syringe to a solution of CuBr • Me_2S complex (4.58 g, 0.0220 mol) in dry THF (50 ml) and anhydrous Me_2S (20 ml) at −4°. The resulting yellow-greenish mixture is stirred for 10 min at −4° and is added dropwise by a solution of (*N*-acyl)oxazolidinone (5.20 g, 0.0148 mol) in dry THF (25 ml) over 60 min at −4°. After being stirred for 90 min at −4°, the reaction mixture is warmed up to room temperature and stirred overnight. The reaction is quenched by the slow addition of saturated aqueous NH_4Cl (70 ml) and stirred for 30 min. The phases are separated, and the aqueous phase is extracted with Et_2O (3 × 50 ml). The combined organic phases are washed with saturated aqueous NH_4Cl (3 × 30 ml), H_2O (3 × 30 ml), and brine (30 ml) and dried over anhydrous $MgSO_4$. The drying agent is filtered, the solvent is evaporated off, and the diastereoisomeric mixture is purified by column chromatography on silica gel (v/v, EtOAc/hexanes : 3/7), followed by recrystallization from EtOAc-hexanes to give optically pure isomers.

3(3*R*),(4*R*)-3-[3′-(4″-Methoxy-2″,6″-dimethylphenyl)butyryl]-4-phenyl-2-oxazolidinone [75%, mp 99.5–101.0°, $[\alpha]_D^{22}$ −88.5 (c 1.07, $CHCl_3$)]
3(3*S*),(4*S*)-3-[3′-(4″-Methoxy-2″,6″-dimethylphenyl)butyryl]-4-phenyl-2-oxazolidinone [85%, mp 104.0–104.5°, $[\alpha]_D^{22}$ +95.6 (c 0.39, $CHCl_3$)]

General Procedure for Bromination. To a cooled solution of 3-[3′-(4″-methoxy-2″,6″-dimethylphenyl)butyryl]-4-phenyl-2-oxazolidinone (3.87 g, 0.0105 mol) in dry CH_2Cl_2 (50 ml) at −78°, dry diisopropylethylamine (DIEA) (2.20 ml, 0.0126 mol) and dibutyl borontriflate solution (1 *M* in CH_2Cl_2, 11.1 ml, 0.0111 mol) are added via a syringe. Meanwhile, in another dry three-necked round bottom flask (100 ml), a suspension of recrystallized *N*-bromosuccinimide (NBS) (2.06 g, 0.0116 mole) in dry CH_2Cl_2 (35 ml) is cooled to −78°. The boron enolate solution is transferred to the NBS suspension at −78° via a cannula. The resulting mixture is stirred at −78° for 2 hr and warmed up to −4°. Then, it is quenched with a $NaHSO_4$ solution (0.5 *M,* 20 ml). The mixture is warmed up to room temperature and stirred for 30 min. The phases are separated, and the aqueous layer is extracted with CH_2Cl_2 (2 × 25 ml). The combined organic phases are washed with 0.5 *M* $NaHSO_4$ (50 ml), 1 *M* $Na_2S_2O_3$ (3 × 50 ml), H_2O (50 ml), and brine (50 ml). The organic layer is then dried over $MgSO_4$, filtered, and evaporated to give the crude bromide as a yellow solid, which is purified further by column chromatography on silica gel [EtOAc/hexanes (v/v): 2/8].

3(2*R*,3*S*), (4*R*)-3-[2′-Bromo-3′-(4″-methoxy-2″,6″-dimethylphenyl)butyryl]-4-phenyl-2-oxazolidinone [85%, mp 172.0–173.5°, $[\alpha]_D^{22}$ −94.0 (c 1.01, $CHCl_3$)]
3(2*S*,3*R*),(4*S*)-3-[2′-Bromo-3′-(4″-methoxy-2″,6″-dimethylphenyl)butyryl]-4-phenyl-2-oxazolidinone [89%, mp 150.0–151.0°, $[\alpha]_D^{22}$ +87.9 (c 1.16, $CHCl_3$)]

General Procedure for Azide Displacement. A solution of 3-[2′-bromo-3′-(4″-methoxy-2″,6″-dimethylphenyl)butyryl]-4-phenyl-2-oxazolidinone (2.20 g, 0.00493 mol) and tetramethylguanidine azide (TMGA) (1.17 g, 0.00740 mol)[38] in dry CH_3CN (22 ml) is stirred under Ar at room temperature, and the reaction is monitored by TLC. After completion, EtOAc (45 ml) is added and the solid filtered. The filtrate is evaporated to dryness and is purified by column chromatography on silica gel [EtOAc/hexanes (v/v): 2/8].

3(2*S*,3*S*),(4*R*)-3-[2′-Azido-3′-(4″-methoxy-2″,6″-dimethylphenyl)butyryl]-4-phenyl-2-oxazolidinone **6** [99%, mp 121.0–121.5°, $[\alpha]_D^{22}$ −48.7 (c 1.24, $CHCl_3$)]

3(2*R*,3*R*),(4*S*)-3-[2′-Azido-3′-(4″-methoxy-2″,6″-dimethylphenyl)butyryl]-4-phenyl-2-oxazolidinone **8** [90%, mp 111.5–112.0°, $[\alpha]_D^{22}$ +49.1 (c 1.24, $CHCl_3$)]

General Procedure for Direct Azidation. To cooled dry THF (15 ml) at −78° is added KHMDS (0.5 *M* in toluene, 9.86 ml, 0.00493 mol) under Ar, followed by a precooled (−78°) solution of 3-[3′-(4″-methoxy-2″,6″-dimethylphenyl)butyryl]-4-phenyl-2-oxazolidinone (1.58 g, 0.00430 mol) in dry THF (12 ml). The resulting yellow solution is kept stirring at −78° for another 30 min. To the just-described stirred solution of potassium enolate at −78° is added, via cannula, a precooled (−78°) solution of 2,4,6-triisopropylsulfonyl azide (1.52 g, 0.00493 mol)[39] in dry THF (18 ml). After 3 min, the reaction is quenched with AcOH (2.2 ml). The reaction flask is immediately immersed into a water bath at 28° for 35 min with stirring and partitioned between CH_2Cl_2 (200 ml) and dilute brine (100 ml). The organic phase is separated, and the aqueous phase is extracted with CH_2Cl_2 (3 × 60 ml). The combined organic phases are washed with saturated $NaHCO_3$ solution, dried over $MgSO_4$ and evaporated to dryness. The crude product is purified by column chromatography on silica gel [EtOAc/hexanes (v/v): 1/9].

3(2*S*,3*R*),(4*R*)-3-[2′-Azido-3′-(4″-methoxy-2″,6″-dimethylphenyl) butyryl]-4-phenyl-2-oxazolidinone **7** [79%, mp 102.5–103.5°, $[\alpha]_D^{22}$ −41.9 (c 1.01, $CHCl_3$)]

3(2*R*,3*S*),(4*S*)-3-[2′-Azido-3′-(4″-methoxy-2″,6″-dimethylphenyl)butyryl]-4-phenyl-2-oxazolidinone **9** [85%, mp 105.5–106.0°, $[\alpha]_D^{23}$ +40.6 (c 1.01, $CHCl_3$)]

General Procedure for the Hydrolysis of 3-[2′-Azido-3′-(4″-methoxy-2″,6″-dimethylphenyl)butyryl]-4-phenyl-2-oxazolidinones. The 3-[2′-azido-3′-(4″-methoxy-2″,6″-dimethylphenyl)butyryl]-4-phenyl-2-oxazolidinone (2.0 g, 0.0049 mol) is dissolved in dry THF (60 ml) and H_2O (20 ml). The mixture is cooled to −4°, and H_2O_2 (30 %, 3.4 ml, 0.0030 mol) is added via a syringe over 5 min. A solution of LiOH (0.44 g of LiOH · H_2O, 0.011 mol in 3 ml H_2O) is then added dropwise over 10 min. The reaction is stirred at 0° for 2 hr and is quenched with Na_2SO_3 (1.3 *M,* 30 ml). The mixture is warmed up to room temperature and stirred

[38] A. J. Papa, *J. Org. Chem.* **31,** 1426 (1966).

[39] R. E. Harmon, G. Wellman, and S. K. Gupta, *J. Org. Chem.* **38,** 11 (1973).

for 30 min. The volatile is removed by rotary evaporation, and the aqueous phase is extracted with CH_2Cl_2 (3 × 40 ml) to remove the chiral auxiliary. The remaining aqueous phase is cooled to 0°, acidified to ca. pH 1.5 with HCl (6 *M*), and extracted with CH_2Cl_2 (3 × 40 ml). The colorless extracts are dried over anhydrous Na_2SO_4, filtered, and concentrated to give an off-white solid.

(2*S*,3*S*)-2-Azido-3-(4′-methoxy-2′,6′-dimethylphenyl)butanoic acid [100%, mp 55.0–57.0°; $[\alpha]_D^{22}$ −17.0 (c 0.82, $CHCl_3$)]
(2*R*,3*S*)-2-Azido-3-(4′-methoxy-2′,6′-dimethylphenyl)butanoic acid [99%, oil, $[\alpha]_D^{23}$ −84.3 (c 1.07, $CHCl_3$)]
(2*R*,3*R*)-2-Azido-3-(4′-methoxy-2′,6′-dimethylphenyl)butanoic acid [100%, mp 77.0–77.5; $[\alpha]_D^{25}$ +16.6 (c 1.10, $CHCl_3$))
(2*S*,3*R*)-2-Azido-3-(4′-methoxy-2′,6′-dimethylphenyl)butanoic acid [100%, oil, $[\alpha]_D^{22}$ + 84.8 (c 1.66, $CHCl_3$)]

General Procedure for the Reduction of Azido Acids. 2-Azido-3-(4′-methoxy-2′,6′-dimethylphenyl)butanoic acid (1.00 g, 0.0038 mol) is dissolved in glacial acetic acid (48 ml) and H_2O (24 ml). After the solution has been bubbled with Ar for 45 min, Pd/C (10%, 0.1 g) is added, and the mixture is bubbled with Ar for another 20 min. The reaction flask is flushed with H_2 (three times), charged with H_2 (36 psi), and shaken for 24 hr. The catalyst is filtered, and the volatile is removed by rotary evaporation. To the remaining aqueous phase is added HCl (6 *M*, 6 ml). It was then evaporated, frozen, and lyophilized to give the amino acid salt as an off-white solid.

(2*S*,3*S*)-2-Amino-3-(4′-methoxy-2′,6′-dimethylphenyl)butanoic acid hydrochloride [100%, mp 109.5–111.5°, $[\alpha]_D^{22}$ −31.9 (c 0.63, CH_3OH)]
(2*R*,3*S*)-2-Amino-3-(4′-methoxy-2′,6′-dimethylphenyl)butanoic acid hydrochloride [99%, mp 118.5–120.0°, $[\alpha]_D^{23}$ −63.7 (c 1.02, CH_3OH)]
(2*R*,3*R*)-2-Amino-3-(4′-methoxy-2′,6′-dimethylphenyl)butanoic acid hydrochloride [100%, oil, $[\alpha]_D^{25}$ +32.1 (c 0.95, CH_3OH)]
(2*S*,3*R*)-2-Amino-3-(4′-methoxy-2′,6′-dimethylphenyl)butanoic acid hydrochloride [100%, mp 110.0–113.0°, $[\alpha]_D^{22}$ +62.5 (c 0.84, CH_3OH)]

General Procedure for the Hydrolysis of Methyl Ethers and Ion-Exchange Purification of the Final Amino Acids [β-Methyl-2′,6′-dimethyltyrosines (TMTs)]. 2-Amino-3-(4′-methoxy-2′,6′-dimethylphenyl)butanoic acid (1.15 g, 0.00485 mol) is dissolved in TFA (50 ml). The solution is cooled to −4°, and thioanisole (3.98 ml, 0.0339 mol) is added and stirred for 10 min. Finally, CF_3SO_3H (6.43 ml, 0.0727 mol) is added via syringe, and the light yellow cloudy solution is stirred at 0 to −4° for 30 min. Volatiles are removed by rotary evaporation. The residue, a dark red-brownish tar, is dissolved in H_2O (100 ml). The solution is loaded on an ion-exchange column (2.5 × 46 cm) with Amberlite IR-120 (H^+) resin (150 g), and the column is washed with H_2O until the eluent is neutral. The amino acid is washed out with NH_4OH solution (5%); this process is monitored by TLC. Fractions containing the product are combined, evaporated to remove excess NH_4OH, frozen, and lyophilized to give an off-white solid.

(2*S*,3*S*)-2-Amino-3-(2′,6′-dimethyl-4′-hydroxyphenyl)butanoic acid **10**[94%, mp 153.0–156.0°, $[\alpha]_D^{22}$ −35.4 (c 0.51, CH_3OH)]
(2*R*,3*S*)-2-Amino-3-(2′,6′-dimethyl-4′-hydroxyphenyl)butanoic acid **11**(96%, mp 175.0–178.0°, $[\alpha]_D^{22}$ −42.8 (c 0.50, CH_3OH)]
(2*R*,3*R*)-2-Amino-3-(2′,6′-dimethyl-4′-hydroxyphenyl)butanoic acid **12**(90%, mp 168.0–170.0°, $[\alpha]_D^{22}$ +35.5 (c 0.71, CH_3OH)]
(2*S*,3*R*)-2-Amino-3-(2′,6′-dimethyl-4′-hydroxyphenyl)butanoic acid **13**[96%, hygroscopic solid, $[\alpha]_D^{22}$ +46.6 (c 0.79, CH_3OH)]

General Procedure for N^α-Boc-TMT Analogues. A solution of the TMT (2.23 g, 0.0100 mol) in a mixture of dioxane (20 ml), H_2O (10 ml), and 1 *N* NaOH (10 ml) is stirred and cooled in an ice-water bath. Di-*tert*-butyl dicarbonate (2.40 g, 0.0110 mol) is added, and stirring is continued at room temperature for 30 min. The solution is concentrated *in vacuo* to about 10 to 15 ml, cooled in an ice-water bath, covered with a layer of EtOAc (30 ml), and acidified with a dilute $KHSO_4$ solution to pH 2–3. The aqueous phase is extracted with EtOAc (3 × 15 ml). The combined organic phases are washed with H_2O (2 × 30 ml) and dried over anhydrous Na_2SO_4. Evaporation *in vacuo* give crude protected amino acids, which are recrystallized from EtOAc and hexanes.[35]

Synthesis of [TMT¹]DPDPE analogues[36]

Cesium Salt of N^α-Boc-D-Pen(pMeBzl). To a suspension of N^α-Boc-D-Pen (pMeBzl) (3.53 g, 0.0100 mol) in water (5 ml) is added $CsHCO_3$ (2.04 g, 0.0105 mol) and EtOH (2 ml). The mixture is stirred until a clear solution forms. Solvents are removed by evaporation, and the residue is lyophilized to give a white solid (4.80 g, 99%).

N^α-Boc-D-Pen-(pMeBzl)-resin.[40] Chloromethylated polystyrene resin (2.89 g, 0.00202 mol) (1% cross-link with divinylbenzene, Peptides International, 0.7 millimol/g) and N^α-Boc-D-Pen(pMeBzl) cesium salt (2.94 g, 0.00606 mol) in anhydrous DMF (18 ml) are placed in a screw-capped vial provided with a magnetic stirring bar. The suspension is stirred overnight at 50°. The resin is filtered, washed throughly with DMF, DMF/H_2O (9 : 1, v/v), DMF, and ethanol, and dried over P_2O_5 *in vacuo.*

General Procedures for Coupling

1. Swell N^α-Boc-D-Pen-(pMeBzl)-resin in CH_2Cl_2 for 2 hr.

[40] B. F. Gisin, *Helv. Chim. Acta.* **56,** 142 (1973).

2. Wash the resin with 50% TFA/CH_2Cl_2 with bubbling under Ar (3 volumes, ~5 min, checking with Kaiser test and seeing if the resin, not the solution, is blue—positive).
3. Wash the resin with 50% TFA/CH_2Cl_2 (3 volumes, ~25 min, checking with Kaiser test; it should be positive!).
4. Wash the resin with CH_2Cl_2 (3 × 3 volumes) and with DMF (2 × 3 volumes).
5. Neutralize with 10% DIEA/CH_2Cl_2 (2 × 3 volumes).
6. Wash with CH_2Cl_2 (4 × 3 volumes) and DMF (4 × 3 volumes).
7. Dissolve the N^α-Boc-amino acid (threefold) and HOBt (threefold) in DMF and add to the resin, wash N^α-Boc-amino acid and HOBt solution with more DMF, and add a minimum amount of DMF.
8. Add DIC (threefold) into the DMF mixture.
9. Bubble the coupling reaction for 2 hr (Kaiser test must be negative after done! If the test is positive, wash out all coupling reagents and do reaction again).
10. Wash the mixture with DMF (4 × 3 volumes).
11. Wash the mixture with CH_2Cl_2 (4 × 3 volumes).
12. Repeat step 2.

Kaiser Test[41]

1. Dissolve ninhydrin (5 g) in ethanol (100 ml).
2. Dissolve liquid phenol (80 g) in ethanol (20 ml).
3. Add a 0.001 *M* aqueous KCN solution (2 ml) to pyridine (98 ml).
4. Sample a few resin beads and wash several times with ethanol.
5. Transfer to a small glass tube and add 2 drops of each of the solutions.
6. Mix well and heat to 120° for 4–6 min. A positive test is indicated by blue resin beads.

Synthesis of [TMT¹]DPDPE. N^α-Boc-D-Pen(pMeBzl)-resin (0.88 g, 0.57 mmol/g, 0.5 mmol) is used as starting material, and the following protected amino acids are added in a stepwise fashion to the growing peptide chain: N^α-Boc-Phe, N^α-Boc-Gly, N^α-Boc-D-Pen(pMeBzl), and optically pure N^α-Boc-TMT. All the N^α-Boc protected amino acids (2 equiv) except for N^α-Boc-TMT are coupled to the growing peptide chain using DIC (2.5 equiv) and HOBt (2.5 equiv) as coupling reagents. N^α-Boc-TMT (1.2 equiv) is added to the growing peptide chain using the BOP reagent (1.44 equiv) and DIEA (1.7 equiv) in 1-methyl-2-pyrrolidinone (NMP) as the solvent. After coupling of the last amino acid, the resin is washed with dichloromethane (6 × 30 ml) and methanol (4 × 35 ml) and is dried by nitrogen gas flow (9 psi) for 10 min. The resin is then stored *in vacuo* for 24 hr.

[41] E. Kaiser, R. L. Colescott, C. D. Bossinger, and P. I. Cook, *Anal. Biochem.* **34,** 595 (1970).

Cleavage. A mixture of the just described peptide resin with liquid HF (approximately 10 ml) and *p*-cresol (0.5 g) and thiocresol (0.5 g) is stirred for 60 min at 0°. The HF is evaporated rapidly by vacuum aspiration at 0° to room temperature over 15 min. The product is washed with anhydrous ether (6 × 30 ml), and the peptide is extracted with glacial acetic acid (8 × 25 ml). The acetic acid fractions are combined and lyophilized to afford a crude linear peptide.

Cyclization and Purification.[42] The crude linear peptide just described is dissolved in a 30 ml of acetonitrile and water mixture (2 : 1, v/v) and is transferred to a syringe. The peptide solution is added to a degassed 0.1 *M* solution of $K_3[Fe(CN)_6]$ (2 equiv of crude linear peptide) at pH 8.3–8.7 at a rate of 0.1 ml/hr via a syringe pump. The pH value is maintained at 8.3–8.7, and the entire process of cyclization is done under the protection of bubbling argon. After all the peptide solution has been transferred to the $K_3[Fe(CN)_6]$ solution, the pH is adjusted to 4 by the addition of acetic acid solution, and the ferro- and excess ferricyanide are removed by stirring the solution with a 9-ml settled volume of Amberlite IR-45 anion-exchange resin (Cl^- form). The mixture is stirred for 1.75 hr, and the anion-exchange resin is filtered and washed by a 50% acetonitrile/water mixture (v/v, 8 × 30 ml). The solution is evaporated down to ca. 100 ml and lyophilized. The residue is dissolved in an acetonitrile and 0.1% trifluoroacetic acid aqueous solution mixture (v/v, 15 : 85) and is purified on a Vydac 218TP1010 C18 RP-HPLC column (25 cm × 1 cm) with a linear gradient elution of 15–75% acetonitrile in 0.1% TFA for 1 min at a flow rate of 3 ml/min. The more lipophilic impurities are washed from the column with 95–100% acetonitrile in 0.1% TFA for 10 min. After equilibrium (11 min, 15% CH_3CN), the column is ready for use again. The UV detector is set at 280 nm during the entire purification process. The major peak is isolated and lyophilized to afford a white powder. The amino acid analysis result for [(2*S*,3*R*)-TMT^1]DPDPE is (2*S*,3*R*)-TMT 0.95 (1.00), Gly 1.04 (1.00), and Phe 1.00 (1.00).

Synthesis of Cyclic Lactam Dynorphin A Analogue[30]*: c[D-Asp^3, Lys^7]DynA(1-11)-NH_2*

The N^α-Boc amino acids (4 equiv) are coupled sequentially to the growing peptide chain using diisoproplycarbodiimide (DIC) and *N*-hydroxybenzotirazole (HOBT) in *N*-methy1-2-pyrrolidine (NMP). The coupling reaction times are 30 min to 1 hr or longer depending on the particular residue. Side chain protecting groups used are 2,4-dichlorobenzyloxycarbonyl or 9-fluorenylmethyoxycarboyl (Fmoc) for Lys, tolylsulfonyl (Tos) for Arg, 2,6-dichlorobenzyl (2,6-Cl_2Bzl) for Tyr, and fluorenylmethyl (OFm) for Asp. Trifluoracetic acid (50% DCM) is used to remove the N^α-Boc-protecting group, and 20% piperidine is used to remove the Fmoc side chain-protecting group for Lys and the OFm group from Asp.

[42] A. Misicka and V. J. Hruby, *Polish J. Chem. Soc.* **68,** 893 (1994).

Diisopropylethylamine (DIEA) is used as a base, and dichloromethane (DCM) and NMP are used as solvents for washing. The cyclic peptides are cyclized on the resin by removing the Fmoc side chain-protecting group from Lys and the OFm group from Asp but leaving the N^{α}-Boc group on the N-terminal residue, and then by forming the lactam ring using BOP (3 equiv) and DIEA (6 equiv) in NMP as coupling reagents. The reaction is allowed to proceed 5–15 hr at room temperature, and repeated if necessary. Then the N-terminal N^{α}-Boc groups are removed and syntheses continued. Each peptide resin is dried *in vacuo,* and the peptide is then cleaved from the resin using liquid anhydrous hydrofluoric acid (HF) in the presence of cresol(10%, w/v) and/or other scavengers for 1 hr at 0°. After removal of HF *in vacuo,* the residue is washed with anhydrous ether and extracted with 30% aqueous acetic acid. The acetic acid solution is lyophilized to give a white residue. The crude peptide is then purified by semipreparative reverse phase HPLC to yield a white powder after lyophilization.

Synthesis of α-Melanotropin Analogue[25]*:*
Ac-Nle^{4}-c[Asp^{5},D-Phe^{7},Lys^{10}] α-MSH(4–10)-NH_2 (MT-II)

The protected peptide resin to the title compound is prepared from 1.5 g of pMBHA resin (0.6 mmol of NH_2/g of resin) by first coupling N^{α}-Boc-Lys (N^{ε}-Fmoc) to the resin to a substitutions level of 0.30 mmol/g of resin. The remaining amino groups are then blocked by acetylation using acetic anhydride. The remaining amino groups are then added to the growing peptide chain by the stepwise addition of N^{α}-Boc-Trp(N^{i}-For), N^{α}-Boc-Arg(N^{g}-Tos), N^{α}-Boc-D-Phe, N^{α}-Boc-His(N^{π}-Bom), N^{α}-Boc-Asp(β-Fom), and N^{α}-Boc-Nle using standard solid-phase synthesis methodology. Each coupling reaction is achieved with a 3-fold excess of DIC and 2.4-fold excess of HOBt. After coupling the last amino acid, the Fmo- and Fmoc-protecting groups are removed by treating the peptide resin with 50% piperidine in DMF for 1 hr. The peptide resin is washed with DMF (3 × 40 ml), DCM (3 × 40 ml), 10% diisopropylethylamine (DIEA) (3 × 40 ml), and DCM (3 × 40 ml), suspended in 15 ml of DMF, and mixed with a 6-fold excess of BOP reagent in the presence of an 8-fold excess of DIEA for 6 hr. The coupling is repeated twice until the resin gives the negative Kaiser test. After cyclization, the N^{α}-Boc—proctecting group is removed, and the amino group is neutralized and acetylated with 25% of acetic anhydride in DCM for 0.5 hr. The finished peptide resin weighs 1.86 g. The peptide resin is cleaved in anhydrous HF in the presence of anisole (~10%) and 1,2-dithioethane (~0.8%) at 0° for 45 min. The residue is washed with anhydrous ether and extracted with 30% acetic acid. The crude peptide is purified by a cation-exchange column with discontinuous NH_4OAc buffer solution followed by gel filtration through Sephadex G-15 using 15% acetic acid as the eluting solvent. The final purification is affected by preparative reverse-phase HPLC on a C18-bonded silica column, eluted with a linear acetonitrile gradient (20–40%) with a constant concentration of trifluoroacetic acid (0.1%, v/v). Separation is monitored at 280 nm. The purification yields a white powder after lyophilization.

Acknowledgment

Research from our laboratory was supported by the grants from the U.S. Public Health Service and the National Institute of Drug Abuse.

[6] Design of Nonpeptides from Peptide Ligands for Peptide Receptors

By VICTOR J. HRUBY, WEI QIU, TORU OKAYAMA, and VADIM A. SOLOSHONOK

Introduction

Peptides as neurotransmitters, neuromodulators, hormones, growth factors, cytokines, etc., are known to influence essentially all vital physiological processes *via* inter- and intracellular communication and signal transduction mediated through various classes of transmembrane receptors.[1,2] During the last two decades, the central role of peptide ligands for the maintenance of human health, in behavior, and in many diseases has been unequivocally proven, and the human genome project will add hundreds or thousands of new polypeptide ligands and targets. Thus a major goal of peptide research has been to elucidate the relationship between the three-dimensional (3D) structure of a peptide and its biological activity so as to develop suitable therapeutic agents in the control or intervention of human health, disease, and dysfunction.

The initial event essential for eliciting (or blocking) a biological response is molecular recognition. In the case of the ligand, the portion of the 3D surface involved in the interaction often is referred to as the *pharmacophore* and may involve either a continuous sequence of amino acids (*sychnologic* organization) or amino acid residues separated from one another throughout the primary structure (*rhegnylogic* organization)[3] but brought together on a surface by the 3D structure of the polypeptide.[1,4] In either case, the peptide backbone serves as a scaffold for the

[1] V. J. Hruby and M. E. Hadley, *in* "Design and Synthesis of Organic Molecules Based on Molecular Recognition" (G. VanBinst, ed.), p. 269. Springer-Verlag, Heidelberg, 1986.
[2] V. J. Hruby, H. I. Yamamura, and F. Porreca, *Ann. N.Y. Acad. Sci.* **757,** 7 (1995).
[3] R. Schwyzer, *Ann. N.Y. Acad. Sci.* **297,** 3 (1977).
[4] V. J. Hruby, *Drug Disc. Today* **2,** 165 (1997).

0076-6879/02 $35.00

key side chain groups involved in the interaction and, in some cases, also serves as a hydrogen bond donor and/or acceptor in molecular recognition (binding). In all cases, the side chain moieties involved directly in the binding are critical for the interaction, and their 3D architecture (topography) and stereoelectronic properties provide the critical complementary shape and chemical properties that favor efficient molecular recognition.

Although native biologically active peptides have a great potential for medical applications, they often need to be modified to overcome certain problems inherent in current drug delivery strategies. The properties include receptor/acceptor selectivity; high potency; stability against proteolytic breakdown; and appropriate biodistribution and bioavailability. These issues often have been addressed by the design of peptidomimetics and nonpeptide ligands.

More success has been obtained with the development of "peptide mimetic" antagonists, but because these antagonists generally come from natural products or company collections, it is not clear that they are truly peptide mimetics. So what is a nonpeptide peptide mimetic? Because this question has developed into an "Alice in Wonderland" quality with almost as many definitions as investigators, we offer a precise but simple definition here. A nonpeptide peptide mimetic is "a designed compound whose pharmacophore stereostructural elements mimic a peptide in 3D space, which mimic the binding, biological activities, selectivities, and structure–activity relationships (SAR) of a natural endogenous peptide ligand or peptide analogue with equivalent or superior bioactivities."[5] Agonists and antagonists would follow the same definition. For compounds that are developed from screening synthetic or natural product libraries and that only mimic some aspects of the peptide ligands, Veber[6] has suggested the use of the term "peptide limetic" to describe such compounds (a limetic is short for ligand mimetic). This is an excellent suggestion, to which we fully subscribe.

The structure-based rational design of nonpeptide ligands for peptide receptors has advanced dramatically over the last two decades (see reviews 7–15). Ultimate

[5] V. J. Hruby and S. Slate, *in* "Advances in Amino Acid Mimetics and Peptidomimetics" (A. Abell, ed.), Vol. 2, p. 191. JAI Press, Greenwich, CT, 1999.

[6] D. F. Veber, *in* "Peptides: Chemistry and Biology" (J. A. Smith and J. E. Rivier, eds.), p. 1. ESCOM Sci., Leiden, 1992.

[7] D. C. Horwell, M. C. Prichard, and J. Raphy, *Adv. Amino Acid Mimet. Peptidomimet.* **2,** 165 (1999).

[8] R. M. Freidinger, *Curr. Opin. Chem. Biol.* **3,** 395 (1999).

[9] J. S. Skotnicki, A. Zask, F. C. Nelson, J. D. Albright, and J. I. Levin, *Ann. N.Y. Acad. Sci.* **878,** 61 (1999).

[10] S. Thaisrivongs and J. W. Strohbach, *Biopolymers* **51,** 51 (1999).

[11] S. Liu, A. M. Crider, C. Tang, B. Ho, M. Ankersen, and C. E. Stidsen, *Curr. Pharm. Des.* **5,** 255 (1999).

[12] S. Krebs and D. Rognan, *Pharm. Acta Helv.* **73,** 173 (1998).

[13] H. Nakanishi and M. Kahn, *in* "Bioorganic Chemistry: Peptides and Proteins" (S. M. Hecht, ed.), p. 395. Oxford Univ. Press, New York, 1998.

goals have been set for this research, including high potency, high receptor selectivity, high stability both *in vitro* and *in vivo,* and high efficacy *in vitro* and *in vivo* for agonists. A systematic stepwise strategy has been developed to accomplish these goals. Once the structure of the natural peptide is knwon, the first phase in this approach involves identifying the key amino acid side chain residues necessary for receptor recognition by single amino acid modifications in the peptide ligand and, in the case of agonist activity, those required for information transduction. Simultaneously, local constraints are built into the ligand so as to constrain the local backbone conformation (ϕ, ψ, and ω angles, Fig. 1) to the most appropriate local mimima in a way that is compatible with molecular recognition. Two widely used methods are novel β-substituted amino acids and/or amide bond replacements. Alternatively, more global constraints are incorporated *via* cyclization to make an appropriate template for all the structural elements that make up the pharmacophore. Compounds with rigid or semirigid conformations are thus produced, and the most active structures are selected for studying conformation-activity relationships.

In addition to the determination of the preferred backbone conformation, which can serve as a template for the bioactive conformation (α helix, reverse turn, β sheet, etc.), it is also very important to examine and determine the preferred side chain topography in χ space (χ^1, χ^2, etc.) in order to provide insight into the topographical requirements for ligand–receptor interactions. The methodology developed includes specific covalent and noncovalent constraints, which can place the constrained side chains at highly preferred conformations [gauche (−), or gauche (+), or *trans*], and careful analysis of the 3D arrangement of critical side chain groups that are critical for bioactivity. These biophysical studies include state-of-the-art nuclear magnetic resonance (NMR) spectroscopy, X-ray crystallography if possible, circular dichroism measurements, and computational methods (molecular mechanics and molecular dynamics calculations) utilizing the biophysical structural parameters as key constraints.

Usually, at this point, one can obtain a quite precise 3D conformation of the pharmacophore. In many cases, such studies can lead to highly potent, selective, and efficacious drug candidates and provide the starting point for the nonpeptide ligands with high stability to biodegradation and good bioavailability, including the ability to cross membrane barriers. Now, the major effort is to find other organic moieties that can replace the peptide scaffold and position the crucial recognition elements correctly in 3D space, and with appropriate dynamic properties. This approach provides a systematic framework for the *de novo* design of nonpeptide ligands as outlined in Fig. 2. For G protein-coupled receptors (GPCRs), it is often

[14] S. H. Rosenberg and S. A. Boyd, *in* "Antihypertensive Drugs" (P. A. Van Zwieten and W. J. Greenlee, eds.), p. 77. Harwood, Amsterdam, 1997.

[15] R. A. Wiley and D. H. Rich, *Med. Res. Rev.* **13,** 327 (1993).

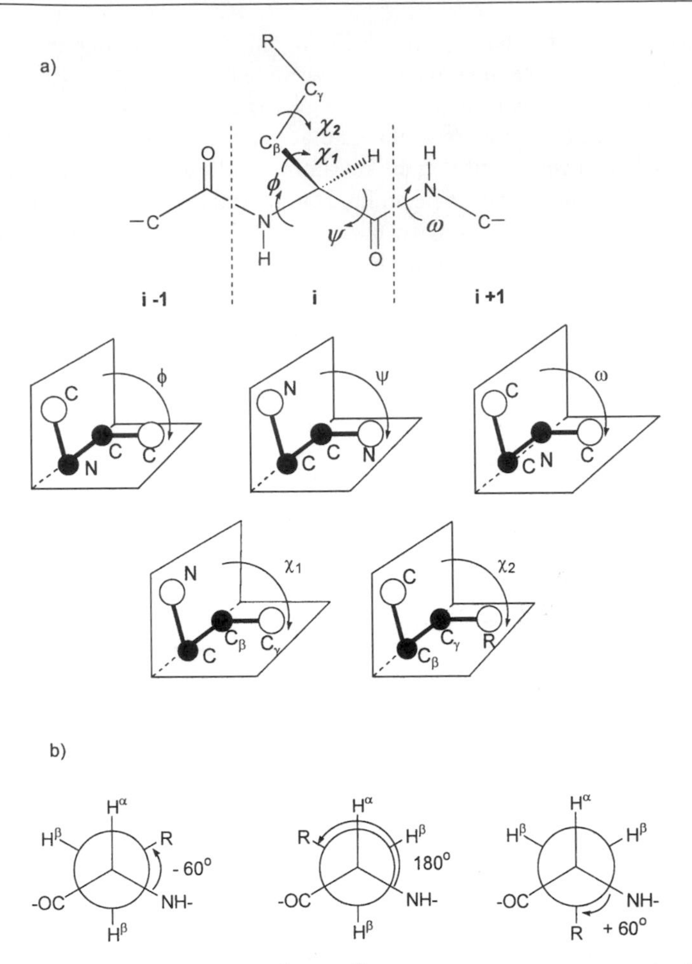

FIG. 1. (a) Definition of the ϕ, ψ, ω, χ^1, and χ^2 torsional angles. (b) Newman projections of the three staggered rotamers in an L-amino acid.

desirable to develop both agonists and antagonists, and in the case of enzymes, usually inhibitors are desired. Studies since the early 1980s have demonstrated that agonists and antagonists have different structure–activity relationships. There still is no universal approach to develop antagonists from agonists for peptide hormones and neurotransmitters, but once a lead has been obtained, there are general approaches that can be used to further develop antagonists.[16] It is worth mentioning here that nonpeptide design is a highly multidisciplinary area that often relies

[16] V. J. Hruby, *in* "Progress in Brain Research" (J. Joosse, R. M. Buijs, and F. J. H. Tilders, eds.), Vol. 92, p. 215. Elsevier Science, Amsterdam, 1992.

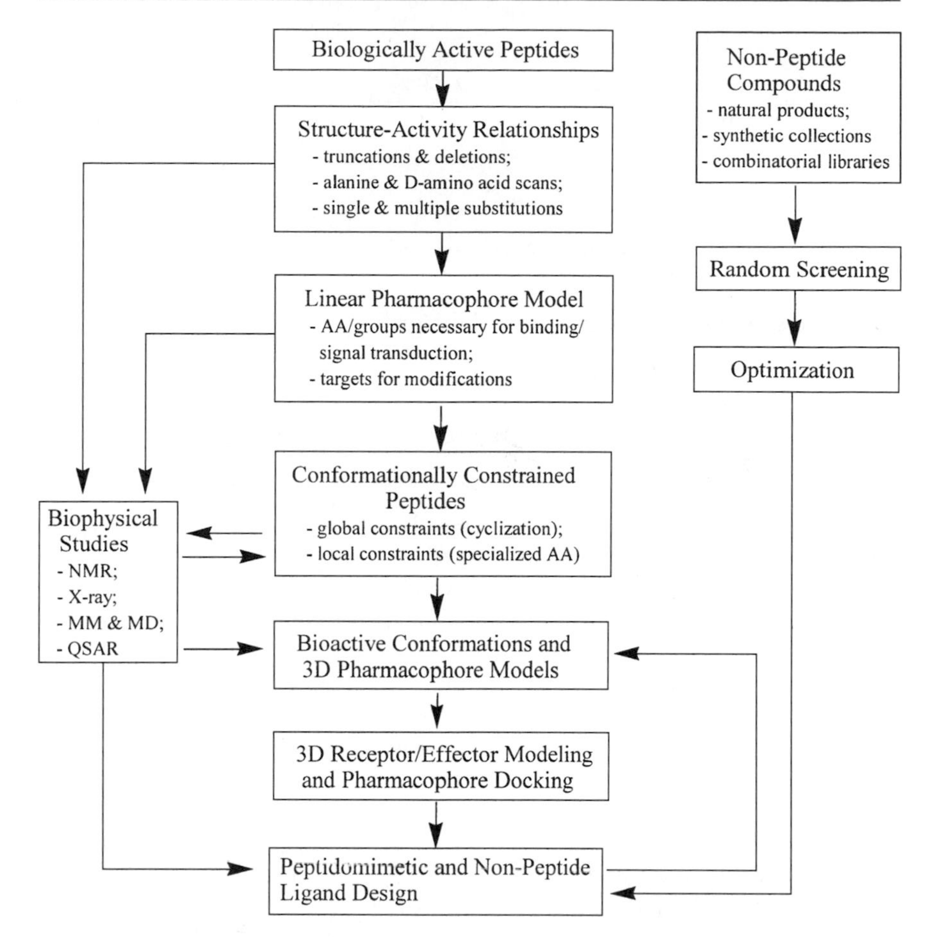

FIG. 2. A *de novo* approach for nonpeptide ligand design.

on (1) new asymmetric synthesis and methodologies for the preparation of novel molecules; (2) high-throughput screening of ligands and libraries of ligands using multiple *in vitro* and *in vivo* biological assays; (3) computational methods; and (4) state-of-the-art biophysical methods for determining structural, conformational, topographical, and dynamic properties of designed ligands (Fig. 2).

In order to develop nonpeptide ligands, efforts are in progress to develop templates and aspects of conformational design that permit assembling of all components necessary for molecular recognition and transduction. Here the proper choice of template that can place the key side chain residue in 3D space is still difficult, and thus only partial success has been achieved in terms of potent and selective ligands. A few examples are presented and discussed in some detail.

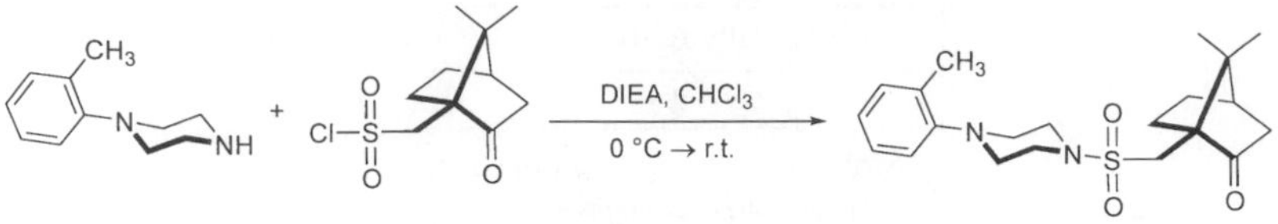

SCHEME 1. Synthesis of oxytocin antagonist L-367,398.

Examples of Nonpeptide Ligand Design*

Oxytocin Antagonist Mimetics

Oxytocin,

$$\text{H-}\overline{\text{Cys-Tyr-Ile-Gln-Asn-Cys}}\text{-Pro-Leu-Gly-NH}_2\ (\text{OT})$$

is a neurohypophyseal hormone physiologically important for its milk-ejecting and uterine-contracting activity in mammals. OT was the first peptide hormone whose primary structure was determined and proved by total synthesis.[17] Intensive SAR studies carried out for four decades have resulted in several hundred analogues of OT.[18,19a]

Compound L-366,948 (Fig. 3), a cyclic hexapeptide, was developed by researchers at Merck based on modification of a screening lead[20] (Fig. 3). It bears some structural resemblance to OT: each molecule contains an aromatic amino acid-isoleucyl dipeptide segment. L-366,948 was found to exhibit high affinity and selectivity for rat uterine OT receptors and antagonized OT induced contraction of the isolated (pA_2, 8.6) and *in situ* (AD_{50}, 0.10 mg/kg iv) rat uterus, but without a useful level of oral bioavailability. Supported by the excellent overlap of structurally similar hydrophobic and hydrophilic elements found in the alignment of the X-ray crystal structure of the simple tolylpiperazine camphor sulfonamide L-367,398 (Scheme 1) with the D-Nal^2-Ile^3 dipeptide portion of the NMR-consistent solution conformation of L-366,948 (Fig. 3), the Merck group has postulated that the tolylpiperazine/spiroindenylpiperidine camphor sulfonamide portion of a camphor-based structural class may serve as a mimetic of the important D-AA^2-Ile^3 dipeptide (AA = aromatic amino acid) in the cyclic hexapeptide class of oxytocin antagonists.[19a,b] This type of alignment also suggested that the affinity-enhancing

[17] V. du Vigneaud, C. Ressler, J. M. Swan, C. W. Roberts, P. G. Katsoyannis, and S. Gordon, *J. Am. Chem. Soc.* **75,** 4879 (1953).

[18] V. J. Hruby and C. W. Smith, *in* "The Peptides: Analysis, Synthesis, Biology" (C. W. Smith, ed.), Vol. 8, p. 77. Academic Press, Orlando, 1987.

[19a] P. D. Williams and D. J. Pettibone, *Curr. Pharm. Design* **2,** 41 (1996).

* For Synthetic Procedures see Appendix.

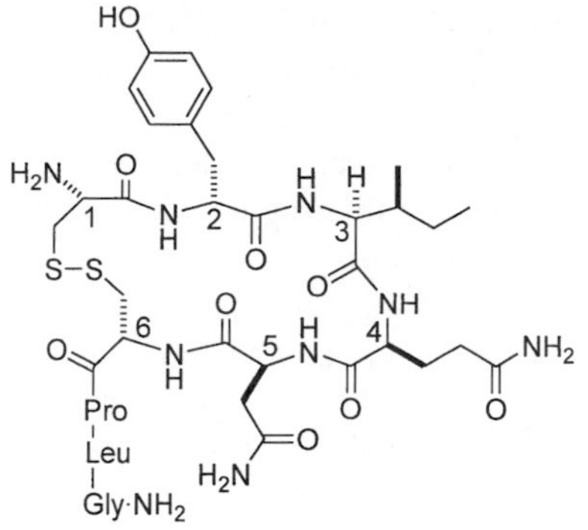

oxytocin

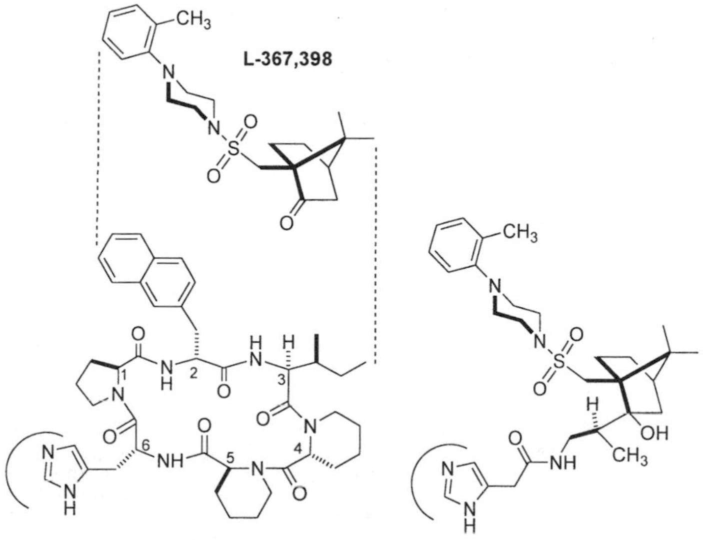

L-367,398

L-366,948

L-367,770

FIG. 3. Schematic representations of the three-dimensional alignment of import functionalities in camphor-based nonpeptide and cyclic hexapeptide OT antagonist: L-367,398 X-ray crystal structure (K_i = 310 nM, rat uterus); L-367,770 low-energy conformer (K_i = 3.2 nM, rat uterus); L-366,948 NMR-consistent solution conformation (K_i = 0.7 nM, rat uterus). From P. D. Williams and D. J. Pettibone, *Curr. Pharm. Design* **2,** 41 (1996). Copyright 1996 by Bentham Science Publishers B. V.

[19b] P. D. Williams, P. S. Anderson, R. G. Ball, M. G. Bock, L. Carroll, S.-H. L. Chiu, B. V. Clineschmidt, J. C. Culberson, J. M. Erb, B. E. Evans, S. L. Fitzpatrick, R. M. Freidinger, M. J. Kaufman, G. F. Lundell, J. S. Murphy, J. M. Pawluczyk, D. S. Perlow, D. J. Pettibone, S. M. Pitzenberger, K. L. Thompson, and D. F. Veber, *J. Med. Chem.* **37,** 565 (1994).

hydrophilic C-2 endo substituent on camphor would be oriented toward the more hydrophilic 5,6-dipeptide region of the cyclic hexapeptide structure. The latter point was illustrated by using L-366,770, a high-affinity analogue in which the preferred rotamer of the C-2 endo substituent on camphor had been established using energy calculations and X-ray crystallography. Aligning a low-energy conformation of L-366,770 with the NMR-consistent solution conformation of L-366,948 produced an excellent overlap of the tolypiperazine camphor sulfonamide and D-Nal^2-Ile^3 dipeptide, and the energetically preferred rotamer of the camphor C-2 endo substituent placed the imidazoleacetyl group directly over the imidazole-containing side chain at position 6 in the cyclic peptide (Fig. 3). In both structural classes the imidazole functional group has proven to be an affinity-enhancing addition. This excellent alignment suggests that the two classes share a similar pharmacophore for binding to the OT receptor and suggests a bright future for the rational design of potent nonpeptide oxytocin antagonists.

TRH Agonist Mimetics (Scheme 2)

Thyrotropin-releasing hormone (TRH) is a tripeptide (pGlu-His-Pro-NH_2, Fig. 4) that displays a variety of biological activities. This hypothalamic peptide functions as a neuroendocrine hormone by increasing the release of thyrotropin-stimulating hormone (TSH), which in turn leads to an elevation of thyroid hormone levels. TRH is a potent neuromodulator that facilitates cholinergic and monoaminergic neurotransmission and is an analeptic agent that also has trophic effects on neurons.[21–23] TRH can also enhance performance in cognitive behavioral models in animals, suggesting its potential to treat cognitive disorders, including those associated with Alzheimer's disease.[24–26] However, endocrine effects, a short half-life, and poor bioavailability are limitations for its use as a drug.

Conformational models have been proposed based on the crystal and solution structures of TRH and TRH peptide analogues.[27,28] It was suggested that the pharmacophore consists of the lactam moiety of the pyroglutamyl residue, the histidine

[20] D. J. Pettibone, B. V. Clineschmidt, E. V. Lis, D. R. Reiss, J. A. Totaro, C. J. Woyden, M. G. Bock, R. M. Freidinger, R. D. Tung, D. F. Veber, P. D. Williams, and R. I. Lowenohn, *J. Pharmacol. Exp. Ther.* **256,** 304 (1991).

[21] G. Yarbrough, *Nature* **263,** 523 (1976).

[22] G. Yarbrough, *Life Sci.* **33,** 111 (1983).

[23] A. Horita, M. A. Carino, and H. Lai, *Annu. Rev. Pharmacol. Toxicol.* **26,** 311 (1986).

[24] S. A. Stwertka, G. P. Vincent, E. R. Gamzu, D. A. MacNeil, and A. Vederese, *Pharmacol. Biochem. Behav.* **41,** 145 (1991).

[25] G. Yarbrough and M. Pomara, *Prog. Neuro-Psychopharmacol. Biol. Psychiat.* **9,** 285 (1985).

[26] M. Miyamoto, N. Yamazaki, A. Nagaoka, and Y. Nigawa, *Ann. N.Y. Acad. Sci.* **553,** 508 (1989).

[27] K. Kamiya, M. Takamoto, Y. Wada, M. Fujino, and M. Nishikawa, *J. Chem. Soc. Chem. Commun.* **438** (1980).

[28] J. Vicar, E. Abillon, F. Toma, F. Piriou, K. Lintner, K. Blaha, P. Fromageot, and S. Fermandjian, *FEBS Lett.* **97,** 275 (1979).

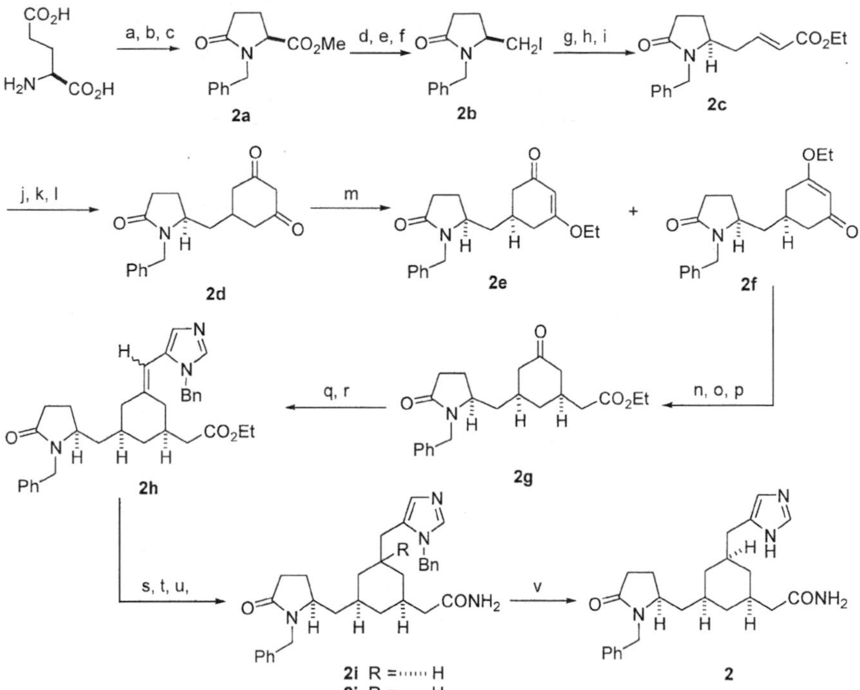

a) PhCHO, $NaBH_4$; b) HCl, H_2O; c) MeOH, H^+; d) $NaBH_4$/MeOH, *t*-BuOH; e) TsCl, DMAP; f) NaI, acetone; g) CH_2=CHMgBr, Li_2CuCl_4; h) O_3/Me_2S; i) Ph_3P=$CHCO_2Et$; j) ethyl acetoacetate, NaOEt; k) KOH, EtOH; l) HCl, H2O; m) *p*-TsOH, EtOH; n) $Me_3SiCH_2CO_2Et$, LDA; o) H_3^+O; p) H_2 Pd/C, 1 atm, 25 °C; q) $(EtO)_2POCH$=$CHNHCH_2Ph$, LDA; r) TOSMIC, $BnNH_2$; s) KOH; t) EtOCOCl, NH_3; u) H_2, catalyst, 1 atm, 25 °C; v) H_2 Pd/C, 50 Psi, 50 °C.

SCHEME 2. Synthesis of TRH agonist **2.**

imidazole ring, and the carboxamide of the terminal prolineamide.[29] Based on this pharmacophore, nonpeptide mimetics have been developed in which a cyclohexane framework replaced the peptide backbone.[30,31] The starting conformation was the

[29] M. L. Moore, "Probing the Thyroliberin Receptor." Washington University, St. Louis, MO, 1978.

[30] G. L. Olson, D. R. Bolin, M. Bonner, M. C. P. Bõs, D. C. Fry, B. J. Graves, M. Hatada, D. E. Hill, M. Kahn, V. S. Madison, V. K. Rusiecki, R. Sarabu, J. Sepinwall, G. P. Vincent, and M. E. Voss, *J. Med. Chem.* **36,** 3040 (1993).

[31] G. L. Olson, H.-C. Cheung, E. Chaing, V. S. Madison, J. Sepinwall, G. P. Vincent, A. Winokur, and K. A. Gary, *J. Med. Chem.* **38,** 2866 (1995).

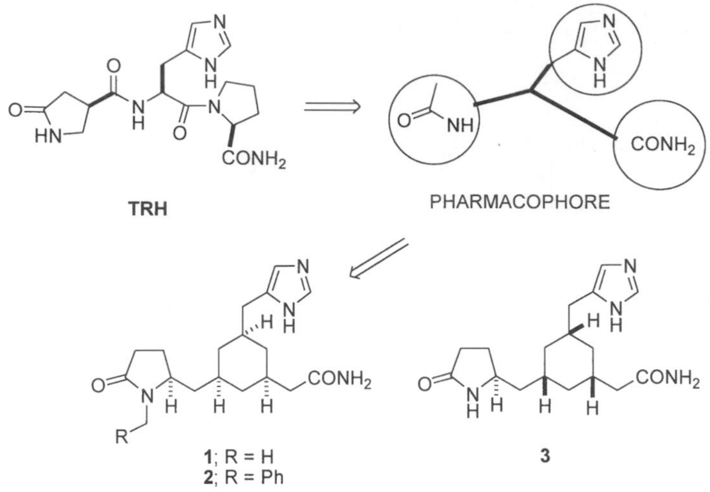

FIG. 4. TRH pharmacophore and design of cyclohexane template for TRH mimetics. Based on G. L. Olson, D. R. Bolin, M.-P. Bonner, M. Bös, C. M. Cook, D. C. Fry, B. J. Graves, M. Hatada, D. E. Hill, M. Kahn, V. S. Madison, V. K. Rusiecki, R. Sarabu, J. Sepinwall, G. P. Vincent, and M. E. Vos, *J. Med. Chem.* **36,** 3039 (1993).

TRH peptide backbone, which approximates a Y-shaped structure, except that in the model the imidazole ring is not locked in hydrogen-bonded interaction of the C-terminal carboxamide as in the crystal, but is oriented in a conformation that is consistent with observations from solution NMR studies (Fig. 4).[32,33] Retaining the three most important pharmacophore groups, the peptide backbone was replaced with a *cis*-1,3,5-trisubstituted cyclohexane scaffold. The cyclohexane ring system retains all of the essential pharmacophore groups but contributes only a nonspecific, hydrophobic character to the mimetic while allowing conformational mobility in the side chain groups (compound **1**, Fig. 4). This scaffold was also intended to make the molecule more like a traditional drug in terms of lipophilicity, improving prospects for oral bioavailability, enzymatic stability, and penetration of the blood–brain barrier. The flexibility of **1** (Fig. 4) is similar to TRH and lies within the volume occupied by the TRH peptide backbone and side chains.

The primary interest in these TRH mimetics was to develop agents to treat cognitive disorders. The most potent of the mimetics in the Morris water maze test, a behavioral model of cognitive impairment, was found to be the *N*-benzyl analogue **2** (Fig. 4). In this test, TRH reduced mean latencies over a dose range of

[32] J. J. Stezowski and E. Eckle, *J. Med. Chem.* **28,** 1654 (1985).

[33] J. Font, "Computer-Assisted Drug Design and the Receptor-Bound Conformation of Thyrotropin Releasing Hormone (TRH); Design and Synthetic Approaches Towards Rigid TRH Analogs." Washington University, St. Louis, MO, 1986.

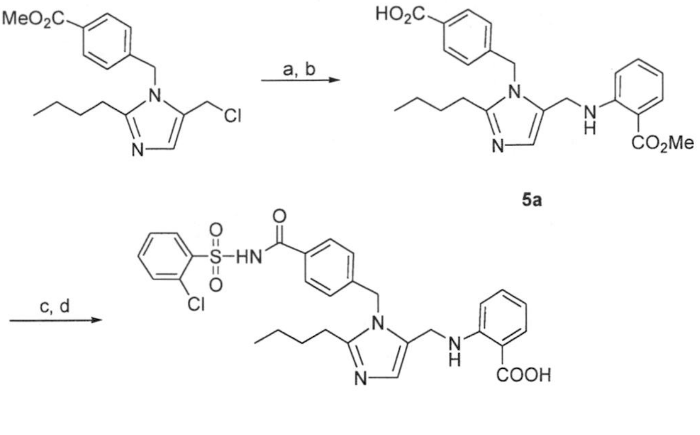

a) methyl 2-aminobenzoate, 2,6-dimethylpyridine, toluene, reflux; b) NaOH, MeOH, H_2O, 50 °C; c) 2-chlorobenzene sulfonamide, EDCI, DMAP, DCM; d) 2N NaOH, MeOH, H_2O, reflux.

SCHEME 3. Synthesis of A II antagonist **5.**

0.003 to 3.0 mg/kg ip, but was inactive by the oral route. Compound **2** was active from 0.0003 to 3.0 mg/kg ip, and 0.0003 to 3.0 mg/kg po.

Inversion of the ring center bearing the imidazolyl methyl group made the compounds about 10-fold less active. In contrast, analogue **3,** in which all three of the ring asymmetric centers are reversed, was equipotent as **2.** This result indicated that the cyclohexane ring serves as a neutral scaffold for the pendant groups because the ring inversion at all three centers, as in **3,** also maintains the 1,3,5-*cis* relationship of the pharmacophoric groups.

While the TRH mimetics exhibited potent activities in the behavioral test, they were unlike TRH in other respects. Some of these analogues were devoid of the TRH-like increase in spontaneous TSH release in cultured rat pituitary cells up to 10^{-5} *M,* whereas TRH caused an increase release of TSH of 300–500% of control with an ED_{50} of 2 n*M*. Furthermore, they did not exhibit binding to high-affinity TRH receptors up to 10^{-5} *M* when [^{3}H]MeTRH was employed as the radioligand.

A II Receptor Antagonists (Scheme 3)

Angiotensin II (A II), H-Asp-Arg-Val-Tyr-Ile-His-Pro-Phe-OH, is produced by the action of angiotensin-converting enzyme (ACE) on angiotensin I. Although ACE inhibitors have been shown to be clinically effective in the treatment of hypertension and congestive heart failure,[34] they also inhibit bradykinin metabolism,

[34] P. Corvol, *Clin. Exp. Hypertens.* **A11** (Suppl. 2), 463 (1989).

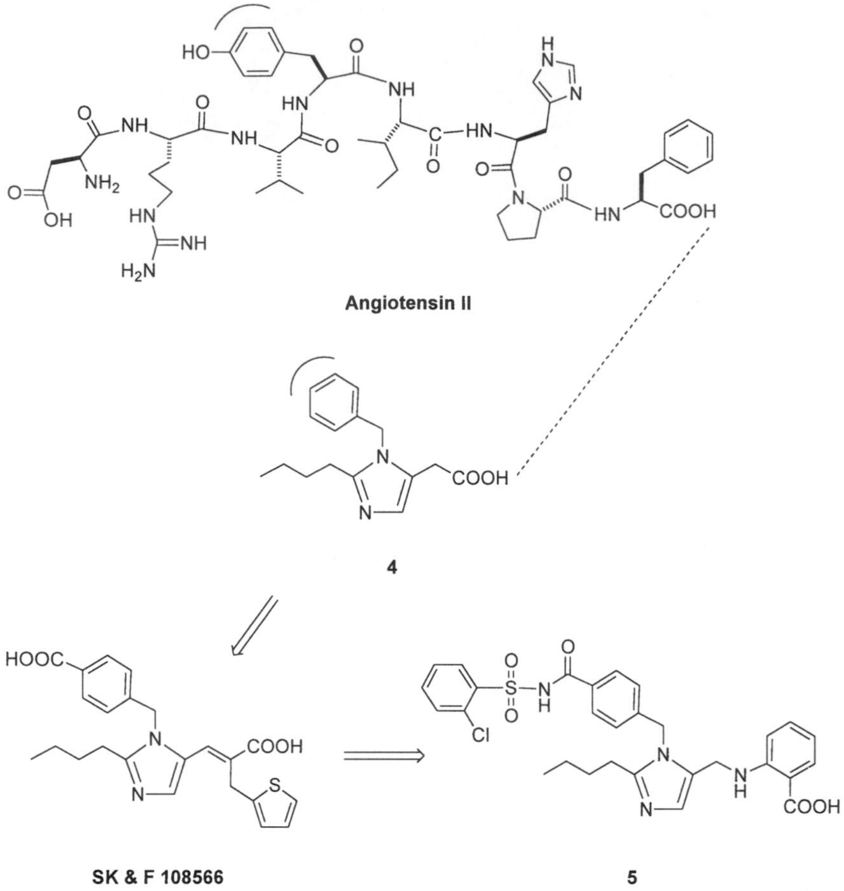

Angiotensin II

4

SK & F 108566

5

FIG. 5. Design of A II receptor antagonists.

which as a result may produce cough and angioedema. Thus an alternate approach of finding selective A II receptor antagonists seems very important.

A SmithKline Beecham group attempted to align imidazole 4 (Fig. 5), a micromolar antagonist with the Fermandjian model[35] for a bioactive conformation of A II.[36] The *N*-benzyl and the carboxyl groups were postulated to align with the Tyr^4 side chain and the C-terminal-COOH group of A II so the carboxyl group

[35] R. R. Smeby and S. Fermandjian, *in* "Chemistry and Biochemistry of Amino Acids, Peptides, and Proteins" (B. Weinstein, ed.), Vol. 5, p. 117. Dekker, New York, 1978.

[36] J. Weinstock, R. M. Keenan, J. Samanen, J. Hempel, J. A. Finkelstein, R. G. Franz, D. E. Gaitanopoulos, G. R. Girard, J. G. Gleason, D. T. Hill, T. M. Morgan, C. E. Peishoff, N. Aiyar, D. P. Brooks, T. A. Fredrickson, E. H. Ohlstein, R. R. J. Ruffolo, E. J. Stack, A. C. Sulpizio, E. F. Weidley, and R. M. Edwards, *J. Med. Chem.* **34,** 1514 (1991).

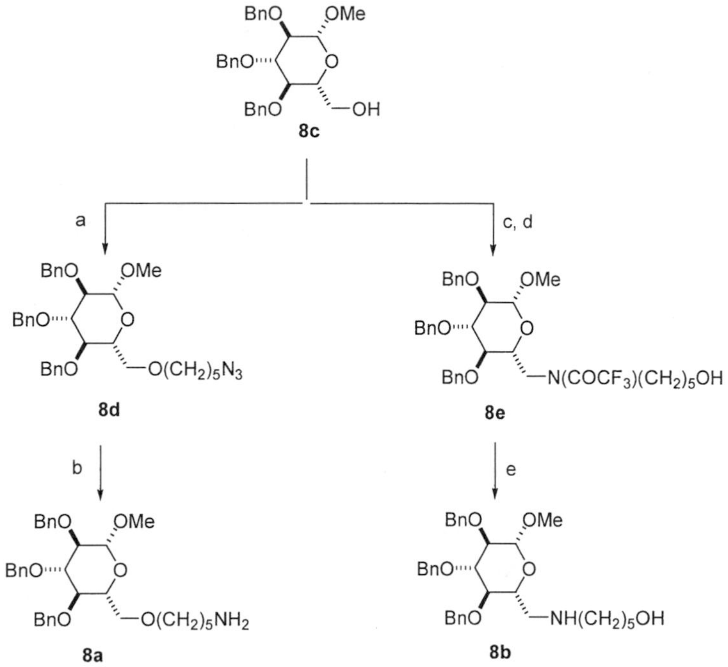

BnO OMe
BnO O
BnO —OH
8c

a | c, d

BnO OMe
BnO O
BnO —$O(CH_2)_5N_3$
8d

BnO OMe
BnO O
BnO —$N(COCF_3)(CH_2)_5OH$
8e

b | e

BnO OMe
BnO O
BnO —$O(CH_2)_5NH_2$
8a

BnO OMe
BnO O
BnO —$NH(CH_2)_5OH$
8b

a) $TfO(CH_2)_5N_3$, NaH, CH_2Cl_2; b) PPh_3, H_2O, THF; c) Tf_2O, 2,6-di-*t*-butyl-4-methylpyridine, CH_2Cl_2, -78 °C; d) $OH(CH_2)_5NHCOCF_3$, 2 equiv NaH, CH_2Cl_2, 24 h; e) 5 M NaOH, EtOH, reflux, 2h.

SCHEME 4. Synthesis of SRIF mimetics **8.**

was extended to better fit the Tyr^4-Phe^8 separation, and an aromatic function was added to mimic the Phe^8 side chain. Modification of the *N*-benzyl functionality gave SK&F 108566, a super antagonist (1.0 n*M*). By examination of this model, it was found that the *n*-butyl group of SK&F 108566 lies in the same space as the side chain of Ile^5 of A II. It was also suggested that the N-C-N imidazole region and the acrylic acid olefin function may mimic peptide amide bonds. A highly potent nonpeptide antagonist (**5** in Fig. 5, IC_{50} 0.8 n*M* on rabbit aorta ring) has been designed and synthesized using the SmithKline Beecham model.[37]

Somatostatin Nonpeptide Mimetics Using a β-D-Glucose Scaffold

The peptide hormone somatostatin (SRIF, Scheme 4, Fig. 6), a cyclic tetradecapeptide, inhibits the release of several hormones, including growth hormone

[37] S. Nouet, P. R. Dodey, M. R. Bondoux, D. Pruneau, J. M. Luccarini, T. Groblewski, R. Larguier, C. Lombard, J. Marie, P. P. Renaut, G. Leclerc, and J. C. Bonnafous, *J. Med. Chem.* **42,** 4572 (1999).

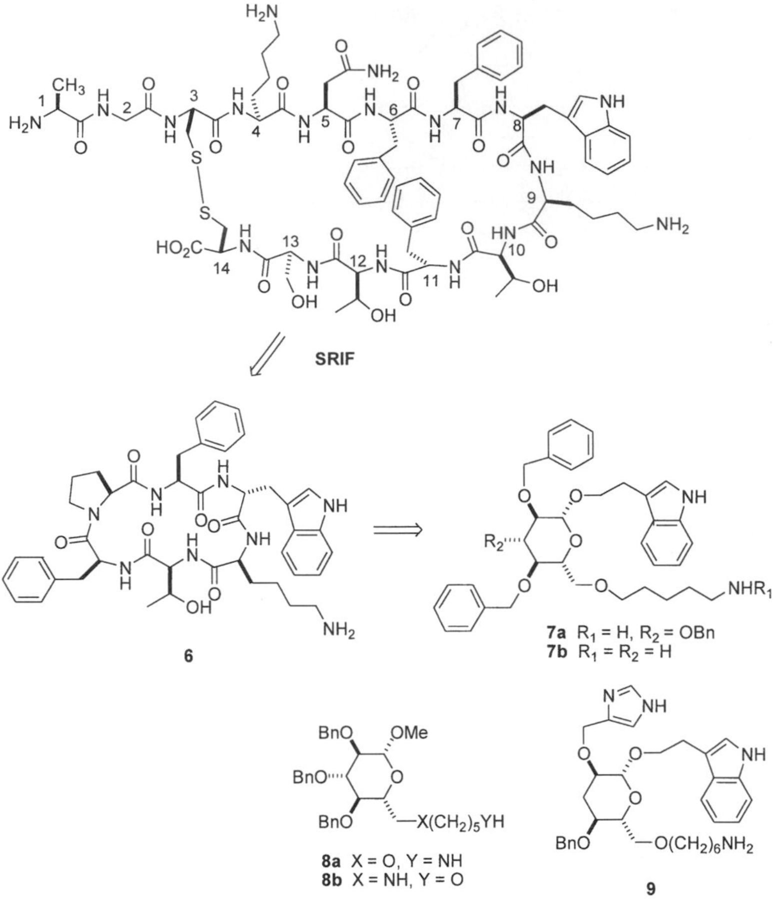

FIG. 6. SRIF and its mimetics.

(GH), glucagon, and insulin.[38] It has a very short biological half-life and does not possess adequate oral bioavailability. SRIF contains a β turn composed of the tetrapeptide Phe-Trp-Lys-Thr.[39,40] This tetrapeptide retains the ability to elicit SRIF-like biological effects as long as the side chains of these four amino acids are constrained in nearly the same orientation as in the bioactive conformation of SRIF. For example, L-363,301 (**6** in Fig. 6) is a potent SRIF agonist.

[38] P. Brazeau, W. Vale, R. Burgus, and R. Guillemin, *Can J. Biochem.* **52,** 1067 (1974).

[39] D. F. Veber, F. W. Holly, W. J. Palaveda, R. F. Nutt, S. J. Bergstand, M. Torchiana, M. S. Glitzer, R. Saperstein, and R. Hirschmann, *Proc. Natl. Acad. Sci. U.S.A.* **75,** 2636 (1978).

[40] B. H. Arison, R. Hirschmann, and D. F. Veber, *Bioorg. Chem.* **7,** 447 (1978).

In search for a nonpeptide scaffold for the SRIF mimetic **6,** D-glucose was found as an attractive option.[41,42] The pyranoside of glucose offers several advantages over other sugars, including its well-defined conformation, the ability to position the required side chains in the equatorial position around the pyranose ring, and the requisite enantiomeric purity of the starting material.

Molecular modeling suggested that in the simple tribenzyl glycoside **7a,** substituents at C-2, C-1, and C-6 could provide appropriately positioned replacements for the critical Phe-Trp-Lys side chains, respectively, and the benzyl group at C-4 may be able to mimic the Phe-Pro dipeptide of **6,** necessary for a favorable hydrophobic interaction with the SRIF receptor (Fig. 6). Because Thr^{10} of SRIF can be replaced by Ala without loss of potency,[43] the side chain group of Thr^{10} was not incorporated into the design of **7.**

It was found that **7a** and **7b** displaced [^{125}I]CGP 23996 from SRIF receptors on membranes from cerebral cortex, pituitary, and AtT-20 cells but with IC_{50} values of 10 and 1.3 μM, respectively. Unexpectedly, analogues **8a** and **8b** lacking the indole side chain also bound to the SRIF receptor (Fig. 6). An analogue containing an imidazole functionality at the 2-position (**9**) was also developed that showed a binding affinity of 1.9 μM. No report on improving these micromolar binding affinities has appeared.

δ-Opioid Selective Nonpeptide Mimetic Agonists

Since the discovery of multiple opioid receptors μ, δ, and κ, tremendous progress has been made in the development of highly selective δ-opioid receptor ligands. This has helped to determine the different functions of opioid receptor types, as well as defining the different structural and conformational properties that are critical to their physiological activities. For example, ligands showing analgesia through the δ receptor may be more advantageous compared to μ or κ compounds in terms of the side effects seen commonly at the other receptors, such as respiratory depression, constipation, and addiction potential.[44] Many conformationally constrained opioid peptide agonists, which can be a template for the *de novo* design, have been synthesized based on endogenous opioid peptides to elucidate crucial pharmacophore and topographical requirements for the receptor binding and

[41] R. Hirchmann, K. C. Nicolaou, S. Pietranico, J. Salvino, E. M. Leahy, P. A. Sprengeler, G. Furst, and A. B. Smith III, *J. Am. Chem. Soc.* **114,** 9217 (1992).

[42] R. Hirschmann, K. C. Nicolaou, S. Pietranico, E. M. Leahy, J. Salvino, B. Arison, M. A. Cichy, P. G. Spoors, W. C. Shakespeare, P. A. Sprengeler, P. Hamley, A. B. Smith III, T. Resine, K. Raynor, L. Maechler, C. Donaldson, W. Vale, R. M. Freidinger, M. R. Cascieri, and C. D. Strader, *J. Am. Chem. Soc.* **115,** 12550 (1993).

[43] W. Vale, J. Rivier, N. Ling, and M. Brown, *Metabolism* **27,** 1391 (1978).

[44] F. Porreca, E. J. Bilsky, and J. Lai, *in* "The Pharmacology of Opioid Peptides" (T. F. Tseng, ed.), p. 219. Harwood Academic, New York, 1995.

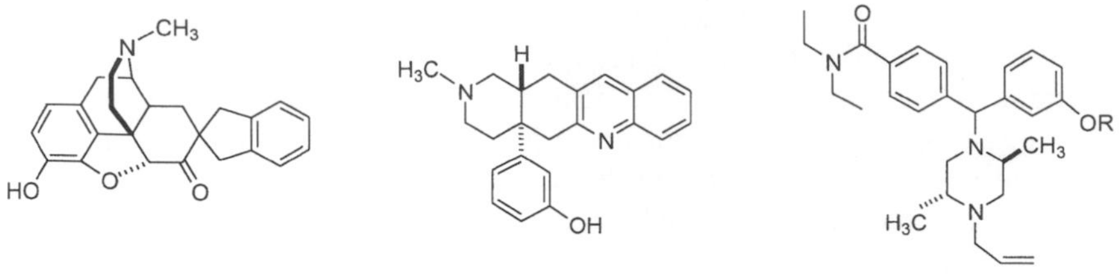

SIOM TAN67 (±)BW373U86 (R= H)
SNC80 (R= CH_3)

FIG. 7. Nonpeptide δ-opioid receptor ligands.

biological activities for the δ-opioid receptors.[45] However, most nonpeptide ligands for the δ-opioid receptor have not been discovered by the rational design methodology, but by random screenings or morphine derivatization. These include SIOM,[46] TAN-67,[47] BW-373U86, and its methoxy analogue (-)-SNC-80[48] (Fig. 7). They may have different ligand–receptor recognition and/or signal transaction mechanisms from the endogenous peptide ligands because they have low activity *in vivo* or different pharmacological profiles. However, SL-3111[49] (Scheme 5) is a novel nonpeptide ligand for the δ-opioid receptor distinctive from others as a nonpeptide peptide mimetic ligand that was developed from a conformational and topographical model of the peptide c[D-Pen^2, D-Pen^5]enkephalin (DPDPE)[50] and its conformationally constrained analogue [(2*S*,3*R*)-TMT^1]DPDPE (TMT = β-methyl-2′,6′-dimethyltyrosine, Fig. 8) and other related analogues.

DPDPE has nanomolar potency at the δ-opioid receptor and high selectivity as a cyclic peptidomimetic of enkephalin. Structure–activity studies demonstrated that Tyr^1, Phe^4, and N^α-amino groups all were critical for the high potency of the ligand at the δ-opioid receptor, whereas the bulky D-Pen residues at position 2 and position 5 were critical for imparting selectivity for the δ vs μ receptor. The 14-membered ring conformation in DPDPE has been well defined through a combination of NMR studies,[51] computational searches,[52] and its X-ray crystal structure.[53] However, the side chain groups of the important pharmacophore of Phe^4 and Tyr^1

[45] V. J. Hruby, F. Al-Obeidi, and W. M. Kazmierski, *Biochem. J.* **268,** 249 (1990).

[46] P. S. Portoghese, S. T. Moe, and A. E. Takemori, *J. Med. Chem.* **36,** 2572 (1993).

[47] T. Suzuki, M. Tsuji, T. Mori, M. Misawa, T. Endoh, and H. Nagase, *Life Sci.* **57,** 155 (1995).

[48] R. J. Knapp, G. Santoro, I. A. De Leon, K. B. Lee, S. A. Edsall, S. Waite, E. Malatynska, E. Varga, S. N. Calderon, K. N. Rice, R. B. Rothman, F. Porreca, W. R. Roeske, and H. I. Yamamura, *J. Pharmacol. Exp. Ther.* **277,** 1284 (1996).

[49] S. Liao, J. Alfaro-Lopez, M. D. Shenderovich, K. Hosohata, J. Lin, X. Li, D. Stropova, P. Davis, K. A. Jernigan, F. Porreca, H. I. Yamamura, and V. J. Hruby, *J. Med. Chem.* **41,** 4767 (1998).

[50] H. I. Mosberg, R. Hurst, V. J. Hruby, K. Gee, H. I. Yamamura, J. J. Galligan, and T. F. Burks, *Proc. Natl. Acad. Sci. U.S.A.* **80,** 5871 (1983).

[51] V. J. Hruby, L.-F. Kao, B. M. Pettitt, and M. Karplus, *J. Am. Chem. Soc.* **110,** 3351 (1988).

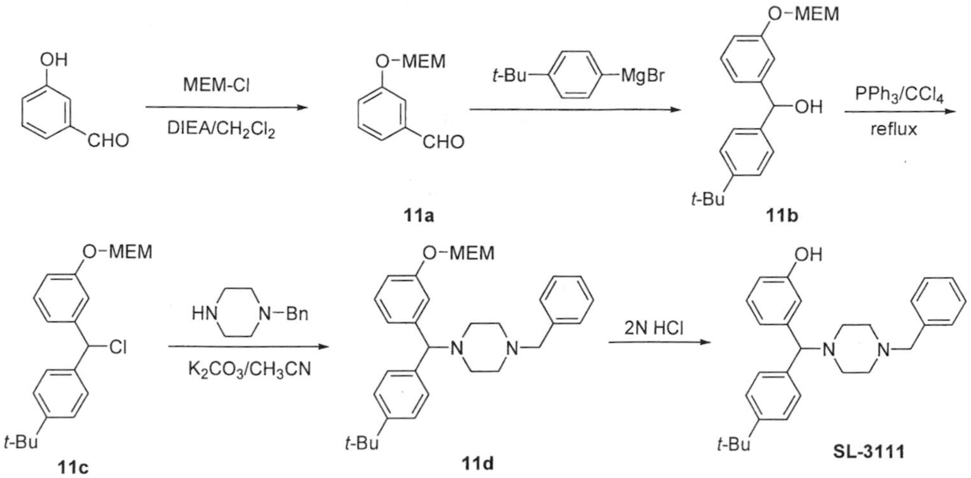

SCHEME 5. Synthesis of nonpeptide δ-opoid ligand SL-3111.

were flexible. To investigate the topographical requirements of the Tyr^1 and Phe^4 side chain residues in DPDPE for interaction with the δ-opioid receptor, analogues of DPDPE incorporated novel topographically constrained amino acid were examined. Substitution of the Phe^4 residue with (2*S*,3*S*)-β-methylphenylalanine[54] showed remarkably high binding selectivity for the δ-opioid receptor over the μ-opioid receptor and suggested that the gauche(−) conformation was the preferred conformation for interaction of the Phe^4 moiety at the δ-opioid receptor.[55] Incorporation of TMT[56] into DPDPE in position 1 was aimed at restricting rotation about both the χ^1 and χ^2 torsional angles of the tyrosine side chain.[57] Among four stereoisomers of TMT, only the (2*S*,3*R*)-containing analogue [(2*S*,3*R*)-TMT^1] DPDPE was one with nanomolar binding affinity and a higher selectivity for the δ receptor than DPDPE itself. Its TMT^1 side chain group was in the *trans* conformation[58] as predicted from the χ^1 and χ^2 energy map of TMT. [(2*S*,

[52] G. V. Nikiforovich, O. Prakash, C. A. Gehrig, and V. J. Hruby, *Int. J. Peptide Protein Res.* **41,** 347 (1993).

[53] J. L. Flippen-Anderson, V. J. Hruby, N. Collins, C. George, and B. Cudney, *J. Am. Chem. Soc.* **116,** 7523 (1994).

[54] F.-D. Lung, G. Li, B.-S. Lou, and V. J. Hruby, *Synth. Commun.* **25,** 57 (1995).

[55] V. J. Hruby, G. Töth, C. A. Gehrig, L.-F. Kao, R. Knapp, G. K. Lui, H. I. Yamanura, T. H. Kramer, P. Davis, and T. F. Burks, *J. Med. Chem.* **34,** 1823 (1991).

[56] X. Qian, K. C. Russell, L. W. Boteju, and V. J. Hruby, *Tetrahedron* **51,** 1033 (1995).

[57] V. J. Hruby, G. Li, C. Haskell-Luevano, and M. D. Shenderovich, *Biopolymers* **43,** 219 (1997).

[58] X. Qian, M. D. Shenderovich, K. E. Kövér, P. Davis, R. Horváth, T. Zalewska, H. I. Yamamura, F. Porreca, and V. J. Hruby, *J. Am. Chem. Soc.* **118,** 7280 (1996).

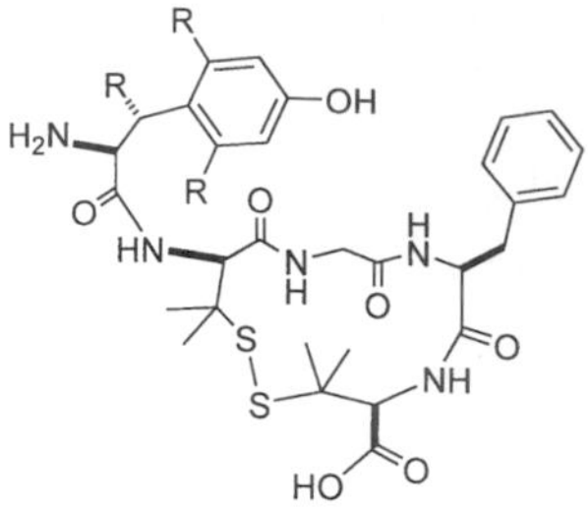

DPDPE (R= H)

[(2*S*,3*R*)TMT[1]]**DPDPE** (R= CH_3)

FIG. 8. Structures of DPDPE and [(2*S*,3*R*)-TMT[1]]DPDPE.

3*R*)-TMT[1]]DPDPE shares the global backbone constrains of DPDPE and was an excellent peptide template for the δ opioid 3D pharmacophore.

Detailed computer-assisted molecular modeling studies using molecular dynamic (MD) simulations with the AMBER force field and a continuum hydration model indicated that there are still several possible conformations available to [(2*S*, 3*R*)-TMT[1]]DPDPE within a 5-kcal/mol energy barrier.[59] Thus, conformational studies comparing [(2*S*,3*R*)-TMT[1]]DPDPE with two potent nonpeptide δ agonists, SIOM and TAN-67, which are potent but less selective agonists for the δ receptor, evolved into a three-dimensional model of the δ-opioid pharmacophore with distances of 7.6 ± 1.5 Å between the tyrosine and the phenyl alanine aromatic side chains (Fig. 9).[60] A computer-assisted template search suggested that a six-membered, cyclohexane-like scaffold could display two *cis*-1,4-disubstituted benzyl-like aromatic rings at the appropriate distances. The hydroxyl group on the phenol ring should be situated in a *meta* position to overlap with the corresponding pharmacophoric element in [(2*S*,3*R*)-TMT[1]]DPDPE. In order to facilitate the synthetic process, 1,4-piperazine was chosen as a six-membered ring scaffold for the first generation of nonpeptide mimetics in which both or either of the amino functions are expected to play a role, although not optimized, as another important element of the pharmacophore. To mimic the hydrophobicity provided by the β-dimethyl groups in the D-Pen residues of DPDPE, an additional variable substituent was added at the benzyl carbon of the phenol side chain. Thus, this

[59] M. D. Shenderovich, S. Liao, X. Qian, and V. J. Hruby, *in* "Peptides: Chemistry, Structure and Biology, Proceedings of the 15th American Peptide Symposium" (J. P. Tam and P. T. P. Kauyama, eds.), p. 404. ESCOM Kluwer Academic, Dordrecht, The Netherlands, 1998.

[60] V. J. Hruby, S. Liao, M. D. Shenderovich, J. Alfaro-Lopez, K. Hosohata, P. Davis, F. Porreca, and H. I. Yamamura, *in* "Peptides: Chemistry, Structure and Biology, Proceedings of the 15th American Peptide Symposium" (J. P. Tam and P. T. P. Kauyama, eds.), p. 144. ESCOM Kluwer Academic, Dordrecht, The Netherlands, 1998.

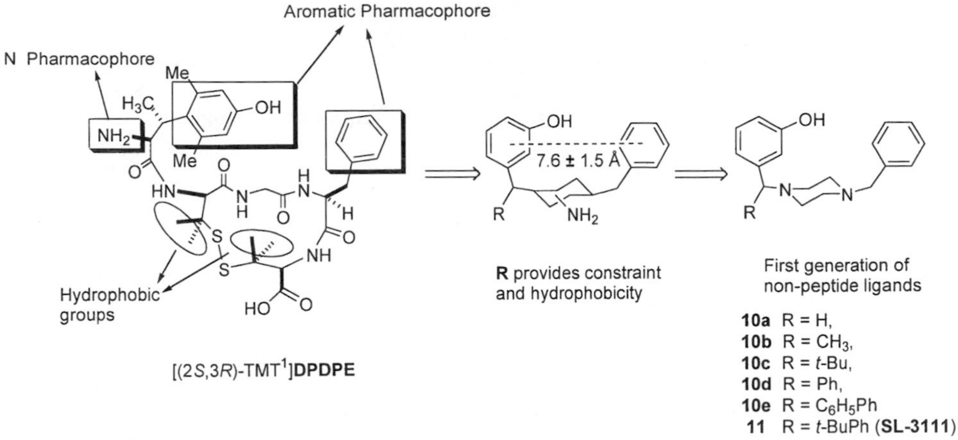

FIG. 9. First-generation nonpeptide mimetic ligands from the peptide lead [(2*S*,3*R*)-TMT^1]DPDPE.

initial series of compounds was designed rationally to explore the roles of the benzyl rings and of the hydrophobic moiety with regard to interaction with the δ-opioid receptor.[49]

Nonpeptide ligands were tested in radioligand-binding assays using $[^3H]$DAMGO[61] for μ-opioid receptors and $[^3H]$[Phe(p-Cl)^4]DPDPE[62] for δ-opioid receptors. The results are shown in Table I. The binding affinity and selectivity for the δ receptor increased with the hydrophobicity and size of the R substituent. Substituting the phenyl group in analogue **10e** with the bulkier *t*-butyl group gave analogue **11,** referred to as SL-3111. SL-3111 is one of the most selective nonpeptide mimetic ligands reported so far for binding preference to the δ-opioid receptor, with a 2000-fold μ/δ selectivity. When the two optically pure enantiomers of SL-3111 were tested, the (-)isomer increased in affinity compared to the racemic mixture, with an IC_{50} value of 4.1 n*M* vs 42 n*M* for the (+)isomer. When tested in the isolated mouse vas deference (MVD, δ receptor) and guinea pig ileum (GPI, μ receptor) bioassay, SL-3111 displayed a 10-fold decrease in potency at the δ-receptor compared to its binding affinity (Table II). To further investigate the role of nitrogen atoms in the piperazine scaffold, analogues in which one of two nitrogens was replaced with CH were synthesized. They revealed that the nitrogen atom linked to the benzyl group in SL-3111 mimics the basic amino group known to be essential in opioid-like drugs,[63] which is thought to imitate the amino terminus on the lead DPDPE.[64]

[61] R. B. Rothman, H. Xu, M. Seggel, A. E. Jacobson, K. C. Rice, G. A. Brine, and F. I. Carroll, *Life Sci.* **48,** 111 (1991).

[62] L. K. Vaughn, R. J. Knapp, G. Toth, Y.-P. Wan, V. J. Hruby, and H. I. Yamamura, *Life Sci.* **45,** 1001 (1989).

TABLE I
BINDING AFFINITY OF δ-SELECTIVE OPIOID LIGANDS[a]

Compound	IC_{50} (nM) [^{3}H]DAMGO (μ)	IC_{50} (nM) [^{3}H][p-ClPhe^4]DPDPE(δ)	Selectivity (μ/δ)
[(2S,3R)-TMT^1] DPDPE	4,300	5.0	860
Analogue **10a** (R = H)	8,100	6400	1.3
Analogue **10b** (R = Me)	780	610	1.3
Analogue **10c** (R = i-Bu)	2,100	420	5.0
Analogue **10d** (R = Ph)	500	34	15
Analogue **10e** (R = C_6H_5Ph)	~27,000	31	~870
Analogue **11** (R = t-BuPh) (SL-3111)	17,000	8.4	2020
(+)-SL-3111	11,000	42	260
(-)-SL-3111	7,700	4.1	1900

[a] Data from S. Liao, J. Alfaro-Lopez, M. D. Shenderovich, K. Hosohata, J. Lin, X. Li, D. Stropova, P. Davis, K. A. Jernigan, F. Porreca, H. I. Yamamura, and V. J. Hruby, *J. Med. Chem.* **41,** 4767 (1998).

Unlike the other nonpeptide ligands, the following evidence revealed that SL-3111 behaves like a peptide ligand at the δ-opioid receptor. The mutant human δ receptor (W284L), in which the Trp at position 284 was mutated to Leu, is known to decrease the affinity of a nonpeptide opioid ligand while maintaining unaltered binding affinities of peptide agonist such as [p-ClPhe^4]DPDPE and deltorphin I.[65] SL-3111 was compared to [p-ClPhe^4]DPDPE and the nonpeptide ligand SNC-80 at both the wild-type cloned human δ-opioid receptor and the W284 mutant receptor. At both receptors, the peptide [p-ClPhe^4]DPDPE exhibited identical binding curves (Table III). The nonpeptide SNC-80 showed significant differences at two receptors in IC_{50} values, whereas SL-3111 showed only threefold differences, which was found to be statistically nonsignificant. Therefore, the binding profile of SL-3111 mimics the peptide ligand but not the nonpeptide ligand, lending credence to the rational design of this nonpeptide mimetic.

Summary/Concluding Remarks

This article concentrated on a few rational approaches that have been considered in the design of nonpeptide ligands from peptide leads. As is evident from

[63] P. S. Farmer, *in* "Drug Design" (E. J. Ariens, ed.), p. 127. Academic Press, New York, 1980.

[64] J. Alfaro-Lopez, T. Okayama, K. Hosohata, P. Davis, F. Porreca, H. I. Yamamura, and V. J. Hruby, *J. Med. Chem.* **42,** 5359 (1999).

[65] R. M. Quock, Y. Hosohata, R. J. Knapp, T. H. Burkey, K. Hosohata, X. Zhang, K. C. Rice, H. Nagase, V. J. Hruby, F. Porreca, W. R. Roseke, and H. I. Yamamura, *Eur. J. Pharmacol.* **326,** 101 (1997).

TABLE II
BIOLOGICAL POTENCIES OF NONPEPTIDE MIMETICS[a]

Compound	EC_{50} (n*M*) GPI (μ)	EC_{50} (n*M*) MVD (δ)	Selectivity (μ/δ)
[(2*S*,3*R*)-TMT[1]]DPDPE	0% at 60 μM (antag., $IC_{50} = 5\mu M$)	1.8	>33,000
SL-3111	39,000	85	460

[a] Data from S. Liao, J. Alfaro-Lopez, M. D. Shenderovich, K. Hosohata, J. Lin, X. Li, D. Stropova, P. Davis, K. A. Jernigan, F. Porreca, H. I. Yamamura, and V. J. Hruby, *J. Med. Chem.* **41,** 4767 (1998).

the cases examined, the *de novo* design of nonpeptides to mimic all of the chemical and biological features of a bioactive peptide is still a considerable challenge, especially if the goal is to obtain an agonist. Although considerable success has been obtained in designing nonpeptide peptidomimetics that will incorporate some aspects of the overall structural and topographical features of a peptide, efforts to design in other important features, such as the proper topographical relationships of three or four critical side chain groups in 3D space and of certain H-bond donating and accepting properties, have been difficult. As a result of these "missing" structural, stereoelectronic, or dynamic properties that the native peptide ligand possesses, perhaps not surprisingly these nonpeptide peptidomimetics often lack some of the critical biological properties that the native ligand, or even a much modified, but still largely "peptidic" ligand, will possess. Because it has been much easier to design potent and selective nonpeptide antagonists for peptide agonist receptors, once one has a lead starting structure, perhaps these results are telling us something regarding the requirements for information transduction resulting from peptide ligand–receptor interactions. For example, because agonists require

TABLE III
EFFECT OF MUTATION OF THE CLONED HUMAN δ-OPIOID RECEPTOR ON THE BINDING AFFINITY (IC_{50}, n*M*) OF δ-SELECTIVE LIGANDS[a]

Ligand	WT[b]	W284L[c]	W284/WT
[(2*S*,3*R*)-TMT[1]]DPDPE	1.53	1.55	1.0
SL-3111	5.51	16.8	3.1
SNC-80	2.85	49.1	17.2

[a] Data from S. Liao, J. Alfaro-Lopez, M. D. Shenderovich, K. Hosohata, J. Lin, X. Li, D. Stropova, P. Davis, K. A. Jernigan, F. Porreca, H. I. Yamamura, and V. J. Hruby, *J. Med. Chem.* **41,** 4767 (1998).
[b] Cloned human δ-opioid receptor.
[c] Mutated human δ-opioid receptor in which Trp-284 was mutated to Leu.

that the ligand–receptor interaction leads to structural changes compatible with information transduction, it might suggest that receptors for relatively small peptide ligands (up to 10 residues) may require the design of nonpeptide peptidomimetics that still retain some key features of the peptide structure, such as H bond-accepting or -donating properties, dynamic behavior of the peptide backbone for appropriate structural changes, and appropriate side chain group conformations and topographies.

Whatever the structural or dynamic features that need to be incorporated into a nonpeptide peptide mimetic, it is hoped that with more effort we will learn how to incorporate these features. Given the importance of the goal, it is certain that synthetic chemists will continue to develop this field.

Synthetic Procedure for 1-(((7-Dimethyl-2-oxobicyclo[2,2,1]heptan-1(S)-yl) methyl)sulfonyl)-4-(2-methylphenyl)piperazine (L-367,398, See Scheme 1)[19b]

To a stirred solution of 1-(2-methylphenyl)piperazine hydrochloride (50.0 g, 235 mmol) and (+)-10-camphorsulfonyl chloride (65.5 g, 260 mmol) at 0° in $CHCl_3$ (1000 ml) is added DIEA (103 ml, 590 mmol) dropwise over 30 min. The solution is stirred at 0° for 1 hr and then at ambient temperature for 3 hr. The solution is extracted with 5% aqueous HCl (2 × 500 ml), water (500 ml), and saturated $NaHCO_3$ (2 × 500 ml). The organic phase is dried ($MgSO_4$) and filtered, and the solvent is removed. The resulting solid is crystallized from MeOH to give the title compound (79 g, 86%). TLC R_f 0.49 (3 : 1 hexanes/EtOAc). FAB MS *m/z* 391 $[M + H]^+$. ^{1}H NMR (300 MHz, $CDCl_3$) δ 0.91 (s, 3H), 1.18 (s, 3H), 1.44 (m, 1H), 1.67 (m, 1H), 1.96 (d, 1H, J = 14 Hz), 2.10 (m, 2H), 2.30 (s, 3H), 2.40 (dt, 1H, J = 14, 3 Hz), 2.57 (m, 1H), 3.00 (m, 4H), 3.40 (d, 1H, J = 16 Hz), 3.45 (m, 4H), 7.0 (m, 2H), 7.2 (m, 2H).

Synthetic Procedures for TRH Agonist 2 (See Scheme 2)[31]

(S)-5-Oxo-1-(phenylmethyl)-2-pyrrolidinecarboxylic Acid Methyl Ester **(2a).** L-Glutamic acid, monosodium salt (250 g, 1.33 mol) is added at room temperature to a solution of NaOH (53.6 g, 1.34 mol) in 550 ml of water. To the resulting solution is added benzaldehyde (142.3 g, 1.34 mol). The mixture is stirred and cooled to 10°, and $NaBH_4$ (15.2 g, 0.409 mol) is added in portions, keeping the temperature at 10–15°. The mixture is stirred for 30 min, and another portion of benzaldehyde (142.3 g, 1.34 mol) is added. After 10 min, a second portion of $NaBH_4$ (3.7 g, 0.10 mol) is added as before. The mixture is then allowed to stir at room temperature overnight. The solution is washed with CH_2Cl_2 (250 ml, discarded) and acidified to pH 3 with 6 *N* HCl. The paste, consisting of *N*-(phenylmethyl)-L-glutamic acid, is diluted with 500 ml of H_2O and heated to reflux overnight. The resulting solution is cooled to room temperature and extracted with $CHCl_3$. The combined extracts are washed with brine, dried over Na_2SO_4,

and concentrated on a rotary evaporator to give 160 g (55% yield) of (*S*)-5-oxo-1-(phenylmethyl)-2-pyrrollidinecarboxylic acid as a white solid. The crude acid (160 g, 0.73 mol) is dissolved in 300 ml of toluene and 550 ml of methanol. To the solution is added 9 ml of concentrated H_2SO_4, and the solution is heated to reflux overnight. The solution is cooled in an ice bath and neutralized to pH 5 with 25% NaOH, followed by adding 50 ml of saturated $NaHCO_3$ to bring the solution to pH 7. Methanol is removed on a rotary evaporator, and the residue is diluted with 500 ml of water and extracted with CH_2Cl_2. The combined extracts are washed with brine and dried over $MgSO_4$. Evaporation of the solvent afforded 132 g of eater **2a** as an oil, which is > 95% pure by NMR. ^1H NMR ($CDCl_3$) δ 2.0–2.6 (m, 4H, CH_2's), 3.68 (s, 3H, CH_3), 3.95(dd, 1H, $J = 4, 9$ Hz, H-4), 4.02 and 5.03 (AB, 2H, $J_{gem} = 16$ Hz, CH_2Ph), 7.2–7.4 (m, 5H, arom).

(S)-5-(Iodomethyl)-1-(phenylmethyl)-2-pyrrolidinone ***(2b).*** A solution of **2a** (132 g, 1.556 mol) in 600 ml of *t*-BuOH is cooled to ca. 18°. $NaBH_4$ (41 g, 1.12 mol) is added in one portion. To the suspension is added 425 ml of methanol over 45 min. The reaction mixture is kept at 15°C during the addition, during which time hydrogen gas evolves. After the addition, the mixture is kept at 20°C until hydrogen evolution has subsided and is allowed to stand at room temperature overnight. Solvents are removed on a rotary evaporator, and the residue is dissolved in 1 liter of water and the pH adjusted to pH 7.5 with 2 *N* HCl. The mixture is extracted with CH_2Cl_2, and the combined extracts are washed with 2 *N* HCl and brine and dried over Na_2SO_4. Evaporation of the solvent affords 96.3 g of crude product, which is recrystallized from toluene to give 89.5 g (77% yield) white solid. mp 82–84° (toluene). ^1H NMR ($CDCl_3$) δ 1.92 and 1.94 (dd, 1H, $J = 6.5$ Hz, OH), 1.96–2.11 (m, 3H, CH_2O and CH), 4.29 and 4.81 (AB, 2H, $J_{gem} = 15$ Hz, CH_2Ph), 7.26–7.35 (m, 5H, phenyl). $[\alpha]^{25}D = +116°$ (c 1.07, MeOH). To a solution of the white solid (143.7 g, 0.70 mol) in 3.5 liter of CH_2Cl_2 are added 4- (dimethylamino)pyridine (94.1 g, 0.77 mol) and *p*-toluenesulfonyl chloride (133.5 g, 0.70 mol). The solution is stirred at room temperature overnight and then washed with cold 1 *N* HCl (600 ml), saturated $NaHCO_3$ (500 ml), and brine and dried over sodium Na_2SO_4. Evaporation of the solvent gives the crude tosylate (244.6 g, 97.2% yield) as a white solid, mp 79–80° (ether), which is refluxed overnight with sodium iodide (305 g, 2.03 mol) in 3 liter of acetone. The suspension is cooled to 10° and filtered. Salts are rinsed with acetone (3 × 250 ml), and acetone washes and the filtrate are concentrated on a rotary evaporator to a thick slurry. CH_2Cl_2 (1.5 liter) is added, and the white precipitate is filtered off and washed with CH_2Cl_2. The filtrate is dried over $MgSO_4$, filtered through a pad of silica gel, and concentrated on a rotary evaporator to give 196.8 g (92% yield crude) of iodide **2b** as an off-white solid. A sample recrystallized from cyclohexane has mp 92–94°. ^1H NMR ($CDCl_3$) δ 1.7–2.75 (m, 4H, CH), 3.18–3.2 (m, 2H, CH_2I), 3.42 (m, 1H, NCH), 3.98 and 4.05(AB, 2H, $J_{gem} = 15$ Hz, CH_2Ph), 7.20 (m, 5H, phenyl). $[\alpha]^{25}D = +12.11°$ (c 0.35, MeOH).

*(E)-4-[2(R)-5-Oxo-1-(phenylmethyl)-2-pyrrolidinyl]-2-butenoic Acid Ethyl Ester (**2c**).* To a solution of crude iodide **2b** (31.5 g, 0.10 mol) in 250 ml of anhydrous THF is added a solution of Li_2CuCl_4 [7.5 ml of a 0.1 *M* solution in THF; prepared from anhydrous LiCl (85 mg) and anhydrous $CuCl_2$ (134.5 mg) in 10 ml of THF]. The solution is cooled to $-78°$, and vinylmagnesium bromide (200 ml, 1 *M* solution in THF) is added over 25 min. After stirring for 1 hr at $-78°$, a second portion of vinylmagnesium bromide (200 ml of a 1 *M* solution) is added as before, and the mixture is stirred at $-78°$ overnight. The cold solution is then poured onto 1.5 kg of ice and acidified with 100 ml of 6 *N* HCl. The mixture is extracted with CH_2Cl_2, washed with saturated $NaHCO_3$ and brine, and dried over $MgSO_4$. The solution is concentrated on a rotary evaporator to give 23.8 g of crude propenylpyrrolidinone as a brown oil. The crude product is chromatographed on 700 g of silica gel, eluting with 1 : 1 EtOAc/hexanes to give 14.2 g (66% yield) of a colorless oil. ^1H NMR ($CDCl_3$) δ 1.7–2.75 (m, 6H, CH_2's), 3.51 (m, 1H, NCH), 4.99 and 5.02 (AB, 2H, $J_{gem} = 15$ Hz, CH_2Ph), 5.0–5.2 (m, 2H, vinyl H), 5.4–5.9 (m, 1H, vinyl H), 7.25–7.40 (m, 5H, phenyl H). $[\alpha]^{25}D = +31.87°$ (c 1.0, MeOH). A solution of the colorless oil (44.8 g, 0.208 mol) in 800 ml of 1 : 1 $MeOH/CH_2Cl_2$ is cooled to $-78°$ and ozonized using a Welsbach ozonizer for 6.5 hr, approximately 1 hr longer than the time required to observe a light blue color of ozone in the solution. Excess ozone is flushed out of the system with oxygen, and the solution is treated with Me_2S (80 ml). The solution is then allowed to come to room temperature and stand overnight. Solvents are removed on a rotary evaporator, and residual liquid is dissolved in 700 ml of CH_2Cl_2, washed with water, and dried over Na_2SO_4. Evaporation of the solvent afforded 43.0 g of crude aldehyde. Chromatography on silica gel (1 kg) eluting with 4% MeOH in CH_2Cl_2 afforded 38 g (84.4% yield) of the pure aldehyde. ^1H NMR ($CDCl_3$) δ 1.60–2.60 (m, 4H, CH_2), 2.77 (one-half of ABX, 1H, $J_{gem} = 18$ Hz, $J_{vic} = 4$ Hz, $COCH_2$), 3.98 (m, 1H, NCH), 4.10 and 4.82 (AB, 2H, $J_{gem} = 15$ Hz, CH_2Ph), 7.15–7.40 (m, 5H, phenyl H), 9.65 (s, 1H, CHO). To a solution of the aldehyde (38 g, 0.175 mol) in 700 ml of toluene is added (carbethoxymethylene)triphenylphosphorane (73.0 g, 0.21 mol), and the mixture is heated to 90° for 5.5 hr. The solvent is removed on a rotary evaporator, and the residue is slurried with 50 ml of EtOAc. The bulk of the precipitate of triphenylphosphine oxide is removed by filtration, and the filter cake is washed with 150 ml of 1 : 1 hexanes/EtOAc. The solution is concentrated on a rotary evaporator, mixed with 200 ml of 3 : 1 hexanes/EtOAc, and allowed to stand at room temperature of 72 hr. The crystalline triphenylphosphine oxide is filtered off, and residual oil (63 g) is chromatographed on 2 kg of silica gel, eluting with EtOAc to give 48 g (95.6% yield) of butenoate **2c** as an oil. ^1H NMR ($CDCl_3$) δ 1.30 (t, $J = 5.5$ Hz, CH_3), 1.60–2.50 (m, 6H, CH_2's), 3.60 (m, 1H, NCH), 3.96 and 4.99 (AB, $J_{gem} = 15$ Hz, CH_2Ph), 4.15 (q, 2H, $J = 5.5$ Hz, OCH_2), 5.85 (d, 1H, $J = 16$ Hz, C-2 vinyl H), 6.74 and 6.80 (dt, 1H, $J = 8, 16$ Hz, C-2 vinyl H), 7.20–7.40 (m, 5 H, phenyl). $[\alpha]^{25}D = -20°$ (c 1.0, MeOH).

(S)-5-([5-Oxo-1-(phenylmethyl)-2-pyrrolidinyl]methyl)-1,3-cyclohexanedione ***(2d).*** Sodium metal (7.82 g, 0.34 g-atom) is dissolved in 1.1 liter of ethanol. Ethyl acetoacetate (44.2 g, 0.34 mol) and butenoate **2c** (66.0 g, 0.23 mol) are added, and the solution is refluxed and stirred overnight. The solvent is removed on a rotary evaporator, and the residue is dissolved in 500 ml of water. The solution is washed with CH_2Cl_2 (discarded), acidified to pH 1 with 6 *N* HCl, and extracted with CH_2Cl_2. The combined extracts are washed with water and dried over Na_2SO_4. The solvent is removed on a rotary evaporator to give 70 g of the intermediate diketo ester as an oil. KOH (128.8 g, 2.3 mol) is dissolved in 2.0 liter of ethanol. To the KOH solution is added 85.5 g (0.23 mol) of the crude diketo ester (from this and a similar run on a smaller scale), and the solution is heated to reflux for 3 h. The ethanol is removed on a rotary evaporator, and to the residue is added 1.2 liter of concentrated HCl. The mixture (gummy residue) is extracted with CH_2Cl_2 (3×300ml), and the combined extracts are washed with water and dried over Na_2SO_4. The solvent is removed on a rotary evaporator to give 60.0 g (87% yield) of cyclohexanedione **2d** as a white solid. mp 159–161°C (EtOAc). ^{1}H NMR ($CDCl_3$) (9 . 1 keto . enol forms) δ 1.38 (ddd, 1H, $J_{vic} = 4, 11$ Hz, $J = 14$ Hz, CH of CH_2), 1.60–2.71 (m, 10H, 4 CH_2, CH of CH_2, CH), 3.37 (s, 10/9 H, CH_2 of keto), 3.46 (m, 1H, NCH), 3.94 and 4.96 (AB, 8/9 H, $J_{gem} = 15$ Hz, CH_2 of enol), 3.95 and 4.99 (AB, 10/9 H, $J_{gem} = 15$ Hz, CH_2 of keto), 5.45 (s, 4/9 H, = CH of enol), 7.19 (br d, 2 H, $J_{ortho} = 7$ Hz, arom), 7.30 (t, 1 H, $J_{ortho} = 7$ Hz, arom), 7.32 (t, 2 H, $J_{ortho} = 7$ Hz, arom).

(5S,1R)-5-[(3-Ethoxy-5-oxo-3-cyclohexen-1-yl)methyl]-1-(phenylmethyl)-2-pyrrolidonone ***(2e)*** *and (5S,1S)-5-[(3-Ethoxy-5-oxo-3-cyclohexen-1-yl)methyl]-1-(phenylmethyl)-2-pyrrolidonone* ***(2f).*** A solution of cyclohexanedione **2d** (60 g, 0.20 mol) and *p*-toluenesulfonic acid monohydrate (3.8 g, 0.02 mol) in 600 ml of ethanol and 1200 ml of toluene is stirred and refluxed for 1.5 hr. The solvent is removed on a rotary evaporator, and the residue is dissolved in CH_2Cl_2. The CH_2Cl_2 solution is washed with saturated $NaHCO_3$ solution and brine and dried over Na_2SO_4. The solvent is removed on a rotary evaporator to give 57.6 g of a crude oil. The crude product is chromatographed on silica gel (800 g) eluting with 2–4% methanol in EtOAc to give a 1 : 1 mixture of diastereomeric products: the (5*S*,1*R*)-ethoxyenone **2e** and the (5*S*,1*S*)-ethoxyenone **2f** (37.5 g, 57% yield). A total of 52.5 g of mixture is chromatographed using HPLC (1 : 24 : 25 MeOH/EtOAc/hexanes) to give 24.3 g (25.0% yield) of the more polar **2e** and 23.9 g (24.6% yield) of the less polar **2f.** (5*S*,1*S*)-ethoxyenone **2f:** mp 92–93°. ^{1}H NMR ($CDCl_3$) δ 1.35 (t, 3H, $J = 6$ Hz, CH_3), 1.50–2.50 (m, 11H, CH_2 and CH), 3.46 (m, 1H, CHN) 3.90 (q, 2H, $J = 6$ Hz, CH_2CH_3), 3.96 and 5.00 (AB, 2H, $J_{gem} = 15$ Hz, CH_2Ph), 5.35 (s, 1H, vinyl H), 7.24–7.34 (m, 5H, phenyl). $[\alpha]^{25}D = +74.61°$ (c 1.0, MeOH).

[1S, 5S,5(2S)]-3-[(Oxo-1-(phenylmethyl)-2-pyrrolidinyl)methyl] cycohexane-acetic acid ethyl ester ***(2g).*** To a solution of diisopropylamine (16.2 g, 160 mmol) in 480 ml of ether is added at −50° a solution of *n*-butyllithium (64.8 ml, 2.5 *M*,

in hexanes, 160 mmol), and the mixture is stirred for 30 min at −20 to −30°. A solution of ethyl(trimethylsilyl)acetate (25.6 g, 160 mmol) in 240 ml of THF is added at −50°, and the mixture is stirred at −30 to −40° for 1 hr. To the solution is added ethoxy enone **2f** (21 g, 64 mmol) in 240 ml of dry THF at −50°. Following the addition, the mixture is allowed to slowly warm over 1 hr to 0°, stirred for 1 hr at 0°, and poured into ice water. The mixture is extracted with CH_2Cl_2, and the solvent is removed on a rotary evaporator. Residual material is dissolved in 330 ml of THF, 330 ml of 6 *N* HCl is added, the mixture is extracted with CH_2Cl_2, and the extracts are washed with saturated $NaHCO_3$ and brine and dried over Na_2SO_4. The solvent is removed on a rotary evaporator, and the residue is chromatographed on silica gel eluting with 1% MeOH in EtOAc to give 12.0 g (51% yield) of enone ester. ^{1}H NMR ($CDCl_3$) δ 1.26 (t, 3H, $J = 7$ Hz, CH_3), 1.25- 2.60 (m, 11H, CH_2 and CH), 3.14, 3.15 (AB, 2H, $J_{gem} = 15$ Hz, CH_2CO), 3.47 (m, 1H, NCH), 3.97, 5.00 (AB, 2H, $J_{gem} = 15$ Hz, CH_2Ph), 4.15 (q, 2H, $J = 7$ Hz, OCH_2), 7.20–7.40 (m, 5H, phenyl). A solution of enone ester (16.4 g, 44.6 mmol) in 500 ml of EtOH is hydrogenated over 10% Pd/C (1.6 g) at 50 psi for 3 hr. The mixture is filtered, and the solvent is removed on a rotary evaporator. The oily product is dissolved in CH_2Cl_2, the solution is dried over Na_2SO_4, and the solvent is removed to give 14.5 g (87.5% yield) **2g** containing ~5% of a *trans* isomer by NMR. The oily product is recrystallized twice from 1 : 1 EtOAc/hexanes to give 7.8 g of **2g** (>97% *cis* isomer). mp 65–70°. ^{1}H NMR ($CDCl_3$) δ 0.95 (m, 1H, axial ring H), 1.25 (t, 3H, $J = 7$ Hz, CH_3), 1.6–2.6 (m, 13H, CH_2 and CH), 3.45 (m, 1H, NCH), 3.97, 4.98 (AB, 2H, $J_{gem} = 15$ Hz, CH_2Ph), 4.12 (q, 2H, $J = 7$ Hz, OCH_2), 7.20–7.40 (m, 5H, phenyl). $[\alpha]^{25}D = +16.88°$ (c 0.936, MeOH).

*(E)-[1S,5S,5(2S)]-5-([5-Oxo-1-(phenylmethyl)-2-pyrrolidinyl]methyl)-3-([1-(phenylmethyl)-1H-imidazol-5-yl]methylene) cycohexaneacetic acid ethyl ester [**2h(E)**] and (Z)-[1S,5S,5(2S)]-5-([5-Oxo-1-(phenylmethyl)-2-pyrrolidinyl] methyl)-3-([1-(phenylmethyl)-1H-imidazol-5-yl]methylene)cycohexaneacetic acid ethyl ester [**2h(Z)**].* To a solution of diisopropylamine (3.6 g, 36 mmol) in 150 ml of ether is added at −30° a solution of *n*-butyllithium in hexanes (14.4 ml, 2.5 *M*, 36 mmol), and the mixture is stirred for 30 min at −30°. A solution of (2-[(phenylmethyl)-amino]ethenyl)phosphonic acid diethyl ester (9.6 g, 36 mmol) in 60 ml of THF is added at −40°. The solution is stirred for 0.5 hr at −15 to −20° and is then cooled to −40°. To the solution is added a solution of **2g** (6.6 g, 18 mol) in 90 ml of THF at −40°. The mixture is stirred for 15 min at −40° and then for 2.5 hr at 0° and subsequently poured into ice water. The organic layer is separated, the aqueous layer is extracted with CH_2Cl_2, and the solvent is removed on a rotary evaporator. The residue is dissolved in 400 ml of CH_2Cl_2, the solution is dried over Na_2SO_4, and the solvent is removed on a rotary evaporator to give the crude imine. ^{1}H NMR ($CDCl_3$) δ 0.65–1.00 (m, 2H, axial ring H), 1.15–1.35 (2t, 3H, CH_3), 3.50 (m, 1H, NCH), 3.95–4.20 (m, 4H, CH_2CO), 4.00, 5.00 (AB, 2H, $J_{gem} = 15$ Hz, CH_2Ph), 5.87, 6.05 (2brd, 1H, vinyl H), 7.20–7.40 (m, 6H, phenyl and imine H),

8034 (br d ca. 0.5H, aldehyde vinyl H), 9.96 [2d, $J = 7.5$ Hz, CHO (*E/Z*)]. The crude imine is dissolved in 210 ml of methanol, and benzylamine (3.84 g, 36 mmol) is added along with 9 g of 3-Å molecular sieves. The mixture is stirred for 1 hr at room temperature, (4-tolylsulfonyl)methyl isocyanide (TOSMIC; 7.2 g, 36 mmol) is added, and the mixture is allowed to stir at room temperature overnight. The mixture is filtered to remove the sieves, and the filtrate is concentrated on a rotary evaporator and then dissolved in 300 ml of CH_2Cl_2. The solution is washed with 2 *N* HCl and saturated $NaHCO_3$, and dried over $MgSO_4$. The solvent is removed on a rotary evaporator, and the residue is chromatographed on silica gel (700 g) eluting with 2.5–5% methanol in CH_2Cl_2 to afford a total of 8.5 g of imidazole ester **2h** as the pure *Z*-isomer **2h(*Z*)** (3.9 g), the *E*-isomer **2h(*E*)** (3.6 g), and mixed fractions (1.0 g). The *E* and *Z* assignments and analytical data are made at the stage of the corresponding amides. For **2h(*Z*),** ^{1}H NMR ($CDCl_3$) δ 0.67 (q, 1H, $J = 12.3$ Hz, axial H of ring CH_2), 1.23 (t, 3H, $J = 7.4$ Hz, CH_3), 1.15–2.28 (m, 12H, 2 CH, 4 CH_2, 2 CH of CH_2), 2.41 (m, 2 H, ring CH_2CO), 2.91 (br d, 1H, $J_{gem} = 12.6$ Hz, equatorial H of ring CH_2), 3.44 (br m, 1H, NCH), 3.94, 4.96 (AB, 2H, $J_{gem} = 14.8$ Hz, CH_2Ph on lactam), 4.11 (q, 2H, $J = 7.4$ Hz, OCH_2), 5.05 (s, 2H, CH_2Ph on imizadole), 5.72 (s, 1H, vinyl H), 6.97, 7.40 (2s, 2H, imidazole H's), 7.0–7.4 (m, 10H, phenyl H's). For **2h(*E*),** ^{1}H NMR ($CDCl_3$) δ 0.66 (br q, 1H, J 11.5 Hz, axial H of ring CH_2), 1.23 (t, 3H, $J = 7.4$ Hz, OCH_3), 1.10–2.60 (m, 12H, 2 CH, 4 CH_2, 2 CH of CH_2), 2.40 (m, 2 H, ring CH_2CO), 2.79 (br d, 1H, $J_{gem} = 12.6$ Hz, equatorial H of ring CH_2), 3.39 (br m, 1H, NCH), 3.93, 4.94 (AB, 2H, $J_{gem} = 14.8$ Hz, CH_2Ph on lactam), 4.10 (q, 2H, $J = 7.4$ Hz, OCH_2), 5.03 (s, 2H, CH_2Ph on imizadole), 5.75 (s, 1H, vinyl H), 6.90, 7.50 (2s, 2H, imidazole H's), 7.0–7.4 (m, 10H, phenyl H's).

[1S,3R,5(2S),5S]-5-([5-Oxo-1-(phenylmethyl)-2-pyrrolidinyl)methyl]-5-[(1H-imidiazol-5-yl)methyl]cyclohexaneacetamide **(2i).** The imidazole ester **2h(*E*)** (7.4 g, 14 mmol) is heated to 75° in 145 ml of ethanol with KOH (4.0 g, 70 mmol) for 30 min. The mixture is concentrated and diluted with 50 ml of water. The solution is washed twice with CH_2Cl_2 and acidified to pH 3.0–3.5 with 6 *N* HCl. The gummy residue is extracted with CH_2Cl_2, dried over Na_2SO_4, and the solvent is removed on a rotary evaporator to give 6 g (86% yield) of the crude acetic acid. The crude acid is dissolved in 325 ml of $CHCl_3$, cooled to −40°, and treated with 6.5 g (60 mmol) of ethyl chloroformate followed by NEt_3 (6.0 g, 60 mmol). The mixture is stirred, allowed slowly warm to 0° for 30 min, and then gaseous ammonia is bubbled into the solution for 20 min at 0°. The white suspension is stirred at room temperature for 2 hr. The mixture is filtered and concentrated. The residue is chromatographed on silica gel, eluting with 5–10% MeOH in CH_2Cl_2 to give 3.78 g [54% overall yield from ester **2h(*E*)**] of a white solid. mp 60–70°. $[\alpha]^{25}D = -28.04°$ (c 1%, MeOH). ^{1}H NMR ($CDCl_3$) δ 0.66 (q, 1H, $J = 12.3$ Hz, axial H of ring CH_2), 1.21 (m, 1H, CH), 1.25 (br m, 1H, CH), 1.48 (t, 1H, $J = 12.3$ Hz, axial H of ring CH_2), 1.53–1.88 (m, 6H, CH_2, 4 CH of CH_2), 2.08 (m, 2H, CH_2CO),

2.34 (m, 1H, CH of CH_2), 2.38 (ddd, 1H, $J_{vic} = 6.8, 9.6$ Hz, $J_{gem} = 17.2$ Hz, CH of ring CH_2CO), 2.49 (ddd, 1H, $J_{vic} = 6.6, 9.6$ Hz, $J_{gem} = 17.2$ Hz, CH of ring CH_2CO), 2.82 (br d, 1H, $J_{gem} = 13.4$ Hz, equatorial H of ring CH_2), 3.42 (m, 1H, NCH), 3.95, 4.94 (AB, 2H, $J_{gem} = 15.3$ Hz, CH_2Ph on lactam), 5.04(s, 2H, CH_2Ph on imizadole), 5.34, 5.41 (2s, 2H, $CONH_2$), 5.76 (s, 1H, vinyl H), 6.90, 7.49 (2s, 2H, imidazole H's), 7.03 (d, 2H, phenyl H's), 7.20 (d, 2H, phenyl H's), 7.22-7.35 (m, 6H, phenyl H's). Irradiation of the vinyl proton at 5.76 ppm results in an observed NOE for the ring proton at C-2 at 2.32 ppm, confirming the assignment of the *E* configuration. HRMS (EI) *m/e* $[M]^+$ calcd for $C_{31}H_{36}N_4O_2$ 496.2838, obsd 496.2823. To a solution of the unsturated imidazole (1.0g, 2.0 mmol) in 100 ml of methanol containing 2% NH_3 is added 10 g of Raney Ni (Aldrich; wet slurry in H_2O, washed twice with H_2O and MeOH). The mixture is hydrogenated at 50 psi in a Parr shaker at room temperature for 48 hr. The catalyst is filtered off, and the methanol is removed on a rotary evaporator to give 620 mg (62% crude yield) of a mixture (88 : 4 : 8 ratio) of **2i,** the more polar 3*S*-isomer **2j,** and unreacted starting material. The crude product is reduced over 10% Pd/C (200 mg) to reduce the remaining starting material. Concentration of the methanol afforded 578 mg of a mixture of **2i, 2j,** and some of the debenzylated compound **2** (see later) in a ratio of 85.8 : 8.6 : 5.6 by reverse-phase HPLC. The product mixture is chromatographed on silica gel eluting with 100 : 20 : 1 $CHCl_3$/MeOH/HOAc. Fractions containing **2i** are combined and concentrated on a rotary evaporator. Water is added followed by NH_4OH until alkaline, and the material is extracted with CH_2Cl_2. The combined extracts are concentrated on a rotary evaporator to afford 380 mg (38%) of pure **2i.** mp 75°C. ^1H NMR ($CDCl_3$) δ 0.36–0.41 (m, 1H, axial ring H), 0.43–0.58 (m, 2H, axial ring H), 3.42 (br s, 1H, NCH), 3.91, 4.93 (AB, 2H, $J_{gem} = 15$ Hz, CH_2Ph on lactam), 5.03(s, 2H, CH_2Ph on imizadole), 5.30 (d, 2H, $J = 12$ Hz, $CONH_2$), 6.80, 7.46 (2s, 2H, imidazole H's), 7.00–7.40 (m, 10H, phenyl H's); HRMS (EI) *m/e* $[M]^+$ calcd for $C_{31}H_{38}N_4O_2$ 498.2995, obsd 498.2990.

[1S,3R,5(2S),5S]-5-([5-Oxo-1-(phenylmethyl)-2-pyrrolidinyl]methyl)-5-[(1H-imidiazol-5-yl)methyl]cyclohexaneacetamide **(2).** A solution of **2i** (360 mg, 0.7 mmol) in 35 ml of MeOH is reduced at 50 psi of hydrogen at 50° over 10% Pd/C (360 mg) in a Fisher–Porter bottle with magnetic stirring overnight. The mixture is cooled and filtered to remove the catalyst, and the solvent is removed on a rotary evaporator. The residue is chromatographed on silica gel (15 g, dry column) eluting with the lower phase of a mixture prepared by shaking $CHCl_3$, MeOH, H_2BD and HOAc in a ratio of 9 : 3 : 1 : 0.6. The fractions are combined, concentrated on a rotary evaporator, diluted with H_2O, made alkaline with NH_4OH, and extracted with CH_2Cl_2. The extracts are dried over Na_2SO_4 and concentrated on a rotary evaporator to afford 146 mg (51% yield) of the monobenzyl amide **2.** ^1H NMR ($CDCl_3$) δ 0.37–0.47 (m, 1H, axial ring H), 0.56–0.68 (m, 2H, axial ring H), 3.44 (m, 1H, NCH), 3.92, 4.95 (AB, 2H, $J_{gem} = 15$ Hz, CH_2Ph), 5.32, 5.41 (br s, 2H,

amide H), 6.74 (s, 1H, imizadole ring H), 7.54 (s, 1H, imizadole ring H), 7.18–7.35 (m, 5H, phenyl); HRMS (EI) calcd for $C_{24}H_{32}N_4O_2$ $[M]^+$ 408.2525, obsd $[M]^+$ 408.2519.

Synthetic Procedures of A II Antagonist 5 (See Scheme 3)[37]

Methyl 2-[(2-Butyl-1-[(4-methoxycarbonyl)phenyl]methyl)-1H-imidazol-5-yl] methyl)-amino]-benzoate **(5a).** Methyl 4-[(2-butyl-5-chloromethyl-1*H*-imidiazol-1-yl) methyl]benzoate hydrochloride (8 g, 22.3 mmol) is suspended in 80 ml of anhydrous toluene. Next, 10.15 g (67.1 mmol) of methyl 2-aminobenzoate and 4.79 g (44.7 mmol) of 2,6-dimethyl pyridine are added. The reaction mixture is refluxed for 8 hr and then poured into cold water. The product is extracted and purified by chromatography with 9 : 1 toluene/2-PrOH to afford 9 g (92%) of an orange oil. ^{1}H NMR ($CDCl_3$) δ 7.91 (d, 2H, $J = 8.3$ Hz), 7.82 (dd, 1H, $J = 1.6$ Hz, 7.9Hz), 7.67 (t, 1H, $J = 4.8$ Hz), 7.28 (m, 1H), 7.04 (s, 1H), 6.92 (d, 2H, $J = 8.3$ Hz), 6.63 (m, 2H), 5.18 (s, 2H), 4.20 (d, 2H, $J = 4.8$ Hz), 3.90 (s, 3H), 3.77 (s, 3H), 2.56 (t, 2H, $J = 8.0$ Hz), 1.70 (m, 2H), 1.30 (m, 2H), 0.87 (t, 3H, $J = 7.3$ Hz). NaOH (0.68 g, 20 mmol) and 10 ml of water are added to a solution of 8.3 g (19.1 mmol) of the orange oil in 80 ml of MeOH. The mixture is heated at 50°C for 3.5 hr and concentrated, and the residue is diluted with 150 ml of water. The aqueous layer is washed with EtOAc and acidified to pH 5 with 1 *N* HCl. The product is extracted with EtOAc. The organic layer is washed with water, dried over $MgSO_4$, and concentrated. The crude product is purified on silica gel with 95 : 5 CH_2Cl_2/MeOH to give 4.8 g of a solid, which is recrystallized from EtOAc, yielding 4.41g (55%) of **3** as a white solid. mp 160°C. ^{1}H NMR ($CDCl_3$) δ 7.92 (d, 2H, $J = 6.6$ Hz), 7.80 (d, 1H, $J = 8.6$ Hz), 7.68 (bt, 1H), 7.30 (t, 1H, $J = 9.0$ Hz), 7.19 (s, 1H), 6.92, (d, 2H, $J = 8.3$ Hz), 6.63 (m, 2H), 5.30 (bs, 1H), 5.22 (s, 2H), 4.20 (d, 2H, $J = 4.9$ Hz), 3.79 (s, 3H), 2.88 (t, 2H, $J = 7.6$ Hz), 1.88 (m, 2H), 1.33 (m, 2H), 0.88 (t, 3H, $J = 7.3$ Hz).

2-[[[2-Butyl-1-[(4-((((2-chlorophenyl)sulfonyl)amino)carbonyl)phenyl) methyl]-1H-imidazol-5-yl]-methyl]amino]benzoic acid **(5).** To a suspension of 2.22 g (5.3 mmol) of **5a** in 55 ml of CH_2Cl_2 are added 1.1 g (5.8 mmol) of 2-chlorobenzenesulfonamide, 0.71 g (5.8 mmol) of DMAP, and 1.11 (5.8 mmol) of EDCI. The mixture is stirred for 3 days at room temperature and concentrated *in vacuo* to afford a residue that is purified by flash chromatography on silica gel using 6 : 4 toluene/2-PrOH as eluent to afford 2.54 g (78%) of the methyl ester: ^{1}H NMR ($CDCl_3$) δ 8.17 (d, 1H), 7.87 (d, 2H), 7.78 (d, 1H), 7.65 (s, 1H), 7.20 (m, 5H), 6.70 (d, 2H), 6.52 (m, 2H), 5.12 (s, 2H), 4.07 (s, 2H), 3.67 (s, 3H), 2.50 (m, 2H), 1.40 (m, 2H), 1.21 (m, 2H), 0.72 (t, 3H). To a solution of 0.8 g (1.4 mmol) of the methyl ester in 20 ml of MeOH is added 0.42 g (10.5 mmol) of NaOH in 2 ml of water. The mixture is heated under reflux for 4 hr. Evaporation of the MeOH afforded a residue that is taken up in water. The pH of the resulting solution is adjusted to 5 with 1 *N* HCl from which the desired product **5** precipitated as a white solid, which is dried *in vacuo* in the presence of P_2O_5 (0.78 g, 88%). mp 198°. ^{1}H NMR (DMSO-d_6) δ 13.20 (bs, 1H), 8.16 (m, 1H), 8.00(d, 1H), 7.86 (d, 2H), 7.80 (d, 1H), 7.46 (m, 4H), 7.34 (t, 1H), 7.04 (d, 2H), 6.65 (m, 2H), 5.49 (s, 2H), 4.46 (s, 2H), 2.80 (t, 2H), 1.42 (m, 2H), 1.21 (m, 2H), 0.76 (t, 3H).

Synthetic Procedures of SRIF Mimetics 8 (see Scheme 4)[42]

*Methyl 2,3,4-tri-o-benzyl-6-o-(5-azidopentyl)-β-D-glucopyranoside **(8d).*** At room temperature, a solution of 5-azido-1-pentanol (0.18 g, 1.40 mmol) and 2, 6-di-*t*-butyl-4-methylpyridine (0.3 g, 1.46 mmol) in CH_2Cl_2 (10 ml) is treated dropwise with triflic anhydride (0.240 ml, 1.43 mmol). After 15 min the mixture is diluted with CH_2Cl_2 (40 ml) and poured into saturated aqueous $NaHCO_3$ (50 ml). The organic phase is washed with brine (2 × 20 ml), dried over $MgSO_4$, filtered, and concentrated, affording a light yellow solid that is used without purification. The alcohol **8c** (0.2 g, 0.429 mmol) and the crude triflate are dissolved in CH_2Cl_2 (2 ml) and treated with NaH (0.025 g, 0.625 mmol). The mixture is stirred for 48 hr, diluted with CH_2Cl_2 (40 ml), and poured into saturated aqueous NH_4Cl (40 ml). The aqueous phase is extracted with CH_2Cl_2 (3 × 20 ml), and the combined organic solutions are washed with brine, dried over $MgSO_4$, filtered, and concentrated *in vacuo*. Flash chromatography (15% EtOAc/hexanes) provided **8d** (0.126 g, 51% yield) as a white solid. mp 69–70° (MeOH/CH_2Cl_2). $[\alpha]^{25}D = +7.7°$ (C 0.75, $CHCl_3$). ^{13}C NMR (125 MHz, $CDCl_3$) δ 23.4, 28.7, 29.2, 51.4, 57.1, 69.7, 71.4, 74.7, 74.8, 75.0, 75.7, 78.0, 82.3, 84.6, 104.7, 127.5, 127.6, 127.8, 127.9, 128.1, 128.3, 128.4, 138.3, 138.5, 138.6. HRMS (FAB, *m*-nitrobenzyl alcohol) calcd for $C_{33}H_{39}N_3O_6$ $[M + Na]^+$ 598.2893, obsd 598.2880.

*Methyl 2,3,4-tri-o-benzyl-6-o-(5-aminopentyl)-β-D-glucopyranoside **(8a).*** Azide **8d** (0.126 g, 0.219 mmol) is dissolved in THF (12 ml) and treated with water (0.096 ml, 5.33 mmol) followed by PPh_3 (0.114 g, 0.44 mmol). The mixture is then heated at 60° for 12 hr, cooled, and concentrated *in vacuo*. Flash chromatography (10% MeOH/ CH_2Cl_2) affords **8a** (87.3 mg, 73% yield) as a white solid. mp 79–80°. $[\alpha]^{25}D = +6.8°$ (c 1.85, $CHCl_3$). ^{1}H NMR ($CDCl_3$) δ 1.35–1.46 (m, 4H); 1.56–1.61 (m, 4H), 2.66 (t, 2H, $J = 6.9$ Hz), 3.40–3.70 (m, 8H), 3.56 (s, 3H), 4.29 (d, 1H, $J = 7.8$ Hz), 4.61 (d, 1H, $J = 10.9$ Hz), 4.70 (d, 1H, $J = 10.9$ Hz), 4.78 (d, 1H, $J = 11.0$ Hz), 4.85 (d, 1H, $J = 10.9$ Hz), 4.91 (d, 1H, $J = 11.0$ Hz), 4.92 (d, 1H, $J = 10.9$ Hz), 7.25–7.35 (m, 15H). ^{13}C NMR ($CDCl_3$) δ 23.5, 29.5, 33.5, 42.0, 57.1, 69.6, 71.6, 74.7, 74.8, 75.0, 77.9, 82.3, 84.6, 104.7, 127.6, 127.7, 127.8, 127.9, 128.1, 128.4, 138.3, 138.5, 138.6. HRMS (FAB, *m*-nitrobenzyl alcohol) *m/z* $[M + Na]^+$ calcd for $C_{33}H_{43}NO_6$ 572.2988, obsd 572.2997.

*Methyl 2,3,4-tri-o-benzyl-6-amino-6-deoxy-6-N-(5-hydroxypentyl)-β- D-glucopyranoside **(8b).*** A stirred solution of **8c** (800 mg, 1.71 mmol) and 2, 6-di-*t*-butyl-4-methylpyridine (632 mg, 3.08 mmol) in CH_2Cl_2 (9 ml) is cooled to

−78° and treated with triflic anhydride (0.345 ml, 2.05 mmol). After 15 min the mixture is warmed to room temperature for 20 min, poured into saturated aqueous $NaHCO_3$ (20 ml), and extracted with EtOAc (50 ml). The organic layer is washed with additional $NaHCO_3$ and brine, dried over $MgSO_4$, filtered, and concentrated *in vacuo*, affording crude triflate, which is used in the next step without further purification. A solution of 5-(trifluoroacetamido)-1-pentanol (1.7 g, 8.6 mmol) in THF (35 ml) is added to a stirred suspension of NaH (855 mg, 21.4 mmol) in THF (60 ml) at 0°. After 10 min the suspension is warmed to room temperature, stirred for 1 hr, and recooled to 0°. A solution of the crude triflate (1.71 mmol) in CH_2Cl_2 (60 ml) is then added, and stirring is continued at 0° for 30 min and at room temperature for 24 hr. The reaction mixture is quenched at 0° with saturated aqueous NH_4Cl and extracted with EtOAc, and the combined organic extracts are washed with water and brine, dried over $MgSO_4$, filtered, and concentrated *in vacuo*. Purification through a small plug of silica gel (30% EtOAc/petroleum ether) gives crude **8e**, which is used immediately in the next step. A stirred solution of the crude **8e** in EtOH (10 ml) is treated with 5 *N* NaOH (3 ml, 15 mmol) at room temperature and then heated at reflux for 2 hr, cooled, and concentrated *in vacuo*. The residue is diluted with CH_2Cl_2 and 2 *N* HCl. The aqueous layer is extracted with CH_2Cl_2 (3 × 50 ml), and the combined organic solutions are washed with brine, dried over $MgSO_4$, filtered, and concentrated *in vacuo*. Recrystallization (EtOAc/ petroleum ether) furnished pure **8b** as a white solid. mp 95–95.5°. $[\alpha]^{25}D = +9.3°$ (*c* 0.15, CH_3CN). ^1H NMR ($CDCl_3$) δ 1.36–1.42 (m, 2H), 1.45–1.51 (m, 2H), 1.53–1.59 (m, 2H), 1.71 (s, 2H), 2.53–2.64 (m, 2H), 2.68 (dd, 1H, J = 12.0, 6.8 Hz), 2.94 (dd, 1H, J = 12.5, 2.1 Hz), 3.36–3.48 (m, 3H), 3.56 (s, 3H), 3.59–3.66 (m, 3H), 4.32 (d, 1H, J = 7.8 Hz), 4.60 (d, 1H, J = 11.0 Hz), 4.70 (d, 1H, J = 11.0 Hz), 4.78 (d, 1H, J = 11.0 Hz), 4.85 (d, 1H, J = 11.0 Hz), 4.90 (d, 1H, J = 7.6 Hz), 4.92 (d, 1H, J = 7.5 Hz), 7.24–7.35 (m, 15H). ^{13}C NMR ($CDCl_3$) δ 23.4, 29.7, 32.5, 49.7, 50.7, 57.2, 62.6, 74.2, 74.7, 75.0, 75.7, 79.7, 82.5, 84.6, 104.7, 127.6, 127.8, 127.9, 128.0, 128.3, 128.4, 138.2, 138.5, 138.6. HRMS (FAB, *m*-nitrobenzyl alcohol) *m/z* $[M + Na]^+$ calcd for $C_{33}H_{43}NO_6$ 550.3168, obsd 550.3179.

Synthetic Procedures of SL-3111 (See Scheme 5)[49]

3-(2-Methoxyethoxy)methoxybenzaldehyde ***(11a).*** Into a stirred solution of 3-hydroxybenzaldehyde (3.7 g, 30 mmol) in 60 ml of CH_2Cl_2 is added diisopropylethylamine (7.6 ml, 43.7 mmol). To the resulting brown solution is added methoxyethoxymethyl chloride (5.0 ml, 43.7 mmol) in 17 ml of CH_2Cl_2 dropwise. The resulting solution is stirred at room temperature for 3 hr and then quenched by the addition of 60 ml of 0.5 *N* HCl. The organic phase is separated, and the aqueous phase is extracted with CH_2Cl_2 (3 × 60 ml). The combined organic phase is washed with 5% Na_2CO_3 (3 × 60 ml) and water (3 × 60 ml) and then dried over $MgSO_4$.

Concentration of the dried solution yields 5.4 g of an orange oil (yield 86%). ^{1}H NMR ($CDCl_3$): δ 9.90 (s, 1H, -CHO), 7.57–7.45 (m, 4H, aromatic protons), 5.33 (s, 2H, -OCH_2O-), 3.86–3.82 (m, 2H, -OCH_2-), 3.58–3.54 (m, 2H, -CH_2O-), 3.38 (s, 3H, -OCH_3). HRMS for $C_{11}H_{15}O_4$, calcd 211.0970 $[M + H]^+$, found 211.0973 $[M + H]^+$.

4-tert-Butyl-3′-(2- methoxyethoxy)methoxybenzhydrol ***(11b).*** Into a stirred solution of *tert*-butylphenylmagnesium bromide (2.0 *M* in ether, 5 eq.) in 10 ml of THF at 0° is added dropwise a solution of **11a** in 5 ml of THF. The resulting solution is allowed to warm to room temperature and is stirred overnight. The reaction mixture is quenched by the addition of 0.5 *N* HCl (30 ml). The product is extracted with ether (3 × 40 ml), and the combined organic phase is washed with brine (60 ml) and water (60 ml) and then dried over $MgSO_4$. Concentration of the solution gave the product, which is purified by flash chromatography (1 : 4 EtOAc/hexanes): colorless oil (yield 85%). ^{1}H NMR ($CDCl_3$) δ ppm 7.36–6.98 (m, 8H, aromatic protons), 5.77 (d, 1H, $J = 3.5$Hz, -CHOH-), 5.24 (s, 2H, -OCH_2O-), 3.82–3.78 (m, 2H, -OCH_2-), 3.55–3.51 (m, 2H, -CH_2O-), 3.35 (s, 3H, -OCH_3), 2.29 (s, 1H, $J = 3.5$Hz, -OH), 1.29[s, 9H, - $C(CH_3)_3$]. HRMS for $C_{21}H_{29}O_4$, calcd 345.2066 $[M + H]^+$, found 345.2014 $[M + H]^+$.

4-tert-Butyl-3′-(2-methoxyethoxy)methoxybenzhydrylchloride ***(11c).*** A mixture of **11b** (2.38 mmol) and triphenylphosphine (3.43 mmol) in 6.0 ml of carbon tetrachloride is refluxed under Ar for 3 hr. A white precipitate is formed. After the mixture cools to room temperature, 10 ml of anhydrous ether is added to the mixture. The solid is filtered off and washed with ether (3 × 10 ml). The combined filtrate is evaporated to dryness and purified by flash chromatography (EtOAc/hexanes): colorless oil (yield 70%). ^{1}H NMR ($CDCl_3$): δ ppm 7.34–7.00 (m, 8H, aromatic protons), 6.06 (s, 1H, -CHCl-), 5.25 (s, 2H, -OCH_2O-), 3.82–3.70 (m, 2H, - OCH_2-), 3.64–3.50 (m, 2H, -CH_2O-), 3.35 (s, 3H, -OCH_3), 1.29 [s, 9H, -$C(CH_3)_3$]. MS for $C_{21}H_{27}O_3Cl$, calcd 362.2 $[M^+]$, found 362.2 $[M^+]$.

1-[4-tert-Butyl-3′-(2-methoxyethoxy)methoxybenzhydryl]-4-benzylpiperazine ***(11d).*** A mixture of 1-benzylpiperazine (0.8 g, 4.8 mmol), potassium carbonate (0.4 g, 2.9 mmol), and **11c** (1.3 mmol) in 6.0 ml of acetonitrile is refluxed under Ar for 2 hr. After cooling to room temperature, the solid is filter off and washed with acetonitrile (3 × 10 ml). Filtrates are combined and concentrated to yield a yellow oil, which is purified by flash chromatography (EtOAc/hexanes containing 1% Et_3N): colorless oil (yield 95%). ^{1}H NMR (CD_3OD) δ ppm 7.28–7.12 (m, 13H, aromatic protons), 5.23 (s, 2H, -OCH_2O-), 4.15 (s, 1H, -CH-), 3.81–3.77 (m, 2H, -OCH_2-), 3.53–3.49 (m, 2H, - CH_2O-), 3.48 (s, 2H,-CH_2Ph), 3.35 (s, 3H, - OCH_3), 2.45 (s, broad, 8H, piperazine protons), 1.25 [s, 9H, - $C(CH_3)_3$].

1-(4-tert-Butyl-3′-hydroxybenzhydryl)-4-benzylpiperazine (SL-3111). A solution of 2 *N* HCl (2 ml/0.1 g) is added into a solution of **11d** in 1 : 1 MeOH/dioxane. The solution is stirred at room temperature for 24 hr, and the volatiles are removed *in vacuo* at room temperature. To the residue is added cold ether to precipitate the product. The white precipitate is filtered off, washed with ether, and dried *in vacuo:* hydrochloride salt, white solid (yield 83%). ^{1}H NMR (CD_3OD): δ 7.73–7.21 (m, 13H), 4.49 (s, 1H, -CHPhOH), 3.73 (s, 2H, -CH_2Ph), 3.73 (s, broad, 4H, piperazine protons), 3.55 (s, 4H, piperazine protons), 1.20 [s, 9H, $C(CH_3)_3$]. HRMS for $C_{28}H_{35}N_2O$, calcd 415.2749 $[M + H]^+$, found 415.2742 $[M + H]^+$. The optically pure enantiomers of SL-3111 are obtained with HPLC separation performed on a Shimadzu liquid chromatograph Model SCL-10A using a chiral column chiralpack-AD (DAICEL) (25 × 0.5 cm) and an isocratic eluent mixture of hexane : isopropanol (97 : 3) as mobile phase with the UV detector set at 254 nm. The concentration of each injection is 1.5 mg/ml to obtain an adequate resolution. (+)-SL-3111: White solid, mp 109–110°; $t_R = 7.2$ min, $[\alpha]_D^{22} = +11.5$ (c 1, $CHCl_3$); (−)-SL-3111: White solid, mp 109–110° $t_R = 8.1$ min, $[\alpha]_D^{22} = -10.6$ (c 1, $CHCl_3$).

Acknowledgments

Supported by grants from the U.S. Public Health Service (USPHS) and the National Institute of Drug Abuse (NIDA). The opinions expressed are those of the authors and do not necessarily reflect those of the USPHS or NIDA.

[7] Strategies for Mapping the Binding Site of the Serotonin 5-HT_{2A} Receptor

By BARBARA J. EBERSOLE and STUART C. SEALFON

Introduction

The objective of structure–activity studies of G protein-coupled receptors (GPCR) is to develop an understanding of the nature and consequences of ligand–receptor interactions at a molecular level. The serotonin 5-HT_{2A} receptor is a member of the GPCR superfamily for which such studies have identified key interactions in the ligand–receptor complexes. One notable group of ligands for this receptor are the serotonergic hallucinogens, such as lysergic acid diethylamide (LSD) and *N,N*-dimethyl 5-HT (bufotenin), which have high affinity for the 5-HT_{2A} receptor. Studying the binding pocket of the receptor and identifying the molecular mechanisms that determine ligand affinity, specificity, and coupling efficiency may help elucidate the basis for the special biological effects of these chemicals.

0076-6879/02 $35.00

An important approach to investigate structure–function relations of GPCRs entails introducing structural perturbations via site-directed mutagenesis and evaluating the resulting receptor phenotype in binding and signal transduction assays. However, determining the phenotype of mutant receptors may not always lead to an unequivocal interpretation concerning the structural perturbation underlying that phenotype. Molecular modeling has facilitated the integration of experimental observations and biophysical data into a mechanistic scheme for receptor structure and function (for review, see Refs. 1 and 2). The recent publication of the crystal structure of rhodopsin[3] improves the accuracy of computational modeling of homologous receptors. Nonetheless, the subtlety and complexity of the pattern of side-chain and main-chain interactions limit the predictive capacity of computational approaches and the structural interpretation of mutagenesis experiments.

In site-directed mutation experiments, a single amino acid substitution is introduced into the receptor, and the resulting effects on receptor function are determined in radioligand-binding and/or coupling assays. It is essential to recognize the difficulties in determining the structural basis for the alteration in receptor function caused by a particular mutation in a GPCR. A mutation may alter the same function by several mechanisms. For example, binding affinity may be decreased because the locus mutated represents a direct site of ligand interaction. However, the same disruption of function could result indirectly from the mutation inducing a structural perturbation of the binding pocket or global receptor conformation that leads indirectly to decreased affinity. Thus the effect of a mutation may derive from a combination of these potential direct and indirect effects.

Insight into the difficulties inherent in deducing how particular mutations have altered receptor function can be gleaned from the mutagenesis studies of prokaryotic proteins of known structure. The functional alterations of site-directed mutants have been correlated with the precise structural alterations observed in solved crystal structures of the mutant proteins (see Refs. 4–6). Mutations are found to alter function indirectly, often through their effects on the backbone and side chains of other residues. The complex mechanisms that may underlie the structural effects of mutations on receptor binding are apparent in a crystallography study of the extracellular domain of the growth hormone receptor. Mutations of the receptor were found to cause large structural rearrangements at distant sites in

[1] J. A. Ballesteros and H. Weinstein, *Methods Neurosci.* **25,** 366 (1995).

[2] I. Visiers, J. A. Ballesteros, and H. Weinstein, *Methods Enzymol.* **343,** [22] 2001 (this volume).

[3] K. Palczewski, T. Kumasaka, T. Hori, C. A. Behnke, H. Motoshima, B. A. Fox, I. Le Trong, D. C. Teller, T. Okada, R. E. Stenkamp, M. Yamamoto, and M. Miyano, *Science* **289,** 739 (2000).

[4] K. P. Wilson, B. A. Malcolm, and B. W. Matthews, *J. Biol. Chem.* **267,** 10842 (1992).

[5] J. A. Bell, W. J. Becktel, U. Sauer, W. A. Baase, and B. W. Matthews, *Biochemistry* **31,** 3590 (1992).

[6] A. R. Poteete, D. P. Sun, H. Nicholson, and B. W. Matthews, *Biochemistry* **30,** 1425 (1991).

the receptor–hormone interface.[7] Thus structural studies in both prokaryotic and eukaryotic proteins indicate that a mutation can cause remarkably large movements in distant side-chain and main-chain contacts.

To overcome the limitations inherent in both site-directed mutagenesis and computational modeling, mutational experiments and computational techniques have been integrated in the study of receptor microdomains.[1,2] The study of structural microdomains facilitates the structural interpretation of the effects of mutations. Strong support for this approach comes from the agreement of our mcirodomain predictions with the crystal structure of rhodopsin. We have defined several conserved GPCR domains and determined their role in receptor function. These microdomains, e.g., a specific helix 2–helix 7 proximity based on double revertant studies[8–10] and an arginine cage motif involved in stabilizing the receptor in an inactive state,[11] have been confirmed by the crystal structure of rhodopsin.[3]

Using this approach to map the binding site of the 5-HT_{2A} receptor, we have determined the pattern of interaction of agonists with a specific helix 5 attachment site[12] and discovered that two specific helix 3 side chains both interact with the amine nitrogen of serotonin.[13] The 5-HT_{2A} receptor, like the other receptors for biogenic amines, has an aspartate residue at a homologous location in the third transmembrane helix (TMH) domain. Site-directed mutagenesis studies with these receptors indicate that, for most ligands, an interaction between the basic nitrogen of the ligand and the carboxyl side chain of the conserved TMH 3 aspartate stabilizes ligand binding. Our collaborators' computational simulations of ligand–receptor complexes of the 5-HT_{2A} receptor using a three-dimensional computational model suggested a complex array of interactions connecting TMH 3 side chains and specific ligands.[2,12] The same charged amino group of 5-hydroxytryptamine (5-HT) that interacts with the TMH 3 aspartate was predicted to form a hydrogen bond with the side chain of a second TMH 3 serine. These structural predictions and the functional role of this microdomain were probed by multiple mutations, simulations, and functional assays. In the molecular model of the receptor, this serine

[7] S. Atwell, M. Ultsch, A. M. De Vos, and J. A. Wells, *Science* **278,** 1125 (1997).

[8] W. Zhou, C. Flanagan, J. A. Ballesteros, K. Konvicka, J. S. Davidson, H. Weinstein, R. P. Millar, and S. C. Sealfon, *Mol. Pharmacol.* **45,** 165 (1994).

[9] S. C. Sealfon, L. Chi, B. J. Ebersole, V. Rodic, D. Zhang, J. A. Ballesteros, and H. Weinstein, *J. Biol. Chem.* **270,** 16683 (1995).

[10] C. A. Flanagan, W. Zhou, L. Chi, T. Yuen, V. Rodic, D. Robertson, M. Johnson, P. Holland, R. P. Millar, H. Weinstein, R. Mitchell, and S. C. Sealfon, *J. Biol. Chem.* **274,** 28880 (1999).

[11] J. Ballesteros, S. Kitanovic, F. Guarnieri, P. Davies, B. J. Fromme, K. Konvicka, L. Chi, R. P. Millar, J. S. Davidson, H. Weinstein, and S. C. Sealfon, *J. Biol. Chem.* **273,** 10445 (1998).

[12] N. Almaula, B. J. Ebersole, J. A. Ballesteros, H. Weinstein, and S. C. Sealfon, *Mol. Pharm.* **50,** 34 (1996).

[13] N. Almaula, B. J. Ebersole, D. Zhang, H. Weinstein, and S. C. Sealfon, *J. Biol. Chem.* **271,** 14672 (1996).

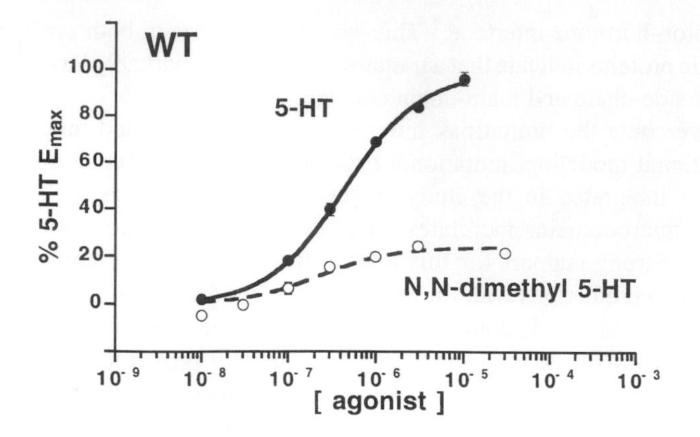

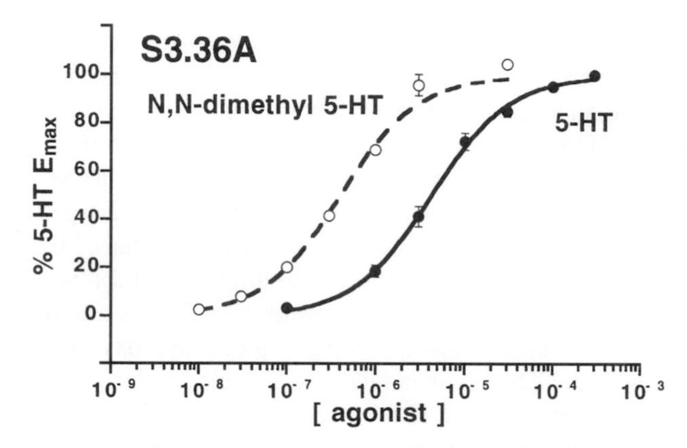

FIG. 1. Accumulation of [^{3}H]IPs in stable cell lines expressing wild-type (WT) and mutant (S3.36A) human 5-HT_{2A} receptors. Data were normalized to the fit E_{max} for 5-HT for each cell line. Note the 20-fold increase in EC_{50} for 5-HT as a result of the mutation, whereas the EC_{50} for *N,N*-dimethyl 5-HT is changed only 4-fold. More striking is the dramatic increase in the intrinsic activity of *N,N*-dimethyl 5-HT at the mutant receptor.

residue is positioned on the same face of the helix as the aspartate. This computational result predicted that the affinity of 5-HT should be affected more by mutation of this serine than would *N,N*-dimethyl 5-HT, which was confirmed by our experimental data (Fig. 1).[12] The role of this microdomain in positioning the ligand and influencing receptor activation was also demonstrated. *N,N*-Dimethyl 5-HT is a partial agonist at the 5-HT_{2A} receptor. However, mutation of the serine site that interacts specifically with the free amine group of 5-HT but not the *N,N*-dimethyl derivative allows the two ligands to fit similarly in the binding pocket and to activate the receptor to a similar degree (Fig. 2).[13]

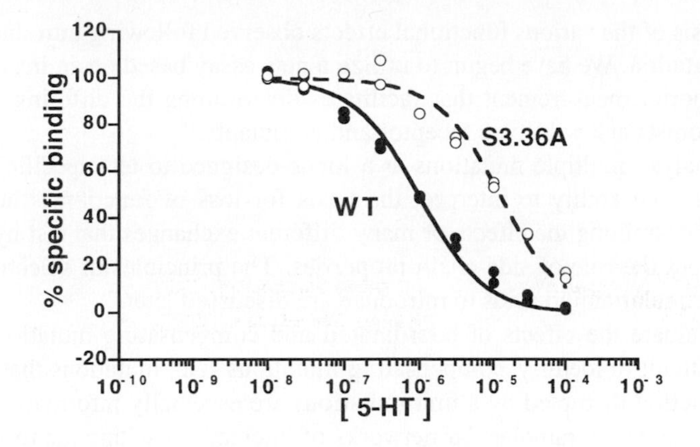

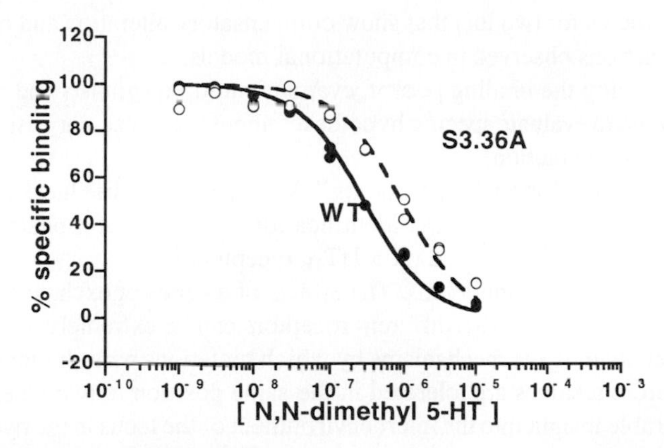

FIG. 2. [^{3}H]ketanserin competiton-binding curves in membranes prepared from COS-1 cells transfected transiently with constructs for wild-type (WT) and mutant (S3.36A) human 5-HT_{2A} receptors. Changes in affinity for 5-HT and *N,N*-dimethyl 5-HT at the mutant receptor parallel the results found in [^{3}H]IPs assays in stably transfected cells (Fig. 1).

Overview of Microdomain Mapping Studies

The following general approach is used in our laboratory.

1. Assay mutant constructs using both radioligand binding and functional assays. Any particular mutation is likely to induce a variety of effects potentially involving direct actions (i.e., interaction of the side chain with ligand) and a variety of indirect effects due to interaction with other receptor loci. Evaluation of the mutant receptor using both ligand binding and functional assays of coupling to signal transduction provides more insight into the

basis of the various functional effects observed following introduction of a mutation. We have begun to utilize a new assay based on an intrinsic gene reporter measurement that facilitates determining the differing effects of agonists at a wild-type receptor and at mutants.[14]

2. Analyze multiple mutations at a locus designed to test specific hypotheses. The ability to interpret the basis for loss of function is augmented by examining the effects of many different exchanges that test hypotheses about the role of side-chain properties. The principles of selection of the particular amino acids to introduce are discussed later.
3. Evaluate the effects of coordinated and compensating mutations. While difficult to identify, compensating mutations (i.e., mutations that restore a function disrupted by a first mutation) are especially informative in identifying the intramolecular networks of interactions within the receptor that have been disrupted,[9,10] Potential sites are identified by searching sequence alignments for two loci that show compensatory alteration and by studying interactions observed in computational models.
4. In studying the binding pocket, evaluate multiple agonists and antagonists selected to evaluate specific hypotheses about the molecular basis of ligand–receptor interaction.
5. Utilize evolutionary "experiments." A number of studies have utilized evolutionary "mutations," e.g., identification of the basis for pharmacological differences of human and rat 5-HT_{1B} receptors.[15]
6. Evaluate parallel mutations. The effects of a series of exchanges at homologous positions in two different receptors can be extremely informative in understanding the mechanisms by which mutations perturb function. When different residues are tolerated in the same position in two receptors, considerable insight into the microenvironment of the locus in the two receptors can be attained. For example, the 5-HT_{2A} and 5-HT_{2C} receptors provide two similar but unique binding sites with which to test conclusions about the specific side-chain interactions with serotonin and other ligands.[12]

Selection of Amino Acid Substitutions

The selection of the amino acid to introduce by mutagenesis must be chosen based on the hypothesis to be tested. Sites that appear interesting should be probed with multiple mutations. In general, mutations should be selected to evaluate the specific physiochemical properties of the side chain, such as charge, hydrogen bonding potential, hydrophobicity, volume, and shape. A chart summarizing the relative position of hydrogen-bond donors and acceptors is found in Figs. 3 and 4,

[14] T. Yuen, W. Zhang, B. J. Ebersole, and S. C. Sealfon, *Methods Enzymol.* **345,** [45] 2001.

[15] D. Oksenberg, S. A. Marsters, B. F. O'Dowd, H. Jin, S. Havlik, S. J. Peroutka, and A. Ashkenazi, *Nature* **360,** 161 (1992).

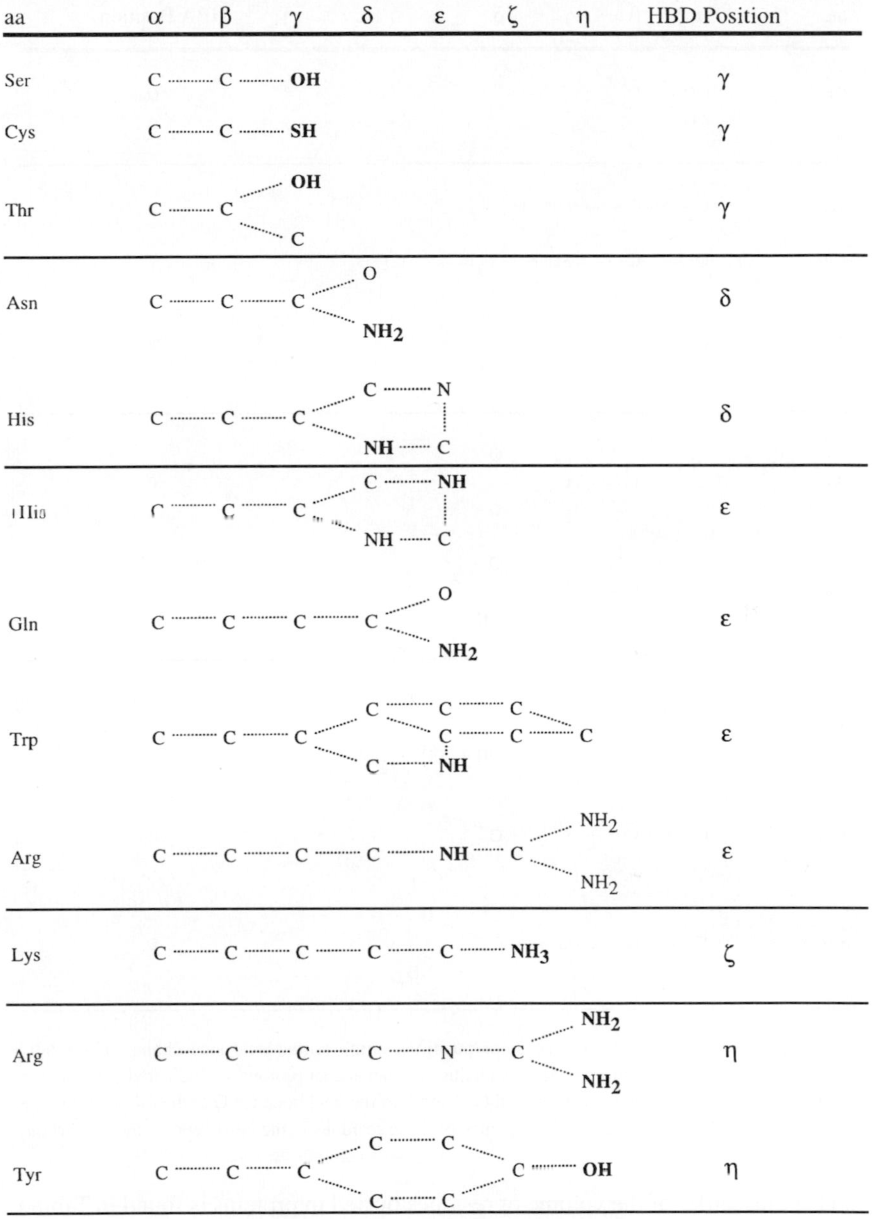

aa	α	β	γ	δ	ε	ζ	η	HBD Position
Ser	C	C	**OH**					γ
Cys	C	C	**SH**					γ
Thr	C	C	**OH** / C					γ
Asn	C	C	C	O / **NH2**				δ
His	C	C	C	C / **NH**	N / C			δ
His	C	C	C	C / NH	**NH** / C			ε
Gln	C	C	C	C	O / **NH2**			ε
Trp	C	C	C	C / C	C / C / **NH**	C / C	C	ε
Arg	C	C	C	C	**NH**	C	NH2 / NH2	ε
Lys	C	C	C	C	C	**NH3**		ζ
Arg	C	C	C	C	N	C	**NH2** / **NH2**	η
Tyr	C	C	C	C / C	C / C	C	**OH**	η

FIG. 3. Identification of H-bond donor positions along amino acid side chains. The relative positions of the H-bond donors of all side chains that can donate protons are indicated.

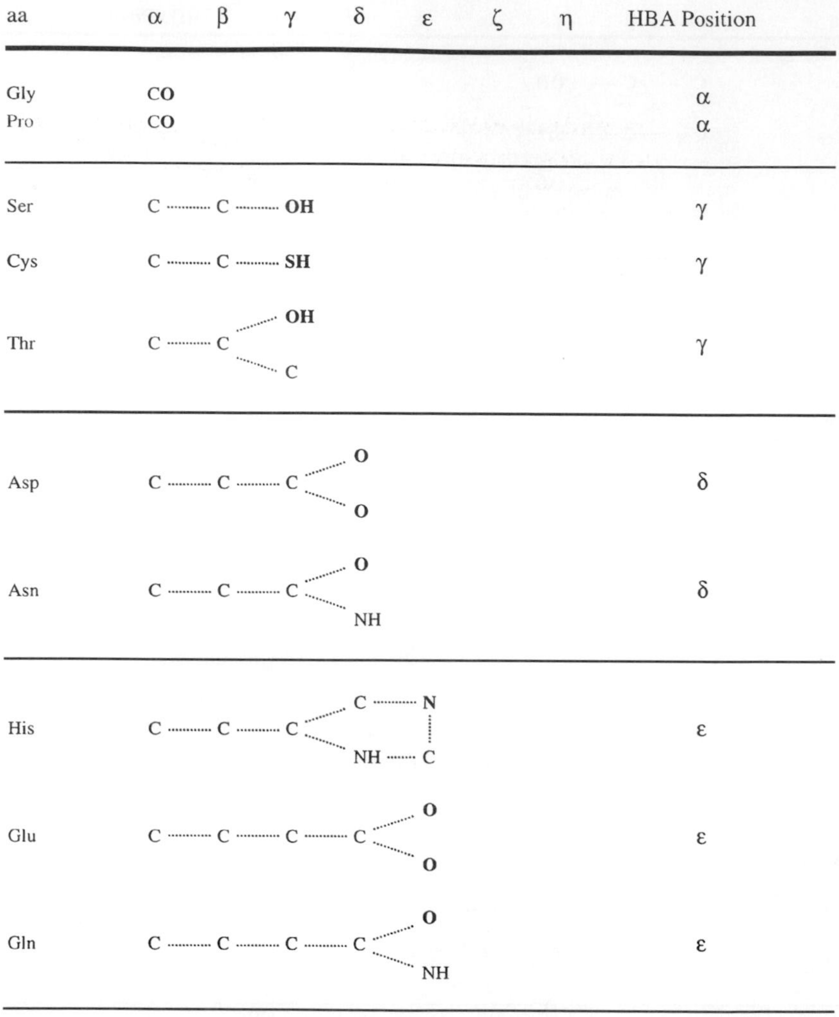

aa	α	β	γ	δ	ε	ζ	η	HBA Position
Gly	CO							α
Pro	CO							α
Ser	C	C	**OH**					γ
Cys	C	C	**SH**					γ
Thr	C	C	**OH** / C					γ
Asp	C	C	C	**O** / **O**				δ
Asn	C	C	C	**O** / NH				δ
His	C	C	C	C / NH	**N** / C			ε
Glu	C	C	C	C	**O** / **O**			ε
Gln	C	C	C	C	**O** / NH			ε

FIG. 4. Identification of H-bond acceptor positions along amino acid side chains. The relative positions of the H-bond acceptors of all side chains that can accept protons are indicated. Note that the α-carbon H-bond acceptor listed for Pro and Gly refers to the backbone C=O at the i-4 position when the residue is present in an α helix. The presence of these residues in the helix renders the i-4 carbonyl accessible for H bounding.

and the rank order of the volume of residues buried in proteins is found in Table I. A common approach followed in the literature is that of Ala scanning where each amino acid is substituted by Ala. Ala substitutions introduce changes in both charge or H-bonding capabilities and in size (Table I), potentially leaving a cavity in comparison to other side chains. While this may be a reasonable initial strategy

TABLE I
RANK ORDER OF VOLUME OF RESIDUES BURIED IN PROTEINS[a]

Gly	66Å	Pro	129Å	Ile	169Å
Ala	92	Asn	135	Lys	171
Ser	99	Val	142	Met	171
Cys-S-S	106	Glu	155	Arg	202
Cys-SH	118	Gln	161	Phe	203
Thr	122	His	167	Tyr	204
Asp	125	Leu	168	Trp	238

[a] Adapted from Chothia.[18]

in many instances, it clearly must be followed up by other substitutions at sites of interest to clarify the basis of the effect found. For example, if Ala at a site of His in the wild-type receptor alters function, additional mutations may be designed to evaluate the likelihood of a H-bond donor or acceptor role. Uncharged His is usually a H-bond donor at the δ position and a H-bond acceptor at the ε carbon. Thus, if His is involved in H bonding, Asn would best mirror its H bonding capacity, and Gln its H-bond accepting capacity (Figs. 3 and 4).

Expression of Mutant Receptors

Receptor mutations are introduced using a commercial kit and following the manufacturer's protocol (Stratagene Quik Change site-directed mutagenesis kit). In order to exclude the possibility of studying the wrong construct, mutations are confirmed by sequencing and reconfirmed following each plasmid purification for transfection. Following generation of the receptor mutants, their properties are characterized by heterologous expression. The choice of expression system and binding and functional assays are adjusted to the goals of the experiment. We routinely assay both ligand binding and signal transduction of the mutant receptors, typically in transiently expressing COS-1 cells.

Expression in Mammalian Cells

Numerous cell lines are available from American Type Culture Collection (Rockville, MD). The choice of the particular cell line depends on the receptor to be studied. It is preferable to avoid cell lines containing endogenous receptors of similar pharmacological profile and to select a cell line with the appropriate machinery for signal transduction of the receptor system under study. The chosen line should be screened for both binding and functional responses to each of the ligands to be used before proceeding with the transfection study. It is possible that functional responses will be observed as a result of the action of the chosen ligands on endogenously expressed receptors with overlapping pharmacological selectivities.

Transient and Stable Cell Lines

Mammalian cells may be transfected transiently or stably with an appropriate eukaryotic expression vector. For example, COS-1 cells have an integrated copy of the early region of SV40 and express the SV40 T antigen. Therefore, vectors that have the SV40 origin of replication, such as pcDNA3 (Invitrogen), are appropriate for transient expression in these cells. Stable transformants incorporate the receptor cDNA into their chromosomal DNA and rely on a selectable marker.

Methods of Transfection

Methods of transfection include the use of $CaPO_4$ precipitation, DEAE-dextran-mediated transfection, lipopolyamine-mediated gene transfer, protoplast fusion, microinjection, electroporation, and viral infection. Currently we are using LIPOFECTAMINE reagent (GIBCO BRL), a lipopolyamine reagent to transfect cells with plasmid DNA. The protocol we follow for transfection with LIPOFECTAMINE transfection is found in a later section.

Characterization of Mutant Receptors

The mutant 5-HT_{2A} constructs are assayed by evaluating concentration–response curves for phosphatidylinositol hydrolysis (Fig. 1) and by radioligand binding (Fig. 2). The protocols we follow for phosphatidylinositol assays and radioligand binding can be found in the detailed methods.

Detailed Methods

Transient Transfection of COS-1 cells

We use LIPOFECTAMINE (GIBCO BRL) to transfect cells with plasmid DNA. The method involves mixing the plasmid and reagent in serum-free medium, allowing DNA–liposome complexes to form, and overlaying cells with this solution.

The method of transfection has a number of variables that need to be optimized for each cell type and plasmid, the most important being (1) ratio of plasmid DNA to lipofectamine reagent; (2) ratio of LIPOFECTAMINE (which is toxic) to cells; (3) time of exposure of cells to DNA–liposome complexes; and (4) number/density of cells at which to carry out the transfection protocol.

In our hands, using COS-1 cells and pcDNA3 encoding receptors for GnRH or 5-HT, high levels of expression are achieved using the following protocol.

1. Cells (3×10^6) are plated (seeded) on a 10-cm tissue culture dish the day before transfection.
2. For each plate, mix 8 μg DNA + 0.8 ml serum-free medium and 48 μl LIPOFECTAMINE + 0.8 ml serum-free medium. Combine the two mixtures and allow complexes to form for 30 min at room temperature.

3. Add serum-free medium to a final volume of 6 ml/plate and overlay cells that have been rinsed twice with serum-free medium. Return the cells to the incubator.
4. Five hours after transfection, the medium is aspirated and replaced with fresh complete growth medium, and the cells are returned to the incubator for 18–24 hr. Alternatively, an equal volume (6 ml) of medium containing 20% fetal bovine serolim (FBS) may be added to the transfection medium; however, not all cell lines will tolerate overnight incubation in the presence of LIPOFECTAMINE reagent.
5. Feed the cells with fresh complete medium the following day (18–24 hr after transfection).
6. Harvest cells for radioligand binding studies 48 to 72 hr after transfection.

Stable Transfection

The method of transfection is the same as for transient transfection except that the cotransfection of two plasmids (or gene and selective marker on same plasmid) is carried out. (COS-1 cells are not suitable for the production of stable cell lines as this line develops resistance to the antibiotics used commonly as selection agents.) The cells are split 72 hr following transfection, at densities that will permit the isolation of colonies arising from single cells. For stable cell lines expressing 5-HT_{2A} receptors, we split HEK293 cells at ratios of 1 : 20 to 1 : 100. Stable tranformants are selected for by growth of the cells in antibiotic containing medium, depending on the selective marker present on the plasmid, e.g., pcDNA3 (Invitrogen) confers resistance to G418. The concentration of G418 required to kill untransfected cells needs to be determined for each cell line.

Measurement of Inositol Phosphate Accumulation

Activation of the receptor in the presence of lithium results in the accumulation of inositol monophosphates, which is the basis of the assay. The measurement of total [^{3}H]inositol phosphates ([^{3}H]IPs) accumulated in the presence of LiCl provides an estimate of the extent of receptor activation. Various methods for the measurement of inositol lipids and phosphates have been described (Ref. 16 and references therein). Currently we are measuring the accumulation of total [^{3}H]IPs by formic acid extraction of cells prelabled with [^{3}H]*myo*-inositol, followed by separation on Dowex columns.

Cells that have been transfected stably or transiently to express the receptor of interest are maintained in appropriate complete growth medium. For transient transfections, we have used COS-1 or HEK293 cells; stable cell lines were developed in HEK293 cells. We find that the most reproducible results are obtained when cells are maintained on a strict schedule of feeding and passaging and are not permitted to become overly confluent. For assays, cells are trypsinized

[16] P. P. Godfrey, *in* "Signal Transduction" (G. Milligan, ed.), p. 105. Oxford Univ. Press, 1992.

when they are about 90% confluent and seeded into 24-well plates at a density of 100,000 cells/well. For assays with stably or transiently transfected HEK293 cells, we found that cells seeded onto poly-D-lysine remained better attached during the assay. [We use either commercial "Biocoat" plates (Becton Dickinson) or prepare plates with poly-D-lysine according to the supplier's instructions.] Coating is not necessary for the very adherent COS-1 cells. Twenty-four hours after seeding, the growth medium is aspirated and cells are washed at least twice with serum-free medium. The cells are then incubated for 18–20 hr in 0.5 ml Dulbecco's modified Eagle's medium (DMEM) that contains 1 μCi/ml [^{3}H]*myo*-inositol. For some cell lines, the use of inositol-free medium may result in better uptake of the [^{3}H]*myo*-inositol. Thorough washing with serum-free medium is necessary to remove traces of serotonin that may be present in FBS and will interfere with subsequent assays.

On the day of the experiment, the loading medium is aspirated and replaced with assay medium (Hank's buffered salt solution containing 20 m*M* HEPES plus 20 m*M* LiCl, pH 7.4 at 37°). HEK293 cells are detached easily, and great care must be taken during media exchanges to avoid dislodging the cells. The cells are incubated for 10–15 min at 37° by floating the plates in a temperature-controlled water bath. Following preincubation, the medium is aspirated and replaced with assay medium that contains the compound of interest at the desired final concentration, and the incubation is continued for a predetermined length of time, usually 15–30 min. (A preliminary experiment to determine the time course of accumulation for each stable cell line should be carried out to ensure that subsequent experiments are being carried out for a time on the linear part of the curve. We found that accumulation in our a stable 5-HT_{2A} wild-type cell line was linear for at least 45 min, whereas in a stable lines expressing a specific mutant 5-HT_{2A} receptor, accumulation was linear for only 20 min. Accordingly, incubations for both lines were carried out for 18 min.) The reaction is stopped by aspiration of the medium and replacement with 1 ml of 10 m*M* formic acid (4°). The plates are stored at 4° for at least 2 hr or up to 72 hr to allow the extraction of water-soluble [^{3}H]IPs before further processing. If desired, the protein content of the cellular material adhering to the plate after extraction can be quantitated as an estimate of cell density, a useful parameter for making comparisons among different experiments.

Total accumulated [^{3}H]IPs are quantitated by ion-exchange chromatography on Dowex AG1-X8 anion-exchange columns (a variation of method developed by Berridge[16a]). We have found that when formic acid is used as the extraction agent, the columns can be reused for at least a year if they are washed with formic acid and stored refrigerated between uses. Columns should not be permitted to totally dry out between uses. If sufficient columns are prepared, 500 samples (twenty 24-well plates) can be procesed in 1 day.

1. Wash columns (1-ml bed volume) with 2.5 ml 3 *M* ammonium formate/0.1 *M* formic acid.

[16a] M. J. Berridge, *Biochem. J.* **212,** 849 (1983).

2. Wash with 10 ml H_2O.
3. Apply samples.
4. Wash columns with 10 ml 60 m*M* sodium formate and 5 m*M* borax.
5. Elute total inositol phosphates into 20-ml scintillation vials with 5 ml 1 *M* ammonium formate and 0.1 *M* formic acid.
6. Add scintillation fluid and count in appropriate liquid scintillation counter. Due to the high ionic strength of the eluate, an appropriate scintillation fluid must be chosen. We find that 15 ml of Universol ES (ICN) is satisfactory to result in a homogeneous gel following vigorous shaking of the sample. (However, if the room temperature is cold, phase separation occurs. This can be reversed by allowing the samples to warm up and reshaking them before they are counted.)

Radioligand-Binding Assays

[^{3}H]ketanserin saturation and competition-binding assays are carried out with membranes prepared from transfected cells. Cells destined for membrane preparation are grown and seeded in the same way, including washes and overnight incubation with serum-free medium, as for measurement of [^{3}H]IPs accumulation (see earlier). Treating these cells according to the same protocol removes variables, such as passage number and growth conditions, that would confound the use of binding parameters to draw inferences regarding the effect of expression levels on experimentally determined functional parameters. COS-1 cells are grown in 100-mm plates; HEK293 cells are grown in 15-cm plates. The plates are washed in phosphate-buffered saline (PBS), and cells are harvested by scraping with a rubber policeman. Cells from one plate are transferred to a 1.5-ml microfuge tube and are pelleted (2 min at 14,000*g*) in a microfuge. After the supernatant has been decanted, the pellets are frozen on dry ice and stored at $-70°$ for up to 6 months until further use.

All membrane preparation procedures are carried out at 4°. Cell pellets are thawed and homogenized in a small volume (5–7 ml) of 50 m*M* Tris–HCl (pH 7.4 at room temperature) in a glass–glass dounce, 8 strokes by hand. The homogenate is diluted to 40 ml with additional buffer and transferred to a 50-ml centrifuge tube. The homogenate is centrifuged at 35,000*g* for 15 min. The supernatant is discarded, and the membrane pellet is transferred to a glass–glass dounce for resuspension in 5 ml buffer by 5–10 strokes by hand. An aliquot of suspension to be used for protein determination can be taken at this time. The homogenate is diluted to the final desired volume and stored on ice until use.

[^{3}H]ketanserin (NEN) is used to label 5-HT_{2A} receptors. Assays tubes contain 30–60 μg membrane protein, [^{3}H]ketanserin (8 concentrations ranging from 0.1 to 8 n*M* for saturation studies and 1.0 n*M* for competition studies), unlabeled ligands (10–13 concentrations for competition studies), and buffer or unlabled 10 μM mianserin (to define nonspecific binding) in a final volume of 0.5 ml.

Other unlabeled ligands such as methysergide may be used to define nonspecific binding; however, we found that with several of the mutations we have examined, the affinity of methysergide is changed drastically, rendering it useless for this purpose. Assays are initiated by the addition of the membrane suspension to tubes containing the other assay components, and incubations are carried out for 30 min at 37°. The assays are terminated by rapid filtration over polyethyleneimine-coated GFC (Whatman) filters and washing with cold buffer. We have used a Brandel 24-well harvester and currently use a Tomtec 96-well harvester that uses a 102 × 258-mm filter mat for this purpose. We have found that the larger surface area offered by the Tomtec permits the filtration of samples containing larger amounts of protein that would clog the filters of a harvester designed to accommodate filter mats of a standard 96-well plate size. The appropriate volume and number of washes should be determined for each harvester type used. Radioactivity remaining on the filter is quantitated by scintillation counting, either by placing individual filter disks in vials or by counting the intact filter mat itself in a scintillation counter (e.g., Wallac Betaplate 1205) designed for that purpose.

Data Analysis

Pharmacological parameters for the inositol phosphate experiments [E_{max} (maximal agonist-induced response) and EC_{50} (concentration of agonist producing 50% of the maximal response)] and binding experiments [K_d (equilibrium dissociation constant of [^{3}H]ketanserin); B_{max} (maximal binding of [^{3}H]ketanserin), and K_i (equilibrium dissociation constant of unlabeled ligands)] are obtained by nonlinear regression. We use either Prism (GraphPad Software), which has appropriate built-in curve-fitting equations, or Kaleidagraph (Synergy Software), which accepts user-defined curve-fitting equations. (For a discussion of the concepts and assumptions implicit in these pharmacological parameters, we suggest that a standard pharmacological reference source, such as Ref. 17, be consulted.) To facilitate comparison of [^{3}H]IPs accumulation experiments, where the dpm among experiments can vary due to loading time, passage number, and so on, we obtain the E_{max} by fitting to raw dpm data and then normalizing all data points to the value of the fit E_{max} for 5-HT for that experiment.

Acknowledgments

The computational studies, on which we based our mutations, were performed in the laboratory of Dr. Harel Weinstein. Figures summarizing the location of hydrogen bonding were originally developed by Dr. Juan A. Ballesteros. The work described is supported by NIH Grant PO1 DA12923.

[17] T. P. Kenakin, "Pharmacologic Analysis of Drug-Receptor Interaction." Lippincott-Raven, New York (1997).

[18] C. Chothia, *Annu. Rev. Biochem.* **53,** 537 (1984).

[8] Use of the Substituted Cysteine Accessibility Method to Study the Structure and Function of G Protein-Coupled Receptors

By JONATHAN A. JAVITCH, LEI SHI, and GEORGE LIAPAKIS

Introduction

Cysteine substitution and covalent modification have been used to study structure–function relationships and the dynamics of protein function in a variety of membrane proteins.[1–6] Moreover, charged, hydrophilic, sulfhydryl reagents have been used to probe systematically the accessibility of substituted cysteines in putative transmembrane segments of a number of proteins. This approach, the substituted cysteine accessibility method (SCAM),[7] has been used to map channel-lining residues in the nicotinic acetylcholine receptor,[8–10] the $GABA_A$ receptor,[11,12] the cystic fibrosis transmembrane conductance regulator,[13] the UhpT transporter,[14] and potassium channels,[15] among others.

We have adapted this approach to map the surface of the binding-site crevice in the dopamine D2 receptor, a member of the G protein-coupled receptor superfamily.[16–22] The binding sites of the D2 receptor and of other receptors in the

[1] A. P. Todd, J. Cong, F. Levinthal, C. Levinthal, and W. L. Hubbell, *Proteins* **6,** 294 (1989).
[2] K. S. Jakes, C. K. Abrams, A. Finkelstein, and S. L. Slatin, *J. Biol. Chem.* **265,** 6984 (1990).
[3] C. Altenbach, T. Marti, H. G. Khorana, and W. L. Hubbell, *Science* **248,** 1088 (1990).
[4] C. L. Careaga and J. J. Falke, *Biophys. J.* **62,** 209 (1992).
[5] A. A. Pakula and M. I. Simon, *Proc. Natl. Acad. Sci. U.S.A.* **89,** 4144 (1992).
[6] K. Jung, H. Jung, J. Wu, G. G. Prive, and H. R. Kaback, *Biochemistry* **32,** 12273 (1993).
[7] A. Karlin and M. H. Akabas, *Methods Enzymol.* **293,** 123 (1998).
[8] M. H. Akabas, D. A. Stauffer, M. Xu, and A. Karlin, *Science* **258,** 307 (1992).
[9] M. H. Akabas and A. Karlin, *Biochemistry* **34,** 12496 (1995).
[10] M. H. Akabas, C. Kaufmann, P. Archdeacon, and A. Karlin, *Neuron* **13,** 919 (1994).
[11] M. Xu, D. F. Covey, and M. H. Akabas, *Biophys. J.* **69,** 1858 (1995).
[12] M. Xu and M. H. Akabas, *J. Biol. Chem.* **268,** 21505 (1993).
[13] M. H. Akabas, C. Kaufmann, T. A. Cook, and P. Archdeacon, *J. Biol. Chem.* **269,** 14865 (1994).
[14] R. T. Yan and P. C. Maloney, *Proc. Natl. Acad. Sci. U.S.A.* **92,** 5973 (1995).
[15] J. M. Pascual, C. C. Shieh, G. E. Kirsch, and A. M. Brown, *Biophys. J.* **69,** 428 (1995).
[16] J. A. Javitch, X. Li, J. Kaback, and A. Karlin, *Proc. Natl. Acad. Sci. U.S.A.* **91,** 10355 (1994).
[17] J. A. Javitch, D. Fu, J. Chen, and A. Karlin, *Neuron* **14,** 825 (1995).
[18] J. A. Javitch, D. Fu, and J. Chen, *Biochemistry* **34,** 16433 (1995).
[19] D. Fu, J. A. Ballesteros, H. Weinstein, J. Chen, and J. A. Javitch, *Biochemistry* **35,** 11278 (1996).
[20] J. A. Javitch, J. A. Ballesteros, H. Weinstein, and J. Chen, *Biochemistry* **37,** 998 (1998).
[21] J. A. Javitch, J. A. Ballesteros, J. Chen, V. Chiappa, and M. M. Simpson, *Biochemistry* **38,** 7961 (1999).
[22] J. A. Javitch, L. Shi, M. M. Simpson, J. Chen, V. Chiappa, I. Visiers, H. Weinstein, and J. A. Ballesteros, *Biochemistry* **39,** 12190 (2000).

0076-6879/02 $35.00

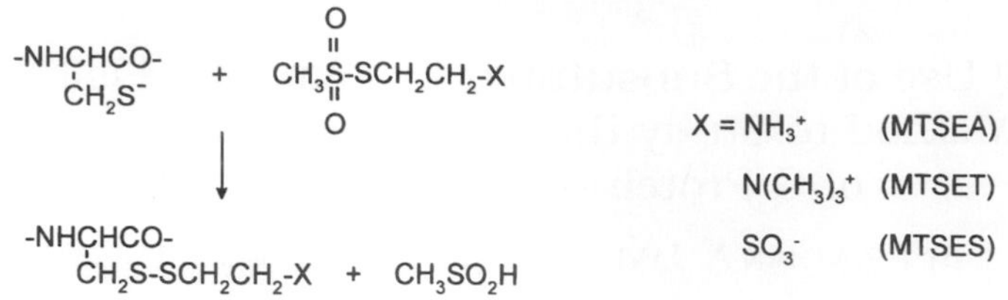

-NHCHCO-(CH_2S^-) + $CH_3S(O_2)$-SCH_2CH_2-X
↓
-NHCHCO-(CH_2S-SCH_2CH_2-X) + CH_3SO_2H

X = NH_3^+ (MTSEA)
$N(CH_3)_3^+$ (MTSET)
SO_3^- (MTSES)

FIG. 1. Structure of methane thiosulfonate derivatives and their reaction with cysteine.

rhodopsin-like subfamily are formed among their seven, mostly hydrophobic, transmembrane segments[23] and are accessible to charged, water-soluble agonists, like dopamine. Thus, for each of these receptors, the binding site is contained within a water-accessible crevice, the binding-site crevice, extending from the extracellular surface of the receptor into the transmembrane domain. The surface of this crevice is formed by residues that can contact specific agonists and/or antagonists and by other residues that may play a structural role and affect binding indirectly.

SCAM provides an approach to map systematically residues on the water-accessible surface of a protein. These residues are identified by substituting them with cysteine and assessing for the reaction of charged, hydrophilic, sulfhydryl reagents with the engineered cysteines. Consecutive residues in putative transmembrane segments are mutated to cysteine, one at a time, and the mutant proteins are expressed in heterologous cells. If ligand binding to a cysteine-substitution mutant is near normal, we assume that the structure of the mutant receptor is similar to that of wild-type receptor and that the substituted cysteine lies in a similar orientation to that of the wild-type residue. In transmembrane segments, the sulfhydryl of a cysteine can face into the binding-site crevice, into the interior of the protein, or into the lipid bilayer; sulfhydryls facing into the water-accessible binding-site crevice should react much faster with charged, hydrophilic, sulfhydryl-specific reagents.

For such polar sulfhydryl-specific reagents, we use derivatives of methane thiosulfonate (MTS): positively charged MTSethylammonium (MTSEA) and MTS-ethyltrimethylammonium (MTSET) and negatively charged MTSethylsulfonate (MTSES) (Fig. 1).[24] These reagents differ somewhat in size with MTSET > MTSES > MTSEA. The largest, MTSET, fits into a cylinder 6 Å in diameter and 10 Å long; thus the reagents are approximately the same size as dopamine. The MTS reagents form mixed disulfides with the cysteine sulfhydryl, covalently linking -SCH_2CH_2X, where X is NH_3^+, $N(CH_3)_3^+$, or SO_3^-. The MTS reagents are specific for cysteine sulfhydryls and do not react with disulfide-bonded cysteines or with other residues.

[23] C. D. Strader, T. M. Fong, M. R. Tota, D. Underwood, and R. A. Dixon, *Annu. Rev. Biochem.* **63,** 101 (1994).

[24] D. A. Stauffer and A. Karlin, *Biochemistry* **33,** 6840 (1994).

Assumptions of SCAM

In order to interpret the results of SCAM, we make a number of assumptions. First, we assume that the highly polar MTS reagents react much faster at the water-accessible surface of the protein than in lipid or in the protein interior. They react with the ionized thiolate (-S^-) more than a billion times faster than with the unionized thiol (-SH).[25] Furthermore, the MTS reagents are very hydrophilic with a relative solubility in water : octanol greater than 2500 : 1.[8,24] Experimental support for the validity of this assumption comes from a study in the aspartate receptor of the accessibility of engineered cysteines to reaction with another hydrophilic, sulfhydryl-specific alkylating reagent.[26] In the $\alpha 2$ helix of the periplasmic domain, a striking correlation was observed between the measured chemical reactivity of each engineered cysteine and the calculated solvent accessibility of the β carbon at the corresponding position in the crystal structure.

We further assume that in transmembrane segments, the access of highly polar reagents to side chains is only through the binding-site crevice, that the addition of -SCH_2CH_2X to a cysteine at the surface of the binding-site crevice is likely to alter binding irreversibly, and, reciprocally, that agonists and antagonists should retard the reaction of the MTS reagents with substituted cysteines that line the binding site.

Another assumption is that the engineered cysteine is an accurate reporter for the water accessibility of the corresponding wild-type residue. If a cysteine-substitution mutant is functional, its overall three-dimensional structure is likely to be similar to the structure of the wild-type receptor. In general, cysteine substitution is remarkably well tolerated (see "Cysteine Substitution"). Nonetheless, local changes at the site of the engineered cysteine could, in principle, alter the accessibility of the residue relative to the accessibility of the wild-type residue. A strength of SCAM results from studying entire transmembrane segments where regular patterns of accessibility can be identified. Given the general consistency of the results obtained in the various receptors and channels studied to date, it is likely that, in most cases, the position of the cysteine residue is similar to that of the wild-type residue it replaced. In several cases, irregular patterns have been observed. We cannot be certain whether the secondary structure in such regions is irregular in the native structure, the protein structure fluctuates in such a region alternately exposing multiple residues, or the cysteine substitution has disrupted the local secondary structure, making the cysteine accessible when the wild-type residue is not.

Detection of Reaction

Reaction can be detected either directly or indirectly by measuring the effect of reaction on a functional property of the protein. Because of the very small quantities of protein produced in most heterologous expression systems, we cannot rely upon

[25] D. D. Roberts, S. D. Lewis, D. P. Ballou, S. T. Olson, and J. A. Shafer, *Biochemistry* **25,** 5595 (1986).
[26] M. A. Danielson, R. B. Bass, and J. J. Falke, *J. Biol. Chem.* **272,** 32878 (1997).

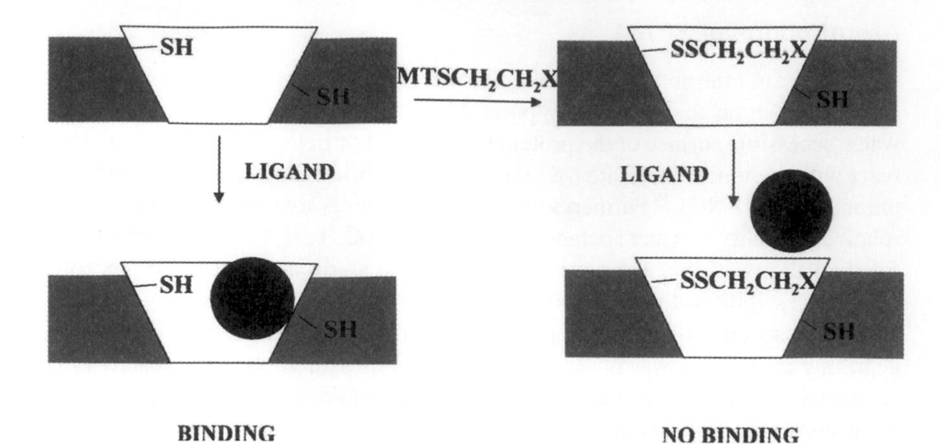

FIG. 2. Schematic representation of the reaction of MTS reagents with a cysteine exposed in the binding-site crevice. The membrane is represented by a shaded rectangle, the binding site crevice is indicated within the plane of the membrane, and the solid oval represents ligand. $-SSCH_2CH_2X$ [where X is NH_3^+, $N(CH_3)_3^+$ or SO_3^-] is linked covalently to the water-accessible cysteine sulfhydryl. In the bound state, represented in the lower left, the ligand is reversibly bound within the binding-site crevice. In the unbound state, represented in the upper left, the binding site is unoccupied. After irreversible reaction with $MTSCH_2CH_2X$, represented in the upper right, ligand binding is altered. The cysteine sulfhydryl-facing lipid or the protein interior reacts vastly more slowly with the charged MTS reagents. Ligand retards the rate of reaction of the MTS reagents with the cysteine sulfhydryl in the binding-site crevice, thereby protecting subsequent ligand binding.

the direct detection of reaction. Instead, we use the irreversible modification of function to assay the reaction. In a receptor, the reaction of an MTS reagent with an engineered cysteine in the binding-site crevice should alter binding irreversibly (Fig. 2). Additionally, reaction with a cysteine near the binding site should be retarded by the presence of bound ligand.

If a MTS reagent has no effect on a mutant, interpretation of the results must be made with caution. The temptation is to infer that the engineered cysteine is inaccessible to the MTS reagents and is therefore not on the water-accessible surface of the protein. While this is the most likely explanation, there are other possibilities. First, electrostatic or steric factors may alter the reactivity of the MTS reagents with a water-accessible residue. Second, while it seems unlikely that a residue forming the surface of the binding-site crevice could be modified covalently by the addition of the charged $-SCH_2CH_2X$ without interfering with binding, such a result is nonetheless possible. In the dopamine D2 receptor, we have observed that the reaction of MTSEA at certain positions has a much greater effect on the binding of particular ligands; e.g., reaction of MTSEA with the highly reactive endogenous cysteine Cys-118 causes a negligible decrease in the affinity of the receptor for particular ligands but a large decrease in its affinity for other ligands.[27] To decrease the likelihood of such a false-negative determination, we typically screen for effects with antagonists from two different structural classes. Nonetheless, these potential complications further demonstrate the importance of systematically mutating to cysteine consecutive residues along an entire transmembrane segment; while mutation of any individual residue might be subject to potential pitfalls due to steric or electrostatic factors or silent reaction, this is unlikely to be a systematic problem affecting the overall pattern of accessibility of multiple residues in a transmembrane segment.

Mechanisms of Altered Binding

The functional effect of the addition of $-SCH_2CH_2X$ to the engineered cysteine could be a result of steric block, electrostatic interaction, or indirect structural changes. Regardless, although we do not know the detailed mechanism of the alterations in binding, an irreversible effect is evidence of reaction and, therefore, of the accessibility of the engineered cysteine. Although reaction usually inhibits binding, it can also potentiate binding. This can be illustrated in the dopamine D2 receptor by the mutation of Asp-108.[17] Mutation to cysteine of this residue at the extracellular end of the third transmembrane segment reduced the affinity of the receptor for antagonist binding about threefold. Reaction of the positively charged MTSEA or MTSET at this position inhibited binding significantly. In contrast, reaction of the negatively-charged MTSES restored the negative charge at this position and shifted the affinity toward that of the wild-type receptor, thereby increasing occupancy and potentiating binding. The fact that reaction can potentiate function necessitates care in experimental design; a potentiation of binding that results from an increase in ligand affinity could be missed by measuring binding at too high a ligand concentration relative to the K_D.

As discussed earlier, reaction with a MTS reagent of a substituted Cys on the water-accessible surface of a receptor, especially if somewhat distant from the binding site, may not produce a change in ligand binding, thereby resulting in a false-negative determination of accessibility. This may be more likely in the case of a neutral antagonist radioligand that is unable to discriminate conformational states of the receptor. In such a case, direct radiolabeled agonist binding or agonist competition of radiolabeled antagonist binding may be a more sensitive indicator of a disruption of receptor structure and thus a more sensitive assay of reaction. Indeed, we have identified an endogenous Cys in the β_2-adrenergic receptor that reacts with MTSEA without altering antagonist binding.[28] We were able to detect reaction in this case by an alteration in the affinity of the agonist isoproterenol in competition with radiolabeled antagonist.

[27] J. A. Javitch, D. Fu, and J. Chen, *Mol. Pharmacol.* **49,** 692 (1996).

[28] G. Liapakis and J. A. Javitch (in preparation).

Endogenous Cysteines

The function of the protein used as the background for SCAM must not be affected by the sulfhydryl reagents. In some cases, endogenous cysteines are not accessible to reaction with the MTS reagents (or reaction produces no functional effect), whereas in other cases, endogenous cysteines are accessible and must first be identified and mutated to other residues. The ideal starting point would be to create a cysteine-less protein with normal expression and function. Such a construct has been possible with the lactose permease,[6] the NhaA-Na^+/H^+ antiporter,[29] and a glutamate transporter,[30] but a dopamine D2 receptor with all five putative transmembrane cysteines simultaneously substituted by other residues expressed too poorly for further study. Nonetheless, in the dopamine D2 receptor, replacement of a single endogenous cysteine (Cys-118) with serine resulted in a 100-fold decrease in the reactivity of the receptor with MTSEA and MTSET.[16] C118S expresses normally and has unaltered binding properties; this mutant was used as the background for further cysteine substitutions.[17–22] When one uses a receptor with endogenous cysteines as a background construct, one must be alert to the possibility that a new engineered Cys might produce its effect, not because the engineered Cys is accessible, but rather because of a mutation-induced alteration of the accessibility of an endogenous Cys, thereby resulting in a false-positive determination of accessibility. Such a scenario has been observed in the serotonin transporter.[31]

Application of the Substituted Cysteine Accessibility Method

In a background of C118S, the mutant dopamine D2 receptor relatively insensitive to MTS reagents, we have used SCAM to determine residues that form the surface of the binding-site crevice. The methods used are detailed later. In summary, cysteines are substituted, one at a time, for residues in a putative transmembrane segment. The mutant receptor is expressed in heterologous cells, and ligand binding is assayed. If the receptor binds ligand, then the effect of reaction with the MTS reagents is determined, always comparing the effects with those on the receptor used as the background for the cysteine substitutions. We then determine the rate of reaction of accessible substituted Cys and test for the ability of ligand to retard the rate of reaction of a MTS reagent at the reactive positions.

Construction of Epitope-Tagged D2 Receptor Construct

The sequence encoding the β_2-adrenergic receptor epitope tagged at the amino terminus with the cleavable influenza-hemagglutinin signal sequence followed by

[29] Y. Olami, A. Rimon, Y. Gerchman, A. Rothman, and E. Padan, *J. Biol. Chem.* **272,** 1761 (1997).
[30] R. P. Seal and S. G. Amara, *Soc. Neurosci. Abstr.* **22,** 1575 (1996).
[31] M. M. Stephan, G. Kamdar, G. Rudnick, and K. M. Y. Penado, *Soc. Neurosci. Abstr.* **25,** 1699 (1999).

the "FLAG" epitope (IBI, New Haven, CT) was a gift from Dr. B. Kobilka.[32] The sequence encoding the epitope tag was excised and ligated in-frame to the D2 receptor cDNA, thereby creating a fusion protein in which the epitope tag leads directly into the sequence of the D2 receptor.[20] The affinity of the epitope-tagged receptor for the antagonists *N*-methylspiperone, YM-09151-2, and sulpiride is unchanged from that of the wild-type receptor. This epitope-tagged D2 receptor fragment was subcloned into the bicistronic expression vector pcin4 (a gift from Dr. S. Rees, Glaxo).[33] With this vector the antibiotic resistance gene is made from an internal ribosomal entry site (IRES), and thus one mRNA encodes both receptor and antibiotic resistance. This ensures that essentially every cell that is antibiotic resistant also expresses receptor, which allows us to work with stably transfected pools of cells and avoid clonal selection (see later).

Site-Directed Mutagenesis

We have generated cysteine mutations using the Altered Sites mutagenesis kit (Promega) or the polymerase chain reaction with Pfu polymerase and with oligonucleotides incorporating a change in a restriction site as well as the desired mutation. Mutants can therefore be identified by restriction mapping, and the mutations are confirmed by DNA sequencing.

Transient and Stable Transfection

HEK 293 cells in DMEM/F12 (1 : 1) with 10% bovine calf serum (BSA; Hyclone) and 293-TSA cells (a clonal line of HEK 293 cells stably expressing the SV40 large T antigen) in DMEM with 10% fetal calf serum are maintained at 37° and 5% CO_2. For transient transfection, 35-mm dishes of 293-TSA cells at 70–80% confluence are transfected with 2 μg of wild-type or mutant D2 receptor cDNA in pcDNAI/Amp (Invitrogen) or pcin4 (see earlier discussion) using 9 μl of LIPOFECTAMINE (GIBCO) and 1 ml of OPTIMEM (GIBCO). Five hours after transfection, the solution is removed and fresh medium is added. Twenty-four hours after transfection the medium is changed. Forty-eight hours after transfection, cells are harvested as described later.

For stable transfection, HEK 293 cells are transfected with D2 receptor cDNA in pcin4 as described earlier. Twenty-four hours after transfection the cells are split to a 100-mm dish, and 700 μg/ml geneticin is added to select for a stably transfected pool of cells. No significant differences in binding affinity or accessibility to the MTS reagents have been detected between receptor from transiently and stably transfected pools of HEK 293 cells. Stable pools are much more convenient for

[32] B. K. Kobilka, *Anal. Biochem.* **231,** 269 (1995).

[33] S. Rees, J. Coote, J. Stables, S. Goodson, S. Harris, and M. G. Lee, *BioTechniques* **20,** 102 (1996).

use in these studies due to the timing necessary in preparing for and performing multiple transient transfections.

Harvesting Cells

Cells are washed with phosphate-buffered saline (PBS; 8.1 mM NaH_2PO_4, 1.5 mM KH_2PO_4, 138 mM NaCl, 2.7 mM KCl, pH 7.2), treated briefly with PBS containing 1 mM EDTA, and then dissociated in PBS. Cells are pelleted at 1000g for 5 min at 4° and are resuspended for binding or treatment with MTS reagents.

N-[^{3}H]Methylspiperone Binding

Whole cells from a 35-mm plate are suspended in 450 μl of buffer A (25 mM HEPES, 140 mM NaCl, 5.4 mM KCl, 1 mM EDTA, 0.006% BSA, pH 7.4). Cells are then diluted 20-fold with buffer A. N-[^{3}H]methylspiperone (Dupont/NEN) binding is performed as described previously.[17] For saturation binding, duplicate polypropylene minitubes contain six different concentrations of N-[^{3}H]methylspiperone between 5 and 800 pM in buffer A with 300 μl of cell suspension in a final volume of 0.5 ml. The mixture is incubated at room temperature for 60 min and is then filtered using a Brandel cell harvester through Whatman 934AH glass fiber filters (Brandel). The filter is washed three times with 1 ml of 10 mM Tris–HCl, 120 mM NaCl, pH 7.4, at 4°. Specific N-[^{3}H]methylspiperone binding is defined as total binding less nonspecific binding in the presence of 1 μM (+)butaclamol (Research Biochemicals). Competition assays are conducted in a final volume of 1 ml with 10 different concentrations of the tested drug and 200 pM N-[^{3}H]methylspiperone. Depending on the level of expression in the various mutants, adjustments in the number of cells per assay tube are made as necessary to prevent depletion of ligand in the case of very high expression or to increase the signal in the case of low expression.

Use of MTS Reagents

At pH 7 and 22°, MTSEA, MTSET, and MTSES (Biotium Inc., Hayward, CA) hydrolyze rapidly with a half-time of 5 to 20 min.[7] At lower pH and lower temperature, hydrolysis is appreciably slower. Stock reagent should be stored desiccated at 4°. A frequently used stock can be kept desiccated at room temperature, and it can be replenished from the 4° stock after appropriate warming to room temperature. The reagents should be weighed but kept dry until immediately before use. The reagents are relatively stable unbuffered in water at 4°; if it is necessary to dissolve the reagents or perform an intermediate dilution in buffer at a physiological pH, this should be done immediately before starting the reaction.

Thiosulfonate reagents can be assayed for activity using the TNB assay as described previously.[7]

Reactions with MTS Reagents

Whole cells from a 35-mm plate are suspended in 400 μl buffer A. Aliquots (45 μl) of cell suspension are incubated with freshly prepared MTS reagents (5 μl) at the stated concentrations at room temperature for 2 min. Cell suspensions are then diluted 16-fold, and 200-μl aliquots are used to assay for N-[^{3}H]methylspiperone (200 pM) binding as described.[19] For screening purposes, to adjust for the intrinsic reactivities of the reagents, we typically screen for effects at 2.5 mM MTSEA, 1 mM MTSET, and 10 mM MTSES. The fractional inhibition is calculated as 1-[(specific binding after MTS reagent)/(specific binding without reagent)] (Table I). We use SPSS for Windows (SPSS, Inc.) to analyze the effects of the MTS reagents by one-way ANOVA with Dunnett's post hoc test ($p < 0.05$).

The second-order rate constant (k) for the reaction of MTSEA with each susceptible mutant is estimated by determining the extent of reaction after a fixed time, 2 min, with four to six concentrations of MTSEA (typically 0.025 to 10 mM) (all in excess over the reactive sulfhydryls). The fraction of initial binding, Y, is fit to $(1-\text{plateau})^* e^{-kct} + \text{plateau}$, where plateau is the fraction of residual binding at saturating concentrations of MTSEA, k is the second-order rate constant (in $M^{-1}\text{sec}^{-1}$), c is the concentration of MTSEA (M), and t is the time (120 sec) (Fig. 3).

Protection

Dissociated cells are incubated in buffer A for 20 min in a 96-well Multiscreen GF/B filter plate (Millipore) at room temperature in the presence or absence of (±)sulpiride and then MTSEA is added, in the continued presence or absence of sulpiride, for 2 min at a concentration chosen to produce approximately 75% of the maximal inhibition of specific N-[^{3}H]methylspiperone binding in the absence of sulpiride. For most mutants, sulpiride is used at a concentration of 10 μM. To compensate for changes in the K_I of particular mutants, sulpiride concentrations are adjusted as necessary. Cells are washed four times for 5 min each and then filtered though the 96-well plates. In the wash buffer, sodium is replaced by choline in order to facilitate the removal of residual sulpiride, the binding affinity of which is sodium dependent. N-[^{3}H]methylspiperone binding to the washed cells is performed in buffer A in the multiscreen plates in a final volume of 0.25 ml. Protection is calculated as 1-[(inhibition in the presence of sulpiride)/(inhibition in the absence of sulpiride)]. The statistical significance of protection by sulpiride is assessed by the paired t test.

If reaction with the sulfhydryl reagents is slowed by antagonist or agonist, we infer that the residue is accessible in the binding-site crevice. Each and every residue that is protected, however, need not contact ligand; ligand could protect residues deeper in the crevice by binding above them and blocking the passage of the MTS reagent from the extracellular medium to the cytoplasmic end of the crevice. In addition, we cannot rule out indirect protection through ligand-mediated

TABLE I

SAMPLE DATA SHOWING ANALYSIS OF THE EFFECTS OF MTS REAGENTS ON *N*-[^{3}H]METHYLSPIPERONE BINDING TO SELECTED CYS SUBSTITUTION MUTANTS IN TM6 OF THE D2 RECEPTOR[a]

	C118S						G380C						H393C					
	NS	Total	EA2.5	EA.25	ET	ES	NS	Total	EA2.5	EA.25	ET	ES	NS	Total	EA2.5	EA.25	ET	ES
×1	280	7189	7228	7600	7695	7433	341	5500	4884	5319	4841	5439	511	5396	2458	3167	4097	4846
×2	318	7050	7066	7393	7382	7431	298	5569	5000	5248	5373	4859	439	5437	2352	3122	3870	4850
×3	224	6852	7294	7582	7417	7064	363	5534	4870	5139	5161	5106	375	5302	2237	2900	3824	4815
mean	280	7031	7196	7525	7498	7309	334	5534	4918	5236	5125	5135	442	5378	2349	3063	3930	4837
spec	—	6751	6916	7245	7218	7029	—	5200	4584	4902	4791	4801	—	4937	1908	2622	3489	4396
MTS resid	—	1.00	1.02	1.07	1.07	1.04	—	1.00	0.88	0.94	0.92	0.92	—	1.00	0.39	0.53	0.71	0.89
MTS inhib	—	—	−0.02	−0.07	−0.07	−0.04	—	—	0.12	0.06	0.08	0.08	—	—	0.61	0.47	0.29	0.11

[a] Sample data from the background construct C118S and from an inaccessible (G380C) and an accessible (H393C) Cys-substitution mutant (in the C118S background) in TM6 of the dopamine D2 receptor. Triplicate values (×1, ×2, ×3) and the mean are shown along with nonspecific binding (NS), total binding (Total), and specific binding (spec). Binding after a 2-min incubation with 2.5 m*M* MTSEA (EA2.5), 0.25 m*M* MTSEA (EA.25), 1 m*M* MTSET (ET), and 10 m*M* MTSES (ES) is shown. The fraction of residual binding (MTS resid) and the fraction of binding inhibited by treatment with the MTS reagents (MTS inhib) are also shown.

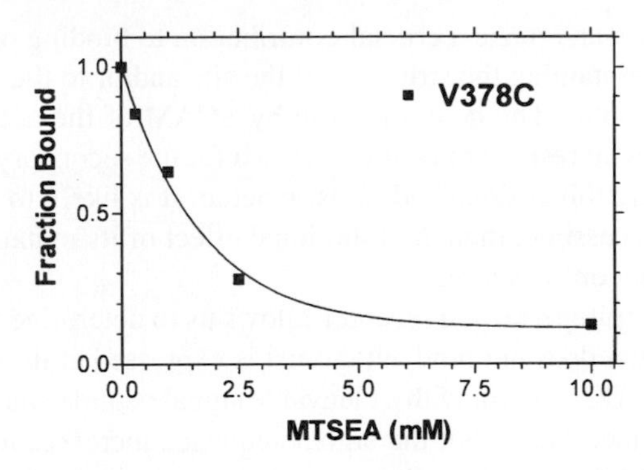

FIG. 3. A representative experiment showing the inhibition of ligand binding by the reaction of MTSEA with a substituted Cys in the dopamine D2 receptor. Data were fit by nonlinear regression using Prism (GraphPad) to a one-phase exponential decay function, $Y = (1\text{-plateu})^{*}e^{-Kx} + \text{plateau}$, where Y is the fraction of initial binding and X is the concentration of MTSEA. The curve starts at 1 and decays to plateau at saturating MTSEA. In this sample fit, $K = 603\ M^{-1}$, plateau $= 0.12$, and $r^2 = 0.99$. Because $K = k^{*}t$ [where k is the second-order rate constant (in $M^{-1}\text{sec}^{-1}$) and t is the time in seconds (120 sec)], $k = K/t = 603/120 = 5\ M^{-1}\text{sec}^{-1}$.

propagated structural rearrangement. We typically use the antagonist sulpiride to screen for protection. This compound is hydrophilic and its binding is sodium dependent, two factors that facilitate its removal and decrease interference of residual drug with the final determination of binding.

Cysteine Substitution

The ability to substitute cysteine residues for other residues and still obtain functional receptor is central to this approach. In the seven transmembrane-spanning segments of the dopamine D2 receptor, 144 of 157 cysteine-substitution mutants bound antagonist with near-normal affinity, i.e., is with a reduction of less than 3-fold compared to that of the background construct, C118S.[17–22] Only 7 mutants had greater than a 5-fold reduction in affinity, and only 4 of these 7 did not bind or bound antagonist with greater than a 10-fold reduction in binding. Three of these 4 mutations were of presumed contact residues. The tolerated substitutions were for hydrophobic residues (alanine, leucine, isoleucine, methionine, and valine), polar residues (asparagine, serine, threonine), neutral residues (proline), acidic residues (aspartate, glutamate), aromatic residues (phenylalanine, tryptophan, tyrosine), and glycine and histidine. Thus, cysteine is a remarkably well-tolerated substitution.

A cysteine-substitution mutant that does not function cannot be studied by the substituted-cysteine-accessibility method (or by traditional site-directed mutagenesis). Residues that cannot be mutated to cysteine either are accessible in the binding site crevice and make a crucial contribution to binding or make a crucial contribution to maintaining the structure of the site and/or to the folding and processing of the receptor. The determination by SCAM of the accessibility of the neighbors of a crucial residue may allow us to infer the secondary structure of the segment containing this residue and, thus, whether it is likely to be accessible as well. If it is not accessible, then the functional effect of its mutation is likely due to an indirect effect on structure.

The use of an epitope-tagged receptor allows us to determine whether the rare mutant receptor that does not bind antagonist is expressed at the cell surface. We have constructed a D2 receptor with a cleavable signal peptide and a FLAG epitope at its N terminus (see Methods); the signal sequence increases the expression of some mutants slightly, and the epitope tag has no effect on the function of the receptor. Even the presence in the membrane of a nonbinding receptor mutant, however, does not prove that the mutated residue contacts ligand, as the residue could still interfere with binding indirectly.

Secondary Structure

To infer a secondary structure, we must assume that if binding to a mutant is not affected by the MTS reagents, then no reaction has occurred and that the side chain at this position is not accessible in the binding-site crevice (but see earlier discussion). In an α-helical structure one would expect the accessible residues to form a continuous stripe when the residues are represented on a helical net. For example, in the third transmembrane segment of the dopamine D2 receptor, the pattern of accessibility is consistent with this transmembrane segment, forming an α helix with a stripe of about 140° facing the binding-site crevice.[17] In contrast, in an antiparallel β strand, one would expect every other residue to be accessible to the reagents.

More complex or irregular patterns of accessibility can be more difficult to interpret, but these findings can also be informative. For example, we have found that cysteines substituted, one at a time, for 10 consecutive residues in the fifth transmembrane segment (TM5) of the dopamine D2 receptor are accessible to MTSEA and MTSET.[18] This pattern of accessibility is not consistent with TM5 being a fixed α helix with one side facing the binding-site crevice. The accessible region of TM5 might be embedded in the membrane but moves rapidly to expose different sets of residues. In such a scenario, at any instant, only a limited set of residues in TM5 might be exposed in the binding-site crevice. Alternatively, the exposed region of TM5, which contains the serines likely to bind agonist, might loop out into the binding-site crevice and be completely accessible to water and thus to MTSEA.

We also observed an unusual pattern of accessibility in the seventh transmembrane segment (TM7) of the dopamine D2 receptor. Again, the overall pattern of

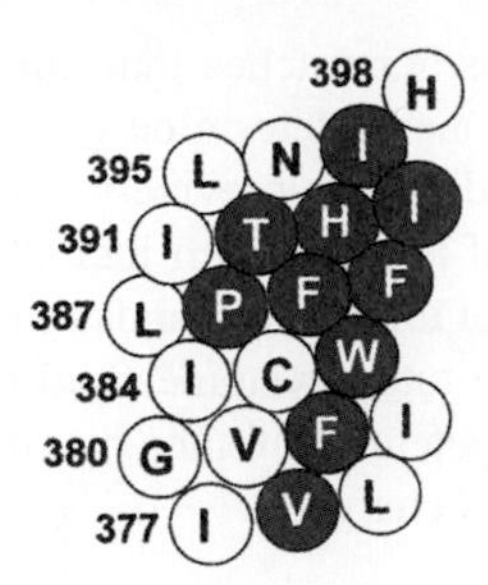

FIG. 4. Helical net representation of the residues in the TM6 segment of the dopamine D2 receptor, summarizing the effects of MTSEA on *N*-[^{3}H]methylspiperone and [^{3}H]YM-09151-2 binding. Reactive residues are filled, and open circles indicate that MTSEA had no significant effect on binding. Adapted from Javitch *et al.*[20]

exposure was not consistent with a simple secondary structure of either α helix or β strand. TM7, however, contains the highly conserved residues Asn-Pro in the middle of the transmembrane segment. In soluble proteins, these residues have been observed to introduce kinks and twists in α helices. In molecular modeling work, we found that the pattern of exposure of the cysteine-substitution mutants to MTSEA can be explained if TM7 is a kinked and twisted α helix.[19] In contrast, accessibility data in TM6 are consistent with a kinked but not twisted α helix[20] (Fig. 4). Thus, these "irregular" patterns of accessibility can lead to new insights or directions for further experimental pursuit.

Distinctions between SCAM and Typical Mutagenesis

A distinction between our approach and that of typical mutagenesis experiments is that we do not rely on the functional effects of a given mutation. The interpretation of the effects of typical mutagenesis experiments requires one to assume that perturbations caused by a mutation, such as changes in binding affinity, are due to local effects at the site of the mutation rather than to indirect effects on protein structure. The validity of this assumption is rarely assessed for individual mutations. The structure of the λ phage receptor maltoporin, however, showed that 50% of the mutated residues that had been implicated in λ phage recognition are actually located in the protein interior.[34] Thus, mutation of these buried residues alters λ phage binding indirectly. Likewise, the crystal structures of several dihydrofolate reductase mutants have demonstrated that a mutation approximately 15 Å from the substrate-binding pocket exerts an effect on catalytic activity through an extended structural perturbation.[35]

[34] T. Schirmer, T. A. Keller, Y. F. Wang, and J. P. Rosenbusch, *Science* **267,** 512 (1995).
[35] K. A. Brown, E. E. Howell, and J. Kraut, *Proc. Natl. Acad. Sci. U.S.A.* **90,** 11753 (1993).

In contrast to mutagenesis approaches that only detect perturbations in protein function, SCAM allows one to determine whether a residue is on the water-accessible surface of the binding-site crevice when the mutant has near-normal function. Other advantages of the approach include the ability to probe binding sites by assessing the ability of different ligands to retard the reaction of the MTS reagents with particular substituted cysteines and the ability to probe the steric constraints and electrostatic potential of sites by comparing the rates of reaction of reagents of varying size and charge.

Electrostatic Potential

Because positively-charged MTSET and negatively-charged MTSES are similar in size, differences in their reactivities with engineered cysteines are likely to be due to differences in the electrostatic potential of the binding-site crevice. For example, in TM3 of the D2 receptor, MTSES did not react with any engineered cysteines more cytoplasmic than Val-111, whereas MTSET reacted with several residues more cytoplasmic than this position.[17] This reflects the negative electrostatic potential deeper in the binding-site crevice. By subsequent applications of positively- and negatively-charged reagents, we can rule out the possibility that reaction has occurred without alteration of function in the case of addition of one but not the other charged moiety. For example, we determined that MTSES did not react with Cys-118 in D2 receptor because subsequent application of MTSET still inhibited binding. If MTSES had reacted silently with Cys-118, it would have prevented the cysteine from subsequent reaction with MTSET.

In contrast, the reactivity of MTSET and MTSES with F110C and V111C, residues located near the extracellular end of TM3, is similar. This indicates that the electrostatic potential near these residues is not as negative as it is below Val-111. The results in TM2 are also consistent with a negative electrostatic potential,[21] but this has been less apparent in the other transmembrane segments studied.[17–20,22] This suggests that Asp-80 and Glu-95 in TM2 and Asp-108 and Asp-114 in TM3 are significant contributors to the negative potential in this region and that the potential is not uniform throughout the crevice at similar depths. This distribution of the field likely helps to orient ligand within the binding-site crevice with the protonated amine toward Asp-114.

Comparison of Reactions with MTSEA and MTSET

When adjusted for the rate constants for their reactions with simple thiols in solution,[24] the reaction of MTSEA with cysteines in the binding-site crevice of the dopamine D2 receptor is typically accelerated approximately 10-fold relative to that of MTSET. MTSEA is smaller than MTSET, and its access to substituted

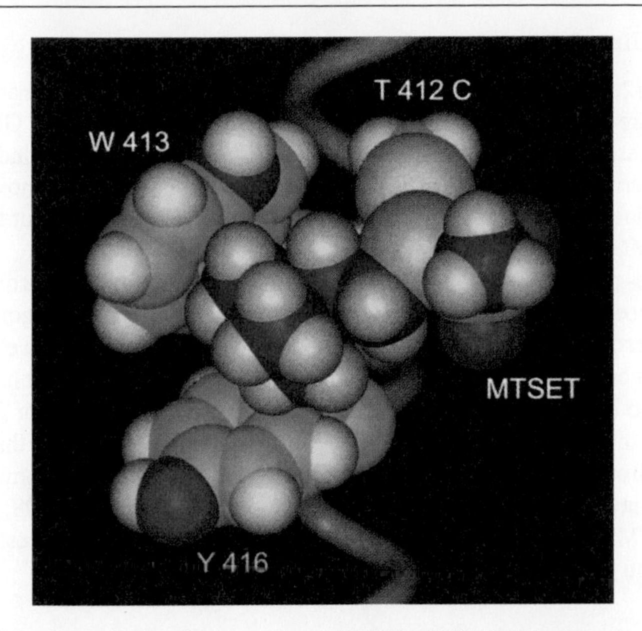

FIG. 5. Molecular model showing interaction of MTSET with TM7 of the dopamine D2 receptor. Trp-413, Tyr-416, and the substituted Cys at position 412 are shown in van der Waals' representations. The potential interaction of the hydrophobic quaternary ammonium group of MTSET with the aromatic side chains is shown in an orientation that accommodates the stereochemical requirements of a bimolecular nucleophilic reaction (SN2) between the substituted Cys and MTSET, thus contributing to the enhanced reaction rate in the T412C mutant. Adapted from Fu *et al.*[19]

cysteines may be less sterically hindered. Moreover, MTSEA, like dopamine, contains an ethylammonium group, and it could be the affinity of this group for the dopamine-binding site that accelerates the reaction of MTSEA relative to that of MTSET. This pattern is broken, however, by T412C in TM7[19] and by F389C and F390C in TM6,[20] which are nearly equireactive with MTSEA and MTSET. Thus, at these positions, the reaction of MTSET is accelerated approximately 10-fold relative to that of MTSEA. For T412C in TM7,[19] we noted that the specific increase may be the result of an interaction of the aromatic side chains of Trp-413 and Tyr-416 with the hydrophobic quaternary ammonium group of MTSET,[36] which would favor the reaction with MTSET at T412C (Fig. 5). Similarly, favorable interactions of the MTSET cation with Trp-386 and other nearby aromatic side chains are likely to be responsible for the increase in the reactivity of MTSET relative to that of MTSEA seen at F389C and F390C.[20]

[36] D. A. Dougherty, *Science* **271,** 163 (1996).

The Binding Site and the Aromatic Cluster

Phe-382, Trp-386, Phe-389, and Phe-390, the four accessible aromatic residues in TM6, are completely conserved within related neurotransmitter GPCRs, and some of these residues have been shown to be important for ligand binding and/or receptor activation.[37,38] Agonist binding may induce coordinated movements of these residues, resulting in a rotational/translational movement about the proline kink in TM6.[20,39]

Ligand–receptor interactions that can be supported by such an "aromatic cluster" have been suggested from structure–activity data for dopamine agonists acting on the D2 receptor.[40] Suggested requirements for potency included an electrostatic interaction between the protonated amine and a negative subsite that has been established to be Asp-114 in TM3,[17,41,42] a hydrogen-bonding group or groups that interact with a serine or serines in TM5,[18,42–45] and an aromatic ring that interacts with a hydrophobic site. This hydrophobic site is likely to be the aromatic cluster in TM6 that extends to the adjacent TM5, where it includes Phe-198, and to the adjacent TM7, where it includes Tyr-416, both of which are also accessible in the binding-site crevice.[18,19]

Conformational Changes Associated with Receptor Activation

In order to further explore the unexpected pattern of accessibility of the substituted-cysteine mutants in TM5 of the D2 receptor, we determined the accessibilities of the aligned residues in the homologous β_2-adrenergic receptor.[46] Surprisingly, the pattern of accessibility of TM5 of the β_2 receptor, unlike TM5 of the D2 receptor, is compatible with a fixed α-helical structure. This may reflect a difference in the structures and/or packing of transmembrane segments in the two receptors, although a major difference in the structures of the two receptors seems unlikely given their sequence similarities, particularly in this region. Alternatively, at rest,

[37] W. Cho, L. P. Taylor, A. Mansour, and H. Akil, *J. Neurochem.* **65,** 2105 (1995).

[38] B. L. Roth, M. Shomam, M. S. Choudhary, and N. Khan, *Mol. Pharmacol.* **52,** 259 (1997).

[39] U. Gether, S. Lin, P. Ghanouni, J. A. Ballesteros, H. Weinstein, and B. K. Kobilka, *EMBO J.* **16,** 6737 (1997).

[40] P. Seeman, *Pharmacol. Rev.* **32,** 229 (1980).

[41] C. D. Strader, I. S. Sigal, M. R. Candelore, E. Rands, W. S. Hill, and R. A. Dixon, *J. Biol. Chem.* **263,** 10267 (1988).

[42] A. Mansour, F. Meng, W. J. H. Meador, L. P. Taylor, O. Civelli, and H. Akil, *Eur. J. Pharmacol.* **227,** 205 (1992).

[43] C. D. Strader, I. S. Sigal, and R. A. Dixon, *Trends Pharmacol. Sci. Suppl.* 26 (1989).

[44] B. A. Cox, R. A. Henningsen, A. Spanoyannis, R. L. Neve, and K. A. Neve, *J. Neurochem.* **59,** 627 (1992).

[45] G. Liapakis, J. A. Ballesteros, S. Papachristou, W. C. Chan, X. Chen, and J. A. Javitch, *J. Biol. Chem.* (2000).

[46] G. Liapakis, D. Fu, and J. A. Javitch (in preparation).

the D2 receptor may be more dynamic than the β_2 receptor, which may be more constrained. Thus, in the absence of ligand, the D2 receptor may undergo sufficient conformational change to alternately expose multiple residues that are not exposed simultaneously. Indeed, a number of antagonists have been found to act as inverse agonists at the D2 receptor, suggesting that there may be significant native constitutive activity of this receptor.[47,48] Movement of the extracellular portion of TM5, which contains the serines likely to bind agonist,[49] might be part of the mechanism of activation of the receptor, and conformational changes of these residues may be associated with receptor activation.

Conformational changes in a protein may result in changes in the accessibility of substituted cysteines as assessed by their rates of reaction with polar sulfhydryl-specific reagents. For example, residues lining the channel of the nicotinic acetylcholine receptor change in accessibility upon activation of the receptor and opening of the channel.[8,10] Similarly, it should be possible to determine changes in the accessibility of residues in GPCRs in different functional states.

To identify activation-induced structural changes in residues forming the surface of the binding site crevice, we sought to determine the relative accessibilities of a series of engineered cysteines in the resting and activated receptor. Agonist cannot be used to activate receptor, however, because the presence of a ligand within the binding-site would interfere with access of the MTSEA to the engineered cysteines. Alternatively, the activated state of the receptor can be achieved by using a constitutively active mutant (CAM) receptor as a background for further cysteine substitution. A CAM receptor is intrinsically active and has a higher affinity for agonist than the wild-type receptor.[50] The high-affinity state for agonist is typically associated with the activated receptor–G protein complex. That agonist affinity is higher in the CAM even in the absence of G protein suggests that the structure of the binding site of the CAM is likely to be similar to that of the agonist-activated, wild-type receptor-binding site (or isomerizes more easily to the active state). Thus, we can compare the resting and active forms of the receptor by determining the accessibility of substituted cysteines in the binding-site crevice in these two states using wild-type receptor and a CAM as background constructs.

We have chosen to pursue initial studies in the β_2-adrenergic receptor because of the availability of a well-characterized CAM[50] (kindly provided by R. Lefkowitz). MTSEA had no effect on the binding of agonist or antagonist to the wild-type β_2 receptor expressed in HEK 293 cells. This suggested that no endogenous cysteines are accessible in the binding site crevice (or that reaction

[47] L. B. Kozell and K. A. Neve, *Mol. Pharmacol.* **52,** 1137 (1997).

[48] D. A. Hall and P. G. Strange, *Br. J. Pharmacol.* **121,** 731 (1997).

[49] C. D. Strader, M. R. Candelore, W. S. Hill, I. S. Sigal, and R. A. Dixon, *J. Biol. Chem.* **264,** 13572 (1989).

[50] P. Samama, S. Cotecchia, T. Costa, and R. J. Lefkowitz, *J. Biol. Chem.* **268,** 4625 (1993).

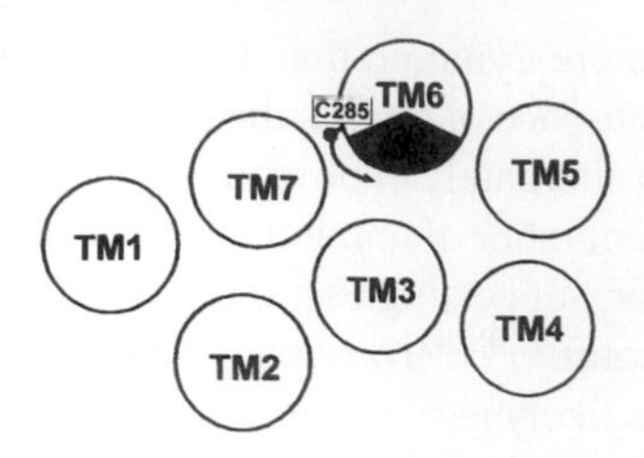

FIG. 6. An illustration of the rotation and/or tilting of the sixth membrane-spanning segment associated with the activation of the β_2 receptor. The indicated rearrangement brings Cys-285 to the margin of the binding-site crevice and allows it to react with MTSEA to inhibit ligand binding. The arrangement of the membrane-spanning segments is based on the projection structure of rhodopsin.[56] The accessible surface of TM6 of the dopamine D2 receptor at the level of the aligned Cys is shaded.[20] Adapted from Javitch *et al.*[51]

took place but was without functional effect). In contrast, in the CAM β_2 receptor, MTSEA inhibited antagonist binding significantly and isoproterenol slowed the rate of reaction of MTSEA.[51] This implies that at least one endogenous cysteine becomes accessible in the binding-site crevice of the CAM β_2 receptor.

We found that Cys-285, in the TM6, is responsible for the inhibitory effect of MTSEA on ligand binding to the CAM.[51] The acquired accessibility of Cys-285 in the CAM may result from a rotation and/or tilting of TM6 associated with activation of the receptor. This rearrangement could bring Cys-285 to the margin of the binding-site crevice where it becomes accessible to MTSEA (Fig. 6). Such a movement of TM6 upon receptor activation is consistent with the results of fluorescence spectroscopy studies in β_2 receptor[39,52] and spin-labeling studies in rhodopsin[53] and suggests that SCAM in a CAM background is a powerful approach for probing conformational change in these receptors.

Structural Bases of Pharmacological Specificity

Conserved features of the sequences of dopamine receptors and of homologous G protein-coupled receptors point to regions, and amino acid residues within these regions, that contribute to their ligand-binding sites. Differences in binding specificities among the catecholamine receptors, however, must stem from their nonconserved residues. Using SCAM, we have identified the residues that form the surface of the water-accessible binding-site crevice in the dopamine D2 receptor. Of ~90 transmembrane residues that differ between the D2 and D4 receptors, only 20 were found to be accessible, and 6 of these 20 are conservative aliphatic substitutions.

[51] J. A. Javitch, D. Fu, G. Liapakis, and J. Chen, *J. Biol. Chem.* **272,** 18546 (1997).

[52] U. Gether, S. Lin, and B. K. Kobilka, *J. Biol. Chem.* **270,** 28268 (1995).

[53] D. L. Farrens, C. Altenbach, K. Yang, W. L. Hubbell, and H. G. Khorana, *Science* **274,** 768 (1996).

We targeted these accessible residues not conserved in the homologous dopamine D4 receptor as candidates for the structural determinants of pharmacological specificity. We reasoned that mutation of these candidate residues in the D2 receptor to the aligned residues in the D4 receptor would generally be well tolerated, based on our previous experience in mutating them to cysteine. In a D2 receptor background, we mutated to the aligned residues in the D4 receptor, individually or in combinations, the 14 accessible, nonconserved residues.[54] We also made the reciprocal mutations in a D4 receptor background. The combined substitution of 4–6 of these residues was sufficient to switch the affinity of the receptors for several chemically distinct D4-selective antagonists by three orders of magnitude in both directions (D2 to D4-like and D4 to D2-like). The mutated residues are in TM2, TM3, and TM7 and form a cluster in the binding-site crevice. Mutation of a single residue in this cluster in TM2 was sufficient to increase the affinity for clozapine to D4-like levels. We can rationalize data in terms of a set of chemical moieties in the ligands interacting with a divergent aromatic microdomain in TM2-TM3-TM7 of the D2 and D4 receptors.

Note

The crystal structure of rhodopsin to 2.8 Å was recently published.[55] This remarkable achievement represents the first high-resolution structure of any GPCR. The impact of the structure is just beginning to be felt, but a preliminary analysis shows that the great majority of structural predictions that we and our collaborators have made based on our SCAM data and on molecular modeling based on the rhodopsin projection structure[56] are validated in the rhodopsin structure. In Fig. 7, to take an isolated example, is TM6 from rhodopsin, showing the aligned positions that we have found to be accessible by SCAM in the D2 receptor. The pattern is consistent with a narrow stripe of α helix at the cytoplasmic, followed more extracellularly by a broader accessible face that results from the proline kink-induced bending of the TM toward the binding-site crevice (compare with Fig. 4).

Sequence analysis of conserved residues and motifs suggests a number of potential structural differences between rhodopsin and biogenic amine receptors. The exact extent of structural similarity between various GPCRs remains to be determined from further analysis and from new high-resolution structures of other GPCRs, but our initial impression from comparison of our SCAM data with the rhodopsin structure is that the similarity in the backbone is remarkably high, even

[54] M. M. Simpson, J. A. Ballesteros, V. Chiappa, J. Chen, M. Suehiro, D. S. Hartman, T. Godel, L. A. Snyder, T. P. Sakmar, and J. A. Javitch, *Mol. Pharmacol.* **56,** 1116 (1999).

[55] K. Palczewski, T. Kumasaka, T. Hori, C. A. Behnke, H. Motoshima, B. A. Fox, I. Le Trong, D. C. Teller, T. Okada, R. E. Stenkamp, M. Yamamoto, and M. Miyano, *Science* **289,** 739 (2000).

[56] G. F. Schertler, C. Villa, and R. Henderson, *Nature* **362,** 770 (1993).

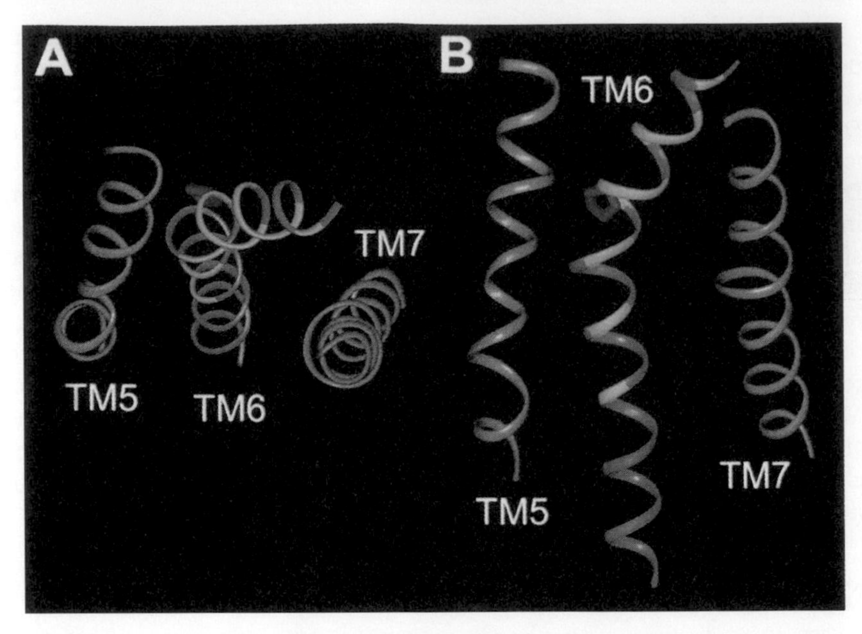

FIG. 7. An extracellular view (A) and a side view from the binding-site crevice (B) of the backbone ribbons from the crystal structure of rhodopsin[55] showing TM6 in the context of the adjacent TM5 and TM7. In TM6, the backbone is shaded cyan at the positions where the aligned residue in the dopamine D2 receptor was found to be accessible in the binding-site crevice. Residues for which the aligned residue was not accessible in the D2 receptor are shaded orange. The side chain of the highly conserved Pro is shown in purple.

in a number of cases where a discrepancy might have been expected based on differences in critical conserved residues.[57]

Acknowledgments

We thank our collaborators Juan Ballesteros and Harel Weinstein for their critical insights into protein structure and molecular modeling of GPCRs. We thank Myles Akabas and Arthur Karlin, our colleagues in the Center for Molecular Recognition, for much helpful discussion and advice. I (J.A.J.) thank the current and previous members of my laboratory for their hard work and important contributions to the work described herein: Wai Chi Chan, Jiayun Chen, Xun Chen, Victor Chiappa, Jasmine Ferrer, Dingyi Fu, Joshua Kaback, Xiochuan Li, Stavros Papachristou, and Merrill Simpson. This work was supported in part by NIH Grants MH01030 and MH54137, by the G. Harold & Leila Y. Mathers Charitable Trust, by the Lebovitz Trust, and by the Lieber Foundation.

[57] J. A. Ballesteros, L. Shi, and J. A. Javitch, *Mol. Pharmacol.* **60,** 1 (2001).

[9] Mass Spectrometric Analysis of G Protein-Coupled Receptors

By DANIEL R. KNAPP, ROSALIE K. CROUCH, LAUREN E. BALL, ANDREW K. GELASCO, and ZSOLT ABLONCZY

Introduction

G protein-coupled receptors (GPCRs) present a challenge for mass spectrometric analysis. GPCRs, as with integral membrane proteins in general, are more difficult to handle in terms of protein chemistry and separations than soluble proteins. Depending on the purpose of the mass spectrometric analysis, the difficulty of analysis can vary greatly. If the need is only to observe a few fragments for identification or to observe modifications on extramembranous domains, appropriate cleavage methods can yield fragment peptides that are as well behaved as fragments from soluble proteins.[1,2] However, if the need is to observe the intact protein or to observe the entire sequence as peptide fragments, e.g., for identification of sites of modification without prior knowledge of their location, GPCRs can present a formidable challenge. Even when the entire sequence of a GPCR can be observed, either as the intact protein or as fragments, the sensitivity of the analysis in terms of required sample amount is usually poorer than for a comparable size soluble protein. Despite these difficulties, GPCRs have yielded to mass spectrometric analysis. This article presents a method for mass spectrometric analysis that permits observation of the entire protein in the case of the prototypical GPCR rhodopsin.[3–5]

Materials and Sources

Rhodopsin in the form of rod outer segment membranes is prepared from commercially obtained bovine retinae (Lawson Co., Lincoln, NE) using the method of McDowell and Kuhn.[6] Reagents are obtained from Sigma/Aldrich (St. Louis, MO) and solvents from Fisher Scientific (Suwanee, GA). Brownlee Aquapore HPLC columns are from Bodman Industries (Aston, PA). Mass spectrometry is done

[1] D. Papac, J. Oatis, R. Crouch, and D. Knapp, *Biochemistry* **32,** 5930 (1993).

[2] M. Roos, V. Soskic, S. Poznanovic, and J. Godovac-Zimmermann, *J. Biol. Chem.* **273,** 924 (1998).

[3] L. E. Ball, J. J. E. Oatis, K. Dharmasiri, M. Busman, J. Wang, L. B. Cowden, A. Galijatovic, N. Chen, R. K. Crouch, and D. R. Knapp, *Protein Sci.* **7,** 758 (1998).

[4] A. Gelasco, R. K. Crouch, and D. R. Knapp, *Biochemistry* **39,** 4907 (2000).

[5] Z. Ablonczy, D. R. Knapp, R. Darrow, D. T. Organisciak, and R. K. Crouch, *Mol. Vis.* **6,** 109 (2000). <http://www.molvis.org/molvis/v6/a15>.

[6] J. H. McDowell and H. Kuhn, *Biochemistry* **16,** 4054 (1977).

0076-6879/02 $35.00

on a Finnigan LCQ instrument (Thermoquest Inc., San Jose, CA) interfaced to a Hewlett Packard 1100 HPLC (Agilent Technologies, Palo Alto, CA).

Analysis Procedure

A membrane sample containing 0.25 mg. rhodopsin (ca. 6 nmol.) is suspended in 200 μl of 1 : 1 (v/v) 1.5 *M* Tris, pH 8.9/1-propanol, reduced, and alkylated by the addition of 5 μl tributylphosphine and 5 μl 4-vinylpyridine. After 1 hr, the sample is delipidated by the addition of 500 μl 95% ethanol and 300 μl hexane. The precipitated protein is isolated by centrifugation, washed with hexane, and dried under vacuum. The protein is dissolved in 400 μl trifluoroacetic acid (TFA) followed by the addition of 180 μl water and 5 μl of a 5 *M* solution of cyanogen bromide in acetonitrile. The resulting solution is allowed to stand at ambient temperature for 18 hr in the dark. The solution is then evaporated to dryness, and the residue is dissolved in 5 μl TFA. Sequential addition of 30 μl acetonitrile, 60 μl 2-propanol, and 4.9 ml water results in 5 ml of sample solution of the same composition as the beginning of the HPLC solvent gradient. This sample is loaded onto a 2.1 × 100-mm C4, C8, or C18 reversed-phase HPLC column equilibrated with 98% solvent A (0.05% aqueous TFA) and 2% solvent B [0.05% TFA in 2 : 1 (v/v) 2-propanol/acetonitrile] at 400 μl/min for 14 min using 2% B to flush the injection loop. The injector is returned to "load" position, the flow is reduced to 200 μl/min, and 2% B is continued for 1 min prior to starting the solvent elution gradient of 2–60% B in 60 min and 60–98% B in 20 min at a flow of 200 μl/min. The HPLC eluant is split, with 90% collected as fractions and 10% directed to the electrospray ionization source of the Finnigan LCQ mass spectrometer. Data are acquired using the following parameters: ESI needle, 4.5 kV; ESI capillary temperature, 200°; ion energy, 45%; isolation window, 2 amu; and scan range, 400–2000 amu. MS data are acquired with repetitive scanning with MS/MS data automatically acquired on the most intense precursor ion in each MS spectrum.

Comments

Two aspects of the just-described procedure are key to the success of the method. First is the reduction and alkylation of cysteine sulfhydryl groups prior to removal of the protein from the membrane lipids. Our prior efforts involving delipidation as the first step in sample preparation were thwarted by irreversible aggregation of the protein, which we suspected might involve the formation of nonnative disulfide linkages. Although there was no proof of such bond formation, use of reduction/alkylation as the first step eliminated the aggregation problem. The second key to success was the method of loading the sample onto the reversed-phase HPLC column. The cyanogen bromide cleavage fragment mixture was loaded onto the column in a solution with the same solvent composition as the initial mobile phase in the gradient (i.e., 2% B). In order to dissolve the fragments in this solvent,

TABLE I
CYANOGEN BROMIDE FRAGMENTS OF BOVINE RHODOPSIN

Fragment	Residues	Mass expected as $(M+H)^{+1}$	Mass observed	Charge state observed	HPLC retention times (min)		
					C4	C8	C18
1	1	144.1	N/D	—	—	—	—
2[a]	2–39	6501.8[b]	6515.7	+1	47.6	48.3	47.8
3	40–44	520.3	520.1	+1	[c]	[c]	25.0
4	45–49	588.4	588.4	+1	42.0	42.1	42.2
5	50–86	4244.1	4243.9	+3	63.5	64.3	64.8
6	67–143	6357.5	6357.5	+4	72.2	72.6	75.0
7	144–155	1374.7	1374.7	+2	32.8	33.8	34.3
8	156–163	862.5	862.5	+1	41.0	41.8	42.1
9	164–183	2160.6	2160.4	+2	43.8	44.7	45.3
10[d]	184–207	3007.3	3007.0	+2	39.8	41.3	40.3
11	208–253	5319.2	5318.5	+3	62.5	63.5	65.6
12	254–257	427.3	427.1	+1	[c]	[c]	26.8
13	258–288	3583.2	3583.0	+2	64.3	65.1	68.0
14	289–308	2199.6	2199.6	+2	50.2	51.1	51.0
15	309	102.1	N/D[e]	—	—	—	—
16	310–317	1097.5	1098.0	+1	18.8	21.3	24.0
17	318–348	3602.3	3602.2	+2	76.3	78.8	89.6

[a] Observed by MALDI-TOF mass spectrometry.
[b] Most abundant glycoform.
[c] Not significantly retained; eluted in or near the injection solvent peak.
[d] Fragment 10 was also observed as +2 ion of the N-terminal pyroglutamate form in approximately equal abundance (retention times: C4, 40.8; C8, 42.6; C18, 41.0).
[e] Fragments that were not detected.

it was necessary to first dissolve them in neat TFA and then successively add the organic and aqueous components. Loading the sample onto the HPLC column in this manner avoids precipitation and aggregation of the peptide fragments at the head of the column, which occurs when such a sample is injected in a strongly solubilizing solvent into the predominantly aqueous solvent at the beginning of a solvent gradient.

Table I[3] summarizes the predicted cyanogen bromide cleavage fragments from rhodopsin along with the observed fragment masses and their HPLC retention behavior on the three columns examined. Figure 1[3] shows chromatograms for the three columns using UV detection at 214 nm. Figure 2[3] shows single ion plots for the molecular ion species detected by mass spectrometry, illustrating that the peptide fragments generally elute as single peaks, although some of the fragments (particularly fragment 6) show some splitting of the chromatographic peak. With the exception of two single residue fragments, these data cover the complete sequence of rhodopsin and as such represents the first complete mass spectrometric

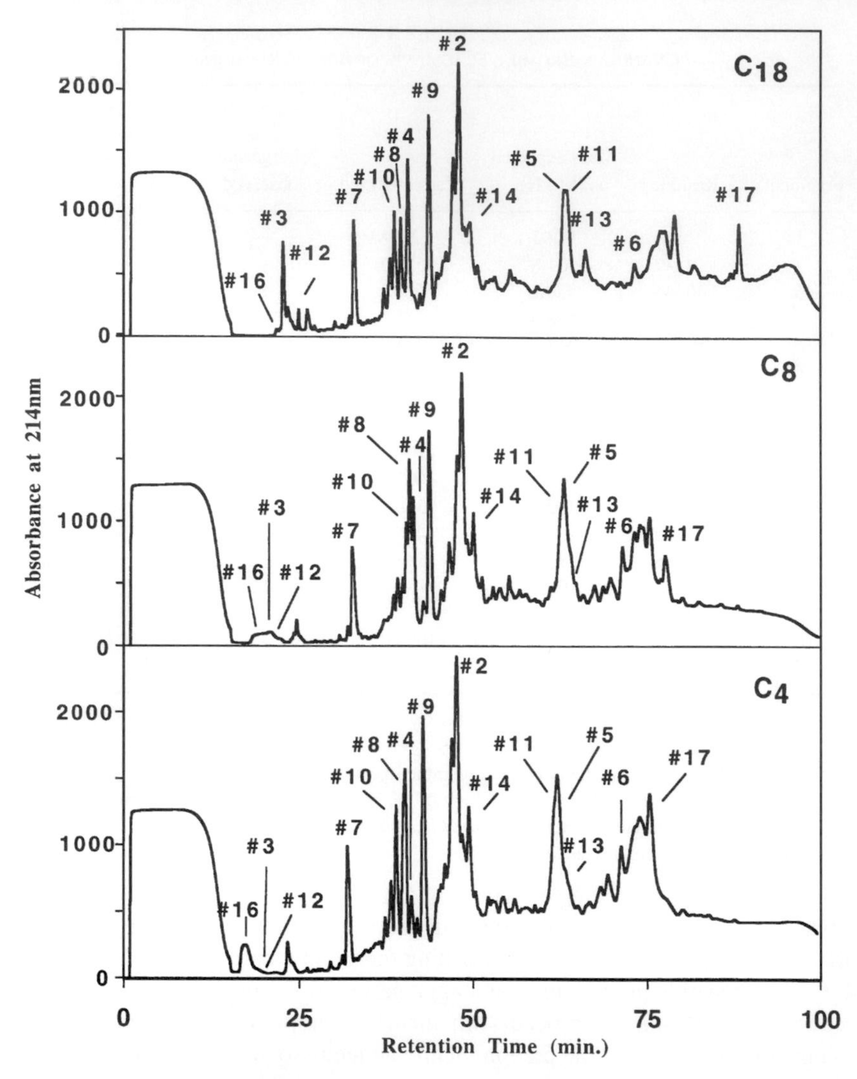

FIG. 1. Chromatograms of HPLC separations of CNBr fragments of rhodopsin on C4, C8, and C18 columns with UV detection at 214 nm. See text for column details and gradient conditions.

mapping of a GPCR.[3] This method has been employed in two subsequent studies[4,5] with minor improvements, which are included in the procedure given in this article. The method has also been applied to the recombinant β-adrenergic receptor, where first attempts yielded coverage of 46% of the receptor sequence.[7] Suboptimal location of native methionines in a receptor sequence will require subsequent secondary cleavage to yield fragments of tractable size for application of the method.

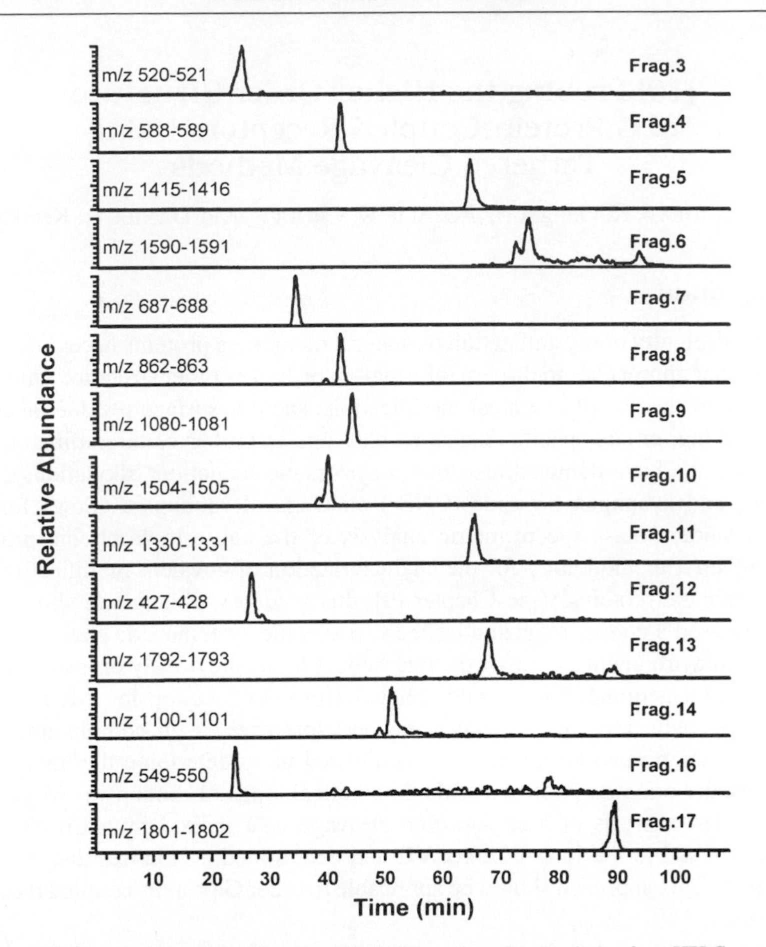

FIG. 2. Selected ion chromatograms from ESI-MS detection of peaks eluting from HPLC separation of CNBr fragments of rhodopsin on a C18 column.

Nonetheless, the method should be applicable to mapping and characterization of covalent modifications on other GPCR proteins.

Acknowledgment

This work was supported by NIH Grants EY08239 (D.R.K.) and EY 04939 (R.K.C.).

[7] K. Dharmasiri, G. M. King, and D. R. Knapp, Proc. 46th ASMS Conf. Mass Spectrom. Allied Topics, p. 1342. Orlando, FL, 1998.

[10] Probing the Higher Order Structure of G Protein-Coupled Receptors Using Tethered Cleavage Methods

By ANDREW K. GELASCO, ROSALIE K. CROUCH, and DANIEL R. KNAPP

Background

The difficulty of crystallization of integral membrane proteins necessitates the use of other approaches to discern information on higher order structure. One such approach is the use of chemical modifications such as surface residue labeling, cross-linking, or site-specific backbone cleavage as probes of three-dimensional structure. We have demonstrated that a systematic reduction, alkylation, delipidation, and cyanogen bromide (CNBr) cleavage of rhodopsin allows for the HPLC–tandem mass spectrometric analysis of the entire hydrophobic protein, opening up a methodology for the characterization of covalent modifications of the membrane proteins[1] (see Chapter [9], this volume). This article shows how we have used a copper reagent attached to a specific cysteine side chain to probe the local environment of one of the interhelical loops previously proposed to undergo conformational changes upon photoactivation of rhodopsin.[2] Modification of a previously developed chemical nuclease/protease,[3–5] in combination with high-resolution mass spectrometry, has allowed us to determine the relative intramolecular orientation of two of the interhelical loops of rhodopsin. We carried out detailed analysis of a site-specific cleavage near helix I by a Cu cleavage agent attached to the loop of helix VII and determined an interresidue distance of ≈10 Å. This approach should be applicable to other G protein-coupled receptor systems.

Materials

All reagents are obtained from Aldrich/Sigma Co. and are used as received without further purification. The phenanthroline derivative, 5-(α-bromoacetamido)-1,10-phenanthroline (OP), is prepared from 5-nitro-1,10-phenanthroline

[1] L. E. Ball, J. E. Oatis, K. Dharmasiri, M. Busman, J. Wang, L. Cowden, A. Galijatovic, N. Chen, R. K. Crouch, and D. R. Knapp, *Protein Sci.* **7,** 758 (1998).

[2] A. Gelasco, R. K. Crouch, and D. R. Knapp, *Biochemistry* **39,** 4907 (2000).

[3] D. S. Sigman, A. Mazumder, and D. A. Perrin, *Chem. Rev.* **93,** 2295 (1993).

[4] D. S. Sigman, M. D. Kuwabara, C.-H. B. Chen, and T. W. Bruice, *Methods Enzymol.* **208,** 414 (1993).

[5] J. Wu, D. M. Perrin, D. S. Sigman, and H. R. Kaback, *Proc. Natl. Acad. Sci. U.S.A.* **92,** 9186 (1995).

0076-6879/02 $35.00

according to a previously published procedure,[6] recrystallized, and characterized by ^{1}H NMR and ESI-MS.

Trypsin, modified sequencing grade, is prepared by dissolving the lyophilized powder as received from Boehringer Mannheim in 1 m*M* HCl at 0.1 μg/μl concentration. *N*-Glycosidase F (PNGase F) is used as obtained from Boehringer Mannheim, 25,000 units/mg in 50 m*M* sodium phosphate, 12.5 m*M* EDTA, 50% glycerol, pH 7.2.

Rod outer segments are prepared from intact bovine retina (Lawson Co., Lincoln, NE) as described previously.[7] Membrane preparations are purified further with a 5 *M* urea wash to remove rhodopsin-associated proteins,[8] leaving a suspension in Tris buffer of nearly pure membrane-bound rhodopsin. Suspensions of the intact membrane preparations are stored at −80° in the dark. All subsequent manipulations are conducted under dim red light or in total darkness until cleavage reagents are removed from the suspension or reactions quenched.

Methods

Cu-Based Cleavage of Intact Rhodopsin

Urea-washed membrane preparations (typically 0.25 mg, 6 nmol rhodopsin) are resuspended in a KP_I buffer (50 m*M* potassium phosphate, 150 m*M* NaCl, pH 7.4) to a concentration of 2 mg/ml. The reagent OP is dissolved in DMF and added to the rhodopsin suspension at 10-fold molar excess, maintaining a total DMF concentration less than 10% of the solution volume. The reaction is allowed to proceed at 4° for 8 hr. OP-modified rhodopsin pellets are isolated by centrifugation (80,000*g*, 20 min, and 4°), washed once with, and resuspended in the KP_I buffer. $CuSO_4$ is added (1.2 equiv./mol of ligand), and the suspension is incubated at room temperature for 15 min. Sodium ascorbate (adjusted to pH 6.5) is added to a final concentration of 100 m*M*, and the mixture is incubated at 37° for 30 min. The cleavage reaction is quenched by the addition of 5 m*M* neocuproine, a quantitative ligand for Cu chelation, the cleaved rhodopsin is isolated by centrifugation, and the pellet is washed three times with dH_2O.

MALDI Mass Spectrometry

Matrix-assisted laser desorption ionization (MALDI) mass spectrometry (MS) is conducted on a Perseptive Biosystems Voyager-DE instrument. Samples of the protein cleavage products are prepared for MALDI-MS by a modification of

[6] D. M. Perrin, A. Mazumder, F. Sadeghi, and D. S. Sigman, *Biochemistry* **33,** 3848 (1994).

[7] J. H. McDowell and H. Kuhn, *Biochemistry* **16,** 4054 (1977).

[8] H. Shichi and R. L. Somers, *J. Biol. Chem.* **253,** 7040 (1978).

published methods.[2,9] The sample, dissolved in formic acid : H_2O : hexafluoroisopropanol (7 : 3 : 2), is diluted 3 : 1 with 50 m*M* sinapinic acid in 70% acetonitrile, 0.1% trifluoroacetic acid (TFA). The MALDI sample is prepared by the addition of 0.4 μl of the matrix solution to the plate that is allowed to dry, and then the 0.4 μl sample/matrix mixture is placed on the dried matrix surface. We have found that this preparation gives significant improvement over the previously published procedure[9] in signal/noise and, to a lesser extent, improvement in peak resolution. Samples of peptide fragments are prepared by dilution in 50 m*M* α-cyano-4-hydroxycinnamic acid in 50% MeOH, 0.1% TFA, which is applied directly to the MALDI sample plate.

Preparation for HPLC–Electrospray Ionization (ESI)–Mass Spectrometry

Membrane-bound fragments are reduced with tributylphosphine and alkylated with 4-vinylpyridine as described previously.[1] A membrane sample containing 0.25 mg rhodopsin (ca. 6 nmol) is suspended in 200 ml of 1 : 1 (v/v) 1.5 *M* Tris, pH 8.9/1-propanol and is reduced and alkylated by the addition of 5 ml tributylphosphine and 5 ml 4-vinylpyridine. After 1 hr, the sample is delipidated by the addition of 500 ml 95% ethanol and 300 ml hexane. The precipitated protein is isolated by centrifugation, washed with hexane, and dried under vacuum. The dried pellet is then subjected to either a trypsin (1.5 μg trypsin in HCl in 25 m*M* NH_4HCO_3, 0.1% *n*-octyl glucoside, pH 8) or a CNBr (in 70% TFA) digest. For the tryptic digest, the supernatant is analyzed by HPLC-MS, as described later. For CNBr cleavage, the protein is dissolved in 400 μl TFA followed by the addition of 180 ml water and 5 μl of a 5 *M* solution of cyanogen bromide in acetonitrile. The resulting solution is allowed to stand at ambient temperature for 18 hr in the dark. The solution is then evaporated to dryness, and the residue is dissolved in 5 ml TFA. The sequential addition of 30 ml acetonitrile, 60 ml 2-propanol, and 4.9 ml water results in 5 ml of sample solution of the same composition as the beginning of the HPLC solvent gradient. This sample is loaded onto a 2.1 × 100-mm C4, C8, or C18 reversed-phase HPLC column equilibrated with 98% solvent A (0.05% aqueous TFA) and 2% solvent B [0.05% TFA in 2 : 1 (v/v) 2-propanol/acetonitrile] at 400 ml/min for 14 min using 2% B to flush the injection loop. The injector is returned to "load" position, the flow is reduced to 200 ml/min, and 2% B is continued for 1 min prior to starting the solvent elution gradient of 2–60% B in 60 min and 60–98% B in 20 min at a flow of 200 ml/min. The HPLC eluant is split with 90% collected as fractions and 10% directed to the electrospray ionization source of the Finnigan LCQ mass spectrometer. Data are acquired using the following parameters: ESI needle, 4.5 kV; ESI capillary temperature, 200°; ion energy, 45%; isolation window, 2 amu; scan range, 400–2000 amu. MS data are acquired with

[9] K. L. Schey, D. I. Papac, D. R. Knapp, and R. K. Crouch, *Biophys. J.* **63** (1992).

repetitive scanning with MS/MS data automatically acquired on the most intense precursor ion in each MS spectrum.

Results

Labeling Cys-316 with 5-(α-Bromoacetamido)-1,10-phenanthroline

We[2] and others[10] have shown that when rhodopsin membranes were incubated at 4°, sulfhydryl modification is limited primarily to Cys-316. After labeling with the OP ligand, the modified uncleaved protein was characterized by both MALDI-MS and HPLC-ESI-MS/MS. MALDI-MS has been demonstrated to be applicable to characterization of the integral membrane proteins (see Chapter 9, this volume), particularly to rhodopsin.[9] The two different sites of glycosylation near the N terminus of rhodopsin at N2 and N15 result in a MALDI mass spectrum containing a broad peak that represents the major glycosylated peptide and a series of shoulders due to the presence of other (smaller) glycosyl side chains. This broad peak shape prevented the use of MALDI-MS as a conclusive method of identifying ligand attachment. However, comparison of MALDI mass spectra of unmodified rhodopsin and ligand-modified rhodopsin, run under identical conditions, shows a mass shift of *m/z* 200. When the rhodopsin/OP ligand reaction was carried out at 23° instead of 4°, the broad peak representing the intact protein was broadened further, appearing as two shoulders, with its average mass value increased *m/z* 500–600 over the unmodified protein. This differential reactivity is consistent with that predicted for the cytoplasmic sulfhydryls Cys-140 and Cys-316.[10,11]

Confirmation of both the extent and the position of modification by OP was obtained through a combination of HPLC-MS/MS and MALDI-MS on CNBr-cleaved, OP-modified rhodopsin. MALDI mass spectra of the peptide mixture resulting from CNBr cleavage of both unmodified and OP-modified rhodopsin indicated that 90–100% of Cys-316 was converted to the 1,10-phenanthroline derivative (Cys-316OP) and that 80–90% of Cys-140 was converted to the 4-vinylpyridyl derivative. HPLC-ESI-MS/MS confirmed the observation of nearly quantitative conversion of Cys-316 to Cys-316OP and confirmed the position of modification through the fragmentation pattern of the CNBr peptide N310-M317.

Single-Site Cu/O_2 Cleavage of Rhodopsin

Following Cu-OP-O_2 cleavage the products were isolated by centrifugation, washed, and subjected to MALDI-MS analysis. Figure 1 shows the MALDI spectrum of Cys-316OP rhodopsin both before (Fig. 1A) and after (Fig. 1B) the

[10] A. D. Albert, A. Watts, P. Spooner, G. Groebner, J. Young, and P. L. Yeagle, *Biochim. Biophys. Acta* **1328,** 74 (1997).

[11] Y. S. Chen and W. L. Hubbell, *Membr. Biochem.* **1,** 107 (1978).

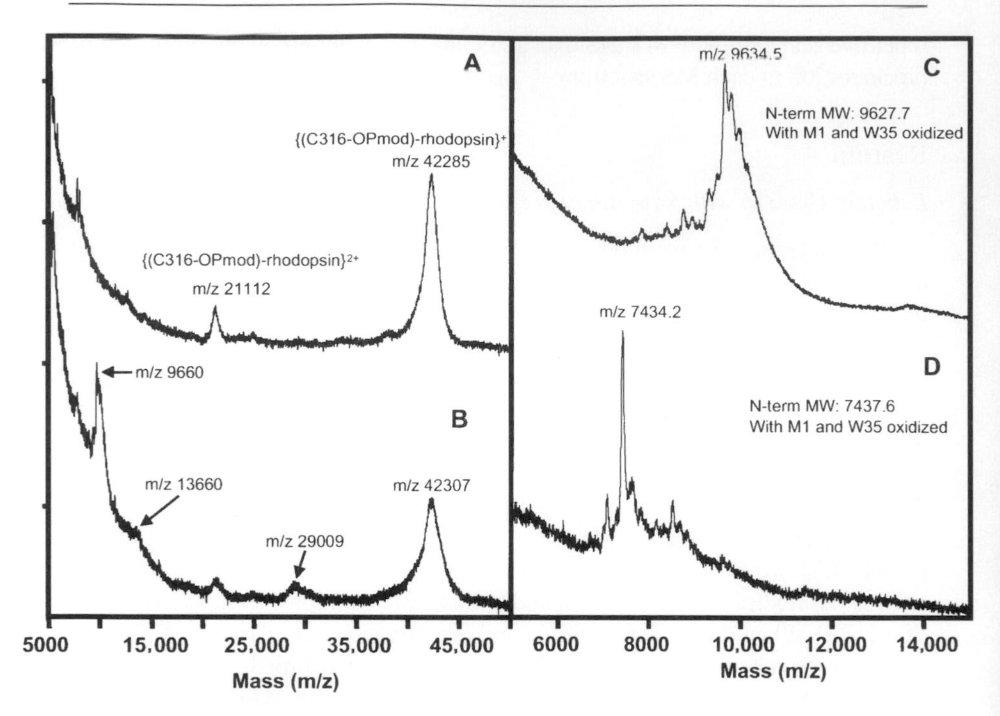

FIG. 1. MALDI-MS of Cys-316OP rhodopsin before (A) and (B) after the cleavage reaction, and MALDI-MS of the major cleavage product before (C) and (D) after treatment with *N*-glycosylase F.

cleavage reaction. Controls using unmodified Cys-316 rhodopsin incubated with Cu and ascorbate and Cys-316OP rhodopsin incubated in ascorbate in the absence of Cu were analyzed, and MALDI spectra were indistinguishable before and after incubation. Figure 1B shows that there are two major products detected by MALDI after the cleavage reaction: intact rhodopsin and the major peak at *m/z* 9640. Broad peaks having masses between 10 and 30 kDa were observed under various conditions, but none had the intensity of either *m/z* 9640 or *m/z* 42,307 peaks.

The shape of the *m/z* 9640 peak suggested that the corresponding fragment peptide contained the N terminus of the protein and was broadened due to inhomogeneous glycosylation. By using the most common glycosylation oligosaccharide modification of this N terminus peptide, the Cu-OP-O_2 cleavage site can be predicted using the known peptide sequence for rhodopsin. Three possible candidate sequences were predicted: M1-V63 (9468.4 Da), M1-Q64 (9596.7 Da), and M1-H65 (9733.8 Da). Reaction of the cleavage products with *N*-glycosidase F, which specifically removes N-linked glycosyl groups, allowed a more specific analysis of the position of cleavage. Analysis by MALDI of the postreaction products showed a shift of the peak observed at *m/z* 9634.5 to *m/z* 7434.2 determined after internal calibration with insulin (*m/z* 5734.6) and thioredoxin (*m/z* 11674.5) (shown in Figs. 1C and 1D). The resulting mass shift is consistent with the loss of the two N-linked glycosyl groups on N2 and N15 and conversion of the asparagine to aspartic acid at these positions. By using the observed mass of the deglycosylated fragment, we predicted that the major cleavage product was the M1-Q64 peptide. While close to the predicted mass, the *m/z* 7434 value observed was 29 Da higher than that calculated for this peptide.

In order to understand the nature of this mass differential, we investigated through LC-MS the possibility of oxidation of M1, which can be oxidized to the sulfoxide or the sulfone, and W35, which can be oxidized at the 2 position on the indole ring. The Cu/O_2-cleaved, deglycosylated rhodopsin mixture was delipidated, pelleted, washed, resuspended in NH_4HCO_3 buffer, and subjected to trypsin cleavage. The reaction was quenched by the addition of TFA to 10% (v/v), the solvents were removed *in vacuo,* and the resulting material was dissolved in the initial mobile phase of the gradient and analyzed by HPLC-ESI-MS. The tryptic peptide M1-K16 was detected in both the native state and in the singly oxidized form separated by 2 min by HPLC. The M1-sulfoxide containing peptide was found in a 2 : 1 excess over the unmodified peptide. No sulfone-containing peptide was detected under the conditions of the experiment. The detection of oxidation at W35 was hindered by the large size of the peptide containing the oxidized tryptophan but could be determined, albeit with less certainty. Assuming oxidation at these two sites results in a peptide of *m/z* 7437.6, within 3.5 Da (0.05% error) of the MALDI peak detected.

Identification of the Specific Cu/O_2 Cleavage Site

The cleavage site was confirmed by HPLC-ESI-MS/MS of the CNBr cleavage products of the Cu/O_2-cleaved protein. The protein was delipidated as described for the trypsin digest and was then subjected to CNBr digestion. CNBr cleavage of rhodopsin yields 17 peptides (see Chapter [9]). Cu/O_2 cleavage at Q64 would result in a cut in the large *m/z* 4242 fragment 5. In order to minimize recovery losses, the reduction and alkylation step prior to delipidation was omitted. The resulting HPLC-MS analysis showed a significant loss of the intact *m/z* 4242 CNBr peptide 5 (L50-M86) and the appearance of two peptides corresponding in mass to two cleaved products of the 36 amino acid fragment. Selected ion chromatograms for the CNBr peptides of the N-terminus cleavage product are shown in Fig. 2. Quantitation of fragment 5, using fragments 3 and 4 from the N terminus and fragment 14 from helix VII as internal standards, indicates a loss of over 95%. Figures 2D and 2F show selected ion chromatograms and corresponding mass spectra for the two fragments resulting from the Cu/O_2 cleavage of the *m/z* 4242 peptide. The N-terminal side of the fragment peptide L50-Q64 was detected

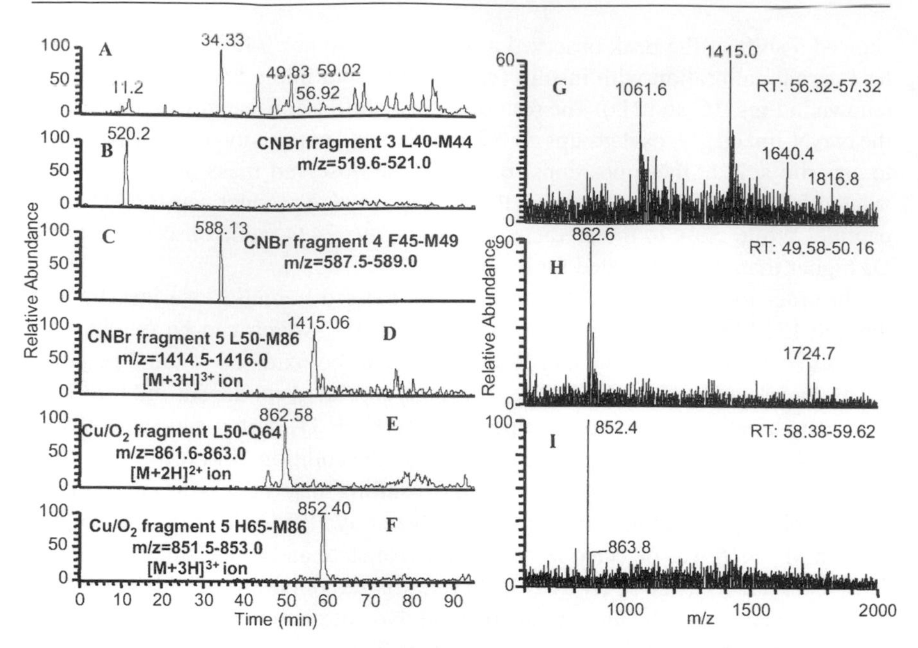

FIG. 2. HPLC-ESI-MS data for the analysis of CNBr fragments of the Cys-316OP Cu/O_2/ascorbate cleavage products of rhodopsin. Total ion current chromatogram (A) and selected ion current chromatograms for CNBr fragments: (B) fragment 3, L40-M44; (C) fragment 4, F45-M49; (D) fragment 5, L50-M86; (E) oxidative cleavage fragment L50-Q64; and (F) oxidative cleavage fragment H65-M86. Mass spectra shown in (G), (H), and (I) are from the peptide peaks found in (D), (E), and (F), respectively.

easily by ESI-MS as both the singly and the doubly charged peptide containing a C-terminal amide (Fig. 2H) with *m/z* values of 862.58 and 1724.7, respectively. This result was consistent with a previously proposed mechanism developed for the cleavage of peptide bonds by the Cu/O_2 phenanthroline reagent.[12,13]

The amide transfer results in only a 1 amu mass shift in the resulting peptide. This shift is observed in both the MS and the MS/MS patterns for the peptide. The nanospray tandem MS of the N-terminal peptide confirmed its identity through the peptide fragmentation pattern. Figure 3 shows a mechanism for the Cu-OP-O_2 cleavage of rhodopsin based on the mass spectrometric evidence observed in these experiments and previously proposed mechanisms based on model peptides. The C-terminal end of the cleaved CNBr peptide (H65-M86) has a predicted mass of 2537.04 for a peptide containing the homoserine lactone C terminus and the

[12] J. Bateman, C. Robert, W. W. Youngblood, J. W. H. Busby, and J. S. Kizer, *J. Biol. Chem.* **260,** 9088 (1985).

[13] A. F. Bradbary and D. G. Smith, *Trends Biochem. Sci.* **16,** 112 (1991).

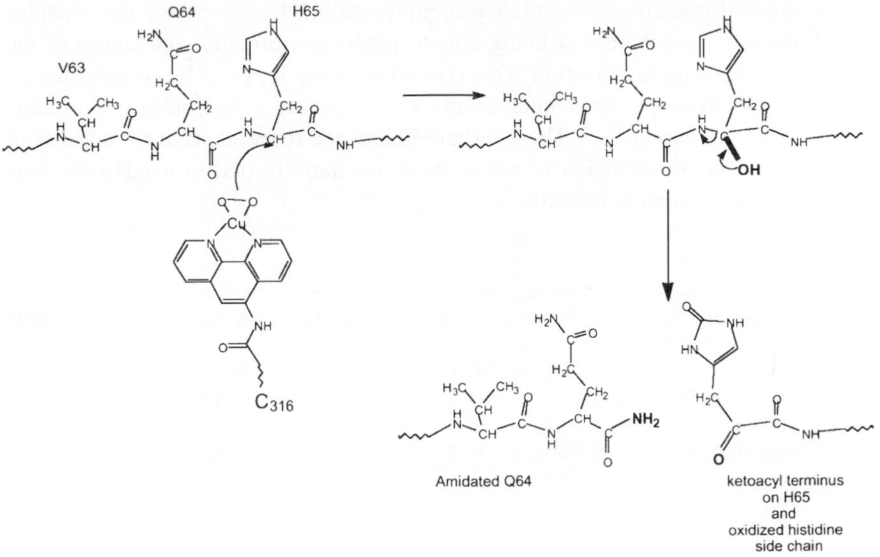

FIG. 3. Proposed mechanism of the copper-mediated oxidative cleavage of the rhodopsin backbone.

ketoacyl group on the N terminus resulting from amide donation to Q64. This peptide should have masses corresponding to the $[M+2H]^{2+}$ and $[M+3H]^{3+}$ ions of 1269.03 and 846.65, respectively. However, we detected a peak eluting 10 min after the L50-Q64 peptide having a *m/z* of 852.38. This peak mass corresponds to that of the $[M+3H]^{3+}$ ion for an oxidation product of the parent peptide or the hydrolyzed lactone C terminus. Unlike the N terminus fragment, this one did not lend itself to MS/MS analysis and additional information regarding this modification was not obtained. Comparison to work with model peptides [14] suggests that oxidation of the N-terminal histidine residue side chain of the H65-M86 cleavage fragment to the 2-oxo derivative is the probable cause of the observed mass shift for the fragment ion.

This work with rhodopsin represents the first use of the tethered chemical protease technique for a G protein-coupled receptor. Previously, it had been demonstrated to work by gel electrophoretic analysis to give a reproducible cleavage pattern in lactose permease[5] and provide information regarding helix packing in this membrane protein. The use of mass spectrometric analysis of the cleavage products has greatly enhanced applicability of the technique because specific cleavage sites, patterns, and mechanisms can be investigated with a high level of precision. In the case of G protein-coupled receptors, this kind of reagent has the potential

[14] M. Khossravi and R. T. Borchardt, *Pharm. Res.* **15,** 1096 (1998).

to be applied toward investigating protein–protein interactions for the receptor itself and has the potential of being able to probe conformational changes of the protein surfaces upon activation. This type of methodology can be complemented by other spectroscopic techniques, such as site-directed spin labeling of proteins demonstrated recently.[15–19] These tethered cleavage reagents have the additional advantage of being applied in the native state, and therefore can be used to examine receptor native state conformations.

[15] Z. Farahbakhsh, K. Ridge, H. Khorana, and W. Hubbell, *Biochemistry* **34,** 8812 (1995).
[16] C. Altenbach, K. Yang, D. L. Farrens, Z. T. Farahbakhsh, H. G. Khorana, and W. L. Hubbell, *Biochemistry* **35,** 12470 (1996).
[17] W. L. Hubbell, A. Gross, R. Langen, and M. A. Lietzow, *Curr. Opin. Struct. Biol.* **8,** 649 (1998).
[18] C. Altenbach, J. Klein-Seetharaman, J. Hwa, H. G. Khorana, and W. L. Hubbell, *Biochemistry* **38,** 7945 (1999).
[19] C. Altenbach, K. Cai, H. G. Khorana, and W. L. Hubbell, *Biochemistry* **38,** 7931 (1999).

[11] Use of Fluorescence Spectroscopy to Study Conformational Changes in the β_2-Adrenoceptor

By BRIAN K. KOBILKA and ULRIK GETHER

Introduction

Transmembrane signaling by G protein-coupled receptors (GPCRs) is the result of ligand-induced changes in the structure of the receptor. These ligand-induced conformational changes lead to changes in interactions between receptors and G proteins. The nature of these conformational changes is poorly understood. Models describing GPCR function have been formulated by inferring the conformational state of GPCRs from their effects on G protein function (GTPase activity or GTPγS binding) or second messenger activation. Studies that directly monitor ligand-induced conformational changes have been limited because of the difficulty of producing and purifying sufficient quantities of functional receptor for biophysical studies. This has been accomplished for rhodopsin and the β_2-adrenergic receptor (β_2AR) (reviewed in Refs. 1, 2). This article describes the general approach for studying ligand-induced conformational changes in wild-type and mutant β_2AR using fluorescence spectroscopy.

[1] U. Gether and B. K. Kobilka, *J. Biol. Chem.* **273,** 17979 (1998).
[2] U. Gether, *Endocr. Rev.* **21,** 90 (2000).

0076-6879/02 $35.00

Production and Purification of β_2AR

Experiments using standard fluorescence spectroscopy to monitor conformational changes in β_2AR labeled with an extrinsic fluorescent probe require 25–1000 pmol of purified receptor per sample depending on the quantum yield of the fluorophore. The number of ligands to be studied and the number of replicates desired determine the total amount of receptor needed for a set of experiments. We typically purify from 2 to 10 nmol of receptor for one set of fluorescence experiments. As much as 50% of the receptor may be lost during the process of labeling the receptor with the fluorescent probe. Therefore, our final yield of labeled receptor is ~1–5 nmol. Considering losses during the various stages of purification, we prefer to start with a cell pellet expressing 10–20 nmol.

Production

β_2AR is produced in insect cells using the baculovirus expression system. We can routinely produce 15–20 nmol of wild-type β_2AR in 1 liter of insect cell culture. Factors that determine the yield of receptor include the quality of the cells and the titer of the virus stock. We use Sf9 cells (obtained from Invitrogen) and have found that these cells retain their ability to produce β_2AR for many passages, up to 6 months of continuous culture. However, we generally start a low-passage cell stock every month.

Preparation of Baculovirus Stocks. SF-hβ_2-6H and mutant DNA constructs are cloned into the baculovirus expression vector pVL1392 and cotransfected with linerarized BaculoGold DNA into Sf9 insect cells using the BaculoGold transfection kit (Pharmingen, San Diego, CA). The resulting virus is harvested after 4–5 days and amplified once before plaque purification (using the method described in Pharmingen instructions). The plaque-purified viruses are amplified several times to obtain 500 ml of a high-titer virus stock (about 1×10^9 pfu). Amplification is accomplished by inoculating insect cells ($1–3 \times 10^6$/ml) with virus at a dilution of 1 : 100 (virus stock : cell culture) and allowing them to incubate until ~50% of the cells are dead. This is repeated until a 1 : 100 dilution of the virus stock produces ~50% cell death by day 3. For the final amplification of our virus stock, we culture for only 2 days. We have observed that baculovirus-infected cells begin to produce proteases after 2 days in culture. Using a virus stock prepared from cultures of more than 2 days can result in proteolysis of receptor protein.

We do not routinely determine a virus titer (plaque-forming units per ml of virus stock) by time-consuming plaque assays. The optimal inoculum for each virus stock is determined by small-scale suspension cultures. Twenty milliliters of $\sim 3 \times 10^6$ cells/ml in 100-ml Erlenmeyer flasks is inoculated with different dilutions of our virus stock and binding is determined after 48 hr in culture. In general, we find that a dilution of from 1 : 100 to 1 : 30 (virus stock : cell culture) is optimal. For high-titer virus stocks, plotting expression as a function of virus inoculum should produce a bell-shaped curve. Expression is reduced by inoculating with too much as well as too little virus.

β_2AR Expression in Large-Scale Insect Cell Culture. Sf9 insect cells are grown in suspension culture in SF 900 II medium (GIBCO, Grand Island, NY) containing 5% fetal calf serum (Gemini, Calabasas, CA) and 0.1 mg/ml gentamicin (GIBCO). For purification, the cells are grown in 1000- to 1200-ml cultures (in 2.8-liter Fernbach flasks). Cells are grown to a density of 3×10^6 cells/ml and are infected with an amount of high-titer virus stock determined from small-scale culture as described earlier. Cells are harvested after 48 hr by centrifugation for 10 min at 5000*g*. The resulting cell pellets are kept at -70° until they are used for purification. We have found that maintaining cultures for more that 48 hr after infection reduces the yield of purified protein, even though the expression of the receptor in membranes may be higher. This is due to the release of proteases by the baculovirus-infected cells during the late stage of infection. Conceivably, these proteases remove the affinity tags that are required for our purification procedure.

Purification

The following procedure requires that the receptor be modified at the amino terminus with a cleavable signal sequence followed by the Flag epitope (Sigma, St. Louis, MO) (Fig. 1) and at the carboxyl terminus with a hexahistidine sequence.[3]

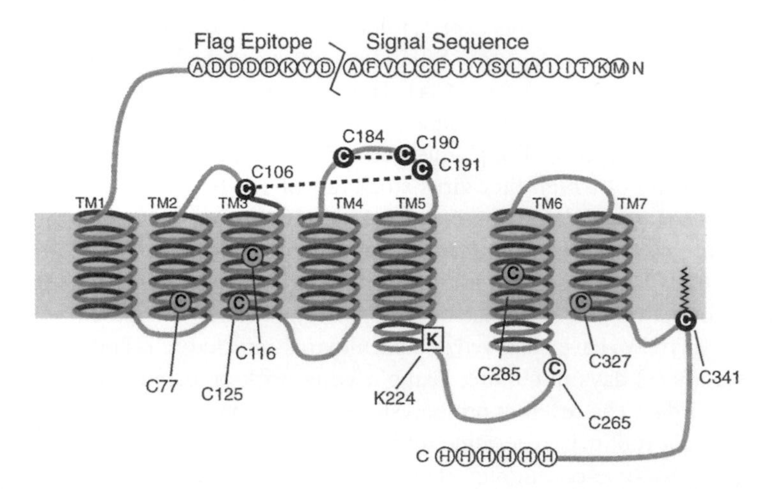

FIG. 1. Diagram of the β_2-adrenergic receptor showing amino- and carboxyl-terminal modifications and sites for labeling with fluorescent probes. The amino terminus of the wild-type human β_2AR has been modified by the addition of a cleavable signal sequence followed by the peptide sequence that is recognized by the M1-Flag antibody. An M1-Flag affinity column is used in the purification procedure. The carboxyl terminus has been modified by the addition of six histidines for nickel affinity chromatography. Several potential labeling sites (cysteines, C, and a lysine, K) are shown.

The cleavable signal sequence improves expression of the β_2AR by up to twofold[4] and is necessary to generate a Flag epitope that can be recognized by the Ca^{2+}-dependent M1 monoclonal antibody used for affinity purification. This antibody only recognizes the epitope at the extreme amino terminus (i.e., not preceded by another amino acid). The M1 affinity column binds the epitope-tagged receptor in the presence of Ca^{2+} and can be eluted in buffer containing EDTA.

Buffers

Protease inhibitors (final concentrations in solutions): 10 μg/ml leupeptin (Boehringer, Mannheim, Germany), 10 μg/ml benzamidine (Sigma)

Lysis buffer: 10 m*M* Tris–HCl buffer (pH 7.5), 10^{-6} *M* alprenolol (a β_2AR antagonist used to stabilize the receptor during solubilization, Sigma) plus protease inhibitors

Solubilization buffer: 20 m*M* Tris–HCl buffer (pH 7.5), 1.0% *n*-dodecyl-β-D-maltoside (DβM) (Anatrace Inc., Maumee, OH), 500 m*M* NaCl, 10^{-6} *M* alprenolol (Sigma) plus protease inhibitors

High-salt buffer: 20 m*M* Tris–HCl (pH 7.5), 500 m*M* NaCl, 0.1% DβM

Low-salt buffer: 20 m*M* Tris–HCl (pH 7.5), 100 m*M* NaCl, 0.1% DβM

No-salt buffer: 20 m*M* Tris–HCl (pH 7.5), 0.1% DβM

HEPES high-salt buffer: 20 m*M* HEPES buffer (pH 7.5), 500 m*M* NaCl, and 0.1 % DβM

Solubilization. Cell pellets from 1–2 liters of baculovirus-infected insect cells are thawed and suspended in lysis buffer (200 ml for a cell pellet from 1 liter of insect cell culture). The cell suspension is centrifuged at 45,000*g* for 30 min, the supernatant is discarded, and the pellets are weighed. Pellets are resuspended in solubilization buffer (10 ml/g of membrane pellet) by douncing (20 strokes with tight pestle) followed by stirring at 4° for 1 hr. Nonsolubilized material is isolated from solubilized protein by centrifugation at 45,000*g* for 30 min.

Nickel-Chelate Chromatography. Imidazole is added to the supernatant to a final concentration of 20 m*M* from a 2.0 *M* stock solution (pH 8.0). Chelating Sepharose fast flow resin (Pharmacia) is charged with nickel (using a 5% solution of nickel sulfate), washed with 10 column volumes of water, and equilibrated in high-salt buffer. Nickel resin is added to the solubilized receptor solution (0.5 ml of resin per gm of lysed cells) and incubated for 1 hr at room temperature, stirring just fast enough to keep the resin in suspension. Nickel resin is isolated by centrifugation for 5 min at 2000*g*. The resin is washed once in batch with 4 column volumes of high-salt buffer, loaded onto a column, and washed with 4 column volumes of high-salt buffer. Elution is done in one-fourth column

[3] B. K. Kobilka, *Anal. Biochem.* **231,** 269 (1995).

[4] X. M. Gaun, T. S. Kobilka, and B. K. Kobilka, *J. Biol. Chem.* **267,** 21995 (1992).

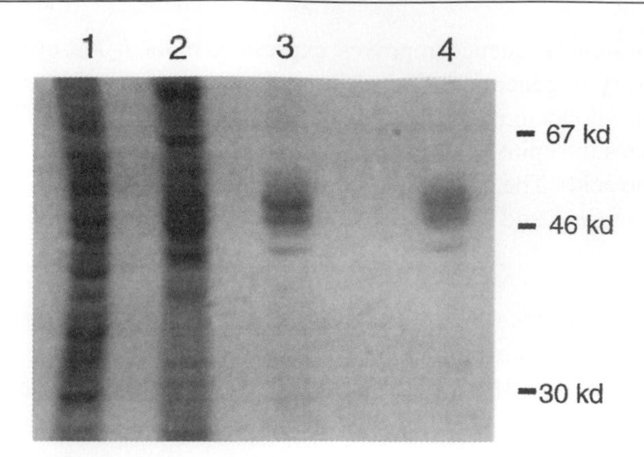

FIG. 2. Polyacrylamide gel electrophoresis of the purification of β_2AR. Lane 1, solubilized Sf9 cell membranes containing 2 pmol of β_2AR; lane 2, 50 pmol of nickel column-purified β_2AR; lane 3, 50 pmol of M1-Flag column-purified β_2AR; and lane 4, 50 pmol of alprenolol column-purified β_2AR.

volume fractions with high-salt buffer containing 200 m*M* imidazole. Fractions are assayed for receptor-binding activity and peak fractions are pooled.

M1-Flag Antibody Chromatography. $CaCl_2$ is added to the pooled nickel-pure fractions to a final concentration of 2.5 m*M*. The pooled fractions are loaded onto an M1 antibody column (Sigma) (0.2 ml resin/nmol of nickel-pure receptor) equilibrated in low-salt buffer and recycled four times by gravity flow. The column is washed with 4 column volumes of low-salt buffer containing 2.5 m*M* $CaCl_2$ and eluted using low-salt buffer containing 1 m*M* EDTA in one-fourth column volume fractions. Fractions are analyzed for receptor-binding activity and peak fractions are pooled.

These two purification steps can produce nearly pure protein (specific activity around 5 nmol/mg of protein) (Fig. 2). GPCRs for which ligand affinity chromatography has not been established can be purified using these two steps alone. However, for β_2AR, approximately half of the protein is nonfunctional.[3]

Alprenolol Affinity Chromatography. To separate the nonfunctional from the functional receptor, we use alprenolol affinity chromatography, which is a standard procedure for the purification of β_2AR. Alprenolol affinity resin can be prepared as described previously.[5] The affinity resin can be reused multiple times. We generally use 1 ml of alprenolol affinity resin for 5–10 nmol of M1-Flag pure β_2AR. Resin that has not been used previously is blocked with 3 column volumes of low-salt buffer containing bovine serum albumin (10 mg/ml). The column is then washed

[5] R. G. Shorr, S. L. Heald, P. W. Jeffs, T. N. Lavin, M. W. Strohsacker, R. J. Lefkowitz, and M. G. Caron, *Proc. Natl. Acad. Sci. U.S.A.* **79,** 2778 (1982).

with 10 column volumes of low-salt buffer. The flag pure receptor is added to the alprenolol affinity resin and is rotated slowly for 1 hr at room temperature and then overnight at 4° (or 3 hr at room temperature). The affinity resin is then loaded into a column and washed six times with 1 column volume of ice-cold high-salt buffer, six times with 1 column volume of ice-cold no-salt buffer, and six times with 1 column volume of ice-cold high-salt buffer. The column is eluted at room temperature in batch mode. The resin is suspended in 3 column volume of high-salt buffer containing 2 m*M* alprenolol for 1 hr. The eluate is recovered by centrifugation at 1000 rpm for 5 min. The elution is repeated once more. The receptor concentration in the eluate is low ($<1\ \mu M$), but can be concentrated by ultrafiltration (Centricon) or nickel chromatography.

Following purification, the quality of the protein is assessed by determining the specific activity, a measure of the number of functional receptor molecules (as measured by ligand binding) per milligram of receptor protein. Soluble binding is performed by diluting the purified receptor to a final concentration of ~10 n*M* in low-salt buffer and adding 10 to 80 μl low-salt buffer and 10 μl of 100 n*M* [^{3}H]dihydroalprenolol (Amersham). The binding reaction is incubated at room temperature for 1 hr. β_2AR with bound ligand is separated from free ligand by gel filtration using 2 ml Sephadex G-50 columns equilibrated with ice-cold low-salt buffer. G-50 columns must be treated to prevent nonspecific binding of receptor protein. We usually use outdated fetal calf serum, but bovine serum albumin (10 mg/ml) can also be used. Protein is determined using the detergent-insensitive Bio-Rad DC protein assay kit (Bio-Rad, Hercules, CA). The purified receptor is also analyzed by 10% SDS–polyacrylamide gel electrophoresis (Fig. 2). Notably, samples should not be boiled before loading onto the gel, as this often causes aggregation of proteins having multiple membrane-spanning domains. The receptor is visualized by standard Coomassie or silver staining.

We have been able to exclude the M1-Flag chromatography in some applications by going directly from the nickel-pure receptor to alprenolol affinity chromatography. A receptor purified using this approach has a specific activity of 3–10 nmol/mg as compared to 10–15 nmol/mg for a receptor purified using nickel, M1-Flag, and alprenolol chromatography. Approximately 5 nmol of purified protein can generally be obtained from a 1-liter culture.

Modifications of the Purification Procedure for Mutant Receptors. We have been able to use the purification scheme described earlier for wild-type as well as mutant receptors in which cysteines have been mutated to limit the number of potential sites for labeling with cysteine-reactive fluorescent probes.[6] We have encountered one mutation requiring a minor modification of the purification procedure. A constitutively active mutant of the β_2AR (CAMβ_2AR) was found to

[6] U. Gether, S. Lin, P. Ghanouni, J. A. Ballesteros, H. Weinstein, and B. K. Kobilka, *EMBO J.* **16,** 6737 (1997).

be biochemically unstable,[7] resulting in considerable denaturation of the protein during purification. To minimize denaturation, we shortened binding to the Alprenolol affinity resin to 2 hr at room temperature and labeled the protein with fluorescent probe immediately after purification.

Labeling of Purified β_2AR with Fluorescent Probes

Fluorescence spectroscopy can be used to study protein structure by monitoring the fluorescence of intrinsic fluorophores, such as tryptophan, or by covalent attachment of fluorescent probes to chemically reactive amino acids such as lysine and cysteine. We have been able to obtain useful structural information studying tryptophan fluorescence in β_2AR[8]; however, this approach has significant limitations. There are eight tryptophans in β_2AR, and the signal from one or two tryptophans that may be sensitive to ligand-induced conformational changes may be diluted by the signal from conformationally silent tryptophans. Mutagenesis could be used to identify the specific tryptophan(s) responsible for ligand-induced changes in tryptophan fluorescence; however, this would require that the majority of tryptophans are not necessary for the folding or function of the protein. A more significant limitation is that most catecholamine derivatives have adsorption spectra that overlap that of tryptophan, making it nearly impossible to study agonist-induced conformational changes. Finally, the quantum yield for tryptophan is low, and studying intrinsic fluorescence requires up to 10 times more protein than needed when the receptor is labeled with extrinsic fluorophores having higher quantum yields.

The receptor protein can be labeled selectively at primary amines (the amino terminus and lysines) or free sulfhydrals (cysteines) with a wide variety of commercially available fluorescent molecules (Molecular Probes, Toronto Research Chemicals). The approach to labeling is similar for these two classes of fluorescent probes with two exceptions. First, buffers containing primary amines (such as Tris buffers) cannot be used when labeling the receptor with amine-reactive fluorophores. During the final step of receptor purification, HEPES buffer can be used in place of Tris or the buffer can be exchanged by a small desalting (G-50 Sephadex) column. Second, most amine-reactive fluorophores are hydrolyzed more readily in aqueous buffers and must therefore be used at higher concentrations than sulfhydral-reactive fluorophores to achieve the same labeling stoichiometry.

There are several factors to consider when choosing a fluorescent probe, including the number and location of the potential sites of labeling on the receptor (the amino terminus, lysines and cysteines); the quantum yield of the probe; the excitation and emission wavelengths of the fluorophore relative to the adsorption

[7] U. Gether, J. A. Ballesteros, R. Seifert, E. Sanders-Bush, H. Weinstein, and B. K. Kobilka, *J. Biol. Chem.* **272,** 2587 (1997).

[8] S. Lin, U. Gether, and B. K. Kobilka, *Biochemistry* **35,** 14445 (1996).

(and possible emission) wavelengths of receptor ligands; the polarity and size of the probe; and the sensitivity of the probe to its molecular environment. These will be discussed in more detail later; however, the choice of fluorophore and the site of labeling must often be determined by trial and error.

Labeling Sites on the Receptor

Cysteines. As discussed previously, lysines and cysteines can be modified covalently with a large number of commercially available fluorescent probes. Some fluorophores found to be useful include fluoresceine, rhodamine, Texas red, Oregon green, NBD, and Cy3. The choice of labeling cysteines or lysines will be determined by the number and location of these amino acids. Generally, lysines will be located in polar regions of the receptor and will be accessible to labeling by polar probes. Cysteines can be found in hydrophobic as well as polar regions of the molecule. Therefore, the accessibility of cysteines will depend on their location in the molecule and on the polarity of the probe. Cysteines that form part of a disulfide bond or are modified covalently (e.g., acylated) will not be labeled unless the protein is specifically treated to reverse these chemical modifications. Most GPCRs will contain more than one reactive cysteine or lysine; however, many of these amino acids may not be located in a region of the receptor that undergoes conformational change. Labeling at these conformationally silent sites will reduce the magnitude of fluorescence changes from conformationally sensitive sites relative to the overall fluorescence of the protein. Therefore, the goal is to target labeling to a few sites that undergo agonist-induced conformational changes. We have observed that fluorescent probes attached to the receptor within or adjacent to transmembrane (TM) domains 3, 5, and 6 are sensitive to ligand-induced conformational changes[6] (Ghanouni and Kobilka, unpublished results; Jensen, Guarnieri, Asmar, Ballesteros, and Gether, submitted).

It is possible that a specific GPCR may already have labeling sites that are located in receptor domains that move on ligand activation. We have been able to observe agonist-induced conformational changes in wild-type β_2AR labeled with two cysteine reactive fluorophores: IANBD [*N,N′*-dimethyl-*N*-(iodoacetyl)-*N′*-(7-nitrobenz-2-oxa-1,3-diazol-4-yl) ethylenediamine; Molecular Probes, Eugene, OR],[6,7,9] and fluorescein maleimide (Molecular Probes) (Ghanouni and Kobilka, unpublished results). While both of these fluorophores are reactive toward cysteines, they report conformational changes from different labeling sites. IANBD is a nonpolar fluorescent probe capable of reacting with cysteines in TM domains. Cysteines in polar domains are also labeled; however, IANBD is strongly quenched by water so IANBD bound to cysteines in polar domains contributes little to the overall fluorescence. β_2AR contains 13 cysteines of which 5 cysteines are not expected to be available for chemical derivatization. In the extracellular loops, four cysteines (Cys-106, Cys-184, Cys-190, and Cys-191) form two disulfide

[9] U. Gether, S. Lin, and B. K. Kobilka, *J. Biol. Chem.* **270,** 28268 (1995).

bridges (Fig. 1),[10–12] and in the intracellular carboxyl-terminal tail, Cys-341 has been shown to be palmitoylated.[13,14] To identify the cysteine(s) responsible for the agonist-induced change in fluorescence and, thus, to establish a system that would allow site-selective incorporation of the IANBD fluorophore, we mutated cysteines in the receptor and generated a series of mutant receptors with one, two, or three cysteines available for chemical derivatization.[6] All these mutants displayed minimal changes in pharmacological properties as compared to the wild type, both with respect to ligand binding and functional coupling to adenylyl cyclase.[6] However, mutation of several cysteine residues led to a reduction in receptor expression.[6] Notably, a mutant receptor with all free cysteines substituted expressed so poorly that purification in sufficient quantities for fluorescence spectroscopy analysis was impossible. Ideally, it should be possible to take out all endogenous cysteines and either reintroduce them one by one or introduce single cysteines in new positions. Unfortunately, this was not possible in β_2AR. Nevertheless, it is possible to obtain site-specific information from a system where it is not possible to remove all cysteines. We were able to show that the fluorescence of IANBD bound to Cys-125 in TM 3 and Cys-285 in TM6 (Fig. 1) is decreased following agonist binding, presumably because conformational changes alter the polarity of the molecular environment around these labeling sites.[6]

In contrast to IANBD, fluorescein is a relatively large, polar molecule that does not readily label cysteines in transmembrane domains. The fluorescence of fluorescein is less sensitive to the polarity of its environment, but is very sensitive to pH. When β_2AR (at a concentration of 1–5 μM) is labeled at a 1 : 1 stoichiometry with fluorescein maleimide, it is possible to restrict labeling to primarily Cys-265 in the carboxyl-terminal region of the third intracellular loop of the β_2AR with fluorescein (Fig. 1). The fluorescence of fluorescein on Cys-265 is decreased by as much as 20% on agonist binding (Ghanouni and Kobilka, unpublished results).

Lysine. β_2AR contains 15 lysines, all of which are in regions of the protein that are likely to be accessible to aqueous solvents. While some selectivity of labeling may be achieved by varying pH, this approach is not very reliable and requires extensive biochemical characterization to identify labeling sites. To overcome this problem we have generated a modified β_2AR in which all of the lysines have been mutated to arginine. This receptor retains wild-type pharmacologic and G protein-coupling properties.[15] It is therefore possible to add back a lysine at a specific site

[10] H. G. Dohlman, M. G. Caron, A. DeBlasi, T. Frielle, and R. J. Lefkowitz, *Biochemistry* **29,** 2335 (1990).

[11] C. M. Fraser, *J. Biol. Chem.* **264,** 9266 (1989).

[12] K. Noda, Y. Saad, R. M. Graham, and S. S. Karnik, *J. Biol. Chem.* **269,** 6743 (1994).

[13] B. Mouillac, M. Caron, H. Bonin, M. Dennis, and M. Bouvier, *J. Biol. Chem.* **267,** 21733 (1992).

[14] B. F. O'Dowd, M. Hnatowich, M. G. Caron, R. J. Lefkowitz, and M. Bouvier, *J. Biol. Chem.* **264,** 7564 (1989).

[15] A. L. Parola, S. Lin, and B. K. Kobilka, *Anal. Biochem.* **254,** 88 (1997).

and achieve site-specific labeling with amine-reactive probes. The amino terminus is labeled under conditions that label lysines. This can, in theory, be prevented by blocking the amino terminus with *N*-hydroxysuccinimide at pH 7, a pH at which most lysines will be protonated. However, in our experience, this is not reliable. We have chosen another approach to remove unwanted amino-terminal labeling. A cleavage site for the tobacco etch virus (TEV, GIBCO BRL) protease is inserted at the amino terminus of β_2AR using recombinant DNA techniques. TEV cleaves between Gln and Gly in the sequence GluAsnLeuTyrPheGln/Gly. After labeling the receptor with an amine-reactive fluorophore, the amino terminus can be removed efficiently with TEV. We have only limited experience with lysine-reactive probes but have found that the fluorescence of a receptor labeled at a single lysine at position 224 in β_2AR (Fig. 1) with rhodamine is increased by up to 6% following agonist binding.

Choice of Fluorophores

In addition to chemical reactivity (cysteine or lysine reactive probes), perhaps the most important consideration in choosing a fuorophore is the spectroscopic properties of the specific agonists and antagonists to be studied. Catecholamines are weakly fluorescent, as are several chemically related synthetic βAR agonists and antagonists, and they are used at high enough concentrations to interfere significantly with fluorescence from intrinsic fluorophores, such as tryptophan, as well as extrinsic fluorophores having similar excitation and emission properties. Moreover, some ligands that are not fluorescent may still interfere with fluorescent measurements if they adsorb either excitation or emission light.

Other properties that may influence the choice of fluorophore are quantum yield, photostability, and sensitivity of the probe to pH and solvent polarity. Photostable fluorophores with high quantum yields will enable more studies to be done with a limited amount of protein. Fluorophores that are bleached readily (not photostable) also require protecting the probe from light during labeling. Many fluorophores are sensitive to changes in pH. This can be a disadvantage if the p*K* of the fluorophore is close to the pH used for experiments. As discussed previously for IANBD, fluorophores that are sensitive to solvent polarity can be used to detect conformational changes in a protein that change the molecular environment around the attached fluorophore.

General Labeling Protocol for β_2AR

Purified β_2AR is incubated with 1- to 10-fold molar excess of a cysteine reactive probe (iodoacetamide or maleimide derivative) or 10- to 100-fold molar excess of a lysine-reactive fluorophore (isothiocyanate or succinimidal ester derivative). Lysine-reactive probes are generally more labile and must be used at higher concentrations to achieve the same labeling stoichiometry as obtained with a cysteine-reactive probe. Lysine-reactive probes are prepared fresh, usually in dimethyl sulfoxide (DMSO). Cysteine-reactive probes are also dissolved in DMSO, but can be stored in aliquots at $-20°$ for several months. The fluorophore stock should be 20–100 times more concentrated than the final concentration to be used in the labeling reaction. β_2AR can tolerate up to 5% DMSO. Care should be taken to protect probes from excessive exposure to light. We generally label 1–1.5 nmol of receptor in a total volume of 100 μl HEPES high-salt buffer by adding up to 5 μl of the concentrated fluorophore stock in DMSO. After adding the fluorophore, the reaction is allowed to proceed for 1 hr on ice, protected from light. The unreacted fluorophore can be quenched by the addition of cysteine to a final concentration of 1 m*M* (for cysteine-reactive fluorophores) or ethanolamine to a final concentration of 1 m*M* (for lysine-reactive fluorophores), followed by 30 min of incubation on ice. We often do not quench unreacted fluorophore because it is removed in the following purification procedure.

In many cases it is possible to remove unreacted fluorophore from labeled protein by a simple desalting procedure using a Sephadex G-50 gel-filtration column (0.5×9 cm). The column is equilibrated with low-salt buffer, and the reaction mixture is applied directly to the column and eluted with low-salt buffer. Elution of the fluorescent receptor can be monitored easily by the visible color or, if necessary, by the frugal use of a weak, hand-held fluorescent light. The labeled receptor can be concentrated in a Centricon-30 filter device (Amicon, Beverly, MA). The eluate containing the labeled receptor is concentrated to approximately 1–5 μM in the Centricon-30 filter device by centrifugation for 45 min at 3000 *g* in a fixed angle rotor (Sorvall SS-34). The labeled receptor is either used directly for the fluorescence spectroscopy analysis or stored on ice. Under these conditions the protein is stable for several days.

We have found that for some fluorophores, such as fluorescein and rhodamine, it is not possible to completely separate unreacted fluorophore from labeled receptor by gel filtration. These fluorophores partition into detergent micelles, which migrate with the receptor on a Sephadex G-50 column. Nickel chromatography is a more effective method of removing unreacted fluorophore. Following the labeling reaction, the receptor is bound to a small ($\sim$200 μl) nickel column by recycling by gravity flow six times. The column is washed with no-salt buffer (3×1 ml), followed by high-salt buffer (3×1 ml). The labeled receptor is eluted in 50-μl fractions with 200 m*M* imidazol in high-salt buffer. Fractions are assayed for fluorescence and peak fractions are pooled. The labeled receptor can be used directly for the fluorescence spectroscopy or stored on ice for several days.

The stoichiometry of labeling can be determined by measuring the fluorophore concentration and the protein concentration. The fluorophore concentration of the labeled protein can be determined spectrophotometrically using the known absorption wavelength and extinction coefficient for the fluorophore. The protein concentration can be determined using the Bio-Rad DC protein assay kit (Bio-Rad), which is insensitive to detergents.

Fluorescence Spectroscopy Analysis

Initial spectroscopic analysis will involve obtaining excitation and emission spectra of the labeled protein. The initial parameters will be chosen based on the known properties of the fluorescent probe. After identifying the optimum excitation wavelength, we obtain emission spectra in the absence and presence of saturating concentrations of several agonists and antagonists, looking for changes in fluorescence intensity and the λ_{max} (the wavelength at which maximum emission intensity is observed). We then perform a time scan, monitoring intensity at λ_{max} as a function of time before and after the addition of ligands. As an example, analysis of β_2AR labeled with IANBD will be described.

Fluorescence spectroscopy is performed at room temperature using a SPEX Fluoromax spectrofluorometer connected to a PC equipped with the Datamax software package. We use the photon-counting mode and generally an excitation and emission bandpass of 4.2 nm. Emission scan experiments are done with 30–50 pmol IANBD-labeled receptor. Usually, 10 μl of receptor is added to 390 μl of low-salt buffer in a 5 × 5-mm quartz cuvette and mixed by pipetting up and down. The excitation wavelength is set at 481 nm, and emission is measured from 490 to 625 nm with an integration time at 0.3 sec/nm.

Similar to emission scans, time scan spectroscopy is performed using 30–50 pmol of labeled receptor. Usually, 10 μl of receptor is added to 490 μl of low-salt buffer in a 5 × 5-mm quartz cuvette. To stabilize the baseline, the mixture is preincubated for at least 10 min in the cuvette before the experiment is started. During the preincubation and the time scan experiment, the mixture is stirred using a 2 × 2-mm magnetic stir bar (Bel-Art Products, Pequannock, NJ). During time scan experiments, the excitation wavelength is fixed at 481 nm, and emission is measured at a wavelength of 525 nm. The time scan is performed routinely over 30 min, and the first addition of ligand is usually done after 5 min. The volume of the added ligands is 1% of the total volume, and fluorescence is corrected for this dilution. The compounds tested in our fluorescence experiments have an absorbance of less than 0.01 at 481 and 525 nm in the concentrations used and therefore have no significant inner filter effects. An example of a time scan on wild-type β_2AR labeled with IANBD is shown in Fig. 3.

A few issues should be emphasized when interpreting data from our fluorescence spectroscopy analysis. It is important to note that the amplitude of the fluorescent change is only a rough indicator of the magnitude of conformational change. For example, we cannot assume that there is a linear correlation between change in fluorescence and magnitude of molecular movements. It is also necessary to ensure that the fluorescent probe, when incorporated into the receptor, does not

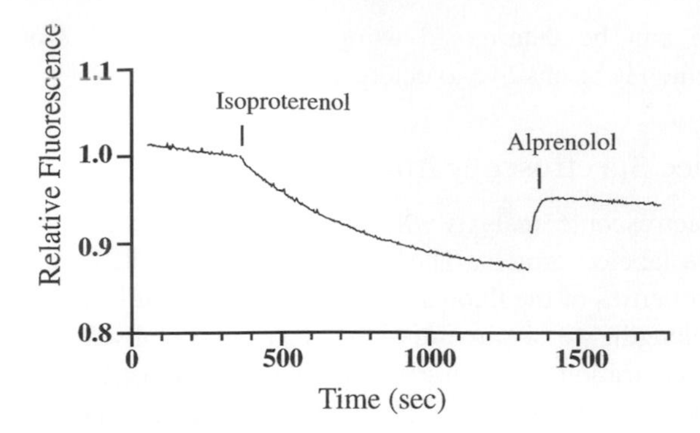

FIG. 3. Time scan of IANBD-labeled β_2AR. Approximately 30 pmol of purified β_2AR labeled with IANBD was excited at 481 nm, and the emission fluorescence intensity at 525 nm was monitored over time. Additions of the agonist isoproterenol (to a final concentration of 20 μM) and the antagonist alprenolol (to a final concentration of 5 μM) are indicated.

interfere with binding of the ligands. In β_2AR, this is highly unlikely. Labeling of the receptor with IANBD does not alter agonist- or antagonist-binding properties,[9] as would be expected if the bound NBD were positioned within the ligand-binding pocket. Results from mutagenesis studies have also provided substantial evidence that amino acids involved in forming the ligand-binding pocket are on a different side of the TM α helix and one to two α-helical turns closer to the extracellular surface relative to Cys-125 and Cys-285.

Concluding Remarks

GPCRs serve as cell surface sensors for a wide variety of biologic stimuli ranging from photons and odorants to hormones, neurotransmitters, and metabolites. The mechanism of GPCR activation has been difficult to study because of the technical challenges in producing and purifying sufficient quantities of protein for structural studies. However, fluorescence spectroscopy can readily detect picomole quantities of a protein labeled with high quantum yield fluorophores. This approach has provided valuable insight into the mechanism of activation for both rhodopsin and β_2AR (reviewed in Gether.[2]) Applying more advanced spectroscopic technology, such as time-resolved fluorescence spectroscopy and single molecule fluorescence spectroscopy, will likely provide even more mechanistic insight from GPCRs labeled with fluorescent probes.

[12] Crystallization of Membrane Proteins *in Cubo*

By PETER NOLLERT, JAVIER NAVARRO, and EHUD M. LANDAU

Introduction

Membrane proteins account for about 25% of the proteins encoded by the genome. They play key roles in a variety of cellular processes, including energy and signal transduction. Comprehensive understanding of the molecular mechanisms of these fundamental biological membrane-associated processes cannot be attained unless the atomic structures of the membrane proteins involved are known. The success of X-ray crystallography, the most widely used method to determine protein structures, depends critically on the availability of well-ordered three-dimensional crystals. Whereas nearly 14,000 structures of soluble macromolecules can be found in the protein databank (http://www.rcsb.org/pdb), the number of high-resolution structures of membrane proteins is still limited to fewer than 30. The two major problems are as follows: (i) Overexpression of membrane proteins is still extremely difficult. (ii) Routine availability of well-ordered crystals has so far been possible with only a few membrane proteins. This article focuses on the second problem: crystallization of membrane proteins. The difficulty in crystallizing membrane proteins arises from their surface duality: membrane proteins possess hydrophilic surfaces, which are exposed to the aqueous medium, and hydrophobic surfaces, which interact with the nonpolar chains of lipids. Because crystallization requires a homogeneous and monodisperse solution of the protein at a relatively high concentration as a starting point, conditions must be found such that both hydrophobic and hydrophilic surfaces of the protein will be solubilized stably when the protein is released from the native membrane. Failure to attain this basic condition might result in rapid aggregation to amorphous species, a process that competes with slow ordered crystal growth.

Starting in the 1970s, it was suggested that detergent solubilization might be useful in overcoming this difficulty.[1,2] This established procedure yields detergent/protein-mixed micelles and renders membrane proteins water soluble so that they may be crystallized analogously to soluble proteins. Indeed, the first reports[3–5] of membrane protein crystals from detergent solution appeared in 1980, and the first high-resolution structure of a membrane protein[6]—a bacterial photosynthetic

[1] A. Helenius and K. Simons, *Biochim. Biophys. Acta* **415,** 29 (1975).
[2] C. Tanford and J. A. Reynolds, *Biochim. Biophys. Acta* **457,** 133 (1976).
[3] R. M. Garavito and J. P. Rosenbusch, *J. Cell. Biol.* **86,** 327 (1980).
[4] R. Henderson and D. Shotton, *J. Mol. Biol.* **139,** 99 (1980).
[5] H. Michel and D. Oesterhelt, *Proc. Natl. Acad. Sci. U.S.A.* **77,** 1283 (1980).
[6] J. Deisenhofer, O. Epp, K. Miki, R. Huber, and H. Michel, *J. Mol. Biol.* **180,** 385 (1984).

0076-6879/02 $35.00

reaction center—was finally published in 1984. Introduction of detergents led initially to a slow progress in the crystallization of membrane proteins, and although in the last few years a dramatic increase in the rate of appearance of new crystal structures took place, major difficulties in the field are still unsolved.

In the past few years, two new concepts for the crystallization of membrane proteins have emerged. The first utilizes genetically engineered Fv antibody fragments that recognize specific epitopes of the protein, thereby facilitating crystallization of the membrane protein.[7] The second is based on the properties of bicontinuous lipidic cubic phases for the incorporation, stabilization, and crystallization of membrane proteins: the *in cubo* crystallization method.[8] We review the practical aspects of the latter.

Rationale: Crystallization *in Cubo*

The basic premise behind the *in cubo* crystallization method is that membrane proteins should crystallize more readily in a lipid bilayer than in a nonbilayer environment, provided that they can be incorporated into an appropriate lipidic matrix, retain their native properties, and diffuse in three dimensions. The lipidic material of choice is a bicontinuous cubic phase of monoacylglycerols and water. Such three-dimensional bilayers have been studied in great detail[9–14] and were shown to provide a stable, structured matrix in which the diffusion of both water-soluble and membrane proteins takes place. We hypothesized that labile membrane proteins, incorporated into continuous lipid bilayers, could be stabilized, diffusing rather freely along the bilayer, similar to the lateral diffusion of lipids.[15,16] On nucleation ("seeding"), this "feeding" may eventually lead to well-ordered crystals.

We demonstrated the feasibility of this idea for the first time by crystallizing the light-driven proton pump bacteriorhodopsin (bR) in a bicontinuous lipidic cubic phase.[8,17] This allowed elucidation of the X-ray structure of bR at ever-higher resolution (from 2.5 to 1.55 Å) from microcrystals grown in a monoolein-based cubic phase.[18–21] We have shown that bR in such crystals retains its biological activity: On photoexcitation it undergoes a photocycle that is indistinguishable

[7] C. Ostermeier, S. Iwata, B. Ludwig, and H. Michel, *Nature Struct. Biol.* **2,** 842 (1995).

[8] E. M. Landau and J. P. Rosenbusch, *Proc. Natl. Acad. Sci. U.S.A.* **93,** 14532 (1996).

[9] L. Rilfors, P. O. Eriksson, G. Arvidson, and G. Lindblom, *Biochemistry* **25,** 7702 (1986).

[10] K. Larsson, *J. Phys. Chem.* **93,** 7304 (1989).

[11] G. Lindblom and L. Rilfors, *Biochim. Biophys. Acta* **988,** 221 (1989).

[12] R. Vargas, P. Mariani, A. Gulik, and V. Luzzati, *J. Mol. Biol.* **225,** 137 (1992).

[13] V. Luzzati, R. Vargas, P. Mariani, A. Gulik, and H. Delacroix, *J. Mol. Biol.* **229,** 540 (1993).

[14] J. Briggs, H. Chung, and M. Caffrey, *J. Phys. II* **6,** 723 (1996).

[15] S. Cribier, A. Gulik, P. Fellmann, R. Vargas, P. F. Devaux, and V. Luzzati, *J. Mol. Biol.* **229,** 517 (1993).

[16] P. O. Eriksson, G. Lindblom, and G. Arvidson, *J. Phys. Chem.* **91,** 846 (1987).

[17] G. Rummel, A. Hardmeyer, C. Widmer, M. L. Chiu, P. Nollert, K. P. Locher, I. Pedruzzi, E. M. Landau, and J. P. Rosenbusch, *J. Struct. Biol.* **121,** 82 (1998).

from that observed for the native purple membrane.[22] Furthermore, crystals grown in lipidic cubic phases have yielded the first high-resolution structures of intermediate states in the photocycle of bR.[23–26] Cubic phase crystallization methodology also proved effective for small molecules and soluble proteins.[17,27] Most importantly, we have successfully crystallized bR directly from the purple membrane in a lipidic cubic phase, avoiding detergent treatment altogether.[28] This development is significant because it allows the crystallization of proteins that are unstable in the presence of detergents. Bacteriorhodopsin is unusual among membrane proteins in that most of its mass is membrane embedded, existing as highly ordered two-dimensional crystalline arrays in purple membranes.[29,30] We generalized the method by applying it successfully to a variety of membrane proteins with different structural characteristics[31]: two photosynthetic reaction centers from *Rhodopseudomonas viridis* (RCvir) and *Rhodobacter sphaeroides* (RCsph), the light-harvesting complex 2 from *Rhodopseudomonas acidophila* (LH2), and halorhodopsin from *Halobacterium salinarum* (hR). Independently, crystals of hR have been obtained from a monoolein-based lipidic cubic phase, leading to the elucidation of its structure to 1.8 Å resolution.[32]

Methods

Principle of Operation

Cubic phases are prepared by mixing lipids with aqueous solutions that contain the protein of choice to be crystallized. The water/lipid ratios are chosen

---

[18] E. Pebay-Peroula, G. Rummel, J. P. Rosenbusch, and E. Landau, *Science* **277,** 1676 (1997).

[19] H. Luecke, H. T. Richter, and J. K. Lanyi, *Science* **280,** 1934 (1998).

[20] H. Belrhali, P. Nollert, A. Royant, C. Menzel, J. P. Rosenbusch, E. M. Landau, and E. Pebay-Peyroula, *Structure* **7,** 909 (1999).

[21] H. Luecke, B. Schobert, H. T. Richter, J. P. Cartailler, and J. K. Lanyi, *J. Mol. Biol.* **291,** 899 (1999).

[22] J. Heberle, G. Büldt, E. Koglin, J. P. Rosenbusch, and E. M. Landau, *J. Mol. Biol.* **281,** 587 (1998).

[23] K. Edman, P. Nollert, A. Royant, H. Belrhali, E. Pebay-Peyroula, J. Hajdu, R. Neutze, and E. M. Landau, *Nature* **401,** 822 (1999).

[24] H. Luecke, B. Schobert, H. T. Richter, J. P. Cartailler, and J. K. Lanyi, *Science* **286,** 255 (1999).

[25] A. Royant, K. Edman, T. Ursby, E. Pebay-Peyroula, E. M. Landau, and R. Neutze, *Nature* **406,** 645 (2000).

[26] H. Luecke, B. Schobert, J. P. Cartailler, H. T. Richter, A. Rosengarth, R. Needleman, and J. K. Lanyi, *J. Mol. Biol.* **300,** 1237 (2000).

[27] E. M. Landau, G. Rummel, S. W. Cowan-Jacob, and J. P. Rosenbusch, *J. Phys. Chem. B* **101,** 1935 (1997).

[28] P. Nollert, A. Royant, E. Pebay-Peyroula, and E. M. Landau, *FEBS Lett.* **457,** 205 (1999).

[29] A. E. Blaurock and W. Stoeckenius, *Nature New Biol.* **233,** 152 (1971).

[30] R. Henderson and P. N. T. Unwin, *Nature* **257,** 28 (1975).

[31] M. L. Chiu, P. Nollert, M. C. Loewen, H. Belrhali, E. Pebay-Peyroula, J. P. Rosenbusch, and E. M. Landau, *Acta Cryst. D* **56,** 781 (2000).

[32] M. Kolbe, H. Besir, L.-O. Essen, and D. Oesterhelt, *Science* **288,** 1390 (2000).

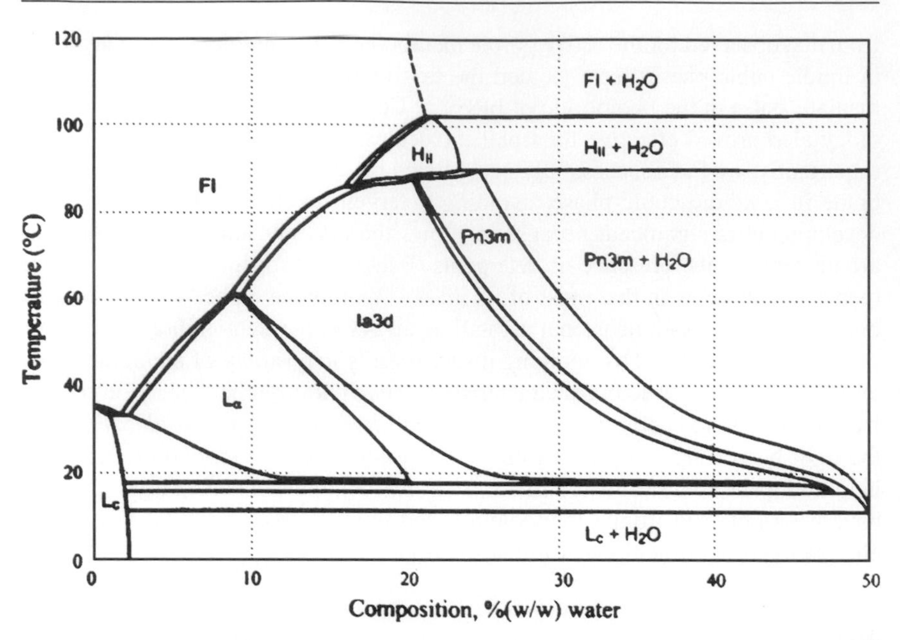

FIG. 1. Lyotropic and thermotropic phase diagram of MO/water mixtures (from Rummel *et al.*,[17] with permission). At room temperature and hydration below 20%, liquid crystalline (Lc) and lamellar (Lα) phases form, whereas lipidic cubic phases (Ia3d and Pn3m) form at increased hydration levels. Note that the Pn3m cubic phase is stable in coexistence with excess water. FI is a fluid isotropic phase, H_{II} is an inverted hexagonal phase.

according to known phase diagrams of the pure lipid water systems (Fig. 1). Cubic phase formation is considered complete once the matrix displays transparency, nonbirefringence, and solid texture. Crystallization is initiated by overlaying precipitating agents on the lipidic cubic phase, as solids or solutions. Crystallization is monitored microscopically, preferably using a polarizing light microscope. Crystals are harvested from the host lipid matrix and subjected to X-ray analysis.

In Cubo *Crystallization Protocols*

Two different protocols for setting up crystallization trials have been devised: (i) the original macro method utilizing glass tubes, aimed at the production of diffraction quality crystals[8,17] and (ii) a micro method using very small amounts of lipid and protein, aimed at preliminary screening of many crystallization conditions.[33] The requirements and performance of both protocols are compared in Table I.

[33] P. Nollert, submitted for publication.

TABLE I
COMPARISON OF THE GLASS TUBE-BASED PROCEDURE AND THE MICRO METHOD

	Glass vial method	Micro method
Volume of a single setup	5–20 μl	0.2 μl
Protein amount in a single setup	17.5–70 μg	0.7 μg
Number of setups from 1 mg protein[a]	57–14	1143[b]
Number of setups/person and day	ca. 48	ca. 500–1000
Applicable observation modes	Dissecting microscope Polarizing microscope	Dissecting microscope Fluorescence microscope Dark-field microscope

[a] Sixty percent lipid in the cubic phase; final protein concentration 3.5 mg/ml.
[b] Including 20% loss of material.

Glass Tube-Based Crystallization. This protocol can be used to reproducibly generate high-quality crystals of bR or other membrane proteins for X-ray diffraction purposes. Crystallization trials are carried out in small glass tubes. Crystal formation and characterization of the lipidic phases, as well as removal of crystals, are monitored by light microscopy with crossed polarizers. The two disadvantages of this method are (1) the use of relatively large amounts of protein and lipid and (2) the numerous and cumbersome handling steps. This protocol is recommended for the production of well-diffracting crystals, after the initial crystallization conditions have been established.

Protocol for Crystallization of bR

1. Purple membrane from *H. salinarum* is solubilized with β-octylglucoside (β-OG). Monomerized bR is purified by gel filtration (Biogel 0.5 A) as described.[34] The protein is concentrated using an Amicon filter (PM-50) to a final concentration of 10 mg/ml in 25 m*M* sodium phosphate, pH 5.6.
2. Dry monoolein (MO) (monooleoyl-*rac*-glycerol) is weighed into a glass tube (3–4 mm inner diameter) and optionally incubated at 37° to melt the MO. The bR solution is added directly to the solid lipid or, after slow cooling to room temperature to the melted MO. The solution/MO ratio is 2/3 (w/w), and the glass tubes are sealed with a strip of sealing film (Parafilm). The components are mixed by repeated centrifugation at 10,000 rpm in a benchtop centrifuge for 5 min in a 20° temperature-controlled microcentrifuge. The tubes are rotated by 180° about their long axis between runs. A purple, transparent, nonbirefringent, and gel-like solid material is obtained at the bottom of the tube.
3. About 0.3 g Sørensen phosphate per 1 g of cubic phase is added as a dry, ground powder to the preparation, either immediately following the addition

[34] B. Lorber and L. J. DeLucas, *FEBS Lett.* **261,** 12 (1990).

of the bR solution in step 2 or alternatively after complete formation of the cubic phase. In the latter case, a further centrifugation run is necessary in order to mix the solid phosphate powder with the preformed lipidic cubic phase. The Sørensen salt mixture is prepared by grinding 94.8 g KH_2PO_4 and 5.2 g $NaH_2PO_4 \cdot H_2O$.

4. The glass tube is sealed again with a strip of Parafilm and is stored in the dark at 20° for several days or weeks. Crystal formation is monitored periodically by microscopic inspection.

Micro Method. This protocol may be used for efficient initial screening of crystallization conditions. Because of the very small size of individual crystallization experiments (ca. 0.2 μl cubic phase) and the use of a semiautomatic dispensing device, it is possible to screen large numbers of conditions. Small amounts of preformed cubic phases may be placed in wells of multiwell plates. We find that manual preparation of crystallization setups in such multiwell plates is practical with plates of ca. 100 individual wells. Solid and/or liquid crystallization agents may be added. The procedure of setting up crystallization experiments using syringes in combination with a semiautomatic dispenser and multiwell plates is described in Nollert[33] (Emerald Biostructures, patent pending).

Protocol

1. Preparation of the lipidic cubic phase using two 250-μl gas-tight syringes coupled *via* a mixer is as described by Chen *et al.*[35] The typical cubic phase is composed of 60% MO and 40% aqueous protein solution.
2. The lipidic cubic phase material is transferred into a microsyringe (10 μl total volume, Hamilton Model 701–26s) for microdispensing purposes. This is done either directly into the syringe barrel, using a needle, or *via* an appropriate adapter. The microsyringe is assembled with a semiautomatic dispenser (Hamilton Model PB600) that allows dispension steps of 0.2 μl each. Refilling is necessary after ca. 50 dispension steps. In order to reduce the buildup of pressure in the syringe, the dispensing needle is shortened to a length of about 1 cm.
3. Prior to dispension of the lipidic material, six microwells (one lane) of a 72-well plate (Nunc) are filled with 1 μl of the precipitant solutions. These solutions may be dried at 60° for 30 min to yield a powder. This procedure resembles crystallization of bR using dry salt as the precipitation agent.[8,17,36] However, solutions may also be used to promote crystallization (see "*In Cubo* Crystallization Screening Conditions").

[35] A. H. Chen, B. Hummel, H. Qiu, and M. Caffrey, *Chem. Phys. Lipids* **95,** 11 (1998).

[36] M. Loewen, M. L. Chiu, C. Widmer, E. M. Landau, J. P. Rosenbusch, and P. Nollert, *in* "G-Protein Coupled Receptors: Methods in Signal Transduction" (T. Haga and G. Berstein, eds.), p. 365. CRC Press, Boca Raton, FL, 1999.

4. Approximately 0.2 mg (corresponding to ca. 0.2 μl) of the preformed cubic phase is added to each microwell by positioning the syringe needle perpendicularly to the well bottom and triggering the dispenser. Lipidic material is extruded through the attached needle and forms an amorphous blob at the bottom of the well.
5. A row of the filled microwells is sealed with clear transparent tape (Crystal Clear, Manco; SealPlate, USA Scientific Inc.). The plates may be stored at different temperatures and inspected using a microscope.

Transfer of the lipidic cubic phase from one syringe into another and use of needles are always accompanied by some loss of material due to dead volumes of the instruments. However, the throughput of this procedure is much higher than that using single glass tubes: More than 1000 individual crystallization experiments at final protein concentration of 3.5 mg/ml can be carried out with 1 mg of protein. In our experience, this number corresponds approximately to the maximum number of setups per person and day. Because of the small volumes used, great care must be taken to prevent dehydration of the crystallization setup.

Harvesting Single Crystals from the Lipidic Cubic Phase Matrix

For X-ray diffraction experiments, single crystals need to be mounted inside a glass capillary or onto nylon cryo loops. The simplest procedure for harvesting crystals is by mechanical manipulation. However, because the crystals are embedded in a highly viscous medium, mechanical manipulation imposes a shear force on the crystals. This in turn might damage the protein crystal, thereby reducing its diffraction quality and increasing mosaicity. To overcome this problem, two harvesting methods have been developed: (i) an enzymatic procedure[37] and (ii) a detergent-solubilization procedure.[21]

Mechanical Harvesting. The simplest isolation method is direct removal, which leaves some remaining matrix lipid attached to the surface of the crystal. Crystals may disintegrate and dissolve after some time.

Protocol

1. A glass tube containing crystals embedded in a lipidic cubic phase (5–20 μl) is broken off with a diamond cutter.
2. A small container (e.g., an hourglass) is filled with 100 μl of an aqueous solution corresponding to the mother liquor (in the case of bR: 20% glycerol in saturated Sørensen phosphate solution). The lipidic cubic phase is removed mechanically with a small spatula and placed into the mother liquor solution.

[37] P. Nollert and E. M. Landau, *Biochem. Soc. Trans.* **26,** 709 (1998).

3. Crystals are cleaned from the surrounding lipidic phase under microscopic inspection using microtools such as needles and small spatulas.
4. An individual crystal with residual adhering lipidic phase may be mounted directly on a cryo loop or first transferred into another solution by pipetting.

As a variation of this method the content of the glass tube may be placed on a dry glass support. On dehydration, the lipidic cubic phase undergoes a phase transition into a lamellar phase that is less viscous. This can be observed easily under a polarizing microscope as a transition from a nonbirefringent to a birefringent material. Crystals can be manipulated as described in steps 3 and 4.

Enzymatic Hydrolysis. This method has been described in detail by Nollert and Landau[37] and yields crystals that do not contain any matrix lipid, as judged by mass spectroscopy.[20]

Protocol

1. The crystallization setup or part thereof is subjected to lipase treatment. In the case of bR, an equal volume of freshly prepared lipase solution is added to the lipidic cubic phase. The lipase used is *Candida rugosa* lipase [EC 3.1.1.3, 890 units/mg of solid (Sigma)] at a concentration of 50 mg lipase/ml in saturated Na/K phosphate buffer. Lipid hydrolysis can be followed microscopically.
2. Incubation for several hours or days converts the lipidic cubic phase into a two-phase system consisting of two immiscible liquids: water/glycerol solution and oleic acid.
3. Crystals are released from the matrix spontaneously and either remain embedded in oleic acid droplets or float freely in solution (Fig. 2). Individual bR crystals may be harvested using micro tools, e.g., microneedles and cryo loops. Crystals are stable in the mother liquor.

Detergent Solubilization. This method is described in Ref. 21.

Protocol

1. A glass tube containing crystals embedded in a lipidic cubic phase (5–20 μl) is broken off with a diamond cutter.
2. A small volume of the lipidic cubic phase is transferred with a small spatula into 100 μl saturated Sørensen phosphate solution containing 0.1% β-OG.
3. Crystals are cleaned mechanically from the surrounding lipidic material. Due to mixing with the detergent, the lipidic cubic phase transforms slowly into a lamellar phase, and crystals can be removed from this solution with a cryo loop.

It is advisable not to expose the crystals to the detergent solution for long periods of time as the crystals disintegrate and the protein is solubilized.

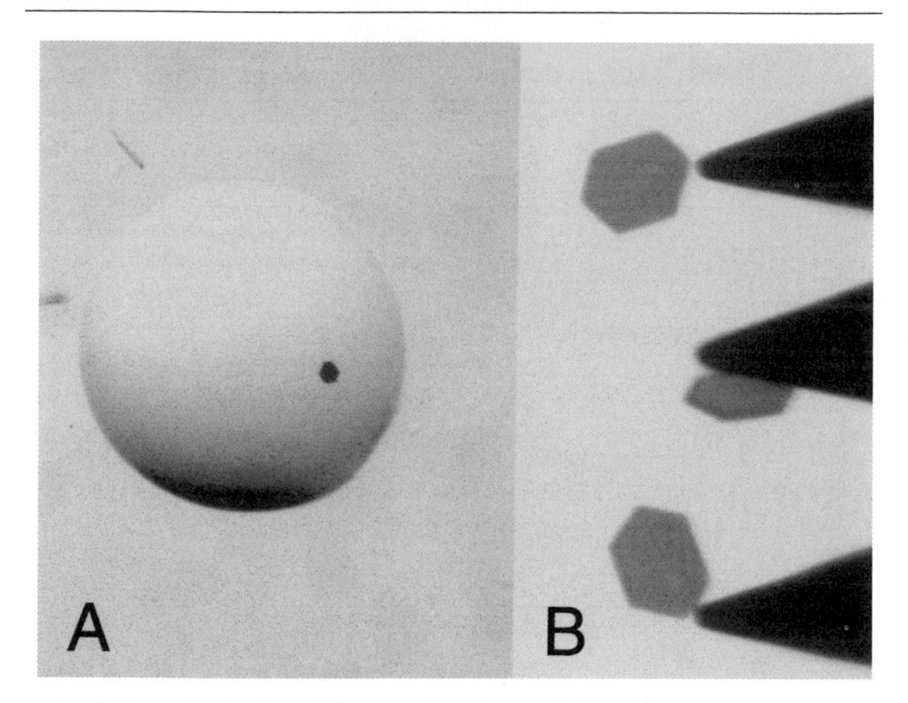

FIG. 2. Enzymatically released bR crystals from the host lipidic cubic phase matrix. Single bR crystals can be handled analogously to crystals grown in solution. (A) bR crystal floating in a drop of water after its release; (B) the crystal is manipulated with a needle in solution; three positions are shown.

In Cubo Crystallization Screening Conditions

Analogous to conventional protein crystallization, the parameters that can be varied are numerous, and usually only a subset of them will be assayed in an actual screening. These parameters include concentration and type of lipidic matrix, detergent, protein, salt, organic solvents, polymers, temperature, pH, and pressure. In the following we present conditions that have yielded membrane protein crystals and discuss parameters with respect to priority and their effect on MO cubic phase stability. Because lipidic cubic phase-mediated membrane protein crystallization is still a novel method, it has not yet been investigated as thoroughly as solution crystallization. The recommendations given herein reflect our limited knowledge to date. Exploration of novel realms is encouraged, as it may unveil yet unexplored territory in the multidimensional crystallization space.

Analysis of Crystallization Conditions. To date, *in cubo* crystallization conditions of five membrane proteins have been reported: bR,[8,17] halorhodopsin,[31,32] photosynthetic reaction centers from *R. viridis* and *R. sphaeroides*, and light-harvesting complex 2 from *R. acidophila*.[31] The crystallization conditions are summarized in Table II.

TABLE II
CRYSTALLIZATION CONDITIONS FOR bR, hR, RCvir, RCsph, AND LH2[a]

Protein	Final protein concentration in cubic phase (mg/ml)	Additive class	Additive
bR	3.0–3.75	Solid salt	0.3 mg Sørensen salt/mg cubic phase
hR	3.3–4.0	Solid salt	0.3 mg Sørensen salt/mg cubic phase
		Solution	4 *M* KCl
			50 m*M* Tris–HCl, pH 7.0
RCvir	9	Solution	750 m*M* sodium acetate
			750 m*M* HEPES, pH 7.5
			37.5 m*M* cadmium sulfate
			2.6% 1,2,3/heptanetriol
RCsph	6	Solution	17% jeffamine M600
			750 m*M* HEPES, pH 7.5
			0.05 mg ammonium sulfate/μl
LH2	6–9	Solution	200 m*M* magnesium chloride
			100 m*M* Tris, pH 8.5
			3.4 *M* 1,6-hexanediol

[a] In all cases, crystallization was carried out in a 60% MO cubic phase at 20°.

Two types of precipitants were used in the *in cubo* crystallization method: solid salt was added to a preformed cubic phase in the cases of bR and hR. In the cases of hR, RCvir, RCsph, and LH2, solutions were added to a preformed cubic phase. It is noteworthy that the latter is possible only because lipidic cubic phases of the bicontinuous type are stable in coexistence with excess water (Fig. 1).

Variable Parameters for Crystallization in Cubo. It is advisable to assay the compatibility of a given lipidic cubic phase with various precipitants before precious protein samples are used in crystallization trials. We infer that the lipidic matrix does not necessarily need to remain in the cubic phase throughout the crystallization process. Therefore, agents that induce lipid phase transitions need not be excluded as long as the transparency of the lipid matrix is maintained, thereby allowing inspection and removal of crystals. Indeed, we have observed that crystals of the photosynthetic reaction center RCvir grow before the lipidic cubic phase is dissolved.[31] Preliminary assays of the stability of the lipidic cubic phase may be performed either with the glass tube method or with the micro method. Figure 3 shows how the addition of ammonium sulfate, the most widely used salt in the crystallization of biological macromolecules, affects the phase behavior of MO. The cubic phase is stable at low salt concentrations; at intermediate concentrations the lipidic material undergoes a phase separation, resulting in a liquid and a transparent lipidic phase; and at high concentrations the lipidic cubic phase becomes turbid. Deterioration of the lipidic cubic phase is not well understood and varies as function of the type of parameter assayed.

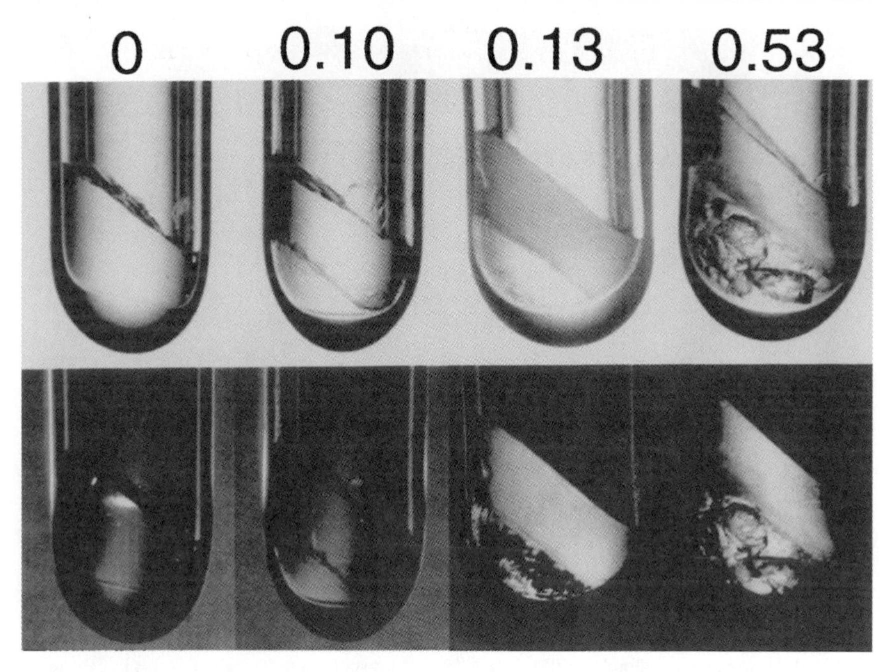

FIG. 3. Stability of a 60% MO/water lipidic cubic phase in the presence of ammonium sulfate. Images of glass tubes with lipidic phases viewed without polarizers (top) and with crossed polarizers (bottom). Numbers indicate the ratio of the mass of added ammonium sulfate to the mass of preformed cubic phase. Addition of a small amount of ammonium sulfate results in the formation of a new liquid phase. Above a threshold ratio of ca. 0.1 g ammonium sulfate/g cubic phase, the lipidic matrix becomes birefringent, indicating that phase transition has taken place.

LIPIDIC MATRIX. Crystals of bR were grown from cubic phases based on two different lipids: MO and monopalmitolein.[8,36] Hexagonal crystals were grown in cubic phases based on MO, whereas rhomboid crystals were grown in monopalmitolein (Fig. 4). Lipids may be used as a solid waxy powder or as a supercooled liquid when mixed with the membrane protein containing solution. Different pathways of hydration and subsequent formation of the cubic phase are presumably followed. Pure MO at room temperature exhibits a lamellar crystal (Lc) phase. The cubic phase is obtained by hydration *via* an intermediate lamellar ($L\alpha$) phase. Depending on the final degree of hydration, either the partially hydrated Ia3d or the highly hydrated Pn3m cubic phase is formed. Alternatively, pure MO may first be melted at temperatures higher than ca. 35° to form a fluid isotropic (FI) phase (Fig. 1). Subsequent slow cooling to room temperature maintains this fluid isotropic phase for at least several minutes. Due to its fluidity, the supercooled liquid MO can be mixed efficiently with the aqueous solution to form a lipidic cubic phase. Although the nature of the transient intermediate phases is unknown, the latter method provides an easy way of mixing and may be used as

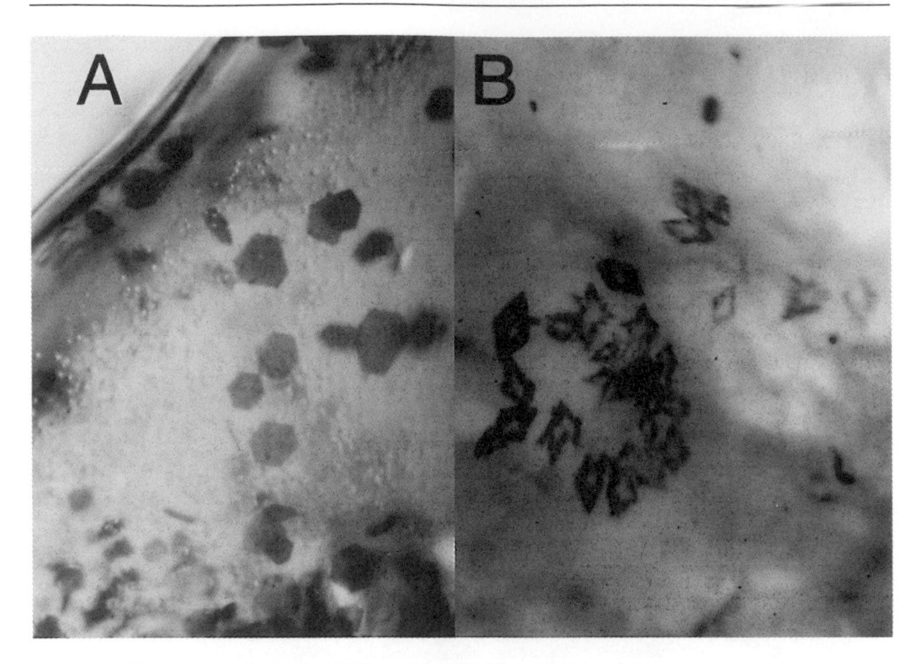

FIG. 4. Hexagonal and rhomboid bR crystals. (A) bR crystallizes as hexagonal plates in monoolein (MO) and (B) as rhomboids in monopalmitolein. Note that rhomboid crystals are extremely thin in the center, which may explain the low diffraction quality of these crystals.

an additional parameter to vary when setting up crystallization experiments. We have not observed any differences in growth or quality of bR crystals using either method.

Although many lipid classes form cubic phases, and some have been used for the generation of bR crystals, we have shown that one lipid, monoolein, at 60% content, can form a matrix for the crystallization of a set of five different membrane proteins.[31] It is therefore recommended to carry out the initial crystallization screening with monoolein and to consider variation of the lipid in the subsequent refinement phase of the crystallization conditions. The lyotropic and thermotropic phase behavior of several monoacylglycerols have been established[14,38,39] and should be consulted when varying the lipid. Most importantly, cubic phases based on monoacylglycerols form readily and spontaneously, as compared to some lipid mixtures that require numerous temperature cycling.[40]

DETERGENT. Both β-OG and dodecyldimethylamine oxide (LDAO) have been employed at low concentrations in combination with the *in cubo* method.[31] High

[38] U. Aota-Nakano, S. J. Li, and M. Yamazaki, *Biochim. Biophys. Acta* **1461,** 96 (1999).

[39] H. Qiu and M. Caffrey, *Biomaterials* **21,** 223 (2000).

[40] B. Tenchov, R. Koynova, and G. Rapp, *Biophys. J.* **75,** 853 (1998).

concentrations of detergent are not compatible with the existence of lipidic cubic phases because of the tendency of detergents to form positively curved entities, whereas bicontinuous cubic phases are composed of negatively curved membranes. Mixtures of detergents with water give rise to specific phase behavior, which may include cubic phases of the micellar type.[41] Therefore, at high concentrations, β-OG and presumably other detergents induce the formation of a lamellar phase.[42] The effect of several detergents on the phase behavior of MO has been investigated,[42,43] and the stability of a MO cubic phase with several detergents is compiled in Landau *et al.*[27] However, due to the complexity of these systems, a comprehensive study is lacking.

Ai and Caffrey[44] have investigated the effect of dodecylmaltoside (DM) on the phase behavior of MO to rationalize its use for membrane protein crystallization. Our experience with DM suggests that it destabilizes MO cubic phases more readily than β-OG. One should therefore avoid its use, at least in the initial screening. Finally, it is noteworthy that detergent plays a minor role, if at all, in the crystallization of bR, as we have demonstrated that the crystallization process can be carried out in the absence of detergent altogether.[28] This approach is feasible in special cases when the membrane protein can be obtained in an enriched form.

PROTEIN. Purification of a membrane protein usually involves detergent solubilization of membranes. As mentioned earlier, large amounts of detergent are not compatible with the existence of the lipidic cubic phase. Rather than the membrane protein itself, the detergent is usually the limiting factor when large amounts of protein are to be incorporated into the cubic phase. The maximum amount of membrane protein that can therefore be incorporated in the cubic matrix depends on the concentration and type of the detergent in the protein solution. This upper limit of detergent can be determined experimentally by mixing various fractions of the protein solution with buffer and cubic phase-forming lipid (e.g., MO) and observing the resulting lipidic phase. Concentrations of membrane proteins used in *in cubo* crystallization are listed in Table II.

SALT. The maximum amounts of salts that may be added to a preformed cubic phase without any significant effect on the optical properties of the cubic phase are listed in Table III.

TEMPERATURE. The stability of the lipidic cubic phase and the membrane protein define the temperature window for the crystallization trials. Typically, MO-based experiments are performed at room temperature or at 20°. It is prudent

[41] P. Sakaya, J. M. Seddon, and R. H. Templer, *J. Phys. II* **4,** 1311 (1994).

[42] B. Angelov, M. Ollivon, and A. Angelova, *Langmuir* **15,** 8225 (1999).

[43] J. Gustafsson, T. Nylander, M. Almgren, and H. Ljusberg-Wahren, *J. Coll. Interf. Sci.* **211,** 326 (1999).

[44] X. Ai and M. Caffrey, *Biophys. J.* **79,** 394 (2000).

TABLE III
STABILITY OF A 60% MO/WATER CUBIC PHASE[a]

Salt	Formula	No. of crystallizations	MW (g/mol)	Compatibility limit (w/w)
Ammonium sulfate	$(NH_4)_2SO_4$	497	32.1	0.2
Sodium chloride	NaCl	261	58.44	1
Magnesium chloride	$MgCl_2*6H_2O$	188	203.3	2
Calcium chloride	$CaCl_2*2H_2O$	103	147.02	0.1
Sodium phosphate (disodium)	Na_2HPO_4	101	141.96	0.3
Sodium phosphate (dihydrogen)	NaH_2PO_4	101	120.0	0.3
Sodium citrate	$Na_3C_6H_5O_7*2H_2O$	95	294.1	0.5
Potassium phosphate (dipotassium)	$K_2 (HPO_4)$	89	174.18	0.2
Potassium phosphate (dihydrogen)	KH_2PO_4	89	136.09	2
Potassium chloride	KCl	53	74.56	1
Magnesium sulfate	$MgSO_4$	31	120.4	0.05
Lithium sulfate	Li_2SO_4	19	109.9	0.13
Manganese(II) chloride	$MnCl_2*4H_2O$	18	197.9	2
Lithium chloride	LiCl	16	42.29	0.33
Ammonium phosphate (diammonium)	$(NH_4)_2HPO_4$	16	132.06	0.5
Zinc sulfate	$ZnSO_4*7H_2O$	14	287.54	0.2
Cadmium(II) chloride	$CdCl_2$	7	183.3	2

[a] Stability was assayed in glass vials following the addition of solid salts. Salts were chosen according to their success rate in promoting the crystallization of soluble proteins as reported in the biological macromolecule crystallization database and the NASA archive for protein crystal growth data, version 2.00 (http://www.bmcd.nist.gov:8080/bmcd/bmcd.html).

to keep this temperature constant, especially during microscopic observation, as even slight variations in temperature may alter the phase of the lipid, which in turn could affect crystallization. Upon increasing the temperature, liquid inclusions, cracks, and turbidity may develop and decrease the visibility in the lipidic phase. Lipidic cubic phases are usually kinetically stable upon cooling: a 60% MO cubic phase (Pn3m) cooled to 5° and kept for days at this temperature does not undergo phase transition to the lamellar crystal (Lc) phase, as would be expected based on its phase diagram. This supercooling phenomenon may be used to one's advantage, e.g., by keeping *in cubo* crystallization trials at 10°.

HYDRATION. As noted previously, 60% MO was used successfully in the crystallization of five different membrane proteins. We therefore recommend not to vary the lipid content in the initial screens. Controlled dehydration of the cubic phase (reduction of the water content by 4%) was used successfully to obtain bR crystals.[45] In this experiment, bR embedded into the MO cubic phase was initially equilibrated with 400 m*M* of a Sørensen phosphate solution; excess solution was subsequently removed and the cubic phase was allowed to dry for several minutes. Loss of water was monitored gravimetrically. BR microcrystals formed within a few days. Because this procedure is very easy to perform, it may be tried routinely before setting up a large number of crystallization experiments.

pH. Because monoacylglycerols are uncharged, pH may be varied widely in *in cubo* crystallization trials. It is advisable to work in a pH range at which the protein function is not altered.

PRESSURE. Pressure may be applied mechanically[46,47] or osmotically. Both have an effect on the phase behavior and unit cell parameters of cubic phases[48] and may thus be used for crystallization purposes. Some lipids form cubic phases only when pressure is applied.[49] It is noteworthy that polymers may be used to increase the pressure by osmotic means. They have been used in the crystallization of RCsph and bR.[31,45]

Crystallization Screening Kits

Several types of commercial screening solutions that may be used in cubic phase crystallization trials are available (Table IV). As pointed out earlier, it is advisable to first assay the effect of these solutions on lipidic cubic phase stability using samples without protein.

Crystal Detection: The Color Issue

So far, only membrane proteins that contain chromophores have been crystallized by the *in cubo* method. Compared to noncolored microcrystals, colored ones are easier to recognize even if the background contains salt crystals, cracks in the cubic phase, and inclusions of liquid droplets—all of which occur frequently in crystallization setups. There is no reason to assume that noncolored proteins cannot be crystallized *in cubo*. In fact, the noncolored soluble protein lysozyme can be crystallized, visualized, and prepared for X-ray diffraction experiments *in cubo*.[27] However, care and expertise are needed in order to distinguish protein crystals and not confuse them with deliberately added salt crystals. Noncolored protein crystals

[45] P. Nollert, H. Qiu, M. Caffrey, J. P. Rosenbusch, and E. M. Landau, *FEBS Lett.*, in press.

[46] C. Cesclik, R. Winter, G. Rapp, and K. Bartels, *Biophys. J.* **68,** 1423 (1995).

[47] R. Winter, J. Erbes, C. Czeslik, and A. Gabke, *J. Phys. Cond. Mat.* **10,** 11499 (1998).

[48] P. Mariani, B. Paci, P. Bosecke, C. Ferrero, M. Lorenzen, and R. Caciuffo, *Phys. Rev. E.* **54,** 5840 (1996).

[49] R. Koynova, B. Tenchov, and G. Rapp, *Biochim. Biophys. Acta Biomembr.* **1326,** 167 (1997).

TABLE IV
SOLUTIONS COMPATIBLE WITH A 60% MO CUBIC PHASE[a]

Screen	No.	Crystallizing agent	Buffer (100 m*M*)	Salt (200 m*M*)
I	2	20% (w/v) PEG-8000	CHES, pH 9.5	—
I	7	10% (w/v) PEG-8000	MES, pH 6.0	$Zn(OAc)_2$
I	9	1.0 *M* $(NH_4)_2HPO_4$	Acetate, pH 4.5	—
I	10	20% (w/v) PEG-2000 MME	Tris, pH 7.0	—
I	11	20% (v/v) 1,4-butanediol	MES, pH 6.0	Li_2SO_4
I	12	20% (w/v) PEG-1000	Imidazole, pH 8.0	$Ca(OAc)_2$
I	16	2.5 *M* NaCl	Na/K phosphate, pH 6.2	—
I	17	30% (w/v) PEG-8000	Acetate, pH 4.5	Li_2SO_4
I	22	10% (v/v) 2-propanol	Tris, pH 7.5	—
I	25	30% (w/v) PEG-400	Tris, pH 8.5	$MgCl_2$
I	37	2.5 *M* NaCl	Imidazole, pH 8.0	—
I	40	10% (v/v) 2-propanol	MES, pH 6.0	$Ca(OAc)_2$
I	42	15% (v/v) ethanol	Tris, pH 7.0	—
I	44	30% (w/v) PEG-400	Acetate, pH 4.5	$Ca(OAc)_2$
I	48	20%(w/v) PEG-1000	Acetate, pH 4.5	$Zn(OAc)_2$
II	1	10% (w/v) PEG-3000	Acetate, pH 4.5	$Zn(OAc)_2$
II	8	10% (w/v) PEG-8000	Na/K phosphate, pH 6.2	NaCl
II	10	1 *M* $(NH_4)_2HPO_4$	Tris, pH 8.5	—
II	11	10% (v/v) 2-propanol	Cacodylate, pH 6.5	$Zn(OAc)_2$
II	12	30% (w/v) PEG-400	Cacodylate, pH 6.5	Li_2SO_4
II	13	15% (v/v) ethanol	Citrate, pH 5.5	Li_2SO_4
II	15	1.26 *M* $(NH_4)SO_4$	HEPES, pH 7.5	—
II	17	2.5 *M* NaCl	Tris, pH 7.0	$Ca(OAc)_2$
II	20	15% (v/v) ethanol	MES, pH 6.0	$Zn(OAc)_2$
II	22	10% (v/v) 2-propanol	Imidazole, pH 8.0	—
II	23	15% (v/v) ethanol	HEPES, pH 7.5	$MgCl_2$
II	27	10% (w/v) PEG-3000	Cacodylate, pH 6.5	$MgCl_2$
II	28	20% (w/v) PEG-8000	MES, pH 6.0	$Ca(OAc)_2$
II	29	1.26 M $(NH_4)_2SO_4$	CHES, pH 9.5	NaCl
II	30	20% (v/v) 1,4-butanediol	Imidazole, pH 8.0	$Zn(OAc)_2$
II	31	1.0 *M* sodium citrate	Tris, pH 7.0	NaCl
II	42	30% (w/v) PEG-400	HEPES, pH 7.5	NaCl
II	43	10% (w/v) PEG-8000	Tris, pH 7.0	$MgCl_2$
II	44	20% (w/v) PEG-1000	Cacodylate, pH 6.5	$MgCl_2$
II	47	2.5 *M* NaCl	Imidazole, pH 8.0	$Zn(OAc)_2$

[a]Solutions are from Emerald Biostructures Wizard screening kit I and II.

may be stained in order to improve their visibility. We have not yet identified a universal stain for such purposes. Alternatively, noncolored proteins may be labeled with a dye either by a chemical reaction[50] or genetically by fusion with a colored

[50] R. P. Haugland, *in* "Handbook of Fluorescent Probes and Research Chemicals," 6th Ed., 2000. http://www.probes.com/

domain, such as green fluorescent protein,[51,52] or a cytochrome domain.[53] For conventional absorption microscopy, the ratio of absorption coefficient/mass of the protein should be larger than 1 cm^2/(mmol Da) if crystals as small as 5 μm are to be identified by their color. When fluorescence tags are used, detection can be performed with fluorescence microscopy. Micro setups are particularly well suited for this observation mode because they consist of thin layers of cubic phase (therefore low background fluorescence) and possess a planar surface.

Summary

Our understaning of lipidic cubic phases for the crystallization of membrane proteins has advanced greatly since the inception of the concept[8] in 1996, and the method is becoming well accepted. Several protocols that allow the efficient screening of crystallization conditions and handling of crystals are presented. State-of-the art micro techniques allow a large number of crystallization conditions to be tested using very small amounts of protein, and diffraction quality crystals can be grown in larger volumes in glass vials. *In cubo* crystallization conditions differ from those employed for detergent-solubilized proteins. Variations comprise the type of lipid matrix, detergent, protein, salt, temperature, hydration, pH, and pressure. Commercially available screening kits may be applied in order to define lead conditions. Once obtained, crystals may be removed from the surrounding cubic phase mechanically, by enzymatic hydrolysis, or by detergent solubilization. We anticipate this set of protocols to be applied successfully to larger, less stable, and noncolored membrane proteins in order to obtain well-diffracting crystals of membrane proteins that have so far evaded crystallization in the detergent-solubilized state.

Acknowledgments

These studies were supported by the following grants: Human Frontier Research Science Organization LT0156/1999-M (to P.N.); Senior Fogarty International Fellowship, NIH F06TW02311, and Robert A. Welch Foundation H-1475 (to J.N.); and Swiss National Science Foundation SPP Biotechnology 5002-37911, -46092, and -55179, and EU Biotechnology PL 970415 (to E.M.L.). We thank Mark L. Chiu, Dirk Neff, Eva Pebay-Peyroula, and Jurg P. Rosenbusch for insightful discussion and Ariane Hardmeyer, Gabriele Rummel, and Christine Widmer for expert technical assistance. The MPL is supported by funds from the HHMI.

[51] L. Kallal and J. L. Benovic, *Trends Pharm. Sci.* **21,** 175 (2000).

[52] G. S. Baird, D. A. Zacharias, and R. Y. Tsien, *Proc. Natl. Acad. Sci. U.S.A.* **96,** 11241 (1999).

[53] G. G. Prive and H. R. Kaback, *J. Bioenerg. Biomembr.* **28,** 29 (1996).

[13] N-Linked Carbohydrates on G Protein-Coupled Receptors: Mapping Sites of Attachment and Determining Functional Roles

By DAVID P. DAVIS and DEBORAH L. SEGALOFF

Introduction

As with other cell surface proteins, G protein-coupled receptors (GPCRs) are typically posttranslationally modified by the attachment of carbohydrates on extracellular sites of the protein. Glycosylation of GPCRs has been found generally to facilitate the proper folding and cell surface expression of the receptors. The degree of dependency on carbohydrates for proper folding and expression, however, can vary from one GPCR to another.

The glycosylation of proteins can be either of two types, N-linked or O-linked, categorized by their method of glycan attachment. N-linked glycosylation refers to the attachment of an oligosaccharide core consisting of $Glc_3Man_9GlcNAc_2$ onto an asparagine residue of a nascent protein. Asparagine residues, which serve as potential substrates for the attachment of N-linked carbohydrates, are found within the consensus sequence Asn-X-Ser/Thr, where X can be any amino acid except proline or aspartate. Not all asparagines within an Asn-X-Ser/Thr sequence, however, are necessarily glycosylated even if they are within a portion of the receptor predicted to be extracellular. Therefore, sites that are glycosylated must be determined experimentally. O-linked glycosylation describes the process of oligosacharide attachment to the hydroxyl moiety of a serine or threonine residue, usually with the core sugar GalNAc. In contrast to N-linked glycosylation, a consensus sequence has not been defined for O-linked glycosylation. Both N- and O-linked carbohydrates are modified as the nascent glycoprotein progresses through the endoplasmic reticulum (ER) and Golgi to the cell surface. In either case, the core oligosaccharide is processed by a series of enzymatic reactions that both trim and add sugar residues to the core saccharide. Depending on their stage of processing, N-linked carbohydrates can be grouped into three classes: high mannose, hybrid, and complex.[1] A high mannose glycoprotein is usually isolated from the ER and represents an ER resident protein or the immature population of a secretory or membrane glycoprotein. Hybrid and complex carbohydrates (characterized by the presence of more than the two core GlcNAc residues) are usually found on glycoproteins that have exited the ER and have been processed by a series of enzymes residing in the Golgi. The degree of processing may include not only varying numbers of GlcNAc residues, but also galactose, glucose, sialyl, and fucosyl residues.

[1] R. Kornfeld and S. Kornfeld, *Annu. Rev. Biochem.* **54,** 631 (1985).

METHODS IN ENZYMOLOGY, VOL. 343
0076-6879/02 $35.00

Because the potential for N-linked glycosylation is readily determined by the presence of the consensus sequence Asn-X-Ser/Thr, GPCRs have typically been evaluated for the presence and function of N-linked carbohydrates. There have been only a few instances where GPCRs have also been examined for O-linked glycosylation as well. Of those, only one GPCR, the V2 vasopressin receptor, has thus far been shown to be both O- and N-glycosylated.[2]

This article focuses primarily on methodologies that can be utilized to determine the sites of N-linked carbohydrates on a GPCR, as well as techniques directed toward elucidating the potential roles of N-linked glycosylation.

Determining the Glycosylation Status of a GPCR

Identifying a Glycosylated GPCR

Several approaches can be utilized to verify the glycosylation status of a GPCR. The most direct method involves monitoring the incorporation of a radiolabeled precursor for N-linked glycosylation into the receptor.[3] This is achieved by incubating cells with [^{14}C]glucosamine and then determining the extent of incorporation of the labeled sugar into the immunoprecipitated receptor.[3] The use of lectins, however, affords a simpler and nonradioactive approach for determining the presence of N-linked carbohydrates on a GPCR. Lectins bind carbohydrates with a high affinity and allow the separation of glycosylated and nonglycosylated proteins from a given sample. Two commonly used lectins include wheat germ agglutinin (WGA) and concanavalin A (Con A). WGA binds preferentially to GlcNAc-enriched glycoproteins, which would be found in complex and hybrid glycans of GPCRs at the cell surface and within the late Golgi. In contrast, ConA binds preferentially to high mannose-containing glycans, which would be found on the immature forms of a GPCR in the ER and early Golgi. Other lectins with differing specificities are also available. The binding of GPCRs to lectins can be ascertained by Western blotting of the receptor with tagged lectins or by testing the absorption of the GPCR to agarose-coupled lectins.

Specific endoglycosidases can also be utilized to determine if a GPCR is glycosylated and the nature of the glycosylation (see Table I). In general, a detergent-soluble extract from cells expressing the receptor is prepared and then incubated in the absence or the presence of a given endoglycosidase. The samples are then resolved by SDS–PAGE, transferred to nitrocellulose, and probed with an antibody specific for the receptor. A decreased molecular mass of the receptor, when treated with endoglycosidase as compared to receptor incubated under identical

[2] H. Sadeghi and M. Birnbaumer, *Glycobiology* **9,** 731 (1999).

[3] U. E. Petaja-Repo, M. Hogue, A. Laperriere, P. Walker, and M. Bouvier, *J. Biol. Chem.* **275,** 13727 (2000).

TABLE I
ENDOGLYCOSIDASES

Enzyme	Alternative name	Specificity	Incubation conditions			
			Enzyme concentration	Time and temperature	Optimal pH	Detergents and buffers
Peptide-*N*-glycosidase F	*N*-Glycosidase F (PNGaseF)	Removes high mannose, hybrid, and complex N-linked glycans via cleavage between the asparagine side chain and the glycan	1–40 units/ml	Overnight at 37°	6–8.5	Active in nonionic detergents and up to 0.2% SDS
Sialidase	Neuraminidase	Removes terminal sialyl residues from O-linked and complex N-linked carbohydrates	20–100 units/mg substrate	2–6 hr at 37°	5–5.5	Active in nonionic detergents
Endo-β-*N*-acetylglucosaminidase H	Endoglycosidase H (Endo H)	Preferential cleavage of high mannose N-linked carbohydrates. Cleaves between core GlcNAc residues	15–40 units/mg substrate	Overnight at 37°	5–6	Active in nonionic detergents. If total protein is less than 100 μg/ml, the SDS concentration should not exceed 0.02% SDS; otherwise active up to a 1 : 1 molar ratio of SDS to total protein
O-Glycan-peptide hydrolase	*O*-Glycosidase	Removes the Gal β(1–3) GalNAc from serine and threonine linkages. Hydrolysis is prevented by the presence of sialic acid residues (which can be removed by neuraminidase)	1–2 milli-unit/mg substrate	Overnight at 37°	6–7.6	Active in nonionic detergents and up to 0.1% SDS. Cl^- ions are inhibitory

conditions but without endoglycosidase, would be indicative of the removal of carbohydrates from the receptor by endoglycosidase. The reliability of this assay depends on the ability of the SDS–PAGE system to resolve small differences in the molecular mass of the GPCR. Therefore, this method works best with proteins that have a high ratio of carbohydrate to protein mass.

O-linked carbohydrates are removed by treating the samples first with neuraminidase, which removes terminal sialic acids, and then with *O*-glycanase. The initial treatment of the sample with neuraminidase is necessary because sialic acids, commonly found on O-linked carbohydrates, inhibit the activity of *O*-glycanase.

All N-linked carbohydrates can be cleaved using PNGaseF. The extent of processing of the N-linked carbohydrates can be assessed further using endoglycosidases specific for N-linked carbohydrates of specific stages of maturation. For example, endoglycosidase H (endo H) cleaves only those N-linked carbohydrates of a high mannose form that would be found on immature receptors within the ER. In contrast, substrates for β-galactosidase, which removes terminal galactose residues, and neuraminidase, which removes terminal sialic acids, are complex and hybrid N-linked carbohydrates found on more mature receptors that have progressed past the ER and *cis*-Golgi. Therefore, the immature precursor forms of a GPCR would be predicted to be sensitive to endo H, but not to neuraminidase or β-galactosidase, whereas the mature form of the receptor on the plasma membrane would be predicted to be insensitive to endo H, but sensitive to neuraminidase or β-galactosidase. Both mature and immature forms, however, should be sensitive to PNGaseF.

An example of results that might be obtained from such an analysis are shown in Fig. 1 for the rat lutropin/choriogonadotropin receptor (rLHR). HEK 293 cells transfected with the cDNA for the rLHR exhibit a broad band of ~81–89 kDa, as well as a band of ~68 kDa. The 81- to 89-kDa complex is shifted to a lower mass after neuraminidase treatment but not after endo H treatment. Conversely, the ~68-kDa band is reduced in mass following endo H treatment but not following neuraminidase treatment. All are reduced to one band of ~59 kDa after PNGaseF treatment. These results are consistent with the identification of the ~81- to 89-kDa complex as the mature rLHR and the ~68-kDa protein as the immature, precursor form of the rLHR.[4,5] Pulse-chase experiments verify the conclusions derived from the differential sensitivity of the rLHR bands to endoglycosidases.[5] The differential sensitivity of a GPCR to endoglycosidases can also provide indirect information with respect to the localization of the protein.[6] For instance, a mutant GPCR retained in the ER would be sensitive to endo H. An example

[4] D. M. Thomas and D. L. Segaloff, *Endocrinology* **135,** 1902 (1994).
[5] R. W. Hipkin, J. Sanchez-Yague, and M. Ascoli, *Mol. Endocrinol.* **6,** 2210 (1992).
[6] T. G. Rozell, H. Wang, X. Liu, and D. L. Segaloff, *Mol. Endocrinol.* **9,** 1727 (1995).

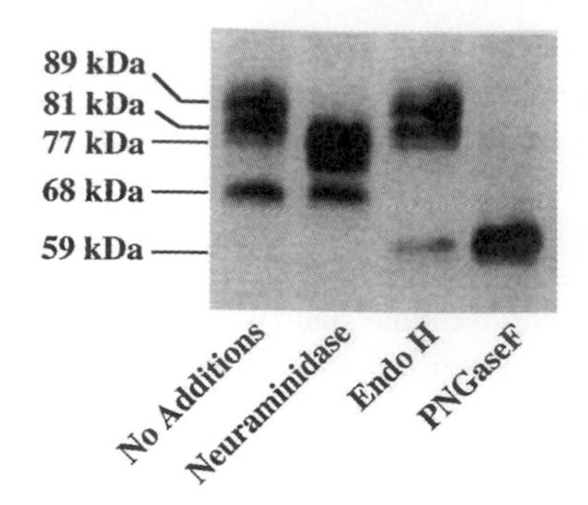

FIG. 1. Effect of glycosidase treatment on rLHR. Detergent-soluble extracts from 293 cells expressing the rLHR (wt) were incubated at 37° with no additions for 18 hr, 300 mU/ml neuraminidase for 2 hr, 300 mU/ml endo H for 18 hr, or 100 U/ml PNGaseF for 18 hr. Samples were electrophoresed on a SDS–polyacrylamide gel under nonreducing conditions, transferred to a PVDF membrane, and probed with a polyclonal antibody to the rLHR. Adapted with permission from D. M. Thomas and D. L. Segaloff, *Endocrinology* **135,** 1902 (1994).

of such a mutant of the rLHR is shown in Fig. 2. Unlike the wild-type rLHR, the rLHR(D397K) mutant is expressed only as the immature, precursor form of the receptor, which is shifted to a lower molecular mass after treatment with endo H. These data are consistent with the intracellular retention of this mutant in the ER.

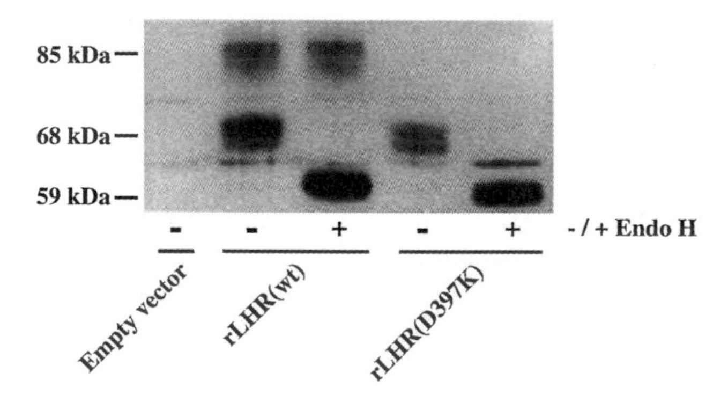

FIG. 2. Expression and endo H sensitivity of wild-type and D397K rLHRs. Clonal 293 cells stably expressing either the wild-type rLHR or rLHR(D397K) were solubilized in detergent and incubated in the absence or presence of 300 mU/ml endo H for 18 hr at 37°. Proteins were resolved on a 7% polyacrylamide gel under reducing conditions, transferred to a PVDF membrane, and probed with an antibody to the rLHR. Adapted with permission from T. G. Rozell, H. Wang, X. Liu, and D. L. Segaloff, *Mol. Endocrinol.* **9,** 1727 (1995).

Determining which Potential N-linked Glycosylation Sites on a GPCR Are Glycosylated

Site-Directed Mutagenesis of Consensus Sequences for N-linked Glycosylation

From the deduced primary sequence of a GPCR, a search for N-linked glycosylation consensus sequences (Asn-X-Ser/Thr) in the predicted extracellular regions would yield those sites that theoretically may contain attached carbohydrates. One approach to determine which sites are actually glycosylated is to purify the receptor, prepare peptides thereof, and analyze the various peptides for glycan attachment. This, however, requires large quantities of receptor and is very labor-intensive. A more commonly used experimental approach, therefore, has been the use of site-directed mutagenesis to selectively disrupt one or more potential sites of glycosylation. In one scheme, the consequences of mutating one site at a time can be studied. Alternatively, or in addition, all but one site can be mutated, leaving one potential site at a time intact. To illustrate the use of mutagenesis to determine sites of N-linked glycosylation, results obtained with the rLHR are presented.[7]

The rLHR contains six potential sites for N-linked glycosylation located at asparagines 77, 152, 173, 269, 277, and 291. As shown in Table II and III, two series of mutants were created by site-directed mutagenesis. In series A, only one or two consensus sequences at a time were disrupted (Table II). In series B, mutants were created in which all but one consensus site were mutated (Table III). Because N-linked carbohydrates are attached to asparagine residues, mutation of the asparagine within the Asn-X-Ser/Thr consensus sequence would obviously prevent the attachment of carbohydrate at that site. To maintain a similarly sized amino acid in its place, asparagine residues are typically mutated to glutamines. It has been shown that the serine or threonine residue within the N-linked glycosylation consensus sequence is essential for hydrogen bonding during the transfer of the precursor oligosaccharide.[1,8] Therefore, mutation of the serine or threonine residue within the consensus sequence can also be used to prevent carbohydrate attachment to the asparagine residue within that consensus site.[7,9,10] In this case, alanine is typically substituted for the serine or threonine.

The rLHR cDNA was mutated using the PCR overlap method originally described by Horton and Ho[11,12] or, for multiple substitutions, the pSELECT method from Promega. HEK 293 cells were transfected with the cDNA for the wild-type or a mutant rLHR and then detergent solubilized. The lysates were resolved by

[7] D. P. Davis, T. G. Rozell, X. Liu, and D. L. Segaloff, *Mol. Endocrinol.* **11,** 550 (1997).
[8] E. Bause and G. Legler, *Biochem. J.* **195,** 639 (1981).
[9] G. Hortin and I. Boime, *J. Biol. Chem.* **255,** 8007 (1980).
[10] M. M. Matzuk and I. Boime, *J. Cell Biol.* **106,** 1049 (1988).
[11] S. N. Ho, H. D. Hunt, R. M. Horton, J. K. Pullen, and L. R. Pease, *Gene* **77,** 51 (1989).
[12] R. M. Horton, H. D. Hunt, S. N. Ho, J. K. Pullen, and L. R. Pease, *Gene* **77,** 61 (1989).

TABLE II
SERIES A rLHR MUTANTS USED TO DETERMINE SITES OF N-LINKED GLYCOSYLATION[a]

Receptor mutant	Consensus sequence(s) disrupted
rLHR(N77Q)	Asn-77
rLHR(N152Q)	Asn-152
rLHR(N77,152Q)	Asn-77 and Asn-152
rLHR(N173Q)	Asn-173
rLHR(T175A)	Asn-173
rLHR(N269Q)	Asn-269
rLHR(N269,277Q)	Asn-269 and Asn-277
rLHR(N291Q)	Asn-291

[a] Adapted with permission from D. P. Davis, T. G. Rozell, X. Liu, and D. L. Segaloff, *Mol. Endocrinol.* **11,** 550 (1997).

SDS–PAGE and the receptor visualized by Western blotting. Using the series A mutants, in which one (or two) sites at a time were disrupted, one would expect to observe a decreased molecular mass (i.e., increased mobility) of the mutant if the carbohydrate was normally attached to the disrupted site(s). Initial experiments with series A mutants, however, showed little change in the mobilities of the mutants as compared to the wild-type rLHR. As mentioned earlier, a decreased mass of a mutant could be difficult to discern if the ratio of carbohydrate to protein mass is relatively small. One method to maximize the contribution of carbohydrate relative to protein mass is to digest the GPCR with a protease that would be predicted to release a receptor fragment containing the potential sites of N-linked carbohydrate. This approach was utilized with the rLHR, where the detergent extracts containing

TABLE III
SERIES B rLHR MUTANTS USED TO DETERMINE SITES OF N-LINKED GLYCOSYLATION[a]

Receptor mutant	Consensus sequence maintained
rLHR(N152,269,277,291Q;T175A)	Asn-77
rLHR(N77,269,277,291Q;T175A)	Asn-152
rLHR(N77,152,269,277,291Q)	Asn-173
rLHR(N77,152,277,291Q;T175A)	Asn-269
rLHR(N77,152,269,291Q;T175A)	Asn-277
rLHR(N77,152,269,277Q;T175A)	Asn-291
rLHR(N77,152,269,277,291Q;T175A)	None

[a] Adapted with permission from D. P. Davis, T. G. Rozell, X. Liu, and D. L. Segaloff, *Mol. Endocrinol.* **11,** 550 (1997).

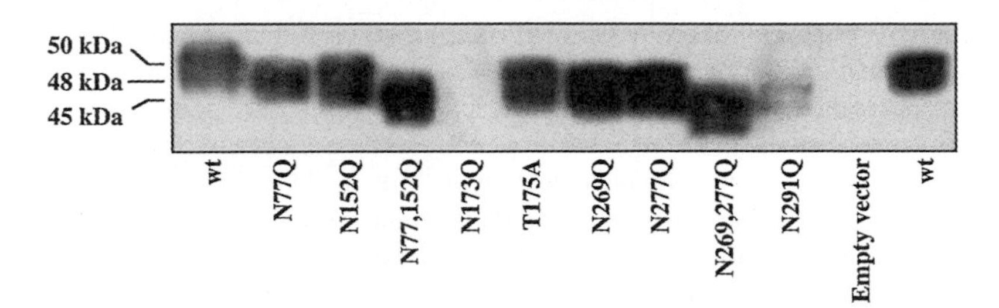

FIG. 3. Identification of rLHR N-linked glycosylation sites using series A mutants in which one or two sites are mutated at a time. HEK 293 cells transiently expressing the empty vector, the wild-type (wt) rLHR, or rLHR mutants containing the indicated substitutions were detergent solubilized. Equal amounts of protein were incubated for 2 hr at 25° in the presence of 50 m*M* *N*-chorosuccinimide. After this incubation, samples were resolved under reducing conditions on a 10% acylamide SDS–PAGE gel, transferred to a PVDF membrane, and probed with an antibody to the rLHR. Adapted with permission from D. P. Davis, T. G. Rozell, X. Liu, and D. L. Segaloff, *Mol. Endocrinol.* **11,** 550 (1997).

receptor were treated with 50 m*M* *N*-chlorosuccinimide (a reagent that cleaves proteins after tryptophan residues) prior to resolving them by SDS–PAGE. The largest fragment predicted to be released by this treatment contains all six potential sites for N-linked glycosylation as well as the epitope recognized by the antibody being used. Figure 3 shows a Western blot of the rLHR in which individual glycosylation consensus sequences were mutated (i.e., the series A mutants shown in Table II) and the receptor was cleaved by *N*-chlorosuccinimide prior to SDS–PAGE. All the mutant receptors that could be detected showed a reduced mass as compared to the wild-type rLHR, suggesting that each of these are normally glycosylated.

Also seen in Fig. 3 is the lack of a detectable receptor when Asn-173 is mutated to glutamine. One interpretation of these data is that the N-linked carbohydrate at Asn-173 is required for the correct folding and stable expression of the rLHR. However, it is also possible that, in addition to potentially preventing the addition of carbohydrate, mutation of a consensus glycosylation site may be disrupting conformation of the polypeptide backbone independent of any effects on N-linked glycosylation. Misfolded receptors would be predicted to be retained in the ER and ultimately degraded, potentially at a faster rate than the wild-type receptor. One way to address whether a given change in GPCR expression (or function) is due to the prevention of attachment of carbohydrate on the asparagine in question versus an alteration in the polypeptide backbone is to instead mutate the serine or threonine residue within the consensus sequence. If decreased expression of the GPCR is observed when an asparagine is mutated, but not when the serine or threonine within the same consensus sequence is mutated, this would confirm that results observed with the asparagine mutation were due to a perturbation of the polypeptide backbone and not to the prevention of the attachment of a carbohydrate at that site. An example of this is seen with the rLHR (see Fig. 3) where mutation of Asn-173 abolishes rLHR expression, but mutation of Thr-175 to Ala has no adverse effect on rLHR expression. Unfortunately, the results remain ambiguous in instances where mutations of either the asparagine or the serine/threonine within a given consensus sequence result in impaired receptor expression.

For the rLHR, results obtained by studying mutants in which only one glycosylation site at a time were disrupted (series A, Table II) suggest that all six sites of the receptor are actually glycosylated (Fig. 3). To further support this conclusion, mutants in series B (in which all but one particular consensus site were disrupted, see Table III) were also studied. As shown in Figs. 1 and 4, both the mature and the immature forms of the wild-type rLHR are converted by PNGaseF treatment to a ~59-kDa protein on SDS gels. However, as shown in Fig. 4, the PNGaseF-treated wild-type rLHR migrates with a larger mass than a fully nonglycosylated mutant of the rLHR (compare lanes 3 and 4). These results suggest that this PNGaseF treatment (15 hr at 37° with 32 U/ml enzyme) or one utilizing even higher concentrations of enzyme does not seem to remove all N-linked carbohydrates from the wild-type rLHR. However, the PNGaseF treatment could efficiently remove N-linked carbohydrates from mutant rLHRs containing fewer potential sites of N-linked glycosylation. Thus, as shown in Fig. 4, there was a reduction in mass observed for each of the series B rLHR mutants after PNGaseF treatment. These results further support the conclusion that all six potential sites for N-linked glycosylation of the rLHR are actually glycosylated.

Assessment of Functional Roles of N-linked Glycosylation on a GPCR

N-linked carbohydrates may affect one or more aspects of GPCR expression and function. For example, the carbohydrates may facilitate the proper folding of the nascent receptor. This may be reflected either in the strict dependence of proper receptor folding on normal carbohydrate content or the less stringent situation where the carbohydrates may aid in the efficiency of the folding process but are not absolutely required. Improper or incomplete folding of a GPCR may, in turn, be reflected by the retention of the receptor in the ER and the possible increased degradation of the receptor. In either case, a decreased expression of the receptor on the cell surface would be expected. In addition to its possible role in receptor folding, N-linked carbohydrates may be required for functional properties of the receptor, such as ligand binding or agonist-stimulated second messenger production.

Creating Nonglycosylated GPCRs and Analyzing Their Expression and Function

The disruption of one or more sites of N-linked glycosylation by site-directed mutagenesis can be utilized to determine the contributions of those oligosaccharide chains to receptor function. Presumably, a panel of receptor constructs where

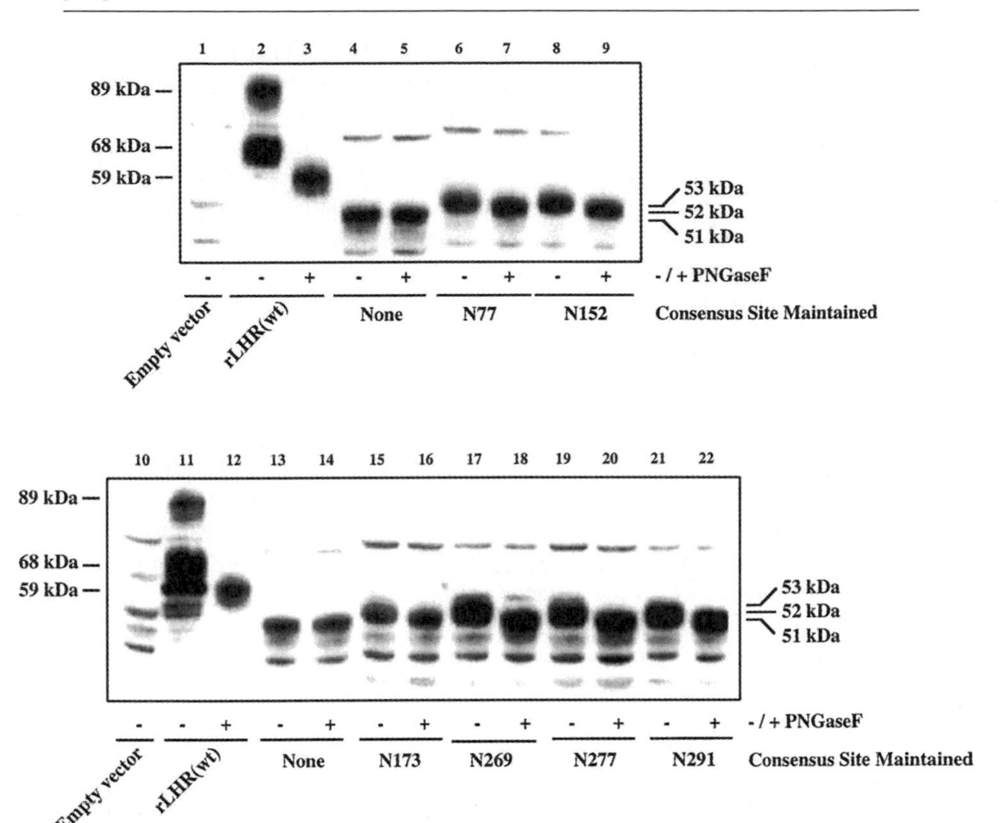

FIG. 4. Identification of rLHR N-linked glycosylation sites using rLHR series B mutants in which only one consensus site was maintained. Nonclonal, stably transfected 293 cells expressing the empty pcDNA1/neo vector (lanes 1 and 10), the wild-type rLHR (lanes 2, 3 and 11, 12), rLHR(N77,152, 269,277,291Q;T175A) (lanes 4, 5 and 13, 14), rLHR(N152,269,277,291Q;T175A) (lanes 6 and 7), rLHR(N77,269,277,291Q;T175A) (lanes 8 and 9), rLHR(N77,152,269,277,291Q) (lanes 15 and 16), rLHR(N77,152,277,291Q;T175A) (lanes 17 and 18), rLHR(N77,152,269,291Q;T175A) (lanes 19 and 20), or rLHR(N77,152,269,277Q;T175A) (lanes 21 and 22) were detergent solubilized. Equal amounts of protein were incubated for 15 hr at 37° in the absence or presence of 32 U/ml PGNase F. After this incubation, samples were resolved under reducing conditions on two 8% SDS–PAGE gels, transferred to PVDF membranes, and probed with an antibody to the rLHR. The N-linked glycosylation sequence that was maintained for a particular mutant is noted below each receptor. Adapted with permission from D. P. Davis, T. G. Rozell, X. Liu, and D. L. Segaloff, *Mol. Endocrinol.* **11,** 550 (1997).

individual consensus sequences for N-linked glycosylation were mutated would have already been created in order to identify the sites actually glycosylated (see earlier discussion). These constructs can then be used to further determine if the receptor is expressed properly at the cell surface and if the cell surface receptor binds ligand with high affinity and stimulates second messenger production in response to agonist.

Site-directed mutagenesis can also be used to create a fully nonglycosylated mutant of a GPCR in which all sites for N-linked glycosylation are mutated. To minimize the number of consensus sequences for N-linked glycosylation that are to be modified by site-directed mutagenesis, it is best to first determine which sites are actually glycosylated (see earlier discussion). Also, by first mutating each site individually, it can be determined if receptor expression is best maintained by mutation of the asparagine residue or the serine/threonine residue within a given consensus sequence. Having obtained these data, a recombinant receptor can be created in which all the sites known to be glycosylated are mutated simultaneously.

It is first necessary to determine that the nonglycosylated mutant receptor is expressed and that it is indeed devoid of any N-linked carbohydrate. These parameters can best be determined by examining the expression and mobility of the receptor on Western blots. If a mutant predicted to be nonglycosylated is not expressed, the results may be indicative of the necessity of N-linked carbohydrates for the proper folding of the receptor. However, it is equally plausible that improper folding may be due to perturbations in the polypeptide backbone as opposed to the absence of carbohydrates. One way to address this possibility is to create alternate mutants in which the asparagines or serines/threonines are substituted with different residues.

Another approach that can be utilized to create a nonglycosylated receptor is to incubate cells expressing the wild-type receptor with tunicamycin. Tunicamycin is an antibiotic that inhibits N-linked glycosylation by blocking the transfer of GlcNAc-1-P from UDP-GlcNAc to the dolichol monophosphate.[13] A number of different homologues of tunicamycin have been identified. Although all inhibit N-linked glycosylation, some also inhibit protein synthesis as well. The concentration of tunicamycin required to inhibit glycosylation depends on the homologue used and its purity. The effects of tunicamycin treatment on total cellular protein synthesis and glycosylation can be ascertained by measuring the incorporation of [^{3}H]alanine and [^{3}H]glucosamine, respectively, into trichloroacetic acid-precipitable material. Unlike a nonglycosylated receptor created by mutagenesis, a nonglycosylated receptor obtained by tunicamycin treatment would only be missing N-linked carbohydrates and would not have any possible additional conformational changes due to mutations in the polypeptide backbone.

A nonglycosylated receptor created by mutagenesis or by tunicamycin treatment would be expected to migrate with a reduced molecular mass on SDS gels comparable to that of the PNGaseF-treated wild-type receptor. In addition, the molecular mass of the nonglycosylated receptor should not be reduced any further by exhaustive treatment of the detergent extract with PNGaseF.

If the nonglycosylated receptor is expressed, it can then be determined if the receptor is present at the plasma membrane. This analysis can be performed by

[13] W. C. Mahoney and D. Duksin, *J. Biol. Chem.* **254,** 6572 (1979).

a number of different approaches, including Western blotting, if it is known that the blots can detect the cell surface mature form(s) of the receptor. Alternatively, immunohistochemistry can be performed, examining receptor expression in intact cells (reflecting cell surface receptor only) versus permeabilized cells (reflecting cell surface and intracellular pools of receptor). Binding assays of ^{125}I-labeled antireceptor antibody or ^{125}I-labeled hormone to intact cells are also commonly used.

If the nonglycosylated receptor is expressed at all at the cell surface then its ligand-binding and signal-transducing properties can be examined further. Because both basal and hormone-stimulated second messenger production can be stoichiometrically dependent on cell surface receptor number[14,15] and because mutant receptors or receptors derived from tunicamycin-treated cells may be expressed at a lower density at the cell surface than wild-type receptors, these analyses require that the numbers of cell surface receptors be taken into account. This can be accomplished by using cells expressing similar numbers of cell surface receptors (i.e., deliberately transiently transfecting cells with decreased amounts of wild-type receptor cDNA or selecting clones stably expressing decreased numbers of cell surface wild-type receptors). Another approach is to normalize data as the response per number of cell surface receptors (or amount of bound hormone to intact cells). This normalization is valid, however, only if it can first be shown that the basal and hormone-stimulated levels of the second messenger are linear over the range of receptor numbers used for both wild-type and mutant receptors.[14,15]

Concluding Remarks

The role of N-linked glycosylation in the expression and function of GPCRs, as well as other cell surface receptors and proteins, has been an area of active investigation for many years. The one functional role of N-linked carbohydrates that may be general to all glycoproteins is their facilitation of the proper folding of the protein. Indeed, two ER chaperone proteins, calnexin and calreticulin, have been shown to facilitate the folding of many glycoproteins through an interaction with nascent glycoproteins primarily through their N-linked carbohydrates.[16] However, the degree to which a given GPCR depends on N-linked carbohydrates for folding may vary. For example, nonglycosylated rLHR is expressed at the cell surface, albeit at very low levels, and is fully functional,[7] suggesting that although the N-linked carbohydrates may facilitate the folding of the rLHR, they are not essential. In contrast, the structurally related rFSHR absolutely requires N-linked carbohydrates in order to fold into a mature conformation.[17] Therefore, the contributions

[14] B. S. Whaley, N. Yuan, L. Birnbaumer, R. B. Clark, and R. Barber, *Mol. Pharmacol.* **45,** 481 (1994).
[15] X. Zhu, S. Gilbert, M. Birnbaumer, and L. Birnbaumer, *Mol. Pharmacol.* **46,** 460 (1994).
[16] L. Ellgaard, M. Molinari, and A. Helenius, *Science* **286,** 1882 (1999).
[17] D. Davis, X. Liu, and D. L. Segaloff, *Mol. Endocrinol.* **9,** 159 (1995).

of N-linked carbohydrates to the expression and function of a given GPCR must be determined directly and cannot necessarily be extrapolated from data on other GPCRs, even those closely related. It is hoped that the overview given in this article will aid investigators in identifying those sites on a GPCR that contain N-linked carbohydrates and in determining the functional roles of N-linked carbohydrates.

Acknowledgment

The studies discussed were supported by NIH Grants HD28970 and HD22196 (to D.L.S.).

[14] Magic Angle Spinning Nuclear Magnetic Resonance of Isotopically Labeled Rhodopsin

By MARKUS EILERS, WEIWEN YING, PHILIP J. REEVES, H. GOBIND KHORANA, and STEVEN O. SMITH

Introduction

G protein-coupled receptors (GPCRs) are an important class of membrane proteins that cells use to sense their surrounding environment and to communicate with other cells. They are activated by intercellular messenger molecules, such as hormones, neurotransmitters, and growth factors, as well as sensory messages, such as light and odorants. The most direct structural data pertaining to the positioning of the seven transmembrane (TM) helices in rhodopsin come from electron density projection maps of Schertler and co-workers.[1–3] These maps show that three of the seven TM helices are oriented roughly perpendicular to the membrane plane and that the remaining four helices are tilted. Based on sequence analyses of the GPCR family, Baldwin has proposed a detailed model of rhodopsin that is consistent with electron diffraction data and the position of polar and conserved residues.[4,5]

In contrast to the lack of structural data, there is a wealth of biochemical and molecular biological data on rhodopsin and other GPCRs. In particular, site-directed mutagenesis has proved extremely valuable for mapping residues that are critical for GPCR structure and function.[6] There are several residues in the

[1] G. F. Schertler, C. Villa, and R. Henderson, *Nature* **362,** 770 (1993).
[2] G. F. Schertler and P. A. Hargrave, *Proc. Natl. Acad. Sci. U.S.A.* **92,** 11578 (1995).
[3] V. M. Unger, P. A. Hargrave, J. M. Baldwin, and G. F. X. Schertler, *Nature* **389,** 203 (1997).
[4] J. M. Baldwin, *EMBO J.* **12,** 1693 (1993).
[5] J. M. Baldwin, G. F. X. Schertler, and V. M. Unger, *J. Mol. Biol.* **272,** 144 (1997).
[6] H. G. Khorana, *Proc. Natl. Acad. Sci. U.S.A.* **90,** 1166 (1993).

0076-6879/02 $35.00

transmembrane helices of the GPCR family that are highly conserved and, therefore, are likely to be important in defining the structure of the receptor or to be involved in the activation mechanism.

Magic angle spinning (MAS) nuclear magnetic resonance (NMR) spectroscopy is well suited to high-resolution studies of the local structure and dynamics of membrane proteins (reviewed in Ref. 7). Much of the work since the early 1990s has, in fact, focused on rhodopsin.[8–15] An advantage of MAS NMR over diffraction and solution NMR methods is that high-resolution structural measurements can be made on proteins in native bilayer environments. MAS experiments rely on the "$3\cos^2\theta - 1$" orientation dependence of the chemical shift and dipolar interactions. Rapid spinning of the sample at an angle of 54.7° (the magic angle) to the external magnetic field averages the chemical shift to its isotropic value and the dipolar coupling to zero, producing high-resolution NMR spectra. The current limitation is that structural measurements must focus on selected regions of interest in the protein (see Fig. 1). This article discusses the design of MAS experiments for targeting selected regions in GPCRs and then describes experimental protocols for isotope labeling of rhodopsin using mammalian HEK293S cells.

MAS NMR Studies of G Protein-Coupled Receptors

MAS NMR measurements require specific labeling with stable isotopes in order to increase the signals of interest over natural abundance background levels. The basic experiment involves specific incorporation of pairs of isotope labels such as ^{13}C..^{13}C, ^{13}C..^{15}N or ^{13}C..^{19}F. Distance and torsion angle measurements between the labeled sites provide high-resolution structural constraints. Selective reintroduction of dipolar interactions in MAS experiments allows the measurement of internuclear distances with resolution on the order of 0.2 Å,[16] whereas torsion angles in bound ligands or prosthetic groups can be determined to within ±10°.[17] Four general areas that can be targeted by MAS NMR are the conformation of the

[7] S. O. Smith, K. Aschheim, and M. Groesbeek, *Quart. Rev. Biophys.* **29,** 395 (1996).

[8] S. O. Smith, I. Palings, M. E. Miley, J. Courtin, H. de Groot, J. Lugtenburg, R. A. Mathies, and R. G. Griffin, *Biochemistry* **29,** 8158 (1990).

[9] S. O. Smith, J. Courtin, H. de Groot, R. Gebhard, and J. Lugtenburg, *Biochemistry* **30,** 7409 (1991).

[10] S. O. Smith, H. de Groot, R. Gebhard, and J. Lugtenburg, *Photochem. Photobiol.* **56,** 1035 (1992).

[11] M. Han, B. S. DeDecker, and S. O. Smith, *Biophys. J.* **65,** 899 (1993).

[12] M. Han and S. O. Smith, *Biochemistry* **34,** 1425 (1995).

[13] M. Eilers, P. J. Reeves, W. W. Ying, H. G. Khorana, and S. O. Smith, *Proc. Natl. Acad. Sci. U.S.A.* **96,** 487 (1999).

[14] A. F. L. Creemers, C. H. W. Klaassen, P. H. M. Bovee-Geurts, R. Kelle, U. Kragl, J. Raap, W. J. De Grip, J. Lugtenburg, and H. J. M. de Groot, *Biochemistry* **38,** 7195 (1999).

[15] P. J. E. Verdegem, P. H. M. Bovee-Geurts, W. J. De Grip, J. Lugtenburg, and H. J. M. de Groot, *Biochemistry* **38,** 11316 (1999).

[16] O. B. Peersen, M. Groesbeek, S. Almoto, and S. O. Smith, *J. Am. Chem. Soc.* **117,** 7228 (1995).

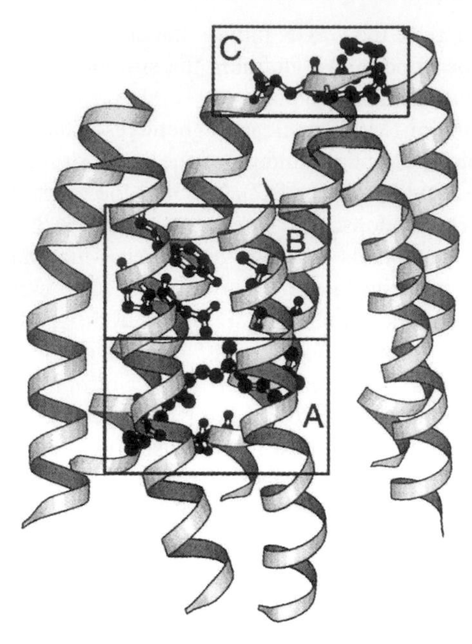

FIG. 1. Model of the transmembrane helices of rhodopsin illustrating selected regions that can be targeted by MAS NMR. Region A contains the retinal-binding pocket. MAS NMR measurements can be used to establish the conformation of the retinal chromophore and the relative position of its Glu-113 counterion. Region B contains the conserved NPXXY sequence of helix 7 and Met-257 of helix 6. Mutations of Met-257 lead to constitutive activity. An interhelical contact between Met-257 and the conserved NPXXY sequence is likely to be important in locking the receptor off in the dark. MAS NMR measurements can be used to establish helix–helix interactions if unique resonances can be resolved or engineered into the receptor. Region C contains the conserved D/ERY sequence at the end of helix 3. Protonation changes of these residues are involved in the activation mechanism of the protein. ^{13}C and ^{15}N chemical shifts are sensitive to protonation states of the D, E, and R functional groups. Coordinates for the rhodopsin model were taken from T. Shieh, M. Han, T. P. Sakmar, and S. O. Smith, *J. Mol. Biol.* **269,** 373 (1997). The model was created in MolScript [P. J. Kraulis, *J. Appl. Crystallog.* **24,** 946 (1991)].

bound ligand, the protonation states of catalytic acids and bases, ligand–protein interactions, and protein–protein contacts (Fig. 1).

Ligand Conformation

Most GPCRs are activated by binding of small molecule ligands, such as amines or peptides. Rhodopsin is unusual in having a covalently bound "ligand" that functions as an antagonist in the 11-*cis* configuration and as an agonist after

[17] X. Feng, P. J. E. Verdegem, Y. K. Lee, D. Sandstrom, M. Eden, P. Boveegeurts, W. J. Degrip, J. Lugtenburg, H. J. M. Degroot, and M. H. Levitt, *J. Am. Chem. Soc.* **119,** 6853 (1997).

isomerization to the all-*trans* configuration. The structure of the protein-bound ligand in all GPCRs is critical for understanding how ligand binding triggers receptor activation. Ligands and prosthetic groups are relatively easy to isotope label[18] and to bind to an unlabeled receptor. As a result, they are typically the first targets in MAS NMR studies of membrane proteins. In rhodopsin, the conformation of the retinal in the vicinity of the C10-C11 = C12-C13 bonds has been determined through measurements of torsion angles[17] and distances[15] using dipolar recoupling techniques.

Protonation States of Key Conserved Residues

Electrostatic interactions are known to be involved in the activation mechanism of GPCRs. For example, rhodopsin and adrenergic receptors are locked in their inactive conformation via a salt bridge between transmembrane helices 3 and 7.[19,20] Two other important residues at the cytoplasmic end of helix 3 are Glu-134 and Arg-135, part of a conserved D/ERY sequence. These residues are critical for receptor activation. Glu-134 is thought to be protonated upon activation, which leads to a change of electrostatic interactions involving Arg-135.[21] In rhodopsin, the E134Q mutant is constitutively active.[22,23]

Isotope labeling of the protein allows the determination of protonation states of acid and base functional groups. The ^{13}C chemical shift of the carboxyl groups of aspartate and glutamate and the side chain ^{15}N chemical shift of histidine, lysine, and arginine are sensitive to the protonation state.[24,25] The key for these experiments is the resolution in chemical shifts or the chemical shift changes between the inactivated and the activated receptor.

The first NMR measurements of isotope-labeled rhodopsin targeted the protonated Schiff base linkage between the retinal chromophore and Lys-296.[13,14] The opsin was labeled with 6-^{15}N-lysine in the mammalian HEK-293S system[13] and in a baculovirus/Sf9 cell expression system.[14] Figure 2 presents the ^{15}N MAS spectrum of ^{15}N-Lys rhodopsin. The resonance at 8.7 ppm corresponds to the 10 free lysines in the protein. The natural abundance ^{15}N resonance from the peptide backbone is observed at 93.6 ppm. Finally, the intense resonance at 156.8 ppm

[18] J. Lou, Q. Tan, E. Karnaukhova, N. Berova, K. Nakanishi, and R. K. Crouch, *Methods Enzymol.* **315,** 219 (2000).

[19] P. R. Robinson, G. B. Cohen, E. A. Zhukovsky, and D. D. Oprian, *Neuron* **9,** 719 (1992).

[20] J. E. Porter and D. M. Perez, *J. Biol. Chem.* **274,** 34535 (1999).

[21] S. Amis, K. Fahmy, K. P. Hofmann, and T. P. Sakmar, *J. Biol. Chem.* **269,** 23879 (1994).

[22] S. Acharya and S. S. Karnik, *J. Biol. Chem.* **271,** 25406 (1996).

[23] J. M. Kim, C. Altenbach, R. L. Thurmond, H. G. Khorana, and W. L. Hubbell, *Proc. Natl. Acad. Sci. U.S.A.* **94,** 14273 (1997).

[24] Z. T. Gu and A. McDermott, *J. Am. Chem. Soc.* **115,** 4282 (1993).

[25] Y. F. Wei, A. C. de Dios, and A. E. McDermott, *J. Am. Chem. Soc.* **121,** 10389 (1999).

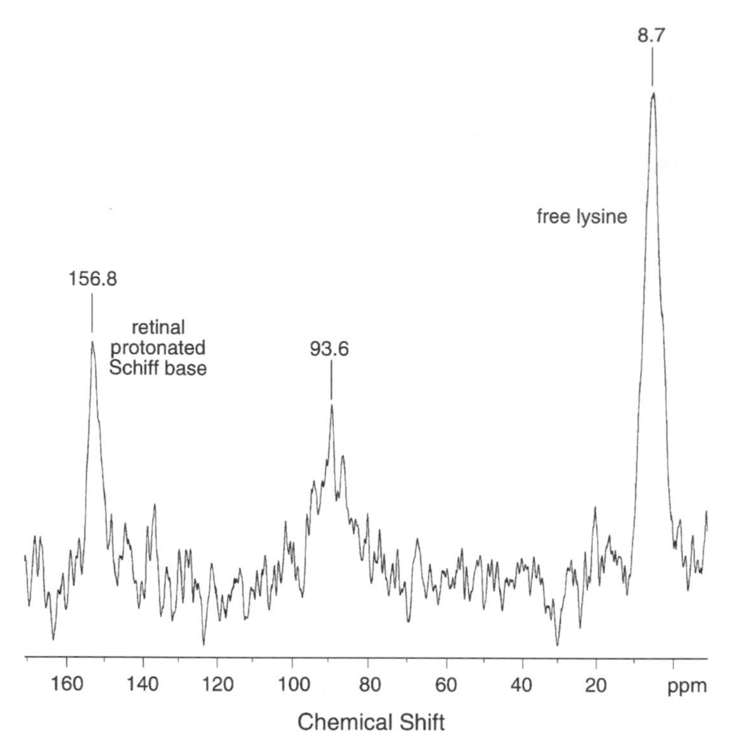

FIG. 2. ^{15}N MAS spectrum of 6.5 mg 6-^{15}N-lysine rhodopsin in DOPC membranes. ^{15}N PSB resonance is observed at 156.8 ppm. ^{15}N signals corresponding to the free lysines and natural backbone of the protein are observed at 8.7 and 93.6 ppm, respectively. The spectrum shown corresponds to 122,880 acquisitions. ^{15}N chemical shifts are reported relative to external 5.6 *M* aqueous $^{15}NH_4Cl$. The spectrum was processed with 20 Hz of exponential line broadening. Taken from M. Eilers, P. J. Reeves, W. W. Ying, H. G. Khorana, and S. O. Smith, *Proc. Natl. Acad. Sci. U.S.A.* **96**, 487 (1999).

is assigned to the ^{15}N resonance of Lys-296, which forms a protonated Schiff base linkage with the retinal. The resonance is well resolved, and its chemical shift correlates with a distance of ~4 Å between the protonated Schiff base and its counterion, Glu-113.[13]

Ligand–Protein Interactions

Ligand binding triggers a structural change of GPCRs to an activated conformation. Ligand-binding sites in GPCRs are as varied as the types of ligands to which they respond. Protein ligands (as in the chemokine receptors) bind to the N terminus or to the extracellular loops, whereas small molecule ligands (as in adrenergic receptors) bind within the transmembrane helices. A combination of isotope labeling of the protein and the ligand allows one to map out specific ligand–protein contacts. The strategy for these studies is generally to make use of unique labels incorporated into the ligand. In rhodopsin, the position of Glu-113 relative to the retinal polyene chain was determined by ^{13}C chemical shift measurements.[11] Specific ligand–protein distance measurements require the resolution of unique labels in both the ligand and the protein. As discussed earlier, unusual electrostatic interactions may lead to chemical shift resolution in the acidic amino acids, Glu and Asp, or the basic amino acids, Arg, Lys, and His. In the case where it is not known which residues form the ligand-binding site in a GPCR, it may be possible to highlight only the most closely packed residues by using double quantum filtering experiments.[26,27] In these experiments, strong dipolar couplings between the ligand and the protein are used to "filter out" the signals from amino acids not in van der Waals contact with the ligand.

Protein–Protein Interactions

Structure–function studies of GPCRs since the early 1990s have shown that the interactions between the transmembrane helices are important for receptor activation.[28,29] Mutations in the transmembrane helices (see Fig. 1 of Ref. 28) are often associated with the constitutive activation of GPCRs. These studies have suggested that specific helix–helix contacts, which are responsible for locking receptors in the inactive state, are disrupted by ligand binding. MAS NMR structural studies of helix–helix contacts are a much greater challenge than studies on ligand conformation or ligand–protein interactions. Isotope labeling of the ligand in these latter cases can provide unique NMR resonances. In contrast, interhelical measurements require two resolved isotope labels within one protein. In these experiments, specific protonated sites may help by providing resolved resonances. An alternative strategy is to engineer labels into unique positions. Cysteine incorporation has been used in the serine receptor to establish the conformational changes in transmembrane helices associated with the binding of serine.[30] In rhodopsin, single and double cysteine mutants on the cytoplasmic loops have been constructed to study loop structures and interactions using EPR spectroscopy. For NMR studies, these cysteine sites can be derivatized with isotopically labeled functional groups to introduce unique labels.[31]

[26] Y. K. Lee, N. D. Kurur, M. Helmle, O. G. Johannessen, N. C. Nielsen, and M. H. Levitt, *Chem. Phys. Lett.* **242,** 304 (1995).

[27] C. M. Rienstra, M. E. Hatcher, L. J. Mueller, B. Q. Sun, S. W. Fesik, and R. G. Griffin, *J. Am. Chem. Soc.* **120,** 10602 (1998).

[28] M. Han, S. O. Smith, and T. P. Sakmar, *Biochemistry* **37,** 8253 (1998).

[29] N. I. Tarasova, W. G. Rice, and C. J. Michejda, *J. Biol. Chem.* **274,** 34911 (1999).

[30] F. Kovacs, O. J. Murphy, B. James, Y. S. Balazs, J. Charette, E. Sicard, and L. K. Thompson, *Biophys. J.* **78,** 66A (2000).

[31] J. Klein-Seetharaman, E. V. Getmanova, M. C. Loewen, P. J. Reeves, and H. G. Khorana, *Proc. Natl. Acad. Sci. U.S.A.* **96,** 13744 (1999).

Protein–protein interactions are also important in the recognition and binding of G proteins to the cytoplasmic loops of the GPCR in the activated conformation. In rhodopsin, the C-terminal tail of the G protein transducin is able to bind to and stabilize the activated metarhodopsin II intermediate alone. This interaction opens up the possibility that G protein–receptor contacts can be established by incorporating unique labels into a synthetic peptide corresponding to the C terminus of transducin.[32]

Isotope Labeling of Rhodopsin

Mammalian cells (e.g., HEK 293 cells) can grow in defined media. This makes it easy to use these cells for isotope labeling of proteins. Defined media contain amino acids, vitamins, inorganic salts, and other compounds (glucose, sodium pyruvate, buffer, and pH indicator) at specific concentrations optimized for cell growth. Amino acids can be replaced by their labeled analogs, and salt concentrations can be adjusted for optimal growth conditions, e.g., the Ca^{2+} concentration for suspension growth. Defined media lacking specific amino acids can be obtained commercially and then supplemented with labeled amino acids in the appropriate concentrations. In addition to defined media, animal serum, which supplies growth factors, lipids, and other essential components, is generally required for the growth of animal cells. Naturally occurring amino acids that would interfere with specific labeling are removed from the serum by dialysis using a 1-kDa cutoff membrane.

Suspension Growth of 293S Cells

The protocol for isotope labeling of rhodopsin is shown schematically in Fig. 3. HEK293S cells have been optimized for growth under suspension conditions. Cell cultures on plates and in spinner flasks are grown in a CO_2 incubator (e.g., Forma Model 3956) where CO_2 (5%), temperature (37°), and humidity can be controlled. Optimum growth is dependent on the Ca^{2+} concentration. High concentrations of Ca^{2+} cause clumping, whereas low concentrations lead to cell death. Therefore, the Ca^{2+} concentration is reduced to 340 μM (50 mg/liter $CaCl_2 \times 2H_2O$) compared to the standard conditions of 1.8 mM (265 mg/liter $CaCl_2 \times 2H_2O$) to minimize clumping, but providing the cells with the essential amount of Ca^{2+} ions. To avoid a concentration shock, the Ca^{2+} concentration is reduced in two steps: the first on plates and the second during transfer to the spinner flasks. For optimal growth, the medium pH is readjusted with $NaHCO_3$ and the cells are fed with glucose at day 5 or 6. The actual growth conditions are described later.

[32] H. E. Hamm, D. Deretic, A. Arendt, P. A. Hargrave, B. Koenig, and K. P. Hofmann, *Science* **241,** 832 (1988).

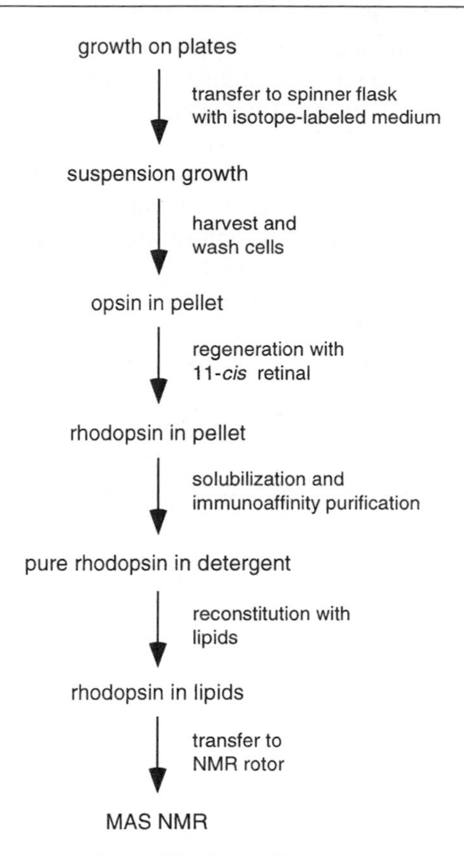

FIG. 3. Scheme of isotope labeling, purification, and NMR sample preparation for rhodopsin.

Materials

The material and sources we have used are described. 293S cells, a suspension-adapted variant of HEK cells (ATCC No. CRL 1573) were provided by J. Nathans (Johns Hopkins School of Medicine). Plasmid pACHEnc was donated by A. Schafferman (Israel Institute for Biological Research). 11-*cis* retinal was a gift from R. Crouch (University of South Carolina and the NEI) or from Hoffmann-La Roche (Switzerland). Dodecyl-β-1-maltoside (DM) and *n*-octyl-β-D-glucoside (OG) can be obtained from Anatrace (Maumee, OH). Protein A-Sepharose was from RepliGen (Cambridge, MA). CNBr-activated Sepharose 4B was from Pharmacia. Custom-made Dulbecco's modified Eagles medium (DMEM) lacking the amino acids of choice, sodium pyruvate, calcium, and sodium bicarbonate was obtained from Atlanta Biologicals (Norcross, GA). Standard DMEM medium HB-Gro was purchased from Irvine Scientific (Santa Ana, CA). Geneticin (G418), trypsin, EDTA, penicillin, and streptomycin were obtained from GIBCO/BRL (Gaithersburg, MD). Aprotinin, benzamidine, leupeptin, and pepstatin A were purchased from Boehringer Mannheim (Indianapolis, IN). Fetal bovine serum (FBS), cell culture grade calcium chloride, Pluronic F-68, phenylmethylsulfonyl fluoride (PMSF), and amino acids were from Sigma (St. Louis, MO) FBS (500 ml) is heat inactivated (56°, 30 min) and dialyzed (1-kDa cutoff) three times against 10 liter of buffer A over 3 days at 4°. Isotope-labeled amino acids (>99% labeled) were from MassTrace (Woburn, MA) and Cambridge Isotope Laboratories (Andover, MA) and used without further purification. The monoclonal antibody 1D4 against the nine C-terminal amino acids of rhodopsin was purchased from the National Cell Culture Center (Minneapolis, MN). The nonapeptide corresponding to the rhodopsin C terminus was synthesized by solid-state methods. 1,2-Dioleoyl-*sn*-glycero-3-phosphocholine (DOPC) was obtained from Avanti (Alabaster, AL).

Buffers. Buffer A: 137 m*M* NaCl, 2.7 m*M* KCl, 1.5 m*M* KH_2PO_4, 8 m*M* Na_2HPO_4, pH 7.2. Buffer B: Buffer A with 4% (w/v) *n*-octyl-β-D-glucoside. Buffer C: Buffer A with 1.46% (w/v) *n*-octyl-β-D-glucoside. Buffer D: 10 m*M* 1,3-bis[tris(hydroxymethyl) methylamino]propane, pH 6.0, with 1.46% (w/v) *n*-octyl-β-D-glucoside. Buffer E: Buffer C + 100 μ*M* nonapeptide. Buffer F: 10 m*M* HEPES, 50 m*M* NaCl, 1 m*M* EDTA, 1 m*M* dithiothreitol (DTT), pH 7.1, with 4% (w/v) *n*-octyl-β-D-glucoside. Buffer G: 50 m*M* Tris, 50 m*M* sodium acetate, pH 7.0.

Growth of Cells in Suspension

The construction of the stable cell line (HEK293S) expressing the opsin gene has been described by Reeves *et al.*[33] The cell lines are stored in liquid nitrogen. Cells are thawed and first grown to confluence in 15-cm culture dishes containing 30 ml of DMEM with the appropriate concentration of geneticin G418 (2 mg/ml).[33] At about 70% confluence, cells are fed with the same medium containing 680 μ*M* Ca^{2+} and grown to confluence.[13] Suspension cultures are set up using 2-liter spinner flasks containing 500 ml custom-made DMEM supplemented immediately before use with dialyzed (1-kDa cutoff) FBS (10%), Pluronic F-68 (0.1%), heparin (50 mg/liter), penicillin (100 units/ml), streptomycin (100 μg/ml), and the missing amino acids (including L-glutamate) at the concentrations used in the original description of DMEM.[34,35] $CaCl_2$ is added at a concentration of 340 μ*M*.[13] Spinner flasks are inoculated using three such culture dishes (6–9 × 10^7 cells) and are incubated at 37° in a humidified incubator. After 5 or 6 days, the culture medium is supplemented further with 6 ml of 20% (w/v) glucose and 4 ml of 8%

[33] P. J. Reeves, R. L. Thurmond, and H. G. Khorana, *Proc. Natl. Acad. Sci. U.S.A.* **93,** 11487 (1996).
[34] R. Dulbecco and G. Freeman, *Virology* **8,** 396 (1959).
[35] J. D. Smith, G. Freeman, M. Vogt, and R. Dulbecco, *Virology* **12,** 185 (1960).

(w/v) $NaHCO_3$.[36] Cells are harvested on day 7 by centrifugation (4000*g*, 10 min, 4°), washed twice with 50 ml buffer A per spinner flask, and resuspended in buffer A (5 ml/g wet pellet) containing protease inhibitors (50 μg/ml aprotinin, 50 μg/ml benzamidine, 50 μg/ml leupeptin, 50 μg/ml pepstatin A, 0.2 m*M* PMSF).

All further manipulations are performed in dim red light. The resuspended cell are incubated with 10 m*M* 11-*cis* retinal in ethanol [250 nmol/g wet pellet; $\varepsilon_{379.5} = 24940\ M^{-1}\text{cm}^{-1}$ (ethanol)[37]], added in two portions, over 4 hr under end-over-end mixing, and pelleted by centrifugation (3600*g*, 10 min, 4°). The cell pellets are solubilized in buffer B (5 ml/g wet pellet) containing protease inhibitors by 2 hr end-over-end mixing at 4°. The amount of rhodopsin can be estimated by dark-light difference UV/VIS spectroscopy ($\varepsilon_{500} = 40,600 M^{-1}\text{cm}^{-1}$ [38]).

Immunoaffinity Purification of Rhodopsin

Rhodopsin is purified by immunoaffinity chromatography with 1D4-Sepharose beads. The monoclonal 1D4 antibody is coupled to CNBr-activated Sepharose 4B according to the manufacturer's (Pharmacia) protocol. The column (10% excess of the total binding capacity is needed) is degassed and equilibrated with 3 bed volumes of buffer B. All further operations are performed in dim red light. The solubilized rhodopsin in buffer B is loaded twice on the column at a flow rate of ≈0.6 ml/min. The column is washed (gravity flow) with 50 bed volumes of buffer C followed by 10 bed volumes of buffer D. Rhodopsin is eluted using buffer E at a reduced flow rate of about 0.05 ml/min. Fractions of 1 bed volume are collected. Labeled rhodopsin prepared in this manner is indistinguishable from rhodopsin purified from rod outer segments (ROS rhodopsin) by all criteria applied.[33] In Fig. 4, the UV/VIS absorption spectrum of the purified HEK293S rhodopsin (A_{280}/A_{500} ratio of 1.65) is compared with rhodopsin purified from ROS. The isotope incorporation efficiency can be determined by mass spectrometry.[13]

Reconstitution of Rhodopsin into Proteoliposomes

Rhodopsin eluted in buffer E is concentrated using concentration devices (Amicon Centricon, 10-kDa cutoff) to 1 mg/ml as determined by UV/VIS spectroscopy ($\varepsilon_{500} = 40,600 M^{-1}\text{cm}^{-1}$ [38]). A 100-fold molar excess of lipid (e.g., DOPC) as a suspension in degassed buffer F is sonicated for 10–15 min. The purified rhodopsin and lipids are then combined and mixed end over end for 1 hr before transferring into dialysis tubing (12- to 14-kDa cutoff). The rhodopsin in mixed micelles is dialyzed against 200 volumes of buffer G at 4°. The buffer is changed every 4 hr over a 24-hr period, and the resulting proteoliposomes are pelleted by

[36] A. Garnier, J. Cote, I. Nadeau, A. Kamen, and B. Massie, *Cytotechnology* **15,** 145 (1994).

[37] R. S. H. Liu and A. E. Asato, *Methods Enzymol.* **88,** 506 (1982).

[38] G. Wald and P. K. Brown, *J. Gen. Physiol.* **37,** 189 (1953).

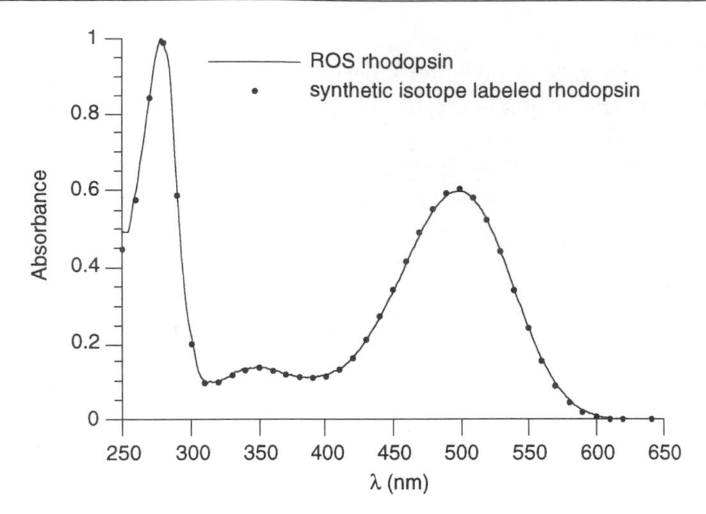

FIG. 4. Absorption spectra of native (line) and synthetic isotope-labeled rhodopsin (dots). Native rhodopsin was obtained from bovine retinas and purified using the 1D4 immunoaffinity column in a manner parallel to the purification of isotope-labeled rhodopsin expressed in HEK293S cells. Spectra were normalized to the absorption at 280 nm. Taken from M. Eilers, P. J. Reeves, W. W. Ying, H. G. Khorana, and S. O. Smith, *Proc. Natl. Acad. Sci. U.S.A.* **96,** 487 (1999).

centrifugation at 100,000*g* for 1 hr at 4° and are loaded into the NMR rotor under dim red light. NMR samples are stored at −80°.

Rhodopsin activity is influenced by the lipid used for reconstitution (for review, see Ref. 39). The nature of the head group, as well as the length and degree of unsaturation of the acyl chains, modulates the conversion of rhodopsin to the activated metarhodopsin II intermediate. We have typically used DOPC for reconstitutions because it is a single lipid species, which allows full rhodopsin activity.

Acknowledgments

We thank Judith Klein-Seetharaman for her critical comments on the manuscript. This work was supported by grants from the National Institutes of Health to S.O.S. (GM 41412) and to H.G.K. (GM 28289).

[39] M. F. Brown, *Chem. Phys. Lipids* **73,** 159 (1994).

[15] Use of Nuclear Magnetic Resonance to Study the Three-Dimensional Structure of Rhodopsin

By PHILIP L. YEAGLE and ARLENE D. ALBERT

Introduction to the Problem

The structure of membrane proteins like G protein-coupled receptors is a largely unsolved problem. This is due to the challenges inherent in the crystallization of membrane proteins. Successful crystallization of membrane proteins is still a relatively rare event and only a few structures of transmembrane proteins have been reported. Therefore, an alternate approach to the structure of membrane proteins would be useful to better understand the function and regulation of G protein-coupled receptors.

Description of Approach and Justification

G protein-coupled receptors are believed to be built around a bundle of seven transmembrane helices. This model suggests that much of the secondary structure of these proteins is contained in the transmembrane helices and the turns that connect them. It can be hypothesized that information about secondary structure such as helices and turns (but not necessarily β sheets) can be obtained from peptides designed to contain the sequences from the protein that putatively code for these secondary structures. A review of the literature provides evidence for this hypothesis.

Small regions of some proteins exhibit the same secondary structure (α helix and β turns) in individual solublized peptides as observed in the corresponding region of the intact protein.[1–7] The stabilization of secondary structures such as β turns by short-range interactions[8] is likely the basis for these observations.

Similar principles have been discovered in examinations of the entire sequences of proteins consisting of helical bundles.[9,10] For example, bacteriorhodopsin is

[1] N. Goudreau *et al., Nature Struct. Biol.* **1,** 898 (1994).

[2] F. J. Blanco and L. Serrano, *Eur. J. Biochem.* **230,** 634 (1994).

[3] M. Adler, M. H. Seto, D. E. Nitecki, J. H. Lin, D. R. Light, and J. Morser, *J. Biol. Chem.* **270,** 23366 (1995).

[4] M. Blumenstein, G. R. Matsueda, S. Timmons, and J. Hawiger, *Biochemistry* **31,** 10692 (1992).

[5] A. P. Campbell, C. McInnes, R. S. Hodges, and B. D. Sykes, *Biochemistry* **34,** 16255 (1995).

[6] J. A. Wilce, D. Salvatore, J. D. Waade, and D. J. Craik, *Eur. J. Biochem.* **262,** 586 (1999).

[7] F. J. S. Cheng and M. Zhang, *Biochem. Biophys. Res. Commun.* **253,** 621 (1998).

[8] A.-S. Yang, B. Hitz, and B. Honig, *J. Mol. Biol.* **259,** 873 (1996).

[9] J. P. L. Cox, P. A. Evans, L. C. Packman, D. H. Williams, and D. N. Woolfson, *J. Mol. Biol.* **234,** 483 (1993).

[10] S. Padmanabhan, M. A. Jimenez, and M. Rico, *Protein Sci.* **8,** 1675 (1999).

METHODS IN ENZYMOLOGY, VOL. 343
0076-6879/02 $35.00

an integral membrane protein, built around a bundle of seven transmembrane helices. Individual peptides were prepared for each of six of the seven helices of bacteriorhodopsin, and five of those six peptides formed helices independent of the remainder of the protein.[11,12] Three loops (connecting helices C and D, D and E, and F and G) of bacteriorhodopsin were individually prepared and shown to form the same loop structure in solution as found in the crystal structure.[13] Nuclear magnetic resonance (NMR) structural studies of small peptides spanning about three-fourths (both helices and loops) of the soluble four-helix bundle, myohemerythrin, show the same secondary structure in solution as those sequences adopt in the intact protein. None of the peptides showed a structure in solution that was different from the native protein.[14] These studies suggest that in proteins consisting of helical bundles, much of the secondary structure of the native protein is retained in small peptides spanning the sequence of the helical bundle.

An alternate approach to the structure of membrane proteins can be built on these observations. In this segmentation approach, the secondary structure of the protein is captured in peptides with sequences of either individual helices of the protein or turns connecting helices. In the latter, the peptide usually exhibits a helix-turn-helix motif. Peptides of this kind can be designed to span the sequence of the protein and overlap each other at the ends. Individual structures are then determined in solution by NMR (or could also be determined by X-ray crystallography). This article emphasizes the use of NMR to determine structure, which in the case of peptides is likely the faster way to get structural information. The elements of secondary structure exhibited by these peptides are then assembled into a structure of the whole protein by superimposing one peptide on the next using the region of overlap in amino acid sequence.

This approach has been used to determine the structure of the G protein-coupled receptor rhodopsin. Details of this approach are described. How this approach can be used to obtain structural information for other G protein-coupled receptors is also described.

Methods

Design of Peptides to Span Sequence

The first step in this segmented approach is to design the set of peptides to span the sequence of the receptor. Several criteria need to be used to design this set properly. An example for rhodopsin is shown in Fig. 1.

[11] K. V. Pervushin, V. Y. Orekhov, A. I. Popov, L. Y. Musina, and A. S. Arseniev, *Eur. J. Biochem.* **219,** 571 (1994).

[12] J. F. Hunt, T. N. Earnest, O. Bousche, K. Kalghatgi, K. Reilly, C. Horvath, K. J. Rothschild, and D. M. Engelman, *Biochemistry* **36,** 15156 (1997).

[13] M. Katragadda, J. L. Alderfer, and P. L. Yeagle, *Biochim. Biophys. Acta* **1466,** 1 (2000).

[14] H. J. Dyson, G. Merutka, J. P. Waltho, R. A. Lerner, and P. E. Wright, *J. Mol. Biol.* **226,** 795 (1992).

MNGTEGPNFYVPFSNKTGVVSPFEAPQYYLAEPWQFSML

QFSMLAAYMFLLIMLG

LIMLGFPINFLTLYVT

TLYVTVQHKKLRTPLNYILLNLAVAD

LAVADLFMVFGGFTTTLY

TTTLYTSLHGYFVFGPTGCNLEGFFATLGGEI

ATLGGEIALWSLVVLAIERYV

AIERYVVCKPMSNFRFGENHAIM

ENHAIMGVAFTWVMALA

VMALACAAPPLVGWSR

LVGWSRYIPEGMQCSCGIDYYTPHEETNNESFVI

ESFVIYMFVVHFIIPLIVIF

LIVIFFCYGQLVFTVKEA

FTVKEAAAQQQESATTQKAEKEVTRMVIIMVIAFL

VIAFLICWLPYAGVA

YAGVAFYIFTHQGSDFGPIFMTIPAF

PAFFAKTSAVYNPVIYIMMNKQFRN

YIMMNKQFRNCMVTTLCCGKNPLGDDEASTTVSKTETSQVAPA

FIG. 1. Primary sequence of rhodopsin. Peptides listed as a sequence on a single line were synthesized and their structures determined in solution by NMR. Overlap regions are underlined. These are the peptides that were actually used in the structure determination of rhodopsin. The overlap regions are smaller than recommended in the text because in the initial stages of the project, the optimal length of overlap was not known. Furthermore, considerable long-range distance information for rhodopsin was known from other experiments, allowing the organization of the protein fragments without optimum overlaps. In the case of another G protein-coupled receptor, where little long-distance information is available from independent experiments on the intact protein, the overlaps recommended in the text will be critical to the successful assembly of the protein structure from its fragments.

1. Overlap regions, one peptide onto the next, must be substantial. Solution structures of individual peptides normally exhibit considerable disorder at the ends of the peptides. Because this approach ultimately requires overlap of the peptides to assemble a construct for the whole protein, the overlap regions must be constructed appropriately. Often the last 3 to 5 residues in a solution structure of a peptide are disordered. A minimum of 4 residues is required in an ordered portion of the structure to perform a superposition of two overlapping peptides. Therefore, the set of peptides spanning the sequence of the protein should overlap each other by about 10 residues.

2. The length of peptides will determine the extent of secondary structure that can be seen in the structure of a particular peptide. In the case of turns, experience with rhodopsin shows that peptides of 15 to 17 amino acids in length will show a turn but nothing else. When the peptide is lengthened to about 23 amino acids, a turn from a protein like rhodopsin shows a helix-turn-helix motif. In this latter case, the helix-turn-helix motif reveals the termination points of the transmembrane helices that connect to the turns and the direction of the helices as they exit the turn.

3. Peptides should encompass a putative turn (loop), a transmembrane helix, a C-terminal domain, or an N-terminal domain to maximize the chance that the

secondary structure will be exhibited by the peptide (turn or helix). A peptide that terminates in the middle of a turn will likely not provide structural information about the turn.

4. Amino acid composition of the peptide will determine solubility. Therefore, the sequence one chooses in a given peptide should be influenced by the desire for solubility of that peptide.

Solvation Problems

Some of the loop peptides from rhodopsin are soluble in water. Peptides for the transmembrane helices of rhodopsin are insoluble in water. Some of the loop peptides of rhodopsin, in the intradiskal face, are insoluble in water, at least at the concentrations required for NMR measurements. (NMR measurements normally require about 1 m*M* concentrations in less than 1 ml of volume.) Thus one must be prepared to work with both hydrophilic peptides and hydrophobic peptides.

Hydrophilic peptides are studied in water. Hydrophobic peptides can be studied in dimethyl sulfoxide (DMSO) (a deuterated version of which must be used in NMR measurements to eliminate confounding resonances from the solvent in the NMR spectrum). Experiments on bacteriorhodopsin have shown that structures in DMSO reported with fidelity on the structure of the intact protein. Three of the loops of bacteriorhodopsin were synthesized and their structures determined in DMSO by NMR (because they were insoluble in water at the concentrations required for NMR experiments). The solution structures were then compared with the crystal structure of the intact protein. Good agreement was obtained between the solution structure of the peptide in DMSO representing the turn and the crystal structure of the intact protein.[13] In another experiment, the structure of the region encompassing the conserved NPxxY sequence of rhodopsin was shown to be helical in both a water-soluble peptide (structure determined in water) with the sequence of the carboxyl-terminal domain of rhodopsin and a hydrophobic peptide (structure determined in DMSO) with the sequence of the seventh transmembrane helix. Therefore, DMSO provides a means of studying relatively hydrophobic peptides by NMR.

The structure of hydrophobic peptides can also be determined in some cases in detergent micelles. A popular detergent for such determinations of structure by NMR is the perdeuterated version of dodecylphosphocholine (DPC). However, it should be noted that a helix like the seventh transmembrane helix of rhodopsin is not solubilized in DPC micelles, presumably because of the highly polar region in the middle around Lys-296.

Structure Determination

Solution structures are initially examined with circular dichroism (CD) if the peptides are soluble in water or in DPC micelles. Because of the high UV cutoff of DMSO, CD cannot be used with that solvent. CD spectra can quickly provide information on whether the peptide is adopting secondary structure in solution, usually either helix or turn. CD is not a very adequate method to determine the quantitative distribution of secondary structure elements in the peptide, as the basis sets for such determinations are proteins, not peptides. Nevertheless, significant negative transitions in the region of 210 to 220 nm are a good indication of the presence of secondary structure in the peptide in solution. Such an observation justifies the attempt to determine the three-dimensional solution structure by NMR.

Determination of solution structures by NMR is described in detail in other monographs[15,16] and will not be described here. A few issues specific to peptide structure determination will be discussed, however.

1. Because these peptides are of a size amenable to solid-phase peptide synthesis, structures are determined exclusively by two-dimensional homonuclear ^{1}H NMR. The cost of synthesis of peptides with stable isotopes such as ^{15}N is prohibitive.
2. Because such peptides have little in the way of tertiary structure, the number of distance constraints per residue that can be obtained is significantly less than that obtained from a larger protein. One typically obtains 10 to 12 unique constraints per residue.
3. Fewer constraints are obtained for the ends of the peptides, which causes the structures to be disordered at the N and C termini. This structural disordering likely reflects conformational motility in those regions.
4. In almost all cases, peptides from rhodopsin showed ordered structures in solution. However, there were a few exceptions. One peptide corresponding to part of the third transmembrane helix did not form a defined structure in solution. The carboxyl-terminal side of the third cytoplasmic loop also did not form a defined structure, although the rest of the loop did. The same observation was made for the third cytoplasmic loop of the PTH receptor.[17] However, those are the only exceptions known for rhodopsin. The vast majority of the peptides made from the rhodopsin sequence form stable structures in solution, mostly helix-turn-helix or helix structures. The same observations were made for the four helix bundle, myohemerythrin, and for the seven helix bundle, bacteriorhodopsin, as described earlier.

Build a Construct

A construct for the whole protein can be assembled from the pieces whose individual structures are determined as described earlier. Because of the design,

[15] J. N. S. Evans, "Biomolecular NMR Spectroscopy." Oxford Univ. Press, Oxford, 1995.

[16] G. C. K. Roberts, "NMR of Macromolecules: A Practical Approach." IRL Press, Oxford, 1993.

[17] D. F. Mierke, M. Royo, M. Pelligrini, H. Sun, and M. Chorev, *J. Am. Chem. Soc.* **118,** 8998 (1996).

the ends of these peptides can be superimposed using the overlapping sequences. This superposition can be done in many programs designed to visualize proteins from pdb files. We have used SYBYL from Tripos to do this and also MacImdad (Molecular Applications Group). Superposition needs to be done with recognition of the disordering that is typically observed at the ends of the peptide. It is best to avoid using the residues that are disordered in the superposition. That is why the overlap suggested for the peptides is so large, as described earlier. The construct eventually consists of the structures of all the peptides with the overlapping sequences. Redundant sequences, mostly in disordered regions, are then removed from the pdb file for a full construct with all the amino acids of the protein.

Simulated Annealing

On this construct, all the distance constraints from the individual structures are written. For the structure determination of rhodopsin, this was done in a mol2 file with SYBYL. This also allows the addition of other distance constraints that can be obtained from independent measurements on the protein. In the case of rhodopsin, a number of experimental long-range distance constraints are available from the literature (see Table I). These are very useful in defining the three-dimensional

TABLE I
EXPERIMENTAL LONG-RANGE CONSTRAINTS FOR DARK-ADAPTED STATE OF RHODOPSIN

	Range (Å)	Ref.
V139C → K248C	12–14	
V139C → E249C	15–20	
V139C → V250C	15–20	
V139C → T251C	12–14	
V139C → R252C	15–20	
H65C → C316	7–10	21,22
C140 → S338	15–21	23
V204 → F276	<5	24
I251 → V138	<13	25
E113 → K296	3	
C140 → C222	2–7	26
K245C → Q312C	2–7	27
S338C → T242C	2–7	28
Various interhelical distances from electron diffraction	X	29
Helix assignment of Baldwin		20

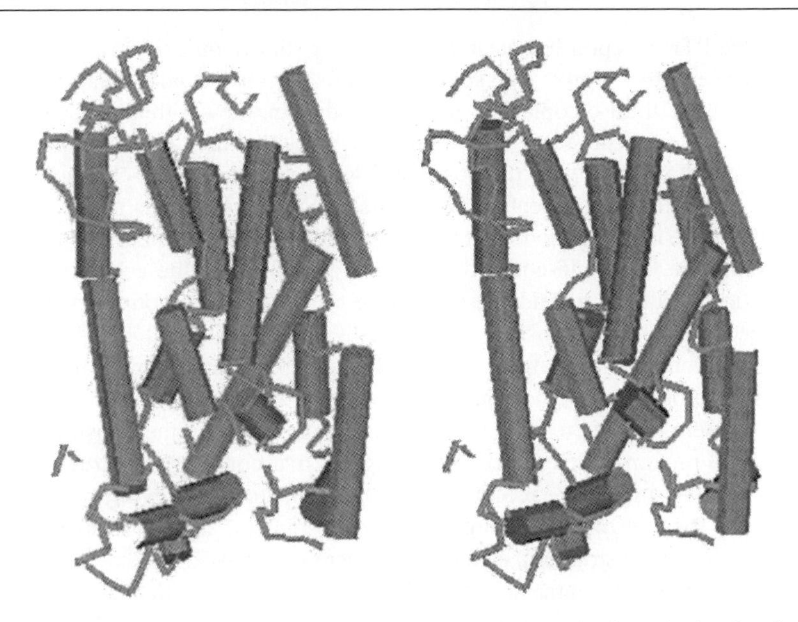

FIG. 2. α carbon map of the three-dimensional structure of rhodopsin, determined as described in the text.

relationship among the elements of secondary structure that are revealed in the individual peptide structure studies.

Simulated annealing is a protocol that allows the simultaneous interplay of all the available experimental distance constraints under conditions of conformational flexibility. On rhodopsin, the protocol for simulated annealing was to "heat" the molecule to 1000° for 1500 fsec and then to "cool" for 1500 fsec to 200°. The energy of the resulting structure was then minimized using SYBYL. Ten consecutive cycles were calculated. All 10 structures are then analyzed for the extent of similarity among them. In the case of a well-defined structure, all the structures obtained from simulated annealing should be reasonably superimposable. The result is a structure that is derived entirely from experimental data. Figure 2 shows the result for rhodopsin, a structure that is consistent with all the available experimental data for dark-adapted rhodopsin (manuscript submitted).

How to Approach the Structure of a New G Protein-Coupled Receptor

Available information suggests that structural information about other G protein-coupled receptors can be obtained by an approach similar to that used successfully with rhodopsin. For example, the structure of the third cytoplasmic loop of the PTH receptor has been determined by studies on a peptide containing the sequence of that loop.[17] The structure of this loop is virtually identical to the structure of the third cytoplasmic loop of rhodopsin,[18] even though there is no significant sequence homology between them. The structure of the seventh transmembrane helix, containing the conserved NPxxY sequence, of the tachykinin receptor as an individual peptide shows a structure very similar to the seventh transmembrane helix of rhodopsin.[19] These results point to similarities in structure among the G protein-coupled receptors, as well as to the ability to obtain information about the secondary structure of other G protein-coupled receptors, as has been done for rhodopsin.

With structural information available for one G protein-coupled receptor, a modified approach can be designed for the partial determination of structure for other G protein-coupled receptors. Key conserved residues in the transmembrane region of G protein-coupled receptors[20] suggest that considerable structural homology may exist from one receptor to another in the transmembrane region. The structure of the transmembrane region of rhodopsin could then be used as a template for assembling structural elements of other G protein-coupled receptors. For example, with a special interest in the cytoplasmic face, one could design peptides encompassing the three cytoplasmic loops and carboxyl-terminal of the receptor of interest and solve their individual structures. These structures could then be substituted for the corresponding structures on rhodopsin to obtain an approximate view of the structure of the cytoplasmic face of the G protein-coupled receptor of interest. The structure of the extracellular face of a G protein-coupled receptor could be similarly constructed from experimental structures of the extracellular loops and the amino terminus.

Because of the homology among G protein-coupled receptors (particularly for the transmembrane region where a significant number of conserved residues are found), homology modeling, using rhodopsin as a template, could be useful to design an initial construct for the receptor. Experimental structures determined

[18] P. L. Yeagle, J. L. Alderfer, and A. D. Albert, *Biochemistry* **36,** 3864 (1997).

[19] J. Berlose, O. Convert, A. Brunissen, G. Chassaing, and S. Lavielle, *FEBS Lett.* **225,** 827 (1994).

[20] J. M. Baldwin, *EMBO J.* **12,** 1693 (1993).

[21] K. Yang *et al., Biochemistry* **35,** 14040 (1996).

[22] D. L. Farrens, C. Altenbach, K. Yang, W. L. Hubbell, and H. G. Khorana, *Science* **274,** 768 (1996).

[23] A. D. Albert, A. Watts, P. J. R. Spooner, G. Grobner, J. Young, and P. L. Yeagle, *Biochim. Biophys. Acta* **1328,** 74 (1997).

[24] H. Yu, M. Kono, T. D. McKee, and D. D. Oprian, *Biochemistry* **34,** 14963 (1995).

[25] S. P. Sheikh, T. A. Zvyaga, O. Lichtarge, T. P. Sakmar, and H. R. Bourne, *Nature* **383,** 347 (1996).

[26] H. Yu, M. Kono, and D. D. Oprian, *Biochemistry* **38,** 12028 (1999).

[27] K. Cai, J. Klein-Seetharaman, J. Hwa, W. L. Hubbell, and H. G. Khorana, *Biochemistry* **38,** 12893 (1999).

[28] K. Cai, R. Langen, W. L. Hubbell, and H. G. Khorana, *Proc. Natl. Acad. Sci. U.S.A.* **94,** 14267 (1997).

[29] J. M. Baldwin, G. F. X. Schertler, and V. M. Unger, *J. Mol. Biol.* **272,** 144 (1997).

as described earlier for loops or other parts of the receptor of interest can be substituted in the homology model to define local structure more accurately. Energy minimization is then usually performed.

Using the simulated annealing approach to determine a complete experimental structure for the whole receptor, as described earlier for rhodopsin, would depend on the availability of sufficient experimental distance constraints. If enough were available, an approach completely analogous to the work on rhodopsin could be used to define experimentally the structure for the whole protein. As in the case of rhodopsin, it is feasible to determine the structures of a series of overlapping peptides that completely span the sequence of the protein. Under favorable circumstances the assembly of these fragments by superimposing their regions of overlap should lead to an approximation of three-dimensional structure. At this point, it will likely be necessary to have some long-range distance information to assist in the organization of the secondary structures into a tertiary structure. As described previously, simulated annealing will then allow the folding of the protein in a manner consistent with all the experimental data.

It should be noted that in cases where the peptides are long, such as in some of the loops or some N terminus or C terminus domains of some G protein-coupled receptors, an alternate approach to preparation of the peptide is required. When peptides get much longer than 50 amino acids, solid-phase synthesis becomes difficult, and as peptides approach 100 residues, two-dimensional homonuclear ^{1}H NMR spectra become difficult to interpret. Such peptides must therefore be produced in an expression system in minimal media to permit stable isotope labeling, such as with ^{15}N. Then multidimensional NMR experiments can be performed, greatly simplifying what would otherwise be very complex two-dimensional NMR data and a difficult structure determination.

Conclusions

Using these techniques, it is now possible to obtain useful structural information on G protein-coupled receptors by a segmentation approach.

[16] Tools for Dissecting Signaling Pathways *in Vivo*: Receptors Activated Solely by Synthetic Ligands

By KIMBERLY SCEARCE-LEVIE, PETER COWARD, CHARLES H. REDFERN, and BRUCE R. CONKLIN

Introduction

The diversity of G protein-coupled receptors (GPCRs) presents a challenge to understanding the connection between a single receptor signaling pathway and a specific physiological or pathological response. Receptors activated solely by synthetic ligands (RASSLs) offer control over the location, timing, and specificity of a G protein signal *in vivo*. These novel, reversible switches for G protein signaling have clarified the role of G_i signaling in cardiac physiology and are now being used to probe sensory transduction and complex neurobehavioral responses. This article summarizes the design of RASSLs and their first use *in vivo*. We supplement a concurrent review[1] on the subject by providing methods for expressing, detecting, and activating RASSLs in a wide variety of tissues.

GPCRs are the largest known family of cell surface receptors, encompassing over 1000 distinct receptors.[2] These receptors can be activated by a variety of natural ligands, including peptide hormones, odorants, photons, biogenic amines, and lipids. Activation of these receptors results in many different physiological responses, including heart rate changes, chemotaxis, cell proliferation, neurotransmission, and hormonal responses.[3] Prolonged stimulation of GPCRs can alter gene transcription and therefore may mediate long-term changes in the biochemistry, physiology, and behavior of an organism.

The same diversity of receptors, ligands, and responses that makes GPCRs biologically important has also complicated the study of their functions *in vivo*. The ability to stimulate a specific GPCR in a particular tissue *in vivo* would be a valuable aid to understanding the resultant changes in signaling and physiology. However, in a whole animal, numerous factors confound this type of study. Although it is feasible to inject a specific GPCR ligand directly into the tissue of interest, there is no way to restrict receptor activation to a specific subpopulation of cells. Furthermore, many GPCRs belong to large families of closely related receptor subtypes and may be activated by similar ligands. For example, the specific agonists and antagonists for many of the serotonin receptor subtypes remain unknown.[4] Finally,

[1] K. Scearce-Levie, P. Coward, C. H. Redfern, and B. R. Conklin, *Trends Pharmacol. Sci.* **22,** 417 (2001).

[2] C. D. Strader, T. M. Fong, M. R. Tota, and D. Underwood, *Annu. Rev. Biochem.* **63,** 101 (1994).

[3] A. M. Spiegel, A. Shenker, and L. S. Weinstein, *Endocr. Rev.* **13,** 536 (1992).

[4] D. Hoyer, D. E. Clarke, J. R. Fozard, P. R. Hartig, G. R. Martin, E. J. Mylecharane, P. R. Saxena, and P. P. A. Humphrey, *Pharmacol. Rev.* **46,** 157 (1994).

METHODS IN ENZYMOLOGY, VOL. 343
0076-6879/02 $35.00

the actions of endogenous ligands can complicate the interpretation of experimental results. In recent years, scientists have developed several "designer" signaling systems that use artificial ligands.[5] These systems, however, have not been able to mimic the complex conformational changes undergone by GPCRs when activated.

One approach to these difficulties has been to develop RASSLs.[6] These genetically engineered receptors are insensitive to their natural, endogenous ligand(s), but can still be fully activated by synthetic, small molecule drugs. Expression of RASSLs in transgenic mice allows us to study GPCR signaling *in vivo*. Using tetracycline-regulated gene expression technology, we can control where and when these RASSLs are expressed. By administering the synthetic drug, we can stimulate a single G protein pathway in a specific tissue quickly and reversibly. This system has already yielded important insights into cardiac function.[6–8] By expressing RASSLs in other tissues, researchers can explore the role of G protein signaling in many physiological and pathophysiological responses.

Construction of a RASSL

To construct our first RASSL, we modified the κ opioid receptor (KOR). Opioid receptors respond to endogenous peptides and signal via the G_i pathway. Because of the importance of this receptor family for pain modulation, the pharmaceutical industry has developed many high-affinity opioid receptor agonists. Small molecule ligands of the KOR are structurally distinct from endogenous peptide ligands, including dynorphin.[9,10] Unlike μ or δ opioid receptor agonists, κ receptor agonists are nonaddictive.[11]

Structure–function studies of the KOR have revealed that the second extracellular loop is critical for the binding of dynorphin and other neuropeptides.[12–14]

[5] A. Bishop, O. Buzko, S. Heyeck-Dumas, I. Jung, B. Kraybill, Y. Liu, K. Shah, S. Ulrich, L. Witucki, F. Yang, C. Zhang, and K. M. Shokat, *Annu. Rev. Biophys. Biomol. Struct.* **29,** 577 (2000).

[6] P. Coward, H. G. Wada, M. S. Falk, S. D. H. Chan, F. Meng, H. Akil, and B. R. Conklin, *Proc. Natl. Acad. Sci. U.S.A.* **95,** 352 (1998).

[7] C. H. Redfern, P. Coward, M. Y. Degtyarev, E. K. Lee, A. T. Kwa, L. Hennighausen, H. Bujard, G. I. Fishman, and B. R. Conklin, *Nature Biotechnol.* **17,** 165 (1999).

[8] C. H. Redfern, M. Y. Degtyarev, A. T. Kwa, N. Salomonis, N. Cotte, T. Nanevicz, N. Fidelman, K. Desai, K. Vranizan, E. K. Lee, P. Coward, N. Shah, J. A. Warrington, G. I. Fishman, D. Bernstein, A. J. Baker, and B. R. Conklin, *Proc. Natl. Acad. Sci. U.S.A.* **97,** 4826 (2000).

[9] G. H. Rimoy, D. M. Wright, N. K. Bhaskar, and P. C. Rubin, *Eur. J. Clin. Pharmacol.* **46,** 203 (1994).

[10] P. A. Reece, A. J. Sedman, S. Rose, D. S. Wright, R. Dawkins, and R. Rajagopalan, *J. Clin. Pharmacol.* **34,** 1126 (1994).

[11] M. J. Millan, *Trends Pharmacol. Sci.* **11,** 70 (1990).

[12] J.-C. Xue, C. Chen, J. Zhu, S. P. Kunapuli, J. K. de Riel, L. Yu, and L.-Y. Liu-Chen, *J. Biol. Chem.* **270,** 12977 (1995).

[13] E. V. Varga, X. Li, D. Stropova, T. Zalewska, R. S. Landsman, R. J. Knapp, E. Malatynska, K. Kawai, A. Mizusura, H. Nagase, S. N. Calderon, K. Rice, V. J. Hruby, W. R. Roeske, and H. I. Yamamura, *Mol. Pharmacol.* **50,** 1619 (1996).

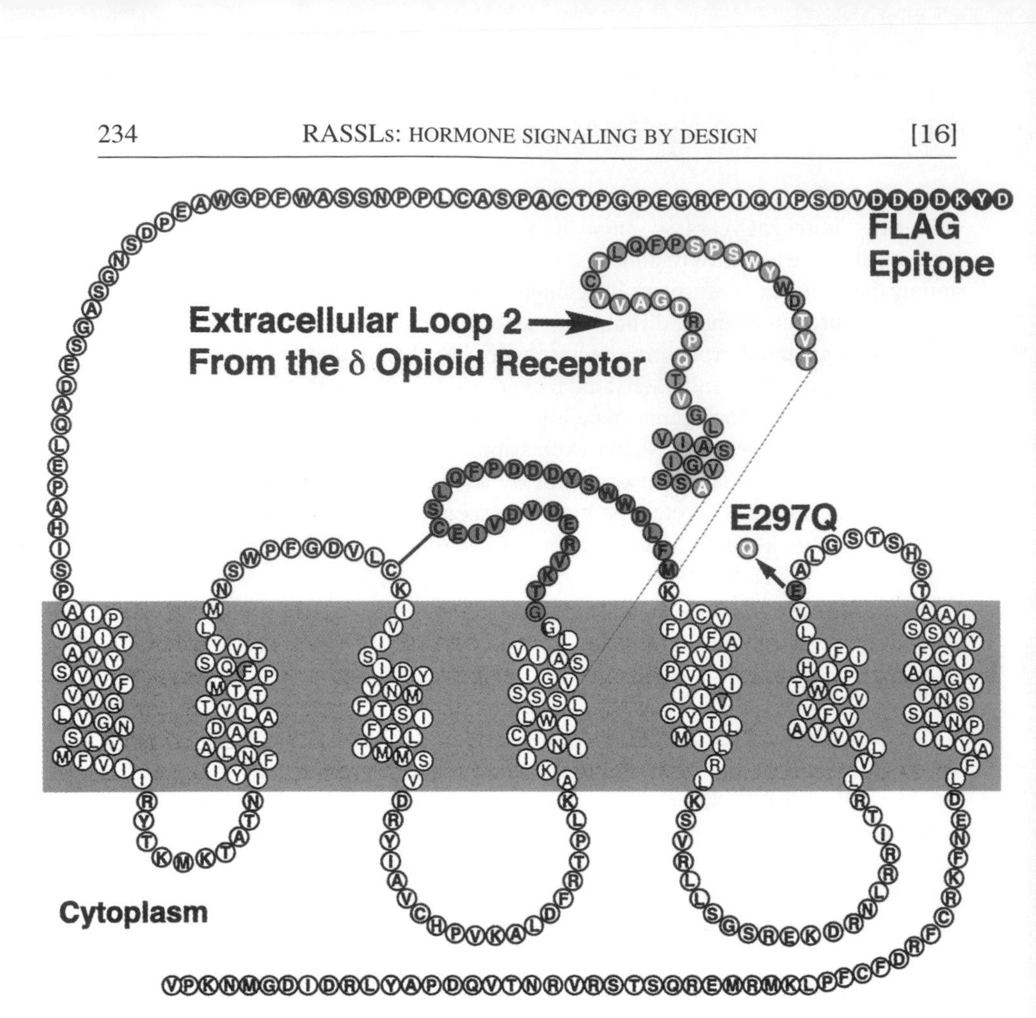

FIG. 1. Construction of an opioid RASSL. Replacing the third extracellular loop of the human KOR with the corresponding sequence from the δ opioid receptor significantly attenuates binding to endogenous peptides, while maintaining affinity for small molecule κ agonists. The specificity of the RASSL for small molecule agonists was enhanced further by mutating the glutamic acid at 297 to glutamine.

However, small molecule agonists have a different binding pocket close to the transmembrane region. This raised the possibility that mutating the extracellular regions of the receptor could interfere with the binding of endogenous, but not synthetic, ligands. To construct our first RASSL, Ro1 (RASSL opioid 1), we substituted the second extracellular loop of the δ opioid receptor for the corresponding portion of the KOR (Fig. 1).[6] This substitution results in low affinity for both

[14] H. Kong, K. Raynor, H. Yano, J. Takeda, G. I. Bell, and T. Reisine, *Proc. Natl. Acad. Sci. U.S.A.* **91,** 8042 (1994).

dynorphin and δ opioid receptor ligands.[5,13–15] Our second RASSL (Ro2) contains all the mutations in Ro1, as well as a substitution of glutamine for glutamic acid 297, located at the junction of transmembrane domain 6 and extracellular loop 3. This residue is thought to contribute to specific opioid peptide binding.[16] Both RASSLs showed reduced affinity for dynorphin, without a significant reduction in the response to spiradoline or other small molecule ligands. The binding affinity of Ro1 for dynorphin was reduced to <0.5%, while Ro2 showed dynorphin binding that is <0.05% of the native KOR.[6]

Despite the changes in the peptide-binding characteristics of Ro1 and Ro2, signaling in response to spiradoline remained intact.[6] This was demonstrated *in vivo* in transfected cells with calcium mobilization and cell proliferation assays. In both cases, spiradoline but not dynorphin induced the expected G_i-mediated response (Fig. 2).

Controlling RASSL Expression *in Vivo*

To take full advantage of the signaling control offered by a RASSL, it was necessary to control the timing and location of RASSL expression *in vivo*. To achieve this, we used the tetracycline (tet) transactivator system developed by Gossen and Bujard.[17] In the tet system, transgene expression is driven by a minimal promoter fused downstream of the tetracycline response element (tetO) from the bacterial *tet* operon (Fig. 3). Expression of the transgene requires the presence of tTA, a transcriptional transactivator protein that binds to tetO. Therefore, the mice must also be transgenic for a second transgene expressing tTA. Tissue specificity comes from the promoter element used to express tTA. Typically, two lines of mice are created: one harboring the tet-O transgene construct and the other harboring the tissue-specific promoter tTA construct and then bred together. Using the tet system to drive expression of the RASSL allows control of where the RASSL is expressed (by choice of tTA line) and when it is expressed (by administration or withdrawal of doxycycline). Two forms of tTA exist: one that binds DNA and activates transgene expression in the presence of doxycycline (the so-called tet-on system) and one that represses expression in the presence of doxycycline (the tet-off system). The tTA (tet-off) system is generally preferred for *in vivo* studies because doxycycline is absent during the experimental period, removing this experimental variable. Doxycycline is known to have many (generally benign) biological effects that are not a problem for breeding mice, but could complicate

[15] F. Meng, Y. Ueda, M. T. Hoversten, R. C. Thompson, L. Taylor, S. J. Watson, and H. Akil, *Eur. J. Pharmacol.* **311,** 285 (1996).

[16] S. A. Hjorth, K. Thirstrup, D. K. Grandy, and T. W. Schwartz, *Mol. Pharmacol.* **47,** 1089 (1995).

[17] M. Gossen and H. Bujard, *Proc. Natl. Acad. Sci. U.S.A.* **89,** 5547 (1992).

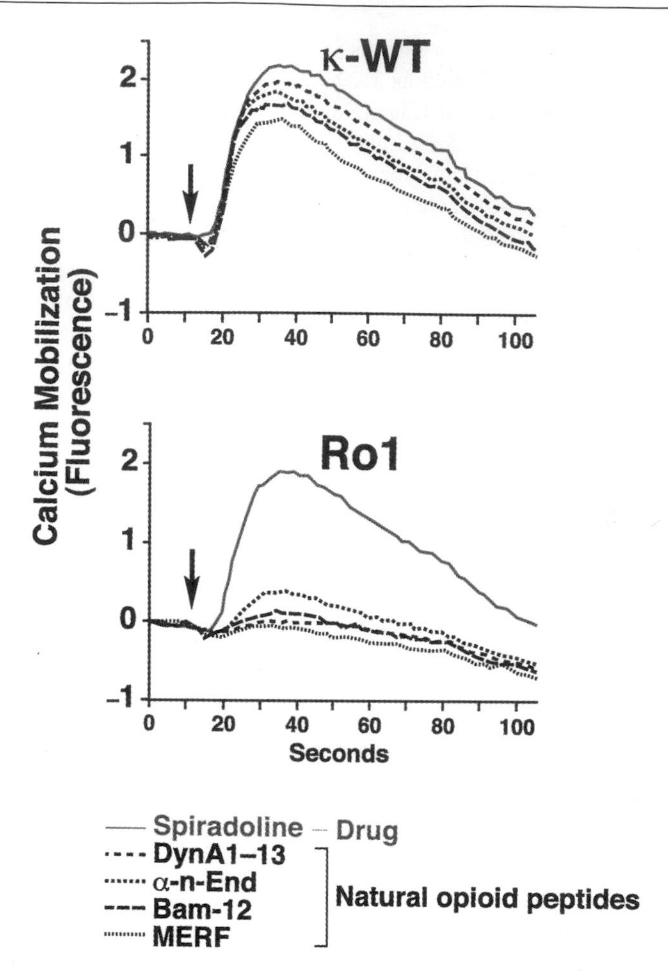

FIG. 2. Prototype RASSL signals specifically in response to the synthetic drug spiradoline. Agonist-mediated changes in intracellular calcium flux were measured with a fluorometric imaging plate reader assay. Curves are sample tracings of actual calcium fluorescence in response to 1 μM doses of agonist. The arrow indicates addition of the agonist.

an experiment if a response only occurs with the application of doxycycline (as in the tet-on system). To date, our collaborators have generated tTA lines that drive the expression of RASSLs in a wide variety of tissues, including heart, brain, salivary gland, liver, kidney, and brown fat (see Table I and more details later). Because our TetO-Ro1 line (available from Jackson Labs, Bar Harbor, ME, www.jax.org) is well characterized and known to be regulated by doxycyline, this line can be used as a reporter line to test expression patterns obtained with new tTA lines.

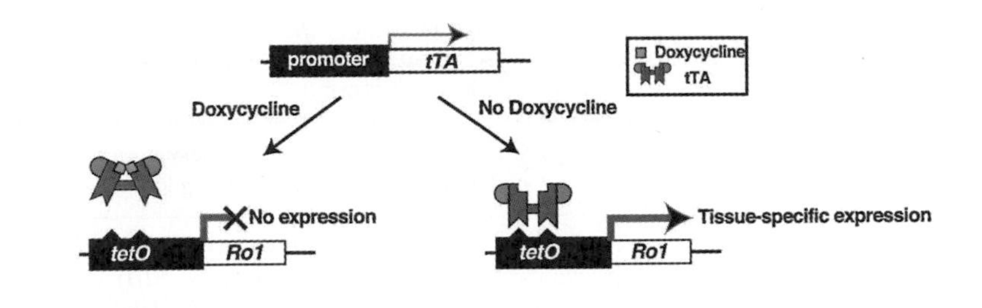

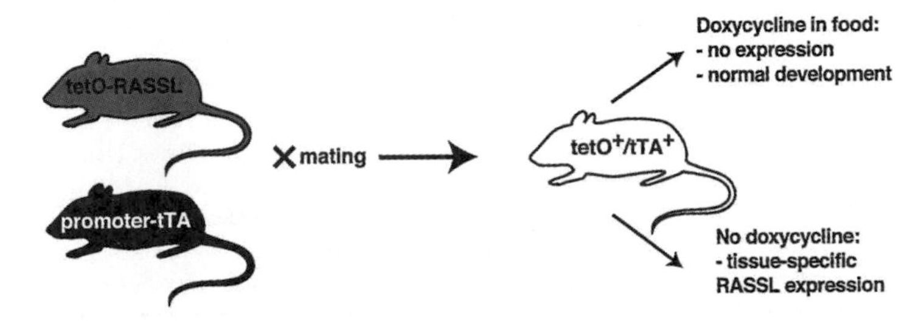

FIG. 3. Tissue-specific gene expression using the tet-tTA system. Two transgenes are required. First, the gene of interest (Ro1, in this case) is placed under the control of the tet operon sequence. Expression of this transgene requires the tetracycline transactivator (tTA). The gene for the tTA can be placed under the control of a tissue-specific promotor to ensure that tTA, and therefore Ro1, is expressed only in a particular tissue. Expression of Ro1 can be suppressed by the addition of doxycycline, which binds to tTA and inactivates it. For gene expression in mice, this system requires two transgenic lines: one with the promoter-tTA transgene and the other with tetO-Ro1. When these mice are mated together, 25% of the offspring will have both transgenes. However, even in these bigenic mice, there will be no Ro1 expression as long as the animals are fed doxycycline. Reprinted with the permission of Nature America, Inc.

We first used the tet system to examine RASSL signaling in the heart. Maximal Ro1 transgene expression in the heart is reached approximately 10 days after withdrawal of doxycyline, when the drug has been washed out of the animal's system completely.[7] Because doxycycline levels can be raised quickly after readministration, the subsequent suppression of transgene expression is rapid and largely dependent on the natural half-life of the protein being expressed. In most cases, transgene expression can be suppressed far more quickly than it can be completely induced. We have observed transgene suppression within 24 hr after doxycycline administration.[8] In contrast, the tet-on system would be expected to have a rapid onset, but a relatively slow suppression of expression *in vivo*.

Doxycycline Administration Protocols for Mice

To avoid expression of the transgene during early development, it is important to maintain mothers on doxycyline during gestation and nursing of litters. Sufficient

TABLE I
PHYSIOLOGICAL EFFECTS OF G_i SIGNALING *in Vivo* USING RASSLs AND THE tTA-tet SYSTEM[a]

Location of tTA expression	Expected effect of G_i signaling	Laboratories using RASSLs
Spinal cord	Analgesia	Iadarola (National Institutes of Health, NIH)
Visual cortex	Abnormal vision	Calloway (Salk Inst.)
Arterial smooth muscle	Muscle contraction	Husain (Univ. Toronto)
Kidney, brown fat	Altered mobilization of fat stores	Kopp (NIH)
Ventral tegmental area, nucleus accumbens	Motivation, addiction	Conklin/Scearce-Levie (Gladstone/UCSF) Nestler (Univ. Texas, Dallas)
Hippocampus	Abnormal learning and memory	Conklin/Scearce-Levie (Gladstone/UCSF)
Astrocytes	Modulation of neuronal activity	McCarthy (Univ. North Carolina)

[a] Preliminary studies communicated with the permission of Michael Iadorola, the National Institutes of Health; Edward Calloway, Salk Institute; Mansoor Husain, University of Toronto; Jeffrey Kopp, the National Institutes of Health; Eric Nestler, University of Texas, Dallas; and Ken McCarthy, University of North Carolina, Chapel Hill.

doxycyline is available to offspring via the placenta or milk to suppress transgene expression.

In Drinking Water. Make a 100× stock solution of 20 mg/ml doxycycline (Sigma, St. Louis, MO) in water. It will take about 2 min of shaking to dissolve the doxycyline. This solution can be frozen in 5-ml aliquots, wrapped in aluminum foil. Add the stock to mouse drinking water bottles for a final concentration of 200 μg/ml. Because doxycycline is light sensitive, use amber-colored bottles (e.g., Wheaton 900 RediPak amber glass packers with caps, 250 ml). It is not necessary to add sucrose to the drinking water. For most purposes, the 200-μg/ml dose is more than adequate to suppress gene expression in peripheral organs and in the brain. Fresh doxycycline should be added to the water weekly; more frequent changes are unnecessary.

In Mouse Chow. Doxycycline-containing food pellets can be obtained by custom order from Bio-Serv (Frenchtown, NJ) with quote # 908-996-2155. The pellets are made by adding 200 mg of doxycycline/kg of regular mouse chow; green food coloring is added to the chow to distinguish it from normal chow.

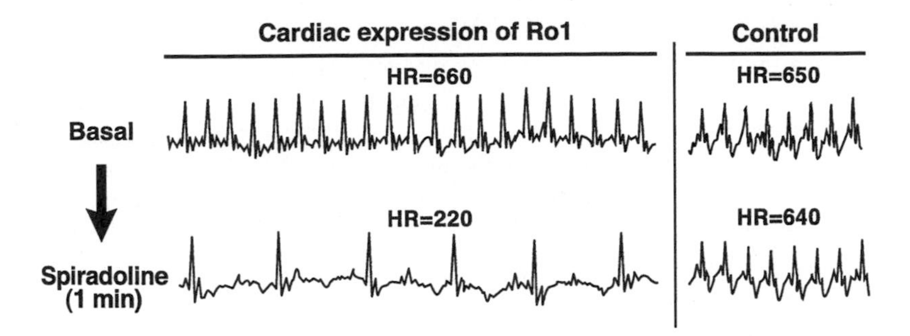

FIG. 4. RASSL-mediated reduction of heart rate (HR). Less than 1 min after spiradoline injection (1×10^{-5} mol/kg), the heart rate of a mouse expressing Ro1 in the heart decreased to one-third of baseline. Spiradoline had no effect on the control mouse (MHC-tTA). Reprinted with the permission of Nature America, Inc.

Doxycycline administration in food is less labor-intensive. Simply replenish the food as needed.

By Intraperitoneal Injection. If transgene expression prevents the animals from eating or drinking normally, it may be necessary to administer an initial dose of doxycycline by intraperitoneal (ip) injection. In a single injection, 10 μg of doxycyline can be administered in 0.5 ml of sterile water.

Controlling RASSL Signaling *in Vivo*

When the receptor is expressed, it should be functionally silent until the administration of spiradoline stimulates the RASSL and activates signaling pathways rapidly and specifically. G_i signaling in heart reduces heart rate.[18] Therefore, when Ro1 is expressed in the heart (using the α-myosin heavy chain promoter), receptor activation can be studied by simply measuring changes in heart rate. Wild-type animals have few KORs in the heart and therefore show no change in heart rate in response to spiradoline administration. In animals expressing Ro1 in the heart, however, heart rate decreases within 30 sec after drug administration (Fig. 4). This brachycardia is reversible within 10–15 min and can be blocked by antagonists.[7] These experiments indicate that a RASSL can be used to modulate a physiological response that requires G_i signaling *in vivo*. Expression of RASSLs in other tissues will enable researchers to regulate many other physiological and behavioral responses.

Signaling of KOR-based RASSLs is controlled by spiradoline administration. Most of the parameters relating to spiradoline administration have been worked out

[18] S. R. Holmer and C. J. Homcy, *Circulation* **84,** 1891 (1991).

in mice expressing Ro1 in the heart. Receptor activation decreases heart rate, which can be measured using implantable telemetry units as described later. It should be noted that repeated drug administration desensitizes the response to spiradoline. This is a characteristic of the desensitization mechanisms native to the KOR. This desensitization demonstates that signaling via a RASSL displays some of the normal physiological characteristics of the native, unengineered KOR. We are currently creating RASSL variants with altered desensitization kinetics to further explore the mechanisms and physiological relevance of receptor desensitization for G_i signaling *in vivo.*

Administration of Spiradoline

Spiradoline (U-62066E) can be purchased from Research Biochemicals International (Natick, MA), which is now a division of Sigma-Aldrich. In mice expressing Ro1 in the heart, the drug can be administered ip in a volume of 10 μl/g body weight. We found that single injections of 0.05–5 mg/kg body weight are sufficient to reduce heart rate in a dose-dependent manner.

Heart Rate Monitoring in Mice

To measure heart rate and wave form, we implanted cardiac telemetry units under sterile conditions into mice expressing Ro1. The mice were anesthetized with Avertin (10 μl/ml). The peritoneum was opened, and a PhysioTel implant (Model TA10EA-F20, Data Sciences International, St. Paul, MN) was inserted; monopolar electrodes were then tunneled subcutaneously across the precordium. Heart rate was monitored with AcqKnowledge III software (BIOPAC Systems, Santa Barbara, CA). For all baseline heart rate measurements, mice were injected with water (10 μl/g, ip) and monitored for 20 min. After treatment with spiradoline, heart rate was monitored for 20 more min. More extended measurements of heart function can be made by sampling heart function at fixed time points.

Eliminating Unwanted Signaling Using KOR Knockout Mice

In a tissue like the heart, where expression of the endogenous KOR is extremely low, the actions of spiradoline at the native receptor are not a major concern. However, for experiments that use RASSLs in KOR-rich tissues, like the brain, experimental results could be complicated by the actions of spiradoline at those receptors. Stimulation of neural KORs results in sedation, an effect that could mask many interesting behavioral responses caused by the action of spiradoline at RASSLs. We therefore used the tet system to express the RASSL in KOR knockout mice[19] (generously provided by Dr. J. Pintar, Rutgers University). With the proper

[19] F. Simonin, O. Valverde, C. Smadja, S. Slowe, I. Kitchen, A. Dierich, M. Le Meur, B. P. Roques, R. Maldonado, and B. L. Kieffer, *EMBO J.* **17,** 886 (1998).

controls, any behavioral or physiological effects of spiradoline administration can then be attributed to the effect of spiradoline at the RASSL, rather than endogenous KORs.

Detecting RASSLs *in Vivo*

For many experiments that use RASSLs to study the role of G_i signaling *in vivo,* it is essential to know precisely where the RASSLs are localized, perhaps even at a subcellular level. For this reason, RASSLs contain an N-terminal FLAG epitope tag, recognized by the commercially available FLAG M1 antibody. We have used the FLAG tag to detect Ro1 and Ro2 expression by a variety of methods: immunoprecipation or Western blotting of homogenized tissue, FACS sorting or ELISA of whole cells, and immunocytochemistry on whole tissue sections. We have not detected any changes in receptor activity resulting from the FLAG tag.

Solubilizing and Homogenizing RASSLs from Whole Tissue

Place 1 ml of homogenization buffer [50 m*M* Tris–HCl, pH 7.4, 1× complete cocktail (Boehringer), 1 m*M* dithiothreitol (DTT); 1 m*M* phenylmethylsulfonyl fluoride (PMSF) in distilled H_2O] in a round-bottom 14-ml tube. Because DTT and PMSF are unstable in water, it is important to use the buffer within 30 min after the addition of these reagents.

Transfer frozen (−90°) tissue into the 14-ml tube. Homogenize with a homogenizer at maximum speed until tissue is fully ground (usually about 30–60 sec).

Pipette the homogenate into an Eppendorf tube and place on ice immediatcly.

Sonicate for 3–5 sec if the homogenate is still too viscous.

Normalize the samples by tissue weight and use approximately 10–50 mg of tissue for solubilization.

Solubilize each sample in an Eppendorf tube under the following conditions: 1× solubilization sample buffer (100 m*M* Tris–HCl, pH 7.4, 1.5 *M* NaCl), 1× Triton-X100, and 1× complete cocktail.

Incubate for 30 min at 4° on a rotator.

Spin for 5 min at 4° at maximum speed.

Transfer supernatant to a new Eppendorf tube.

Immunoprecipitation of FLAG-Tagged RASSLs

Add the following to the supernatant of solubilized sample (use 50 ng of a FLAG-tagged protein): 2 μl of 1 *M* $CaCl_2$, 2 μl of FLAG M1-antibody (1 μg/μl; Sigma), and 15 μl of 50% suspension protein-A agarose beads (Sigma).

Incubate overnight at 4° on a rotator.

Spin down beads for 1 min at 4° at 3000 rpm.
Remove supernatant with a pipette or vacuum (being careful not to disturb beads) and discard.
Add 1 ml of TBS-T Ca^{2+} wash buffer (1× TBS; 0.1% Tween 20; 1 m*M* $CaCl_2$) to beads. Invert tube several times and then spin for 1 min at room temperature at 3000 rpm.
Discard supernatant.
Repeat wash.
Add 50 μl of 1× electrophoresis sample buffer [2× Tris–glycine SDS sample buffer (Novex) and 5% β-mercaptoethanol, diluted to 1× with ddH_2O] to beads. Vortex and boil sample for 5 min.
Vortex again. Spin for 2 min at room temperature at 5000 rpm.
Using flat tips, transfer supernatant to new tube or gel. Pipette carefully to avoid disturbing the beads.

Immunoblotting FLAG-Tagged RASSLs

Remove precast 10% Tris–glycine gel (Novex) from refrigerator and warm to room temperature.
Prepare controls and/or standards and adjust to the same volume as samples:
Standard:
5 μl Kaleidoscope prestained standard (Bio-Rad)
5 ng FLAG-tagged protein
40 μl of 1× sample buffer
Remove precast gel from bag. Remove tape from bottom of gel. Place gel into apparatus (Novex: X Cll II mini-cell). Remove comb. Fill the inside chamber with running buffer (1× Tris/glycine/SDS buffer; Bio-Rad). Using a pipette rinse wells with buffer. Make sure that there is no leakage from the inside to the outside chamber. Fill the outside chamber with running buffer until the buffer covers the opening on the bottom of the gel.
Using flat tips, load 45–50 μl of each sample into well. Fill blank wells with buffer to minimize curving of bands along the sides of the gel.
Run gel at 125 constant volts for 90 min or until blue dye front has exited gel.
While gel is running, prewet nitrocellulose, filter paper, and blotting pads in blotting buffer (1× blotting buffer and 25× Tris–glycine transfer buffer for blotting, Novex; 20% methanol).
Remove gel from apparatus. Using a razor blade, slice the lower opening of the gel and remove gel pieces in the slot. Pop open plastic plates with a knife. Remove the top portion (1 cm) of the gel.
Assemble gel transfer sandwhich as follows: (facing + electrode) top pad, filter paper, nitrocellulose, gel, filter paper, and (facing − electrode) bottom pad.

Fill the inside of the transfer chamber with blotting buffer and the outside with water.

Transfer for 1–2 hr at 30 constant volts.

Disassemble apparatus. Place nitrocellulose in container with blocking solution (50 ml of 5% milk in TBS). Incubate for 0.5–3.0 hr at room temperature, with gentle shaking.

Discard blocking solution. Rinse nitrocellulose for 5 min in TBS.

Place nitrocellulose in container with FLAG M1 antibody directly conjugated to horseradish peroxidase (Chromaprobe Inc.; custom conjugation). Incubate for 1–3 hr at room temperature, with gentle shaking. Cover container with foil.

Incubation mix:
1 m*M* $CaCl_2$
1% BSA
0.25 μg/ml M1-HRP
in 1× TBS and 0.1% Tween 20

Rinse blot with TBS-T Ca^{2+} wash buffer. Then, wash twice for 15 min each.

Remove nitrocellulose from buffer and place on plastic wrap.

From the Amersham ECL kit, combine 4 ml of solution 1 with solution 2.

Immediately pour mixture onto the surface of the nitrocellulose. Allow reaction to proceed for exactly 1 min at room temperature.

Pour off developing mixture. Enclose nitrocellulose in plastic wrap and place a fluorescent marker next to nitrocellulose. Transfer to a film cassette.

In dark room, place a piece of film on top of plastic wrap. Expose for 1 min at room temperature and develop film. If signal is low, increase exposure time.

Detection of FLAG-Tagged RASSLs in Suspended Cells by Flow Cytometry

This method is used for detecting RASSLs on lymphocytes of transgenic mice expressing Ro1. With modifications, it can be used to detect RASSLs on a variety of suspended cells. We use a similar protocol to monitor RASSL expression on cultured mammalian cell lines (HEK 293 and Rat1a).

Cut mouse tail with sharp razor.

Collect 200 μl of blood in 15-ml Falcon conical tube containing 12 ml of fresh phosphate-buffered saline (PBS) and heparin (1–2 units/ml).

Invert tube several times and store on ice while cutting other tails.

Spin Falcon tubes in centrifuge at 1000–1200 rpm for 5 min.

Pour off supernatant (careful, the pellet is not tight) and resuspend in 3 ml of 1× lysis buffer (10× lysis buffer: 80.2 g NH_4Cl, 8.4 g $NaHCO_3$, 3.7 g EDTA in 1 liter distilled H_2O).

Let sample stand in lysis buffer for 5–10 min at room temperature.
Spin for 3 min at 1000 rpm.
Resuspend in medium (Dulbecco's modified Eagle's medium + 2% fetal calf serum) and spin again for 3 min at 1000 rpm.
Resuspend in antibody staining solution (for each reaction: 1 ml of medium solution plus 1 μg FLAG M1 antibody. It requires custom conjugation to phycoerythrin by Molecular Probes, Eugene, OR). Keep protected from light. Shake gently on ice for 20–30 min.
Pellet cells by spinning; wash in 2 ml of medium.
Pellet cells again; wash in 1% paraformaldehyde in PBS.
Pellet cells again; resuspend in 500 μl of 1% paraformaldehyde/PBS in a 15-ml Falcon tube (2054).
Cover with aluminum foil and refrigerate until time for FACS analysis.

In theory, immunolabeling of the same FLAG epitope can be used to localize the RASSL *in vivo* by using immunohistological techniques on tissue sections. However, the FLAG M1 antibody is notoriously difficult to use on tissue sections. We are making efforts to improve labeling specificity for the FLAG tag in whole tissue. However, we do have some alternative methods that can be used to visualize RASSL expression in tissue sections.

β-Galactosidase Staining

An indirect approach to detecting RASSL expression in tissue section relies on the coinjection of tetO-lacZ along with tetO-RASSL DNA into transgenic mouse lines. This allows both the RASSL and the lacZ to integrate stably into the genome at the same location.[20] β-Galactosidase activity can then be used to indicate the expression of lacZ. Although it is not possible to estimate the precise expression level of the RASSL from the intensity of β-galactosidase staining, β-galactosidase is useful for rapid, clear visualization of transgene expression. Typically, we inject three times the amount of RASSL construct relative to the quantity of lacZ. This ratio results in cointegration of the constructs in $>90\%$ of our founder lines.

We have used the following protocol for a variety of fresh frozen tissue samples, including heart, brain, liver, and whole embryo.

Cut 10-μm cryostat sections from fresh-frozen tissues onto pap-penned slides. Immerse immediately in cold formaldehyde/glutaraldehyde for 5 min and then rinse in ddH_2O for 1 min.
Let section dry completely onto slide.

[20] R. R. Behringer, T. M. Ryan, M. P. Reilly, T. Asakura, R. D. Palmiter, R. L. Brinster, and T. M. Townes, *Science* **245,** 971 (1989).

Rinse with 1× PBS.

Apply final X-Gal solution (dilute 40X stock of 40 mg/ml X-Gal in DMSO into a solution of 5 m*M* potassium ferricyanide crystalline, 5 m*M* potassium ferricyanide trihydrate, 2 m*M* MgCl in PBS) to sections and incubate at 37° for 30 min to 24 hr. Check sections under microscope regularly; stop reaction when deep blue staining is observed.

Rinse with PBS.

Wash twice for 2 min each with distilled H_2O.

Counterstain for 3 min with nuclear fast red.

Wash twice for 2 min each with distilled H_2O.

Coverslip with Gel Mount (Biomeda).

RASSLs Tagged with Green Fluorescent Protein

The immunological and histological techniques just described typically require that the tissue be fixed or lysed before processing. Sometimes, though, it is useful to visualize the location of the RASSL in living cells. For this purpose, we have tagged our Ro2 RASSL with green fluorescent protein (emerald GFP from Packard). We placed the GFP tag on the N terminus of the RASSL to avoid a potential interference between GFP and the cellular sorting of the RASSL via the cytoplasmic surface of the receptor. GFP is located after the signal sequence and the FLAG tag, but before the coding region of the receptor. This GFP–RASSL fusion protein can be expressed in mammalian cell lines. In preliminary studies of HEK 293 and Rat 1a cells, the fusion protein was sorted to the membrane correctly and signaled normally. One advantage that GFP has over β-galactosidase is that in very large cells, like neurons, the location of β-galactosidase staining might not correspond directly to the location of the receptor. Portions of the cells where protein is translated will stain positive for β-galactosidase, but neurons may transport the receptor protein to distal regions that do not have enough β-galactosidase for detection.

Currently, we are using GFP-Ro2 to monitor receptor internalization after agonist stimulation. We can now use this protein to assay for localization of the RASSL both *in vitro* and *in vivo* in transgenic mice.

Applications of RASSL Technology

Acute activation of RASSLs expressed in discrete tissues or cell types will allow researchers to probe the role of G_i signaling in the normal function of those cells. Correspondingly, overexpression or extended activation of RASSLs can be used to explore disease models of pathologies related to hyperactive G protein signaling.

Tissue-Specific Expression and Activation of RASSLs

RASSLs have been expressed successfully in a variety of tissues and organs of transgenic mice, including heart, liver, salivary glands, smooth muscle, adipose tissue, and specific brain regions (see Table I). This offers several opportunities for new research programs. Our laboratory is particularly interested in using RASSLs to study the control of electrically active tissues by G protein signaling.[7] We have already demonstrated that RASSL activation in the heart can slow heart rate. A number of mouse lines that express the RASSL in specific brain regions are being used to study how the activation of G_i signaling in a given brain nucleus can affect the activity of downstream neural circuits.

Long-Term Expression and Activation of RASSLs

RASSLs can also be used to probe the role of G protein signaling in a number of long-term biological changes. RASSLs that are overexpressed or activated chronically will yield changes in gene expression and long-term adaptive shifts in the cells expressing the receptors. Long-term changes in cellular function underlie such important, yet poorly understood, biological functions as neuronal plasticity, cytoskeletal remodeling, apoptosis, proliferation, and differentiation. By using RASSLs to induce chronic shifts in G protein signaling, scientists can better understand how G protein signaling mediates both adaptive and pathological changes in cell function.

Reversible Models of Disease States

The ability to modulate both RASSL expression (by the tetracycline-inducible expression system) and signaling (by administration of spiradoline) offers the opportunity to create reversible models of disease caused by abnormal G protein signaling.

Our laboratory has developed a mouse model of dilated cardiomyopathy by overexpressing Ro1 in the heart.[8] Mice that express the RASSL for more than 3 weeks begin to develop abnormal heart function, including decreased contractility and wide QRS complexes on electrocardiograms, even in the absence of synthetic ligand. The hearts of these animals have enlarged ventricles and elevated levels of fibrosis and collagen. This phenotype appears to be caused by increased basal G_i signaling due to the overexpression of Ro1, as cardiomyopathy occurs without the addition of ligand and it can be blocked by suppressing Ro1 expression. Similarly, blocking signaling by administering antagonists to the receptor (norbinaltorphimine dihydrochloride) or inhibitors of G_i (pertussis toxin) can prevent the development of the phenotype or partially rescue it, depending on the time of administration (N. Cotte, T. Nanevicz, and B. Conklin, unpublished observations). This work suggests that hyperactive G_i signaling can disrupt heart function and lead to dilated cardiomyopathy.

Potentially, RASSLs could be used to study other diseases that may be related to hyperactive G_i signaling. In the brain, a number of pathologies may be caused by signaling abnormalities. For instance, seizures may be induced by the asynchronous activation of G proteins in a particular population of neurons.[21] Dementias and neurodegenerative disorders have been associated with abnormal G protein signaling.[22] Psychiatric disorders, such as schizophrenia, may be related to hyperactive G protein signaling through dopamine receptors.[23] Other diseases that may be tied to abnormal G protein signaling include osteoporosis, vasospasm, and immune disorders such as lupus and Crohn's disease.

Gene Expression Fingerprints for Signaling Pathways and Disease States

RASSLs can be combined with cDNA microarray technologies to monitor the changes in gene expression induced by G protein signaling. This type of study will allow researchers to create gene expression "fingerprints" that identify genes whose expression is regulated by G protein activation in different tissues or at different times in development. This approach is a potentially powerful means for improving our understanding of both normal biological function and pathological states. Data from gene expression experiments based on Ro1-induced cardiomyopathy can be viewed in sample data provided with a new bioinformatics software program developed in the Conklin laboratory called GenMAPP (www.GenMAPP.org).

Future Directions: New RASSLs

Existing RASSLs will allow study of the effects of G_i signaling in specific tissues under specific circumstances. Of course, it would be even more valuable to extend these studies to include other G protein signals, like G_s and G_q. New RASSLs can be developed by using many of the same principles used to develop the existing G_i-coupled RASSLs. The ideal RASSL activator should be a small molecule drug that is readily available from commercial sources. Its binding site should be well characterized. It must be highly specific for a single receptor subtype to minimize effects of the drug at non-RASSL receptors. Ideally, this could be a synthetic system in which the ligand would not activate endogenous receptors. Failing that, the ligand should not cause significant side effects, and the effects of activation of its natural receptor should be relatively minor in humans or rodents. Because small molecule ligands developed by the pharmaceutical industry are approved for human use and become freely available to researchers, there will be new opportunities to develop RASSLs that signal through other G protein pathways.

[21] N. G. Bowery, K. Parry, A. Boehrer, P. Mathivet, C. Marescaux, and R. Bernasconi, *Neuropharmacology* **38,** 1691 (1999).

[22] A. J. Berger, A. C. Hart, and J. M. Kaplan, *J. Neurosci* **18,** 2871 (1998).

[23] A. Cravchik, D. R. Sibley, and P. V. Gejman, *J. Biol. Chem.* **271,** 26013 (1996).

The combination of RASSL technology with the tet system provides a reversible molecular switch that allows researchers to control the timing, location, and specificity of G protein signaling *in vivo*. With the completion of the Human Genome Project, all GPCRs will soon be identified. We can then focus on the significant challenge of understanding how this diverse family of proteins modulates physiological processes. Recent findings revealing the significance of dimerization,[24] protein–protein interactions,[25] and alternative splicing of GPCRs[26] suggest that the signaling and modulation of these proteins are even more complex than previously believed. RASSLs provide a new tool to help study these processes *in vivo* and provide specific functional data for a variety of physiologically important signaling pathways.

[24] H. Möhler and J.-M. Fritschy, *Trends Pharmacol. Sci.* **20,** 87 (1999).

[25] Y. Tang, L. A. Hu, W. E. Miller, N. Ringstad, R. A. Hall, J. A. Pitcher, P. DeCamilli, and R. J. Lefkowitz, *Proc. Natl. Acad. Sci. U.S.A.* **96,** 12559 (1999).

[26] G. J. Kilpatrick, F. M. Dautzenberg, G. R. Martin, and R. M. Eglen, *Trends Pharmacol. Sci.* **20,** 294 (1999).

[17] Analysis of Structure–Function from Expression of G Protein-Coupled Receptor Fragments

By SADASHIVA S. KARNIK

The family of receptors coupled to heterotrimeric guanyl nucleotide-binding proteins (G proteins) consists of transmembrane receptors that transduce signals in response to light, odorants, hormones, neurotransmitters, and a variety of intracellular signals of unknown nature.[1,2] G protein-coupled receptors (GPCRs) share a common structural motif consisting of seven transmembrane (7TM) α-helical segments separated by four extracellular segments and four cytoplasmic segments. The extracellular segments predominantly play a role in the folding and the assembly of the receptor. In a variety of peptide hormone receptors the extracellular domain is also involved in hormone binding. As a consequence of this role, an N-terminal hormone binding domain distinct from the 7TM structure is evolved in some receptors, which is structurally distinct and functionally autonomous. In the majority of GPCRs, the ligand pocket is formed by the TM domain. This structure is fundamental to signal generation in GPCRs. The intracellular regions interact

[1] T. H. Ji, M. Grossmann, and I. Ji, *J. Biol. Chem.* **273,** 17299 (1998).

[2] U. Gether and B. K. Kobilka, *J. Biol. Chem.* **273,** 17979 (1998).

0076-6879/02 $35.00

with G proteins and other cytoplasmic proteins in mediating the signals. Mutagenesis studies have played a major role in defining the structure and the molecular mechanisms that govern GPCR function. Structure–function analysis from expression of GPCR fragments has become a powerful tool in the study of GPCRs and other membrane proteins. The approach opens up new insights into tertiary structure analysis and molecular mechanisms governing GPCR functions. Promise of developing new tools for regulating specific GPCR functions has emerged in the study of several GPCRs.

Cytoplasmic Domain Fragments

Mutagenesis of the cytoplasmic loops (CL) in GPCRs suggests a prominent role for the CL-3 in G protein activation; however, other cytoplasmic loops also play a role. Peptides derived from CL-2, CL-3, and CL-4 in a variety of GPCRs have been shown to block the receptor-specific interaction of G proteins.[3–7] However, the exact role of each cytoplasmic loop in the specificity of G protein binding, stimulation of GDP release and GTP uptake in the receptor-bound G protein, or dissociation of G protein subunits from the receptor is not very well understood. Advances in protein engineering have demonstrated that it is possible to separate individual domains of multidomain proteins preserving the domain function. The G protein activating function in GPCRs is accomplished by the cytoplasic domain consisting of discontinuous loops connecting TM helices. Therefore, engineering a soluble G protein-activating domain of a GPCR would require extensive reengineering of the loops. This article describes the construction of TAPER (transducin activating protein engineered from rhodopsin) through a series of minigenes for expressing in *Escherchia coli.*

Design of TAPER Constructs

TAPER I contains four cytoplasmic segments of bovine rhodopsin (see Fig. 1)—loop I (Thr^{58}–Asn^{73}), loop II (Lys^{141}–His^{152}), loop III (Lys^{231}–Arg^{252}), and loop IV (Asn^{310}–Cys^{322})—and the octapeptide epitope (Thr^{340}–Ala^{348}) for the monoclonal antibody 1D4. Because the conserved ERY sequence located at the cytoplasmic end of TM3 appears to play an important role in rhodopsin function, the segment Ile^{133}–Cys^{140} was designed to be a part of loop II in all TAPER constructs that contain this loop. Two of the cysteines, Cys^{140} and Cys^{322} were retained,

[3] B. König, A. Arendt, J. H. McDowell, M. Kahlert, P. A. Hargrave, and K. P. Hofmann, *Proc. Natl. Acad. Sci. U.S.A.* **86,** 6878 (1989).

[4] A. H. Cheung, C. H. Ruey-Ruey, M. P. Granziano, and C. D. Strader, *FEBS Lett.* **279,** 277 (1991).

[5] T. Okamoto, Y. Murayama, Y. Hayashi, M. Inagaki, E. Ogata, and I. Nishimoto, *Cell* **67,** 723 (1991).

[6] T. Okamoto and I. Nishimoto, *J. Biol. Chem.* **267,** 8342 (1991).

[7] H. M. Dalman and R. R. Neubig, *J. Biol. Chem.* **266,** 11025 (1991).

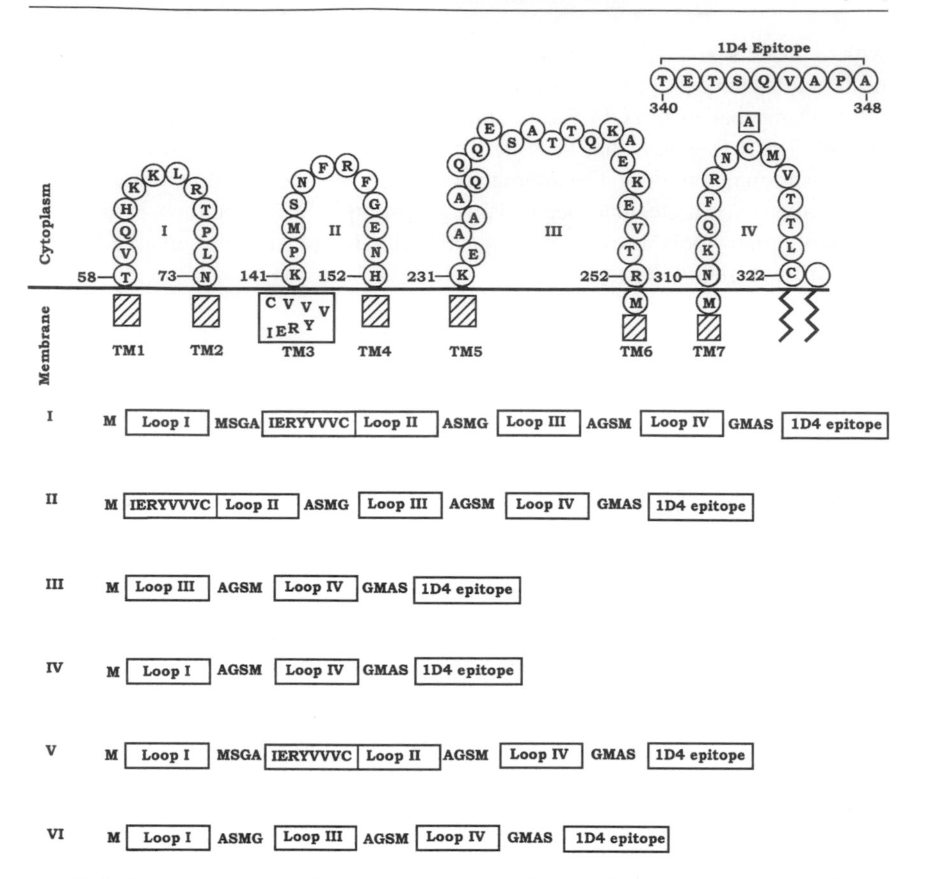

FIG. 1. Schematic representation of the cytoplasmic domain of bovine rhodopsin and TAPER polypeptides constructed. The residue numbering is according to bovine opsin. The boundaries of transmembrane helices and loops are chosen arbitrarily based on recent structure–function studies. The ERY sequence was retained as an extension of transmembrane helix 3, which may be important for activating transducin. Palmitoylated Cys-322 and Cys-323 are indicated. The 1D4 epitope was placed at the C-terminal end of all TAPER constructs. The tetrapeptide linker sequence is indicated.

whereas Cys^{316} was replaced with an Ala. A tetrapeptide linker consisting of amino acids Met, Ser, Gly, and Ala was used in different sequence because these amino acids occur with nearly equal preference in different protein secondary structural elements. Therefore the linker sequence is less likely to influence the conformation assumed by the loops. Because the functional contribution of different loops to various steps of the GTPase cycle of transducin is not clearly established, a series of TAPER constructs that differ in the loop composition were also constructed. They are shown in Fig. 1.

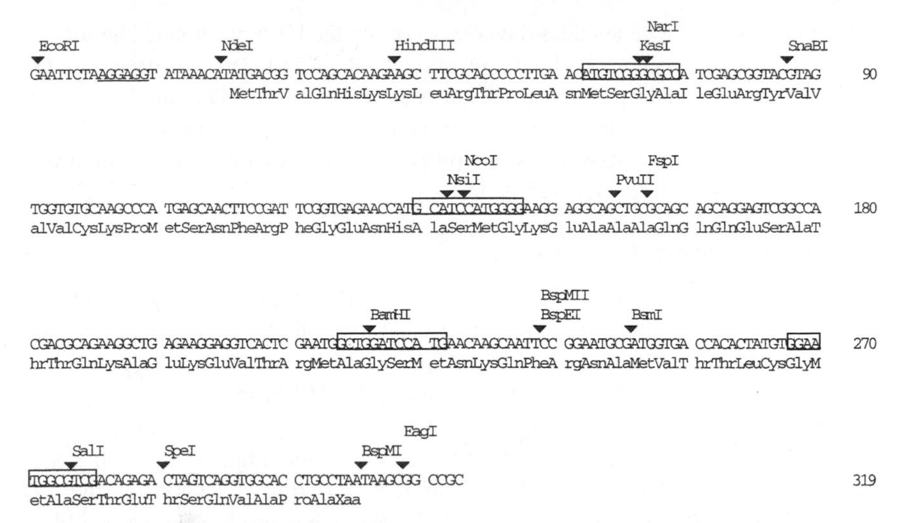

FIG. 2. The synthetic gene encoding TAPER I. The gene was assembled from synthetic oligonucleotides and characterized as described in Ferretti *et al.*[8] The remaining TAPER constructs contain the identical sequence except for the segment that is missing. Segments encoding tetrapeptide linkers are shown in boxes and the ribosome binding site is underlined.

Minigene Expression

A synthetic gene encoding TAPER I is shown in Fig. 2. Minigenes encoding different TAPER polypeptides were constructed by the methods developed by Ferretti and colleagues,[8] which is described in detail elsewhere in this volume [19]. To facilitate efficient expression in *E. coli,* a synthetic ribosome-binding site was introduced upstream of the initiator Met codon in each minigene. Combinations of the cytoplasmic loops in each TAPER are illustrated in Fig. 1. For example, TAPER I contains all four cytoplasmic loops, whereas TAPER II contains only loops II, III, and IV, which are thought to be important for G protein coupling, and TAPER III contains loops III and IV. An octapeptide epitope for the monoclonal antibody, 1D4 is present at the C-terminal end of all TAPER constructs. This facilitated easy detection of the TAPER by immunoblotting, as well as purification by immunoaffinity chromatography.[9,10] Expression of these minigenes utilizes the bacteriophage T7 RNA polymerase to drive transcription of the TAPER minigene.[11] The *E. coli* strain, BL21(DE3), is a lysogen bearing the bacteriophage

[8] L. Ferretti, S. S. Karnik, H. G. Khorana, M. Nassal, and D. D. Oprian, *Proc. Natl. Acad. Sci. U.S.A.* **83,** 599 (1986).

[9] R. S. Molday and D. McKenzie, *Biochemistry* **22,** 653 (1983).

[10] S. S. Karnik and H. G. Khorana, *J. Biol. Chem.* **265,** 17520 (1990).

T7 RNA polymerase gene under the control of the lac UV5 promoter. The second component of this system is the T7 promoter, which is located upstream of the TAPER minigene in the pT7-5 expression vector. Expression is initiated by the addition of 5 m*M* nonmetabolizable inducer isopropylthio-β-D-galactoside (IPTG). Addition of IPTG derepresses T7 RNA polymerase expression in turn, which will activate expression of the TAPER gene.

Induction of Minigene Expression

A single colony of the BL21(DE3)/pT7-5-TAPER transformant is used to inoculate 5 ml LB medium containing 50 μg/ml ampicillin. The overnight culture (37°) is then diluted 1000-fold in 10 ml of enriched medium (2% tryptone, 1% yeast extract, 0.5% NaCl, 0.2% glycerol, and 50 m*M* KH_2PO_4, pH 7.2) containing ampicillin and grown overnight at 30° with shaking at 150–200 rpm. This culture is used to inoculate 1 liter of rich medium containing ampicillin in 2-liter Erlenmeyer flasks, maintaining shaking at 150–200 rpm at 30°. When the OD 650 nm reaches 0.3, IPTG (30 μM) and chloramphenicol (1 μg/ml) are added. Chlorophenicol addition improves the expression of TAPER sometimes, but does not inhibit expression anytime. About 5 hr after the addition of the inducer, bacteria are collected by centrifugation at 4°. The cells are washed with ice-cold TES buffer [50 m*M* Tris (pH 8.0), 10 m*M* EDTA, 2 m*M* dithiothreitol (DTT), 0.1 m*M* phenylmethylsulfonyl fluoride (PMSF) and 10 mg/ml benzamidine]. Cell pellets are frozen by immersion in liquid nitrogen and kept frozen at −80°.

Lysis of E. coli

About 5 g (wet weight) of a fully thawed cell pellet is suspended in 20 ml of TES buffer containing 10% sucrose and 10 mg/ml lysozyme (Sigma). Mix well and place on ice for 1 hr. Add 150 ml 50 m*M* Tris (pH 8.0) buffer, containing 10 m*M* $MgCl_2$ and 10 m*M* 2-mercaptoethanol. Mix well on ice, and sonicate using a probe sonicator until the viscosity is reduced. Add 500 μl DNase I (1 mg/ml stock) and incubate for 1 hr at 37°. Place on ice, add NaCl to a final concentration of 200 m*M*, mix for 30 min, and spin at 10,000*g* for 10 min. The supernatant is spun again at 50,000 *g* for 1 hr. Add CHAPS [10% (w/v) stock in water] to a final concentration of 0.5% (w/v), and spin at 50,000 *g* for 1 hr. TAPER is recovered in the supernatant as indicated by the immunoblot analysis shown in Fig. 3A.

Purification of TAPER Peptides

Further purification is carried out by immunoaffinity chromatography on 1D4-Sepharose. The 50K supernatant is mixed with 5 ml of 1D4-Sepharose (50%

[11] S. Tabor, *in* "Current Protocols in Molecular Biology" (K. Janssen, series ed.), p. 16.2.1. Current Protocols; Green Publishing Associates, Inc., and John Wiley & Sons, Inc., 1987.

A. 1D4-Sepharose chromatography of TAPER

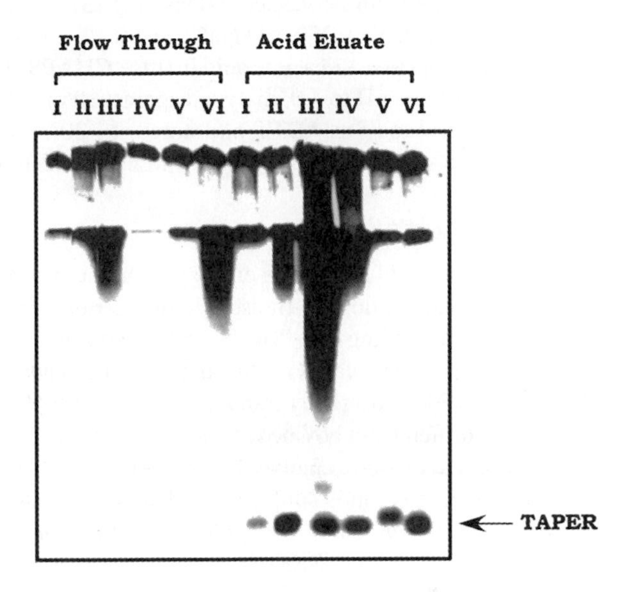

B. Purified TAPER

FIG. 3. (A) Immunoaffinity purification of TAPER polypeptides from *E. coli* lysates. The cleared lysate prepared from lysed *E. coli* BL21(DE3) cells expressing various TAPER constructs was subjected to immunoaffinity chromatography on 1D4-Sepharose as described in the text. Sample fractions (10 μl) were resolved by SDS–PAGE on a 14% polyacrylamide gel. The resolved proteins were transferred to nitrocellulose for Western blot analysis with the 1D4 monoclonal antibody as the primary antibody and the alkaline phosphatase-conjugated antimouse IgG as the secondary antibody. Protein bands were visualized by reaction with NBT and BCIP according to directions from the manufacturer (Promega). All sample buffers contained β-mercaptoethanol. (B) Purified TAPER samples (10 μl) after dialysis were resolved on a 14% polyacrylamide gel and immunobloted as described earlier. Note that the resolution obtained on this gel does not distinguish differences in the size of TAPER polypeptides.

slurry prepared as described earlier) with an end-to-end mixing at 4° overnight. The Sepharose beads are recovered by a gentle spin at 3000 rpm on a refrigerated tabletop centrifuge. The resin is washed extensively (3 × 20 bed volumes each) with 50 m*M* Tris–Cl (pH 6.5%), 200 m*M* NaCl, and 0.1% CHAPS. Bound TAPER is eluted from the column in 0.1 *M* acetic acid in 0.1% CHAPS. The eluate is mixed with Tris base to adjust the pH to 7.0, dialyzed against 10 m*M* Tris buffer (pH 7.5) containing 20% glycerol and 5 m*M* DTT, and stored at −20°. Protein concentration is determined, and the samples are analyzed by immunoblotting shown in Fig. 3B.

Functional Assay for TAPER

Two different functional assays are employed. In the first assay, the ability of TAPER to interfere with rhodopsin–transducin interaction is measured by following the light-dependent binding of [^{35}S]GTPγS by transducin. In the second assay, the ability of TAPER to stimulate GTPase activity of transducin is measured by following [γ-^{32}P]GTP hydrolysis by transducin.

Transducin is purified from bovine retinae according to the procedure of Fung *et al.*[12] and is subjected to ion-exchange chromatography on a DE-52 column. The eluate from the DE-52 column is subjected to dialysis against 10 m*M* Tris buffer (pH 7.5) containing 50% glycerol, 2 m*M* $MgCl_2$, and 1 m*M* DTT and is stored at −20°. The assay mixture for [^{35}S]GTPγS binding to transducin consists of 2 n*M* spectroscopically pure rhodopsin solubilized in 0.1% dodecyl maltoside, 2 μM transducin, 4 μM [^{35}S]GTPγS in the assay buffer [10 m*M* Tris buffer (pH 7.5), 100 m*M* NaCl, 5 m*M* $MgCl_2$, 2 m*M* DTT, 0.01% dodecyl maltoside]. Purified TAPER is added to 100 n*M* final concentration. The reaction is initiated in the dark by the addition of [^{35}S]GTPγS; after 1 min of incubation (20°), the assay mixture is illuminated at >495-nm light. Samples are withdrawn at 1-min intervals and filtered through nitrocellulose. Filters are washed five times with 5 ml of 10 m*M* Tris buffer (pH 7.5), 100 m*M* NaCl, 5 m*M* $MgCl_2$, and 2 m*M* DTT. Radioactivity retained on the filter is measured by scintillation counting.

The assay mix for the measurement of GTPase activity of transducin consists of either 100 n*M* purified TAPER or 0.01 n*M* light-activated pure rhodopsin in 0.1% dodecyl maltoside, 1.2 n*M* transducin, 1 μM [γ-^{32}P]GTP in the assay buffer [10 m*M* Tris buffer (pH 7.5), 100 m*M* NaCl, 5 m*M* $MgCl_2$, 2 m*M* DTT, 0.01% dodecyl maltoside]. The assay is initiated by the addition of [γ-^{32}P]GTP and is followed for 20 min at 20°. The γ-$^{32}P_i$ released during the assay is extracted into a molybdic acid complex and estimated as described earlier.[10]

Table I shows the activity of various TAPER constructs in the two assays. All TAPER constructs, except TAPER IV, were effective in inhibiting transducin activation by the activated rhodopsin. However, none of the TAPER constructs were

[12] B. K. K. Fung, J. B. Hurley, and L. Stryer, *Proc. Natl. Acad. Sci. U.S.A.* **78,** 152 (1981).

TABLE I
FUNCTIONAL ACTIVITY OF PURIFIED TAPER PEPTIDES

	Relative transducin activation	
	Rhodopsin-dependent [^{35}S]GTPγS binding	[γ-^{32}P]GTP hydrolysis
Control	1.0[a]	1.0[b]
TAPER I	0.32	<0.01
TAPER II	0.38	<0.1
TAPER III	0.55	<0.001
TAPER IV	1.0	<0.001
TAPER V	0.68	<0.001
TAPER VI	0.51	<0.001

[a] Light-activated rhodopsin activated 270 ± 15 mol transducin/mol rhodopsin. This value was used as 1.0 to calculate the relative activity of each of the TAPERS (50-fold molar excess in the reaction).

[b] Light-activated rhodopsin stimulated 40 ± 5 mol/mol GTP hydrolysis. This value was used as 1.0 to calculate the GTPase-activating potential of each TAPER construct (100 n*M* final GTP concentration). Basal GTPase activity of transducin in our assay system was 4 ± 2mol/mol.

very effective in the direct activation of transducin. Successful design and construction of a more effective transducin-activating protein domain, starting from the cytoplasmic loops of rhodopsin, could be useful for undertaking structural studies. However, the observation from TAPER studies is consistent with the hypothesis that critical contacts between GPCR and G proteins may involve residues within TM domain. The movement of TM helices may actually be necessary to bring critical residues to the receptor–G protein interface to initiate the biochemical events involved in nucleotide exchange.

Previous studies employing purified, reconstituted G proteins and receptor-derived peptides or toxins have demonstrated that specific peptides are capable of stimulating G proteins.[4–6] For example, peptides derived from the carboxyl region of the third cytoplasmic loop of β-adrenergic receptor, M4-Muscatine receptor, and α_{2a}-adrenergic receptor mimic the receptor, can directly activate G-proteins *in vitro*. However, most GPCR-derived peptides are effective in blocking receptor G protein interaction through competition.[7,13,14] It has been proposed that this form of antagonism may provide a model for development of a class of receptor

[13] L. M. Luttrell, S. Cotecchia, J. Ostrowski, H. Kendall, and R. J. Lefkowitz, *Science* **259,** 1453 (1993).

[14] B. E. Hawes, L. M. Lutterell, S. T. Exum, and R. J. Leftkoitz, *J. Biol. Chem.* **269,** 15776 (1994).

antagonists that specifically blocks the receptor G protein interaction. Studies of Luttrell *et al.*[13] suggest that the cellular expression of cytoplasmic loops of both α_{1B}-adrenergic receptor and D_{1A} dopamine receptor results in specific antagonism of the receptor G protein interaction in intact cells. Several possible mechanisms could be responsible for the observed antagonism. Competition between the loop peptides and the activated agonist–receptor complex is most likely, as the inhibition is overcome by an increase in receptor density. Receptor specificity of inhibition, i.e. α_{1B}-loop peptide, is more effective in blocking the α_{1B}-adrenergic receptor signal than any other receptor that uses G_q/11 suggesting that the mechanism of antagonism may be more complicated than simple receptor antagonism. The heterogeneity of G protein heterodynes and intermolecular interactions of CL-3 with other intracellular regions of the receptor may have significant impact on the specificity observed.

Transmembrane Domain Fragments

Many different membrane proteins have been shown to be capable of assembly in functional form from two or more protein fragments.[15–25] Examples include bacteriorhodopsin,[15–17] β_2-adrenergic receptor,[18] muscarinic acetylcholine receptor,[19,20] lactose permease,[21] rhodopsin,[22–24] and the yeast α-factor receptor.[25] The assembled functional unit casually referred to as a "split receptor" could be a valuable research tool in many studies.

The approach of using polypeptide fragments to study the mechanism of integral membrane protein folding and assembly was pioneered during investigations on bacteriorhodopsin, a seven transmembrane helical light-transducing proton pump.[15] Proteolysed polypeptide fragments containing one or more transmembrane segments were found to be able to insert into lipid vesicles, assemble with their complementary partners, and form a chromophore with all-*trans* retinal and a light-driven proton pump. A two-step model for folding of this family of proteins envisaged is depicted in Fig. 4, in which each of the transmembrane α helices

[15] K. S. Huang, H. Bayley, M. J. Liao, E. London, and H. G. Khorana, *J. Biol. Chem.* **256,** 3802 (1981).
[16] T. W. Kahn and D. M. Engelman, *Biochemistry* **31,** 6144 (1992).
[17] O. K. Hansen, M. Pompejus, and H. J. Fritz, *Biol. Chem. Hoppe-Seyler* **375,** 715 (1994).
[18] B. K. Kobilka, T. S. Kobila, K. Daniel, J. W. Regan, M. G. Caron, and R. J. Lefkowitz, *Science* **240,** 1310 (1988).
[19] R. Maggio, Z. Vogel, and J. Wess, *FEBS Lett.* **319,** 195 (1993).
[20] T. Schoneberg, J. Liu, and J. Wess, *J. Biol. Chem.* **270,** 18000 (1995).
[21] E. Bibi and H. R. Kaback, *Proc. Natl. Acad. Sci. U.S.A.* **87,** 4325 (1990).
[22] K. D. Ridge, S. S. J. Lee, and L. L. Yao, *Proc. Natl. Acad. Sci. U.S.A.* **92,** 3204 (1995).
[23] H. Yu, M. Kono, T. D. Melcee, and D. D. Oprian, *Biochemistry* **34,** 14963 (1995).
[24] K. D. Ridge, S. S. J. Lee, and N. G. Abdulaev, *J. Biol. Chem.* **271,** 7860 (1996).
[25] N. P. Martin, L. M. Leavitt, C. M. Sommers, and M. E. Dumont, *Biochemistry* **38,** 682 (1999).

1. Cotransfection of minigenes encoding each fragment

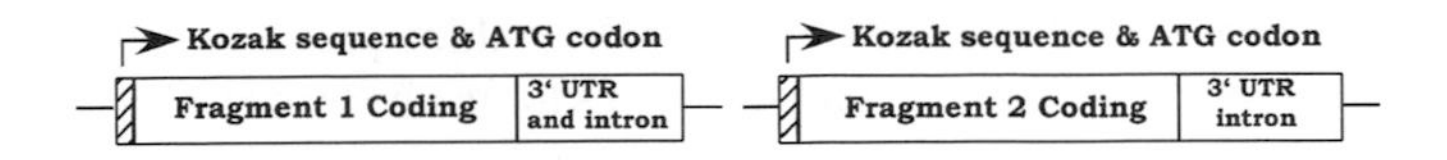

2. Fragments fold independently in membrane

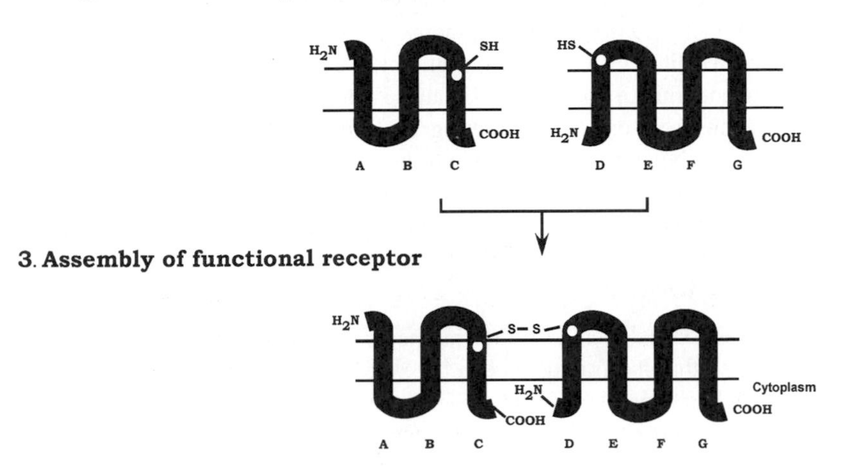

FIG. 4. Schematic representation of assembly of the functional receptor from coexpressed complementary polypeptide fragments. The highly conserved disulfide linkage in the extracellular domain shown is expected to be formed after the assembly of fragments to form the functional receptor.

fold independently of each other, insert into the membrane, and following which assemble functional transmembrane bundle without major rearrangement. Complementary fragment assembly into functional entities has been demonstrated unequivocally in a number of polytopic transmembrane proteins since then. These studies have demonstrated that each of the polypeptide fragments appears capable of endoplasmic reticulum translocation and topogenesis. Singly expressed fragments are not degraded readily by cellular proteases suggesting that each helical segment is an independent folding domain. The experiments also suggest that complementary segments associate through a side-to-side interaction within the membrane.

General Principles

The approach of using fragments of the receptor polypeptide needs demonstration that the cotransfection of gene fragments generates a functional split receptor that folds into a native conformation as judged by ligand selectivity. In rhodopsin, fragments generated by proteolytic cleavage remain associated in the membrane and after solubilization in detergent solution.[14–22] Recombinant expression studies indicate that coexpression of two or three fragments allows the formation of noncovalently assembled GPCRs, which exhibit properties similar to the wild-type receptor. For a detailed treatment of this topic, the reader is referred to Refs. 21–25.

Choosing Fragments for Expression

Split receptors exhibit functional activity comparable to that of the wild type when the split site did not disrupt segments known to be important for function. For instance, a split in the C-terminal segment of EF loop led to a ~75% decrease in G protein coupling in both rhodopsin and β-adrenergic receptor. A normal precaution that must be exercised in selecting a split site is to carefully restrict to those regions of the polypeptide chain shown to be tolerant of mutagenic changes. For example, a split that results in a discontinuous TM helix or loop region that is important for helical alignment may not be assembled easily into a stable split receptor.

Expression of Fragments

For expression, in each case the fragments are encoded on separate expression vectors so that the fragments can be expressed individually or they can be coexpressed. Each fragment is cloned into the vector as a cassette containing a Kozak consensus ribosome-binding sequence, CCACC, immediately 5′ of the initiator methionine codon, ATG, followed by the coding sequence of the fragment placed in frame with the ATG codon and ending with one or two nonsense codons. Semiconfluent COS1 cells (~2–3 × 10^6/60-mm dish) grown in Dulbeco's minimal essential medium (DMEM) containing 10% bovine calf serum are transfected by the lipofectamine method per the manufacturer's recommendation. *In situ* functional studies are carried out 48 hr after transfection. Transfected cells are grown for 72 hr for the purpose of maximizing the protein level for ligand binding or functional reconstitution studies. Purification of the split receptor constructs can follow solubilization with either 1% (w/v) CHAPS or β-D-dodecyl maltoside. The details of solubilization and purification steps need careful standardization for each system. The cellular expression of polypeptide fragments was examined by protein immunoblotting of whole cell detergent extracts with fragment-specific monoclonal antibodies. In general, the yield of the split receptor is ~30% compared to the yield of the wild-type receptor in the same expression system.

Application

The split receptor approach could be a general, rapid, and easy assay for the routine determination of tertiary contacts in single chain, polytopic, integral membrane proteins. Split receptors used in combination with techniques such as Cys cross linking and site-directed reporter attachment studies would prove extremely useful. The approach is to use disulfide cross-linking of site-directed Cys mutations to demonstrate the proximal location of residue side chains within the protein. Cys residues can be engineered into split receptor constructs. Cross-linking can be detected readily by a mobility shift on SDS–PAGE. This approach has been used successfully in rhodopsin.[22–24] Traditionally, the disulfide cross-linking method has found limited application because it is restricted to proteins composed of multiple subunits. With split receptors, however, the limitation can be overcome. Cys cross-linking demonstrates that the two candidate residues are capable of coming close together enough to form a covalent bond in the protein.[26] It is important to consider other interpretations of the observation in terms of protein structure and dynamics of protein conformations. A major factor for consideration is the trapping of random conformational fluctuation in the protein. For this reason, it is very important to functionally characterize the cross-linked proteins. In addition, analysis of the global data set obtained from a systematic Cys scan of the domain for a pattern of self-consistency can thwart such anomalies.

Studies on functional heterodimerization and protein complementation in GPCR function elude to an entirely novel application of the split receptor approach.[27,28] It should now be possible to design GPCR mutants and fragments that are able to interfere with the specific aspects of wild-type receptor function or folding. The experimental design and general principles will be identical to those described earlier for split receptors.

Thus, the construction of minigenes to express carefully designed GPCR fragments could be extremely valuable in analysis of the tertiary structure of GPCR, elucidation of complex GPCR signaling networks, and development of novel *in vivo* modifiers of GPCR functions. Therefore, this approach should be very valuable.

Acknowledgments

This work was supported in part by National Institute of Health Grants EY9704, HL57470 and an Established Investigator Award from the American Heart Association. I am grateful to Thomas Boyle, Shreeta Acharya, and Yasser Saad for the experimental work; and John Boros, Jingli Zhang, and Robin Lewis for assistance in the preparation of the manuscript.

[26] J. J. Falke and D. E. Koshland, Jr., *Science* **237,** 1596 (1987).
[27] Z. Zhu and J. Wess, *Biochemistry* **37,** 15773 (1998).
[28] B. A. Jordan and L. A. Devi, *Nature* **399,** 697 (1999).

[18] Construction and Analysis of Function of G Protein-Coupled Receptor–G Protein Fusion Proteins

By GRAEME MILLIGAN

Introduction

The bulk of studies on the interactions between G protein-coupled receptors (GPCRs) and the α subunits of different heterotrimeric G proteins utilize their coexpression into appropriate cell systems. However, this poses problems for detailed quantitative analysis on the effectiveness and/or specificity of these interactions, as it is rarely possible to confirm equal levels of expression of the individual G proteins or an identical cellular distribution of the GPCR and G protein. In recent times, fusion proteins in which the N terminus of a G protein α subunit is attached directly to the C-terminal tail of the GPCR have become popular constructs in the analysis of ligand regulation of G protein activation. Although they are clearly unnatural entities, a number of groups have commented on and reviewed many of the benefits of the unique characteristics of such fusion proteins.[1–3] These include the defined 1:1 stoichiometry of GPCR and G protein defined by the fusion, the equivalent physical proximity of the protein partners following the expression of fusion proteins consisting of different G proteins fused to the same GPCR, and, perhaps most importantly for quantitative analysis, the capacity of agonist ligands to stimulate guanine nucleotide exchange and hydrolysis by the G protein partner of the fusion protein. In all cases reported to date, the agonist-stimulated GTPase activity of such fusion proteins has characteristics of an enzyme that behaves in accordance with the formalisms derived by Michaelis and Menten. Therefore, simple enzyme kinetic analysis of the stimulated GTPase activity provides quantitative means of analysis of features as distinct as ligand efficacy[4] and the effects of point mutations within either the GPCR[5] or the G protein.[6,7] These features should also mean that such GPCR-G protein fusions will be useful for quantitative analysis of the effects of allosteric modulators and of other proteins known to interact with GPCR or G protein.

[1] D. Colquhoun, *Br. J. Pharmacol.* **125,** 924 (1998).
[2] R. Seifert, K. Wenzel-Seifert, and B. K. Kobilka, *Trends Pharmacol. Sci.* **20,** 383 (1999).
[3] G. Milligan, *Trends Pharmacol. Sci.* **21,** 24 (2000).
[4] A. Wise, I. C. Carr, D. A. Groarke, and G. Milligan, *FEBS Lett.* **419,** 141 (1997).
[5] R. J. Ward and G. Milligan, *FEBS Lett.* **462,** 459 (1999).
[6] V. N. Jackson, D. S. Bahia, and G. Milligan, *Mol. Pharmacol.* **55,** 195 (1999).
[7] E. Kellett, I. C. Carr, and G. Milligan, *Mol. Pharmacol.* **56,** 684 (1999).

METHODS IN ENZYMOLOGY, VOL. 343
0076-6879/02 $35.00

Construction of GPCR-G Protein Fusion Proteins

In all examples reported to date, the 3′ end of a GPCR DNA or cDNA has been modified, generally via polymerase chain reaction (PCR), such that the stop codon is removed or modified and an appropriate restriction enzyme site added. Equivalent PCR to introduce the same (or a mutually compatible) restriction site adjacent to codon 1 or 2 of the G protein cDNA then allows ligation and in-frame linkage to produce a chimeric DNA anticipated to encode a single open reading frame capable of producing the fusion protein. The strategy used to generate a fusion protein between the human $5HT_{1A}$ receptor and the α subunit of G_{i1}[7] is displayed in Fig. 1 as a representative example, although a range of basic molecular biological approaches can be employed equally effectively. No obvious consensus has been reached about whether it is most appropriate to remove codon 1 (the initiator methionine) from the G protein sequence or indeed on the relevance of disruption of the sequence of the extreme C-terminal tail of the GPCR during the construction process. In most cases, pragmatism appears to have prevailed with ease of construction and cloning being the prevailing factors in the final sequence of the chimeric DNA. This has resulted in fusion proteins being generated in which a few amino acids have been eliminated from the C-terminal sequence of the GPCR, such as the originally reported β_2-adrenoceptor–$G_s\alpha$ fusion protein,[8] or additional amino acids being added to the sequence at the interface between GPCR and G protein. For example, the addition of *Bam*H1 restriction sites to ligate together the GPCR and G protein DNAs results in introduction of a new Gly-Ser dipeptide between the two proteins (see Fig. 1).

Except for studies that have addressed issues of whether the coupling efficiency between a GPCR and a G protein is enhanced by greater physical proximity,[9,10] little published data have examined the importance or otherwise of the length of the GPCR C-terminal tail, and thus the linker length between transmembrane region VII of the GPCR and the G protein, in the production of functionally active and appropriately folded GPCR-G protein fusion proteins. However, as a fusion protein between the human A_1 adenosine receptor and the α subunit of G_{i1} in which green fluorescent protein was placed between the GPCR and the G protein produced larger agonist-mediated signals than a form in which the GPCR and the G protein were linked directly to each other,[11] this is clearly an issue deserving detailed analysis.

[8] B. Bertin, M. Friessmuth, R. Jockers, A. D. Strosberg, and S. Marullo, *Proc. Natl. Acad. Sci. U.S.A.* **91,** 8827 (1994).

[9] K. Wenzel-Seifert, T. W. Lee, R. Seifert, and B. K. Kobilka, *Biochem. J.* **334,** 519 (1998).

[10] R. Seifert, K. Wenzel-Seifert, U. Gether, U. V. T. Lam, and B. K. Kobilka, *Eur. J. Biochem.* **260,** 661 (1999).

[11] N. Bevan, T. Palmer, T. Drmota, A. Wise, J. Coote, G. Milligan, and S. Rees, *FEBS Lett.* **462,** 61 (1999).

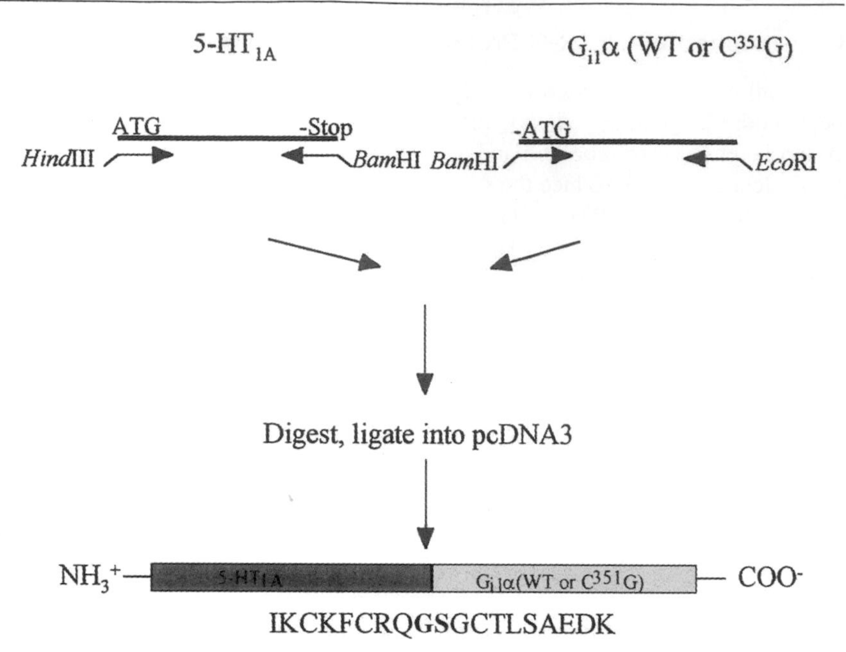

FIG. 1. Construction of human 5-HT_{1A} receptor–$G_{i1}\alpha$ fusion proteins. Fusion protein cDNAs were constructed between the human $5HT_{1A}$-receptor and either wild-type (cys^{351}) (C) or a pertussis toxin-resistant (gly^{351})(G) form of rat $G_{i1}\alpha$ [see E. Kellett, I. C. Carr, and G. Milligan, *Mol. Pharmacol.* **56,** 684 (1999) for further details]. The human $5HT_{1A}$ receptor clone (EMBL/GenBank accession number X13556) in pSP64 was digested with *Xba*I/*Bam*HI, and the resulting 1.5kb fragment was ligated to pcDNA3. To obtain the open reading frame (ORF) of 1.3 kb, PCR was carried out using the following primers to introduce a *Hind*III restriction site at the 5′ end and to remove the stop codon (-Stop) and introduce a *Bam*HI restriction site at the 3′ end, respectively: 5′ CTG<u>AAGCTT</u>ATGGATGTGCTCAGCCCTGGTC 3′; 5′ CTG<u>GGATCC</u>CTGGCGGCAGAAGTTACACTTAATG3′ (restriction enzyme sites underlined). The PCR fragment was digested with *Hind*III and *Bam*HI and ligated into pcDNA3 to make the plasmid p5HT. To link the $G_{i1}\alpha$ wild-type (Cys^{351})cDNA (EMBL/GenBank accession number X13556) to the $5HT_{1A}$ receptor sequence, PCR was carried out on $G_{i1}\alpha$ to produce compatible restriction sites and to alter the start codon (-ATG). The oligonucleotides used to do this were 5′ CTG<u>GGATCC</u>GGCTGCACACTGAGCGCTGAG3′ at the 5′ end and 5′ GA<u>GAATTC</u>TTAGAAAGAGACCACAGTC3′ at the 3′ end. Plasmid p5HT was digested with *Bam*HI/*Eco*RI, as was the $G_{i1}\alpha$ PCR fragment, and the two were ligated to give the plasmid p5HTGi1. To construct the $5HT_{1A}$ -(gly^{351})$G_{i1}\alpha$ fusion plasmid, (gly^{351})$G_{i1}\alpha$ in pBS was digested with *Sac*II/*Eco*RI, and the 730-bp fragment was used to replace the corresponding fragment in p5HTGi1. The constructs were then sequenced to verify the DNA sequence. This strategy resulted in the addition of two amino acids (in bold) as a linker, which are not present in either the isolated GPCR (solid bar) or the G protein (open bar).

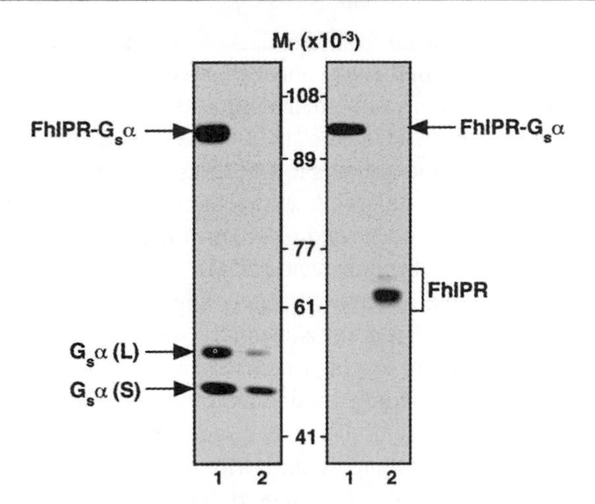

FIG. 2. Expression of FLAG-IP prostanoid receptor and FLAG-IP prostanoid receptor–G protein fusion proteins: immunological detection. Membranes from clones expressing FLAG-human IP prostanoid receptor (FhIPR)-$G_s\alpha$ (1) or FhIPR (2) were resolved by SDS–PAGE and immunoblotted with either antiserum CS, which identifies the C-terminal decapeptide of forms of $G_s\alpha$ (left), or an anti-FLAG antibody (right) as described in text. Reproduced with permission from C. W. Fong and G. Milligan, *Biochem. J.* **342,** 457 (1999). Copyright the Biochemical Journal.

Detecting Expression of GPCR-G Protein Fusion Proteins

Following either transient or stable expression of a GPCR-G protein fusion protein DNA, a number of approaches are available to confirm and quantitate appropriate expression. Immunoblotting studies are important and can be particularly helpful if the GPCR of the fusion protein construct has additionally been N-terminally epitope tagged. Because many of the widely available, commercial, anti-G protein α subunit antisera are directed to the extreme C-terminal of the protein, then specific identification of polypeptides of equivalent and appropriate size with both antiepitope tag and G protein antisera following SDS–PAGE confirms expression of the full-length fusion protein[8,9,12–15] (Fig. 2). A wide range of other studies have used either only an anti-GPCR antiserum[11] or antisera directed to either the G protein C terminus[16,17] or an internal epitope in the G protein[7,18–20] and

[12] C. W. Fong and G. Milligan, *Biochem. J.* **342,** 457 (1999).

[13] R. Seifert, K. Wenzel-Seifert, T. W. Lee, U. Gether, E. Sanders-Bush, and B. K. Kobilka, *J. Biol. Chem.* **273,** 5109 (1998).

[14] K. Wenzel-Seifert, J. M. Arthur, H.-Y. Liu, and R. Seifert, *J. Biol. Chem.* **274,** 33259 (1999).

[15] T. M. Loisel, H. Ansanay, L. Adam, S. Marullo, R. Seifert, M. Lagace, and M. Bouvier, *J. Biol. Chem.* **274,** 31014 (1999).

[16] A. Wise, M. Sheehan, S. Rees, M. Lee, and G. Milligan, *Biochemistry* **38,** 2272 (1999).

[17] Y. Wang, R. T. Windh, C. A. Chen, and D. R. Manning, *J. Biol. Chem.* **274,** 37435 (1999).

[18] A. Wise and G. Milligan, *J. Biol. Chem.* **272,** 24673 (1997).

surmized full-length expression from the size and functionality of the expressed protein. In the absence of appropriate antisera, reverse transcriptase-polymerase chain reaction has been employed, either with a forward primer designed to anneal to the sequence of the GPCR and the reverse primer to sequence from the G protein[12] or with appropriate primers to amplify segments of the G protein.[14] In the former situation, appropriate amplification confirms transcription of a chimeric RNA, whereas the latter has been used to confirm that the correct G protein sequence was indeed present in the transfected cDNA. Apart from examples in which a proteolytic cleavage site has been deliberately engineered into the construct between the GPCR and G protein to address specific issues,[10,15] there is little available evidence that GPCR-G protein fusion proteins are more susceptible to proteolytic degradation than the corresponding, individual protein partners.

Quantitation of expression of GPCR-G protein fusion proteins is based routinely on saturation ligand-binding studies with appropriate antagonist ligands. Although it might be assumed that the single polypeptide nature of GPCR-G protein fusion proteins would define that there should be a binding site with a single affinity for both agonist and antagonist ligands, this is not the case.[13] As such, where quantitation of expression level studies must be performed with agonist ligands, equivalent provisos on the presence of both high- and low-affinity-binding sites, as would be inherent with coexpression of the isolated GPCR and G protein, must be considered, which can lead to underestimation of expression levels.[14] Assuming that B_{max} levels can be obtained, then the 1:1 stiochiometry of GPCR and G protein now defines the levels of expression of the G protein. This information is integral to effective enzyme kinetic analysis.

Examining the Functionality of GPCR-G Protein Fusion Proteins

Guanine Nucleotide Exchange Studies

Activation of heterotrimeric G proteins requires GPCR-induced release of GDP from the G protein α subunit and its subsequent replacement by GTP.[21] The active state is terminated by hydrolysis of the terminal phosphate of GTP by the intrinsic GTPase activity of the G protein α subunit to restore the bound nucleotide to GDP. Because the GPCR-induced release of GDP is well appreciated to be the rate-limiting step of this cycle,[21] then monitoring GTPase activity and its stimulation by agonist ligands can provide a direct monitor of the rate and extent of G protein activation (Fig. 3) (see Refs. 22–24 for practical details). This approach

[19] A. Wise, I. C. Carr, and G. Milligan, *Biochem. J.* **325,** 17 (1997).

[20] M. Waldhoer, A. Wise, G. Milligan, M. Freissmuth, and C. Nanoff, *J. Biol. Chem.* **274,** 30571 (1999).

[21] A. Gilman, *Annu. Rev. Biochem.* **56,** 615 (1987).

[22] P. Gierschik, T. Bouillon, and K. H. Jakobs, *Methods Enzymol.* **237,** 13 (1994).

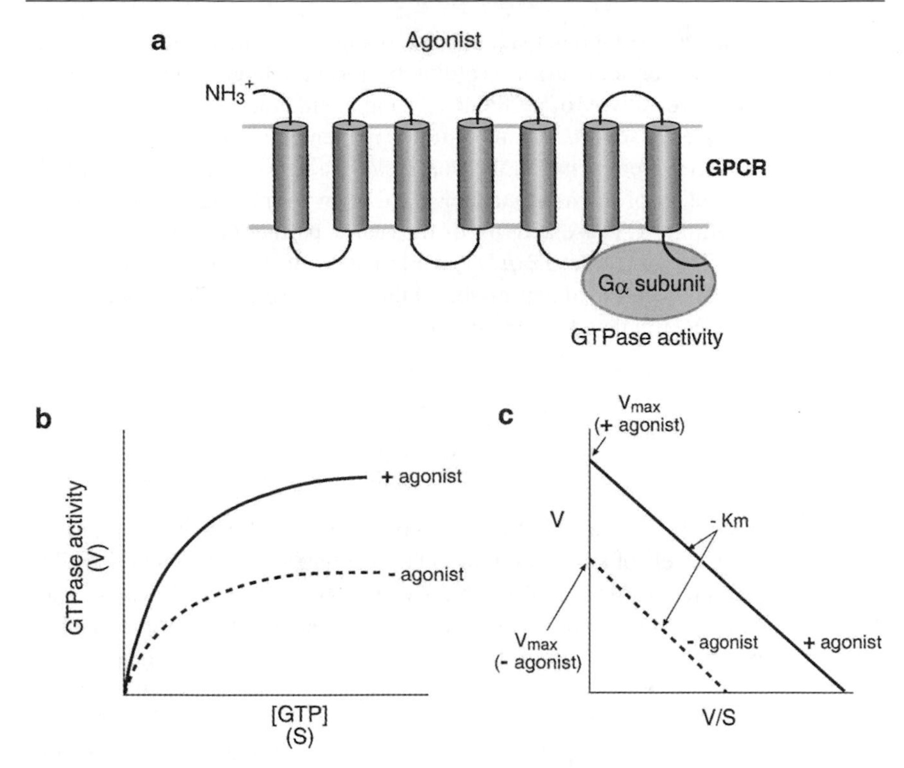

FIG. 3. Analysis of ligand activation of a GPCR-G protein fusion protein. (a) Physical linkage of the N terminus of a G protein α subunit to the C-terminal tail of a GPCR from which the stop codon has been removed allows expression of a single polypeptide encoding the functions of both proteins. Functionality and appropriate folding and processing of the two elements of the fusion protein are demonstrated by its capacity to bind radiolabeled ligands and the ability of appropriate agonists to stimulate guanine nucleotide exchange and hydrolysis (GTPase activity) by the G protein element of the fusion. (b) GPCR-G protein fusion proteins function as agonist-activated enzymes. Following expression of the fusion protein and membrane preparation, high-affinity GTPase activity (V) can be measured at a range of substrate (GTP) concentrations (S). The presence of agonist stimulates V with variation of S in a hyperbolic fashion, as anticipated for an enzyme displaying Michaelis–Menten kinetics. (c) Plotting of data such as b as a Eadie–Hofstee transformation allows extrapolation of enzyme activity to V_{max} and assessment of K_m for GTP. The stimulation of GTPase activity at V_{max} by a maximally effective concentration of agonist can then be converted directly to the agonist-induced turnover number for GTP (i.e., mol GTP hydrolyzed per unit time per mol fusion protein) if expression levels of the construct are measured in parallel. A partial agonist increases V_{max} to a lesser extent than a full agonist, and comparisons of the effectiveness of two ligands at V_{max} provide direct measures of their relative intrinsic activity. Reprinted from G. Milligan, *Trends Pharmacol. Sci.* **21,** 24 (2000), with permission from Elsevier Science.

has provided the most comprehensive means to apply enzyme kinetics to the activation of GPCR-G protein fusion proteins by receptor ligands and to explore potential constitutive activity of a GPCR.[3] Using a wide range of concentrations of GTP as substrate, Wise *et al.*[19] demonstrated in membranes of cells transfected to express a fusion protein between the porcine α_{2A}-adrenoceptor and a pertussis toxin-resistant mutant of the α subunit of G_{i1} that an agonist ligand was able to stimulate high-affinity GTPase activity of the fusion partner G protein. The estimated K_m for GTP was some 0.3 μM. Extrapolation of data to V_{max} in concert with knowledge of the levels of expression of the fusion protein allows calculation of turnover of GTP stimulated by agonist binding to the GCPR as

$$\text{Turnover number (min}^{-1}) = \frac{\begin{array}{c}\text{stimulated high-affinity GTPase activity at } V_{max} \\ \text{(pmol min}^{-1}\text{ mg protein}^{-1})\end{array}}{\begin{array}{c}\text{fusion protein expression level} \\ \text{(pmol mg protein}^{-1})\end{array}}$$

Addition of a variety of agonist ligands at maximally effective concentrations resulted in varying levels of stimulation of GTPase activity and, when measured at V_{max}, provided direct estimates of the relative intrinsic activity of these ligands to produce conformational alterations in the GPCR required to promote guanine nucleotide exchange by the G protein. Such an approach has also been applied to studies on $5HT_{1A}$ receptor-containing fusion proteins[7,25] and the human formyl peptide receptor.[14] Because all of the widely expressed G_i family α subunits comprise 354 or 355 amino acids, then construction of a series of equivalent fusion proteins in which a range of G_i family members are linked to the same GPCR provides an effective means to measure activation selectivity of GPCRs for distinct but highly related G proteins. For the porcine α_{2A}-adrenoceptor, a rank order for activation by epinephrine of $G_{i1} > G_{i3} \geq G_{i2} = G_{o1}$ has been recorded with a greater than threefold range of turnover numbers between G_{i1} and G_{o1}.[5] Interestingly, although the rank order of relative intrinsic activity of a series of agonists at the α_{2A}-adrenoceptor to activate G_{i1} or G_{o1} was unaltered when compared to epinephrine, a series of partial agonists displayed higher relative intrinsic activity to stimulate G_{o1}. A similar characteristic has been noted following coexpression of an α_{2A}-adrenoceptor and these G proteins.[27] In contrast, no statistically

[23] I. Mullaney, *in* "Signal Transduction: A Practical Approach" (G. Milligan, ed.), 2nd Ed., p. 73. Oxford Univ. Press, Oxford, 1999.

[24] A. Wise, *in* "Signal Transduction: A Practical Approach" (G. Milligan, ed.), 2nd Ed., p. 103. Oxford Univ. Press, Oxford, 1999.

[25] G. Milligan, E. Kellett, V. Dubreuil, E. Jacoby, C. Dacquet, and M. Spedding, *Neuropharmacology,* in press.

[26] Deleted in proof.

[27] Q. Yang and S. M. Lanier, *Mol. Pharmacol.* **56,** 651 (1999).

significant differences in agonist-stimulated GTPase turnover numbers of G_{i1}, G_{i2}, and G_{i3} were reported for fusion proteins incorporating the human formyl peptide receptor.[14]

Inverse agonism of a range of ligands has also been explored for the β_2-adrenoceptor[13] and the $5HT_{1A}$[7] and formyl peptide[14] receptors by their capacity to reduce the constitutive GTPase activity imbued to the G protein elements of these fusion proteins by their partner GPCRs.

Although not as appropriate as GTPase activity measurements for detailed studies, the widespread use of ligand regulation of the binding of [^{35}S]GTPγS as a monitor of G protein activation (see Refs. 23 and 28 for methodological details) reflects both the technical ease of the approach and its capacity for significant levels of throughput. As such, it is not surprising that this approach has been applied to GPCR-G protein fusion proteins.[14,16,17,29] These experiments have largely been used simply to demonstrate the capacity of ligands to active the GPCR-linked G protein. However, they have also been used to monitor ligand efficacy.[14,29]

G Protein Regulation of Effectors

A widely held view of agonist activation of G proteins is that this results in dissociation of GPCR and G protein allowing the G protein subunits to modulate the activity of a variety of enzymes and ion channels. Clearly, full physical separation of GPCR and G protein cannot result from the activation of a GPCR-G protein fusion protein and thus a potential scenario was that such proteins would fail to regulate downstream effectors. As expression of a β_2-adrenoceptor–$G_s\alpha$ fusion protein in cells that genetically lack endogenous expression of $G_s\alpha$ allows agonist-mediated stimulation of adenylyl cyclase activity,[8] then such regulation is clearly not prohibited. Furthermore, when expressed in insect Sf9 cells, the effectiveness of fusion proteins between the β_2-adrenoceptor and either long and short splice variants of $G_s\alpha$ is different and this can be correlated with the relative GTPase rates of the two fusion proteins.[13] Effective inhibition of adenylyl cyclase by ligands at the $5HT_{1A}$ receptor is also achieved in a manner that is unaffected by pertussis toxin treatment when this GPCR is linked to a pertussis toxin-insensitive variant of $G_{i1}\alpha$.[7] Many GPCRs are able to produce elevation of intracellular [Ca^{2+}] when coexpressed with the so-called "universal adapter" G protein $G_{15}\alpha$/$G_{16}\alpha$. Thus, not surprisingly, agonist occupation of an α_{2B}-adrenoceptor–$G_{15}\alpha$ fusion protein causes elevation of intracellular [Ca^{2+}] when expressed in CHO cells[30] and fusion proteins between either the α_{1B}-adrenoceptor or the thyrotropin-releasing hormone receptor and the phosphoinositidase C-linked G protein, $G_{11}\alpha$ do likewise

[28] T. Wieland and K. H. Jakobs, *Methods Enzymol.* **237,** 3 (1994).

[29] D. S. Dupuis, S. Tardif, T. Wurch, F. C. Colpaert, and P. J. Pauwels, *Neuropharmacology* **38,** 1035 (1999).

[30] P. J. Pauwels, S. Tardif, F. Finana, T. Wurch, and F. C. Colpaert, *J. Neurochem.* **74,** 375 (2000).

following expression in cells derived from a $G_q\alpha + G_{11}\alpha$ knockout mouse[31] (Fig. 4). It thus appears that a wide range of effectors can be accessed directly by the G protein element of such fusion proteins. This is also true for G protein regulation of ion channels. Expression of a fusion protein between the m_2 muscarinic acetylcholine receptor and the α subunit of G_z allowed agonist regulation of coexpressed G protein-activated K^+ (GIRK) channels in a pertussis toxin-insensitive manner.[32] Because GIRK channels are regulated physiologically via the G protein β/γ complex released from pertussis toxin-sensitive members of the G_i/G_o family,[33] then these results indicated that the m_2 muscarinic receptor–$G_z\alpha$ fusion protein was activated by agonist but also that it must have the capacity to bind and release β/γ complex upon agonist stimulation.[32] The capacity of GPCR-G protein α subunit fusion proteins to interact with the β/γ complex has been examined since the initial report on their use appeared.[8] Concern about this issue may stem from knowledge that the N-terminal region of the G protein α subunit is a key interaction site for β/γ and that this is linked directly to the GPCR in fusion constructs. However, as well as the direct functional analysis noted earlier, cotransfection of cells with fusion proteins containing the α_{2A}-adrenoceptor with various forms of either G_{i1} or G_{o1} along with the β_1/γ_2 complex results in higher levels of agonist-stimulated GTPase activity per mole of fusion protein than for the fusion protein without β_1/γ_2.[19] Such results also clearly indicate functional interactions with the β/γ complex.

Regulation of downstream effectors by the G protein partner of GPCR-G protein fusion proteins has not been observed in all studies, however, even when the linked G protein clearly becomes activated by agonist binding to the GPCR. Following stable expression in Rat-1 fibroblasts of a fusion protein between the porcine α_{2A}-adrenoceptor and a variant of $G_{i1}\alpha$, which was rendered insensitive to the actions of pertussis toxin by replacement of Cys^{351} by Gly, the addition of receptor agonist produced a strong stimulation of high-affinity GTPase activity and both inhibition of forskolin-amplified adenylyl cyclase activity and activation of ERK MAP kinases in untreated cells.[34] However, following pertussis toxin treatment of the cells to eliminate possible contacts of the GPCR with endogenous G_i, agonist-stimulated GTPase activity was reduced substantially and regulation of both adenylyl cyclase and ERK MAP kinase was blocked completely. Such results are consistent with the GPCR of the fusion protein being able to access and activate the endogenously expressed pool of G_i family G proteins, as well as the linked G protein of the fusion protein. However, these results also indicated that all the agonist-induced modulation of downstream effector proceeded via the

[31] P. A. Stevens, J. Pediani, J. J. Carillo, and G. Milligan, *J. Biol. Chem.* **276** (in press).

[32] D. Vorobiov, A. Kanti Bera, T. Keren-Raifman, R. Barzilai, and N. Dascal, *J. Biol. Chem.* **275,** 4166 (2000).

[33] J. L. Leaney, G. Milligan, and A. Tinker, *J. Biol. Chem.* **275,** 921 (2000).

[34] A. R. Burt, M. Sautel, M. A. Wilson, S. Rees, A. Wise, and G. Milligan, *J. Biol. Chem.* **273,** 10367 (1998).

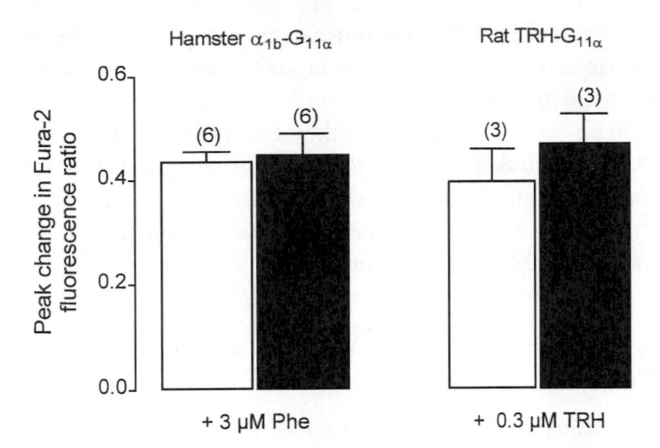

FIG. 4. Agonist-induced $[Ca^{2+}]_i$ elevation by fusion proteins containing $G_{11}\alpha$. A fibroblast cell line (EF88) [J. Mao, H. Yuan, W. Xie, M. I. Simon, and D. Wu, *J. Biol. Chem.* **273,** 27118 (1998) and R. Yu and P. M. Hinkle, *J. Biol. Chem.* **274,** 15745 (1999)], which was derived from embryos of mice in which the α subunits of both G_q and G_{11} had been knocked out by targeted gene disruption, was grown in Dulbecco's modified Eagle's medium (DMEM) supplemented with 10% (v/v) heat-inactivated fetal bovine serum and L-glutamine (1 m*M*) in a 95% air and 5% CO_2 atmosphere at 37°. For transfection experiments, cells were plated onto glass coverslips (22-mm diameter, grade 0 thickness). After a 24-hr growth period, they were transfected with cDNAs encoding either a hamster α_{1b}-adrenoceptor mouse $G_{11}\alpha$ fusion protein or an equivalent construct with the long isoform of the rat thyrotropin-releasing hormone receptor-1, using lipofectAMINE. (1) All transfection solutions were made up in sterile centrifuge tubes. (2) For each coverslip, 1–3 μg of each cDNA was diluted in 100 μl of OPTIMEM-1. (3) Five microliters of stock lipofectAMINE reagent was diluted 1 : 20 using OPTIMEM-1 (final volume = 100 μl). (4) The diluted lipofectAMINE reagent was then combined with the DNA/OPTIMEM-1 mixture, mixed gently, and incubated at room temperature for 40–45 min to allow DNA–liposome complexes to form. (5) After complex formation, an aliquot (800 μl/coverslip) of OPTIMEM-1 was added to the tube containing the DNA–liposome complexes and mixed gently. (6) Cells were washed twice with OPTIMEM-1 solution and then incubated with the diluted transfection complex solution (1 ml volume) at 37° for 3 hr. (7) After 3 hr, cells were washed twice with OPTIMEM-1 and then cultured in DMEM growth medium for a further 24 hr prior to Fura-2 experimentation. Transfected cells growing on coverslips were loaded with the Ca^{2+}-sensitive dye Fura-2 by incubation (15–20 min, 37°) in culture medium containing the membrane-permeant acetoxymethylester form of the dye (1.0 μM). A rise in $[Ca^{2+}]_i$ causes a corresponding rise in the Fura-2 fluorescence ratio recorded from cells loaded with this dye, which allows receptor-mediated changes in $[Ca^{2+}]_i$ to be monitored using standard, microspectrofluorimetric techniques [G. Grynkiewicz, M. Poenie, and R. Y. Tsien, *J. Biol. Chem.* **265,** 3440 (1985)]. Fura-2 fluorescence ratios were recorded (excitation wavelengths: 340 and 380 nm) at 4 Hz from single cells at room temperature. Data were digitized and recorded directly to a computer disk using hardware and software (Version 5.2) obtained from Cairn Research (Faversham, Kent, UK). Phenylephrine (Phe)(3 μM) or thyrotropin-releasing hormone (TRH) (0.3 μM) was used as an agonist. Agonist-evoked $[Ca^{2+}]_i$ signals were quantified by peak height (i.e., difference between the baseline resting ratio level and that attained at the peak response). In some experiments (dark bars), after an initial 24-hr transfection/growth period, transfected cells were treated with pertussis toxin by incubating the cells in growth medium containing this toxin (25 ng ml^{-1}) for 24 hr.

endogenous G protein pool.[34] A m_2 muscarinic receptor–$G_Z\alpha$ fusion protein has also been reported able to activate endogenous G_i family G proteins,[32] although in this case, downstream function was also produced directly by the fusion protein.

To date, there are no reports that other coexpressed GPCRs can assess and activate the G protein linked to a GPCR. However, given the current interest in the possibility of heterodimerization between distinct but related GPCRs, use of the fusion protein strategy in such a manner may be amenable to monitoring the close proximity of two GPCRs in native membranes.

Measuring the Effects of Point Mutations in GPCRs and G Proteins

A frequent concern with studies on the effects of point mutations in GPCRs or G proteins on the effectiveness of their interactions is that the mutation can alter their levels of expression or cellular targeting and distribution. The GPCR-G protein fusion strategy ensures that the stoichiometry of the two polypeptides will not alter and, providing the mutations do not compromise antagonist binding, then expression levels can be monitored as for the wild-type protein. For GPCRs, this has been used so far only to demonstrate that a Asp^{79} Asn mutation in the porcine α_{2A}-adrenoceptor has equal effects on the capacity of epinephrine to activate $G_{i1}\alpha$, $G_{i2}\alpha$, and $G_{i3}\alpha$.[5] These studies were undertaken because although when expressed in pituitary AtT20 cells the wild-type α_{2A}-adrenoceptor can regulate each of adenylyl cyclase activity and both K^+ and Ca^{2+} currents via pertussis toxin-sensitive G proteins, the Asp^{79} Asn version of the receptor selectively lost the capacity to regulate K^+ current.[35] Such observations could have been interpreted as a selective loss of coupling efficiency for one G_i family G protein member expressed by these cells, but the GPCR-G protein fusion protein results indicate this not to be case.

A very much wider range of studies have examined the effects of mutations in the G protein element of such fusion proteins. One approach has used the fact that if the role of agonist binding to a GPCR is to promote the rate of release of GDP from a G protein, then, as a corollary, increasing concentrations of GDP should increase the dissociation rate of agonist from GPCR. Using fusion proteins between the A_1-adenosine receptor and either wild-type or a pertussis toxin-resistant Cys^{351} Gly mutant of G_{i1}, Waldhoer *et al.*[20] showed that although the association rate of the agonist $(-)$-N^6-3[^{125}I](iodo-4-hydroxyphenylisopropyl)adenosine ([^{125}I]HPIA) to the two fusion proteins was not different, the dissociation rate was considerably slower from the fusion protein containing the wild-type G_{i1} sequence. As agonist dissociation would be expected to be slower from a high-affinity GPCR-G protein complex than from one with lower affinity, such results are consistent with the Cys^{351} Gly mutation reducing the affinity of A_1 adenosine receptor–$G_{i1}\alpha$ interactions. Furthermore, when equilibrium [^{125}I]HPIA-binding

[35] A. Surprenant, D. A. Horstman, H. Akbarali, and L. E. Limbird, *Science* **257,** 977 (1992).

experiments were performed in the presence of increasing concentrations of GDP, the IC_{50} for GDP was lower at the fusion protein containing the Cys^{351}Gly mutation, again consistent with a lower affinity of interactions between these two proteins when Cys^{351} was replaced by Gly.[20]

Part of the reason for selecting Cys^{351} for alteration was this is the site for pertussis toxin-catalyzed ADP-ribosylation in $G_{i1}\alpha$, and because pertussis toxin treatment abolishes functional interactions between GPCRs and G_i family members, this is clearly a key contact site.[36,37] Analysis of mutations at this site has been performed in some detail, as it has been useful to generate pertussis toxin-insensitive forms of G_i family proteins to overcome the lack of cell systems that do not express such G proteins.[38–40] When Bahia *et al.*[41] systematically mutated Cys^{351} of $G_{i1}\alpha$ to all other amino acids and then monitored their activation by an agonist at a coexpressed α_{2A}-adrenoceptor, a spectrum of activity was obtained in which the presence of charged amino acids at this position largely prevented functional interactions while effective coupling was restored as the *n*-octanol/H_2O partition coefficient of the amino acid increased. Because some of the mutated forms of $G_{i1}\alpha$ did not express very effectively, Carr *et al.*[42] used the GPCR-G protein fusion protein approach to measure the capacity of epinephrine to stimulate wild type and Cys^{351}Gly $G_{i1}\alpha$. Stimulated activity of Cys^{351}Gly $G_{i1}\alpha$ measured at V_{max} was only 50% of that of the wild-type G protein.[42] Even more dramatically, when a Cys^{351}Ile $G_{i1}\alpha$ containing fusion protein was added to the analysis and the effects of partial agonist ligands were studied, all displayed increased intrinsic activity relative to epinephrine in the order Ile > Cys > Gly.[6] An interpretation of this is that a better interface between GPCR and G protein provided by more hydrophobic amino acids at residue[351] of the G protein allows partial agonists to more effectively stabilize a conformational state suitable for G protein activation. In contrast, more efficacious agonists are better able to overcome a suboptimal GPCR-G protein contact interface. A prediction that can be extended from this concept is that agonist-independent (or constitutive) signal transduction between a GPCR and G protein would be limited if the GPCR/G protein interface was made less hydrophobic. At least for interactions between the $5HT_{1A}$ receptor and $G_{i1}\alpha$, direct support for this is available. A fusion protein between the $5HT_{1A}$ receptor and Ile^{351} $G_{i1}\alpha$ has constitutive activity, which is inhibited in a concentration-dependent manner by the inverse agonist spiperone. In contrast, an equivalent fusion protein containing Gly^{351} $G_{i1}\alpha$ has no detectable constitutive activity and is unaffected by spiperone.[7]

[36] H. R. Bourne, *Curr. Opin. Cell Biol.* **9,** 134 (1997).

[37] E. Kostenis, F. Y. Zeng, and J. Wess, *Life Sci.* **64,** 355 (1999).

[38] A. Wise, M. A. Watson-Koken, S. Rees, M. Lee, and G. Milligan, *Biochem. J.* **321,** 721 (1997).

[39] S. E. Senogles, *J. Biol. Chem.* **269,** 23120 (1994).

[40] T. W. Hunt, R. C. Carroll, and E. G. Peralta, *J. Biol. Chem.* **269,** 29565 (1994).

[41] D. S. Bahia, A. Wise, F. Fanelli, M. Lee, S. Rees, and G. Milligan, *Biochemistry* **37,** 11555 (1998).

[42] I. C. Carr, A. R. Burt, V. N. Jackson, J. Wright, A. Wise, S. Rees, and G. Milligan, *FEBS Lett.* **428,** 17 (1998).

GPCR-G Protein Fusions Containing Unnatural and Chimeric G Proteins

As the proximity of the protein partners in GPCR-G protein fusions enhances the effectiveness of their coupling, it would be extremely useful if such an approach allowed GPCRs that couple to members of the G_s and G_q families of G proteins to generate easily measurable agonist stimulation of GTPase activity or [^{35}S]GTPγS binding. Historically, this has been much more difficult to observe in coexpression studies than for G_i family proteins. This has been achieved for β_2-adrenoceptor–$G_s\alpha$ fusions expressed in insect Sf9 cells[2,13] and for a fusion between the IP prostanoid receptor and $G_s\alpha$ when expressed stably in HEK293 cells.[12] However, as chimeric G proteins, which have the extreme C-terminal regions switched to allow GPCR recognition to be channeled to an unnatural second messenger end point, have been widely used to address issues of both academic interest and in industrial ligand screening programs,[43] these have also been incorporated into GPCR-G protein fusion proteins. Fusion of the IP prostanoid receptor to a chimeric G protein in which the last six amino acids of $G_{i1}\alpha$ were replaced with the equivalent sequence from $G_s\alpha$ produced a system in which agonist-stimulated GTPase activity was substantially greater than from equivalent levels of a fusion between this GPCR and full-length $G_s\alpha$.[12] As analysis of fusion proteins between the yeast mating factor GPCR and chimeric yeast-mammalian G proteins had indicated that the predominant role for the G protein α subunit C terminus was no longer noted following the production of fusion proteins,[44] then a fusion protein between the IP prostanoid receptor and full-length $G_{i1}\alpha$ was also produced.[12] This, however, failed to be activated in an agonist-dependent manner[12] indicating that simple proximity resulting from GPCR-G protein fusion is insufficient to overcome the natural selectivity of GPCR-G protein pairs.

Conclusions

Although clearly unnatural entities, fusion proteins between GPCRs and G protein α subunits have provided the means to address a number of features of interactions between these proteins in a quantitative manner that would not have been easy to achieve with coexpression of the individual polypeptides (Table 1). It is anticipated that as well as extensions to the types of applications they have currently been used for, they will provide valuable means to explore the quantitative effects of a range of other proteins found to interact with either GPCRs or G proteins on information transfer between these partners.

[43] G. Milligan and S. Rees, *Trends Pharmacol. Sci.* **20,** 118 (1999).
[44] R. Medici, E. Bianchi, G. Di Segni, and G. P. Tocchini-Valentini, *EMBO J.* **16,** 7241 (1997).

TABLE I

RECEPTOR–G PROTEIN PAIRINGS THAT HAVE BEEN USED AS FUSION PROTEINS[a]

Receptor	G protein	Refs.[c]
β_2-AR	G_s	a–g
IP prostanoid	G_s, G_{i1}, G_{i1}-G_s chimera	h
α_{2A}-AR	G_{i1}, G_{i2}, G_{i3}, G_{o1}	i–q
α_{2B}-AR	G_{15}	r
$5HT_{1A}$	G_{i1}, G_{o1}	s–v
A_1-adenosine	G_{i1}, G_{i2}, G_{i3}, G_{o1}	w–y
Formylpeptide	G_{i1}, G_{i2}, G_{i3}	z
M_2 muscarinic	G_z	aa
α_{1b}-AR	G_{11}	bb
TRHR-1[b]	G_{11}	cc
Ste 2	Gpal, Gpal-G_s chimera	dd
edg-2	G_{i1}	ee

[a] No attempt was made to record specific point mutations introduced into either GPCR or G protein in the specific pairings studied.

[b] Thyrotropin-releasing hormone receptor-1.

[c] Key to references: a. B. Bertin, M. Friessmuth, R. Jockers, A. D. Strosberg, and S. Marullo, *Proc. Natl. Acad. Sci. U.S.A.* **91,** 8827 (1994). b. B. Bertin, A. D. Strosberg, and S. Marullo, *Int. J. Cancer* **71,** 1029 (1997). c. R. Seifert, K. Wenzel-Seifert, T. W. Lee, U. Gether, E. Sanders-Bush, and B. K. Kobilka, *J. Biol. Chem.* **273,** 5109 (1998). d. K. Wenzel-Seifert, T. W. Lee, R. Seifert, and B. K. Kobilka, *Biochem. J.* **334,** 519 (1998). e. R. Seifert, K. Wenzel-Seifert, U. Gether, U. V. T. Lam, and B. K. Kobilka, *Eur. J. Biochem.* **260,** 661 (1999). f. R. Seifert, U. Gether, K. Wenzel-Seifert, U. and B. K. Kobilka, *Mol. Pharmacol.* **56,** 348 (1999). g. T. M. Loisel, H. Ansanay, L. Adam, S. Marullo, R. Seifert, M. Lagace, and M. Bouvier, *J. Biol. Chem.* **274,** 31014 (1999). h. C. W. Fong and G. Milligan, *Biochem. J.* **342,** 457 (1999). i. A. Wise, I. C. Carr, and G. Milligan, *Biochem. J.* **325,** 17 (1997). j. A. Wise, I. C. Carr, D. A. Groarke, and G. Milligan, *FEBS Lett.* **419,** 141 (1997). k. A. Wise and G. Milligan, *J. Biol. Chem.* **272,** 24673 (1997). l. M. Sautel and G. Milligan, *FEBS Lett.* **436,** 46 (1998). m. A. R. Burt, M. Sautel, M. A. Wilson, S. Rees, A. Wise, and G. Milligan, *J. Biol. Chem.* **273,** 10367 (1998). n. I. C. Carr, A. R. Burt, V. N. Jackson, J. Wright, A. Wise, S. Rees, and G. Milligan, *FEBS Lett.* **428,** 17 (1998). o. V. N. Jackson, D. S. Bahia, and G. Milligan, *Mol. Pharmacol.* **55,** 195 (1999). p. R. J. Ward and G. Milligan, *FEBS Lett.* **462,** 459 (1999). q. A. Cavalli, K. Druey, and G. Milligan, *J. Biol. Chem.* **275,** 23693 (2001). r. P. J. Pauwels, S. Tardif, F. Finana, T. Wurch, and F. C. Colpaert, *J. Neurochem.* **74,** 375 (2000). s. E. Kellett, I. C. Carr, and G. Milligan, *Mol. Pharmacol.* **56,** 684 (1999). t. Y. Wang, R. T. Windh, C. A. Chen, and D. R. Manning, *J. Biol. Chem.* **274,** 37435 (1999). u. D. S. Dupuis, S. Tardif, T. Wurch, F. C. Colpaert, and P. J. Pauwels, *Neuropharmacology* **38,** 1035 (1999). v. G. Milligan, E. Kellett, V. Dubreuil, E. Jacoby, C. Dacquet, and M. Spedding, *Neuropharmacology.* w. A. Wise, M. Sheehan, S. Rees, M. Lee, and G. Milligan, *Biochemistry* **38,** 2272 (1999). x. M. Waldhoer, A. Wise, G. Milligan, M. Freissmuth, and C. Nanoff, *J. Biol. Chem.* **274,** 30571 (1999). y. N. Bevan, T. Palmer, T. Drmota, A. Wise, J. Coote, G. Milligan, and S. Rees, *FEBS Lett.* **462,** 61 (1999). z. K. Wenzel-Seifert, J. M. Arthur, H.-Y. Liu, and R. Seifert, *J. Biol. Chem.* **274**, 33259 (1999). aa. D. Vorobiov, A. Kanti Bera, T. Keren-Raifman, R. Barzilai, and N. Dascal, *J. Biol. Chem.* **275,** 4166 (2000). bb. P. A. Stevens *et al., J. Biol. Chem.* (2001), in press. cc. J. Pediani, P. A. Stevens, and G. Milligan, unpublished observations. dd. R. Medici, E. Bianchi, G. Di Segni, and G. P. Tocchini-Valentini, *EMBO J.* **16,** 7241 (1997). ee. M. S. Beer, J. A. Stanton, K. Salim, M. Rigby, R. P. Heavens, D. Smith, and G. McAllister, *Ann. N.Y. Acad. Sci.* **905,** 118 (2000).

[19] Synthetic Gene Technology: Applications to Ancestral Gene Reconstruction and Structure–Function Studies of Receptors

By BELINDA S. W. CHANG, MANIJA A. KAZMI, and THOMAS P. SAKMAR

Introduction

The use of synthetic gene technology offers several major advantages for the study of G protein-coupled receptor (GPCR) structure and function,[1] and incorporation of the polymerase chain reaction (PCR) into gene synthesis strategies means that genes of over 1 kb can now be synthesized efficiently and economically in a matter of weeks.[2–4] In creating the gene of interest, the degeneracy of the genetic code can be used to yield nucleotide sequences that have useful properties, such as large numbers of endonuclease restriction sites, optimized primer sites for the PCR and sequencing, and desired levels of GC content and codon bias. Moreover, in certain cases, such as in studies of ancestral protein function, the easy and efficient creation of a gene *de novo* is critical.

Traditionally, studies of GPCRs have relied heavily on mutagenesis techniques to identify residues important for structure and function. While a number of methods are available for site-directed mutagenesis, the use of a properly designed synthetic gene offers many advantages, particularly where extensive mutagenesis is planned. The large number of restriction enzymes now available allows as many as 40 unique restriction sites to be introduced for a synthetic gene of 1 kb in length. In addition, as many optimized PCR primer sites as desired can be incorporated.

A novel and important application of synthetic gene technology in studies of GPCRs is for the recreation of ancestral receptors.[5] The superfamily of GPCRs consists of a large number of related seven transmembrane proteins of highly varied function.[6] Recreating "fossil" receptors for functional studies in the laboratory can provide important insights into the constraints that shaped molecular structure and function in this diverse superfamily that are difficult to attain using more traditional mutagenesis methods. Except in cases where the ancestral receptor of interest is

[1] C. J. L. Carruthers and T. P. Sakmar, *Methods Neurosci.* **25,** 322 (1995).

[2] B. S. W. Chang and T. P. Sakmar, in preparation.

[3] C. Withers-Martinez, E. P. Carpenter, F. Hackett, B. Ely, M. Sajid, M. Grainger, and M. J. Blackman, *Protein Eng.* **12,** 1113 (1999).

[4] W. P. C. Stemmer, A. Crameri, K. D. Ha, T. M. Brennan, and H. L. Heyneker, *Gene* **164,** 49 (1995).

[5] B. S. W. Chang and M. J. Donoghue, *Trends Ecol. Evol.* **15,** 109 (2000).

[6] W. C. Probst, L. A. Snyder, D. I. Schuster, J. Brosius, and S. C. Sealfon, *DNA Cell Biol.* **11,** 1 (1992).

METHODS IN ENZYMOLOGY, VOL. 343
0076-6879/02 $35.00

quite similar to extant sequences, synthetic gene technology is essential for this kind of study.

In addition, in studying receptor function via heterologous expression of mutant receptors, it is often useful to create chimeric receptors in which putative functional domains are exchanged. For example, the transfer of a cytoplasmic loop sequence from one pharmacological receptor subtype to another is one approach to study the specificity of ligand-dependent G protein activation. A synthetic gene for one receptor subtype can be engineered readily to produce a chimeric construct by exchanging a portion of a functionally different receptor subtype, or even of a reconstructed ancestral receptor. The use of synthetic genes allows domain exchanges without the potential limitations of naturally occurring restriction endonuclease cleavage sites.

This article presents a summary of methods and applications of synthetic gene technology, with an emphasis on synthesizing ancestral genes. Algorithms for inferring ancestral sequences and general considerations for designing synthetic genes are discussed in detail. Detailed laboratory procedures for the various steps in gene synthesis, from oligonucleotide preparation to cassette mutagenesis, are also given.

Applications of Synthetic Gene Technology

Many studies of GPCR function have relied on restriction fragment replacement (cassette mutagenesis) methods, which can be facilitated greatly by the use of a synthetic gene. Site-directed point mutations are introduced easily into cloned DNA by a variety of mismatch primer methods, some of which employ PCR. However, site-directed cassette mutagenesis can be employed successfully to introduce extensive alterations of the nucleotide sequence within a particular gene segment. Such an approach may be useful, for example, in structure–function studies of discrete receptor domains, such as the cytoplasmic loops of a GPCR.[7–9] A long stretch of amino acid residues can be replaced easily by a random sequence, by a homologous sequence from a related protein, or by a portion of an ancestral gene.

A new and promising approach to studies of gene function that also relies heavily on synthetic gene technology is for the study of ancestral genes. Here, the use of synthetic genes is not only useful and convenient, but in many cases where the gene has diverged significantly from existing molecules, absolutely essential. The only other option, which would be to start with an existing sequence and use mutagenesis to incorporate all substitutions in the ancestral sequence, rapidly becomes infeasible for sequences that have even relatively small levels

[7] O. Moro, J. Lameh, P. Högger, and W. Sadée, *J. Biol. Chem.* **268,** 22273 (1993).

[8] R. R. Franke, T. P. Sakmar, R. M. Graham, and H. G. Khorana, *J. Biol. Chem.* **267,** 14767 (1992).

[9] R. R. Franke, B. König, T. P. Sakmar, H. G. Khorana, and K. P. Hofmann, *Science* **250,** 123 (1990).

of divergence. Both of these approaches, and how they can incorporate synthetic gene technology, are discussed below.

Mutagenesis by Restriction Fragment Replacement (Cassette Mutagenesis)

Mutagenesis by restriction fragment replacement was first demonstrated in the naturally occurring gene of bacteriorhodopsin.[10] This mutagenesis strategy was possible because of the fortuitous natural placement of unique restriction sites. Cassette mutagenesis in this case involved replacement of a restriction fragment by a synthetic duplex counterpart in order to introduce the desired codon alteration(s). However, the incorporation of large numbers of unique and evenly spaced restriction sites in a carefully designed synthetic gene increases the convenience and usefulness of this method. In addition, cloning is "directional" and screening is not generally necessary for identification of a desired recombinant transformant.

Cassette mutagenesis is also of more general utility than most forms of site-directed mismatch primer mutagenesis because of the ease of producing defined mutations at multiple sites within a domain to yield deletions, extensive substitutions, domain swaps, or the construction of chimeric genes. In cases where initial mutagenesis experiments that target a particular amino acid did not prove informative, an alternative strategy is to prepare large numbers of mutations in a particular putative domain if an adequate functional screening method can be devised. Combinatorial cassette mutagenesis can be employed in such situations.[11–16] The general strategy of combinatorial cassette mutagenesis is to perform restriction fragment replacement with a set of synthetic duplexes that should provide codons for each of the 20 amino acids at one or more positions within the duplex.[17] This can be accomplished by synthesizing the noncoding (top) strand of the duplex with equal mixtures of all four bases in the first two positions of a codon, and with equal mixtures of guanine and cytosine at the third position. Inosine is inserted at each of the randomized base positions in the bottom strand because it is able to pair with each of the four natural bases. The heterogeneous top strand oligonucleotides and the bottom strand oligonucleotide are annealed and the resulting duplex is ligated into an appropriate vector. Bacterial transformation essentially produces a library of mutants that can be cloned or studied batchwise depending on particular circumstances.

[10] K.-M. Lo, S. S. Jones, N. R. Hackett, and H. G. Khorana, *Proc. Natl. Acad. Sci. U.S.A.* **91,** 2285 (1984).

[11] K. Poindexter, R. Jerzy, and R. B. Gayle, *Nucleic Acids Res.* **19,** 1899 (1991).

[12] W. C. Chan and T. Ferenci, *J. Bacteriol.* **175,** 858 (1993).

[13] J. C. Hu, N. E. Newell, B. Tidor, and R. T. Sauer, *Protein Sci.* **2,** 1072 (1993).

[14] L. M. Gregoret and R. T. Sauer, *Proc. Natl. Acad. Sci. U.S.A.* **90,** 4246 (1993).

[15] A. P. Arkin and D. C. Youvan, *Proc. Natl. Acad. Sci. U.S.A.* **89,** 7811 (1992).

[16] S. Delagrave, E. R. Goldman, and D. C. Youvan, *Protein Eng.* **6,** 327 (1993).

[17] J. F. Reidhaar-Olson and R. T. Sauer, *Science* **241,** 53 (1988).

Synthetic Ancestral Genes

A novel approach to the study of molecular structure and function that is complementary to mutagenesis methods is the study of ancestral gene function.[5,18,19] Often mutagenesis studies may be limited by the number of mutants that can reasonably be made and screened in the laboratory. *A priori* hypotheses that are critical in guiding the position and identity of amino acids to mutate may be difficult to formulate, especially for receptors about which little structural information is known. The lack of appropriate hypotheses usually means one must be prepared to screen large numbers of mutants. In addition, random mutations may result in either nonfunctional receptors or receptors with wild-type function (in the particular assay used), for reasons which can be difficult to determine *a posteriori*, rendering the results uninformative in terms of receptor structure–function.

The superfamily of GPCRs is an ideal system in which to use ancestral genes to study molecular structure–function. In the molecular evolutionary radiation of these genes, negative selection has already screened out all the mutations that result in nonfunctional receptors. Moreover, by focusing attention on ancestral receptors where key functional shifts are thought to have occurred (Fig. 1), important amino acid substitutions that form the basis of those functional shifts may be isolated. Essentially, reconstructing ancestral proteins allows the identification of mutations that caused major shifts in protein function, in the background in which these changes originally occurred. Reconstructing critical mutations in the ancestral background may avoid the problem of creating nonfunctional or improperly folded/expressed proteins by mutating extant molecules. Note that this approach also provides very specific *a priori* hypotheses about the nature of the amino acid substitutions, and the particular function affected.

Studies of ancestral gene function are made possible by advances in the statistical estimation of ancestral sequences, which are summarized briefly here. These methods are critical to the success of this approach. Although moving from the statistical inference of ancestral proteins to their actual synthesis in the laboratory is a major step, it can provide unique information about the context in which adaptive replacements may have occurred and other structure–function information difficult to obtain using more traditional molecular methods.

Ancestral Gene Inference Methods

In an ideal situation, one would start from a well-supported molecular phylogeny, from which the ancestral protein sequence of interest would be determined

[18] T. M. Jermann, J. G. Opitz, J. Stackhouse, and S. A. Benner, *Nature* **374,** 57 (1995).

[19] U. M. Chandrasekharan, S. Sanker, M. J. Glynias, S. S. Karnik, and A. Husain, *Science* **271,** 502 (1996).

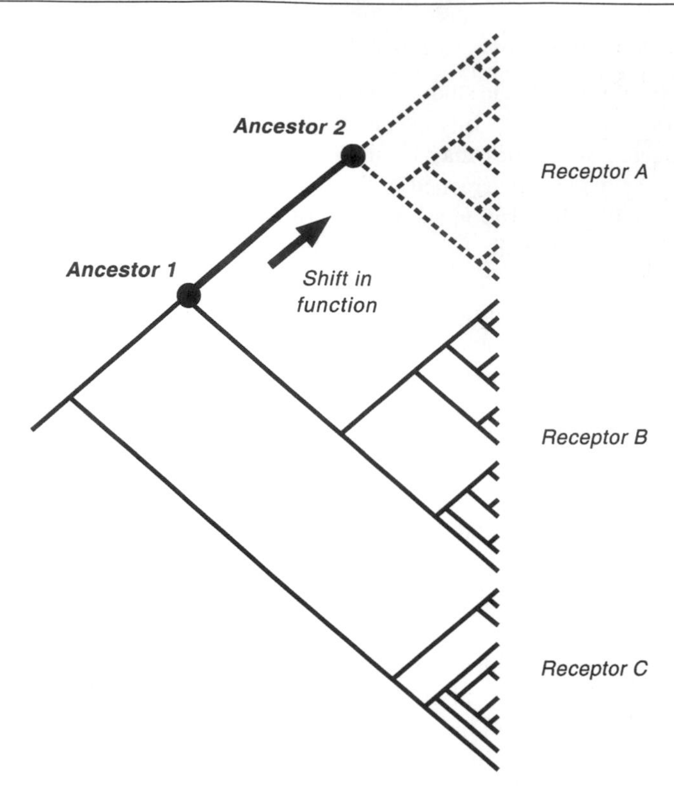

FIG. 1. Ancestral genes in studies of receptor structure and function. In the phylogeny depicted of a superfamily of related receptor genes, receptor A has evolved a function different from that of receptors B and C. Receptors with this new function are indicated by dashed lines. As an alternative to mutagenesis studies to discover the amino acid substitutions that underlie this shift in function, ancestral genes relevant to this functional transition can be synthesized and assayed instead (ancestors Nos. 1 and 2). This method offers the advantage of assaying mutations in the background in which they originally occurred and avoids problems of misfolded or nonfunctional receptors, which is often the case in structure–function studies of extant sequences. See text for details.

unambiguously. However, given that a typical protein is composed of hundreds of amino acid sites, all of whose ancestral states must be inferred, it is extremely rare that all sites can be reconstructed unambiguously. Moreover, often the most interesting evolutionary changes in biochemical function do not occur at extremely low levels of sequence divergence, where ancestral states are inferred more easily. This may be a problem particularly when divergent sequences are combined with varying rates of evolution.[20]

[20] D. Schluter, *Nature* **377,** 108 (1995).

A comprehensive review of phylogenetic methods, or methods for inferring ancestral states, can be found elsewhere.[21,22] This chapter highlights methodological considerations that are most directly relevant when the goal is to proceed to reconstruct sequences in the laboratory. One source of error in inferring ancestral states is lack of resolution of the tree itself, or the existence of a variety of plausible trees. If this problem cannot be overcome, it would be necessary to thoroughly explore the sensitivity of the inferred ancestral sequences to alternative resolutions of poorly supported nodes in the tree.[23,24] Even when the data strongly support a single tree, there may still be ambiguity in inferences of ancestral states at particular internal nodes. Strategies to proceed in the face of such ambiguity are presented below.

Parsimony methods (implemented in programs such as PAUP*[25]) evaluate phylogenetic relationships and ancestral state assignments based on the amount of evolutionary change along the branches of the tree. Specifically, trees or ancestral states that require the fewest changes are preferred.[21,26] While precisely what parsimony assumes about rates of change may be controversial,[22,27,28] it is clear from simulation studies that it can fail under certain circumstances, such as when rates of change are highly unequal along different branches.[29,30] It is difficult to correct for multiple substitutions at a site in an explicit model of evolution, or to take branch lengths into consideration in most parsimony analyses. These properties of parsimony are problematic from the standpoint of reconstructing ancestral states.[23,31] Finally, in practice it is clear that equivocal assessments of ancestral states are common using parsimony. Even a small percentage of such ambiguities scattered along an entire sequence could yield a large number of permutations and combinations, which might then require the examination of a large number of proteins in the laboratory. For this reason it is important to consider alternative strategies to narrow down the possibilities.

[21] D. L. Swofford, G. J. Olsen, P. J. Waddell, and D. M. Hillis, *in* "Molecular Systematics" (D. M. Hillis, C. Moritz, and B. K. Mable, eds.), 2nd Ed., p. 407. Sinauer, Sunderland, MA, 1996.
[22] J. Felsenstein, *Annu. Rev. Genet.* **22,** 521 (1988).
[23] J. Zhang and M. Nei, *J. Mol. Evol.* **44,** S139 (1997).
[24] M. J. Donoghue and D. D. Ackerly, *Philos. Trans. R. Soc. London. Ser. B* **351,** 1241 (1996).
[25] D. L. Swofford, "PAUP*, Phylogenetic Analysis Using Parsimony (*and Other Methods)," Version 4.0. Sinauer, Sunderland, MA, 1999.
[26] W. P. Maddison, *Syst. Zool.* **40,** 304 (1991).
[27] P. O. Lewis, *in* "Molecular Systematics of Plants II: DNA Sequencing" (P. S. Soltis, D. E. Soltis, and J. J. Doyle, eds.), p. 132. Kluwer Boston, 1998.
[28] Z. Yang, *J. Mol. Evol.* **42,** 294 (1996).
[29] J. P. Huelsenbeck, *Syst. Biol.* **46,** 69 (1997).
[30] J. Felsenstein, *Syst. Zool.* **27,** 401 (1978).
[31] Z. Yang, S. Kumar, and M. Nei, *Genetics* **141,** 1641 (1995).

Likelihood phylogenetic methods (implemented in programs such as PHYLIP,[32] MOLPHY,[33] PAML,[34] and NHML[35]) use as an optimality criterion a likelihood score, calculated according to a specified model of evolution.[36] This likelihood score represents the probability of observing the sequence data, given a particular tree topology and model of evolution, and is maximized in reconstructing phylogenetic relationships and ancestral sequences. In reconstructing ancestral states, likelihood methods offer several advantages over parsimony.[27,31,37] Likelihood methods not only use an explicit model of evolution, they also make use of additional information ignored by parsimony contained in branch lengths. An explicit model allows the incorporation of knowledge of the mechanisms and constraints acting on coding sequences, as well as the possibility of comparing the performance of different models, ultimately resulting in the development of more realistic models.[38] Finally, one of the frequently cited drawbacks of likelihood methods, namely the computationally intensive calculations required for most phylogenetic analyses, is not an issue for ancestral reconstructions because most of the computational burden is due to calculating the likelihood of different tree topologies, and not the calculations of the ancestral states themselves. Thus once a reliable tree has been obtained, in most cases it is entirely feasible to use likelihood methods to reconstruct ancestors.[27]

These properties of likelihood are especially important for ancestral state reconstruction, as sites that have ambiguous ancestral state assignments under parsimony can then be explored under different models in likelihood.[23,31] Likelihood methods not only offer the opportunity to assess the comparative fit of various models to the sequence data at hand, they also provide information about specific probabilities associated with particular ancestral reconstructions. This information can be extremely useful in narrowing down to one or a few reconstructions for the purpose of designing ancestral proteins for synthesis in the laboratory.

If differences in ancestral reconstructions depend on the choice of the evolutionary model, then choosing a realistic model is critical. Not surprisingly, this has been the focus of recent developments in phylogenetic methods. Likelihood models describe molecular evolution at three different levels: nucleotide, amino acid, and codon. The simplest nucleotide model, Jukes–Cantor,[39] assumes equal base

[32] J. Felsenstein, "PHYLIP: Phylogeny Inference Package," Version 3.4. University of Washington, Seattle, WA (1991).

[33] J. Adachi and M. Hasegawa, "MOLPHY" Version 2.2. Institute of Statistical Mechanics, Tokyo, Japan, 1994.

[34] Z. Yang, *Comput. Appl. Biosci.* **13,** 555 (1997).

[35] N. Galtier and M. Gouy, *Mol. Biol. Evol.* **15,** 871 (1998).

[36] J. Felsenstein, *J. Mol. Evol.* **17,** 368 (1981).

[37] J. M. Koshi and R. A. Goldstein, *J. Mol. Evol.* **42,** 313 (1996).

[38] N. Goldman, *J. Mol. Evol.* **36,** 345 (1993).

[39] T. H. Jukes and C. R. Cantor, *in* "Mammalian Protein Metabolism" (H. N. Munro, ed.), p. 21. Academic Press, New York, 1969.

frequencies and equal rates of transitions and transversions, which is clearly not realistic for most data sets. More complex nucleotide models incorporate parameters, such as those allowing unequal base frequencies,[36] transition/transversion bias,[40] among-site rate heterogeneity,[41] or nonstationary base composition.[35] However, even these models, and models such as the GTR model,[42] which allows unequal number of substitutions among all the different classes of nucleotides in the rate matrix, fail to take into account codon position or amino acid information.

Amino acid models tend to be even more parameter rich because they involve 20 states instead of only 4, as with nucleotide data. The simplest of these models is the Poisson model, which assumes equal amino acid frequencies and equal rates of substitution among all the amino acids.[43] Both of these assumptions are problematic, as amino acids are known to occur at very different frequencies, and rates of substitution among classes of amino acids can be highly variable due to functional and structural constraints at the protein level. Amino acid models have also been developed that incorporate parameters allowing unequal amino acid frequencies[44] and among-site rate heterogeneity,[41] in addition to a GTR model for amino acids, which allows for unequal numbers of substitutions in the rate matrix for all the different classes of amino acids.[34] However, it is not always necessary to estimate the substitution rate matrix from the sequence data at hand. Rate matrices have been calculated for a number of data sets, including those of Dayhoff[45,46] and Jones[47,48] for globular proteins, mitochondrial transmembrane proteins,[49] and rhodopsin proteins.[50] Use of rate matrices derived from other data sets (if appropriate) allows for the reduction in the number of parameters in the model of evolution. A significant advantage of amino acid models is that they avoid many of the problems associated with modeling evolution at the nucleotide level, such as base compositional bias. However, in these models, all nucleotide-encoded information is lost, including those potentially relevant to the task of ancestral reconstruction.

Codon-based models of molecular evolution are among the most recent developments and have the advantage of incorporating information on both nucleotide and amino acid levels. The original codon-based models assumed equal nonsynonymous to synonymous rate ratios among sites and lineages.[51,52] Subsequent

[40] M. Kimura, *J. Mol. Evol.* **16,** 111 (1980).
[41] Z. Yang, *J. Mol. Evol.* **39,** 306 (1994).
[42] Z. Yang, *J. Mol. Evol.* **39,** 105 (1994).
[43] M. J. Bishop and A. E. Friday, *Proc. R. Soc. Lon. Ser. B* **226,** 271 (1985).
[44] M. Hasegawa and M. Fujiwara, *Mol. Phylogenet. Evol.* **2,** 1 (1993).
[45] M. O. Dayhoff, R. M. Schwartz, and B. C. Orcutt, *in* "Atlas of Protein Sequence and Structure" (M. O. Dayhoff, ed.), p. 345. National Biomedical Research Foundation, Washington, DC, 1978.
[46] H. Kishino, T. Miyata, and M. Hasegawa, *J. Mol. Evol.* **31,** 151 (1990).
[47] D. T. Jones, W. R. Taylor, and J. M. Thornton, *Comp. Appl. Biosci.* **8,** 275 (1992).
[48] Y. Cao, J. Adachi, T. Yano, and M. Hasegawa, *Mol. Biol. Evol.* **11,** 593 (1994).
[49] J. Adachi and M. Hasegawa, *J. Mol. Evol.* **42,** 459 (1996).
[50] B. S. Chang and M. J. Donoghue, in preparation.

models have allowed that ratio to vary across lineages or sites in the protein,[53,54] or even allowed the incorporation of unequal frequencies of different types of nonsynonymous substitutions based on the nature of the various amino acids.[55]

Given the diversity of models now available, the choice of model for use in phylogenetic analysis and ancestral inference is critical. An inappropriate model of evolution can lead to inconsistency in the likelihood analysis and convergence to an incorrect result.[21,28,56] Ancestral inference methods are particularly sensitive to model choice. The possibility of an incorrect result can be reduced by selecting a model of evolution that displays a good fit to the sequence data at hand.

Likelihood ratio tests can be used to compare two models of evolution that are nested with respect to one another in order to determine whether the more complex model fits the sequence data significantly better than the simpler model.[36,57,58] For nested models, a more complex model (H_1) will contain all the parameters of the original model (H_0), as well as additional parameters. If the models are not nested, they cannot be compared directly using a likelihood ratio test, and other methods, such as the generation of the distribution of the test statistic using Monte Carlo simulation, must be used.[38] For nested models, a more complex model (H_1), with additional parameters, should fit data better than a simpler model (H_0), as judged by the likelihood score, or the natural logarithm of the likelihood, of each model (L_0, L_1). If H_0 is correct, this difference in fit to data can be approximated by a χ^2 distribution, with degrees of freedom (*df*) equal to the difference in number of parameters between the two models[59]:

$$2(L_1 - L_0) = \chi^2[df]$$

However, if the observed difference is greater than the χ^2 critical value, then the simpler model (H_0) will be rejected and the more complex model (H_1) will be the preferred model. In other words, in this case the more complex model fits the data even better than would be expected because of its additional parameters relative to the simpler model.

Although the exact details of the ancestral reconstruction methods such as model choice will differ according to the particular data set used, inference of an ancestral rod opsin gene is presented here as an example.[50] While different methods

[51] N. Goldman and Z. Yang, *Mol. Biol. Evol.* **11,** 725 (1994).
[52] S. V. Muse and B. S. Gaut, *Mol. Biol. Evol.* **11,** 715 (1994).
[53] Z. Yang, *Mol. Biol. Evol.* **15,** 568 (1998).
[54] R. Nielsen and Z. Yang, *Genetics* **148,** 929 (1998).
[55] Z. Yang, R. Nielsen, and M. Hasegawa, *Mol. Biol. Evol.* **15,** 1600 (1998).
[56] J. P. Huelsenbeck, *Syst. Biol.* **47,** 519 (1998).
[57] J. P. Huelsenbeck and B. Rannala, *Science* **276,** 227 (1997).
[58] Z. Yang, N. Goldman, and A. Friday, *Mol. Biol. Evol.* **11,** 316 (1994).
[59] W. C. Navidi, G. A. Churchill, and A. von Haeseler, *Mol. Biol. Evol.* **8,** 128 (1991).

often infer the same ancestral reconstructions, occasionally parsimony methods may yield more than one most parsimonious reconstruction, as is illustrated here (Table I). Such ambiguity can be problematic if the aim is to reconstruct a protein in the laboratory. Although a few such ambiguous sites can be dealt with by synthesizing all possible combinations, this approach becomes intractable with even relatively few ambiguous sites. Parsimony methods alone offer no means of deciding among several most parsimonious reconstructions. In contrast, with likelihood methods it is possible to compare the fit to data of different models using likelihood ratio tests, if these are nested models. Furthermore, for each model, information can be obtained about the relative probabilities of different amino acid reconstructions in the form of marginal posterior probabilities.[31] This information can provide a basis for choosing to create one or a few inferred sequences in the laboratory.

Table I shows a portion of the ancestral protein that gave rise to the rhodopsin family of genes inferred from a phylogeny of vertebrate rhodopsin and other visual pigment sequences using both parsimony and likelihood methods. For this data set, based on pairwise nested comparisons using likelihood ratio tests, the HKY + Γ model (for nucleotides) and the GTR + Γ model (for amino acids) were chosen as the most appropriate for maximum likelihood ancestral reconstruction. Two simpler models, which show a significantly poorer fit to the data, JC and Poisson, are shown for comparison. In addition, maximum parsimony reconstructions are also given.

For many sites there is good correspondence between parsimony and likelihood inferences, and also among models of evolution utilizing different information (e.g., nucleotides *versus* amino acids). However, this is not the case for all sites. In parsimony, ambiguity can result from different reconstructions generated from amino acids *versus* nucleotides (site 138), or it can result from more than one most parsimonious state assignment (sites 137, 139, 150). Choices can sometimes be made among these using likelihood methods, by choosing the model with the best fit to the data, and the reconstruction with the highest posterior probability under this model (site 137). However, this is not always the case. Furthermore, strongly violating the assumptions of a model or using a model that poorly fits the data (regardless of whether it is the best fit of the models compared) may result in spurious inferences. For example, site 139 is reconstructed as an Ile under the (in this case, oversimplified) Poisson amino acid model, whereas it is reconstructed as a Val under all other likelihood models. In addition, for a given model, ancestral reconstruction with the highest marginal probability may not be correct, especially if that probability is not high and other reconstructions have comparable posterior probabilities. Finally, different likelihood models that make use of different information, such as amino acids *versus* nucleotides, can yield different reconstructions (site 150), and in such cases it may be desirable to test both possibilities in the laboratory.

TABLE I
RECONSTRUCTED ROD OPSIN ANCESTOR[a]

	Parsimony[c]		Likelihood[d]					
			Amino acids			Nucleotides[e]		
Site[b]	AAS	BPS[e]	Poisson	GTR + Γ	Post. Prob's	JC	HKY + Γ	Post. Prob's
119	L	L	L	L	0.999	L	L	0.913
120	G	G	G	G	1.000	G	G	0.952
121	G	G	G	G	1.000	G	G	0.663
122	E	E	E	E	1.000	E	E	0.667
123	V	V	V	V	0.937	V	V	0.334
124	A	A	A	A	0.973	A	A	0.699
125	L	L	L	L	1.000	L	L	0.874
126	W	W	W	W	1.000	W	W	1.000
127	S	S	S	S	1.000	S	S	0.877
128	L	L	L	L	1.000	L	L	0.905
129	V	V	V	V	1.000	V	V	0.979
130	V	V	V	V	1.000	V	V	0.944
131	L	L	L	L	1.000	L	L	0.995
132	A	A	A	A	1.000	A	A	0.974
133	I	I	I	I	0.999	I	I	0.984
134	E	E	E	E	1.000	E	E	0.996
135	R	R	R	R	1.000	R	R	0.592
136	Y	Y	Y	Y	1.000	Y	Y	0.888
137	***I/V***	I	I	I	0.909	I	I	0.749
138	V	G	V	V	1.000	V	V	0.626
139	***I/V***	***I/V***	I	V	0.509	V	V	0.600
140	C	C	C	C	1.000	C	C	0.998
141	K	K	K	K	1.000	K	K	0.909
142	P	P	P	P	1.000	P	P	0.919
143	M	M	M	M	1.000	M	M	1.000
144	G	G	G	G	0.996	G	G	0.995
145	N	N	N	N	1.000	N	N	0.998
146	F	F	F	F	1.000	F	F	0.956
147	R	R	R	R	0.999	R	R	0.944
148	F	F	F	F	1.000	F	F	0.982
149	G	G	G	G	0.997	G	G	0.677
150	***S/D/G/N***	S	S	S	0.957	***D***	***D***	0.579
151	T	T	T	T	0.967	T	T	0.897
152	H	H	H	H	1.000	H	H	0.645

[a] From Chang and Donoghue.[50]

[b] Amino acid reconstructions are for amino acid sites 119–152, numbered according to bovine rhodopsin.

[c] Parsimony reconstructions were done in PAUP*[25] using unweighted analyses for amino acids (AAS), and 2-to-1 T_V/T_S weighting for nucleotides (BPS). Discrepancies among reconstructions are highlighted in bold italics.

TABLE II
SYNTHETIC GPCR GENES

Gene	Length (bp)	Reference
Ancestral rod opsin	1114	Chang *et al.*, in preparation
Budgerigar UV pigment	1153	Sakmar *et al.*, in preparation
Rhodopsin	1048	Ferretti *et al.*[60]
Red cone pigment	1130	Oprian *et al.*[63]
Green cone pigment	1130	Oprian *et al.*[63]
Blue cone pigment	1080	Oprian *et al.*[61]
D4-2 dopamine receptor	1164	Chio *et al.*[101]
D4-2 dopamine receptor	1170	Kazmi *et al.*[62]
D4-4 dopamine receptor	1266	Kazmi *et al.*[62]
D4-7 dopamine receptor	1410	Kazmi *et al.*[62]
Glucagon receptor	1472	Carruthers and Sakmar[1]
Galanin type 3 receptor	1104	Kolakowski *et al.* (GenBank AF042785)
Angiotensin receptor	1131	Noda *et al.*[102]
Calcitonin receptor	1493	Nussenzveig *et al.*[103]
Serotonin 5HT3 receptor	1906	D. S. Johnson (GenBank U59673)

Gene Synthesis Methods

Synthetic Gene Design

In designing a synthetic gene, the ultimate goals are to facilitate subsequent mutagenesis and chimeric gene studies, while achieving high levels of *in vitro* expression of the designed gene. Ensuring ease and flexibility of genetic manipulation can be done in two ways. Incorporating large numbers of unique restriction sites is required for studies using replacement of restriction fragments, or cassette mutagenesis, whereas PCR-based mutagenesis methods are facilitated by the incorporation of appropriately optimized primer sites. Levels of protein expression can be optimized by adjusting GC content and codon bias. These and other considerations are outlined briefly.

The choice of restriction endonuclease sites to be considered in the design of a synthetic gene includes the following criteria: (1) reliability and availability, (2) high activity and freedom from any exonuclease activity, (3) a recognition sequence of five or more nucleotides, and (4) the generation of staggered rather

[d] Likelihood reconstructions were done in PAML[34] using the GTR + Γ model,[34] the Poisson model[44] for amino acids, and the HKY85+Γ model[100] and the Jukes-Cantor model[39] for nucleotides. Marginal posterior probabilities are given for each amino acid reconstruction for the best-fitting models, GTR + Γ and HKY85 + Γ.

[e] Nucleotide reconstructions were translated to amino acids for the purposes of comparison with the amino acid reconstructions.

than blunt ends. Because of the degeneracy of the universal genetic code, a very large number of potential nucleotide sequences can encode a given amino acid sequence. This potential variability in nucleotide sequence generates a large number of potential restriction maps.

The traditional approach for synthetic gene design was to begin with the native DNA sequence and restriction map, retain all potentially useful restriction sites, and then attempt to add new sites in the intervening sequences.[60] This approach, however, is not general. Manual approaches were used to reverse translate restriction endonuclease recognition sequences in order to consider the locations of all possible restriction sites in a particular amino acid sequence.[61] More recently, this general approach has been facilitated greatly by the availability of sequence analysis software packages such as LaserGene (DNAStar, Inc.), which allow the identification of all of the potential restriction sites within a putative gene. This can be accomplished by starting with the amino acid sequence and using a reverse translation algorithm to create a degenerate nucleotide sequence from which potential restriction sites may be identified.

In order to reduce the number of restriction sites identified in the degenerate sequence, it is preferable to limit the file of restriction enzymes to those with recognition sequences (palindromic or interrupted palindromic) of at least five bases that generate cohesive ends of two or more nucleotides. Methylase-sensitive enzymes should be avoided, but in some cases, nucleotides outside of the endonuclease recognition sequence can be altered to remove the methylase recognition sequence. Sites for enzymes generating blunt ends can be used if long gaps are present after all enzymes generating staggered ends are considered. However, blunt-end cutters should not be juxtaposed in the restriction map. Enzymes that generate identical cohesive overhangs should also not be juxtaposed. Unique restriction sites can be ensured by removing other potential sites from the sequence. Convenient cloning sites should be chosen for each end of the gene, as well as for gene construction via ligation of the long oligonucleotides should PCR methods fail. For example, a number of genes have been synthesized with an *Eco*RI site at the 5′ end and a *Not*I site at the 3′ end for ease of cloning.[1,2,61–63]

In order to optimize gene construction using PCR-based methods, to ensure the success of PCR-based mutagenesis methods, and to facilitate later construction of chimeric genes by "swapping" of PCR fragments, it is necessary to give careful consideration to the design of appropriately placed PCR primer sites. These primers should be designed with standard considerations in mind, such as minimizing hairpins, primer duplexes, mispriming, and optimizing the melting temperature.

[60] L. Ferretti, S. S. Karnik, H. G. Khorana, M. Nassal, and D. D. Oprian, *Proc. Natl. Acad. Sci. U.S.A.* **83,** 599 (1986).

[61] T. P. Sakmar and H. G. Khorana, *Nucleic Acids Res.* **16,** 6361 (1988).

[62] M. A. Kazmi, L. A. Snyder, A. M. Cypess, S. G. Graber, and T. P. Sakmar, *Biochemistry* **38,** 3734 (2000).

[63] D. D. Oprian, A. B. Asenjo, N. Lee, and S. L. Pelletier, *Biochemistry* **30,** 11367 (1991).

In addition, primers that will be used together should be designed so as to minimize possible primer–dimers. All these can be done via many programs available for this purpose, such as Vector NTI (InforMax, Inc.).

After defining the nucleotide sequence that corresponds to the desired restriction map and optimized primer sites, a majority of the gene sequence may still remain undefined. In some cases, the natural sequence can be retained. However, other factors, such as codon usage and GC content, may also be considered in order to optimize expression levels and ease of molecular genetic manipulation.

Codon usage bias can be optimized for a particular expression system in order to achieve desired expression levels.[64] In expression systems such as those using *Escherichia coli,* rare codons (e.g., the AGA codon for Arg) have been shown to cause translational problems most likely due to limited tRNA availability, resulting in misincorporations, frameshifts (leading to truncated proteins), and overall reduced translational efficiency.[65–67] These rare codons and over represented codon pairs, which have also been shown to slow translation,[68] should generally be avoided in designing synthetic genes. Optimizing codon usage frequencies can result in much higher expression levels. However, it may sometimes be appropriate to deliberately incorporate unpreferred codons in order to slow the translation of signal sequences so that cellular membrane translocation systems are not saturated. This was the case for the expression in *E. coli* of the gene for the light-driven proton pump bacteriorhodopsin from *Halobacterium halobium.*[69]

The synthetic gene design process also allows for the reduction of GC content if desired. Stretches of four or more guanines or cytosines can be avoided where possible to minimize potential difficulties in oligonucleotide synthesis, PCR, and DNA sequencing. For example, the GC content in dopamine receptor variants was reduced from 74.2 to 49.4% in the synthetic genes.[62] In addition, some investigators have found it useful to place a mammalian translation initiation consensus sequence immediately preceding the initiation methionine codon.[60,61,70]

Finally, the user-defined DNA sequence should be translated to confirm the correct amino acid sequence. This is important because the amino acid sequence translated from a degenerate codon sequence will not always match the original. For example, with sixfold degenerate amino acids such as serine, four of the codons form TCN and two others AGY. The two types reduce to the single degenerate codon WSN. If this degenerate codon is translated, it will be assigned an unknown

[64] P. M. Sharp, E. Cowe, D. G. Higgins, D. C. Shields, K. H. Wolfe, and F. Wright, *Nucleic Acids Res.* **16,** 8207 (1988).

[65] J. F. Kane, *Curr. Opin. Biotechnol.* **6,** 494 (1995).

[66] M. D. Forman, R. F. Stack, P. S. Masters, C. R. Hauer, and S. M. Baxter, *Protein Sci.* **7,** 500 (1998).

[67] A. Deana, R. Ehrlich, and C. Reiss, *Nucleic Acids Res.* **26,** 4778 (1998).

[68] B. Irwin, J. D. Heck, and G. W. Hatfield, *J. Biol. Chem.* **270,** 22801 (1995).

[69] S. S. Karnik, M. Nassal, T. Doi, E. Jay, V. Sgaramella, and H. G. Khorana, *J. Biol. Chem.* **262,** 9255 (1987).

[70] M. Kozak, *Nucleic Acids Res.* **12,** 857 (1984).

amino acid X, as WSN can expand to any of the following: TCN (Ser), ACN (Thr), AGY (Ser), AGR (Arg), TGY (Cys), TGA (Ter), or TGG (Trp).

Synthetic Gene Construction

Methods of synthetic gene construction rely on overlapping synthesized oligonucleotides of varying lengths. In earlier studies, these were extended using T7 DNA polymerase and were ligated together to form a complete gene.[71] The incorporation of PCR, especially the use of heat-stable enzymes such as *Pfu* that have additional proofreading functions and higher processivity than *Taq,* has not only made synthetic gene construction much faster, but also easier and more economical.[2–4,72–76] It is now entirely feasible to have a complete gene synthesized and expressed in a matter of weeks.

Because gene construction using PCR has rendered unnecessary purification steps to ensure full-length synthesized oligonucleotide template and requires very little starting material, it has become feasible to synthesize genes using longer oligonucleotide fragments of up to 300 bases. Longer oligonucleotides offer several advantages in addition to rendering gene design more straightforward. Short oligonucleotides require on the order of 50–100 overlapping fragments, and it may become difficult to optimize them all for PCR. More importantly, regions of overlap are required to be about 20 bp in length, regardless of oligonucleotide length. Therefore, fewer longer oligonucleotides are required to cover the entire synthetic gene, minimizing both the total amount of overlap required and the total number of bases synthesized. This strategy is thus much more economical and the oligonucleotides can be synthesized more quickly.

In order to demonstrate the procedure for synthetic gene construction, a recently constructed ancestral rod opsin gene is presented here as an example (Fig. 2).[2] The entire gene (1114 bp) was constructed from five long synthetic oligonucleotides that were amplified and assembled using a stepwise PCR procedure (Fig. 3). Although it may be feasible to combine all the reactions in one PCR step,[4] a stepwise procedure was chosen primarily to facilitate troubleshooting should any of the PCRs fail. The five overlapping fragments were synthesized as single-stranded oligonucleotides on an Applied Biosystems oligonucleotide synthesizer. In the first

[71] D. D. Moore, *in* "Current Protocols in Molecular Biology" (F. M. Ausubel, R. Brent, R. E. Kingston, D. D. Moore, J. A. Smith, J. G. Seidman, and K. Struhl, eds.), p. 8.2.8. Green Publishing Associates and Wiley-Interscience, New York, 1994.

[72] R. W. Graham, T. Atkinson, D. G. Kilburn, R. C. Miller, and R. A. J. Warren, *Nucleic Acids Res.* **21,** 4923 (1993).

[73] D. R. Casimiro, P. E. Wright, and H. J. Dyson, *Structure* **5,** 1407 (1997).

[74] E. K. Jaffe, M. Volin, C. R. Bronson-Mullins, R. L. Dunbrack, J. Kervinen, J. Martins, J. F. Quinlan, M. H. Sazinsky, E. M. Steinhouse, and A. T. Yeung, *J. Biol. Chem.* **275,** 2619 (2000).

[75] C. Prodromou and L. H. Pearl, *Protein Eng.* **5,** 827 (1992).

[76] A. N. Vallejo, R. J. Pogulis, and L. R. Pease, *in* "PCR Primer: A Laboratory Manual" (C. D. G. Dieffenbach, ed.), p. 603. Cold Spring Harbor Laboratory Press, New York, 1995.

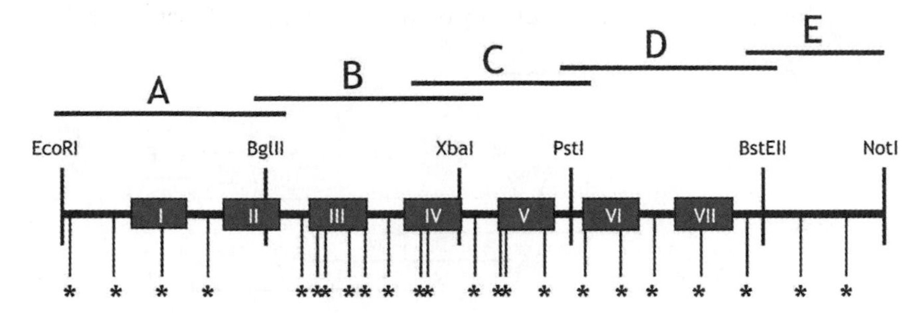

FIG. 2. Gene synthesis by the stepwise PCR method. This method involves the synthesis of long overlapping oligonucleotides of 200–300 bp, followed by several PCRs to assemble the gene in a stepwise manner from the synthetic oligonucleotides. Once the complete gene has been obtained in this manner, it is then cloned into the appropriate expression vector. A schematic is presented for the synthesis of the rod ancestral opsin gene as an example. This gene is 1114 bp in length and was designed with 29 unique restriction sites (*) and 10 PCR primer sites at the ends of each long oligonucleotide. The seven transmembrane segments are depicted as black rectangles and are labeled I–VII. Five gene fragments were prepared using an ABI oligonucleotide synthesizer: fragment A, *Eco*RI to *Bgl*II (277 bases); fragment B, *Bgl*II to *Xba*I (300 bases); fragment C, *Xba*I to *Pst*I (221 bases); fragment D, *Pst*I to *Bst*EII (308 bases); and fragment E, *Bst*EII to *Not*I (159 bases). Adjacent fragments were designed to overlap by at least 20 nucleotides. Detailed procedures for gene synthesis by this method are presented in the text.

round, each synthesized oligonucleotide was converted into a duplex and amplified in a PCR reaction containing flanking primers of 25–30 bp. In the second round of PCR, the resulting duplex PCR products were joined together pairwise (AB, BC, CD, and DE) in separate PCR reactions. The third round of PCR started the elongation of the gene through stepwise PCR reactions. For example, AB and BC were spliced together to give fragment ABC. In the fourth round, fragments ABC and CD were spliced together to give ABCD. In the fifth and final round, ABCD was spliced together with DE to give the full-length gene ABCDE. The products of each round of PCR were separated on low-melt agarose gels and purified using a Qiaex II kit (Qiagen) or used directly in the next round of PCR. The final product was cloned directly into pCR-Blunt (Invitrogen), a vector designed specially for direct cloning of blunt-end PCR products generated using *Pfu* polymerase. Several recombinants were sequenced using flanking and internal sequencing primers.

Experimental Procedures

Oligonucleotide Synthesis

Automated oligonucleotide synthesis can be carried out easily on oligonucleotide synthesizers with commercially available solvents and reagents. The most commonly used chemistry involves the phosphite triester approach using protected β-cyanoethyl phosphoramidite nucleosides. The fully protected 3′-terminal phosphoramidite of the oligonucleotide is coupled to a solid support, such as control

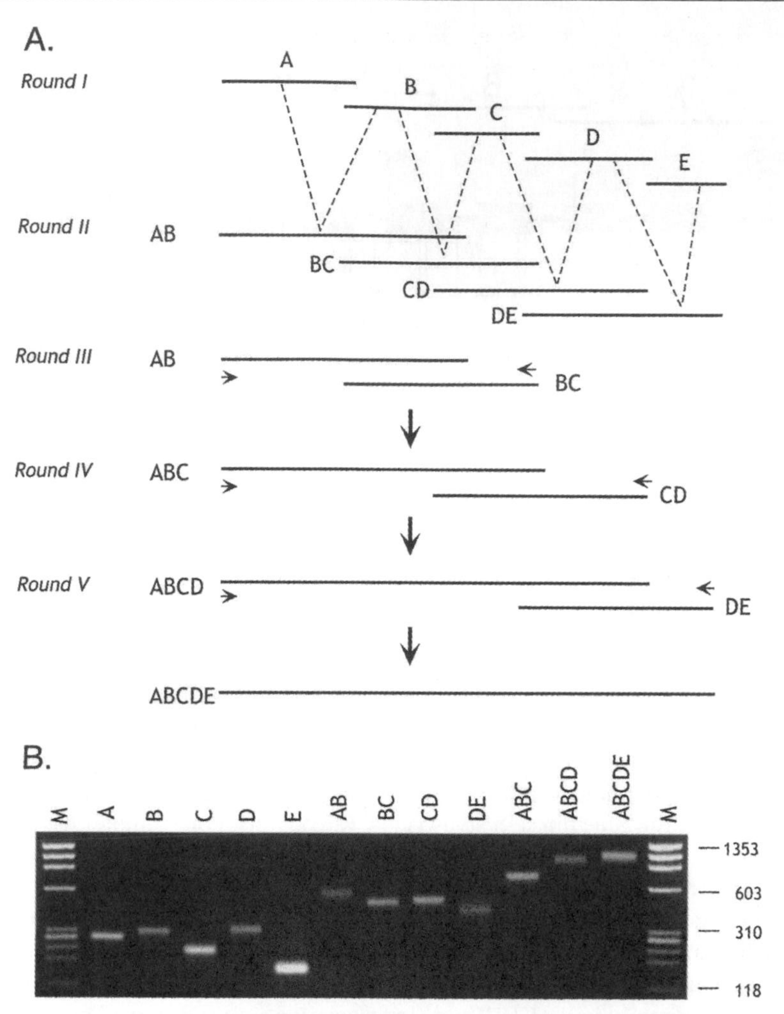

FIG. 3. Synthesis of the rod ancestral opsin gene by stepwise PCR. (A) Assembly of the five long overlapping oligonucleotide fragments (A–E) is accomplished by five rounds of PCR. In the first round, synthesized oligonucleotide fragments are amplified using primers at each respective end. In the second round, adjacent pairs of oligonucleotide fragments are sewn together; in subsequent rounds, these joined fragments are concatenated until the entire gene is amplified. (B) Agarose gel showing the results of each PCR (3% agarose run in 1xTAE buffer and stained with ethidium bromide; ladders on either end are 0.5 ng ϕX-*Hae*III digest). Lanes A thru E are from round I, AB thru DE from round II, ABC from round III, ABCD from round IV, and ABCDE from round V.

pore glass or polystyrene.[77] After protic acid treatment to remove the 5′-protecting group, the fully protected incoming phosphoramidite is activated by tetrazole so that a phosphite triester bond is formed at high efficiency. The small amount of unreacted 5′-hydroxyl of the first nucleoside is capped by a quantitative reaction with acetic anhydride in the presence of 1-methylimidazole. Finally, the newly formed internucleotide linkage is converted from a phosphite triester to a more stable phosphate triester by oxidation with iodine where water is the oxygen donor. The 5′-hydroxyl of the dinucleotide can now be deprotected with acid treatment to complete a cycle. The cycle is repeated until the full-length oligonucleotide is obtained. Thus, the oligonucleotide is elongated from 3′ to 5′. Cleavage from the support and removal of phosphate and exocyclic amine protecting groups is achieved by treatment with concentrated ammonium hydroxide.

For synthesis of the ancestral rod opsin gene, automated oligonucleotide synthesis is performed on an Applied Biosystems Model 392 DNA synthesizer. Phosphoramidite chemistry is employed using 40-pmol synthesis scales and standard cycle routines.

Each synthetic oligonucleotide is cleaved automatically from the solid support after removal of the terminal 5′-hydroxyl protecting group. Each oligonucleotide solution is transferred into a screw-top vial. After the addition of 2 ml of fresh concentrated ammonium hydroxide, the vial is capped tightly and heated at 55° for at least 5 hr. Each fully deprotected oligonucleotide is dried by vacuum centrifugation in a polypropylene tube, and the pellets are dissolved in 50 μl of TE (10 m*M* Tris–HCl, 1 m*M* EDTA, pH 7.4). An ultraviolet spectrum is measured from 310 to 210 nm after making the proper dilution in TE. The yield in total absorbance units at 260 nm is calculated. The crude oligonucleotides are subjected to PCR amplification as described next.

Stepwise PCR

Oligonucleotides corresponding to the overlapping fragments A–E (277, 300, 221, 308, and 159 bp, respectively) are synthesized (Fig. 2). Adjacent oligonucleotides have overlaps of 20–40 bp. A total of 10 primers of 21–24 bp corresponding to the 5′ and 3′ region of each long oligonucleotide are synthesized. Primer pairs a1/a2,b1/b2,c1/c2,d1/d2, and e1/e2 flank fragments A–E, respectively. These primer pairs and their corresponding crude oligonucleotide templates are used in a PCR to amplify the five gene fragments (Fig. 3). In round I synthesis, 100 pmol of each template oligonucleotide is added to 50 μl PCR mixture [20 m*M* Tris–HCl, pH 8.0, 2 m*M* $MgCl_2$, 10 m*M* KCl, 6 m*M* $(NH_4)_2SO_4$, 0.1% Triton X-100, 10 μg/ml bovine serum albumin (BSA), 0.4 m*M* each dNTP, 2.5 U of *Pfu* polymerase, and 1 μM each flanking primer]. The PCR program consists of one

[77] M. D. Matteucci and M. H. Caruthers, *J. Am. Chem. Soc.* **103,** 3185 (1981).

denaturation step at 94° for 45 sec, followed by 25 cycles at 94° for 45 sec, 58° for 1 min, 72° for 2 min, and a final incubation at 72° for 10 min. The fragments are separated on 2% NuSieve GTG agarose (FMC BioProducts) and purified using Qiaex II purification kits (Qiagen). These PCR purification conditions are used in each subsequent round of PCR.

For round II PCR, products A–E from round I are diluted 1 : 500, and 1 μl of each adjacent pair is used a template. Primer pairs a1/b2,b1/c2,c1/d2, and d1/e2 flank fragments AB, BC, CD, and DE, respectively. For round III, products AB and BC are diluted 1 : 500, and 1 μl of each is used as a template for primers a1 and c2. The resulting product ABC and product CD from round II are diluted 1 : 500, and 1 μl of each is used as a template with primers a1 and d2 in round IV to generate fragment ABCD. In the fifth and final round of PCR, products ABCD and DE from round II are diluted 1 : 500, and 1 μl of each is used a template with primers a1 and e2 to amplify the entire gene.

Cloning of PCR Product

The final PCR product (ABCDE) is cloned directly into pCR-Blunt (Invitrogen). In a 10-μl reaction, 1 μl of the purified product is mixed with 25 ng of vector and ligated under the manufacturer's guidelines. The ligated material is transformed into TOP-10 cells (Invitrogen) following standard protocols, and the recombinant transformants are plated onto LB plates containing 50 μg/ml kanamycin. Several recombinant clones are streak purified, and plasmid DNA is purified from each using Qiagen miniprep kits. Each clone is sequenced with two internal and two flanking primers. Alternatively, the final PCR product may be digested with the restriction enzymes engineered at the 5′ and 3′ end of the gene and ligated into an expression vector using directional cloning. Sequence errors that are due to depurination during oligonucleotide synthesis or due to misincorporations during PCR are expected to be distributed randomly among different clones and can be corrected easily by combining error-free fragments by restriction cloning. Any remaining errors can be repaired using the QuikChange kit (Stratagene).

Cassette Mutagenesis

Site-directed mutagenesis of the synthetic gene is accomplished by synthesizing a pair of complementary oligonucleotides to form a duplex containing the desired codon alteration and the appropriate cohesive-terminal overhangs. After purification and annealing as described previously, the 5′ end nonphosphorylated synthetic duplex is ligated into the plasmid /gene DNA fragment linearized with the appropriated restriction endonucleases. Alternatively, for longer oligonucleotide replacements, the insert can be synthesized single stranded, amplified using PCR, and then cloned into the synthetic gene using the appropriate restriction sites.

Expression of Synthetic Genes

As discussed previously, one of the advantages of the use of synthetic genes is that they can be transferred easily among a variety of vectors and that codon usage can be optimized where relevant to achieve maximal levels of expression. Synthetic GPCR genes will generally be expressed in mammalian cells in tissue culture where pharmacological and cellular physiological effects can be correlated with structural changes introduced by mutation. In the case of the synthetic gene for bovine rhodopsin, large quantities of the opsin apoprotein can be produced in monkey kidney cells by transfection where transcription is under the control of the human adenovirus major-late promoter[78] or in stable cell lines.[79] Apoprotein in the plasma membrane can be regenerated with the chromophore 11-*cis*-retinal to form rhodopsin. The recombinant rhodopsin can be solubilized with detergent treatment and purified using an affinity adsorption method.[8,78]

Conclusions

Because of their advantages for mutagenesis studies, synthetic GPCRs have been expressed in a variety of heterologous expression systems. For example, synthetic genes have been expressed in *E. coli*,[69] in monkey kidney cells in tissue culture,[60] in insect Sf9 cells,[80] and in yeast.[80–82] In visual pigment structure–function studies, synthetic receptor genes for the rhodopsin and for the human blue, green, and red cone pigment genes have been expressed in mammalian cells and purified from cell extracts after reconstitution with 11-*cis*-retinal chromophore.[63] Purified site-directed mutant pigments have been studied by a variety of bio chemical[83–86] and biophysical techniques.[87–92] These studies have led to a greater

[78] D. D. Oprian, R. S. Molday, R. J. Kaufman, and H. G. Khorana, *Proc. Natl. Acad. Sci. U.S.A.* **84,** 8874 (1987).

[79] P. J. Reeves, R. L. Thurmond, and H. G. Khorana, *Proc. Natl. Acad. Sci. U.S.A.* **93,** 11487 (1996).

[80] M. S. Urdea, J. P. Merryweather, G. T. Mullenbach, D. Coit, U. Heberlein, P. Valenzuela, and P. J. Barr, *Proc. Natl. Acad. Sci. U.S.A.* **80,** 7461 (1983).

[81] T. Tokunaga, S. Iwai, H. Gomi, K. Kodama, E. Ohtsuka, M. Ikehara, O. Chisaka, and K. Matsubara, *Gene* **39,** 117 (1985).

[82] T. Tanaka, S. Kimura, and Y. Ota, *Nucleic Acids Res.* **15,** 3178 (1987).

[83] G. B. Cohen, T. Yang, P. R. Robinson, and D. D. Oprian, *Biochemistry* **32,** 6111 (1993).

[84] T. A. Zvyaga, K. C. Min, M. Beck, and T. P. Sakmar, *J. Biol. Chem.* **268,** 4661 (1993).

[85] T. P. Sakmar, R. R. Franke, and H. G. Khorana, *Proc. Natl. Acad. Sci. U.S.A.* **86,** 8309 (1989).

[86] K. Fahmy and T. P. Sakmar, *Biochemistry* **32,** 7229 (1993).

[87] S. W. Lin, T. P. Sakmar, R. R. Franke, H. G. Khorana, and R. A. Mathies, *Biochemistry* **31,** 5105 (1992).

[88] K. Fahmy, F. Jäger, M. Beck, T. A. Zvyaga, T. P. Sakmar, and F. Siebert, *Proc. Natl. Acad. Sci. U.S.A.* **90,** 10206 (1993).

[89] J. F. Resek, Z. T. Farahbakhsh, W. L. Hubbell, and H. G. Khorana, *Biochemistry* **32,** 12025 (1993).

[90] J. W. Lewis, I. Szundi, W. Y. Fu, T. P. Sakmar, and D. S. Kliger, *Biochemistry* **39,** 599 (2000).

understanding of the mechanism of wavelength regulation by visual pigments[84,93–98] and of the mechanism of rhodopsin–transducin interaction.[9,99]

However, despite advances in mutagenesis studies using synthetic gene technology, these studies are still limited by the requirements of some *a priori* knowledge of protein structure and function in order to decide which mutants to test. A new and promising approach that can shed light on this problem, and requires little prior information, is the study of ancestral genes. This approach has been made possible in recent years by advances in statistical methods in ancestral sequence inference, and the obvious benefits of synthetic gene technology for bringing these ancestral genes into the laboratory are just beginning to be exploited.

In conclusion, gene synthesis should be considered when extensive long-term structure–function studies are planned or when ancestral genes are the subject of study. The initial investment in gene design and oligonucleotide synthesis increases the ease and flexibility of later DNA manipulation. Improved economical automated DNA synthesis, PCR techniques, and the availability of a large number of quality restriction endonucleases have combined to make gene synthesis rapid and efficient for nearly all molecular biology laboratories.

Acknowledgments

We thank Michael Donoghue for discussions concerning ancestral inference and Wing-Yee Fu for technical assistance on the design and synthesis of the Budgerigar UV pigment. Support for this work was provided in part by the Howard Hughes Medical Institute, the National Institutes of Health (DK54718, EY07138), and the Allene Reuss Memorial Trust.

---

[91] O. P. Ernst, C. K. Meyer, E. P. Marin, P. Henklein, W. Y. Fu, T. P. Sakmar, and K. P. Hofmann, *J. Biol. Chem.* **275,** 1937 (2000).

[92] M. Eilers, P. J. Reeves, W. Ying, H. G. Khorana, and S. O. Smith, *Proc. Natl. Acad. Sci. U.S.A.* **96,** 487 (1999).

[93] T. Chan, M. Lee, and T. P. Sakmar, *J. Biol. Chem.* **267,** 9478 (1992).

[94] Z. Wang, A. B. Asenjo, and D. D. Oprian, *Biochemistry* **32,** 2125 (1993).

[95] E. A. Zhukovsky and D. D. Oprian, *Science* **246,** 928 (1989).

[96] T. P. Sakmar, R. R. Franke, and H. G. Khorana, *Proc. Natl. Acad. Sci. U.S.A.* **88,** 3079 (1991).

[97] S. W. Lin, G. G. Kochendoerfer, K. S. Carroll, D. Wang, R. A. Mathies, and T. P. Sakmar, *J. Biol. Chem.* **273,** 24583 (1998).

[98] G. G. Kochendoerfer, S. W. Lin, T. P. Sakmar, and R. A. Mathies, *Trends Biochem. Sci.* **24,** 300 (1999).

[99] K. Fahmy and T. P. Sakmar, *Biochemistry* **32,** 9165 (1993).

[100] M. Hasegawa, H. Kishino, and T. Yano, *J. Mol. Evol.* **22,** 672 (1985).

[101] C. L. Chio, R. F. Drong, D. T. Riley, G. S. Gill, J. L. Slightom, and R. M. Huff, *J. Biol. Chem.* **269,** 11813 (1994).

[102] K. Noda, Y. Saad, A. Kinoshita, T. P. Boyle, R. M. Graham, A. Husain, and S. S. Karnik, *J. Biol. Chem.* **270,** 2284 (1995).

[103] D. R. Nussenzveig, C. N. Thaw, and M. C. Gershengorn, *J. Biol. Chem.* **269,** 28123 (1994).

[20] Considerations in the Design and Use of Chimeric G Protein-Coupled Receptors

By JÜRGEN WESS

Introduction

Systematic functional analysis of chimeric G protein-coupled receptors (GPCRs) has provided a wealth of information about the structural elements determining specific aspects of GPCR function.[1-4] During the past decade, this strategy has been applied with great success to delineate GPCR domains involved in ligand binding, G protein coupling, regulation of receptor activity (including processes such as phosphorylation, desensitization, internalization, or downregulation), receptor assembly, and receptor trafficking.[1-4]

In most cases, chimeric receptor studies are designed as gain-of-function studies, as data derived from such experiments can be interpreted in a rather straightforward manner. This approach examines whether the substitution of a distinct region from a "donor receptor" into an "acceptor receptor" results in the transfer of a specific functional property that is characteristic for the donor receptor. In contrast, loss-of-function hybrid receptor studies analyze whether replacement of a specific region in an acceptor receptor with the homologous region from a donor receptor leads to the loss of a functional property that is specific for the acceptor receptor. Data derived from loss-of-function studies need to be interpreted with particular caution because it is often difficult to exclude the possibility that the loss of the specific receptor function under investigation is caused by improper folding of the analyzed mutant receptor proteins. However, loss-of-function mutagenesis experiments can nevertheless provide useful information, particularly when applied in combination with a gain-of-function mutagenesis approach.

This article describes methods that can be employed to generate hybrid GPCRs and that are useful for the proper analysis of such mutant receptors. All examples discussed in this article are taken from our own work involving the mutational analysis of different members of the muscarinic acetylcholine receptor family (M_1–M_5). These receptors are ideally suited for hybrid receptor studies, as the five muscarinic receptor subtypes show distinct pharmacological and functional properties, while sharing a high degree of sequence homology.[3-6]

[1] H. G. Dohlman, J. Thorner, M. G. Caron, and R. J. Lefkowitz, *Annu. Rev. Biochem.* **60,** 653 (1991).
[2] S. Suryanarayana and B. K. Kobilka, *Methods Neurosci.* **25,** 61.
[3] J. Wess, *Crit. Rev. Neurobiol.* **10,** 69 (1996).
[4] J. Wess, *Pharmacol. Ther.* **80,** 231 (1998).
[5] E. C. Hulme, N. J. M. Birdsall, and N. J. Buckley, *Annu. Rev. Pharmacol. Toxicol.* **30,** 633 (1990).

0076-6879/02 $35.00

General Considerations

As indicated in the introduction, the analysis of hybrid receptors is very useful for identifying structural domains that determine one or more functional properties that are characteristic for one of the two parent receptors. However, for obvious reasons, a chimeric receptor strategy does not allow the delineation of regions or residues that have a similar functional role in both parent receptors.

Correctly folded hybrid receptors are usually obtained with relative ease when the two parent receptors show a high degree of sequence homology. For this reason, most hybrid receptor studies have been carried out using individual members of a specific GPCR subfamily such as adrenergic[1,2] or muscarinic acetylcholine[3] receptor families. Improper receptor folding frequently presents a problem when domain replacements are carried out between members of different subclasses of GPCRs that do not share a high degree of sequence identity (e.g., muscarinic acetylcholine and dopamine receptors[7]).

The likelihood of obtaining functional hybrid receptors is also strongly dependent on the site(s) within the receptor proteins at which the two sequences are joined ("chimeric junctions"). As a general rule, chimeric junctions within the transmembrane (TM) domains are usually less well tolerated than in extracellular or cytoplasmic receptor regions. This phenomenon is most likely due to the fact that specific interactions between residues located on different TM helices (TM I–VII) are required for the proper assembly of the TM receptor core.

Initially, the construction of chimeric receptors requires that the sequences of the two parent receptors be aligned, either manually (this can be done when the sequence homology between the two wild-type receptors is very high) or by using commercially available computer software. If possible, it is advantageous to join the two parent receptors at sites characterized by a high degree of sequence identity. This precaution will reduce the likelihood that the newly created chimeric junctions will cause major folding deficits. This latter issue becomes even more relevant when the hybrid receptors to be generated contain more than one chimeric junction (see later).

The decision as to which specific receptor regions to replace or to exchange between two wild-type receptors depends on which functional aspects of receptor function are to be examined. For example, if the goal is to map structural aspects of the ligand-binding site of a specific GPCR, the extracellular domains and the exofacial segments of the TM helices should represent the primary targets. However, if the aim of the planned study is to analyze structural elements involved in receptor/G protein coupling or regulation of receptor activity (including processes such as phosphorylation, sequestration, or downregulation), the primary focus

[6] M. P. Caulfield, *Pharmacol. Ther.* **58,** 319 (1993).

[7] J. Wess, unpublished results (1992).

should be on the potential involvement of the intracellular receptor domains and adjacent TM segments.

Typically, the application of a hybrid receptor approach initially involves the analysis of constructs harboring relatively large substitutions to allow for a rough mapping of functionally important receptor domains. In the next step, progressively smaller substitutions need to be carried out to define the structural elements that specify the receptor function under investigation in greater detail. In several cases, systematic application of this approach has led to the identification of single amino acids that play key roles in the function of a particular GPCR (see, e.g., Ref. 8).

Construction of Chimeric Receptors

Several different approaches can be employed to construct hybrid GPCRs. A simple way of generating chimeric receptor constructs involves the use of restriction sites that are conserved among the two parent receptors. If present, such sites offer a rather convenient approach to an initial mapping of functionally important receptor domains. However, such conserved restriction sites are usually relatively rare, even in rather closely related GPCRs of a specific receptor subfamily. Moreover, an obvious disadvantage of this strategy is that the precise location of the chimeric junctions is not chosen based on receptor topography but is dictated by where in the receptor sequences the conserved restriction sites are located.

Another strategy that has been used occasionally to generate hybrid GPCRs is the adapter-mediated ligation of restriction fragments (reviewed in Ref. 2). This approach also requires conveniently located restrictions sites in both receptors (these sites, however, need not be identical) and suffers from similar disadvantages as the "restriction fragment swapping strategy" described in the previous paragraph.

The advent of polymerase chain reaction (PCR) technology has greatly facilitated the construction of hybrid GPCR genes. Most importantly, PCR-based mutagenesis techniques can be employed to generate chimeric junctions at any desired point within the receptor sequences. Several examples illustrating how PCR-based mutagenesis methods can be used to construct hybrid GPCRs are outlined in Figs. 1–3.

To avoid time-consuming subcloning steps, we routinely design PCR mutagenesis strategies in a fashion that the PCR fragments containing the desired mutations can be inserted directly into a mammalian expression vector. In the past, we have carried out most of our GPCR mutagenesis work using pcD- or pcD-PS-based expression plasmids.[9,10] Because of its small size, the pcD (pcD-PS) expression vector has proven particularly versatile for assembling the final mutant GPCR

[8] K. Blüml, E. Mutschler, and J. Wess, *J. Biol. Chem.* **269,** 11537 (1994).

[9] H. Okayama and P. A. Berg, *Mol. Cell. Biol.* **3,** 280 (1983).

[10] T. I. Bonner, A. C. Young, M. R. Brann, and N. J. Buckley, *Neuron* **1,** 403 (1988).

constructs without the need for further subcloning steps. Moreover, transfection of COS-7 cells with pcD-based receptor expression plasmids generally results in high GPCR expression levels that usually exceed levels that can be obtained with most commercially available mammalian expression vectors.[11]

Generation of Hybrid GPCRs with One Chimeric Junction

Figure 1 depicts a general strategy that can be used to introduce a single chimeric junction into a receptor protein. This strategy, which involves an approach that is also referred to as "recombinant PCR,"[12] requires the use of four PCR primers (oligonucleotides; primers A–D) and the two receptor cDNAs as PCR templates. The two "internal" PCR primers (primers B and C) are chimeric in nature because they contain sequences derived from both parent receptors. In contrast the two "external" PCR primers are receptor specific (primer A: receptor 1; primer D: receptor 2).

Primer A is a sense primer located upstream of a restriction site X in receptor 1 that is useful for the assembly of the final chimeric construct. Site X should be located close to the chimeric junction, as this will reduce the size of the PCR product that needs to be generated and sequenced. This site does not need to be unique as long as it is useful for generating a restriction fragment that can be used (e.g., in a three-piece ligation) to assemble the final chimeric construct. Primer D is an antisense primer that corresponds to the C-terminal segment of receptor 2 and includes a stop codon and a restriction site (Y) that is useful for cloning purposes. For example, site Y can be a unique restriction site located in the 3′-untranslated region of receptor 1 or a site that is present in the polylinker sequence of the expression vector. We routinely add at least five extra bases at the 5′ end of primer D to improve the efficiency of digestion by enzyme Y.

The mutagenic oligonucleotides, primers B and C, are complementary to each other. The 3′ and 5′ portions of primer B consist of receptor 1 and 2 sequences, respectively. Conversely, the 3′ and 5′ portions of primer C contain sequences derived from receptors 2 and 1, respectively. We usually get good results with hybrid oligonucleotides that are about 40 bases long containing 20 bases from each of the two receptors.

In the first step of the recombinant PCR protocol, two separate PCR reactions are carried out with primers A/B (template: receptor 1 DNA) and primers C/D (template: receptor 2 DNA), using standard PCR conditions that are appropriate for the desired amplification. To reduce the number of misincorporations by the thermostable DNA polymerase, it is recommended that at least 100 ng of template DNA be used. The two resulting PCR products (1 and 2) overlap in the sequence that corresponds to that of the mutagenic PCR primers. The two overlapping primary

[11] J. Wess, unpublished results.

[12] R. Higuchi, *in* "PCR Technology" (H. A. Erlich, ed.), p. 61. Stockton Press, New York, 1989.

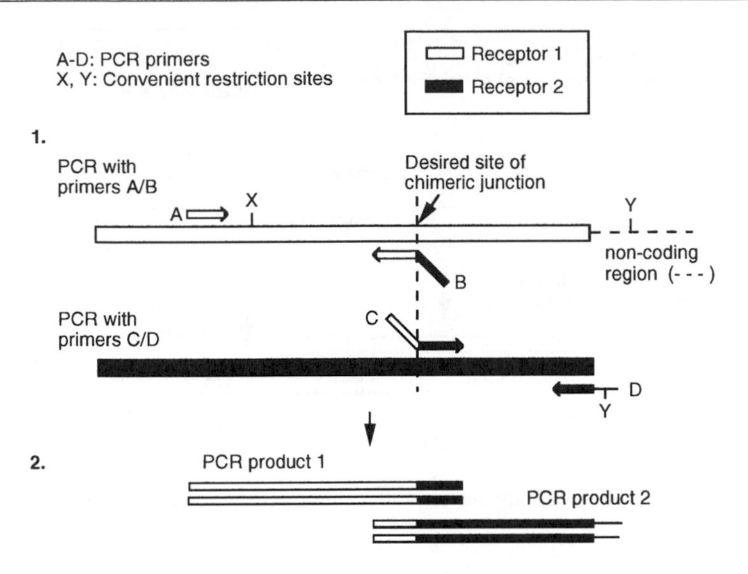

3. Isolate PCR products 1 and 2, mix, denature, and renature

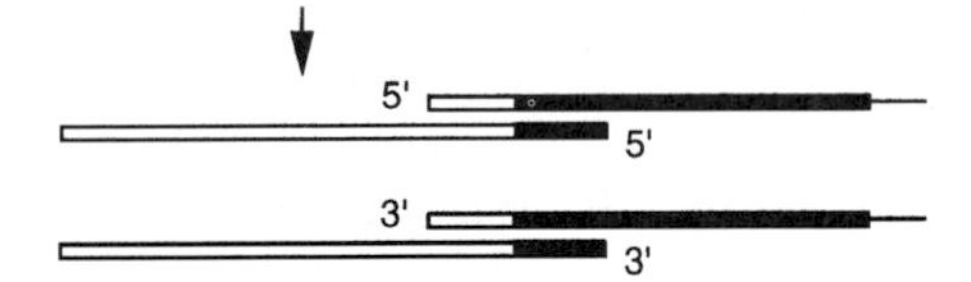

4. Extend 3' ends/PCR with primers A/D

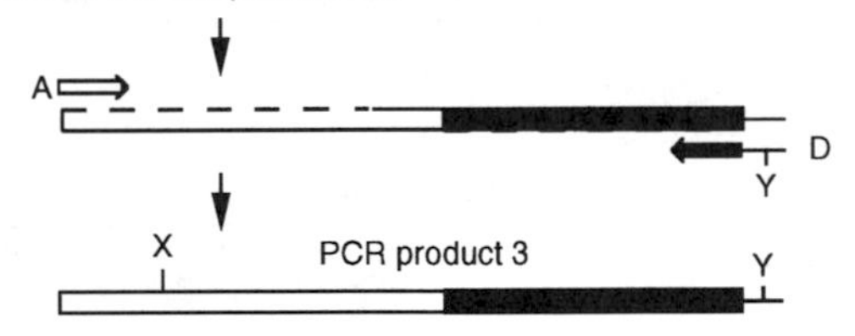

5. Cut PCR product 3 with X and Y; replace X-Y segment in receptor 1 with PCR product 3

FIG. 1. A general PCR strategy for generating hybrid GPCRs with a single chimeric junction. See text for details.

PCR products are then isolated, mixed, and used as templates (about 50 ng each) for a final PCR reaction using the two outside primers (primers A and D). The resulting recombinant PCR product (product 3) is isolated, digested with enzymes X and Y, and then used to replace fragment X–Y in the receptor 1 expression plasmid by employing standard molecular biological techniques.

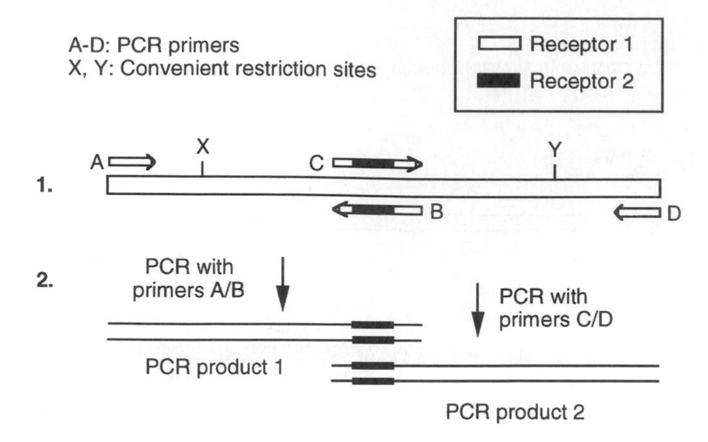

3. Isolate PCR products 1 and 2, mix, denature, and renature

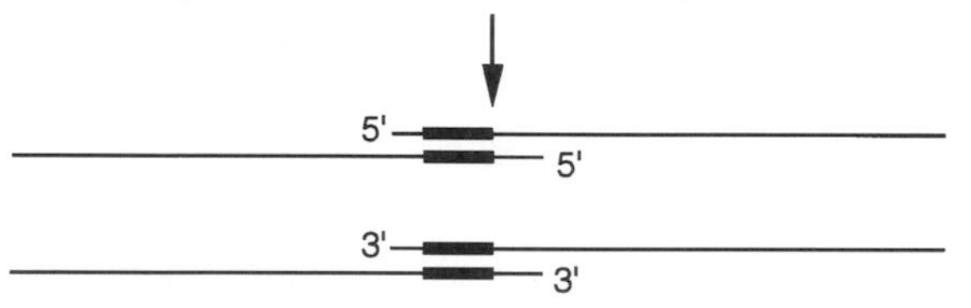

4. Extend 3' ends/PCR with primers A/D

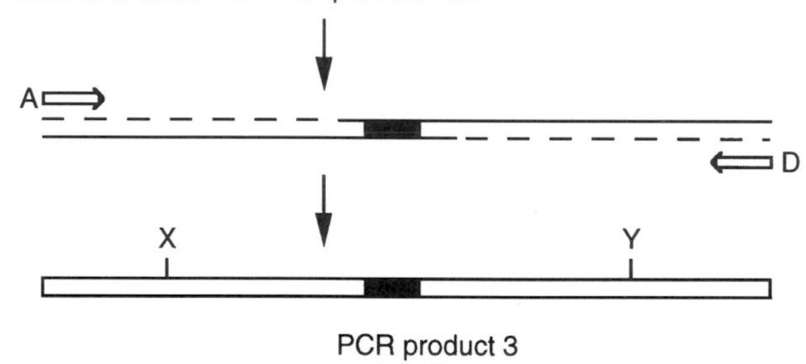

5. Cut PCR product 3 with X and Y; replace X-Y segment in receptor 1 with PCR product 3

FIG. 2. A PCR strategy for generating hybrid GPCRs with small internal replacements (example). See text for details.

Generation of Chimeric GPCRs with Small Internal Replacements

Modified versions of the recombinant PCR protocol outlined in Fig. 1 can be employed to generate mutant GPCRs with two chimeric junctions in which an internal segment of receptor 1 is replaced with the corresponding segment derived from receptor 2. Figure 2 illustrates a multistep PCR protocol that can be used to generate small internal substitutions (about 40 bp or less). The general strategy is similar to that depicted in Fig. 1, except that receptor 1 DNA serves as the template for both initial PCR reactions. The mutagenic oligonucleotides, primers B and C, are complementary to each other. The 3′ and 5′ ends of primers B and C contain about 15–18 bases of receptor 1 sequence. The sequences of the central portions of the two primers are derived from receptor 2. The length of this central mutagenic segment is dictated by the specific sequence replacement that needs to be generated. For practical reasons related to the total possible length of primers B and C, the central receptor 2 sequence normally cannot be longer than about 40 bases. All subsequent steps required to generate the final chimeric construct are essentially identical to those described in the previous paragraph.

If a useful restriction site is located close to the chimeric junction, it may be possible to incorporate this site into the 5′ end of a chimeric oligonucleotide. In this case, a simple single-step PCR procedure can be employed to generate the desired hybrid receptor.

Generation of Chimeric GPCRs with Large Internal Replacements

Recombinant PCR strategies can also be used to generate hybrid GPCRs carrying large internal substitutions. Figure 3 provides an example for a strategy that can be applied if the receptor region to be substituted is flanked by two useful restriction sites. These sites need not be unique as long as they are useful for generating restriction fragments that can be joined in a three-piece ligation to assemble the final chimeric construct. The strategy outlined in Fig. 3 requires the synthesis of two hybrid oligonucleotides, primers A and B. At their 3′ ends, the two primers contain about 20 bases of receptor 2 sequence, allowing them to anneal to the receptor 2 DNA template. The 5′ portions of primers A and B consist of receptor 1 sequence and include the receptor 1 sequences between the chimeric junctions and the two restrictions sites (X, Y), the X and Y sites, and several additional bases to allow for the efficient cleavage of the resulting PCR product by the two restriction enzymes. Based on our experience, some restriction enzymes (e.g., *Bst*XI) require at least 10 extra nucleotides for efficient cleavage, which can be added to the 5′ end of the PCR primer. If this is not possible because of the overall length of the mutagenic oligonucleotide, the PCR fragment can be extended via another round of PCR using a newly designed PCR primer that overlaps with the end of the PCR product that contains the digestion-resistant restriction site.

The method depicted in Fig. 3 represents a very straightforward mutagenesis strategy. However, in most cases, the specific receptor 1 domain that needs to be replaced is not flanked by conveniently located restrictions sites at either end that can be incorporated into the mutagenic oligonucleotides. In this situation, the basic protocol shown in Fig. 3 can be subjected to various modifications. One approach involves the initial use of site-directed mutagenesis to introduce one or two silent

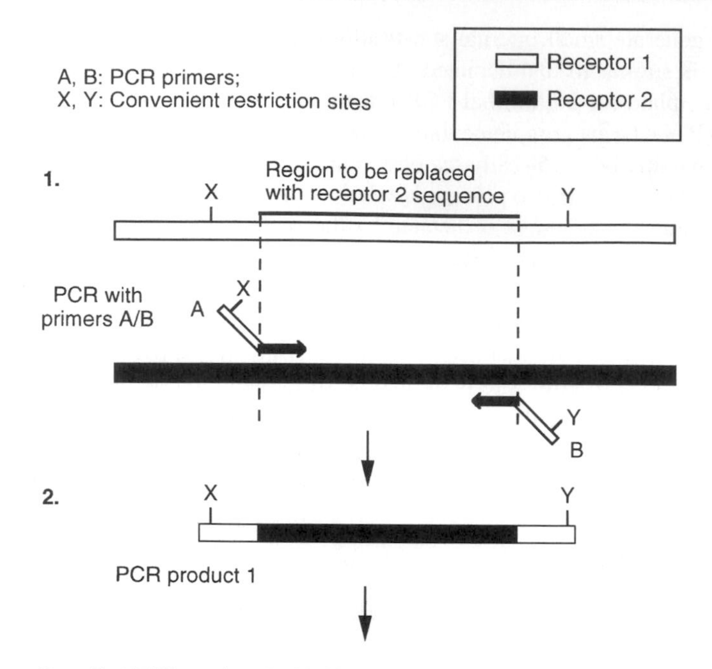

FIG. 3. A PCR strategy for generating hybrid GPCRs with large internal replacements (example). See text for details.

restriction sites into receptor 1 that flank the chimeric junctions. Alternatively, one can apply modified versions of the multistep PCR protocol outlined in Figs. 1 and 2 to generate the desired hybrid receptors.

PCR Conditions

Several precautions can be taken to reduce the frequency of undesired mutations in the final PCR product. Such measures include the use of the "hot start" technique (preheating samples at 94° for 5 min prior to the addition of the thermostable DNA polymerase) and avoiding PCR cycle numbers >30. Moreover, the use of thermostable DNA polymerases endowed with proofreading capacity (e.g., Vent DNA polymerase from New England Biolabs, Inc., or *Pfu* DNA polymerase from Stratagene) is also useful for reducing PCR error rates. However, as reported by other investigators,[2] we have observed that these high-fidelity polymerases frequently fail to generate the desired PCR products under conditions where *Taq* DNA polymerase (Perkin-Elmer/Applied Biosystems) is highly efficient in amplifying the target sequence. The likelihood of failure associated with the use of the high-fidelity thermostable DNA polymerases to generate the desired PCR product is usually directly correlated with the number of mismatches between the PCR primers and the DNA target sequence. A particular mutagenesis strategy should therefore be revised only after *Taq* polymerase fails to amplify the target sequence.

Typically, PCR reaction mixtures consist of the oligonucleotide primers (final concentration: 1 μM each), PCR buffer (10 mM Tris–HCl, pH 8.3, 50 mM KC1, 1.5 mM $MgCl_2$), deoxynucleotides (dATP, dGTP, dCTP, dTTP, 200 μM each), DNA template (about 100 ng), Amplitaq DNA polymerase (2.5 units per reaction; Perkin-Elmer/Applied Biosystems), and sterile water in a total reaction volume of 100 μl. We generally perform PCR amplifications in a Perkin–Elmer thermocycler using the following standard conditions (30 cycles total): denaturation at 94° for 1 min, primer annealing at 42–55° for 2 min, and primer extension at 72° for 2 min. The last PCR cycle is followed by a 10-min incubation at 72° to ensure that that primer extension is complete. It is recommended to choose the highest possible annealing temperature, as this will reduce the likelihood of misannealing of the PCR primers. However, if the number of mismatches between the mutagenic oligonucleotides and the receptor template is rather high, successful PCR amplification usually requires a relatively low annealing temperature ($<50°$).

To verify the success of PCR amplification, an aliquot of the PCR reaction mixture (5–10 μl) is subjected to agarose gel electrophoresis. The remaining PCR product is then purified, digested with the appropriate restriction enzymes, and used to assemble the desired chimeric construct employing standard molecular biological techniques. All PCR-derived receptor regions need to be sequenced to verify the identity of the chimeric constructs and to exclude the presence of unwanted PCR errors.

Expression of Hybrid Receptors in Mammalian Cells

To study the functional consequences of receptor domain replacements, we routinely express wild-type and mutant GPCR constructs in COS-7 cells. A DEAE/dextran transfection method that routinely gives good receptor expression levels is given in protocol 1. Immunocytochemical studies with epitope-tagged muscarinic receptor constructs have shown that this protocol results in a transfection efficiency of about 10–20%.[13] The transient COS-7 cell expression system offers the great advantage that functional data can be obtained within several days after the desired receptor constructs have been generated. In contrast, the creation of mammalian cell lines that stably express the mutant receptors under investigation is a considerably more time-consuming process. However, the availability of stable cell lines can be advantageous under conditions where the rather high number

[13] T. Schöneberg and J. Wess, unpublished results (1995).

of receptors present in a small subpopulation of cells (as is the case with transiently transfected COS-7 cells) is problematic. For example, receptor-mediated inhibition of adenylyl cyclase is often difficult to study in a transient expression system and usually requires the derivation of stable cell lines to allow for accurate measurements. However, it is recommended to always carry out an initial set of studies in a transient expression system (e.g., in COS-7 cells). If a specific hybrid (mutant) GPCR fails to display ligand-binding activity in the transient expression system, it will also fail to do so in stable cell lines. Similarly, when a particular mutant receptor is poorly expressed in transiently transfected COS-7 cells (low number of radioligand-binding sites), it is usually difficult to generate stable cell lines that express the receptor at sufficiently high levels. It may therefore be wise to exclude mutant receptors that display obvious deficits in folding and/or stability in the COS-7 cell system from further studies that require the production of stable cell lines.

Protocol 1: Transfection of COS-7 Cells with Receptor DNA

1. Grow COS-7 cells to be transfected at 37° in a humidified 5% CO_2 incubator to about 70% confluence [growth media: Dulbecco's modified Eagle's medium (DMEM) supplemented with 10% fetal bovine serum (FBS), 100 IU/ml penicillin G, and 100 μg/ml streptomycin].
2. One day before the transfection, split cells into 100-mm dishes at a density of about 1×10^6 cells/dish (a 70% confluent 175-cm^2 tissue culture flask yields about 12–14 100-mm dishes).
3. Transfect cells about 18–24 hr after seeding into 100-mm dishes.
4. Immediately before transfections, prepare the following mixture (amounts are given per 100-mm dish): 4 μg receptor DNA, 1 μg G protein α subunit DNA (optional; see text for details), 850 μl phosphate-buffered saline (PBS) with Ca^{2+} and Mg^{2+}, and 55 μl DEAE dextran (10 mg/ml in PBS). Mix well and let mixture sit at room temperature for 30–60 min.
5. Wash cells once with PBS containing Ca^{2+} and Mg^{2+} and slowly add the DNAdextran mixture along the side of the dish.
6. Tilt the dish to spread the transfection mixture evenly and incubate at 37° (5% CO_2) for 2 hr. Tilt the plate carefully every 20 min.
7. Add 7 ml of DMEM containing 10% FBS and chloroquine (80–100 μM final concentration) and incubate for 3–5 hr (37°, 5% CO_2).
8. Aspirate the chloroquine-containing medium and add fresh DMEM containing 10% FBS.
9. For radioligand-binding studies, culture the cells for an additional 2–3 days. For second messenger assays (e.g., measurement of receptor-mediated changes in inositol phosphate or cAMP levels), process the cells as described in Protocol 2.

Ligand-Binding Studies

For many different classes of GPCRs, high-affinity radioligands are available that allow for a quick and accurate determination of the number of functionally competent (properly folded) receptors. To study the expression of hybrid or other mutant muscarinic receptors, we routinely use the muscarinic antagonist *N*-[^{3}H]methylscopolamine ([^{3}H]NMS; 79.5 Ci/mmol; NEN Life Science Products) as a radioligand. [^{3}H]NMS is an excellent radioligand that displays high binding affinity for all five muscarinic receptor subtypes (M_1–M_5)[14] and shows very low nonspecific binding activity when used at standard concentrations (<2 n*M*).

To assess muscarinic receptor expression levels, we carry out [^{3}H]NMS-binding assays using membrane homogenates (about 10–50 μg membrane protein per tube) prepared from transfected COS-7 cells (see Protocol 1). The protein content of membrane preparations is determined by using the Bio-Rad protein assay kit. Incubation buffer consists of 25 m*M* sodium phosphate (pH 7.4) containing 5 m*M* $MgCl_2$ (total volume per tube: 1 ml). In [^{3}H]NMS saturation-binding experiments, we usually use six different concentrations of the radioligand (25–800 p*M*). However, it may be necessary to employ higher [^{3}H]NMS concentrations in the case of hybrid or mutant muscarinic receptors that show considerably reduced [^{3}H]NMS-binding affinities. Binding affinities of agonists or other "cold" ligands are determined in competition-binding assays. In these experiments, membranes are incubated with 10 different concentrations of the competing ligand and a fixed concentration of [^{3}H]NMS (usually 200 p*M*). Incubations are carried out for 3 hr at 22° to achieve equilibrium-binding conditions. Nonspecific binding is determined in the presence of 1 μM atropine. Bound radioactivity is separated from free by filtration through GF/C filters.

Ligand-binding affinities (equilibrium dissociation constants) are calculated from the saturation isotherms and the competition-binding curves using nonlinear least-squares curve-fitting procedures. For the analysis of saturation-binding data, we have routinely used the computer program LIGAND.[15] For the evaluation of competition-binding data, we commonly use the program KALEIDAGRAPH (Synergy Software).

Adjustment of Receptor Densities

A meaningful comparison of the functional profiles of wild-type and hybrid GPCRs requires that the mutant receptors are expressed at levels comparable to those found with the wild-type receptors. This issue is of particular relevance for quantitative studies analyzing receptor/G protein coupling and receptor-mediated

[14] F. Dörje, J. Wess, G. Lambrecht, R. Tacke, E. Mutschler, and M. R. Brann, *J. Pharmacol. Exp. Ther.* **256,** 727 (1991).

[15] P. J. Munson and D. Rodbard, *Anal. Biochem.* **107,** 220 (1980).

activation or inhibition of downstream signal transduction pathways. It is well known that the magnitude and kinetics of these latter responses strongly depend on overall receptor densities. Unfortunately, hybrid or other mutant GPCRs are frequently expressed at levels that are considerably lower than the corresponding wild-type receptor levels. This phenomenon can be due to a variety of factors, including reduced receptor stability and/or impaired trafficking of mutant receptors to the cell surface.

However, use of the COS-7 cell expression system allows the application of a simple strategy by which wild-type receptor levels can be reduced progressively to levels found with specific mutant receptors, thus allowing for a meaningful analysis of mutant GPCR phenotypes. This can be achieved by simply reducing the amount of transfected wild-type receptor DNA. However, it is essential that the total amount of transfected plasmid DNA (4 μg) is kept constant by adding the appropriate amount of vector DNA (pcD-PS). An example illustrating this approach is depicted in Fig. 4. Figure 4 demonstrates that the expression levels of the rat M_3

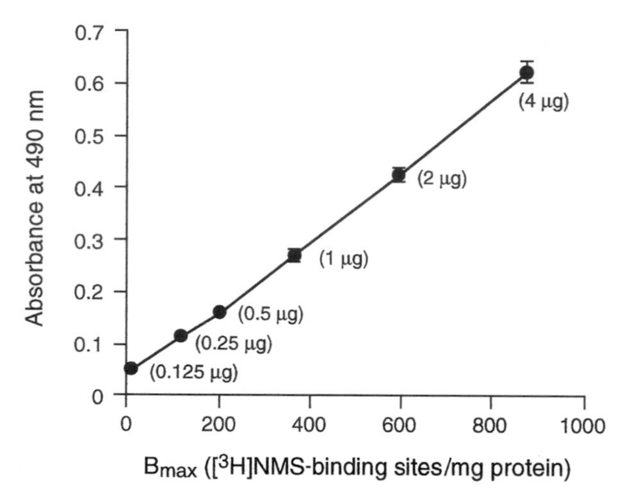

FIG. 4. Relationship among amount of transfected receptor DNA, maximum number of radioligand-binding sites, and signal detected in a cell surface ELISA. COS-7 cells grown in 100-mm dishes were transfected with different amounts (0.125–4 μg) of a mammalian expression plasmid coding for rat M_3 muscarinic receptor containing an N-terminal HA epitope tag (Rm3pCD-N-HA). The transfected receptor DNA was supplemented with vector DNA (pcD-PS) to keep the total amount of transfected plasmid DNA constant at 4 μg. B_{max} values were determined in [^{3}H]NMS saturation-binding studies as described in the text. To obtain a measure of receptor cell surface expression that is independent of ligand-binding activity, aliquots of transfected cells were subjected to the cell surface ELISA detailed in Protocol 3. This assay is carried out with intact COS-7 cells and is based on the detection of the extracellular HA epitope tag by the 12CA5 monoclonal antibody (see text for details). ELISA readings (absorbance at 490 nm; means $\pm$ SEM) were plotted against mean B_{max} values. These studies reveal a linear relationship between B_{max} values and the magnitude of the ELISA signals (adapted from Ref. 16).

muscarinic receptor (as determined in [^{3}H]NMS saturating-binding studies) can be reduced gradually by the transfection of COS-7 cells with progressively smaller amounts of receptor DNA.[16]

Functional Analysis of Hybrid Receptors in COS-7 Cells

Once conditions have been established that allow for the expression of wild-type and mutant receptors at comparable levels, experiments can be carried out to study the specific receptor function of interest (ligand binding and G protein coupling profiles, receptor phosphorylation, desensitization, internalization, or downregulation, etc.). During the past decade, for example, we have generated and analyzed a large number of M_2/M_3 hybrid muscarinic receptors to identify residues that are critical for the muscarinic receptor-mediated activation of G proteins of the $G_{q/11}$ family (this group of G proteins, in contrast to G proteins of the $G_{i/o}$ class, cannot be inactivated by treatment with pertussis toxin[17,18]). The M_3 muscarinic receptor, but not the M_2 subtype, is known to productively couple to $G_{q/11}$-type G proteins, whereas the M_2 muscarinic receptor preferentially interacts with G proteins of the $G_{i/o}$ family.[3–6] Therefore, we systematically substituted distinct M_3 receptor domains (or single M_3 receptor residues) into the M_2 receptor background and analyzed the resulting hybrid receptors for their ability to gain coupling to G_q (G_{11}). Activation of this class of G proteins is known to result in the breakdown of phosphoinositide (PI) lipids mediated by the activation of specific subtypes of phospholipase Cβ.[17,18] As a measure of G_q (G_{11}) activation, many investigators therefore examine the receptor-mediated generation of inositol phosphates (this response should remain unaffected by pretreatment of cells with pertussis toxin). We use the following protocol routinely to measure receptor-mediated PI hydrolysis in a quick and reliable fashion in transfected COS-7 cells.

Protocol 2: Receptor-Mediated PI Hydrolysis Studied with Transfected COS-7 Cells

1. About 18–24 hr after transfections (see Protocol 1), split cells into six-well dishes (approximately 0.4×10^6 cells per well) and add 3μCi/ml [*myo*-^{3}H]inositol (20 Ci/mmol) (NEN Life Science Products).
2. Label cells for 24–36 hr (37°, 5% CO_2).

[16] T. Schöneberg, J. Liu, and J. Wess, *J. Biol. Chem.* **270,** 18000 (1995).

[17] A. M. Spiegel, *in* "G proteins" (A. M. Spiegel, T. L. Z. Jones, W. F. Simonds, and L. S. Weinstein, eds.), Molecular Biology Intelligence Unit, p. 18, R. G. Landes Company, Austin, Texas, 1994.

[18] S. Rens-Domiano and H. E. Hamm, *FASEB J.* **9,** 1059 (1995).

3. After the labeling period, wash cells once with Hanks' balanced salt solution containing 20 m*M* HEPES and add 1 ml of the same media containing 10 m*M* LiCl to each well.

4. Shake cells gently on an orbital shaker for 20 min at room temperature.

5. Add the appropriate amount of agonist and/or antagonist ligand and incubate for 1 hr at 37°.

6. Aspirate the ligand-containing medium and add 750 μl of ice-cold 20 m*M* formic acid.

7. Incubate for 40 min at 4°.

8. Neutralize with 250 μl of 60 m*M* ammonium hydroxide.

9. Collect cell extracts and isolate the inositol monophosphate fraction via anion-exchange chromatography as originally described by Berridge *et al.*[19]

Coexpression of Hybrid Receptors with Promiscuous G Proteins

Proper analysis of functional data obtained with mutant or hybrid GPCRs requires that these receptors adopt a fold that is similar to that of the corresponding wild-type receptors. This is of particular relevance when mutant receptors are studied that have lost or fail to acquire the specific receptor function under investigation. Obviously, the complete loss of radioligand-binding activity is indicative of a major folding defect. However, it is possible that the ligand-binding domain of a specific mutant GPCR is properly arranged while its intracellular G protein-binding surface is misfolded.

In previous studies, for example, we have identified a series of hybrid M_2/M_3 muscarinic receptors that can bind muscarinic ligands normally but are unable to stimulate the hydrolysis of PI lipids (see, e.g., Ref. 20). To be able to interpret these data properly, one needs to demonstrate that such mutant receptors are principally capable of interacting with G proteins. We therefore coexpressed hybrid M_2/M_3 muscarinic receptors that failed to mediate agonist-dependent stimulation of PI hydrolysis with $G\alpha_{15}$ (mouse),[21] a G protein α subunit that is not expressed endogenously by COS-7 cells (see Protocol 1). This G protein, like its human homologue $G\alpha_{16}$, can be activated by almost all GPCRs, including the M_2 muscarinic receptor,[22] resulting in the breakdown of PI lipids mediated by the activation of distinct isoforms of phospholipase $C\beta$.[23] Coexpression studies with hybrid

[19] M. J. Berridge, M. C. Dawson, C. P. Downes, J. P. Heslop, and R. F. Irvine, *Biochem. J.* **212,** 473 (1983).

[20] E. Kostenis, B. R. Conklin, and J. Wess, *Biochemistry* **36,** 1487 (1997).

[21] T. M. Wilkie, P. A. Scherle, M. P. Strathmann, V. Z. Slepak, and M. I. Simon, *Proc. Natl. Acad. Sci. U.S.A.* **88,** 10049 (1991).

[22] S. Offermanns and M. I. Simon, *J. Biol. Chem.* **270,** 15175 (1995).

[23] C. H. Lee, D. Park, D. Wu, S. G. Rhee, and M. I. Simon, *J. Biol. Chem.* **267,** 16044 (1992).

M_2/M_3 muscarinic receptors showed that analysis of receptor-mediated activation of G_{15} (or G_{16}) represents a useful strategy to verify the proper overall fold of the intracellular receptor surface.[20]

Quantitation of Cell Surface Receptor Expression in the Absence of Ligand-Binding Activity

In many cases, transfection of mammalian cells with hybrid GPCR constructs does not lead to detectable levels of radioligand-binding activity. As discussed earlier, this phenomenon is most likely to occur when the two parent receptors show a low degree of sequence homology, the chimeric junctions are located within the TM helices, or when the TM receptor core contains TM helices derived from both parent receptors. By definition, hybrid receptors that lack ligand-binding activity are misfolded (see earlier discussion), assuming that both wild-type receptors are able to bind the employed radioligand with high affinity. Improper folding frequently results in intracellular retention of the mutant GPCRs.[24,25] To examine whether binding-defective hybrid receptors (or mutant GPCRs in general) are transported to the cell surface, various techniques can be employed.

To study the subcellular distribution of wild-type and mutant GPCRs, we and others have used epitope-tagged GPCR constructs. We have found that addition of the nine amino acid hemagglutinin (HA) epitope tag (YPYDVPDYA)[26] to the N terminus of a variety of GPCRs has little effect on receptor function.[16,25] The presence of the N-terminal HA tag allows the analysis of cell surface expression of wild-type and mutant GPCRs by using the 12CA5 mouse monoclonal antibody (Roche Molecular Biochemicals), which specifically recognizes the HA sequence. If the mutant receptors are present on the cell surface, the epitope tag protrudes into the extracellular space and can be recognized by the 12CA5 antibody using intact cells. Using this strategy, cell surface receptors can be visualized by immunofluorescence microscopic techniques.[16,25] Alternatively, one can apply an ELISA strategy that is based on a similar immunological approach but allows quantitation of surface receptor densities. An ELISA protocol that we have successfully applied to different classes of GPCRs is given in Protocol 3.[16] Importantly, this ELISA procedure allows the determination of cell surface receptor densities in the absence of ligand-binding activity. In many cases, such information is useful to interpret the results of studies that are designed as "loss-of-function studies." Moreover, such

[24] C.-H. Sung, B. G. Schneider, N. Agarwal, D. S. Papermaster, and J. Nathans, *Proc. Natl. Acad. Sci. U.S.A.* **88,** 8840 (1991).

[25] T. Schöneberg, J. Yun, D. Wenkert, and J. Wess, *EMBO J.* **15,** 1283 (1996).

[26] P. A. Kolodziej and R. A. Young, *Methods Enzymol.* **194,** 508 (1991).

knowledge is also useful to properly interpret the phenotypes of nonfunctional hybrid (mutant) receptors and to elucidate the molecular mechanisms involved in the trafficking of GPCRs to the cell surface.

Protocol 3: Quantitation of Cell Surface Receptor Densities via ELISA

1. Transfect COS-7 cells with 4 μg of plasmid DNA coding for a GPCR containing an HA epitope tag at its N terminus (see Protocol 1).
2. About 18–24 hr after transfections, split cells into 96-well plates (about 2.5×10^4 cells/well) and culture in a 5% CO_2 incubator at 37° for 48 hr.
3. Fix cells with 4% formaldehyde (Sigma) in PBS for 30 min at room temperature and wash cells twice with PBS (180 μl/well).
4. Add 180 μl 10% FBS in DMEM to each well and incubate at 37° for 1 hr.
5. Add 100 μl of 12CA5 monoclonal antibody (Roche Molecular Biochemicals) (20 μg/ml in DMEM with 10% FBS) to each well and incubate at 37° for 2 hr.
6. Wash plates four times with PBS (180 μl/well).
7. Add 100 μl of goat antimouse IgG antibody conjugated with horseradish peroxidase (Amersham Life Science) (1:1000 dilution in DMEM with 10% FBS) and incubate at 37° for 1 hr.
8. Wash plates four times with PBS (180 μl/well).
9. Add 100 μl of "color development solution" [2.5 m*M* H_2O_2 and 2.5 m*M* *o*-phenylenediamine (Sigma) in 0.1 *M* phosphate-citrate buffer, pH 5.0] and incubate at room temperature for 15–30 min.
10. Add 50 μl of 1 *M* H_2SO_4 containing 0.05 *M* Na_2SO_3 to terminate color development.
11. Measure optical densities at 490 and 630 nm (background) using the BioKinetics reader (EL 312, Bio Tek Instruments, Inc.) or an equivalent apparatus from another manufacturer.

To correlate ELISA signals with actual amounts of cell surface receptors, the wild-type M_3 muscarinic receptor containing an N-terminal HA epitope tag was expressed at different densities (B_{max}, determined in [^{3}H]NMS saturation-binding studies). This was achieved by stepwise reduction of the amount of transfected receptor DNA (Fig. 4). In parallel, aliquots of transfected COS-7 cells were subjected to the ELISA protocol (Protocol 3). This analysis showed that the ELISA signals (absorbance at 490 nm) were directly correlated with the number of radioligand-binding sites (B_{max}) (Fig. 4). A similar relationship has been observed for other classes of GPCRs, such as the V2 vasopressin receptor.[27] The strategy described

[27] T. Schöneberg, V. Sandig, J. Wess, T. Gudermann, and G. Schultz, *J. Clin. Invest.* **100,** 1547 (1997).

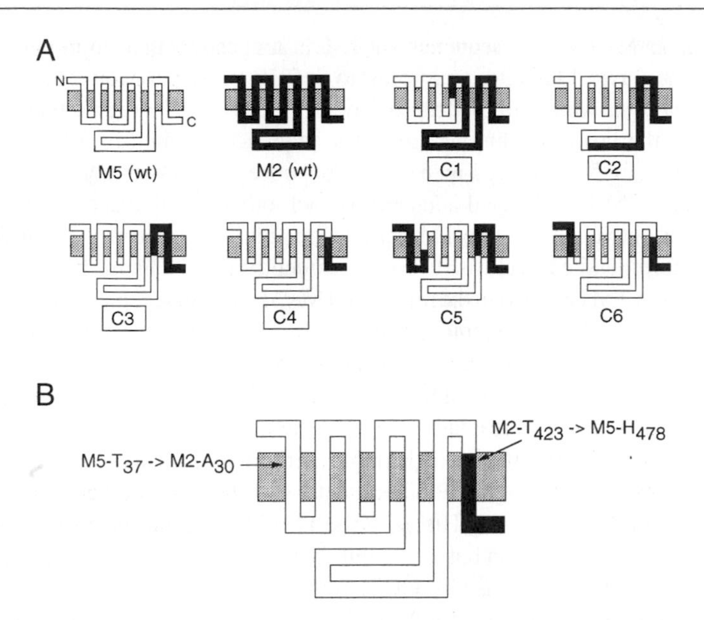

FIG. 5. Structure of hybrid M_2/M_5 muscarinic receptors. (A) In contrast to the two wild-type receptors and the C5 and C6 hybrid constructs, the C1–C4 mutant receptors (shown boxed) are unable to bind significant amounts of $[^3H]$NMS when expressed in COS-7 cells.[29,30] (B) Either of the two indicated amino acid substitutions (M_5-Thr_{37}→M_2-Ala_{30} or M_2-Thr_{423}→M_5-His_{478}) restores $[^3H]$NMS-binding activity to the C4 hybrid construct.[30] The precise composition of the hybrid receptors is given in Refs. 29 and 30 (adapted from Ref. 30).

here should be generally useful to examine whether binding-defective mutant GPCRs are properly transported to the cell surface.

Potential Usefulness of Misfolded Hybrid Receptors to Study Mechanisms of GPCR Assembly

Several studies have shown that functionally inactive hybrid receptors can serve as useful tools to gain insight into GPCR structure.[28–31] For example, while analyzing a series of M_2/M_5 hybrid muscarinic receptors, we identified several mutant receptors (C1–C4 in Fig. 5) that were unable to bind significant amounts of muscarinic radioligands, such as $[^3H]$NMS.[29,30] A common structural feature shared by the C1–C4 hybrid receptors is that they contain M_2 receptor sequence in

[28] S. Suryanarayana, M. von Zastrow, and B. K. Kobilka, *J. Biol. Chem.* **267,** 21991 (1992).
[29] Z. Pittel and J. Wess, *Mol. Pharmacol.* **45,** 61 (1994).
[30] J. Liu, T. Schöneberg, M. van Rhee, and J. Wess, *J. Biol. Chem.* **270,** 19532 (1995).
[31] T. Mizobe, M. Maze, V. Lam, S. Suryanarayana, and B. K. Kobilka, *J. Biol. Chem.* **271,** 2387 (1996).

TM VII and M_5 receptor sequence in TM I, suggesting that improper packing between these two TM helices may lead to deficits in receptor folding. Consistent with this notion, we found that replacement in C1–C4 of a region that included TM I with the corresponding M_2 receptor sequence resulted in fully functional receptors (e.g., C5 and C6 in Fig. 5).[29,30] This observation strongly suggested that TM VII and TM I are located adjacent to each other and that specific molecular interactions between these two TM helices are required for proper GPCR folding.

In a more detailed analysis,[30] we subsequently targeted specific amino acids located within TM I and VII in the misfolded M_2/M_5 hybrid receptors. Strikingly, all misfolded M_2/M_5 hybrid receptors could be functionally rescued by introduction of a single point mutation into either TM I (M_5-$Thr_{37}\rightarrow M_2$-Ala_{30}) or TM VII (M_2-$Thr_{423}\rightarrow M_5$-His_{478}) (shown for C4 in Fig. 5).[30] These data are consistent with the concept that these two residues are located at the TM I/TM VII interface where they are engaged in structurally important interhelical interactions. Helical wheel models[32,33] of the TM core of GPCRs indicate that the likely location of M_5-Thr_{37} (M_2-Ala_{30}) and M_2-Thr_{423} (M_5-His_{478}) at the TMI/TM VII interface is in agreement with a clockwise orientation but is absolutely incompatible with an anticlockwise arrangement of TM I–VII (as viewed from the extracellular space). These studies provide an instructive example for how mutational analysis of misfolded chimeric or other mutant GPCRs can help elucidate details of GPCR structure.

Concluding Remarks

Studies with hybrid GPCRs have greatly advanced our knowledge of the structural elements involved in many aspects of GPCR function. However, as outlined in this article, a meaningful interpretation of hybrid receptor data requires that experimental conditions are carefully chosen, such that differences in receptor numbers or improper receptor folding or localization do not interfere with proper data analysis. Given the extraordinary size of the GPCR superfamily and the ease with which chimeric GPCRs can be created by the use of modern PCR technology, hybrid GPCRs will continue to serve as important tools for researches in the GPCR field for many years to come.

Acknowledgments

The contributions by Drs. T. Schöneberg and E. Kostenis to the development and optimization of several of the protocols described in this article are gratefully acknowledged.

[32] J. M. Baldwin, *EMBO J.* **12,** 1693 (1993).
[33] J. M. Baldwin, G. F. X. Schertler, and V. M. Unger, *J. Mol. Biol.* **272,** 144 (1997).

[21] Strategies for Modeling the Interactions of Transmembrane Helices of G Protein-Coupled Receptors by Geometric Complementarity Using the GRAMM Computer Algorithm

By ILYA A. VAKSER and SULIN JIANG

Introduction

A significant number of proteins in a genome are membrane proteins (20–30%[1]). The existing direct structural information on membrane proteins derived from experimental techniques (X-ray crystallography, nuclear magnetic resonance spectroscopy and others) is not enough to draw conclusions on many aspects of their structure and function. In sharp contrast with soluble proteins, where more than 13,000 three-dimensional (3D) structures are presently known,[2] only a handful of membrane protein structures have been determined. Moreover, only one of these structures, published very recently, is a G protein-coupled receptor (GPCR) protein (rhodopsin[3]). At the same time, more than a 1000 GPCR sequences are presently known.[4] This emphasizes the importance of modeling approaches to structural studies of GPCRs. The problem with modeling GPCR structure is that many of the GPCRs have the degree of sequence similarity to rhodopsin lower than the one needed to use the 3D structure of rhodopsin as the modeling template (GPCRDB, http://www.gpcr.org). Although all of them have the same structural motif—a seven transmembrane (TM) helix bundle—the rest of the structure, including the relative orientation of the TM helices, is largely unknown. Thus, methods not based on structural templates (*ab initio* methods) are important for GPCR modeling. There have been numerous attempts to model the GPCR structures (see, e.g., a review of Ballesteros and Weinstein[5]). Generally, there are four aspects in GPCR modeling: (1) identification of TM segments in the sequence, (2) 3D modeling of TM helices, (3) packing of TM helices in a bundle, and (4) modeling of intra- and extracellular loops. These modeling steps are, to a certain degree, similar to the structure prediction for soluble proteins. The low accuracy of soluble protein modeling, in cases of low homology or no homology to known structures, is

[1] E. Wallin and G. Heijne, *Protein Sci.* **7,** 1029 (1998).

[2] H. M. Berman, J. Westbrook, Z. Feng, G. Gilliland, T. N. Bhat, H. Weissig, I. N. Shindyalov, and P. E. Bourne, *Nucleic Acids Res.* **28,** 235 (2000).

[3] K. Palczewski, T. Kumasaka, T. Hori, C. A. Behnke, H. Motoshima, B. A. Fox, I. Le Trong, D. C. Teller, T. Okada, R. E. Stenkamp, M. Yamamoto, and M. Miyano, *Science* **289,** 739 (2000).

[4] F. Horn and G. Vriend, *J. Mol. Mod.* **76,** 464 (1998).

[5] J. A. Ballesteros and H. Weinstein, *in* "Methods in Neurosciences," Vol. 25, p. 366. Academic Press, New York, 1995.

METHODS IN ENZYMOLOGY, VOL. 343
0076-6879/02 $35.00

well known.[6,7] Thus, the expectations that in cases of no strong homology to rhodopsin, GPCR modeling alone can deliver the atomic resolution precision, are premature. The solution to this problem is to complement the modeling by a system of experimentally derived constraints.[8]

Because of the physicochemical environment similarity, it is reasonable to assume the similarity of TM packing principles in all membrane proteins, including GPCRs. Thus, to develop methods for GPCR modeling, one can use the knowledge derived from other membrane proteins. Similarly, the prediction methods that work for non-GPCR membrane proteins should work for GPCRs.

The TM helix bundle is the core of the GPCR structure. The packing of TM helices is the basis of the GPCR function (signal transduction through the GPCR structure). It is also structurally important for the folding of intra- and extracellular loops. The helices in the bundle are tightly packed.[9] The packing has a number of common features in different membrane proteins.[10] Factors determining helices interactions—steric complementarity, electrostatic interactions, and hydrogen bonding—appear to be highly correlated with the interaction energy so that they can be used separately to predict the packing. Methods for helix packing prediction, which are based on electrostatic interactions[11] and hydrogen bonding,[12] have been developed. This article presents our method of modeling TM helix packing based on steric interactions alone. The method may be used to predict the low-resolution structure of the helix bundle. The method is independent from other physicochemical and experimental considerations. Thus, it can be used in combination with other computational and experimental techniques to obtain a higher resolution structure. The method has been developed and validated on several known structures of TM helix bundles, with the assumption that packing principles for all TM bundles, including those of GPCRs, are similar.

Matching Procedure

Similarity of Packing and Docking

The packing of two protein fragments (helices) is similar to the docking of two separate molecules. Thus, the methods developed for protein docking are applicable to the prediction of the helix bundle structure. Although the bundle

[6] C. A. Orengo, J. E. Bray, T. Hubbard, L. LoConte, and I. Sillitoe, *Proteins Suppl.* **3,** 149 (1999).
[7] A. G. Murzin, *Proteins Suppl.* **3,** 88 (1999).
[8] P. Herzyk and R. E. Hubbard, *Biophys. J.* **69,** 2419 (1995).
[9] M. Eilers, S. C. Shekar, T. Shieh, S. O. Smith, and P. J. Fleming, *Proc. Natl. Acad. Sci. U.S.A.* **97,** 5796 (2000).
[10] J. U. Bowie, *J. Mol. Biol.* **272,** 780 (1997).
[11] M. Suwa, T. Hirokawa, and S. Mitaku, *Proteins* **22,** 363 (1995).
[12] I. D. Pogozheva, A. L. Lomize, and H. I. Mosberg, *Biophys. J.* **70,** 1963 (1997).

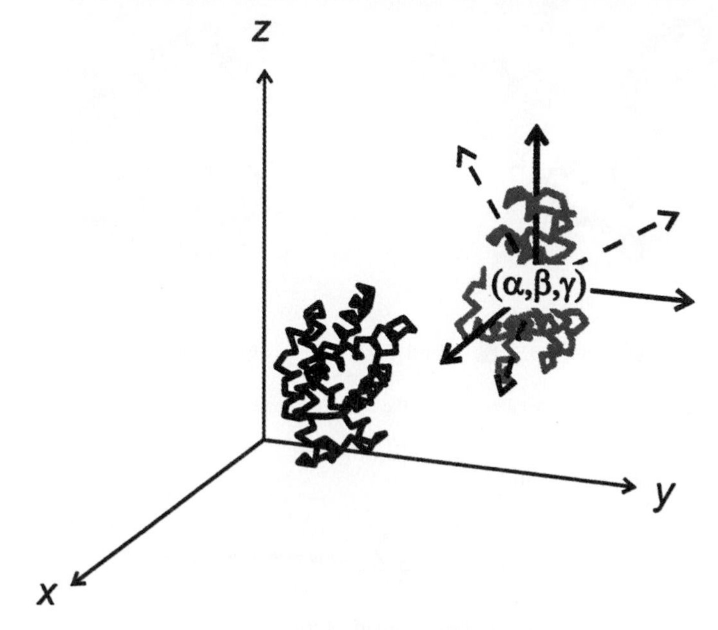

FIG. 1. The system of docking coordinates.

involves several helices, the analysis of known transmembrane structures suggests that the bundle can be approximated by a system of binary helix–helix interactions.[13] The main principle of a binary complex prediction (docking) is illustrated in Fig. 1. One fragment is considered stationary and the other fragment is translated in three dimensions and rotated by three angles, in search for the best fit between the fragments.

Docking Algorithm

We developed a geometric recognition algorithm for protein docking.[14] It predicts the structure of a complex by maximizing surface overlap between the two molecular images. The images are digital representations of molecular shape that distinguish between the surface and the interior of a molecule. The images are obtained by projecting the 3D atomic structures of the molecules A and B on an $N \times N \times N$ grid so that functions $a_{l,m,n}$ and $b_{l,m,n}$ $(l,m,n = 1, \ldots, N)$ represent molecule A and B, respectively (Fig. 2). The algorithm is based on the correlation

[13] S. Jiang and I. A. Vakser, *Proteins* **40,** 429 (2000).

[14] E. Katchalski-Katzir, I. Shariv, M. Eisenstein, A. A. Friesem, C. Aflalo, and I. A. Vakser, *Proc. Natl. Acad. Sci. U.S.A.* **89,** 2195 (1992).

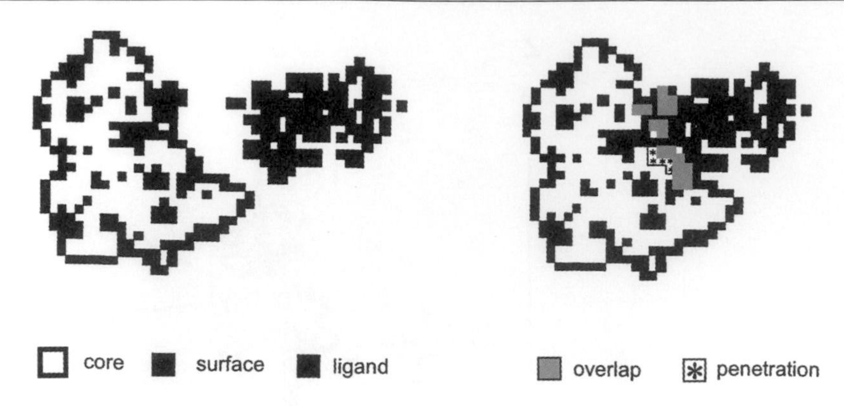

FIG. 2. Cross section through the 3D representation of the molecules.

between the functions a and b:

$$C_{x,y,z} = \sum_{l=1}^{N} \sum_{m=1}^{N} \sum_{n=1}^{N} a_{l,m,n} b_{l+x,m+y,n+z}$$

where x, y, and z are the number of grid steps by which molecule B is shifted with respect to molecule A in each dimension. The correlation function is calculated rapidly using fast Fourier transformation. The procedure is repeated for all combinations of angles α, β, and γ, which determine the molecule B orientation. Correlation peaks found by the procedure indicate geometric match and thus represent a potential complex. The procedure is equivalent to the full six-dimensional search (three translations and three rotations of the molecule B) but is much faster by design, and the computation time is only moderately dependent on molecular size. Since our paper on the FFT application to protein docking,[14] there have been many reports by other groups on successful application of this methodology to the docking problem.[15–23]

The practical application of the algorithm to independently determined protein structures faced major problems. The two most significant problems were the

[15] R. W. Harrison, I. V. Kourinov, and L. C. Andrews, *Protein Eng.* **7,** 359 (1994).

[16] L. F. Ten Eyck, J. Mandell, V. A. Roberts, and M. E. Pique, *in* "ACM/IEEE Supercomputing Conference," San Diego, CA, 1995.

[17] F. Ackermann, G. Herrmann, F. Kummert, S. Posch, G. Sagerer, and D. Schromburg, *in* "Intelligent Systems for Molecular Biology" (C. Rawlings, D. Clark, R. Altman, L. Hunter, T. Lengauer, and S. Wodak, eds.), p. 3. AAAI Press, Menlo Park, CA, 1995.

[18] M. Meyer, P. Wilson, and D. Schomburg, *J. Mol. Biol.* **264,** 199 (1996).

[19] N. S. Blom and J. Sygusch, *Proteins* **27,** 493 (1997).

[20] J. M. Friedman, *Protein Eng.* **10,** 851 (1997).

[21] H. A. Gabb, R. M. Jackson, and M. J. E. Sternberg, *J. Mol. Biol.* **272,** 106 (1997).

[22] A. A. Bliznyuk and J. E. Gready, *J. Comput. Chem.* **20,** 983 (1999).

[23] D. W. Ritchie and G. J. L. Kemp, *Proteins* **39,** 178 (2000).

structural inaccuracy of the proteins and the multiple-minima problem (multiple false-positive matches). These factors interfered with the correct prediction of the structure of the complex.

Low-Resolution Docking

The problems of structural inaccuracy and multiplicity of false-positive matches were addressed in the low-resolution rigid body docking approach.[24–26] Low-resolution docking is a transition to larger grid steps, which is equivalent to the elimination of smaller structural features. Its utility is based on the assumption that surface complementarity involves structural elements of different size. After the elimination of smaller (atom-size) elements, the presence of larger structural motifs is still sufficient for the recognition (with lower precision).

Overlap of the low-resolution molecular images was shown to be equivalent to the intermolecular energy calculated with a long-range step function potential[27]:

$$F = \sum_{i,j} E(r_{ij}), \qquad E(r_{ij}) = \begin{cases} U, & 0 < r_{ij} \leq R \\ -1, & R < r_{ij} \leq 2R \\ 0, & r_{ij} > 2R \end{cases}$$

where E is the energy step function, U is the height of the repulsion part of the potential, R is the range of the potential (the grid step), and r_{ij} is the distance between atoms i and j. From the point of view of intermolecular energy, low-resolution recognition is based on the smoothing of the energy landscape that reveals the funnel in the energy landscape.[28,29] An application of an alternative energy-smoothing approach to helix–helix interaction was described by Pappu *et al.*[30]

Figure 3a illustrates the problem in rigid body docking with "accurate" potentials, which approximate the Lennard–Jones potentials. Even an "ideal" search procedure will give a wrong answer because conformational changes in the complementarity region will move the global minimum of energy to a different place. The longer ranges of the potential (in our definition) smooth the energy profile, making smaller conformational changes negligible. The global minimum remains in place. However, its width increases. Thus, precision of the predicted ligand position becomes lower (Fig. 3b). The low-resolution docking approach eliminates the multiple-minima problem in protein interactions, including helix–helix interactions, by consolidating low-energy matches in the bottom of the smoothed energy funnel (Fig. 4). An important consequence of elimination of high-resolution

[24] I. A. Vakser, *Protein Eng.* **8,** 371 (1995).

[25] I. A. Vakser, *Biopolymers* **39,** 455 (1996).

[26] I. A. Vakser and G. V. Nikiforovich, *in* "Methods in Protein Structure Analysis" (M. Z. Atassi and E. Appella, eds.), p. 505. Plenum Press, New York, 1995.

[27] I. A. Vakser, *Protein Eng.* **9,** 37 (1996).

[28] C.-J. Tsai, S. Kumar, B. Ma, and R. Nussinov, *Protein Sci.* **8,** 1181 (1999).

[29] K. A. Dill, *Protein Sci.* **8,** 1166 (1999).

[30] R. V. Pappu, G. R. Marshall, and J. W. Ponder, *Nature Struct. Biol.* **6,** 50 (1999).

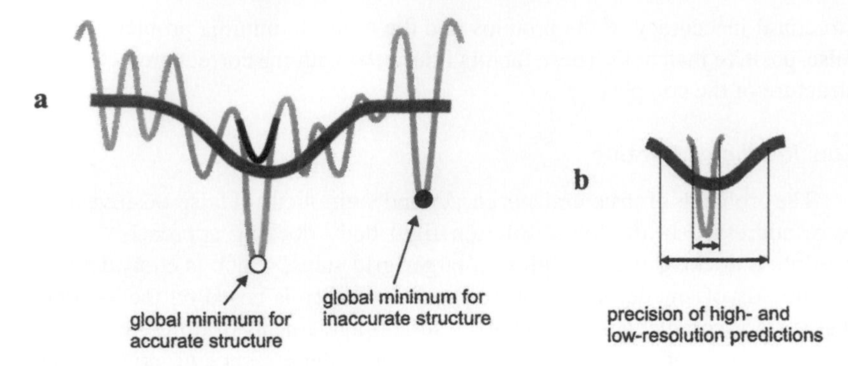

FIG. 3. Schematic intermolecular energy profiles in high- and low-resolution docking. In high-resolution docking (light gray thin curve), when the accurate structure is substituted by the inaccurate/uncomplexed structure (the change in the energy profile is shown in black), the global energy minimum moves to another location. The energy profile in low-resolution docking (dark gray bold curve) reflects only the underlying funnel-like energy landscape (a). The smoothing of energy profile corresponds to wider energy minima and, thus, to lower precision in their prediction (b).

recognition details is that a single "exact" match splits into multiple "approximate" matches[24] (e.g., multiple ligand orientations within the same receptor site). This effect creates a possibility of predicting correct contact residues on one molecule, with often incorrect contact residues on the other molecule.

The methods just described are implemented in our program GRAMM (http://reco3.musc.edu). It has been installed in more than 500 academic and industrial sites, where it is used for protein docking and packing helices in TM bundles.

Results

Docking of Cocrystallized Helices: Role of Steric Interactions

GRAMM was applied at high resolution (1.7 Å grid step) to recreate the 3D structure of the L subunit of the photosynthetic reaction center 1prc.[31] The practical utility of a high-resolution rigid body docking for TM bundle modeling is limited because of structural uncertainties, including conformational changes, in the helices. However, because GRAMM is based on geometric complementarity only, it is possible to reveal the role of steric interactions in the packing of TM helices. A significant role of steric interactions would allow us to explore geometry-based low-resolution and/or flexible matching techniques for the prediction of unknown GPCR structures.

The 1prc structure is shown in Fig. 5a. Visual examination of the L subunit showed tight packing of helix pairs 1–2, 2–3, and 4–5. Helices from these pairs were separated and redocked. The docking results were compared with the experimental

[31] J. Deisenhofer, O. Epp, I. Sinning, and H. Michel, *J. Mol. Biol.* **246,** 429 (1995).

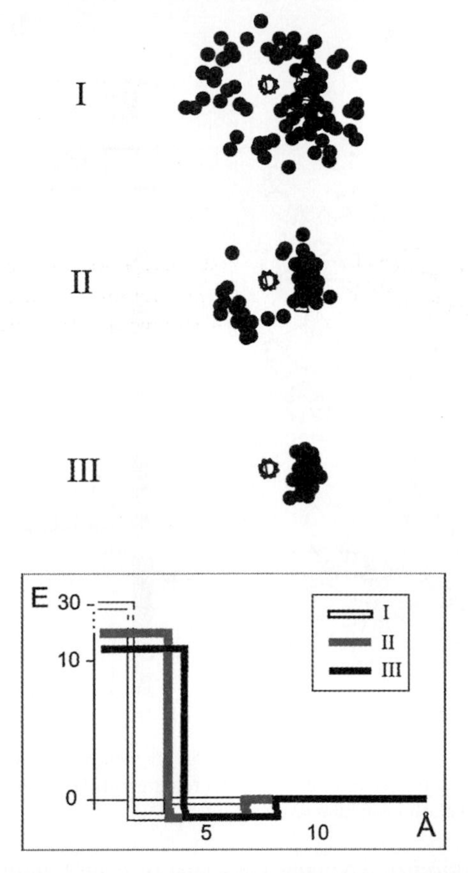

FIG. 4. Low-energy configurations of a helix–helix complex. The first 100 lowest energy positions of helix 2 in complex with helix 1 from the L subunit of the photosynthetic reaction center. The pair of helices is shown in the crystallographically determined configuration. Black circles represent the predicted positions of the helix 2 center of gravity. The potential function was varied from short (with characteristic sizes similar to Lennard–Jones potentials) to longer ranges, corresponding to lines I–III in the lower panel. Positions predicted with longer ranges of the potential consolidate in the area of the helix–helix energy minimum (the position of helix 2 in the experimental structure).

structure. Results (Fig. 5b) showed that in all three cases the correct prediction scored first in the list of all matches. This confirmed our assumption that steric interactions are strongly correlated with the interhelical interaction energy, and thus can be used to predict the structure of the helix bundle.

Problems with High-Resolution Docking

There are four major problems in the application of high-resolution docking to TM bundle prediction.

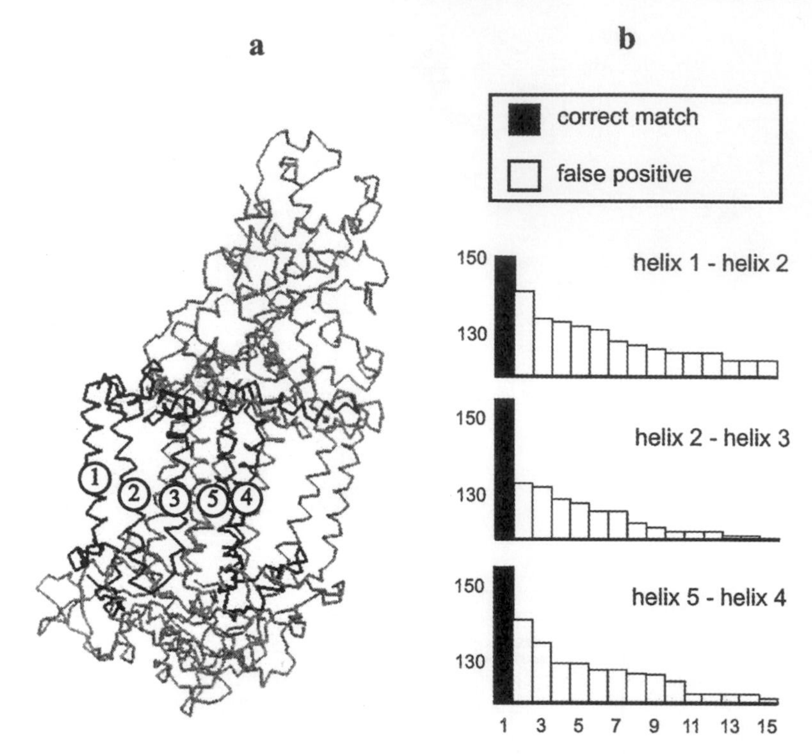

FIG. 5. An overview of the results of interhelical packing studies of the photosynthetic reaction center. The helices tested (a) are part of the L subunit (dark gray). The histograms (b) present sorted scores of different interhelical configurations.

Side Chain Flexibility. The concave shape of deep binding pockets in many proteins restricts conformational freedom of the receptor side chains, thus making these pockets relatively rigid. Because of this, rigid body high-resolution docking procedures may be successful in predicting protein complexes that involve such structures.[32] However, in structures with no such restrictive fold motifs, the side chains have more conformational freedom. Such side chain flexibility prevents the high-resolution prediction of the complexes.[33,34] TM helices clearly belong to the second type of structures. In a helix A–helix B complex, conformation of the side chains on helix A are determined not only by the helix A structure, but also by the position and the structure of helix B. Thus matching of individually

[32] N. C. J. Strynadka, M. Eisenstein, E. Katchalski-Katzir, B. K. Shoichet, I. D. Kuntz, R. Abagyan, M. Totrov, J. Janin, J. Cherfils, F. Zimmerman, A. Olson, B. Duncan, M. Rao, and R. Jackson, M. Sternberg, and M. N. G. James, *Nature Struct. Biol.* **3,** 233 (1996).

[33] J. S. Dixon, *Proteins Suppl.* **1,** 198 (1997).

[34] I. A. Vakser, *Proteins Suppl.* **1,** 226 (1997).

determined structures of A and B, by a high-resolution rigid body procedure, will fail. The practical applicability of the procedures that search through the side chain conformations to a blind, unconstrained prediction of a helix–helix complex is presently unclear. In the low-resolution docking approach, differences in the side chain conformations become indistinguishable. This allows one to ignore the side chain flexibility at the low-resolution stage of matching.

Helix End Point Uncertainty. The existing methods for prediction of the membrane-spanning fragments have a limited accuracy of plus or minus several residues.[35] This adds significantly to the structural uncertainty of the TM helices to be matched. This uncertainty is much less important at low resolution.

Loops Influence. Although the TM helices are tightly packed in the bundle, it can be assumed that this packing, to some extent, is influenced by the loops outside the membrane. Thus, even the "ideal" high-resolution docking procedure, applied to the TM helices separated from the loops, potentially may be inadequate. The degree of the loops influence on the helix bundle packing is unknown. Low-resolution docking is much less sensitive to such an influence.

Multiple-Minima Problem. This is a "classical" problem in protein structure prediction and protein docking. The objective function to be minimized in a structure prediction procedure (the energy of the system) has a complicated form, with many "local" minima that make it difficult to find the "global" minimum—the minimum corresponding to the native structure. In the docking of TM helices, this problem appears as the multiplicity of false-positive fits. Low-resolution docking is equivalent to docking with the extended ranges of the atom–atom potentials. The longer potential ranges smooth the energy function and thus largely eliminate the multiple-minima problem.

These four problems are generally solved by the low-resolution docking approach. However, low-resolution predictions have low precision. Transition from the low-resolution prediction to the high-resolution one is crucial to the success of TM bundle modeling.

Low-Resolution Helix Recognition

Low-resolution docking is based on the funnel-like intermolecular energy landscape.[28] From the geometrical point of view, it is based on spatial complementarity of the macrostructural (larger than atom size) features. In globular proteins, the largest recognition factors are determined by the protein fold. This corresponds to the lowest docking resolution of ~7 Å.[27,36] In helices, obviously, there are no recognition factors larger than the side chains, which corresponds to the lowest docking resolution of ~4 Å.[27] The existence of low-resolution recognition

[35] G. V. Nikiforovich, *Protein Eng.* **11,** 279 (1998).

[36] I. A. Vakser, *Protein Eng.* **9,** 741 (1996).

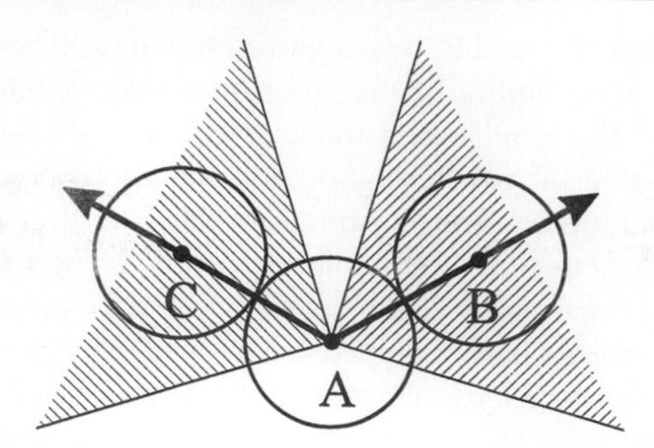

FIG. 6. Definition of the helix–helix interface. View along the helical axis shows helix A interacting with helices B and C. The hatched 90° sectors are the interfaces in helix A.

in TM helices suggests that the residues are arranged along the helix in the way that maximizes the helix–helix contact area.

To test this hypothesis, we calculated the average length of side chains in the interface and the noninterface areas in known structures of integral membrane proteins. Structures used in the analysis were bacteriorhodopsin (2brd[37]), photosynthetic reaction center (1prc[31]), cytochrome c oxidase (1occ[38]), potassium channel (1bl8[39]), and cytochrome bc1 (1bcc[40]). Well packed helix–helix interfaces were selected based on three criteria: (1) the area of the interface between the helices had to be $\geq$200 Å^2, (2) the angle between the axes of the helices had to be $\leq$30°, and (3) the difference between distances separating the tips of the helices had to be $\leq$6 Å. Fifty-six TM helices satisfied these criteria and were selected for the analysis. Definition of the helix–helix interfaces is shown in Fig. 6. Residues with the C^{α} atoms inside the 90° sectors that cover the interacting helices were defined as the interface ones and all other residues were defined as noninterface ones. The conformation-independent side chain length was assigned to all 20 types of residues, according to Levitt.[41] The side chain length was averaged for interface and noninterface areas on each TM helix. Data shown in Fig. 7 indicate that, on average, side chains at the interfaces are shorter than in noninterface areas.[13] This point is illustrated in Fig. 8. This example shows the shorter side chains at the interfaces. Results suggest that, in general, the transmembrane helices, viewed along

[37] N. Grigorieff, T. A. Ceska, K. H. Downing, J. M. Baldwin, and R. Henderson, *J. Mol. Biol.* **259,** 393 (1996).

[38] T. Tsukihara, H. Aoyama, E. Yamashita, R. Tomizaki, H. Yamaguchi, K. Shinzawa-Itoh, R. Nakashima, R. Yaono, and S. Yoshikawa, *Science* **272,** 1136 (1996).

[39] D. A. Doyle, J. M. Cabral, A. K. Pfuetzner, J. M. Gulbis, S. L. Cohen, B. T. Chait, and R. MacKinnon, *Science* **280,** 69 (1998).

[40] Z. Zhang, L. Huang, V. M. Shulmeister, Y. I. Chi, K. K. Kim, L. W. Hung, A. R. Crofts, E. A. Berry, and S. H. Kim, *Nature* **392,** 677 (1998).

[41] M. Levitt, *J. Mol. Biol.* **104,** 59 (1976).

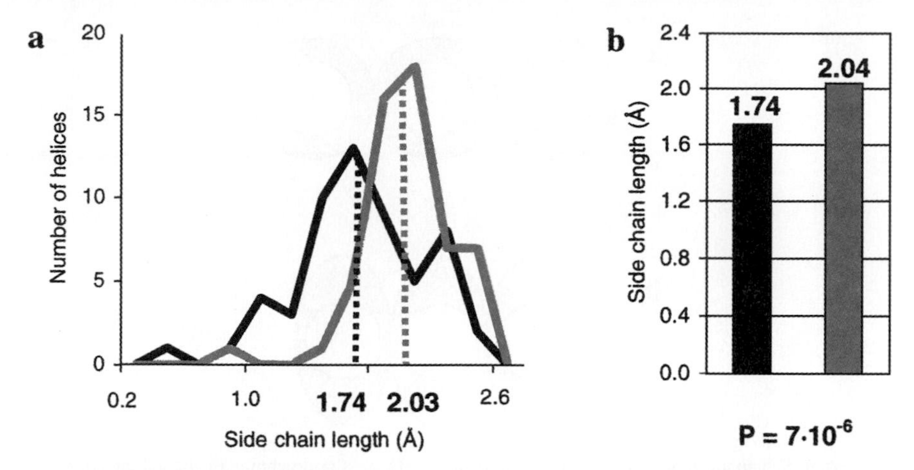

FIG. 7. Conformation-independent statistics of the side chain length. Results for interface and noninterface areas are in black and gray, respectively. Distributions of helices according to the side chain length are shown as histograms with the intervals of 0.2 Å; medians are in bold (a). The difference between the mean values (b) is statistically significant; P value is 7×10^{-6}.

their axes, look like ellipses rather than circles (Fig. 9). This effect increases the surface of the contact between the helices and creates a tighter helix–helix packing. This structural effect, in general, provides the basis for low-resolution helix–helix recognition. However, helix–helix docking, even at low resolution, involves structural recognition factors of different size and nature. Thus, a systematic search algorithm is needed to predict the low-resolution match.

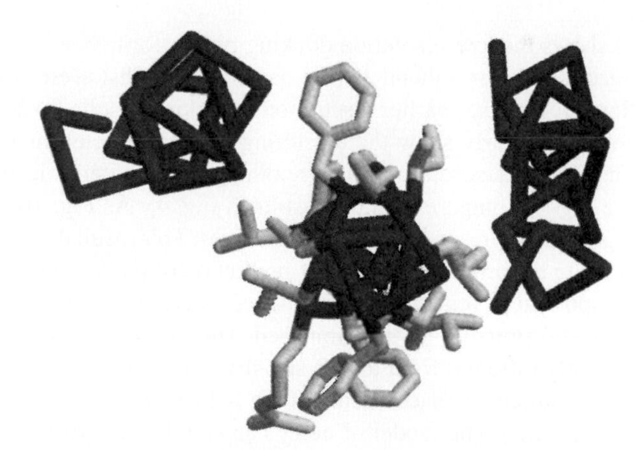

FIG. 8. Example of side chain distribution in interacting helices. Helices are from 1prc, subunit L, residues 117–136, 229–248, and 172–191.

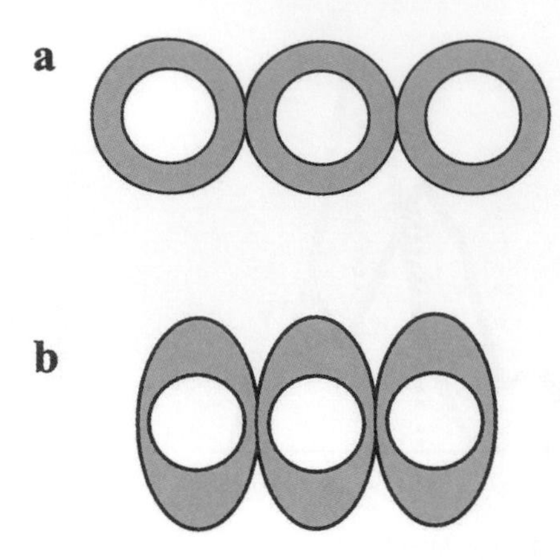

FIG. 9. Schematic representation of the helix shape. Helices (side chains in shaded areas), when viewed along the axes, generally look not like circles (a) but rather like ellipses (b).

Low-Resolution Docking of Modeled Helices

To assess GRAMM applicability to low-resolution GPCR modeling, it was tested on models of TM helices in cases where the crystal structure of the bundle was known. The backbone of the TM segments was put in the α-helical conformation, and the conformations of the side chains were optimized systematically (models were kindly provided by G. V. Nikiforovich). For comparison, we also docked the crystallographically determined structures of these helices at the same low resolution. The docking results were compared with the experimental structures of the bundles.

Figure 10 shows the low-resolution docking results for helices from the photosynthetic reaction center subunit L (the same helices that were used for the high-resolution docking, see earlier discussion). Docking results for helices from the crystal structure clearly show the clustering of the predicted matches at the helix–helix interfaces, according to the smoothing of the intermolecular energy landscape (see Fig. 3; compare with the results in Fig. 4). As Fig. 10 shows, this clustering, in general, is preserved in modeled helices. This result demonstrates the possibility of using the modeled helices for prediction of the low-resolution helix bundle. It is important to reiterate that the helices were modeled from sequence only, with no crystal structure information used. The structure of the modeled helices is significantly different from the crystal structure. This shows that GRAMM can tolerate large structural inaccuracies in the helices. Figure 11 illustrates the degree of this tolerance. The model of helix 1 is significantly different from the crystal structure of the same helix. A most prominent structural discrepancy, in

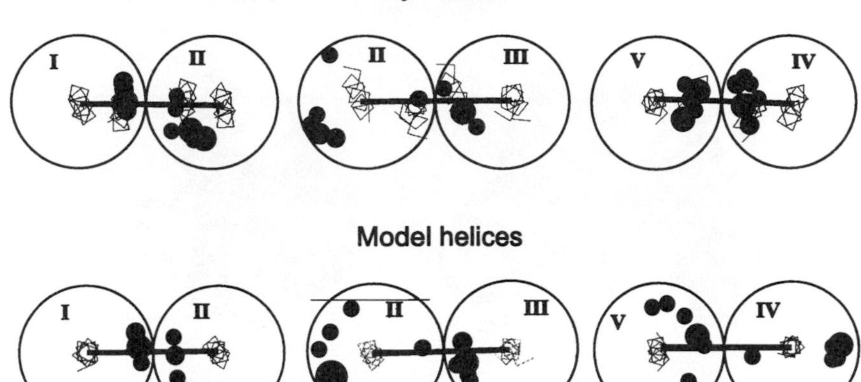

FIG. 10. Low-resolution docking results for helices 1–5 from the L subunit of the photosynthetic reaction center. Helices from the crystal structure are shown in the cocrystallized configuration. Filled spheres are the predicted 10 lowest energy positions of the docked helix center of mass. Larger spheres correspond to more than one match. Docking results are shown in large circles. For a helix X – helix Y pair, circle X shows the predicted positions of helix Y around helix X, and circle Y shows the predicted positions of helix X around helix Y. The centers of mass of helices X and Y, in the experimental position, are connected by a line. Both model and crystallographic structures of helices are docked at resolution 4.1 Å.

the middle of the helix–helix interface, is the extended conformation of the Met side chain. A high-resolution rigid body matching of these helices would make positioning of helix 2 at the correct interface impossible. However, low-resolution docking unequivocally predicts the correct position of helix 2, which corresponds to a pronounced cluster of matches.

Discussion and Future Directions

Transition from Low Resolution to High Resolution

Studies of GPCR function require the atomic-level resolution of GPCR models. Transition from low-resolution to high-resolution helix–helix structure predictions is an important step in atomic resolution precision. Such a transition can be achieved by transformation of the smoothed intermolecular energy landscape to the "regular" one (so called "backtracking"), which is determined by a "conventional" force field (see Fig. 3). Within the framework of GRAMM ideology, this corresponds to a gradual reduction of the grid step from ~4 Å to less than 2 Å (Fig. 12). This backtracking protocol has to take into account side chain flexibility,

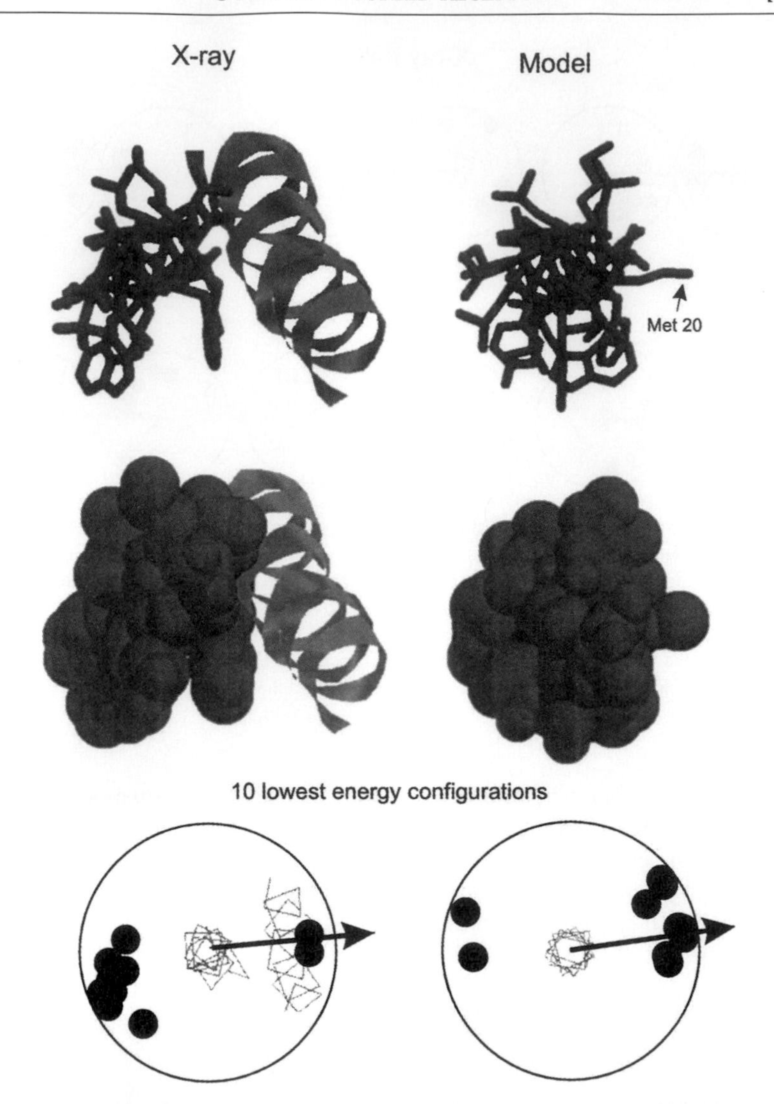

FIG. 11. Example of GRAMM tolerance to structural inaccuracies at low resolution. Helix 1 and helix 2 (ribbon) from bacteriorhodopsin are docked at 4.1 Å resolution. The helix structures, on the left, are from the crystal structure and, on the right, are models. Conformations of the side chains in the model are significantly different from those in the crystal structure. The representation of the lowest energy docking results is analogous to the one in Fig. 10.

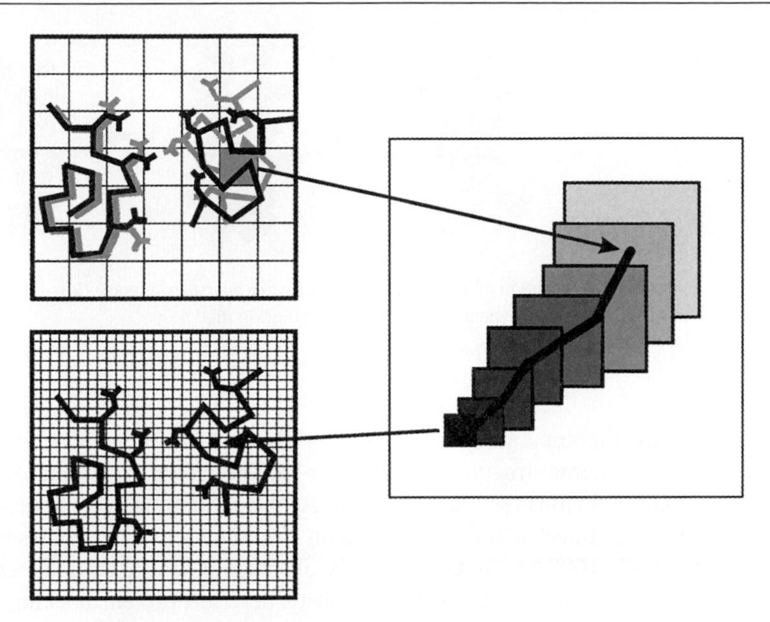

FIG. 12. Schematic illustration of the principles of the transition from low-resolution to high-resolution structures of complexes. (Left) Two-dimensional cross sections through low-resolution (top) and high-resolution (bottom) 3D grids. In the low-resolution complex, structures in black are inaccurate (or native) structures. For illustrative purposes, they are overlapped with structures with proper conformations in the complex (in gray). Shaded squares correspond to the predicted position of the structure's center of mass (the global minimum of the energy function). (Right) The backtracking trajectory from the low-resolution to the high-resolution prediction by a gradual reduction of the grid step. The final point of the trajectory is not necessarily located within the initial low-resolution cube of the grid (light gray square).

with the conformational search determined by the current grid resolution. This would allow us to avoid the oversampling of the conformational space and, thus, will make the backtracking computationally feasible. Development of the backtracking procedure is currently in progress in our group.

Modeling of GPCRs as Constrained Optimization

It is important to point out that, realistically, the atomic resolution model of a GPCR can be obtained only by utilizing all available information regarding this structure. The generation of the low-resolution model and the transition to the high-resolution model can be viewed as constrained optimization (Fig. 13). The structural constraints may come from the homology to a known 3D structure, from physicochemical preferences, and from experimental data (e.g., mutational analysis, cross-linking).

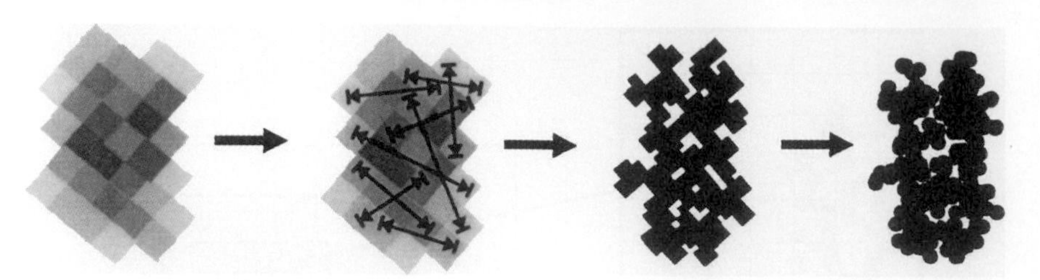

FIG. 13. Constrained optimization of the GPCR structure from a sparse grid image (low-resolution structure) to a fine-grid image and, eventually, to the high-resolution model.

Conclusion

The progress in comparative modeling of globular proteins is based on the large number of available structural templates (>13,000 known 3D structures, the number of nonhomologous templates >1000). At the same time, the importance of GPCR modeling studies that are not based on structural templates comes from the fact that out of >1000 known GPCRs, only one such potential template exists (rhodopsin). We took advantage of the similarity between protein docking and helix–helix packing and applied our protein docking procedure GRAMM to the matching of helices in the TM bundle. GRAMM is based on the search for the steric complementarity at variable resolution. We showed that the steric interactions between TM helices are correlated with the helix–helix interaction energy to the degree that the steric interactions alone, in general, can be used to predict helix packing. We demonstrated the ability of GRAMM to predict the low-resolution match between the helices based on the smoothing of the energy landscape. We determined the existence of large-scale structural recognition factors in TM helices that facilitate the low-resolution match. GRAMM was shown to tolerate significant structural inaccuracies in the modeled helices. This structural tolerance provides an opportunity to generate a low-resolution model of the TM helix bundle from sequence only. At the same time, realistically, the atomic resolution GPCR model can be obtained by a combination of this and other approaches, which should take advantage of all existing information about the structure of that GPCR.

Acknowledgments

The models of individual helices were provided by Gregory Nikiforovich. The authors acknowledge the encouragement of Garland Marshall and Dan Knapp for GRAMM application to the GPCR modeling. The project is supported by NIH GM6188901, NSF DBI-9808093, and South Carolina Commission on Higher Education.

[22] Three-Dimensional Representations of G Protein-Coupled Receptor Structures and Mechanisms

By IRACHE VISIERS, JUAN A. BALLESTEROS, and HAREL WEINSTEIN

Introduction

Nearly 1000 sequences likely to encode G protein-coupled receptors (GPCR) are currently available, in both publicly accessible and in proprietary databases, and the number continues to grow rapidly.[1] This wealth of information about GPCRs will continue to grow and be enhanced with the refinement of the human genome and the sequencing of other mammalian genomes. The great biological importance of GPCRs suggested by these data is also reflected in the enormous interest they engender in the pharmaceutical industry as the targets of nearly 50% of all existing medications.[2] This explains both the interest in GPCRs evident in most areas of biomedical research and the specific clamor for structural information about them at a detailed molecular level.

G protein-coupled receptors have been grouped into five somewhat distinct families: one resembling rhodopsin, another identified with the secretin receptor, a class related to the metabotropic glutamate receptor, another to the fungal pheromone receptor, and finally a class of cAMP receptors.[3,4] The recent breakthrough in determining the crystal structure of rhodopsin[5] has confirmed the general expectation that GPCRs are composed of seven helical transmembrane segments connected by intracellular and extracellular loop segments, as well as the expected topology of an extracellular N terminus and an intracellular C terminus. Although no significant sequence conservation or identity has been found among receptors in the different families, they seem nevertheless to share the composition of seven stretches of hydrophobic amino acids (AAs), connected by segments with greater variability and polarity. This similarity leads to the common (but as yet unverified) assumption that they all share a rhodopsin-like topology, if not structure. As described in this article, the construction, evaluation, and use of the three-dimensional (3D) molecular models of GPCRs from the various families

[1] A. Bairoch and R. Apweiler, *Nucleic Acids Res.* **28,** 45 (2000).

[2] T. Gudermann, B. Nurnberg, and G. Schultz, *J. Mol. Med* **73,** 51 (1995).

[3] F. Horn, J. Weare, M. W. Beukers, S. Horsch, A. Bairoch, W. Chen, O. Edvardsen, F. Campagne, and G. Vriend, *Nucleic Acids Res.* **26,** 275 (1998).

[4] J. Bockaert and J. P. Pin, *EMBO J.* **18,** 1723 (1999).

[5] K. Palczewski, T. Kumasaka, T. Hori, C. A. Behnke, H. Motoshima, B. A. Fox, I. LeTrong, D. C. Teller, T. Okada, R. E. Stenkamp, M. Yamamoto, and M. Miyano, *Science* **289,** 739 (2000).

0076-6879/02 $35.00

reflect the general current understanding of their architecture. This understanding is implemented in the 3D receptor models based on a variety of direct experimental data, as well as results from various approaches from computational genomics, biophysics, and bioinformatics.

One important aim of molecular modeling for membrane proteins such as the GPCRs is to provide a coherent structural context for the interpretation and integration of the abundant data collected in experimental structure–function studies. As illustrated in detail later, the molecular models can incorporate directly and consistently all types of structure-related information obtained experimentally and are, in turn, continuously refined by such data. This modeling process not only provides a complete three-dimensional description of specific receptor proteins for which structures are not available from experiment, but it also yields generalizable templates for families of these proteins and details about various subtypes. Moreover, as described in the final sections, 3D molecular models can be developed to address the characteristics of different states of the receptor molecule that relate to its function and can be used in computational simulations of functional mechanisms. Such simulations can reveal novel aspects of the receptor mechanisms based on the dynamic properties of the proteins revealed by such computational studies. Together, the inferences from both computational modeling and simulation serve as mechanistic working hypotheses for new and more focused experiments.

Development of a Secondary Structure Map

Sequence Alignments and Comparisons

Both when a specific structural template for the modeled protein is available and when the model is based on less definitive structural information, the first step in the construction of a molecular model requires an analysis of the shared characteristics of the sequences of cognate proteins. Such an analysis is based most commonly on multiple sequence alignments constructed so as to maximize similarities identifiable at every AA site. A good variety of computational algorithms are available to perform this task and connect it to analysis and evaluation methods. These approaches have been reviewed and documented extensively[6–8] and will not be described further here.

For implementation of the alignment in the modeling of membrane proteins such as GPCRs, a manual refinement has been recommended that takes into

[6] J. D. Thompson, D. G. Higgins, and T. J. Gibson, *Nucleic Acids Res.* **22,** 4673 (1994).

[7] R. F. Doolittle (ed.), *Methods Enzymol.* **266** (1996).

[8] R. Durbin, S. R. Eddy, A. Krogh, and G. Mitchison, "Biological Sequence Analysis," Cambridge Univ. Press, Cambridge, 1998.

account structural criteria[9,10] and now the structure of rhodopsin as a template.[5] Such refinements enrich the predictive power of sequence comparisons by providing inferences about secondary structure based on criteria such as the low probability of finding insertions and deletions in helices.[11] Because gaps may be inserted in the predicted transmembrane helices (TMHs) by an automated sequence alignment, the refinement (often performed manually) is used to relocate such mismatches toward the loop segments of the sequence, without significant loss of homology. The relocation of gaps to the loop regions is especially advisable where nonconserved Pro of Gly residues occur because the high flexibility conferred by a Gly or Pro residue is often found to have a specific structural role that would have dictated conservation. For example, conserved Gly or Pro in loops is suggestive of specific structures such as turns. A nonconserved occurrence of these residues indicates a structurally permissive site where gaps can be accommodated.

Specific structural criteria that will be reflected in the sequence alignments of rhodopsin-like GPCRs can now be derived from the analysis of the new crystal structure of rhodopsin.[5] Other alignment clues are gathered from the functional roles of residues and groups of residues (i.e., sequence motifs), identified as illustrated later, from structure–function studies. Such sequence motifs often translate into *structural motifs* (SM) that have specific functional roles [i.e., *functional microdomains* (FM) in receptor activation, ligand binding, and/or maintenance of the proper fold of the receptor]. These SM/FM groups can be used to refine the sequence alignment and yield more meaningful alignments for structure prediction purposes.

Sequence comparisons from the refined alignments are subsequently used to identify the likely determinants for structural commonalty of the GPCRs as well as the basis for divergent functional properties of subtypes within a family or among different families. The two basic assumptions underlying the extraction of such structural information from a set of aligned sequences of GPCRs are (i) that they all share a structural framework and (ii) that highly conserved residues are essential for the structural and/or functional integrity of the receptor. Consequently, AA that present a high degree of conservation are proposed to face the protein interior. In membrane proteins, these are most often polar residues that appear in the transmembrane (TM) regions. Residues at positions exhibiting high variability through evolution are considered to face the lipid chains. Most often these are hydrophobic residues. A low degree of conservation can also identify residues responsible for receptor subtype differences, and these are predicted to face the protein interior.

[9] J. A. Ballesteros and H. Weinstein, *Methods Neurosci.* **25,** 366 (1995).

[10] R. Sanchez and A. Sali, *Curr. Opin. Struct. Biol.* **7,** 206 (1997).

[11] S. Pascarella and P. Argos, *J. Mol. Biol.* **224,** 461 (1992).

A refined analysis of conservation from sequence alignments is obtained from the identification of amino acid positions that exhibit evolutionary constraints in terms of discrete physicochemical properties such as hydrophobicity (hdp), local volume, charge, and aromaticity. These properties are chosen (see Ballesteros and Weinstein[9]) because they can be related to the orientation of an AA with respect to the membrane bilayer or protein interior, and the *periodicity* of appearance of the property in a stretch of sequence can be related to the secondary structure of that sequence segment (e.g., to reflect the presence of a helix and its boundaries). The application and validation of this approach are illustrated later. Notably, the approach applies in general to membrane proteins (e.g., see Norregaard *et al.*[12]) and yields useful insight into the secondary structure of this family as described later.

Generic Numbering Scheme for GPCR Sequences

The numbering scheme used throughout has been devised as described elsewhere[9] to make possible comparisons among different GPCRs. It is based on the positional analogy in the sequence that implies spatial correspondence in the structure. The numbering is composed of a digit identifying the transmembrane segment (from 1 to 7) and a number associated with a position in the segment relative to the *most conserved position* in that helix, which is assigned the identifier 50. Other positions are thus numbered relative to this conserved residue with numbers increasing toward the C terminus and decreasing toward the N terminus.

Criteria for Secondary Structure Analysis from Conservation Patterns

The most commonly used approach for predicting transmembrane helical domains from GPCR sequences draws on the calculation of hydrophobicity profiles (for some general illustrations, see Ballesteros and Weinstein,[9] Baldwin,[13] Donnelly *et al.*,[14] Cserzo *et al.*,[15] but other useful approaches are available as well for parsing of the primary protein structure, as illustrated in the following subsections).

The Hydrophobicity Profile as a Parsing Criterion. In the hydrophobicity profile approach,[16] a sequence is scanned with a fixed window size and the average hydrophobicity is calculated within that window. The plot of average hydrophobicity against sequence number identifies regions with high overall hydrophobic

[12] L. Norregaard, I. Visiers, C. Loland, J. A. Ballesteros, H. Weinstein, and U. Gether, *Biochemistry* **39,** 15836 (2000).

[13] J. M. Baldwin, *Curr. Opin. Cell Biol.* **6,** 180 (1994).

[14] D. Donnelly, J. B. Findlay, and T. L. Blundell, *Receptors Channels* **2,** 61 (1994).

[15] M. Cserzo, J. M. Bernassau, I. Simon, and B. Maigret, *J. Mol. Biol.* **243,** 388 (1994).

[16] J. Kyte and R. F. Doolittle, *J. Mol. Biol.* **57,** 105 (1982).

character such as expected of TMH domains. A hydrophobic region identified in the plot is predicted as a TMH if it spans at least 18 AA, a length considered to represent the minimum number of residues needed to traverse the cell membrane in an α-helical conformation. TMH boundaries predicted by the hydrophobicity profile must represent minimal TMH boundaries, as the helices could actually extend into the more polar lipid head group region of the membrane (see later). The mean hdp—which in the commonly used scale varies between 4.5 (highly hydrophobic) and -4.5 (highly hydrophylic)—is calculated first at every position in the alignment, and then the average mean hdp is calculated for a given segment of 12 residues by shifting the 12 residue window along the alignment. An hdp plot for GPCR alignment can thus be obtained in which each transmembrane domain is identifiable as a peak of high hydrophobicity (average $\gg 0$) that spans a number of residues sufficient for the segment to cross the lipid domain of the membrane. Hydrophilic segments (average hdp < 0) flank such a TM stretch of high hdp.

Once the TM domains have been identified from the hdp plot, the next steps involve the refinement of the boundaries of each TM segment and of the secondary structure it incorporates.

Occurrence of Insertions and Deletions in Aligned Sequences. As discussed in the beginning of this section, the use of insertions and deletions to determine helix boundaries is based on the low probability of their occurrence in helical segments[11] where they would disrupt structural and functional patterns. This criterion overlaps the hdp plot, as boundaries identified from the insertion/deletions in the sequence alignment most likely represent the maximal boundaries for each transmembrane segment, whereas the hdp plot represents the minimal TM fragment.

Occurrence of Pro Residues within Putative Transmembrane Helix Segments. Proline residues disrupt α helices because they lack the H bonds to the preceding turn and instead clash sterically with the carbonyl at position (i–4) in an α helix. Due to this characteristic distortion produced by a Pro residue in an α helix,[17,18] when single Pro residues appear within TMHs, they are highly conserved.[19] Analysis of GPCR sequences has shown that single Pro residues that are not conserved are common at the ends of putative TMHs (see Javitch *et al.*[20]). However, analysis of known protein structures has shown that such single and nonconserved Pro residues can occupy the first three positions at the N terminus and at most the last position at the C terminus without distorting the helix significantly.[21,22] Consequently, the occurrence of such nonconserved Pro residues can be used to indicate the ends of TMHs.

[17] R. Sankararamakrishnan and S. Vishveshwara, *Int. J. Pept. Protein Res.* **39,** 356 (1992).
[18] I. Visiers, B. Braunheim, and H. Weinstein, *Protein Eng.* **13,** 603 (2000).
[19] K. A. Williams and C. M. Deber, *Biochemistry* **30,** 8919 (1991).
[20] J. A. Javitch, L. Shi, M. M. Simpson, J. Chen, V. Chiappa, I. Visiers, H. Weinstein, and J. A. Ballesteros, *Biochemistry* **39**(40), 12190 (2000).
[21] M. W. MacArthur and J. M. Thornton, *J. Mol. Biol.* **218,** 397 (1991).
[22] K. G. Strehlow, A. D. Robertson, and R. L. Baldwin, *Biochemistry* **30,** 5810 (1991).

Prediction of TMH Segments from the α-Helix Periodicity of Property Profiles Quantified with Fourier Transform Analysis. The prediction of secondary structure in GPCR sequences can be based on the periodicity exhibited by the conservation pattern of properties.[23] This periodicity is a discriminant indicator for secondary structure because an α-helix periodicity (3.6 AA/turn) is very different from that of a β strand (2AA), so that α helix or β strand-like periodicity of conservation identified in a given segment can be interpreted as a high probability that the segment will adopt the corresponding secondary structure.[9]

The periodicity of the conservation is quantified with Fourier transform (FT) analysis. Analysis of a property profile over a window of size N produces a power spectrum $P(\omega)$ that reveals all existing periodicities ω.[24,25] If the sequence contained within the window N adopts an α-helical conformation, a peak in the power spectrum should appear around 105°, the angle between adjacent side chains for an α helix viewed down its axis.[26] An α-helix periodicity (AP) calculated from $P(\omega)$ describes the extent of the periodicity in the helical region compared to that over the entire spectrum:

$$P(\omega) = \left[\sum_{j=1}^{N}(U_j - \overline{U})\sin(j\omega)\right]^2 + \left[\sum_{j=1}^{N}(U_j - \overline{U})\cos(j\omega)\right]^2 \tag{1}$$

$$AP = \frac{(1/30)\int_{90}^{120} P(\omega)d\omega}{(1/180)\int_{0}^{180} P(\omega)d\omega} \tag{2}$$

A value of AP > 2 was suggested as significant indication of an α-helical character based on an analysis of the known structure of the PRC,[27] although parameters such as the window size and number of sequences can affect the AP value. Because AP measures the extent of α periodicity, a plot of AP versus residue number can be used to identify the TM segments[27] and to infer their boundaries.[28] The plotted values are calculated by scanning the sequence alignment with a constant window size N. The sequence segment at which the AP value drops significantly indicates the N or C terminus of the TMH. In a previous review,[9] we illustrated in detail the application of this procedure for the variability profile in the determination of the putative ends of TMH using an alignment of dopaminergic, serotonergic, and adrenergic GPCRs. Plots obtained for each of the seven TMHs clearly identify each of them as a region of increased AP values within the AP plot, indicating the

[23] D. Donnelly, M. S. Johnson, T. L. Blundell, and J. Saunders, *FEBS Lett.* **251,** 109 (1989).
[24] D. Eisenberg, R. M. Weiss, and T. C. Terwilliger, *Proc. Natl. Acad. Sci. U.S.A.* **81,** 140 (1984).
[25] H. Komiya, T. O. Yeates, D. C. Rees, J. P. Allen, and G. Feher, *Proc. Natl. Acad. Sci. U.S.A.* **85,** 9012 (1988).
[26] D. Donnelly, J. P. Overington, and T. L. Blundell, *Protein Eng.* **7,** 645 (1994).
[27] D. C. Rees, H. Komiya, T. O. Yeates, J. P. Allen, and G. Feher, *Annu. Rev. Biochem.* **58,** 607 (1989).
[28] D. Donnelly and R. J. Cogdell, *Protein Eng.* **6,** 629 (1993).

presence of helix periodicity within these regions. The TMH boundaries predicted by these plots are determined for the first 12 AA segment at which the AP changes from a basal value to one indicating significant helix periodicity and then back again. The N and C terminus of the predicted TMH segment is the first AA within the window that displays, respectively, the first and last peak in the AP. According to this criterion, Fig. 8 in Ballesteros and Weinstein[9] indicates that the boundaries for TMH2 would be residue 2.37 as the N terminus and 2.59 as the C terminus. The same study for TMH7 suggests 7.38 as the N terminus of the segment and 7.71 as the C terminus (see Fig. 8 in Ballesteros and Weinstein[9]). It is noteworthy that factors such as the presence of prolines and subtype selective residues can disrupt the helical periodicity of a certain property. The presence of prolines in α helices disrupts the helical periodicity because it induces a face twist of the two helical portions preceding and following the proline inducing an offset in the periodicities.[9,29–31] This local perturbation results in a reduction of the AP value calculated for windows that include proline kink regions. The subtype selectivity disruption is illustrated by TMH2, which exibits a subtype-specific distribution at position 2.60 and 2.64 so that the C terminus identified at 2.59 fails to include the strip of subtype selective residues at 2.60 and 2.64.

Based on the special considerations related to local distortion and to possible artifacts in the calculation of AP plots, the prediction of TMH boundaries by the Fourier transform method alone must be considered incomplete. Like other such methods that are based on the analysis of sequence properties, it must be augmented whenever possible with inferences drawn from other methods. Many of these methods are refined by the increasing availability of complete 3D structures that offer new influences and generalizable guides. A more secure prediction of TMH ends, which is essential both for the construction of complete 3D molecular models that include the loop segments and for structure–function inferences, requires a convergence of results from several different methods.

Prediction of Transmembrane Helix Segments from the α-Helix Periodicity Identified on Surface Patches in a Helical Net Representation. This method is analogous to the Fourier transform method discussed earlier because both measure helical periodicity of a property. However, it draws on inferences from analysis of several noncontiguous segments of sequence and is strengthened by the inspection of graphical representations. Because α helices have a periodicity of 3.6 AA/turn, individual residues separated by 3 or 4 AAs and sharing a given property (such as hydrophobicity or conservation) would define a continuous patch on the helical surface whose limits are predicted to be the TMH boundaries. The analysis of helical patches is best undertaken on graphical representations of helical nets, which

[29] D. Fu, J. A. Ballesteros, H. Weinstein, J. Chen, and J. A. Javitch, *Biochemistry* **35,** 11278 (1996).
[30] K. Konvicka, F. Guarnieri, J. A. Ballesteros, and H. Weinstein, *Biophys. J.* **75,** 601 (1998).
[31] M. S. P. Sansom and H. Weinstein, *Trends Pharmacol. Sci.* **21,** 445 (2000).

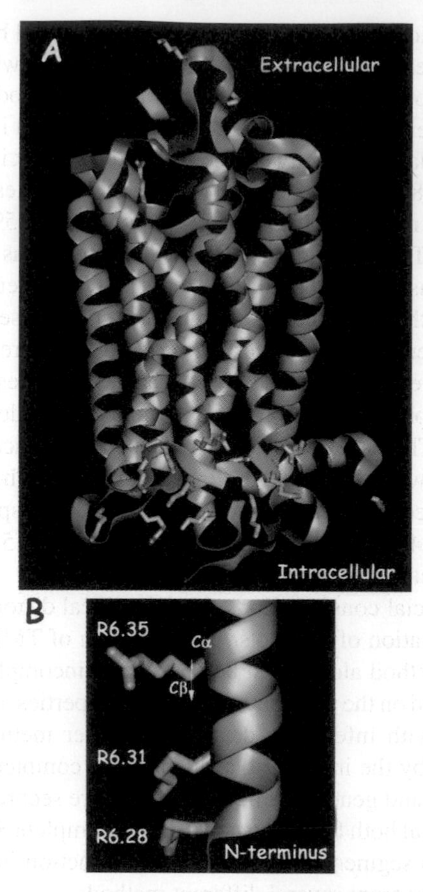

FIG. 1. (A) Positions of Arg and Lys in rhodopsin. The structure is the 2.5-Å resolution crystal structure.[5] Positions containing Arg/Lys are highlighted. Note the prevalence of such residues at the cytoplasmic end. (B) Detail of the structure in A showing the cytoplasmic part of TMH6 in rhodopsin. Note the direction of the Arg or Lys side chain: the $C_\alpha \rightarrow C_\beta$ bond is always oriented toward the N terminus so that the long side chain of Arg or Lys can extend over nearly two turns toward the N terminus on the cytoplasmic side.

are a correct two-dimensional representation of positions on a three-dimensional cylinder defined by the Cα atoms of an α helix (e.g., see Fig. 2).

Use of Arg and Lys Positions to Define the Cytoplasmic Ends of Transmembrane Segments. Commonly predicted to lie outside the TMH by the hydrophobicity profile,[32] the Arg and Lys residues, positioned predominantly at the cytoplasmic ends of bacteriorhodopsin (BR),[33] the photoreaction center (PRC),[34] and rhodopsin[5] (Fig. 1A) were actually found to belong to the TMH domains in the

[32] P. Cronet, C. Sander, and G. Vriend, *Protein Eng.* **6,** 59 (1993).

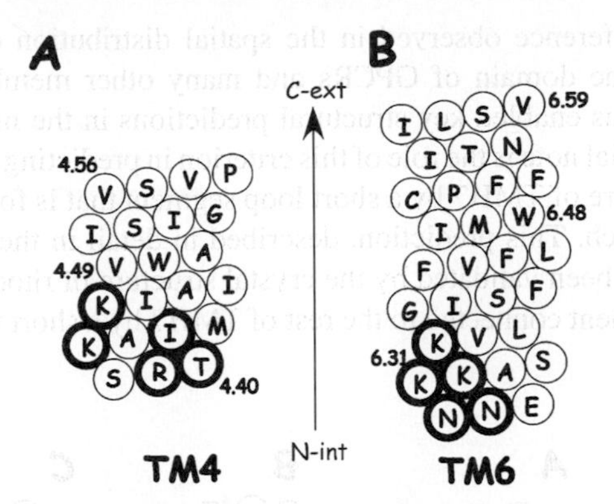

FIG. 2. Helical net representations of TMH4 (A) and TMH6 (B) of the $5HT_{2C}$ receptor show that Arg/Lys at the cytoplasmic ends continue the hydrophobic patch of the TMH core and lie inside the TM region of the helix. TMH ends were obtained from the crystal structure of rhodopsin.[5] Positions containing nonconserved Arg/Lys in an alignment of GPCR sequences are highlighted with a thicker circle; note that these polar residues actually belong to the TMH, not the loop.

available structures of these proteins. The hypothesis that these residues anchor the TMH to the membrane through ionic interactions with phospholipid head groups[35] would explain the cytoplasmic localization of the Arg/Lys, as it is known that the inner leaflet of cell membranes is richer in negatively charged phospholipids than the outer one. For use as a criterion, it is reasonable to predict that any Arg/Lys at the cytoplasmic ends of a TMH that appear in the sequence with an α-helical periodicity would actually continue the hydrophobic patch of the TMH core and lie inside the TM region of the helix.

This criterion was suggested to refine the prediction of TMHs[35] and was shown to be stronger for helices with an intracellular N terminus (N_{int}) and an extracellular C terminus (C_{ext}).[9] The reason is that in an α helix the $C\alpha \rightarrow C\beta$ bond is always oriented toward the N terminus (Fig 1B). Therefore, the long side chain of Arg or Lys is oriented toward the N-intracellular terminus and can extend over two turns toward the cytoplasm to reach the phospholipid head groups. Such is the case of TMH4 and TMH6 in rhodopsin, both of which have nonconserved arginines and lysines in the cytoplasmic segment of the helix.[5] TMH4 displays nonconserved R and K at positions 4.39, 4.40, 4.41, 4.43, and 4.45 (Fig. 2A) and TMH6 at positions 6.28, 6.31, and 6.35 (Fig. 2B). The helix end criterion based on R and K conservation is illustrated in Fig. 1B, in which the structure of the cytoplasmic part of TMH6 of rhodopsin is shown with its three positively charged residues in the accessible side of the helix.

[33] E. Pebay-Peyroula, G. Rummel, J. P. Rosenbusch, and E. M. Landau, *Science* **277,** 1676 (1997).

[34] J. Deisenhofer, H. Michel, T. O. Yeates, H. Komiya, D. C. Rees, J. P. Allen, and G. Feher, *Proc. Natl. Acad. Sci. U.S.A.* **84,** 6438 (1987).

[35] J. A. Ballesteros and H. Weinstein, *Biophys. J.* **62,** 107 (1992).

The clear preference observed in the spatial distribution of the Arg/Lys in the transmembrane domain of GPCRs and many other membrane proteins, as discussed here, has enabled key structural predictions in the modeling of GPCR structure. Of special note is the role of this criterion in predicting the interruption of the helical structure of TMH7 by a short loop segment that is followed by another short helical stretch. This prediction, described in detail in the illustration given later (Fig. 3), has been validated by the crystal structure of rhodopsin[5] showing a short helical segment connected to the rest of TMH7 by a short flexible hinge (see Figs. 3D and 3E).

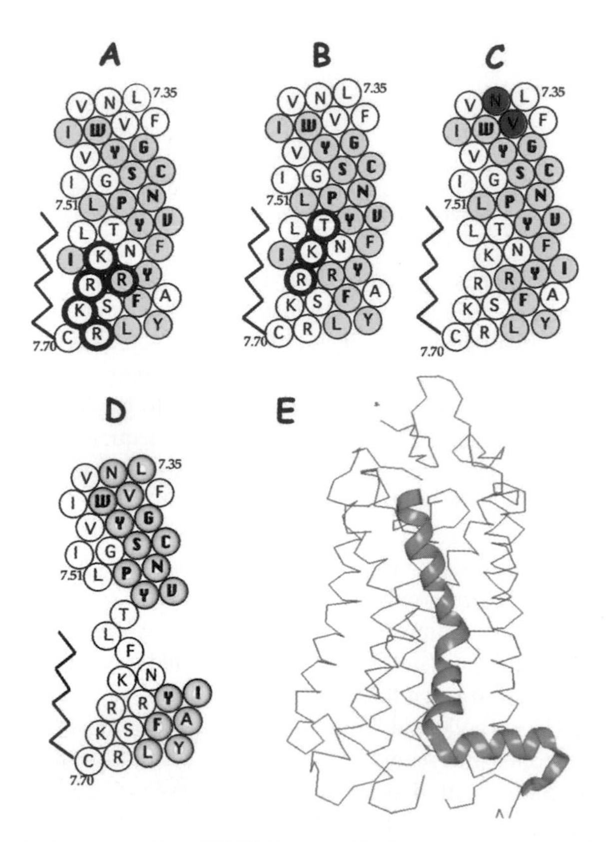

FIG. 3. Helical net representation of TMH7 in a model of the $5HT_{2C}$ receptor. (A) Yellow circles indicate conserved hydrophobicity (σ_{hdp} < 1.2). (B) Filled circles represent residues with σ_{vol}<22 Å (C) Yellow circles indicate residues with a CI larger than 0.7. Red circles indicate residues predicted to face the protein interior by criteria other than a high CI. (D) Consensus prediction for the helix faces. Gold circles indicate residues predicted to face the protein interior; white circles indicate residues facing the membrane. (E) TMH7 in the crystal structure of rhodopsin.[5] Each fragment of the helix corresponds to the helical fragments predicted with the methods described in the text. Note the flexible hinge region that separates the two helical stretches. Each of these observed stretches corresponds to the predicted patch in D.

Determination of Protein–Lipid and Protein–Protein Interfaces

Modeling the interactions that determine the tertiary structure of the GPCRs involves the prediction of the orientation of each TMH relative to the lipid and the protein interior. This entails predicting the residues that lie in the helix–helix or the helix–lipid interface for each TMH. All methods for predicting the orientation of TMHs toward the lipid or other TMH interfaces rely on the same conservation or polarity criteria that served in the definition of the TMH segments described earlier. Based on the conservation criterion, the face of the TMH containing more conserved residues is considered to face the protein interior. The second criterion for defining the orientation of each TMH in the helix bundle is based on the complementarity of the surface polarity between a TMH and the membrane lipids. In the hydrophobic core of the membrane, the more apolar face of the TMH is predicted to be oriented toward the membrane lipids, whereas for the region containing the Arg/Lys described in the previous section, the more hydrophilic face (containing the patch of basic residues) of the TMH cytoplasmic extension is oriented toward the membrane at the level of the phospholipid head groups.

The following methods represent a quantitative approach based on the same criteria we reviewed previously.[9] This quantitative approach is designed to identify sequence positions where the variability appears to be constrained in evolution and to relate these to spatial orientation properties. In attempting to quantify the degree of conservation for a given property, the evolutionary constraints are interpreted in terms of a conserved physicochemical property. On this basis, we can distinguish the conservation of hydrophobicity and volume and the conservation of other properties (e.g., aromaticity, charge, hydrogen bonding capacity). The conservation of a specific AA at a given position is discussed in a subsequent section.

A common approach in quantifying these conservation properties for the purpose of structure prediction is the representation of the property under study on a helical net. This makes it possible to identify patches of AA that share a particular property. Individual residues separated by three to four AAs, and predicted from their properties to be likely to face a given environment, will then appear as continuous property-sharing patches on a helical net representation (see earlier discussion). The orientation of such patches defines the predicted interface of a TMH according to the shared property, e.g., the hydrophobic interface to the lipid can be identified from such a hydrophobic patch.

Quantification of the Conservation of Hydrophobicity and Volume

This analysis is performed based on the hdp scale of Kyte–Doolittle[16] and on the volumes derived for each standard AA from known protein and peptide structures.[36] A statistical analysis is applied for both the hdp and the volume,

[36] C. Chothia, *Annu. Rev. Biochem.* **53,** 537 (1984).

computing the arithmethic mean and the standard deviation at each AA position. Because a particular sequence alignment is incomplete and thus may not represent fully the variability of any given locus, calculations are based on the different residues present at a given AA position in the alignment rather than on the number of occurrences of each of them.[23] For example, what is registered is that both Ala and Tyr are present at a given position in the sequence, regardless of how many times they are found to occur at that position in the sequence alignment. The extent of conservation of hdp or volume is quantified through the standard deviation. The calculated mean value is taken as the putative molecular determinant for positions that have a low standard deviation (σ).

Identification of a conserved position in terms of volume or hdp is based on an empirical physicochemical criterion rather than the statistical significance of the conservation. The numerical value for the criterion is based on an analysis of 10 globin structures that showed that core residues with equivalent environments, mutations were restricted to a volume change of one methyl group.[37] Therefore, the volume is considered to be conserved if the standard deviation is smaller than the volume of a methyl group (22 Å^3), i.e., σ(Vol)<22 Å^3. Similarly, analysis of the conservation of hdp character can be done by considering subsets of AA instead of a hdp scale,[37,38] thus precluding a similar threshold as in the analysis of volume conservation described earlier. Because σ_{vol}<22 Å^3 represents a 13% deviation over the volume range (66–238), the hdp is considered to be conserved for a deviation not exceeding 13% over the range (−4.5, 4.5), i.e., $\sigma_{\text{hdp}} < 1.2$.

Quantification of the Conservation of Aromaticity, Charge, and Hydrogen-Bonding Capabilities

Analysis is based on the percentage occurrence at a given locus of a subset of AA types that display a particular property, e.g., conservation of aromatic character in a particular position. Consequently, percentage aromaticity is quantified as the sum of percentage presence at that position for a Trp residue, a Phe, or a Tyr. The threshold value is chosen arbitrarily as 70%.

Charge and H-bonding capabilities are calculated in analogous fashion. Notably, H-bonding capabilities are classified according to their position in the side chain. For example (not inclusive), Asn and Asp are H-bond donors and acceptors at position δ along the side chain, whereas Gln, His, Trp, and Arg are H-bond acceptors at position ε along the side chain and Gly and Pro are H-bond acceptors at position α along their side chain. Thus, the conservation criterion is sensitive not only to the conservation of H-bonding capability, but also to its particular spatial type.

[37] D. Bordo and P. Argos, *J. Mol. Biol.* **211,** 975 (1990).

[38] J. Overington, M. S. Johnson, A. Sali, and T. L. Blundell, *Proc. R. Soc. Lond. (Biol.)* **241,** 132 (1990).

Quantification of the Conservation at a Given Position and the Definition of a "Conservation Index"

The conservation index (CI) has been quantified as the number of different AA at a given position, but such a definition assumes that all 20AA are equally distinct, thus overlooking the significance of the physicochemical implications of the AA substitutions discussed above. A different procedure for the calculation of the conservation index that takes into account the relations among the 20 natural AA has been considered to yield better modeling criteria.[9] The procedure relates to the well-known measurement of the quantitative significance of AA substitutions in terms of pairwise substitution probabilities or mutation matrices. Such matrices are based on alignments of similar sequences[39] or on inferences from protein pairs that have similar structure but only low (or no) sequence similarity.[40] The matrices express the probability of finding a mutation of a given residue—termed AA_1 (e.g., Ala)—to another residue termed AA_2 (e.g., Glu), at a given locus. This probability is denoted $P(AA_1 \rightarrow AA_2)$. The probability values are calculated from a multiple sequence alignment analysis of homologous proteins belonging to the same protein family. The statistics are improved by combining results from a number of such alignments, corresponding to a set of cognate protein families. A variety of mutation matrices can be used for such calculations.[39–43] One example is the mutation matrix calculated specifically for AA located within α-helices.[39] The use of mutation matrices for the calculation of the conservation index (CI) requires the estimation of the probability for the presence of a set of different AA at a given position $P(AA_1, AA_2, \ldots, AA_k)$, from a set of pairwise substitution probabilities $\{P(AA_1 \rightarrow AA_2), P(AA_1 \rightarrow AA_3), \ldots\}$ also denoted as $(P_{1,2}, P_{1,3}, \ldots P_{i,j})$. There is a well-defined mathematical formula for such calculation in the theory of polytopes.[44] This formulation is used in our quantification of the conservation index, CI, in the following manner.

For a set of different residues $\{AA_1, AA_2, \ldots, AA_k\}$ considered as points in an euclidean space, mutation probabilities $P(AA_i \rightarrow AA_j)$ $(P_{i,j})$ are distances among these points, as represented in the following scheme:

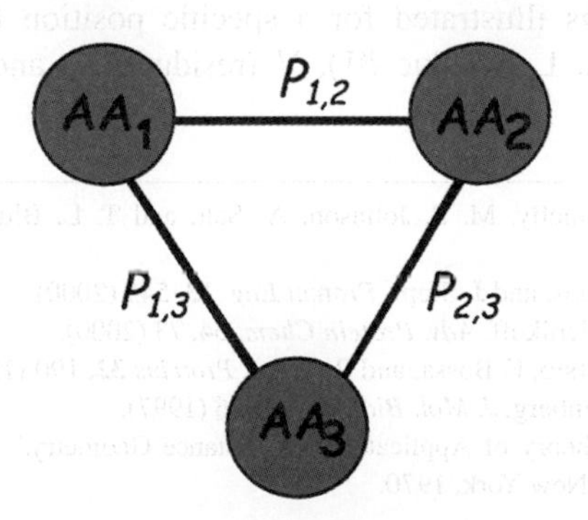

$P_{1,2}$ is the probability computed for the mutation of residue AA_1 into residue AA_2. $P_{1,3}$ is the probability computed for the mutation of residue AA_1 into residue AA_3, and $P_{2,3}$ is the probability computed for the mutation of residue AA_2 into residue AA_3. If the triangular inequality $P_{1,3} > P_{1,2} + P_{2,3}$ is fulfilled for every set of three residues taken from the mutation matrices, then for N residues $\{AA_1, AA_2, AA_3, \ldots, AA_N\}$ the total number of distances or probabilities is $[N^*(N-1)]/2$.

Under these conditions, a polytope, defined as the closed geometrical object in a $N-1$ dimensional space defined by these $[N^*(N-1)]/2$ distances, exists so that the volume of the space contained within it is the minimal possible. This volume estimates the conservation at that position, given the set of amino acids present and their respective pairwise mutation probabilities.

For the above example, three AAs define the distances that delimit a triangle in 2-dimensional space (the polytope). In the present case, the area of the triangle is a quantitative measurement of the conservation at that position. In general, the "volume" associated—or space contained—of the polytope is a quantification of the conservation. A "volume" equal to 1 (a point) corresponds to 100% conservation. Similarly, a line quantifies the conservation finding that only two AAs occur at this locus by its length (the distance $P_{i,j}$). The volume of the polytope in a general case from pairwise $P_{i,j}$ mutation matrices is defined below. This volume $V(N_{AA})$ defines the conservation index at a given AA site.

$$V\{AA_1, AA_2, \ldots, AA_N\} = \sqrt{\frac{(-1)^N}{2^{N-1} * [(N-1)!]^2} * \begin{vmatrix} 0 & 1 & 1 & \cdot & 1 \\ 1 & 1 & P_{1,2}^2 & \cdot & P_{(1,N)}^2 \\ 1 & P_{(2,1)}^2 & 1 & \cdot & P_{(2,N)}^2 \\ \cdot & \cdot & \cdot & \cdot & \cdot \\ 1 & P_{(N,1)}^2 & P_{(N,2)}^2 & \cdot & 1 \end{vmatrix}} \quad (3)$$

Since this $V(N_{AA})$ volume corresponds to the probability attached to performing the necessary mutations, the actual values vary between 1 and 0 for maximal conservation and minimal conservation, respectively.

The calculation is illustrated for a specific position that has three different AA ($N=3$), e.g., L (residue #1), V (residue #2), and I (residue #3). The

[39] J. Overington, D. Donnelly, M. S. Johnson, A. Sali, and T. L. Blundell, *Protein Sci.* **1,** 216 (1992).

[40] A. Prlic, F. S. Domingues, and J. Sippl, *Protein Eng.* **13,** 545 (2000).

[41] S. Henikoff and J. G. Henikoff, *Adv. Protein Chem.* **54,** 73 (2000).

[42] S. Pascarella, R. De Persio, F. Bossa, and P. Argos, *Proteins* **32,** 190 (1998).

[43] D. W. Rice and D. Eisenberg, *J. Mol. Biol.* **267,** 1026 (1997).

[44] L. M. Blumenthal, "Theory of Applications of Distance Geometry." Oxford Univ. Press, 1953. Reprinted by Chelsea, New York, 1970.

substitution probability table for residues in α helices yields the following substitution probabilities[39]:

$$\begin{aligned} P_{L,V} &= 0.11 \\ P_{L,I} &= 0.076 \\ P_{V,L} &= 0.133 \\ P_{V,I} &= 0.112 \\ P_{I,L} &= 0.139 \\ P_{I,V} &= 0.178 \end{aligned} \tag{4}$$

With these values the volume (for $N = 3$) is calculated as

$$V(3) = \sqrt{\frac{(-1)^3}{2^{3-1} * [(3-1)!]^2} \begin{vmatrix} 0 & 1 & 1 & 1 \\ 1 & 1 & 0.11^2 & 0.076^2 \\ 1 & 0.133^2 & 1 & 0.112^2 \\ 1 & 0.139^2 & 0.178^2 & 1 \end{vmatrix}} = 0.426 \tag{5}$$

A refined formulation of this polytope-based approach to the calculation of the conservation index has been described recently (L. Shi *et al., Biochemistry* (2001), in press). This formulation resolves difficulties with the triangular inequality for the probability values, and addresses the issue of the frequency of appearance of a particular amino acid in the sequence alignment. A new term was added to the conservation index calculation that differentiates among cases in which the same total number of different amino acids appears at a given position in the alignment, but with different frequencies. For example, a position at which 3 different amino acids appear with frequencies 98 : 1 : 1 compared to frequencies of 30 : 30 : 40 for the same three AAs.

Quantification of Criteria for the Prediction of TMH–Lipid Interfaces

Residue positions that are highly variable through evolution yet conserve a high hydrophobic character are proposed to face the lipid chains. The criterion combines the energy-related consideration for the preference of highly hydrophobic AA for the lipid milieu, with the evolutionary consideration that relates low conservation requirements with exposure to nonspecific interactions in the lipid. This hypothesis has been proved quantitatively in the protein family of the photoreaction center through the calculation of AA substitution matrices for lipid-exposed residues.[45] In quantitative terms, residues pointing outward from the transmembrane bundle, toward the lipid phase, will have a low CI, low hdp standard deviation (σ_{hdp}),

[45] D. Donnelly, J. P. Overington, S. V. Ruffle, J. H. A. Nugent, and T. L. Blundell, *Protein Sci.* **2,** 55 (1993).

and high hdp mean value. This leads to a quantitative criterion by defining the probability of an AA to lie in a TMH–lipid interface as

$$P(\text{TMH} \rightarrow \text{Lipid}) = (1 - \text{CI})^{*}[\text{mean(hdp)} + 4.5]^{3}/\sigma_{\text{hdp}} \qquad (6)$$

A CI lower than 0.7 restricts this quantitative analysis to poorly conserved positions. Note that in Eq. (6) the range of hydrophobicity values for the 20 natural AAs (−4.5, 4.5) has been rescaled to (0 to 9) by adding 4.5 in order to avoid negative probabilities.

For modeling purposes, it is more useful to select those sites that are strongly predicted to face the lipids by use of this criterion. The function proposed in Eq. (6) is very discriminant when used with a threshold value of 6 for the probability function.

Illustration of Concerted Application of Prediction Methods: Helix Boundaries and Interfaces for TMH7 in GPCRs

Application of the methods outlined earlier is illustrated for the seventh segment of GPCRs in the rhodopsin-like family. The analysis is carried out in the context of the sequence of TMH7 in the serotonin receptor $5HT_{2C}$.

Prediction of TMH Boundaries

According to the hydrophobicity profile,[46] N and C termini of TMH7 are residues 7.34 and 7.56, respectively. However, the entire segment up to 7.70 exhibits a conservation index larger than 0.7 on a common face of the TMH (Fig. 3C). This is consistent with a continuous TMH up to 7.70. Thus, the analysis yields a strong prediction for the cytoplasmic TMH boundary at the palmytolated Cys 7.70 (notably, this locus is strongly predicted to face the lipid milieu, as its aliphatic component is chemically identical to the lipid chains). The prediction is reinforced by the disruption of TMH periodicity by the distribution of R/K residues after C7.70; the break is sharp and clear, as a continuous stretch of 5 AA sites at which R/K residues occur in other GPCRs follows the predicted TMH end.

The extracellular boundary predicted from the conservation index would have the helix start at 7.40 rather than 7.34 as predicted from the hdp profile. The reason may be that the role of the extracellular portion of TMH7 in ligand binding, as suggested for position 7.39 based on mutagenesis studies (e.g., see Smolyar and Osman[47]), points to subtype selective variability in this position of the TM segment. As mentioned earlier, subtype selective positions produce a low conservation value and may distort artificially the patch of residues predicted to face the interior

[46] D. Julius, A. B. MacDermott, R. Axel, and T. M. Jessell, *Science* **241,** 558 (1988).
[47] A. Smolyar and R. Osman, *Mol. Pharmacol.* **44,** 882 (1993).

of the protein (for a recent example, see the discussion about the extracellular end of TMH4 in the dopamine receptors[20]). The criterion of conservation of the hdp character provides additional support for the predicted TMH boundary at the extracellular side: V7.37 has high mean hydrophobicity conserved at this locus ($\sigma_{hdp} < 1.2$, Fig. 3A). Notably, other conservation criteria, including conservation of volume (Fig. 3B, filled circles represent residues with σ_{vol}<22 Å), support the determination of TMH ends from the other approaches (e.g., note the high conservation of aromatic character for residues Y7.60 and F7.64).

The helical stretch predicted for TMH7 is interrupted: insertions and deletions in the sequence alignment can be found consistently at 7.54 and 7.57 in a number of GPCR and are thus a strong prediction of nonhelical character. Nonconserved prolines appear at positions 7.54, 7.58, and 7.62. Interestingly, positions with nonconserved prolines are four residues apart, defining a continuous patch on the TMH surface (Fig. 3B). Because the patch defined by prolines 7.54, 7.58, and 7.62 also belongs to the Arg/Lys motif (Fig. 3A), all three positions are predicted to be on the TMH–lipid interface. This suggests that the presence of these prolines involves specific structural constraints that are more likely related to α-helical character than to loop structure. Given the conservation of the pattern and its constraints, conformational effects of the disruption of helical structure in the TMH7 segment are likely to have special significance for the orientation and flexibility of the entire TMH7 domain. Notably, both the conservation pattern and the analysis of the Arg/Lys motif would support the prediction of a continuous TMH7 extending into the cytoplasmic side to Cys7.70 (Fig. 3A). However, reconciling a continuous TMH7 to Cys7.70 with the presence of both insertion/deletions and Pro residues within the same segment requires the turn 7.54–7.58 of TMH7 to be a flexible hinge[48] (Fig. 3D). The cytoplasmic extension of TMH7 after this hinge is predicted to be helical. Notably, it could still form a continuous TMH with the core of TMH7 at some point during the activation process of the GPCR. It is a most noteworthy confirmation of the remarkable predictive power of the tools and their application described here that the crystal structure of rhodopsin confirms the presence of just such a hinge and a helical cytoplasmic extension. In the crystal structure, the helical extension is formed by residues 7.58 to 7.70 and is connected to the rest of the helix by a flexible hinge formed by residues 7.55 to 7.57 (Fig. 3E). This short helix is in the membrane plane, with the axis parallel to the membrane plane. Experimental data support a functional role, most likely through significant conformational changes, for this helical extension predicted by the modeling and observed in rhodopsin. The rearrangement has been implicated in G protein coupling in the context of the full receptor.[30,49–51] The rearrangement

[48] J. A. Ballesteros, Doctoral Thesis, Mount Sinai School of Medicine, 1997.

[49] R. J. Lefkowitz and M. G. Caron, *J. Biol. Chem.* **263,** 4993 (1988).

[50] M. G. Caron, *Curr. Opin. Cell Biol.* **1,** 159 (1989).

is such that palmytolation of Cys7.70 seems to be important for GPCR–G protein interaction.[52,53] This inference is based on observations that mutation of the palmytolated Cys 7.70 to Gly produces an uncoupled β_2-adrenergic GPCR,[52] and dithiothreitol treatment, which removes the palmitoyl, leads to significant changes in the FTIR spectrum of this segment on rhodopsin activation. The FTIR frequencies for the segment were assigned to helical conformations,[54] and spin-labeling experiments in rhodopsin have shown a significant movement between residues 1.60 and 7.63 on rhodopsin activation.[55]

Prediction of TMH–Lipid and TMH–Protein Interfaces for TMH7

According to the criteria outlined earlier, residues with a high CI are predicted to face the interior of the protein. The helical net representation of such residues (Fig. 3C, yellow-filled circles) clearly shows high conservation on one face of the helix defined by residues 7.40, 7.42, 7.43, 7.45, 7.46, 7.49, 7.50, 7.52, and 7.53. Some special considerations are noteworthy: residue N7.36 is predicted to face the protein interior (Fig. 3C) due to its high polarity, although it is not conserved. Position 7.39, which was identified in a previous section as being related to the selectivity of the receptor for different ligands, is also predicted to face the protein interior (Fig. 3C). Those two positions are contiguous with the patch of conserved residues in the helical net representation, supporting their assignment to the protein interface. Finally, positions 7.41 and 7.51 are part of the nonconserved face of the helix and have high hydrophobicity so that they are predicted to face the phospholipid environment.

A helical net representation of the final predictions made for TMH7 is presented in Fig. 3D. The predictions are validated by calculation of the solvent-accessible surface area (SASA) for TMH7 in the rhodopsin crystal structure. The SASA is defined as the surface accessible to a probe of 1.4 Å that rolls over the protein surface. The calculated fraction of the total surface that is accessible to such a probe is shown in Fig. 4 for each residue in TMH7 of rhodopsin. The pattern of accessibility to the probe follows a clear helical periodicity such that residues 7.33 (286), 7.34 (287), 7.37 (290), 7.41 (294), 7.44 (297), 7.47 (300), 7.48 (301), and 7.51 (304) are mostly accessible to the probe (number in parentheses indicates the rhodopsin

[51] B. Kobilka, *Annu. Rev. Neurosci.* **15,** 87 (1992).

[52] B. F. O'Dowd, M. Hnatowich, M. G. Caron, R. J. Lefkowitz, and M. Bouvier, *J. Biol. Chem.* **264,** 7564 (1989).

[53] D. I. Papac, K. R. Thornburg, E. E. Bullesbach, R. K. Crouch, and D. R. Knapp, *J. Biol. Chem.* **267,** 16889 (1992).

[54] U. M. Ganter, T. Charitopoulos, N. Virmaux, and F. Siebert, *Photochem. Photobiol.* **56,** 57 (1992).

[55] K. Yang, D. L. Farrens, C. Altenbach, Z. T. Farahbakhsh, W. L. Hubbell, and H. G. Khorana, *Biochemistry* **35,** 14040 (1996).

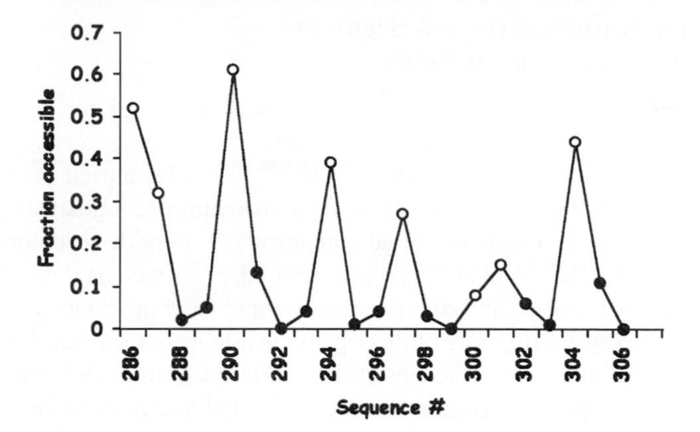

FIG. 4. Solvent-accessible surface area of residues in TMH7 of rhodopsin, calculated from the crystal structure.[5] Values represent the fraction of the total surface accessible to a spherical probe with a 1.4 Å radius (see text).

sequence number of the locus). On helical net representation, Fig. 5 compares residues accessible to the probe according to SASA calculations for TMH7 of rhodopsin and the pattern of accessibility predicted as described in this section for TMH7 of the $5HT_{2C}$. Interestingly, the accessibility pattern for rhodopsin (Fig. 5A) coincides with the residues predicted to face the lipid environment in $5HT_{2C}$ (Fig. 5B).

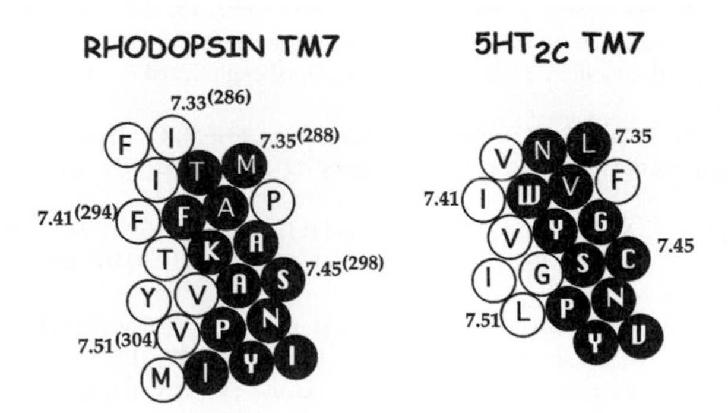

FIG. 5. Relation between calculated surface accessibility and TMH packing in the transmembrane helix bundle. Comparison of helical net representations of TMH7 in rhodopsin (left) and a model of the $5HT_{2C}$ receptor (right). (Left) Solvent-accessible residues in TMH7 of rhodopsin (see Fig. 4) are shown in black circles. (Right) The protein-facing patch predicted for TMH7 of the $5HT_{2C}$ receptor, from analysis of the conservation pattern of biophysical properties (see text), is composed of black circles.

Criteria for Bundling Helical Segments into a Transmembrane Domain

Although some modeling algorithms have been proposed for the automatic construction of the transmembrane bundle from energy-based criteria,[56–58] the cryomicroscopy difraction map of rhodopsin[13,59–61] has been used extensively as a template to model the bundling of the seven transmembrane segments of GPCRs. These have included models of opioid receptors,[62–64] peptide receptors, including TRH,[65,66] GnRH,[68,69] and CCK,[70] cannabinoids,[67] as well as rhodopsin,[71] the secretin receptor family,[72] and adrenergic receptors,[73] among others. Now, the coordinates of the crystal structure of rhodopsin,[5] available in the PDB[74] (1F88), are likely to be used similarly as the new structural template to model other GPCRs. However, in developing a general method for helix–helix packing in the transmembrane bundle of GPCRs, a note of caution is necessary regarding the assumption that all GPCRs share a common template. In general, the expectation of structural similarity among proteins is modulated by two major factors. First, it depends on the extent of sequence similarity (identity; homology) in the superfamily, with 50% having been proposed as a threshold for structurally homologous proteins based on

[56] P. Herzyk and R. E. Hubbard, *Biophys. J.* **69,** 2419 (1995).

[57] P. A. H. Herzyk and R. E. Hubbard, *J. Mol. Biol.* **281,** 741 (1998).

[58] I. D. Pogozheva, A. L. Lomize, and H. I. Mosberg, *Biophys. J.* **72,** 1963 (1997).

[59] V. M. Unger and G. F. Schertler, *Biophys. J.* **68,** 1776 (1995).

[60] V. M. Unger, P. A. Hargrave, J. M. Baldwin, and G. F. Schertler, *Nature* **389,** 203 (1997).

[61] J. M. Baldwin, G. F. X. Schertler, and V. M. Unger, *J. Mol. Biol.* **272,** 144 (1997).

[62] I. D. Pogozheva, A. L. Lomize, and H. I. Mosberg, *Biophys. J.* **75,** 612 (1998).

[63] C. M. Topham, L. Mouledous, and J. C. Meunier, *Protein Eng.* **13,** 477 (2000).

[64] M. Filizola, L. Laakkonen, and G. H. Loew, *Protein Eng.* **12,** 927 (1999).

[65] L. J. Laakkonen, F. Guarnieri, J. H. Perlman, M. C. Gershengorn, and R. Osman, *Biochemistry* **35,** 7651 (1996).

[66] R. Osman, A.-O. Colson, J. H. Perlman, L. J. Laakkonen, and M. C. Gershengorn, *in* "Structure-Function Analysis of G Protein-Coupled Receptors" (J. Wess, ed.), p. 59. Wiley-Liss, New York, 1999.

[67] R. D. Bramblett, A. M. Panu, J. A. Ballesteros, and P. H. Reggio, *Life Sci.* **56,** 3205 (1995).

[68] W. Zhou, C. Flanagan, J. A. Ballesteros, K. Konvicka, J. S. Davidson, H. Weinstein, R. P. Millar, and S. C. Sealfon, *Mol. Pharm.* **45,** 165 (1994).

[69] J. Ballesteros, S. Kitanovic, F. Guarnieri, P. Davies, B. J. Fromme, K. Konvicka, L. Chi, R. P. Millar, J. S. Davidson, H. Weinstein, and S. C. Sealfon, *J. Biol. Chem.* **273,** 10445 (1998).

[70] P. Gouldson, P. Legoux, C. Carillon, B. Delpech, G. Le Fur, P. Ferrara, and D. Shire, *Eur. J. Pharmacol.* **400,** 185 (2000).

[71] S. W. Lin, Y. Imamoto, Y. Fukada, Y. Shichida, T. Yoshizawa, and R. A. Mathies, *Biochemistry* **33,** 2151 (1994).

[72] D. Donnelly, *FEBS Lett.* **409,** 431 (1997).

[73] F. Fanelli, C. Menziani, A. Scheer, S. Cotecchia, and P. G. DeBenedetti, *Int. J. Quantum Chem.* **73,** 71 (1999).

[74] H. M. Berman, J. Westbrook, Z. Feng, G. Gilliland, T. M. Bhat, H. Weissig, I. N. Shindyalov, and P. E. Bourne, *Nucleic Acids Res.* **28,** 235 (2000).

criteria developed for water-soluble proteins[75] (see, however Sali and Kuriyan,[76] Sanchez and Sali,[77,78] and Ortiz *et al.*[79]). Second, it depends on the extent to which the presence of particular AA substitutions can be expected to entail significant structural modifications, such as the long-range structural effects of Pro in TMHs discussed later, or the introduction of new sites for constraining TMH–TMH interactions. Consequently, the process of TM bundle construction cannot be reduced to the structural mimicry of rhodopsin. The correct modeling of structural characteristics for specific members of the rhodopsin-like family, as well as structural differences expected among members of different GPCR families, is certain to require the consideration of specific modeling criteria such as outlined later.

Definition of the Relation Between Surface Patches and Helix Orientation

The predictions of TMHs interfaces provide a guiding criterion for bundling. The extent and orientation of the lipid exposure predicted for each TMH and the position and orientation of the predicted TMH–protein interface must be consistent with the final 3D model. Thus, the predicted TMH–lipid and TMH–protein interfaces for each transmembrane segment become constraints for modeling the helix packing arrangement of the seven TMHs. These constraints incorporate both subtype-specific and family-specific criteria described earlier.

Effect of Proline Kink Distortion

As described in a previous section, the structural perturbation introduced by the proline kink rearranges the positions of the segments preceding and following the proline locus. The magnitude of the proline kink can vary,[17,18,31] thus determining the extent of the deviation of the helical portions before and after the proline kink from a continuous helix axis. Consequently, the position and conformation of the proline kink deserve special attention in any effort to obtain a realistic bundle for a specific GPCR. This has been illustrated repeatedly with specific examples for various GPCRs (e.g., see Ballesteros and Weinstein,[9] Fu *et al.*,[29] Konvicka *et al.*,[30] and Gether *et al.*[80]). The distortion introduced by the 100% conserved P6.50 in the sixth helix in rhodopsin is an interesting illustration of the structural effect of Pro in a helix in GPCRs[5]; it emphasizes the importance that modeling an appropriate kink can have in the bundling process. Figure 6 shows that in rhodopsin the high kink induced by the proline in TMH6 bends the helix toward the neighboring

[75] C. Chothia and A. M. Lesk, *EMBO J.* **5,** 823 (1986).
[76] A. Sali and J. Kuriyan, *Trends Cell Biol.* **9,** M20 (1999).
[77] R. Sanchez and A. Sali, *Proteins* Suppl. 50 (1997).
[78] R. Sanchez and A. Sali, *Proc. Natl. Acad. Sci. U.S.A.* **95,** 13597 (1998).
[79] A. R. Ortiz, A. Kolinski, and J. Skolnick, *J. Mol. Biol.* **277,** 419 (1998).
[80] U. Gether, J. A. Ballesteros, R. Seifert, E. Sanders-Bush, H. Weinstein, and B. K. Kobilka, *J. Biol. Chem.* **272,** 2587 (1997).

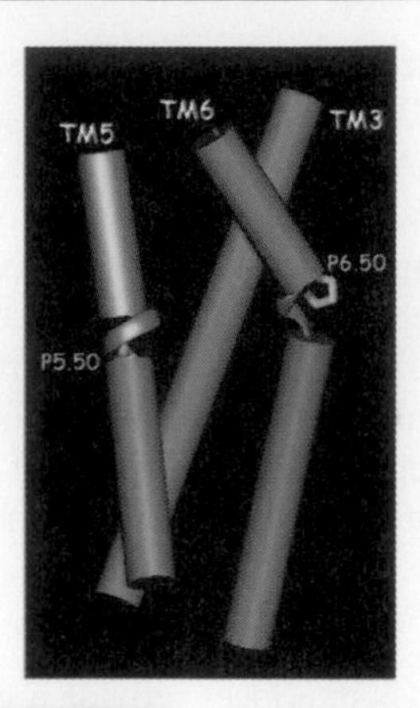

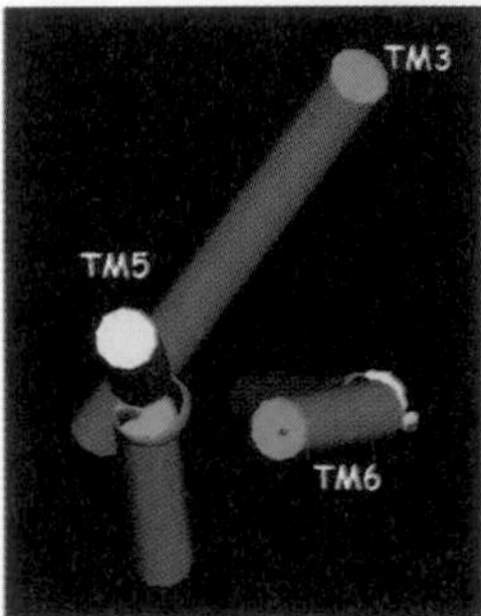

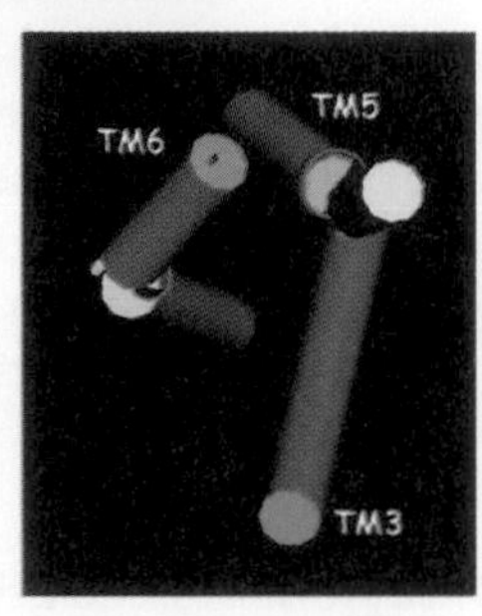

FIG. 6. Representation of the TMH3/TMH5/TMH6 bundle in the crystal structure of rhodopsin[5] shown in three different perspectives: The top panel shows a perspective parallel to the membrane; the bottom panels show two views from the extracellular side. Note that the high bend induced by proline in TMH6 strongly affects its interaction with neighboring helices. This indicates the importance of modeling a correct proline kink to obtain a reliable helix bundle.

TMH5. In contrast, the proline kink in TMH5 is much smaller and has less effect on interactions with the neighboring transmembrane helices. Figure 6 illustrates how the interaction between TMH3 and TMH6 at the cytoplasmic part and between TMH5 and TMH6 at the extracellular side of the receptor would not be possible if TMH6 were constructed as an ideal α helix.

Clearly, modeling proline-containing helices as ideal helices may hinder important features of the receptor structure, decreasing the quality and reliability of the resulting 3D model. However, *de novo* structural modeling of proline kink (PK) regions is currently problematic due to the intrinsic flexibility of the proline kink. Sankararamakrishnan and Vishveshwara[17] suggested specific backbone dihedral angles for modeling PKs. Use of such initial values for the PK is preferable over the approach taken by some in GPCR modeling to start by considering the

PK-containing TMHs as ideal α helices and relying on energy minimizations to produce the appropriate conformations. The latter approach is most likely to produce artificial local distortions in the minimization process, resulting in structures that differ significantly from known PK geometries. However, the use of reasonable initial values for the PK is advantageous because the intrinsic flexibility of the PKs will allow those starting values to be modified through the modeling process so as to optimize tertiary structure interactions.

In determining the structural effect of the PK, it is necessary to take into account the modulation of its magnitude by residues surrounding the proline. For example, the presence of Thr at position (i-1) from Pro produces a larger kink than the one caused by Pro alone, as has been demonstrated for connexins.[81] Thr at position (i-1) from the (i) position of the proline is capable of forming a hydrogen bond through its hydroxyl group to the carbonyl of the backbone of residue (i-5) from the proline. This hydrogen bond stabilizes the helix in a highly kinked conformation and is also likely to modify the directionality of the bend (data not shown) measured by the wobble angle.[18] This is likely to be a general structural motif applicable to transmembrane proteins, including GPCRs and channels, as illustrated recently.[31,81]

When N or D is present at position (i-1) from the (i) position of the proline, the "NP" or "DP" motif has a highly pronounced proline kink.[30] Such a motif is conserved in TMH7 of rhodopsin-like GPCRs and other families. The importance of the perturbation produced in the α-helix structure by the NP/DP motif relates to the observation that an ideal α helix or a regular PK cannot fulfill all the structural constraints imposed by interactions between residues in TMH7 and surrounding TMHs. As discussed earlier, these structural constraints were inferred from various experiments[68,82–84] and must be observed simultaneously by the models. For the specific structural role of the conserved NP/DP motif in TMH7, Konvicka *et al.*[30] suggested that the apparent structural incompatibilities among the various experimental constraints are resolved by the significant perturbation in the helicity of the TMH7. The extent of the perturbation produced by the NP/DP motif was inferred from structural data for soluble proteins. It is remarkable that the PK observed in the structure of rhodopsin around proline 7.50[5] is not only very large, but also exhibits a pronounced face shift that causes a tightening of the helix at that turn. This tightening enables N7.49 to reach D2.50, although N7.49 would be facing the lipids if the PK produced by P7.50 were a regular kink.

[81] Y. Ri, J. A. Ballesteros, C. K. Abrams, S. Oh, V. K. Verselis, H. Weinstein, and T. A. Bargiello, *Biophys. J.* **76,** 2887 (1999).

[82] J. Liu, T. Schoneberg, M. van Rhee, and J. Wess, *J. Biol. Chem.* **270,** 19532 (1995).

[83] J. H. Perlman, A. O. Colson, W. Wang, K. Bence, R. Osman, and M. C. Gershengorn, *J. Biol. Chem.* **272,** 11937 (1997).

[84] S. C. Sealfon, L. Chi, B. J. Ebersole, V. Rodic, D. Zhang, J. A. Ballesteros, and H. Weinstein, *J. Biol. Chem.* **270,** 16683 (1995).

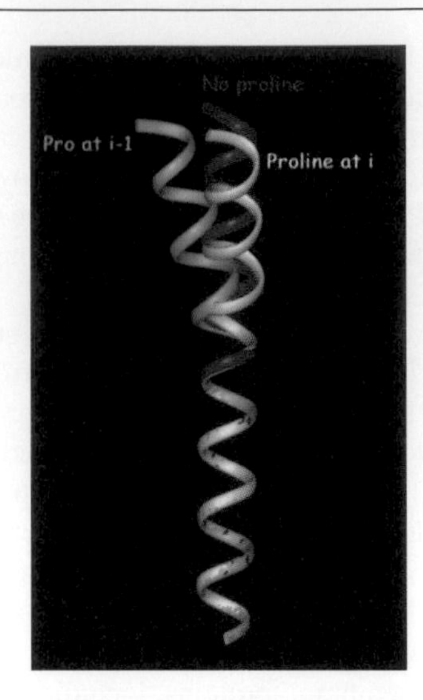

FIG. 7. Proline residues in two contiguous positions in a helix bend it toward different areas of the 3D space. The red ribbon represents a straight ideal helix. The yellow ribbon represents the helix with proline at position (i) in which the proline kink (PK) causes the upper portion to bend out of the plane of the page; the green ribbon shows a helix with proline at position (i-1) with respect to the yellow helix; note that the PK causes a bend in the plane of the figure.

The occurrence and specific position of prolines in the transmembrane segments of different GPCRs and different GPCR families are two of the most striking characteristics that differentiate members of GPCR families. As discussed in the literature (e.g., see Ballesteros and Weinstein,[9] Wess,[85] and Sansom and Weinstein[31]), the special structural and dynamic properties of the prolines may be responsible for some of the key functional and pharmacological differences among GPCRs. For example, rhodopsin differs from neurotransmitter receptors by the presence of a proline in TMH1 (1.48) and at the top of TMH7 (7.38), whereas the P2.59 characteristic of neurotransmitters is absent in rhodopsin. Interestingly, chemokine receptors have a proline at position 2.58 (rather than 2.59), and this difference is likely to cause a kink in the helix in a direction that is different from that produced by P2.59 (Fig. 7 shows the different effect that proline residues in two contiguous positions in a helix can have on the directionality of the kink).

This difference in kinking suggests that differences are likely to exist in the structural and functional involvement of the extracellular segment of TMH2 in rhodopsin compared to neurotransmitter GPCRs and to chemokine receptors. The distortion of Pro residues in TM segments and the modulation of the structural

[85] J. Wess (ed.), "Structure-Function Analysis of G Protein-Coupled Receptors." Wiley-Liss, New York, 1999.

perturbation produced by the PK are therefore likely to play key structural roles in differentiating GPCRs in different families. Until more structural information provides new insights and criteria, use of the rhodopsin structure[5] as a universal model for GPCRs should be viewed with great caution.

Assignment of Side-Chain Conformation in the TM Bundle

The allocation of appropriate torsional angles[86–88] to residues in TMHs is an important element in 3D modeling of GPCRs because inappropriate rotamers may affect both the validity of helix-packing interactions and the identification of functional mechanisms. Modeling the correct side chain conformation is facilitated greatly by the availability of side chain rotamer libraries specific for protein families and elements of secondary structure.[89,90] For example, an α-helix environment significantly restricts the values of the X_1 angle for all AAs to values around $-60°$ and $180°$. The reason is the steric overlap between the side chain atoms in position γ of residue (i) and the carbonyl oxygen of the third preceding residue (N-terminal to i), denoted here as $C=O_{i\text{-}3}$. The statistical analysis of X_1 distribution in helical enviroments in high-resolution crystal structures clearly supports this generalization, particularly for aromatic side chains. Additional interactions between their side chains and the helix backbone that give rise to preferential X_1 rotamers have been identified before for specific sets of residues.[9] Briefly summarized, they include the following examples.

1. Ser, Thr, and Cys. These residues are capable of H bonding their side chain to the carbonyl at position (i-4) or (i-3). This introduces a conformational constraint for the X_1 rotamer to $-60°$ (g+) and $60°$ (g−), with (g+) being the most abundant. Because this constraint arises from favorable interactions rather than steric hindrances, these residues could also adopt other X_1 rotamer values, e.g., $X_1 = 180°$ for Ser and Cys, although such values appear with much lower frequency. Thr residues are generally restricted to $X_1 = -60°$ due to the C_β-branched character of the side chain, as explained later. Figure 8 illustrates such a hydrogen bond for the case of Thr 3.33(118) in the rhodopsin crystal structure.[5] Ser and Thr may also cause a special bend in the helix in the $X_1 = 60$ (g−) due to the hydrogen bond with $C=O_{i\text{-}4}$.[91]
2. Val/Ile/Thr. In an α helix, C_β-branched residues are restricted to one rotamer, as suggested by the relative rotamer populations observed in an analysis of

[86] N. L. Summers and M. Karplus, *Methods Enzymol.* **202,** 156 (1991).
[87] L. Piela, G. Nemethy, and H. A. Scheraga, *Biopolymers* **26,** 1273 (1987).
[88] J. W. Ponder and F. M. Richards, *J. Mol. Biol.* **193,** 775 (1987).
[89] M. J. McGregor, S. A. Islam, and M. J. Sternberg, *J. Mol. Biol.* **198,** 295 (1987).
[90] M. J. Sternberg and M. J. Zvelebil, *Eur. J. Cancer* **26,** 1163 (1990).
[91] J. A. Ballesteros, X. Deupi, M. Olivella, E. E. Haaksma, and L. Pardo, *Biophys. J.* **79,** 2754 (2000).

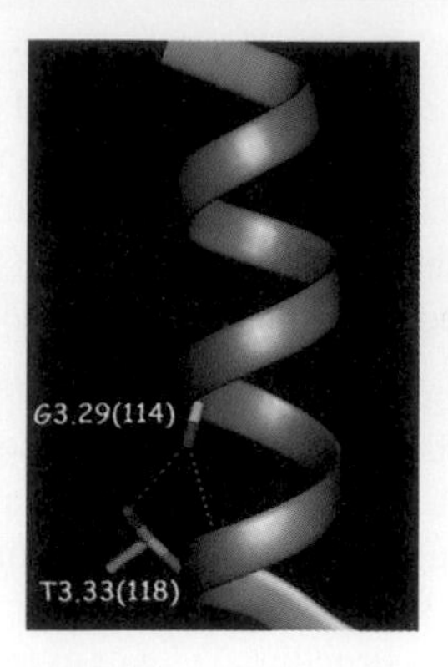

FIG. 8. In an α helix, the side chain of Thr or Ser at position (i) makes a hydrogen bond with the carbonyl oxygen of the residue at position (i-4): illustration of the hydrogen bond between T3.33(188) in the rhodopsin crystal structure[5] and residue G3.29 (114) at position (i-4) from it (the number in parentheses is the sequence number of the residue in rhodopsin).

structures deposited in the PDB. The reason for this preference is the steric clash of a C_γ atom at $X_1 = 60°$ with the $C=O_{i\text{-}3}$. Because C_β-branched AA have two atoms at position γ, the only rotamer that avoids the steric clash at 60° is the one that places the two γ atoms at X_1 values of $-60°$ (and 180°; for reasons of nomenclature, this rotamer corresponds to $X_1 = 60°$ for Ile and Thr, and $X_1 = 180°$ for Val).

Incorporation of Constraints from Experimentally Derived Segment–Segment Interactions

The complex process of modeling the transmembrane bundle of the TMH[9] based mainly on physicochemical properties and biophysical principles (e.g., see Herzyk and Hubbard[57] and Pogozheva *et al.*[58,62]) can gain much from the incorporation of experimentally derived constrains. This is well illustrated in complete modeling studies of specific GPCRs,[9] including models for TRH receptors,[65,66] adrenergic receptors,[73] opiod receptors,[64,92] and various hormone receptors.[69,68,93] The constraints are obtained from experiments designed to explore the structural properties of GPCRs and structure–function relations. A voluminous literature reports these results (see review chapters this volume), and we illustrate only briefly the utility of such data for the modeling process. As this literature grows quickly and is augmented by complete 3D structures of GPCRs, the new and more powerful constraints that will continue to emerge can be used in the same manner, not

[92] D. Strahs and H. Weinstein, *Protein Eng.* **10,** 1019 (1997).
[93] F. Fanelli, *J. Mol. Biol.* **296,** 1333 (2000).

only to obtain consensus structures for GPCRs, but also to develop increasingly accurate models for very specific members of the various GPCR families based on differentiating constraints.

Specific TMH–TMH interactions in GPCRs have been proposed based on spin labeling,[94,55] fluorescence labeling,[95] Cys cross-linking,[94,96,97] engineered Zn-binding sites,[98–100] and double revertant mutant constructs.[68,82,84,101,102] Constraints resulting from such studies outlined in the following subsections are illustrated in Table I.

Distances and Relative Positions Obtained from Spin Labeling and Cys Cross-Linking Experiments in Rhodopsin. The cytoplasmic portions of TMH3 and TMH6 in rhodopsin were shown to be in spatial proximity and to move away upon activation.[94] The pattern of distances between residues 3.54 (139) and 6.30 (247) to 6.35 (2.55) and the subsequent changes observed upon activation led to the proposal that rhodopsin activation involves a movement of TMH6 away from TMH3 and a rotation of TMH6 in an anticlockwise direction viewed from the extracellular side.[94] Similar studies on rhodopsin by the laboratories of Hubbell and Khorana have identified significant movement between residues H1.60(65)C and C7.63(316), which were found to be closer than 10 Å in the ground state, but significantly farther away upon activation.[55] Notably, very similar inferences about the displacement of the cytoplasmic ends of TMH6 and TMH3 relative to each other were obtained from fluorescence experiments in β-adrenergic receptors.[95]

Spin-labeling experiments were complemented by the analysis of Cys–Cys cross-linking patterns between the C3.54 in TMH3 and a Cys placed sequentially at positions 6.30 to 6.35 in TMH6.[94] Cross-linking resulted in rhodopsin inactivation, consistent with findings from spin-labeling experiments showing that TMH6 and TMH3 move away from each other upon activation. That C3.54 could cross-link to all substituted Cys from position 6.30 to 6.34, involving a whole turn of TMH6, is inconsistent with the helical conformation of the region.

Another set of Cys cross-linking studies on rhodopsin[96,97] provided results that are consistent with an α-helical conformation of both TMH5 and TMH6.

Structural and Mechanistic Insights from Engineered Zn^{2+}-Binding Sites. The structures of many soluble Zn^{2+}-binding proteins are known from X-ray crystallography and, therefore, the geometry of the interactions between Zn^{2+} and the

[94] D. L. Farrens, C. Altenbach, K. Yang, W. L. Hubbell, and H. G. Khorana, *Science* **274,** 768 (1996).

[95] U. Gether, S. Lin, P. Ghanouni, J. A. Ballesteros, H. Weinstein, and B. K. Kobilka, *EMBO J.* **16,** 6737 (1997).

[96] H. Yu, M. Kono, T. D. McKee, and D. D. Oprian, *Biochemistry* **34,** 14963 (1995).

[97] M. Kono, H. Yu, and D. D. Oprian, *Biophys. J.* **70,** 395 (1996).

[98] C. E. Elling, S. M. Nielsen, and T. W. Schwartz, *Nature* **374,** 74 (1995).

[99] C. E. Elling and T. W. Schwartz, *EMBO J.* **15,** 6213 (1996).

[100] S. P. Sheikh, T. A. Zvyaga, O. Lichtarge, T. P. Sakmar, and H. R. Bourne, *Nature* **383,** 347 (1996).

[101] V. R. Rao, G. B. Cohen, and D. D. Oprian, *Nature* **367,** 639 (1994).

[102] M. Han, S. W. Lin, M. Minkova, S. O. Smith, and T. P. Sakmar, *J. Biol. Chem.* **271,** 32337 (1996).

TABLE I

EXPERIMENTALLY DERIVED INFORMATION ABOUT TMH–TMH INTERACTIONS THAT SERVE AS SPATIAL CONSTRAINTS

TMH–TMH	AA ←→ AA[a]	Method[b]	GPCR[c]	State[d]
H1-H7	T1.39A ←→ H7.36T	2rm	MUSC	R
	H1.60C ←→ C7.63	spin	RHO	
H2-H7	D2.50 ←→ N7.49	2rm	5HT2a, GnRH	
	K7.43 ←→ G2.57G	2rm	RHO	R*
H2-H3	Y2.64H ←→ H3.28	Zn-His	NK1	R
H3-H5	N3.29H ←→ E5.35H	Zn-His	NK1	R
	D3.32 ←→ S5.42	Lig	DOP/ADR	
	D3.32 ←→ ST5.43	Lig	5HT/DOP/ADR	
	D3.32 ←→ S5.46	Lig	5HT2a/DOP/ADR	
H3-H6	V3.54C ←→ E6.30C	Cys-Cys	RHO	R, not R*
	V3.54C ←→ E6.31C	Cys-Cys	RHO	R, not R*
	V3.54C ←→ E6.32C	Cys-Cys	RHO	R, not R*
	V3.54C ←→ E6.33C	Cys-Cys	RHO	R, not R*
	V3.54C ←→ E6.34C	Cys-Cys	RHO	R, not R*
	V3.53H ←→ T6.34H	Zn-His	RHO	R, not R*
	V3.53H ←→ T6.34H	Zn-His	$\beta 2$	R, not R*
	V3.53H ←→ T6.34H	Zn-His	PTHR	R, not R*
	V3.53H/K3.56H ←→ T6.34H	Zn-His	RHO	R, not R*
	G3.36L ←→ F6.44A	2rm	RHO	R
	D3.32 ←→ N6.55	Lig	$\beta 2$	
	D3.32 ←→ F6.52	Lig	5HT2a/ADR	
H3-H7	K7.43 ←→ E3.28	Lig	RHO	R
	D3.32 ←→ W7.40	Lig	$\alpha 2$	R
	D3.32 ←→ N7.39	Lig	5HT1a/$\beta 2$	R
	D3.32 ←→ N7.39	Zn-His	$\beta 2$	R*
H5-H6	E5.35H ←→ Y6.59H	Zn-His	NK1	R
	H5.39 ←→ Y6.59H	Zn-His	NK1	R
	N5.35C ←→ F6.55C	Cys-Cys	RHO	
	V5.39C ←→ F6.55C	Cys-Cys	RHO	
	V5.39C ←→ A6.56C	Cys-Cys	RHO	
	V5.39C ←→ F6.59C	Cys-Cys	RHO	
	V5.39C ←→ T6.60C	Cys-Cys	RHO	
	I5.40C ←→ F6.55C	Cys-Cys	RHO	
	I5.40C ←→ F6.56C	Cys-Cys	RHO	
	I5.40C ←→ F6.59C	Cys-Cys	RHO	
	Y5.58C ←→ V6.37C	Cys-Cys	RHO	
	G5.59C ←→ V6.37C	Cys-Cys	RHO	

[a] Interactions are indicated by the residue identifier, the AA of the corresponding receptor, and, when appropriate, the mutant.

[b] The experimental technique used to identify the constraint. Codes for experimental techniques are Cys-Cys (Cys cross-linking), Zn-His (engineered Zn-binding sites), spin (spin labeling of Cys residues), 2rm (double revertant mutants), and Lig (ligand-binding residues).

[c] The receptor in which the constraint was determined.

[d] The conformational state of the receptor [active (R*) or inactive (R)].

coordinating residues is well characterized.[12] Consequently, engineered or natural Zn^{2+}-binding sites in membrane proteins have been used to obtain important distance constrains between the transmembrane segments. For example, engineered Zn^{2+}- and Cu^{2+}-binding sites between TMH3 and TMH7 at the extracellular part of the transmembrane bundle have been shown to involve residues 3.32 and 7.39 in the β_2-adrenergic receptor[103] and the tachykinin NK2 receptor.[104] Similarly, Zn^{2+}-binding sites designed to estimate distances between the cytoplasmic portions of TMH3 and TMH6 in rhodopsin,[100] as well as in the β_2-adrenergic and parathyroid hormone (PTH) receptor,[105] indicated that V3.53H was close to T6.34H and K3.56. Because occupancy of the metal-binding site by Zn^{2+} was found to markedly impair the ability of each receptor to mediate ligand-dependent activation of G protein, binding of Zn^{2+} was considered to prevent the movement of TMH6 relative to TMH3 that was characterized from spectroscopic experiments discussed in the previous section. Thus, insights from this engineered Zn^{2+}-binding site connect a structural constraint with a functional determinant in the mechanism of receptor activation.

Structural Constraints Identified from Correlated Mutations and Evolutionary Revertant Mutants. An important step in the construction of the helix bundle is the identification of TM contacts. To this end, useful information is obtained from the identification of possible evolutionary correlated mutations (ERM). Such correlations are observable from the analysis of sequence alignments and have been attributed to the evolutionary pressure to preserve protein structure and function against mutational drift (e.g., see Ballesteros and Weinstein,[9] Gobel *et al.*,[106] and Neher[107]). This pressure constrains the nature of permissible mutations in areas of intramolecular 3D contact and in regions of functional importance. Conversely, mutations that are tolerated only in tandem identify sites that are spatially adjacent.[106,108–111] Residues that display a coordinated AA substitution pattern are thus predicted to interact among themselves. Such reasoning has led to the identification of adjacencies between transmembrane segments 2 and 7 in GPCRs[68,84] involving conserved AA at 2.50 and 7.49. These appear to be an example of evolutionary correlation because when D2.50 is replaced by N2.50

[103] C. E. Elling, K. Thirstrup, B. Holst, and T. W. Schwartz, *Proc. Natl. Acad. Sci. U.S.A.* **96,** 12322 (1999).

[104] B. Holst, C. E. Elling, and T. W. Schwartz, *Mol. Pharmacol.* **58,** 263 (2000).

[105] S. P. Sheikh, J. P. Vilardarga, T. J. Baranski, O. Lichtarge, T. Iiri, E. C. Meng, R. A. Nissenson, and H. R. Bourne, *J. Biol. Chem.* **274,** 17033 (1999).

[106] U. Gobel, C. Sander, R. Schneider, and A. Valencia, *Proteins* **18,** 309 (1994).

[107] E. Neher, *Proc. Natl. Acad. Sci. U.S.A.* **91,** 98 (1994).

[108] F. Pazos, M. Helmer-Citterich, G. Ausiello, and A. Valencia, *J. Mol. Biol.* **271,** 511 (1997).

[109] V. De Filippis, C. Sander, and G. Vriend, *Protein Eng.* **7,** 1203 (1994).

[110] F. Pazos, O. Olmea, and A. Valencia, *Comput. Appl. Biosci.* **13,** 319 (1997).

[111] I. N. Shindyalov, N. A. Kolchanov, and C. Sander, *Protein Eng.* **7,** 349 (1994).

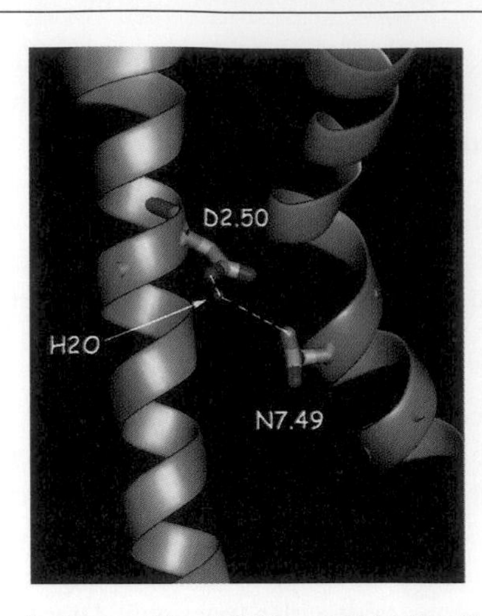

FIG. 9. Spatial proximity between residues D2.50 and N7.49 identified by analysis of correlated mutations (see text) was verified in the crystal structure of rhodopsin,[5] where a water molecule mediates the interaction between D2.50 and N7.49.

(as in the GnRHR), residue N7.49 is replaced by D7.49. The restoration of function in a double-revertant mutant in the gonadotropin-releasing hormone receptor (GnRHR) showed that the pair N2.50/D7.50 has the same ligand-binding properties as the D2.50/N7.50 pair found in most GPCRs, leading to the inference of a shared microenvironment and hence adjacency of TMH2 and TMH7 in that region.[68,84] Analogous results were obtained for the β_2-adrenergic receptor[14] and more recently for TRHR[83] and CCKR.[112] Structural implications of the hypothesis are verified in the crystal structure of rhodopsin where such proximity is found in the microenvironment of the 2.50 and 7.49 loci, with the two TM segments bridged by hydrogen bonds through a water molecule (Fig. 9). It is necessary to point out here that such an interaction is possible only if TMH7 is highly distorted in the region of the conserved P7.50, as discussed in a previous section. The distortion is caused by the NP motif created by N7.49, which modulates both the size and the shape of the kink induced in the helix by P7.50.

Other revertant mutants have also been described, e.g., K7.43-D2.57-E3.28 in rhodopsin.[101,113] Residue F7.38 has been proposed to face TMH3 and/or TMH6 based on revertant chimeras in adrenergic receptors.[114] The effect of the structural perturbation produced by the H7.36(423)T mutation in muscarinic receptors was

[112] D. Donnelly, S. Maudsley, J. P. Gent, R. N. Moser, C. R. Hurrell, and J. B. Findlay, *Biochem. J.* **339,** 55 (1999).

[113] D. D. Oprian, *J. Bioenerg. Biomembr.* **24,** 211 (1992).

[114] T. Mizobe, M. Maze, V. Lam, S. Suryanarayana, and B. K. Kobilka, *J. Biol. Chem.* **271,** 2387 (1996).

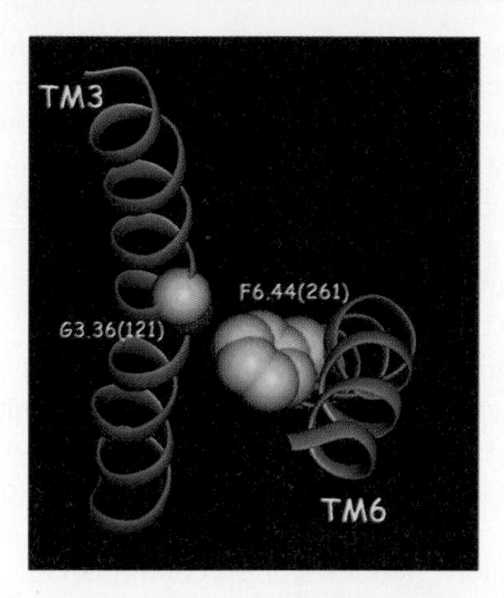

FIG. 10. Spatial proximity between residues 3.36 and 6.44 identified by analysis of correlated mutations (see text) was verified in the crystal structure of rhodopsin,[5] as illustrated in the figure for residues G3.36(121) and F6.44(261). Note the spatial complementarity between the bulky phenylalanine and the small glycine residue.

reversed by the subsequent mutation of T1.39(37) A,[82] suggesting that H7.36(423) and T1.39(37) are in spatial proximity. Another structural revertant was observed in rhodopsin with the G3.36(121)L and F6.44(261) A mutations,[102] and the proximity of these loci is verified by the crystal structure of rhodopsin[5] (Fig. 10). All these examples illustrate the significant predictive power of the revertant mutant criterion in the identification of spatial adjacencies.

Modeling of Loop Regions That Connect Transmembrane Helical Segments

Complete 3D models of GPCRs must include the loop regions, both extracellular and intracellular, because of their important role in the functional mechanisms of these proteins. This is not an easy task because homology modeling methods are not very successful for such short segments, especially in the complete absence of valid structural templates.[115] Many attempts at loop modeling have simply ignored these concerns, but the outcomes are not likely to be found valid. More reliable methods involve a structure–prediction component and some form of energy-based steps, e.g., by extensive structure annealing[116] or energy-based criteria.[115] Note that the following procedural steps pertain to work on relatively short loop stretches connecting the transmembrane segments, as the more extensive segments, such as the third intracellular loop in monoamine GPCRs, may require the application of

[115] A. Fiser, R. K. Do, and A. Sali, *Protein Sci.* **9,** 1753 (2000).
[116] R. Sankararamakrishnan and H. Weinstein, *Biophys J.* **79,** 2331 (2000).

the more commonly used modeling techniques designed for medium-sized soluble proteins. The applicable modeling steps are best used in combinations that are likely to increase the validity of results from these as yet imperfect approaches.

a. Multiple sequence alignment[117–119] used to identify any level of sequence homology among corresponding loop segments in the various GPCRs. Using neural pattern recognition (e.g., the method of Rost and Sander[120,121]), the propensities at the aligned positions can be averaged and used in predictions of secondary structure.

b. Search for homologous sequences of various lengths in the PDB[74] that may have homologous structures. Larger segments can be "synthesized" from superpositions and additions of short stretches with defined structure, with extensive energy-based structure optimization (see point d) serving for annealing and feasibility checks.

c. Modeling of short loops using conformation selection based on statistical trends[122–124] combined with energy minimization (by extensive molecular dynamics and/or Monte Carlo methods) to model backbone conformation,[125,126] including complete conformational searches of connecting segments of the loops.

d. For short loop segments and for loop segments connecting model structures obtained from secondary structure prediction of homology modeling, data on the conformational space can be obtained with methods for conformational searches (e.g., see Guarnieri and Weinstein[125]) and specific methods for loop modeling.[127] Thus, loop searches can be performed in data bases from known protein structures, based on initial fragment conformations obtained from calculations with restricted torsion angles and side chain conformations restricted to "allowed" values from statistical considerations and most populated conformations.[125] These must then be followed by energy-based conformational scans to refine the fragment of structures and identify incorrect selections. The computational methods in the CHARMM package[128] and Monte Carlo methods, such as Conformational Memories,[125] can serve in these explorations.

[117] P. Argos and M. Vingron, *Methods Enzymol.* **183,** 352 (1990).

[118] G. J. Barton, *Acta Crystallogr. D. Biol. Crystallogr.* **54,** 1139 (1998).

[119] M. Vihinen, *Methods Enzymol.* **183,** 447 (1990).

[120] B. Rost and C. Sander, *J. Mol. Biol.* **232,** 584 (1993).

[121] B. Rost and C. Sander, *Annu. Rev. Biophys. Biomol. Struct.* **25,** 113 (1996).

[122] S. Salzberg and S. Cost, *J. Mol. Biol.* **227,** 371 (1992).

[123] S. A. Benner and D. Gerloff, *Adv. Enzyme Regul.* **31,** 121 (1991).

[124] J. M. Levin, B. Robson, and J. Garnier, *FEBS Lett.* **205,** 303 (1986).

[125] F. Guarnieri and H. Weinstein, *J. Am. Chem. Soc.* **118,** 5580 (1996).

[126] I. Visiers, S. Hassan, and H. Weinstein, *Protein Eng.* **14,** in press (2001).

[127] L. E. Donate, S. D. Rufino, L. H. Canard, and T. L. Blundell, *Protein Sci.* **5,** 2600 (1996).

[128] B. R. Brooks, R. E. Bruccoleri, B. D. Olafson, D. J. States, S. Swaminathan, and M. Karplus, *J. Comp. Chem.* **4,** 187 (1983).

Conserved Structural Motifs and Functional Microdomains Play a Key Role in the Refinement of Receptor Models

A powerful approach in the process of refining initial molecular models of GPCRs has evolved from the strong cooperation between experimental explorations of structure–function relations and computational simulation approaches (e.g., see Sealfon *et al.*,[84] Ballesteros *et al.*,[69] and Ebersole and Sealfon[129]). This approach is based on parsing the receptor sequence into groups of residues that correspond to "microdomains." The microdomains are characterized by specific structural motifs not necessarily contiguous but exhibit a very high degree of conservation. Such microdomains are most likely to have a specific role in the function of the proteins. The identification of such structural and functional microdomains is based on sequence comparisons and on inferences from biophysical principles and experimental verification. Some specific examples discussed later illustrate the application of these approaches based on (1) sequence analysis for correlated mutations and mutagenesis[68,84,130]; (2) biophysical principles and mutagenesis[69]; and (3) the substituted cysteine accessibility method (SCAM) and computational simulation.[131]

The elucidation of structural and functional details of such microdomains plays a key role in the refinement of molecular models of GPCRs because the functional implications, when verified experimentally, can offer solid information about specific structural relationships in different parts of the receptor molecules. The accuracy and predictive power of such information are higher than the considerations that have produced the initial receptor model.

In most cases, the functional microdomains participate in receptor mechanisms through conformational changes in backbone or side chain interactions. These changes are propagated into large segments of the receptor protein, thus revealing sets of overlapping structural constraints that can be used in refining the model. The best described structural motifs (SM) characterized as functional microdomains include the ligand-binding site (e.g., see Wang *et al.*,[132] Almaula *et al.*,[133,134] Choudhary *et al.*,[135,136] Roth *et al.*,[137] and Javitch *et al.*[138]), the "aromatic cluster"

[129] B. J. Ebersole and S. C. Sealfon, *Method. Enzymol.* **343,** [7] 2001 (this volume).

[130] C. Flanagan, W. Zhou, L. Chi, T. Yuen, V. Rodic, D. Robertson, M. Johnson, P. Holland, R. Millar, H. Weinstein, R. Mitchell, and S. Sealfon, *J. Biol. Chem.* **274,** 28880 (1999).

[131] J. A. Javitch, J. A. Ballesteros, H. Weinstein, and J. Chen, *Biochemistry* **37,** 998 (1998).

[132] C. D. Wang, T. K. Gallaher, and J. C. Shih, *Mol. Pharmacol.* **43,** 931 (1993).

[133] N. Almaula, B. J. Ebersole, D. Zhang, H. Weinstein, and S. C. Sealfon, *J. Biol. Chem.* **271,** 14672 (1996).

[134] N. Almaula, B. J. Ebersole, J. A. Ballesteros, H. Weinstein, and S. C. Sealfon, *Mol. Pharmacol.* **50,** 34 (1996).

[135] M. S. Choudhary, S. Craigo, and B. L. Roth, *Mol. Pharmacol.* **43,** 755 (1993).

[136] M. S. Choudhary, N. Sachs, A. Uluer, R. A. Glennon, R. B. Westkaemper, and B. L. Roth, *Mol. Pharmacol.* **47,** 450 (1995).

in TMH6,[131,139] and the "arginine cage" comprising residues in the cytoplasmic end of TMH3[69] and E6.30.[48,139,140] As illustrated briefly later, structural constraints inferred from the properties of microdomains enhanced the modeling of GPCRs in the important context of their functional mechanisms. In particular, they were shown to provide steric constrains that were essential for refining the interfaces of various TMH-specific interactions among them and positions of TMHs in the transmembrane bundle.[69,131,133,134,138]

Modeling Criteria Based on the Ligand-Binding Pocket

The identification of ligand–receptor interactions imposes some distance and conformational constrains among the residues that constitute the binding pocket microdomain. Such information is highly valuable in the process of modeling the TM bundle and in the positioning of the loops relative to the TM domain. For example, while the binding site in receptors for opsins, monoamines (serotonin, norepinephrine, dopamine), and canabinoids seems to be located in the transmembrane core, binding of the endogenous ligands of peptide receptors, including chemokines, GnRH, TRH, and opioids, involves interactions with residues in the extracellular loops. It is worth mentioning that the binding site for small nonpeptide ligands of the same peptide receptors may overlap only partially with the endogenous peptide binding site and is likely to involve primarily the transmembrane segment, e.g., for chemokine receptors (e.g., see Bockaert and Pin[4]). Moreover, ligand receptor interactions are both ligand and receptor dependent so that different ligands may bind differently in the same receptor and the same ligand may bind differently in different receptors. To illustrate constraints and modeling criteria based on ligand–receptor interactions, the examples provided later focus on monoamine receptor binding sites; for a comprehensive review of ligand-binding domains for other GPCRs see Gether *et al.*[141]

The combination of data from mutagenesis, structure–activity studies, and direct structural elucidation identifies the binding site of endogenous ligands of rhodopsin-like monoamine receptors as being formed mainly by residues in TMHs 3, 5, and 6. The aspartate conserved at position 3.32 of these receptors has been shown to interact electrostatically with the positively charged nitrogen of the ligand in the β_2-adrenergic receptor[142] and in the receptors for dopamine,[143]

[137] B. L. Roth, M. Shoham, M. S. Choudhary, and N. Khan, *Mol. Pharmacol.* **52,** 259 (1997).

[138] J. A. Javitch, D. Fu, J. Chen, and A. Karlin, *Neuron* **14,** 825 (1995).

[139] I. Visiers, J. A. Ballesteros, and H. Weinstein, *Biophys. J.* **78,** 68A:394 (2000).

[140] S. C. Sealfon, B. J. Ebersole, S. Dracheva, J. A. Ballesteros, and H. Weinstein, *Soc. Neurosci. Abstr.* **24,** 773 (1998).

[141] U. Gether, *Endocr. Rev.* **21,** 90 (2000).

[142] C. D. Strader, T. Gaffney, E. E. Sugg, M. R. Candelore, R. Keys, A. A. Patchett, and R. A. Dixon, *J. Biol. Chem.* **266,** 5 (1991).

[143] A. Mansour, F. Meng, W. J. H. Meador, L. P. Taylor, O. Civelli, and H. Akil, *Eur. J. Pharmacol.* **227,** 205 (1992).

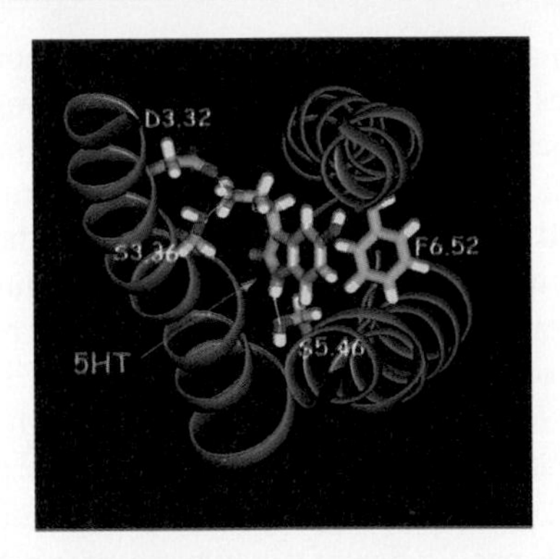

FIG. 11. Serotonin (5HT) in the binding pocket of the $5HT_{2C}$ receptor. The main interactions involve (i) the protonated amine in 5HT and D3.32 and S3.36, (ii) the indole ring of 5HT and F6.52, and (iii) the indole nitrogen of 5HT and S5.46.

serotonin,[132,144] histamine,[145] and acetylcholine.[146] Four residues away from this D3.32, and thus on the same face of the TMH, the S3.36 in the $5HT_{2A}$ receptor was shown to provide a second point of interaction with primary and secondary amines.[133] Additional interactions for 5HT in the binding pocket were inferred from mutagenesis and modeling studies to include S5.46[134] and F6.52, which is likely to interact through an aromatic–aromatic interaction with the indole ring in serotonin.[135,137] The catechol ring in β_2-adrenergic receptor ligands who also shown to interact with F6.52,[147] and two hydrogen bonds between the catechol hydroxyl groups and two serines (S5.43 and S5.46) in TMH5 have been described in the β_2-adrenergic receptor,[148] as well as a hydrogen-bonding interaction with agonist containing a meta-OH.[149] Another residue in TMH6, N6.55 seems to be involved in a hydrogen-bonding interaction with the β-hydroxyl of epinephrine.[150] In addition, N7.39 provides for a hydrogen-bonding residue to interact with the ligand in the β_2-adrenergic receptor. Figure 11 illustrates the main interactions of serotonin in the binding pocket of the $5HT_{2C}$ receptor.

Despite the great variety of receptors and ligands, the findings described earlier identify a spatially compact binding microdomain for this family of receptors.

[144] B. Y. Ho, A. Karschin, T. Branchek, N. Davidson, and H. A. Lester, *FEBS Lett.* **312,** 259 (1992).

[145] I. Gantz, J. DelValle, L. D. Wang, T. Tashiro, G. Munzert, Y. J. Guo, Y. Konda, and T. Yamada, *J. Biol. Chem.* **267,** 20840 (1992).

[146] T. A. Spalding, N. J. Birdsall, C. A. Curtis, and E. C. Hulme, *J. Biol. Chem.* **269,** 4092 (1994).

[147] M. R. Tota and C. D. Strader, *J. Biol. Chem.* **265,** 16891 (1990).

[148] C. D. Strader, M. R. Candelore, W. S. Hill, R. A. Dixon, and I. S. Sigal, *J. Biol. Chem.* **264,** 16470 (1989).

[149] G. Liapakis, J. A. Ballesteros, S. Papachristou, W. C. Chan, X. Chen, and J. A. Javitch, *J. Biol. Chem.* (2000).

[150] K. Wieland, H. M. Zuurmond, C. Krasel, A. P. Ijzerman, and M. J. Lohse, *Proc. Natl. Acad. Sci. U.S.A.* **93,** 9276 (1996).

The spatial distribution of the sites that interact simultaneously with the relatively small ligand constitutes a "distance geometry" connecting the helices that bear the contributing residues. This matrix of spatial criteria must be taken into consideration by the specific models of GPCR families. The refinement of models for individual subtypes will incorporate in a similar manner the known pharmacological differences between closely related receptors. An illustration of such a case is the selectivity of antagonists for D2 vs D4 dopamine receptors. As described elsewhere,[151] a cluster of six residues distributed in TMH2, TMH3, and TMH7 (positions 2.60, 2.61, 2.64, 3.28, 3.29, and 7.35), which differ in the two dopamine receptor subtypes, was identified as a functional motif for the discriminant ligand binding. When interchanged between D2 and D4 dopamine receptors, the motif interconverts the ligand specificities of the two receptor subtypes.

The importance of structural criteria based on the binding microdomain for the correct modeling of structural details in GPCRs is emphasized by findings showing that differences in the orientation of ligands in the binding pocket of the $5HT_{2A}$ receptor[152] can account for the differences in their measured pharmacological efficacy.[129] Accurate placement in modeled GPCRs of the SM/FM elements is thus seen to be essential for the construction of valid models that can be used to obtain mechanistic inferences.

Modeling Criteria Based on the "Aromatic Cluster" in TMH6

Structural constraints related to the motif formed by the conserved aromatic residues in TMH6 (F6.44, W6.48, F6.51, and F6.52)[131] further refine the model, based on the relation of this SM to the binding microdomain. Thus, the sensitivity of the receptor activation mechanism to the position of the ligand suggested by our findings for the $5HT_{2A}R$[129,139,152] indicates that at least one of the elements in the binding microdomain must be sensitive to variations in the orientation of ligands in the binding pocket. One of the components of the aromatic cluster in TMH6, the conserved F6.52, is accessible in the binding site[131] and is likely to serve as such a "sensor." Through the interactions of the ligand with F6.52, the structural properties of the aromatic cluster allow it to respond through concerted conformational rearrangements of the aromatic side chains, like a "toggle switch," to promote agonist-mediated receptor activation. The concerted conformational changes propagated in this cluster on activation can transmit the binding stimulus further toward the cytoplasmic side of the protein. This dynamic hypothesis is supported by the reported conformational change in the orientation of W6.48 in rhodopsin from an orientation "perpendicular" to the membrane in the

[151] M. M. Simpson, J. A. Ballesteros, V. Chiappa, J. Chen, M. Suehiro, D. S. Hartman, T. Godel, L. A. Snyder, T. P. Sakmar, and J. A. Javitch, *Mol. Pharmacol.* **56,** 1116 (1999).

[152] N. Almaula, B. J. Ebersole, D. Zhang, H. Weinstein, and S. Sealfon, *J. Biol. Chem.* **271,** 14672 (1996).

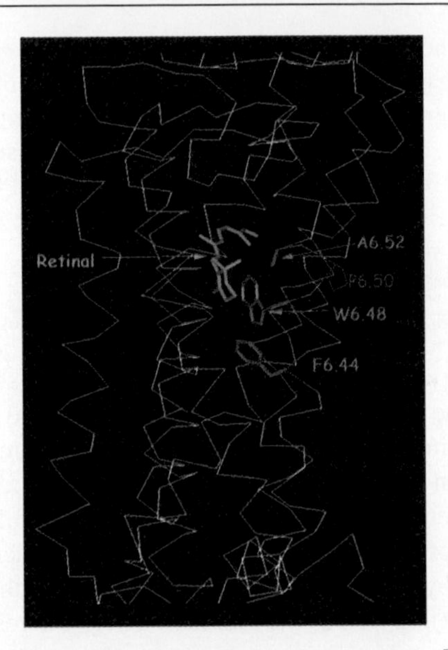

FIG. 12. The strategic position of retinal in rhodopsin crystal structure[5] relative to residue W6.48 locks this residue side chain in orientation perpendicular to the plane of the membrane. Earlier spectroscopic data suggested that this perpendicular orientation corresponds to the inactive state of the receptor.[153]

inactive form to a "parallel" orientation in activated rhodopsin.[153] In serotonin $5HT_{2A}R$, our findings point to the interaction between the aromatic moieties of F6.52 and 5HT as a trigger for the conformational change in the highly conserved W6.48.[139]

Interestingly, rhodopsin with the retinal attached covalently to the receptor does not conserve an aromatic ring at position 6.52, while conserving P6.50 and W6.48 (Fig. 12). Inspection of the crystal structure of the inactive form of rhodopsin[5] reveals that a direct hydrophobic interaction between W6.48 and the retinal ring (Fig. 12) keeps W6.48 in the perpendicular orientation suggested from spectroscopy.[153] In rhodopsin, the ligand is already buried deep in the membrane where it interacts directly with W6.48. Here, the "sensor," which in neurotransmitter receptors is the aromatic ring at position 6.52, is imbedded in the "ligand." The rearrangement triggered in the neurotransmitter receptor by the interaction of the ligand with F6.52 is likely to be produced in rhodopsin by the conformational change of the retinal after isomerization around the double bond (i.e., by releasing the W6.48 from its perpendicular position shown in Fig. 12 to its "planar" conformation). This prediction should be verified in the structure of an activated

[153] S. W. Lin and T. P. Sakmar, *Biochemistry* **35,** 11149 (1996).

form of rhodopsin, where W6.48 should have "escaped" the constraint of the retinal and have the ring positioned in parallel to the membrane plane. The steric and orientational congruency of the elements in the aromatic cluster motif, to the binding microdomain, constitutes a powerful modeling constraint. This constraint has been a part of our 3D models of the neurotransmitter GPCRs (e.g., see Gether *et al.*,[95] Javitch *et al.*,[131] and Visiers *et al.*[139]) even before the crystal structure[5] verified it.

Modeling Criteria Based on the DRY Motif at the Cytoplasmic End of TMH3

Analysis of conservation patterns of the sequence motifs surrounding the highly conserved R3.50 at the cytoplasmic end of TMH3 suggested that the arginine side chain is constrained in the inactive receptor state. The constraints are imposed by an ionic interaction with the neighboring conserved residue D3.49 and by the conserved steric properties of the type of residues found at 3.54.[69] The positioning of this constrained motif in the initial model of the helix packing between TMH3 and TMH6 is guided by the position of D3.32 and F6.52 with respect to the ligand (see earlier discussion), as well as the interaction between residues 3.53–3.54 and 6.30–6.34.[94] In the serotonin receptor, this model brings the conserved R3.50 in close proximity to the conserved E6.30, and computational modeling of the interaction indicates formation of an ionic bond between these two residues.[48] The conservation pattern of E6.30 in neurotransmitter receptors parallels the 100% conservation of R3.50 and D3.49, thus supporting the proposed interactions. Consequently, R3.50 was proposed to be restrained in the inactive state of the receptor by two acidic groups: E6.30 and D3.49.[48,140] This structural hypothesis based on the analysis of local sequence conservation patterns in the microdomain surrounding the conserved DRY motif has been validated by the crystal structure of rhodopsin in which the side chain of R3.50 is "caged" through an electrostatic interaction with D3.49 and E6.30 as predicted (Fig. 13). Interaction between

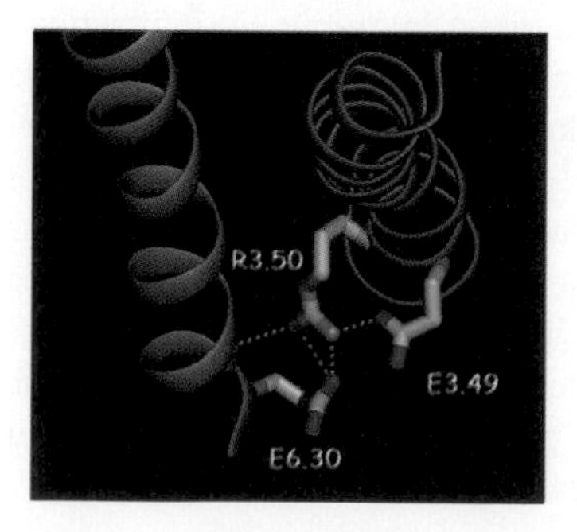

FIG. 13. The conserved "arginine cage" in the crystal structure of rhodopsin.[5] The side chain of R3.50 interacts with D3.49 and E6.30.

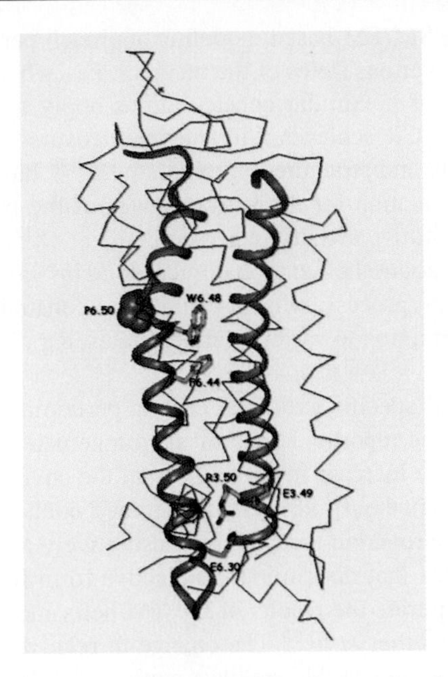

FIG. 14. The interaction between TMH3 and TMH6 at the cytoplasmic side in the inactive state is likely to depend on the ionic interaction between E6.30 and R3.50. Such an interaction exists in rhodopsin crystal structure[5] (see Fig. 13 and text for functional implications).

the cytoplasmic ends of TMH3 and TMH6, which has been demonstrated by a variety of methods (see earlier sections), is therefore likely to depend on this ionic interaction between E6.30 and R3.50 observed experimentally for rhodopsin[94,100] (Fig. 14) and to be sensitive to the protonation state of the carboxylic groups at positions 3.49 and 6.30 in rhodopsin-like GPCRs. These protonation states are likely to be different in different states of the receptor (e.g., inactive vs ligand activated) and therefore produce structural rearrangements in the receptor due to changes in the status of the "arginine cage."

Modeling Various Functional States of the Receptor

The identification of structural motifs in GPCRs that can be implicated mechanistically into functional microdomains offers a perspective on the cascade of conformational rearrangements in the GPCR molecule that relate the inactive form to the activated state of the receptor. In particular, the receptor modeling approach based on SM/FM parsing makes possible the construction of separate models for inactive and active states of the receptor. The discussion of the DRY motif and "arginine cage" structures in the previous section provides one example of the manner in which the SM/FM-based modeling approach permits a more accurate representation of the various states of the receptor, i.e., when complexed with the ligand or when free of it. Similar considerations apply to the modeling of the state in which the GPCR achieves a ligand-free activated form ("constitutively active"). Mutagenesis and structure–function studies[129] have produced a wealth of experimental information for the various states of the protein, including different levels of constitutive activity (e.g., see Perez *et al.*[154] and Gether *et al.*[80]). Specific information about the functional microdomains can be used in the initial stages of the modeling process, whereas additional constraints can be identified from information pertaining to various defined states, e.g., in a constitutive active mutant.

Some examples of specific structural criteria pertaining to different states of the receptor include the reported structural rearrangements associated with agonist activation [e.g., the increase in the polarity of the environment around C3.44 (125) and C6.47 (285) in the β_2-adrenergic receptor[95] or changes in the orientation of TMH6 of the β_2-adrenergic receptor in constitutively active mutants[155]]. The structural determinants that discriminate the active form from the inactive form of the receptor incorporate the results of the TM helix movements described by Farrens *et al.*[94] and Gether *et al.*[95] The change in relative position of the cytoplasmic ends of TMH3 and TMH6 entails a motion of TMH6 outward and clockwise from the intracellular side around a flexible hinge created by the conserved P6.50.[9,95,156] This P6.50 residue is conserved among all GPCRs, and through the rearrangements described in a previous section can provide the connection between the conformational changes that happen in the binding microdomain to the third intracellular loop (IL3) where the key interactions with the G protein occur. As discussed in detail earlier, the special local flexibility of the PK confers a dynamic behavior to the proline-containing helix that propagates the functional mechanisms. This appears to be a generalized mechanism in membrane proteins (see Ballesteros and Weinstein[35], Sansom and Weinstein[31]) and was demonstrated experimentally for connexin-32 where structural distortion of the α helix by a conserved proline was found to play a key role in the voltage-dependent gating mechanism.[81]

It is not surprising, therefore, to find that an absolutely conserved proline (P6.50) is likely to play a central role in the structural rearrangement of rhodopsin-like GPCRs on transition from the inactive to the activated state. The nature of the rearrangement identifies key differences between structures of the

[154] D. M. Perez, J. Hwa, R. Gaivin, M. Mathur, F. Brown, and R. M. Graham, *Mol. Pharmacol.* **49,** 112 (1996).

[155] J. A. Javitch, D. Fu, G. Liapakis, and J. Chen, *J. Biol. Chem.* **272,** 18546 (1997).

[156] X. Luo, D. Zhang, and H. Weinstein, *Protein Eng.* **12**(7), 1441 (1994).

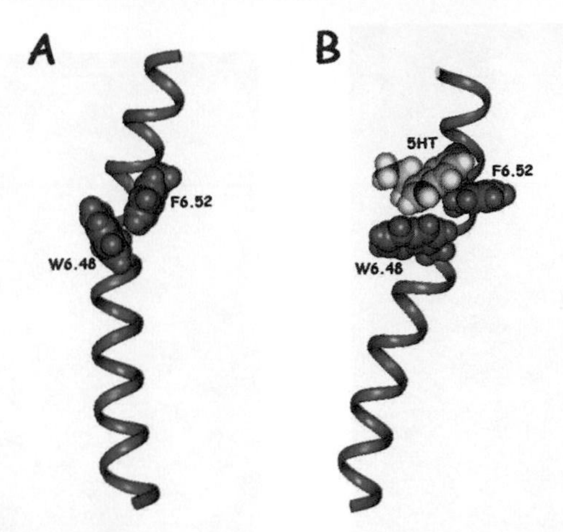

FIG. 15. Illustration of the expected conformational change of the aromatic cluster in TMH6 on ligand binding, which is likely to involve the proline kink of the conserved P6.50 in the motion of TMH6. (A) Model of TMH6 in the $5HT_{2C}$ receptor in the inactive state. Note that the helix is highly kinked, and W6.48 adopts an orientation perpendicular to the membrane.[153] (B) Active state represented by the agonist-bound $5HT_{2C}$ receptor. Note that the helix is minimally kinked and W6.48 is parallel to the membrane.[153]

active and inactive state models of these GPCRs that relate to the geometry of the proline kink at P6.50. Interestingly, the aromatic cluster of residues in TMH6 (including, as described in the previous section, residues F6.44, W6.48, F6.51, and F6.52) straddles the absolutely conserved proline P6.50. Computational simulations of the dynamic properties of GPCRs, using Monte Carlo and molecular dynamics methods, show[157] that the side chain conformation of the residues comprising the aromatic cluster are interrelated dynamically and that depending on the presence of a ligand in the binding pocket, the "toggle switch" mechanism described in the previous section will rearrange these conformations.[139] Because of the position of the conserved P6.50 in the middle of this "toggle switch," the rearrangement also determines the preferred size and geometry of the kink produced by P6.50. Thus, when the ligand is bound to the receptor, the activated form of the receptor is triggered by the interaction with F6.52 that induces the rearrangement in the W6.48 to an orientation parallel to the membrane (Fig. 15). This causes a drastic decrease in the proline kink of TMH6. The active state of the receptor molecule is thus characterized by a nearly straight TMH6 (in contrast to the kinked conformation observed in the inactive form of the rhodopsin crystal structure[5]). This reduction in the bend angle of TMH6 causes the movement of the cytoplasmic

[157] I. Visiers and H. Weinstein, unpublished (2000).

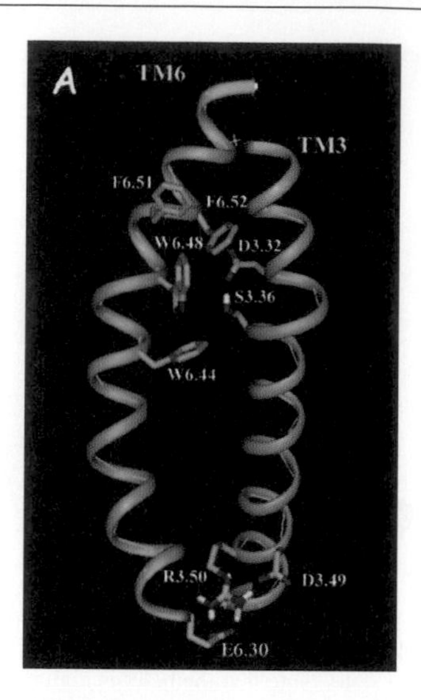

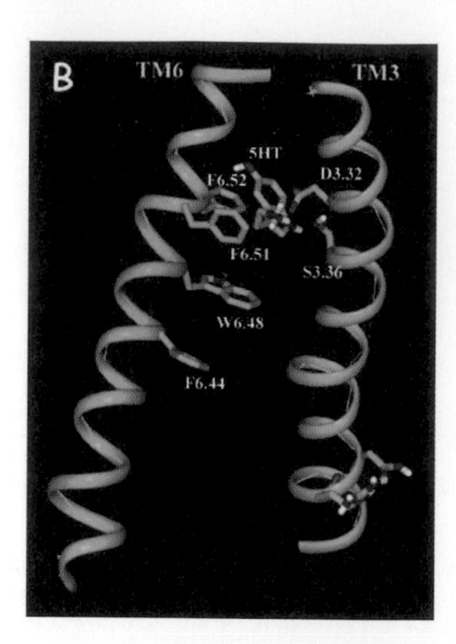

FIG. 16. Changes in the proline kink of P6.50 result in the observed increase in the distance between the cytoplasmic segments of TMH3 and TMH6. (A) Model of the inactive state of the TMH3 and TMH6 in the $5HT_{2C}$ receptor. Note the perpendicular orientation of W6.48 relative to the membrane plane, the strongly kinked TMH6, and the R3.50 caged by D3.49 and E6.30. (B) The active state (ligand is bound) model, characterized by the parallel orientation of W6.48 relative to the membrane plane, a minimally kinked TMH6, and an increased distance between TMH3 and TMH6 cytoplasmic portions that eliminates the interaction between R3.50 and E6.30.

ends of TMH6 away of TMH3 (Fig. 16) observed experimentally.[94] Therefore, in the active state model of the GPCR, the E6.30–R3.50 interaction described in the previous section is removed, which explains both the increased distance between the ends of TMH3 and TMH6 (Fig. 16B) and the observed increase in solvent accessibility and side chain mobility at the TMH6 cytoplasmic boundaries of rhodopsin.[158] These structural rearrangements constitute key elements in the construction of discriminant molecular models representing the different states of GPCRs.

Concluding Remarks

The complexities identified in this article of molecular modeling methods and validation approaches reflect the special challenge presented by the characteristics of flexible, allosteric proteins such as GPCRs. These proteins have evolved

[158] C. Altenbach, K. Yang, D. L. Farrens, Z. T. Farahbakhsh, H. G. Khorana, and W. L. Hubbell, *Biochemistry* **35,** 12470 (1996).

to carry out functions based on a cascade of structural changes that transduce ligand-binding signals from one region of the molecule into another. Their functional mechanisms relate structural changes to a modulation of their affinity for the various proteins with which they interact in cellular signal transduction mechanisms. The dynamic properties of the molecular structures of GPCRs therefore constitute an important element of the emerging understanding of their structure–function relations. Consequently, even the availability of 3D structures of GPCRs for one state of the protein or another (such as the breakthrough structure of inactive rhodopsin published recently[5]) is unlikely to supersede soon the need for 3D molecular models of a very large number of different GPCRs in their various families. As illustrated briefly in the previous section, these models, and the large number of mutant constructs that can be derived from them rapidly, can be queried computationally for the dynamic aspects of their properties and functions. These computational simulations can be expanded to include interactions with G proteins and other cellular factors. In turn, as the 3D models evolve and are refined by comparisons to biological data and to structures obtained directly from crystallography and/or nuclear magnetic resonance, the methods that serve in their construction are validated and/or refined as well. Importantly, these refined methods and approaches validated by specific experimental data can serve in the modeling of a variety of other membrane proteins, such as channels and transporters. As the analysis presented in this article indicates, the use of the methods is justified by the many structural attributes that these proteins share with GPCRs by virtue of their position in the cell, the nature of their activation mechanisms, and the sensitivity to specific ligands.

Acknowledgments

This work was supported by NIH Grants DA-09083, DA-12923, and DA-00060 (to HW). Computational support was provided by the Cornell Supercomputer Facility and the Advanced Scientific Computing Laboratory at the Frederick Cancer Research Facility of the National Cancer Institute (Laboratory for Mathematical Biology). I.V. and J.A.B. contributed equally to this article.

[23] Reconstitution of G Protein-Coupled Receptors with Recombinant G Protein α and $\beta\gamma$ Subunits

By WILLIAM E. MCINTIRE, CHANG-SEON MYUNG, GAVIN MACCLEERY, QI WANG, and JAMES C. GARRISON

Introduction

The interaction of a G protein-coupled receptor (GPCR) with its cognate heterotrimeric G protein is the initial step in the generation of signals that regulate cellular function. Current understanding of this signaling pathway shows it to be surprisingly complex, with large families of proteins comprising receptors, G proteins, and effectors[1–4] and with most cell types expressing multiple isoforms of each category. There appear to be over 1000 GPCR,[1] 23 different α subunits (including splice variants), 5 β subunits, and 12 γ subunits known from their cDNA clones.[5] Many receptors are known to signal through several G proteins, and, in turn, many effectors receive signals from different isoforms of both α and $\beta\gamma$ subunits of G proteins.[3,4,6] While it is possible to reconstitute the major functions of this signaling system with pure receptors and G protein heterotrimers alone, it is now clear that many accessory proteins regulate the system in the intact cell, adding additional signaling specificity. Some of the important accessory proteins include regulators of G protein signaling (RGS proteins),[7–9] activators of G protein signaling (AGS proteins),[10] and receptor kinases.[11,12]

Nevertheless, a basic issue in cell signaling is identification of those receptors that interact with specific isoforms of the multiple G protein α and $\beta\gamma$ subunits. In addition, at the molecular level, it is important to understand which domains of the G protein α and $\beta\gamma$ subunits interact with the intracellular loops

[1] C. D. Strader, T. M. Fong, M. R. Tota, D. Underwood, and R. A. F. Dixon, *Annu. Rev. Biochem.* **63,** 101 (1994).

[2] E. J. Neer, *Cell* **80,** 249 (1995).

[3] D. E. Clapham and E. J. Neer, *Annu. Rev. Pharmacol. Toxicol.* **37,** 167 (1997).

[4] H. E. Hamm, *J. Biol. Chem.* **273,** 669 (1998).

[5] G. B. Downes and N. Gautam, *Genomics* **62,** 544 (1999).

[6] R. K. Sunahara, C. W. Dessauer, and A. G. Gilman, *Annu. Rev. Pharmacol. Toxicol.* **36,** 461 (1996).

[7] H. G. Dohlman and J. Thorner, *J. Biol. Chem.* **272,** 3871 (1997).

[8] D. M. Berman and A. G. Gilman, *J. Biol. Chem.* **273,** 1269 (1998).

[9] J. R. Hepler, *Trends Pharm. Sci.* **20,** 376 (1999).

[10] A. Takesono, M. J. Cismowski, C. Ribas, M. Bernard, P. Chung, S. Hazard III, E. Duzic, and S. M. Lanier, *J. Biol. Chem.* **274,** 33202 (1999).

[11] J. P. Pitcher, J. Inglese, J. B. Higgins, J. L. Arriza, P. J. Casey, C. Kim, J. L. Benovic, M. M. Kwatra, M. G. Caron, and R. J. Lefkowitz, *Science* **257,** 1264 (1992).

[12] J. G. Krupnick and J. L. Benovic, *Annu. Rev. Pharm. Toxicol.* **38,** 289 (1999).

0076-6879/02 $35.00

and transmembrane helices of receptors. Thus, it is useful to have methods to study the receptor:G protein interaction with recombinant proteins. To date, many techniques have been developed to examine this question, which use intact cells, isolated plasma membranes, and/or purified receptors and G proteins. These methods include the following.

Determination of Receptor-Heterotrimer Specificity

Stimulating receptors in isolated plasma membranes with a specific agonist and then identifying the activated α subunits by photolabeling with [^{32}P]azido-GTP can determine the α subunit(s) activated by the receptor. The radiolabeled α subunits can be extracted from the membrane and immunoprecipitated with specific antibodies. This approach has the major advantages of causing minimal disruption of the receptor:α:$\beta\gamma$ interaction and that activation of the receptor targets its cognate α subunit.[13] The method has the disadvantages that it does not identify the particular $\beta\gamma$ subunit supporting receptor coupling and that it cannot be used easily to examine the function of mutant α and/or $\beta\gamma$ dimers.

Knockout of Specific Proteins

Antisense mRNA methods have been developed to selectively remove or reduce the concentration of specific G protein α and $\beta\gamma$ subunits in cells. Injection of antisense mRNA into the nucleus followed by the assay of receptor function in single cells can provide very detailed information about the interaction of receptors and G proteins.[14–16] Disadvantages of this method include a difficulty in proving that a particular protein is completely removed from the membrane of the cell and that it cannot be used easily to examine the function of mutant α and/or $\beta\gamma$ dimers. In addition, the large number of receptors and G protein isoforms makes it possible that redundancy will complicate interpretation of data. Studies of patients with mutations in certain G protein α subunits or mice with specific genes in the G protein signaling pathway knocked out have provided significant information about the role of G protein α subunits in signaling.[17]

Transfection of Cultured Cells

Transfection of mammalian cells with expression vectors containing powerful promoters and a cDNA encoding specific members of the signaling pathway is widely used to determine function. This method has the advantage that it is easy

[13] T. Gudermann, F. Kalkbrenner, and G. Schultz, *Annu. Rev. Pharm. Toxicol.* **36,** 429 (1996).

[14] C. Kleuss, J. Hescheler, C. Ewel, W. Rosenthal, G. Schultz, and B. Wittig, *Nature* **353,** 43 (1991).

[15] C. Kleuss, H. Scherubl, J. Hescheler, G. Schultz, and B. Wittig, *Nature* **358,** 424 (1992).

[16] C. Kleuss, H. Scherubl, J. Hescheler, G. Schultz, and B. Wittig, *Science* **259,** 832 (1993).

[17] A. Farfel, H. Bourne, and T. Iiri, *N. Eng. J. Med.* **340,** 1012 (1999).

to perform and can be applied to many types of biological questions.[18] It has the disadvantages that the native pathways of the cell are left intact and functioning. Thus, clear interpretation of results may be difficult. Moreover, because there is a large amount of similarity among the isoforms of the G protein α and $\beta\gamma$ subunits, significant overexpression of proteins can obscure the specificity inherent in the native systems.

Reconsitution of Pure Receptors and G Proteins into Synthetic Lipid Vesicles

Whether used with native or recombinant proteins, reconstitution of the G protein signaling pathway with pure proteins provides a flexible experimental system that can be used to answer a large variety of experimental questions.[19–21] Moreover, because pure proteins are used, the system is defined precisely and clear answers are obtained. Two major disadvantages of this approach are that purification and reconstitution of receptors and G proteins are complicated tasks and that such systems may miss the contribution of known and/or unknown proteins that provide additional specificity in the intact cell.

Reconstitution of G Proteins into Membranes Expressing Recombinant Receptors

It is also possible to reconstitute cell membranes isolated from cultured cells or Sf9 insect cells overexpressing recombinant receptors with pure, recombinant G protein α and $\beta\gamma$ subunits and examine the specificity of the receptor:G protein interaction.[22] This approach provides a flexible assay system for examining the interaction of receptors and G proteins, combines some of the advantages of the methods just described, and eliminates certain disadvantages.

The major advantages of this approach are that (a) the first steps in receptor: G protein interaction can be measured; (b) receptors are expressed and inserted in a native membrane, thus the problems of purification and reconstitution of receptors into synthetic lipid vesicles are eliminated. (c) Either cultured mammalian cells or Sf9 insect cells may be used in these experiments. (d) If Sf9 cells are used, they contain few mammalian receptors and the process of viral infection reduces the expression of native insect cell proteins greatly.[23] Thus, membranes of Sf9 cells

[18] J. Sambrook, E. F. Fritsch, and T. Maniatis, "Molecular Cloning: A Laboratory Manual." Cold Spring Harbor Laboratory Press, Cold Spring Harbor, NY, 1989.

[19] R. C. Rubenstein, M. E. Linder, and E. M. Ross, *Biochemistry* **30,** 10770 (1991).

[20] G. H. Biddlecome, G. Berstein, and E. M. Ross, *J. Biol. Chem.* **271,** 7999 (1996).

[21] R. A. Figler, M. A. Lindorfer, S. G. Graber, J. C. Garrison, and J. Linden, *Biochemistry* **36,** 16288 (1997).

[22] R. A. Figler, S. G. Graber, M. A. Lindorfer, H. Yasuda, J. Linden, and J. C. Garrison, *Mol. Pharm.* **50,** 1587 (1996).

infected with a virus expressing a receptor contain a minimum of potentially interfering proteins. Treating the membranes with urea can further reduce interference from endogenous G proteins. (e) Specific mutations in receptors or G proteins can be studied by constructing cDNAs or a recombinant baculovirus to express the modified protein of interest. (f) It is possible to examine the stoichiometry of the receptor:α:$\beta\gamma$ interaction with precision. Thus, questions related to differing affinities can be answered.

This article outlines the use of recombinant receptors and G proteins overexpressed in baculovirus-infected Sf9 insect cells to examine receptor:α:$\beta\gamma$ interactions. Protocols for the construction of recombinant baculoviruses for receptors and G proteins, procedures for infection of Sf9 cells and harvesting the proteins, and purification of membranes and G proteins have been described in detail elsewhere[24,25] and will be outlined only briefly. Procedures for the reconstitution of G proteins into Sf9 cell membranes are fully presented. The utility of the procedures is demonstrated using two assays of receptor:G protein interaction: (1) reconstitution of the high-affinity agonist-binding conformation of the receptor[22] and (2) reconstitution of receptor-driven GTPγS binding to the G protein α subunit.[26] Each assay depends on the interaction of both α and $\beta\gamma$ subunits with the receptor and can be used to probe differences in receptor activation of the many different isoforms of the α and $\beta\gamma$ subunits.[22,26]

Description of Baculovirus Techniques

General Considerations

The high levels of protein expression typically achieved with the baculovirus expression system are due to the powerful promoter of the polyhedrin gene from the *Autographica californica* nuclear polyhedrosis virus (AcMNPV). This promoter regulates the gene encoding the polyhedrin protein that encapsulates the virus and can usurp over 60% of the protein synthetic machinery of infected insect cells at certain stages post infection.[23,27] If the cDNA for another protein is placed downstream of the polyhedrin promoter, a high level of expression of the foreign protein will occur in insect cells infected with the virus.[23,27] A full description of these techniques is outside the scope of this article, but very complete explanations of the procedures can be found in two laboratory manuals.[23,28]

[23] D. R. O'Reilly, L. K. Miller, and V. A. Luckow, "Baculovirus Expression Vectors: A Laboratory Manual." Freeman, New York, 1992.

[24] S. G. Graber, R. A. Figler, and J. C. Garrison, *Methods Enzymol.* **237,** 212 (1994).

[25] S. G. Graber, M. A. Lindorfer, and J. C. Garrison, *Methods Neurosci.* **29,** 207 (1996).

[26] H. Yasuda, M. A. Lindorfer, K. A. Woodfork, J. E. Fletcher, and J. C. Garrison, *J. Biol. Chem.* **271,** 18588 (1996).

[27] L. K. Miller, *Annu. Rev. Micro.* **42,** 177 (1988).

[28] M. D. Summers and G. E. Smith, Texas Agric. Exp. Stat. Bull. **1555** (1987).

Subcloning Strategy for Receptors or G Protein α and $\beta\gamma$ Subunits

The large size of the baculovirus genome (~130 kbp) makes direct manipulations of the DNA impractical, thus recombinant viruses are produced by cotransfection of Sf9 cells with wild-type viral DNA and a transfer vector containing the foreign cDNA sequence. Transfer vectors contain the polyhedrin promoter adjacent to a multiple cloning site flanked by wild-type viral sequences. Homologous recombination in cultured insect cells substitutes the foreign cDNA sequence for wild-type polyhedrin sequences, producing a recombinant virus. Construction of recombinant baculoviruses can be accomplished using the baculovirus transfer vectors pVL-1392 and pVL-1393 originally described by Dr. Max Summers of Texas A&M University[28] or the Bac to Bac system sold by Life Technologies. The generation and isolation of recombinant baculoviruses have been simplified greatly by the baculovirus kits offered by commercial vendors. These vectors are available commercially in baculovirus kits from a number of vendors (PharMingen, Strategene, InVitrogen, and others). Our procedures for producing and purifying recombinant baculoviruses using PharMingen's BaculoGold kit have been described in detail.[25] A major advantage of the BaculoGold kit is that it provides a wild-type viral DNA containing a lethal mutation that is rescued by homologous recombination with the BaculoGold transfer vector. Thus, only recombinant viruses replicate, nearly eliminating the wild-type viral placques. This strategy makes identification and isolation of the recombinant virus very efficient.[25] Detailed descriptions of the subcloning strategies for the preparation of recombinant viruses encoding specific G protein α, β, or γ subunits have been described.[29–32] The preparation of transfer vectors for A_1-adenosine and ET_B receptors has also been described.[22,32] Given the amount of effort involved in producing recombinant viruses and then recombinant proteins, it is important to sequence all completed transfer vectors to determine that the subcloned cDNA will generate the desired primary sequence of the expressed protein.

Sf9 Cell Culture

Maintenance of Sf9 Cells

Spodoptera frugiperda (Sf9) cells may be obtained from the American Tissue Culture Collection (ATCC No.: CRL 1711). Cells are maintained in logarithmic growth in TNM-FH medium, which is Grace's medium supplemented with yeastolate (3.33 g/liter) and lactalbumin hydrolysate (3.33 g/liter), containing 10% fetal

[29] S. G. Graber, R. A. Figler, and J. C. Garrison, *J. Biol. Chem.* **267,** 1271 (1992).
[30] S. G. Graber, R. A. Figler, V. K. Kalman-Maltese, J. D. Robishaw, and J. C. Garrison, *J. Biol. Chem.* **267,** 13123 (1992).
[31] H. Yasuda, M. A. Lindorfer, C. Myung, and J. C. Garrison, *J. Biol. Chem.* **273,** 21958 (1998).
[32] M. A. Lindorfer, C. Myung, H. Yasuda, Y. Savino, R. Khazan, and J. C. Garrison, *J. Biol. Chem.* **273,** 34429 (1998).

bovine serum.[28] We routinely use antibiotics in both maintenance cultures and infections. Cultures are supplemented with antibiotics as follows. A stock solution containing 10 U/ml of penicillin G, 10 μg/ml streptomycin sulfate, and 25 μg/ml of amphotericin B is made up in Grace's medium, and 1 ml of antibiotic solution is added to each 100 ml of culture medium. Cells are maintained at 27° as 75-ml suspension cultures in spinner flasks stirred at a rate of 60 rpm in an atmosphere of 50% O_2/50% N_2 or 50% O_2/50% air. For normal passage, spinners are seeded at a density of 0.8–1.2 $\times$ 10^6 cells/ml. The doubling time for these cells is typically 18–22 hr. To prevent the accumulation of toxic by-products, every 3 days the cells are pelleted by gentle centrifugation (120 *g* for 6 min) and resuspended in fresh medium. Because the health of the cells is essential for high levels of protein expression, care should be taken to ensure that cell viability remains greater than 95% and the cell borders appear cleanly rounded when examined with a microscope. To test, aliquots from each spinner are diluted 1:1 with 0.4% (w/v) trypan blue in 0.85% (w/v) normal saline and counted using a Neubauer hemocytometer.

General Procedures for Infection and Harvest of Sf9 Cells

Sf9 cells may be infected in culture volumes ranging from 75 ml to 3 liters, depending on the amount of final product required. A 75-ml culture will yield about 1–1.25 g (wet weight) of infected cells, whereas a 3-liter culture will yield 40–45 g (wet weight) of cells. For infections, the appropriate number of cells is spun at 120*g* at room temperature for 7 min with the centrifuge brake off. Spent medium is removed, and the cell pellet is resuspended gently in the appropriate viral stock using a multiplicity of infection (MOI) of 1–5 placque-forming units (pfu) per cell. In most instances, we use a MOI of 3. A small volume of additional medium may be used to aid in cell suspension, but the cell density should be at least 1 $\times$ 10^7 cells/ml during the infection process. After 1 hr at 27°, infected cells are transferred back to spinner flasks and suspended at 2.5–3.0 $\times$ 10^6 cells/ml with fresh medium. Infected cells are harvested between 48 and 72 hr, depending on their appearance and density, usually when trypan blue staining reaches 20–25% of the total number. Because infectivity and protein expression levels vary depending on culture conditions and the particular recombinant virus used, a time course of protein expression should be carried out with each virus to accurately determine the infection period yielding the highest level of expression. This will also allow determination of the optimum multiplicity of infection and assess the degree of proteolysis that may occur as the infection progresses. Infection of a Sf9 cell culture with two or more different viruses should be expected to reduce the expression of each individual protein. Infected cells may be harvested by centrifugation at 150*g* for 7 min at 4° with the centrifuge brake off. The supernatant is aspirated, and the cell pellet is washed three times in ice-cold insect cell phosphate-buffered saline (7.3 m*M* NaH_2PO_4, pH 6.2, 58 m*M* KCl, 47 m*M* NaCl, 5.0 m*M* $CaCl_2$). Following this rinsing step, the infected cell pellets are resuspended in a buffer selected as optimal for the next step of purification (see later). Cell pellets can be stored frozen at −70° or used immediately as desired.

Preparation of Membranes Overexpressing G Protein-Coupled Receptors

General Procedures

The examples provided in this article have been performed using the rat β_1-adrenergic receptor, the endothelin ET_B receptor, and the bovine A_1 adenosine receptor. Procedures for making Sf9 membranes expressing the A_1 adenosine receptor and treating them with urea are described as an example. Membranes containing the other two receptors are prepared using similar protocols, but with slightly different buffer compositions (see later). Sf9 insect cells are infected with a high-titer stock of baculovirus coding for the bovine A_1 adenosine receptor at a MOI of 3, and the infection is continued for 48–60 hr.[22] Cell pellets are washed three times with insect cell phosphate-buffered saline (7.3 m*M* NaH_2PO_4, pH 6.2, 58 m*M* KCl, 47 m*M* NaCl, 5.0 m*M* $CaCl_2$), resuspended in about 1 ml/g wet weight of cell pellet in membrane homogenization buffer 25 m*M* HEPES, pH 7.5, 1% (v/v) glycerol, 100 m*M* NaCl, 1 μ*M* adenosine, 17 μg/ml phenylmethylsulfonyl fluoride (PMSF), 20 μg/ml benzamidine, and 2 μg/ml each of aprotinin, leupeptin, and pepstatin A, and stored frozen at −70°. To prepare membranes, cell pellets are thawed in 15× their wet weight of ice-cold membrane homogenization buffer and burst by N_2 cavitation (600 psi, 20 min) on ice. Cavitated cells are centrifuged at 4° for 10 min at 750*g* to remove unbroken nuclei and cell debris. The supernatant from the low-speed spin is centrifuged at 4° for 30 min at 28,000*g* to obtain the membrane pellet. The supernatant from this spin is discarded, and the membranes are washed twice in membrane homogenization buffer by resuspension and centrifugation, resuspended at a concentration of about 5 mg protein/ml, snap frozen in liquid nitrogen, and stored at −70°. If desired, the membranes can be treated with urea to remove endogenous G proteins before freezing (see later). We have used a slightly different homogenization buffer for the β_1-adrenergic receptor. This buffer contains 20 m*M* HEPES, pH 7.5, 2 m*M* $MgCl_2$, 1 m*M* EDTA, 17 μg/ml PMSF, and 2 μg/ml of aprotinin and leupeptin.

Treatment of Membranes with Urea

Endogenous Sf9 cell G proteins may couple with receptors overexpressed in the membranes and complicate experiments in which purified, recombinant G proteins are reconstituted into the membranes. It is also possible to overexpress receptors in the membranes of mammalian cells such as COS-7 or HEK-293 cells and reconstitute purified G proteins into these membranes. If mammalian cell membranes are used, the endogenous G proteins may interfere with the interpretation of the results.

Urea has been used in experiments with rod outer segments and other membranes to remove endogenous G proteins,[33–35] and a similar protocol can be used to reduce the level of endogenous G proteins in Sf9 or mammalian cell membranes.[32] Thus, it is possible to treat the cell membranes overexpressing recombinant receptors with urea and obtain a membrane preparation containing a reduced complement of endogenous proteins. This membrane will contain high levels of functional receptors and is an excellent background for reconstitution experiments.

Pelleted Sf9 cell membranes containing the A_1 adenosine receptor (see earlier discussion) are weighed and resuspended with a Teflon pestle homogenizer in the original homogenization buffer (HEPES, NaCl, glycerol, and protease inhibitors) containing 7 *M* urea. The amount of buffer used in the resuspension is based on the wet weight of the membrane pellet to avoid measuring the protein concentration at this step. We add about 50 μl buffer per milligram of wet weight of the pellet (20 mg pellet/ml buffer). Then the mixture is incubated on ice for 60 min.[32] After the 60-min incubation, the urea concentration is diluted to 4 *M* urea with the homogenization buffer and the suspension is centrifuged to pellet the membranes. The force required varies with the composition of the buffer used for the urea incubation. We have used 28,000 to 142,000*g* for 30 min to pellet the membranes. If necessary, the membranes can be treated twice with urea. In this case, the first treatment with 7 *M* urea is shortened to 30 min, the membranes are pelleted in 4 *M* urea as described earlier, and the treatment is repeated for an additional 30 min. The urea-stripped membranes are washed twice with membrane homogenization buffer by centrifugation at 28,000*g* for 30 min. After the washes, the membranes are resuspended in this buffer at about 2 mg/ml membrane protein, frozen, and stored at $-70°$. Urea treatment may remove about 50% of the protein in the membrane preparation. To determine that urea treatment did not harm the overexpressed receptor, it is useful to measure the B_{max} and K_d values of the receptors before and after urea treatment. These measurements have been made for recombinant ET_B receptors,[32] M_1 muscarinic receptors,[32] and 5-HT_{1a} receptors.[35] Binding data indicate that urea treatment does not change the affinity for ligand greatly. Most importantly, the receptors in these urea-treated membranes couple well to recombinant G proteins reconstituted into the membranes.[32,35]

Preparation of G Protein α and $\beta\gamma$ Subunits

Expression and Purification of Recombinant G Protein α Subunits

Recombinant G protein α subunits can be purified using a number of protocols. Members of the G_i family (G_{i1}, G_{i2}, G_{i3}, and G_o) can be purified from baculovirus-infected Sf9 cells in bulk using three sequential columns (DEAE, hydroxyapatite,

[33] H. Shichi and R. L. Somers, *J. Biol. Chem.* **253,** 7040 (1978).

[34] S. K. Debburman, P. Kunapuli, J. L. Benovic, and M. M. Hosey, *Mol. Pharm.* **47,** 224 (1995).

[35] J. L. Hartman and J. K. Northup, *J. Biol. Chem.* **271,** 22591 (1996).

and MonoP) as described.[24,25] This protocol can produce 10–20 mg of pure $G_i\alpha$ subunit from a 50-g pellet of Sf9 cells.[25] Members of the G_q, G_s, and $G_{12/13}$ families are purified more readily using their affinity for the $\beta\gamma$ subunit as initially described by Kozasa.[36,37] If desired, this method can also be used to purify members of the G_i family. In this approach, Sf9 cells are coinfected with three viruses encoding the G α of interest, a β subunit (usually, the β_1 subunit), and a γ subunit (usually, the γ_2 subunit). The viruses for one or both of the β and γ subunits are engineered to express one or more epitope tags (hexahistidine and/or FLAG) at their N terminus. The α:$\beta\gamma$ complex is expressed in the cells, extracted from the plasma membranes with detergent, and passed over an affinity column, which binds the epitope tag on the β or γ subunit (or both). The α:$\beta\gamma$ complex on the column is washed to remove contaminants and then highly pure α subunits can be eluted from the column by activating the α subunit with AlF_4^- to release it from the $\beta\gamma$ subunit.[36,37] This strategy ensures that native and active α subunits are obtained. Detailed protocols for the purification of the G_q α subunit have been published.[20,32] Purification of the G_s and G_{13} α subunits is accomplished with similar protocols.[37]

Expression and Purification of Recombinant $\beta\gamma$ Complexes

Sf9 cells are coinfected with the appropriate β and γ baculoviruses at a multiplicity of infection of 3 each and harvested 48–60 hr after infection. The construction of our panel of recombinant baculoviruses encoding various β and γ subunits has been described.[26,30–32] The $\beta\gamma$ dimers expressed by the Sf9 cells are extracted with detergents (usually Genapol C-100) and purified as described.[25] The major purification step in this procedure is affinity purification on a recombinant $G_i\alpha$ subunit column. This column is critical in obtaining highly active preparations of $\beta\gamma$ dimers because only those recombinant dimers folded properly and modified posttranslationally will bind to the column with high affinity and be released easily with AlF_4^-.[25] Detailed protocols for construction of an α subunit affinity column and its use in purification of G protein $\beta\gamma$ dimers have been described.[25]

It is also possible to prepare recombinant $\beta\gamma$ dimers using an modification of the triple infection protocol described by Kozasa to purify G_q or G_s.[37] In this approach, Sf9 cells are coinfected with a virus expressing the $G_i\alpha$ subunit containing a hexahistidine tag within the molecule and the viruses for the desired $\beta\gamma$ subunits. The heterotrimer is extracted from the cells with detergent and applied to a Ni^{2+}-NTA column. The entire heterotrimer is bound to the column, washed as needed, and the $\beta\gamma$ dimer eluted with AlF_4^- as described.[37]

[36] T. Kozasa and A. G. Gilman, *J. Biol. Chem.* **270,** 1734 (1995).

[37] T. Kozasa, *in* "G Proteins, Techniques for Analysis" (D. R. Manning, ed.), p. 23. CRC Press, Boca Raton, FL, 1999.

Reconsitution of G Proteins into Sf9 Cell Membranes Expressing Recombinant Receptors

G protein α and $\beta\gamma$ subunits can be reconstituted into membranes using a variety of methods. However, all methods depend on the preference of G proteins to leave solution and incorporate into plasma membranes. A small amount of detergent (Genapol, cholate, or CHAPS) often improves the amount of protein inserted in the membranes. We use slightly different protocols to reconstitute G proteins depending on the assay used to measure coupling to the receptor.

Reconstitution of Membranes for Recovery of High Affinity Binding Conformation of the Receptor

The following protocol provides an example of the reconstitution of G proteins with the recombinant A_1 adenosine receptor prior to performing the high-affinity, agonist-binding assay. Prior to reconstitution, concentrated stocks of the purified preparations of G protein α and $\beta\gamma$ subunits are diluted in specific buffers as follows. The $G_i\alpha$ subunit is diluted 10-fold from a stock of ~2 mg/ml in a buffer containing 50 m*M* HEPES, pH 8.0, 10 m*M* $MgCl_2$, 1 m*M* EGTA, 0.1% (w/v) bovine serum albumin (BSA), 5 μM GDP, and 1 m*M* dithiothreitol (DTT). This is the working stock of the α subunit (about 200 μg/ml) and is stored frozen in small, single-use aliquots. The concentrated stock of the $\beta\gamma$ subunits contains about 200–600 μg/ml of protein in 20 m*M* HEPES, pH 8.0, 200 m*M* NaCl, 49 m*M* $MgCl_2$, 1 m*M* EDTA, 0.6% (w/v) CHAPS, 5 μM GDP, 3 m*M* DTT, 10 m*M* NaF, and 30 μM $AlCl_3$. At this point, the stock α and $\beta\gamma$ subunits each are diluted about fivefold into the following buffers. The α subunit is diluted into membrane-binding buffer containing 10 m*M* HEPES, pH 7.4, 5 m*M* $MgCl_2$, 1 m*M* EDTA, and 1 m*M* DTT. The $\beta\gamma$ subunit is diluted into 20 m*M* HEPES, pH 7.5, 150 m*M* NaCl, 1 m*M* $MgCl_2$, and 0.04% (w/v) CHAPS. Before adding the membranes to the G proteins, the desired amounts of α and $\beta\gamma$ subunits are added to individual tubes in small volumes and brought to a final volume of 12 μl in 10 m*M* HEPES, pH 7.4, 5 m*M* $MgCl_2$, 1 m*M* EDTA, and 1 m*M* DTT. Because the final incubation volume will be 50 μl (see later), the concentrations of α and $\beta\gamma$ subunits at this stage are diluted to be fourfold higher than the intended concentration in the reconstitution. It is important that the volume of G proteins added to this 12 μl is small (1–2 μl). If the volume becomes too large, it may be difficult to keep the detergents in the α and $\beta\gamma$ preparations at low enough levels in the final assay (see later). To perform the reconstitution, frozen Sf9 cell membranes overexpressing the A_1 adenosine receptor are thawed, pelleted at 12,000*g*, and resuspended to a concentration of about 0.5 mg protein/ml in reconstitution buffer containing 5 m*M* HEPES, pH 7.4, 100 m*M* NaCl, 5 m*M* $MgCl_2$, 1 m*M* EDTA, 1 m*M* DTT, 500 n*M* GDP, and 0.04% (w/v) CHAPS. Then 10-μl aliquots of membranes are added to each tube containing the G protein α and $\beta\gamma$ subunits, and the volume is increased to a total of 50 μl with membrane-binding buffer (see later). Experiments have shown that the partition of G protein subunits into the membrane is very dependent on the final concentration of CHAPS (or cholate) in the reconstitution buffer. Partition of G proteins into the membrane is constant (at about 25–30%) when the CHAPS concentrations ranges between 0.005–0.04% (w/v) and decreases when the CHAPS concentration becomes higher than 0.05% CHAPS.[26] Therefore, care should be taken to hold the final CHAPS concentration in the reconstitution mix slightly below 0.04%. The reconstitution mixture is incubated for 30 min on ice. Following the 30-min incubation, the reconstituted membranes are added to the agonist-binding assay (see the high-affinity binding assay later).

Reconstitution of G Proteins for the GTPγS-Binding Assay

Reconstitution of the α and $\beta\gamma$ subunits into receptor-containing Sf9 cell membranes prior to the assay for receptor-stimulated GTPγS binding is performed according to the principles given earlier with the following modifications. To perform the reconstitution, frozen Sf9 cell membranes overexpressing the A_1 adenosine receptor are thawed and an aliquot containing about 30 μg protein is pelleted at 12,000*g* and resuspended to a concentration of about 60 μg protein/ml in reconstitution buffer containing 25 m*M* HEPES, pH 7.4, 100 m*M* NaCl, 5 m*M* $MgCl_2$, 1 m*M* EDTA, 1 m*M* DTT, 500 n*M* GDP, 14 U/ml adenosine deaminase, and 0.1% BSA. The resuspended membranes are typically brought to a volume of 450 μl. The next step is to incubate the membranes with 1 m*M* AMP-PNP for 15 min to occupy all ATP-binding sites with the analogue. This treatment helps reduce background [^{35}S]GTPγS binding. To begin the reconstitution, purified preparations of G protein α and $\beta\gamma$ subunits are thawed, and generally 1 μl of the desired α subunit and 1–2 μl of the desired $\beta\gamma$ dimer are added directly to the resuspended membranes with gentle vortexing. The mixture is incubated for 30 min on ice to complete reconstitution. Aliquots of the reconstituted membranes are removed and taken directly to the GTPγS-binding assay as described later. Reconstitution of G protein α and $\beta\gamma$ subunits into membranes containing the β_1-adrenergic receptor is performed using a similar protocol with the exception that no adenosine deaminase is included in the reconstitution buffer.

Assays of Receptor–G Protein Interaction using Recovery of High-Affinity, Agonist Ligand Binding

Introduction

Some receptors undergo a large affinity shift for agonists in the presence of GTPγS, reflecting uncoupling of the receptor from the G protein heterotrimer. Thus, if a receptor with this property is under study, it is possible to measure the interaction of the receptor and G protein α:$\beta\gamma$ subunits by measuring the recovery of the high-affinity, agonist-binding conformation of the receptor. This is the case for the bovine A_1 adenosine receptor, and the assay described later takes advantage of the finding that the overexpression of receptors in membranes often yields an excess of receptors over the available G proteins. Treatment of the membranes with urea can denature or remove G proteins and increases the fraction of receptors in the low-affinity, uncoupled state. Thus, prior to reconstitution, the majority of receptors are not coupled to G proteins and their affinity for agonists is reduced.[22] The high-affinity state of the receptor can be recovered by reconstituting the receptors with appropriate amounts of the α and $\beta\gamma$ subunits with which they interact. An important point is that recovery of the high-affinity, agonist-binding state requires both α and $\beta\gamma$ subunits and can be used to measure the ability of the G protein subunits to interact with each other and/or the receptor.[22] Thus, the assay provides a convenient way of measuring the ability of the receptor to interact with G proteins at equilibrium. The recovery of the high-affinity state can be measured using a radiolabeled agonist at a concentration near the K_D of the high-affinity state (see Fig. 1).

An example of a protocol used to measure the recovery of the high-affinity, agonist-binding conformation of the receptor is described later. Results of the experiment are illustrated in Fig. 1. Membranes are reconstituted with 40 nM of purified $G_{i1}\alpha$ and the $\beta_1\gamma_2$ dimer (or vehicle) as described earlier. To perform

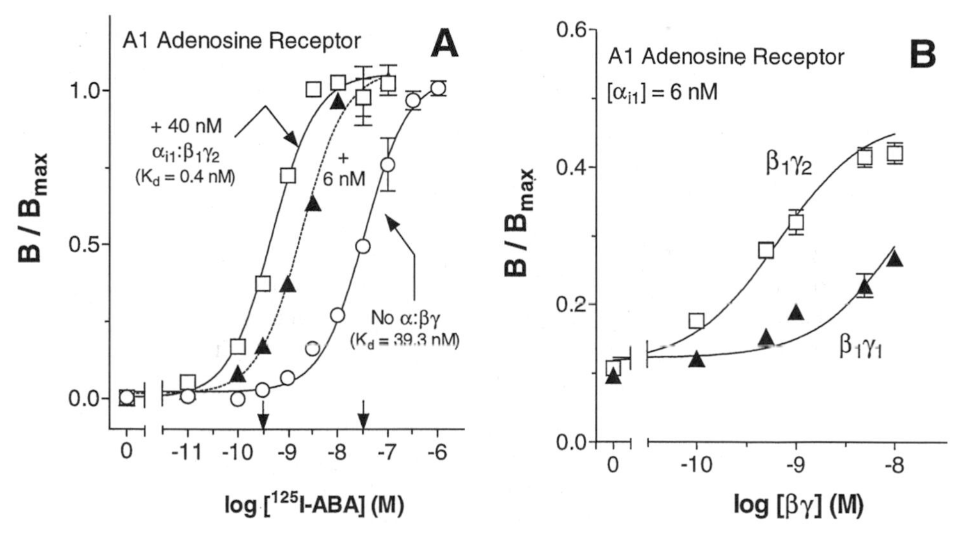

FIG. 1. (A) The effect of reconstituting 6 and 40 nM $G_{i1}\alpha:\beta_1\gamma_2$ into Sf9 cell membranes expressing the A_1 adenosine receptor on the affinity of the receptor for the agonist ligand $[^{125}I]$ABA. The reconstitution protocol and the binding assay were performed as described in the text. The effect of the heterotrimer on the recovery of high-affinity binding is most dramatic between 0.3 and 30 nM (arrows). (B) Differences in the coupling of the $\beta_1\gamma_1$ and $\beta_1\gamma_2$ dimers to the A_1 adenosine receptor as measured with 0.4 nM $[^{125}I]$ABA and 6 nM $G_{i1}\alpha$ subunit.

the binding assay, a membrane-binding buffer is prepared containing 10 mM HEPES, pH 7.4, 5 mM $MgCl_2$, 1 mM EDTA, 5 U/ml of adenosine deaminase (to remove endogenous adenosine), and 2 nM [^{125}I]aminobenzyladenosine (about 250,000 cpm/50 μl). Fifty microliters of this buffer is added to the 50 μl of the reconstituted membrane mixture prepared as described previously. In the experiment shown in Fig. 1A, various concentrations of cold I-ABA were also added to each tube to obtain the desired final concentration of I-ABA (ranging from 1 to 1000 nM). Naturally, this is not necessary in most experiments where the ligand concentration is held constant (see Fig. 1B). The agonist ligand, [^{125}I]ABA, is prepared by iodination of the parent compound to theoretical specific activity (2200 Ci/mmol).[22] The final mixture (100 μl in each tube) is incubated for 3 hr at 25°. The assay is terminated by filtration over Whatman GF/C glass fiber filters. The filters are rinsed three times with 4 ml of ice-cold 10 mM Tris–Cl, pH 7.4, 5 mM $MgCl_2$, and counted in a γ counter. For each reconstitution protocol, specific binding and nonspecific binding in the presence of 10 μM R-PIA are measured in duplicate. Nonspecific binding of [^{125}I]ABA to Sf9 cell membranes is minimal.[22]

The ability of the G protein heterotrimer to increase the fraction of receptors in the high-affinity, agonist-binding conformation is illustrated in Fig. 1A. As outlined earlier, the affinity of the A_1 adenosine receptor for [^{125}I]ABA is very dependent on the receptor coupling to the heterotrimer. Note that the uncoupled receptor has a K_D for the ligand of about 39 nM. This value corresponds well to the 20 nM affinity measured in intact membranes in the presence of GTPγS.[22] When an excess of heterotrimer (40 nM) is reconstituted into the membrane and the receptor couples to the G protein, the affinity shifts about 100-fold to the left, giving a K_d of 0.4 nM. Thus, if one uses a ligand concentration ranging from 0.3 to 30 nM (arrowheads in Fig. 1A), coupling of the G protein heterotrimer causes a marked increase in [^{125}I]ABA binding. Inspection of Fig. 1 also shows that the magnitude of the signal will be reduced by using either too high or too low an agonist concentration. The increase in agonist binding obtained in such an experiment is large enough to be measured readily using either ^{3}H- or ^{125}I-labeled ligands. By measuring the total number of receptors in the membranes with an antagonist and comparing this value with the B_{max} found with the agonist, it is possible to calculate the percentage of coupled receptors. In this experiment using 40 nM α:$\beta\gamma$, about 90% of the receptors coupled to the reconstituted G protein.

Figure 1A also shows that the amount of heterotrimer reconstituted into the membrane determines the position of the curve on the X axis (ligand concentration). Note that when 6 nM α:$\beta\gamma$ is reconstituted, the curve is intermediate between the fully coupled and the uncoupled receptors (dotted lines and closed triangles). Thus, by varying the concentration of either the α and/or the $\beta\gamma$ subunits appropriately and using a low concentration of ligand, the assay can be used to measure the ability of the receptor to couple to a particular α:$\beta\gamma$ combination. For example, to make the assay dependent on the α subunit, a high concentration of $\beta\gamma$ dimer could be reconstituted into the membrane and the amount of α subunit varied. In Fig. 1B, the protocol made the assay dependent on the concentration of the $\beta\gamma$ dimer by reconstituting 6 n*M* $G_i\alpha$ subunit into membranes containign 0.4 n*M* of the A_1 adenosine receptor (receptor:α ratio of 15) and varying the concentration of two different $\beta\gamma$ dimers. Note that the $\beta_1\gamma_2$ couples to the receptor more effectively than the $\beta_1\gamma_1$ dimer. By varying the concentration and type of α subunit or $\beta\gamma$ dimers used in the reconstitution, a variety of interactions between the receptor and G proteins can be examined.

Formal Analysis of Data

The ternary complex and extended ternary complex models describe the binding constants that define the equilibrium between receptors and G proteins in the G protein cycle.[21,38] A clear understanding of the kinetic reactions in the G protein cycle is important for a proper interpretation of experiments such as those presented in Fig. 1. A full numerical analysis of binding data to define the number of receptors in the high- and low-affinity states is beyond the scope of this article. The interested reader is referred to one of the many books on this subject.[39] However, for the experiment presented in Fig. 1, it is important to realize that the recovery of the high-affinity, agonist-binding state of the receptor measures at least two interactions: the affinity between the α and $\beta\gamma$ subunit ($K_{D\alpha:\beta\gamma}$) and the affinity of the receptor for the heterotrimer ($K_{Drec:\alpha:\beta\gamma}$).[21,38] Thus, measurement of these two affinities is needed to fully interpret data such as that shown in Fig. 1.

Assay of Receptor-Stimulated GTP-γ-S binding to α subunits

Introduction

The major feature of this assay is that it measures a fundamental property of the receptor:G protein interaction: the ability of the receptor to cause release of GDP from the α subunit and initiate binding of GTP to the α subunit. As with recovery of the high-affinity, agonist-binding state of the receptor, both α and $\beta\gamma$ subunits of the G proteins are required for interaction.[26] Thus, the assay can be used to measure the receptor:α interaction, the receptor:$\beta\gamma$ interaction, or the α:$\beta\gamma$ interaction. Neither a large shift in receptor affinity for agonists on coupling nor radiolabeled agonist ligands are required for this assay.

Examples are provided later using this assay with the recombinant at β_1-adrenergic receptor or the bovine A_1 adenosine receptor. To perform the assay, the following protocol is used. Sf9 cell membranes expressing recombinant A_1

[38] P. Samama, S. Cotecchia, T. Costa, and R. J. Lefkowitz, *J. Biol. Chem.* **268,** 4625 (1993).

[39] H. Motulsky, "Intuitive Biostatistics, Oxford Univ. Press, New York, 1995.

adenosine receptors (~2–4 pmol/mg membrane protein) are treated with AMP-PNP and reconstituted with the desired G protein α and $\beta\gamma$ subunits as described earlier. Typically, membranes are reconstituted with 5 nM of the $G_{i1}\alpha$ subunit (containing 4 μM GDP) and 10 nM of the $\beta_1\gamma_2$ dimer by incubation for 30 min on ice in a buffer containing 25 mM HEPES, pH 7.4, 100 mM NaCl, 5 mM $MgCl_2$, 1 mM EDTA, 1 mM DTT, and 0.1% (w/v) BSA. The final GDP concentration in the assay (provided in part by the α subunit) is about 550 nM, and the final CHAPS concentration (provided mostly by the $\beta\gamma$ subunit) is kept below 0.005% (w/v).The final incubation volume is typically 500 μl, and the G protein α and $\beta\gamma$ subunits are diluted about 500-fold directly into the assay mixture. In these experiments, the receptor:$G_i\alpha$ ratio ranges from 1:1 to 1:30, depending on the experiment. Following reconstitution, about 450 μl of the membrane suspension is transferred to a 25° water bath for a 10-minute preincubation. At 10 min of incubation, 50 μl of carrier-free [^{35}S]GTPγS (about 7×10^6 cpm) is added to the tube, bringing the final volume to 500 μl. At this point the 500-μl incubation mixture is split into two aliquots: one of 210 μl (for the control) and the other of 290 μl (for the agonist). Addition of the [^{35}S]GTPγS establishes the zero time point for the time course. A baseline rate of [^{35}S]GTPγS binding is established by taking samples over the 16-min incubation period from the control tube (see Fig. 2 for timing of the removal of samples). At 8 min after the zero time point, the A_1 receptor agonist, *n*-(2-phenylisopropyl)adenosine (R-PIA), is added to a final concentration of 100 nM to the other tube. The receptor-activated time course is established by removing 30-μl aliquots every 60 sec. All samples are filtered through nitrocellulose filters (Millipore, HAWP-025). The filters are washed three

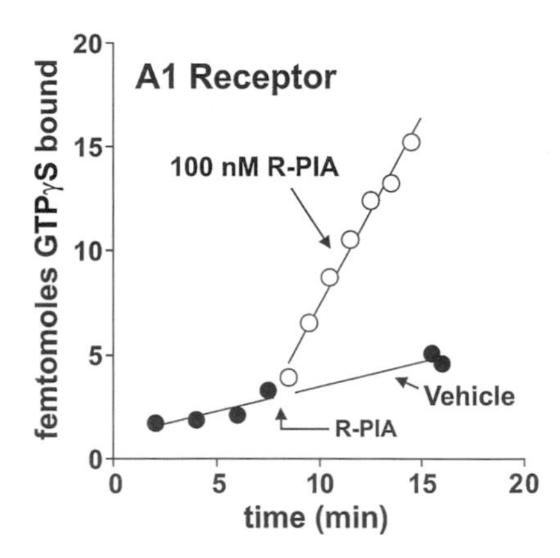

FIG. 2. A time course of agonist-dependent binding of [^{35}S]GTP binding to the $G_{i1}\alpha$ subunit in Sf9 cell membranes expressing the A_1-adenosine receptor and reconstituted with 2.5 nM $G_{i1}\alpha$ and 10 nM $\beta_1\gamma_2$. The receptor was stimulated with 100 nM R-PIA at 8 min. Samples were withdrawn and filtered as described in the text.

times with 4 ml of an ice-cold buffer containing 10 m*M* Tris, pH 7.4, and 5 m*M* $MgCl_2$ and counted by liquid scintillation counting.

A typical time course experiment is presented in Fig. 2. Note that R-PIA activation of the A_1 adenosine receptor drives a seven- to eightfold increase in the rate of GTPγS binding to the $G_i\alpha$ subunit (open versus closed circles). The increased rate is relatively linear over the period between 8 and 15 min. In some experiments, the rate of agonist-stimulated GTP binding falls off after about 5 min of stimulation. Thus, one needs to ensure that the stimulated rate is linear in protocols in which only one time point is taken after agonist addition. The amount of GTPγS bound to the $G_i\alpha$ subunit by receptor activation is about 13 fmol. This value represents about 9% of the total G_i added to the membrane in the reconstitution step. Because only 25–30% of the G protein added to the Sf9 membrane is incorporated into the membrane during the reconstitution protocol,[26] approximately 25% of the α subunit incorporated in the membrane-bound GTP when activated by the receptor in this experiment.

Concentration of GTPγS in the Assay

The concentrations of both GDP and GTPγS are important for obtaining an optimal signal in the receptor activation assay. A number of factors can affect the basal rate of GTP binding and the magnitude of the signal obtained. These issues are discussed in this section. GDP is needed to help the receptor couple to the heterotrimer in the receptor:α GDP:$\beta\gamma$ basal state, and GTP is needed as the "substrate" for the receptor-activated α subunit. Moreover, the amount of [^{35}S] GTPγS bound determines the magnitude of the signal observed. Data in Fig. 3 present experiments performed to determine the optimal concentrations of these guanine nucleotides in the assay. The experiments were performed using the A_1 adenosine receptor and the $G_{i1}\alpha$:$\beta_1\gamma_2$ heterotrimer via the protocol shown in Fig. 2 with one time point taken after 7 min of stimulation. Figure 3A shows that both the level of background [^{35}S]GTPγS binding and the magnitude of the receptor activated signal are dependent on GDP concentration. Concentrations of GDP below 500 n*M* lead to an increased background and a smaller signal. Surprisingly, GDP levels greater than 500 n*M* did not change the background or signal. Therefore, about 500 n*M* GDP has been used as an optimal concentratio of GDP in most of our experiments. Figure 3B illustrates the dependence of [^{35}S]GTPγS binding on the GTPγS concentration in the assay. This experiment was performed using exactly the same assay conditions as in Fig. 3A except that the GTPγS concentration was varied. Note that the signal increases significantly over GTPγS concentrations ranging between 0.3 and 50 n*M*. In our experience, about 5–10 n*M* GTPγS represents a workable compromise between the amount of radioactivity added to the assay and the magnitude of the signal (see dotted lines). In most experiments, the [^{35}S]GTPγS is used carrier free.

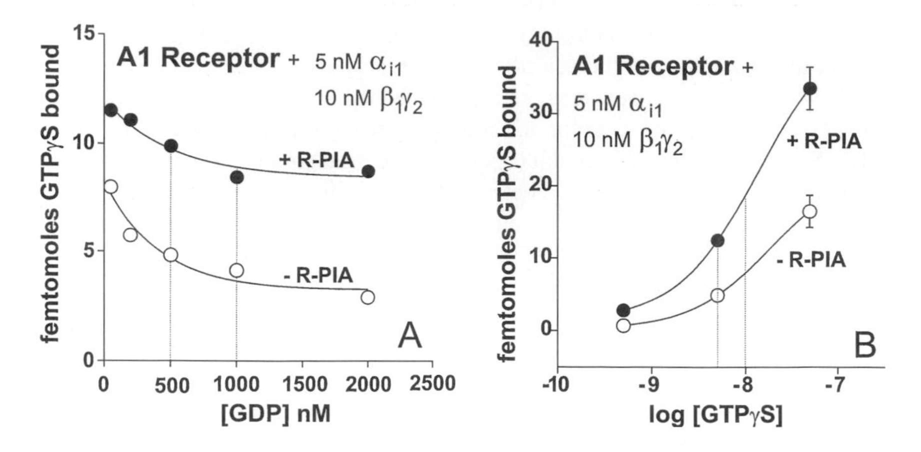

FIG. 3. (A) Effects of GDP concentration on the amount of $[^{35}S]$GTP bound to the $G_{i1}\alpha$ subunit in Sf9 cell membranes expressing the A_1 adenosine receptor. The experiment was performed as described in the text. The receptor was stimulated with 100 n*M* R-PIA, and the time course was stopped 7 min after agonist addition. (B) Effect of GTPγS concentration on the magnitude of the signal obtained in Sf9 cell membranes expressing the A_1 adenosine receptor. The experiment was performed as in A except that the GTPγS concentration was varied and the GDP concentration was 500 n*M*. Dotted lines in each figure indicate the range of GDP or GTPγS concentrations used routinely in these assays.

Constitutive Activity of Receptors and Effects of Urea

Some recombinant receptors are constitutively active when overexpressed. The A_1 adenosine receptor has this property[40] and has the additional complication that the agonist adenosine is released continuously by membrane preparations.[22] Thus, the "basal" state of this receptor in the protocols described here is actually an activated state. These issues are illustrated in Figs. 4A and 4B. Note from Fig. 4A that the rate of GTPγS binding to membranes in which G poteins were *not* reconstituted is very low (closed squares), as is the rate of GTPγS binding to the $G_i\alpha$ subunit alone (open squares). However, as the membranes are reconstituted with increasing amounts of the $G_{i1}\alpha$ subunit in the presence of 10 n*M* $\beta_1\gamma_2$, the basal rate of GTPγS binding increases with the concentration of α subunit added. Note that the rate increases about twofold as the concentration of α subunit is increased from 1.25 to 5 n*M*. Thus, when these experiments are performed with a constitutively active receptor, an increase in the basal rate of GTPγS binding should be expected as more receptors couple to the heterotrimer. Logically, if the elevated basal rate of GTP binding is due to an activated receptor, an antagonist would be expected to reduce the rate. Figure 4B shows an experiment in which 2.5 n*M* G_{i1} and 10 n*M* $\beta_1\gamma_2$ were reconstituted into the membranes. Note that the basal rate can be reduced about twofold by including 5 μM of the A_1 adenosine receptor antagonist

[40] A. L. Tucker, L. Jia, D. Holeton, A. J. Taylor, and J. Linden, *Biochem. J.* **352,** 203 (2000).

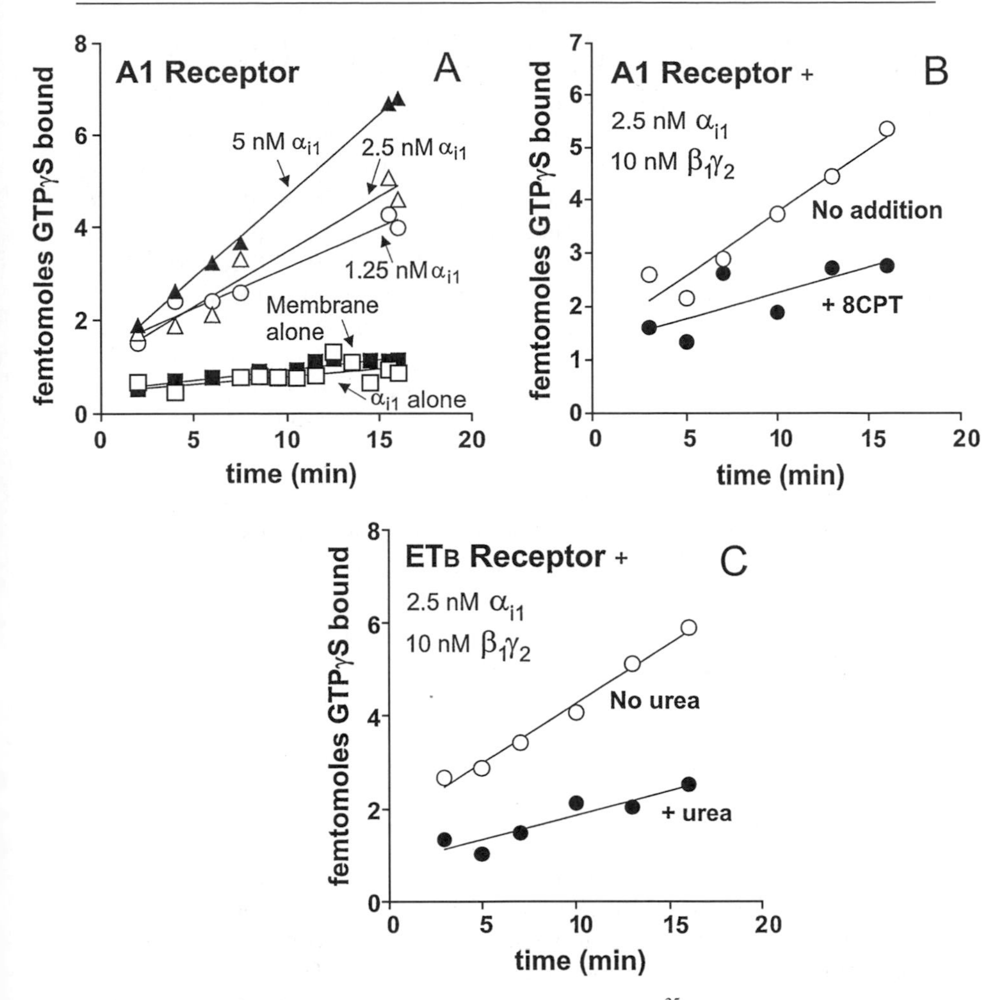

FIG. 4. (A) Illustration of factors that affect the basal rate of $[^{35}S]GTP\gamma S$ binding to Sf9 cell membranes expressing the A_1 adenosine receptor. $\beta_1\gamma_2$ (10 n*M*) was reconstituted with 1.25, 2.5, or 5 n*M* concentrations of the $G_{i1}\alpha$ subunit. No $\beta\gamma$ was added to the membranes alone (closed squares) or the $G_{i1}\alpha$ alone (open squares). (B) The effect of 5 μM of the A_1 adenosine receptor antagonist 8-CPT on the basal rate of $[^{35}S]GTP\gamma S$ binding to the $G_{i1}\alpha$ subunit in membranes containing the A_1 adenosine receptor. (C) The effect of urea stripping on the basal rate of $[^{35}S]GTP\gamma S$ binding to the $G_{i1}\alpha$ subunit in membranes containing the endothelin ET_B receptor. No agonist was added, and the experiment was performed as described in the text.

8-cyclopentaltheophylline (8-CPT) in the incubation (open circles versus closed circles). These data prove that the receptor is, indeed, constitutively active and/or stimulated by endogenous adenosine. With the A_1 adenosine receptor, some of this problem can be minimized by including adenosine deaminase in the incubation to metabolize the adenosine. Fortunately, most other receptor preparations do not release endogenous agonists, but the possibility of constituitive activity should always be considered if a high baseline is observed.

As noted earlier, it is possible to treat isolated plasma membranes with urea to remove or denature the endogenous G proteins prior to reconstitution of the G proteins. In our experience, urea treatment will reduce background GTPγS binding and increase the magnitude of the receptor-activated GTPγS signal. Figure 4C illustrates both a decrease in background binding and a reduced rate of GTP binding following treatment of membranes expressing the endothelin B receptor (ET_B) with urea. Note that the amount of background binding of the ET_B receptor is reduced about twofold by urea treatment. Moreover, the rate of GTPγS binding obtained when 2.5 nM G_{i1} and 10 nM $\beta_1\gamma_2$ are reconstituted into the membranes is reduced about threefold. Overall, treating the membranes with urea provides more consistent data and larger receptor-activated signals in these protocols.

Specificity of the Assay

An important issue with reconstitution experiments is to determine if the specificity expected from intact cell experiments is retained. Data in Fig. 5 illustrate the specificity obtained in Sf9 membranes expressing the A_1 adenosine receptor reconstituted with 2.5 nM G_i or $G_s\alpha$ subunits and 10 nM of the $\beta_1\gamma_2$ dimer. Note from Fig. 5A that the agonist, R-PIA, causes eightfold increase in the rate of GTPγS binding to the $G_i\alpha$ subunit (open versus closed circles), whereas Fig. 5B shows that reconstitution of 5 nM of the $G_s\alpha$ subunit into the the A_1 adenosine receptor containing membranes yields no effect (triangles). This indicates that, as expected, the $G_s\alpha$ subunit does not couple to the A_1 adenosine receptor. A similar set of experiments using Sf9 cell membranes containing the β_1-adrenergic receptor

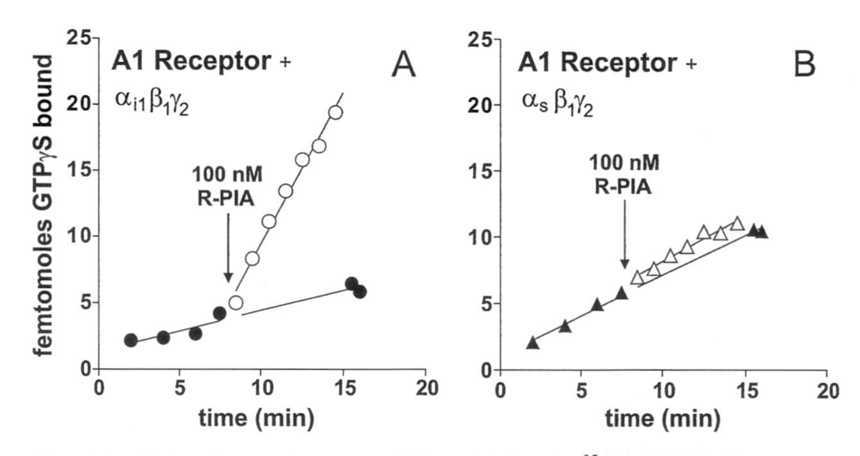

FIG. 5. Specificity of reconstitution. (A) Effect of R-PIA on [^{35}S]GTPγS binding to the $G_{i1}\alpha$ subunit reconstituted into A_1 adenosine membranes with 10 nM $\beta_1\gamma_2$. (B) Effect of 100 nM R-PIA on [^{35}S]GTPγS binding to the $G_s\alpha$ subunit reconstituted into A_1 adenosine membranes with 10 nM $\beta_1\gamma_2$.

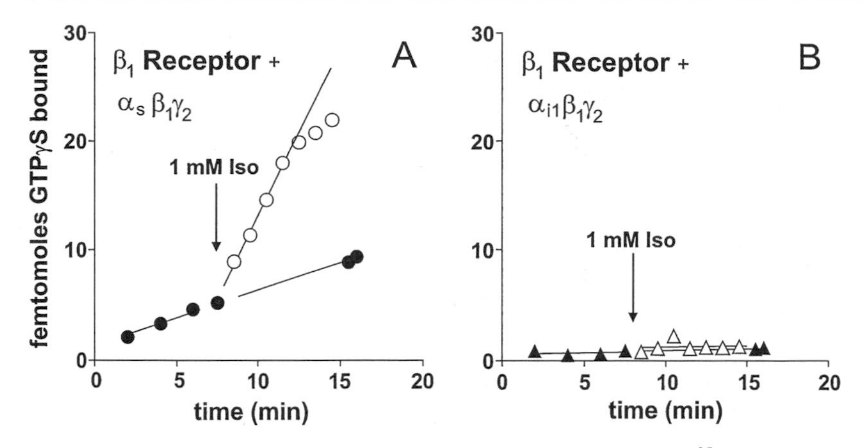

FIG. 6. Specificity of reconstitution. (A) Effect of 1 mM isoproterenol (iso) on [^{35}S]GTPγS binding to the $G_s\alpha$ subunit reconstituted into β_1-adrenergic receptor containing membranes with 10 nM $\beta_1\gamma_2$. (B) Effect of 1 mM isoproterenol (iso) on [^{35}S]GTPγS binding to the $G_{i1}\alpha$ subunit reconstituted into A_1 adenosine receptor containing membranes with 10 nM $\beta_1\gamma_2$.

is shown in Fig. 6. A marked response to 1 mM isoproterenol (open versus closed circles) occurs when 5 nM $G_s\alpha$ and 10 nM of the $\beta_1\gamma_2$ dimer are reconstituted into the membranes (Fig. 6A). Importantly, if 5 nM $G_i\alpha$ is reconstituted with the $\beta\gamma$ dimer, no stimulation of GTPγS binding is observed following the addition of isoproterenol (Fig. 6B). As described earlier, this experiment demonstrates the expected preference of the β_1-adrenergic receptor for the $G_s\alpha$ subunit. Similar specificity has been obtained when reconstituting M_1 muscarinic receptors with G_q and $G_i\alpha$ subunits. The M_1 muscarinic receptor couples to G_q but not to G_i.[32]

Use of the Assay to Examine Selectivity of Coupling to α or $\beta\gamma$ Subunits

The concentration of G protein α or $\beta\gamma$ subunits in the reconstitutions can be varied to make the response dependent on the α or the $\beta\gamma$ subunits. By comparing experiments performed with different amounts of α subunits or $\beta\gamma$ dimers, the investigator can obtain the apparent affinity of a given receptor for different heterotrimers. Data in Fig. 7 illustrate how the rate of GTPγS binding varies as the concentration of the α subunit is increased from 0 to 5 nM. Sf9 membranes containing the A_1 receptor were reconstituted with 10 nM of the $\beta_1\gamma_2$ dimer and the indicated concentrations of the $G_i\alpha$ subunit between 0 and 5 nM and stimulated with 100 nM of the agonist R-PIA. Note how the rate of GTPγS binding increases with increasing concentrations of the α subunit. By using a large number of concentrations of the α subunit, the efficiency of coupling can be determined as an EC_{50} value and compared among different receptors and heterotrimers. Thus, this protocol can be used to measure the ability of various receptors to couple to defined heterotrimers.

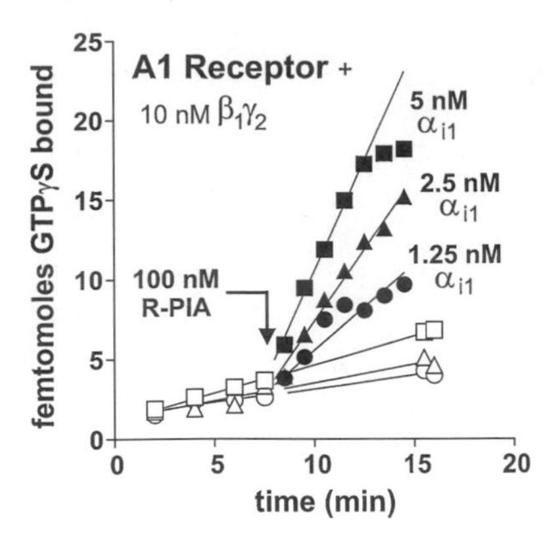

FIG. 7. Effect of the amount of the $G_{i1}\alpha$ subunit reconstituted into A_1 adenosine receptor containing membranes with 10 nM $\beta_1\gamma_2$ on the magnitude of the signal. The experiment was performed as described in the text.

Similar experiments can be performed to determine the EC_{50} values for different $\beta\gamma$ dimers. Data in Fig. 8 illustrate an experiment performed with the A_1 adenosine receptor and the $G_{i1}\alpha$ subunit. In this case, the concentration of $G_i\alpha$ was held constant at 5 nM and three concentrations of $\beta_1\gamma_2$ (0.5, 5, and 10 nM) were reconstituted into the membranes. When stimulated with 100 nM R-PIA, the rate of GTPγS binding was linear with each concentration of $\beta\gamma$ dimer added (the 0 time corresponds to agonist addition). Thus, the assay is dependent on the concentration of the $\beta\gamma$ dimer. If the experiment is repeated with 10 concentrations of

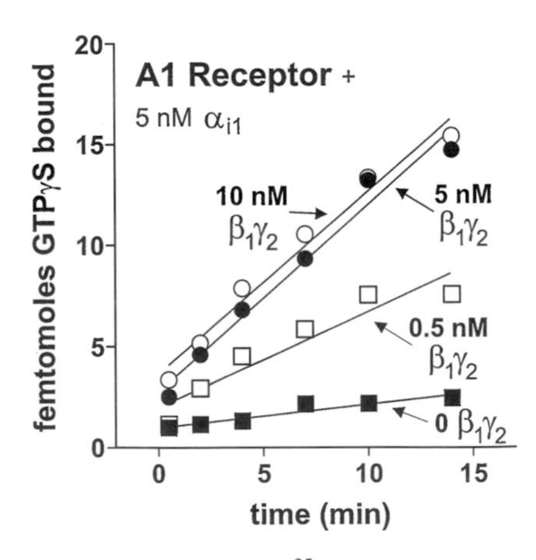

FIG. 8. Effect of varying $\beta\gamma$ concentrations on [^{35}S]GTPγS binding to 5 nM of the $G_{i1}\alpha$ subunit reconstituted into A_1 adenosine membranes. The experiment was performed as described in the text and in Fig. 2. Zero time is the time that 100 nM R-PIA was added.

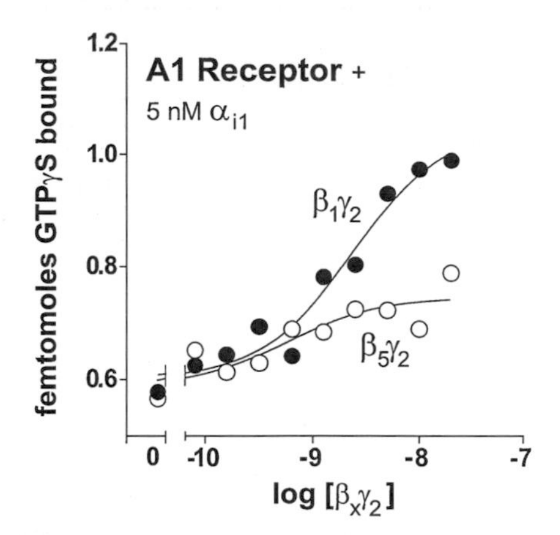

FIG. 9. Effect of R-PIA on [^{35}S]GTPγS binding to 5 n*M* of the $G_{i1}\alpha$ subunit reconstituted into A_1 adenosine receptor containing membranes with various concentrations of $\beta_1\gamma_2$ or $\beta_5\gamma_2$. The experiment was performed as described in the text.

$\beta\gamma$ dimer ranging between 0 and 20 n*M* and stopped at the 7-min time point after R-PIA addition, data shown in Fig. 9 are obtained. This experiment indicates that the receptor : $G_{i1}\alpha : \beta_1\gamma_2$ interaction has a EC_{50} value of 2.4 n*M* (closed circles). If the $\beta_5\gamma_2$ dimer is used in place of the $\beta_1\gamma_2$ dimer, a much lower activation is observed (open circles). This result would be expected considering the weak coupling between the $G_i\alpha$ subunit and the $\beta_5\gamma_2$ dimer.[32] A series of such experiments can provide considerable information about the interaction of a variety of different receptors with G protein heterotrimers of defined composition.

Summary

The methods outlined in this article describe experiments that can probe the first steps in receptor:G protein interaction using defined, recombinant receptors and G proteins. The protocols have the advantages that the receptors are inserted properly in a cell membrane and that the investigator has complete control of the proteins reconstituted with the receptor. Specific mutations in the receptors or G proteins are studied easily and the protocols allow precise examination of the stoichiometry of the receptor:α:$\beta\gamma$ interaction.

Acknowledgments

This work was supported by NIH Grant DK-19952. The authors thank Drs. Elliott Ross and Joel Linden for the baculoviruses encoding the β_1-adrenergic receptor and the A_1 adenosine receptor, respectively.

[24] Cell-Free Membrane Desensitization Assay for G Protein-Coupled Receptors

By MARY HUNZICKER-DUNN and LUTZ BIRNBAUMER

Principle

Exposure of most heptaspanning guanine (G) nucleotide-binding protein-coupled receptors to saturating concentrations of agonist promotes an attenuation or desensitization of receptor-dependent effector activity.[1] Homologous, receptor-specific desensitization usually results from an uncoupling of the agonist-activated receptor from its cognate G protein. This is an extremely well-studied phenomenon, with over 600 references in the last 10 years. The great majority of reports on receptor desensitization have utilized either an intact cellular model or purified proteins reconstituted with lipids to evaluate the mechanisms of receptor desensitization. From these studies, a compelling series of events appear to be responsible for G protein-coupled receptor desensitization.

For the prototypical β_2-adrenergic receptor, desensitization in response to saturating agonist is believed to be triggered by phosphorylation of the agonist-activated receptor by a G protein-coupled receptor kinase (GRK).[2] Activation of at least GRK2 and GRK3 is regulated in part by their interaction with G protein $\beta\gamma$ subunits (generated on receptor-stimulated G protein activation), which promote translocation of the cytoplasmic GRKs to the plasma membrane.[3] In response to receptor phosphorylation, the clathrin adaptor protein[4] β-arrestin is recruited from the cytoplasm,[5] binds with high affinity to the phosphorylated receptor, and quenches receptor signaling to G proteins.[6] β-Arrestin binding to the β_2-adrenergic receptor additionally promotes receptor sequestration away from the cells surface[2]. Receptor sequestration is reported to be independent of receptor uncoupling from G proteins,[1] although both events generally require the binding of an arrestin to the phosphorylated receptor.[6–8] Following sequestration and internalization, the

[1] W. P. Hausdorff, M. G. Caron, and R. J. Lefkowitz, *FASEB. J.* **4,** 2881 (1990).

[2] R. J. Lefkowitz, *J. Biol. Chem.* **273,** 18677 (1998).

[3] C. V. Carmen and J. L. Benovic, *Curr. Opin. Neurobiol.* **8,** 335 (1999).

[4] F.-T. Lin, K. M. Krueger, H. E. Kendall, Y. Daaka, Z. L. Fredericks, J. A. Pitcher, and R. J. Lefkowitz, *J. Biol. Chem.* **272,** 31051 (1997).

[5] S. S. G. Ferguson, L. S. Barak, J. Zhang, and M. G. Caron, *Can. J. Physiol. Pharmacol.* **74,** 1094 (1996).

[6] M. J. Lohse, S. Andexinger, J. Pitcher, S. Trukawinski, J. Codina, J.-P. Faure, M. G. Caron, and R. J. Lefkowitz, *J. Biol. Chem.* **267,** 8558 (1992).

[7] O. B. Goodman, J. G. Krupnick, F. Santini, V. V. Gurevich, R. B. Penn, A. W. Gagnon, J. H. Keen, and J. L. Benovic, *Nature* **383,** 447 (1997).

[8] U. Wilden, S. W. Hall, and H. Kuhn, *Proc. Natl. Acad. Sci. U.S.A.* **83,** 1174 (1986).

0076-6879/02 $35.00

β_2-adrenergic receptor returns to the surface of the cell in a dephosphorylated, activatable state.[9]

Desensitization of at least some G protein-coupled receptors can also be evaluated using a cell-free membrane model, which was developed and initially used to evaluate desensitization of the luteinizing hormone/chorionic gonadotropin (LH/CG) receptor.[10] For the LH/CG receptor, this *in vitro* membrane model faithfully mimics the time course and extent of receptor-specific LH/CG receptor desensitization seen in ovarian follicles.[10–13] Utilizing this model, we have demonstrated that LH/CG receptor desensitization is mediated by the binding of membrane-delimited β-arrestin to the third intracellular loop of the receptor.[14,15] We have shown that this β-arrestin is released from its membrane-docking site upon activation of ADP-ribosylation factor (ARF)6[16] by the activated LH/CG receptor. Activation of ARF6 appears to comprise the GTP requirement[17–19] for LH/CG receptor desensitization.[16]

The utility of this membrane model to evaluate receptor desensitization requires that all of the enzymes, other proteins, and cofactors needed to mediate receptor desensitization be present and activatable in the membrane preparation. This membrane model offers the advantage of accessibility to added reagents. Rather than needing to transfect cells, purified proteins, peptides, antibodies, or cell-impermeable activators or inhibitors of various enzymes can be added directly to the membranes to evaluate their effect on receptor desensitization. That the effect of the addition is on the receptor and not on the G protein or adenylyl cyclase can be tested readily by evaluating AlF- or cholera toxin (for G protein directed)- or forskolin (for adenylyl cyclase directed)-stimulated adenylyl cyclase activities. Another advantage of this cell-free membrane model is that it offers the ability to investigate receptor uncoupling from G proteins in the absence of receptor sequestration, as the latter event cannot be studied in this cell-free membrane

[9] J. Zhang, L. S. Barak, K. E. Winkler, M. G. Caron, and S. S. G. Ferguson, *J. Biol. Chem.* **272,** 27005 (1997).

[10] J. Bockaert, M. Hunzicker-Dunn, and L. Birnbaumer, *J. Biol. Chem.* **251,** 2653 (1976).

[11] M. Hunzicker-Dunn, *Biol. Repro.* **24,** 279 (1981).

[12] M. L. G. Lamm and M. Hunzicker-Dunn, *Mol. Endocrinol.* **8,** 1537 (1994).

[13] J. M. Marsh, T. M. Mills, and W. J. Lemaire, *Biochim. Biophys. Acta* **304,** 197 (1973).

[14] S. Mukherjee, K. Palczewski, J. L. Benovic, V. V. Gurevich, and M. Hunzicker-Dunn, *Proc. Natl. Acad. Sci. U.S.A.* **96,** 493 (1999).

[15] S. Mukherjee, K. Palczewski, V. V. Gurevich, and M. Hunzicker-Dunn, *J. Biol. Chem.* **274,** 12984 (1999).

[16] S. Mukherjee, V. V. Gurevich, J. C. R. Jones, J. E. Casanova, S. R. Frank, M.-F. Bader, R. A. Kahn, K. Palczewski, K. Aktories, and M. Hunzicker-Dunn, *Proc. Natl. Acad. Sci. U.S.A.* **97,** 5901 (2000).

[17] R. C. Ekstrom, E. M. Carney, M. L. G. Lamm, and M. Hunzicker-Dunn, *J. Biol. Chem.* **267,** 22183 (1992).

[18] E. Ezra and Y. Salomon, *J. Biol. Chem.* **255,** 653 (1980).

[19] E. Ezra and Y. Salomon, *J. Biol. Chem.* **256,** 5377 (1981).

model. Additionally, the membrane model offers the advantage of convenience. Purified membrane preparations are stable at low temperatures ($<-70°$) for at least 6 months.

This cell-free membrane model therefore offers an alternate method to evaluate the steps of G protein-coupled receptor desensitization. However, while its utility beyond the LH/CG receptor has not been examined extensively, there are reports of cell-free membrane receptor desensitization for the glucagon receptor,[20,21] vasopressin receptor,[22] follicle-stimulating hormone receptor,[23] and the β-adrenergic receptor.[24] Cell-free membrane models have also been used to evaluate the mechanisms of heterologous desensitization.[25,26]

Membrane Preparation

Generally, gentle homogenization protocols to yield a plasma membrane preparation are required.[27] The membranes need to be purified away from nuclei, largely due to contaminating nucleotides and enzymes. Following homogenization of tissue (1 g/20 ml) in a glass/glass homogenizer (Blaessig Glass Co., Rochester, NY; approximately seven strokes with loose pestle and four strokes with tight pestle), the homogenate is filtered successively through one, two, and three layers of gauze to remove cell debris, layered over a cushion of 45% (wt/wt) sucrose, 1 m*M* EDTA, 10 m*M* Tris–HCI (pH 7.2; 30 ml homogenate per 5-ml cushion), and then centrifuged at 60,000*g* for 60 min at 4°. The interface of membranes on top of the sucrose cushion is aspirated off, diluted 10-fold with 10 m*M* Tris–HCI, pH 7.2, and centrifuged at 20,000*g* for 30 min. The pellet (membranes) is resuspended in 10 m*M* Tris–HCI, pH 7.2, to a protein concentration of 3–5 mg/ml. Membranes are then quick frozen in a dry ice–acetone bath and stored in aliquots at $-70°$ until further use. Membrane adenylyl cyclase activity varies somewhat from batch to batch of membranes and is not stable to freezing and rethawing.

Critical Reagents

In order to determine the nucleotide dependence of desensitization, GTP-free reagents are required. ATP and cAMP are both readily available from commercial sources free of contaminating GTP. The creatine phosphate/creatine kinase

[20] H. Attramadal, L. Eikvar, and V. Hansson, *Endocrinology* **123,** 1060 (1988).

[21] R. Iyengar, P. W. Mintz, T. L. Swartz, and L. Birnbaumer, *J. Biol. Chem.* **255,** 11875 (1980).

[22] C. Roy, G. Guillon, and S. Jard, *Biochem. Biophys. Res. Commun.* **72,** 1265 (1976).

[23] J. Sanchez-Yague, R. W. Hipkin, and M. Ascoli, *Endocrinology* **132,** 1007 (1993).

[24] W. B. Anderson and C. J. Jaworski, *J. Biol. Chem.* **254,** 4596 (1979).

[25] M. W. Kunkel, J. Friedman, S. Shenolikar, and O. H. Clark, *FASEB. J.* **3,** 2067 (1989).

[26] H.-L. Lai, T. H. Yang, R. O. Messing, Y.-H. Ching, S.-C. Lin, and Y. Chern, *J. Biol. Chem.* **272,** 4970 (1997).

[27] R. C. Ekstrom and M. Hunzicker-Dunn, *Endocrinology* **124,** 2470 (1989b).

ATP-regenerating system used in the adenylyl cyclase assay, however, needs to be purified of contaminating nucleotides, as described in Iyengar *et al.*[21] Contaminating nucleotides are removed from creatine phosphate by adsorption to activated charcoal [2.67 g creatine phosphate in 10 ml water plus 50 mg Norit A, mix 10 min at 4°, discard charcoal pellet collected by centrifugation (2500 rpm, 10 min at 4°), filter supernatant through Whatmann #1]. Creatine phosphokinase (50 mg in 1 ml water) is passed over a 10-ml column of Sephadex G-25, collecting 1-ml fractions. Peak protein fractions are pooled, mixed with creatine phosphate, neutralized to pH 7.2, and frozen in aliquots.

Procedure

The membrane receptor desensitization reaction generally consists of a two-stage incubation (Fig. 1A). In stage 1, membranes are incubated (40 min, 30°) under conditions that do not promote receptor activation and thus do not lead to receptor desensitization (in the absence of agonist) or under conditions that promote receptor activation and therefore lead to receptor desensitization (in the presence of agonist). Stage 2 is contiguous with stage 1 and consists of a 5-min adenylyl cyclase assay to evaluate receptor activity resulting from stage 1 incubations. As shown in Fig. 1B, where hCG was used as the agonist with ovarian

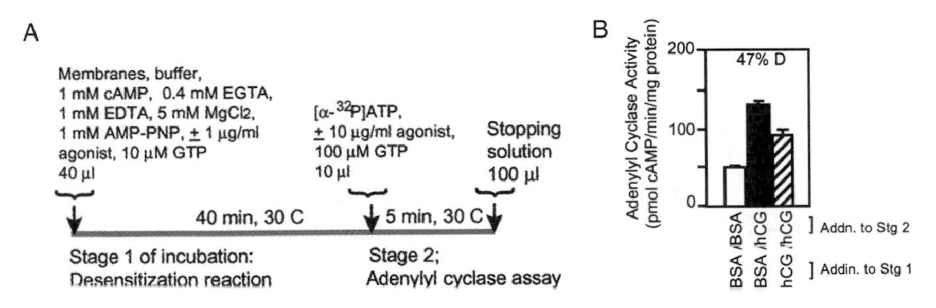

FIG. 1. Schematic of the two-stage reaction to measure receptor desensitization. (A) The composition and volumes of the reaction mix and times of incubation for the two stages of the reaction. From J. Bockaert, M. Hunzicker-Dunn, and L. Birnbaumer, *J. Biol. Chem.* **251,** 2653 (1976). (B) An example of the results obtained from a two-stage reaction using ovarian follicular membranes (30 μg membrane protein) and hCG as the agonist. Buffer in stage 1 is 25 m*M* 1,3-bis[tris (hydroxymethyl)methylamino]propane, pH 7.2. Stage 1 also contains an ATP-regenerating system, consisting of 0.2 mg/ml creatine phosphokinase and 20 m*M* phosphocreatine. Stopping solution consists of 40 m*M* ATP, 10 m*M* cAMP, and 1% sodium dodecyl sulfate. Percentage of desensitization (% D) represents the reduction in hCG-stimulated adenylyl cyclase activity with BSA in stage 1 (BSA/hCG) when incubations contained hCG in stage 1 (hCG/hCG) minus basal (BSA/BSA) activity. From R. M. Rajagopalan-Gupta, S. Mukherjee, X. Zhu, Y.-K. Ho, H. Hamm, M. Birnbaumer, L. Birnbaumer, and M. Hunzicker-Dunn, *Endocrinology* **140,** 1612 (1999).

follicular membranes, samples incubated in the absence of agonist in stage 1 and then in the absence or presence of agonist in stage 2 measure basal (BSA/BSA) or maximal agonist-dependent (BSA/hCG) adenylyl cyclase activities, respectively. Samples incubated in the presence of agonist in stages 1 and 2 (hCG/hCG) measure receptor desensitization. Potential modulators of receptor desensitization are tested in the stage 1 reaction, or in a preincubation; the ultimate response is determined in the stage 2 adenylyl cyclase assay. While reagents may vary in stage 1 reactions, optimal conditions to measure adenylyl cyclase activity are required for stage 2 reactions. These optimal conditions vary with the cell and the receptor. In general, however, the reaction requires millimolar concentrations of ATP, 5–10 μCi [α-^{32}P]ATP (when detecting adenylyl cyclase activity as the conversion of [α-^{32}P]ATP to cAMP), a phosphodiesterase inhibitor to prevent degradation of synthesized cAMP (usually 1 m*M* cAMP), optimal concentrations of GTP to support receptor-dependent G protein activation (generally 100 μM), optimal concentrations of $MgCl_2$, and chelation of calcium (to prevent nonspecific inhibition of adenylyl cyclase).

Use of this membrane model has revealed a number of characteristics of agonist-dependent LH/CG receptor desensitization. While some of these characteristics can be demonstrated readily using intact cells (e.g., dose and time dependence of receptor desensitization), some of the characteristics can only be shown using a cell-free model (e.g., GTP dependence of receptor desensitization). Some examples of the utility of this membrane model to evaluate LH/CG receptor desensitization are described.

Varying the incubation time and hCG concentrations in stage 1 show that LH/CG receptor desensitization increases with time of incubation and concentration of hCG (Fig. 2).[10,12] Unlike the β_2-adrenergic receptor, LH/CG receptor

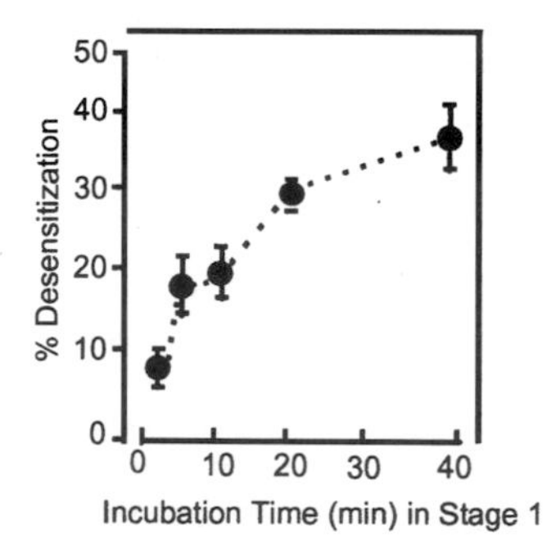

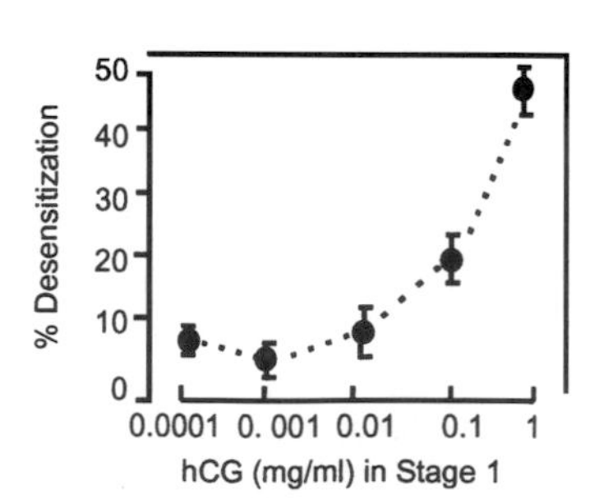

FIG. 2. Effect of incubation time and hCG concentrations in stage 1 on percentage desensitization of LH/CG receptor-stimulated adenylyl cyclase activity. From M. L. G. Lamm and M. Hunzicker-Dunn, *Mol. Endocrinol.* **8,** 1537 (1994).

desensitization is not accompanied by a right-hand shift in the dose–response curve for agonist.[10,28]

LH/CG receptor desensitization is dependent on the concentration of GTP in stage 1 (Fig. 3A). Demonstration of the submicromolar GTP requirement for receptor desensitization requires not only the generation of nucleotide-free regents and a membrane preparation free of contaminating nucleotides, but also the presence of millimolar concentrations of a second nucleotide to protect GTP from hydrolysis, such as adenylylimidodiphosphate (AMP-PNP) (Fig. 3B) or CTP, UTP, or ATP.[29] AMP-PNP is the nucleotide of choice because it is poorly converted into other nucleotides, the results of which could confound the interpretation of results. However, once a GTP dependence has been demonstrated using AMP-PNP, any of these nucleotides will protect GTP from hydrolysis and support receptor desensitization. That LH/CG receptor desensitization occurs when CTP or UTP alone (in the absence of ATP, AMP-PNP, or cAMP) is added to the stage 1 reaction indicates that the product of adenylyl cyclase (cAMP) is not required to promote LH/CG receptor desensitization. Isoproterenol-stimulated desensitization of the β-adrenergic receptor in membranes prepared from rat kidney cells was also been shown to be GTP dependent.[24]

LH/CG receptor desensitization (Fig. 3C, hatched bars) is reversed by the addition of the G protein inactivator GDPβS. For these studies, membranes were subjected to a three-stage incubation. Stage 1 was a standard desensitization reaction; membranes were then diluted fivefold with 10 mM Tris–HCl, pH 7.2, pelleted by centrifugation, and subjected to a new stage 2 incubation consisting of incubation in the presence of 100 μM GTP or various concentrations of GDPβS at 30° for 40 min.[17] Membranes were then diluted and pelleted again by centrifugation and subjected to the standard 5-min adenylyl cyclase assay (in the presence of optimal GTP concentrations). The reversibility of receptor desensitization by a nucleotide that inactivates G proteins has not been reported for other G protein-coupled receptors but is consistent with the requirement for LH/CG receptor desensitization of submicromolar GTP concentrations.[17–19] These results suggest that the GTP requirement for LH/CG receptor desensitization is either unique for this receptor or that this nucleotide requirement for receptor desensitization of other G protein-coupled receptors is awaiting discovery.

LH/CG receptor desensitization is enhanced markedly by ethanol, whereas ethanol does not affect basal, full hCG-, or fluoride-stimulated adenylyl cyclase activities (Fig. 4A). While maximal agonist-dependent LH/CG receptor desensitization in the presence of ethanol retains its requirement for GTP (Fig. 4B), some desensitization is now observed in the absence of added GTP. The simplest explanation of these potentiating effects of ethanol on LH/CG receptor desensitization

[28] R. J. Lefkowitz, J. M. Stadel, and M. G. Caron, *Annu. Rev. Biochem.* **52,** 159 (1983).

[29] R. C. Ekstrom and M. Hunzicker-Dunn, *Endocrinology* **125,** 2470 (1989a).

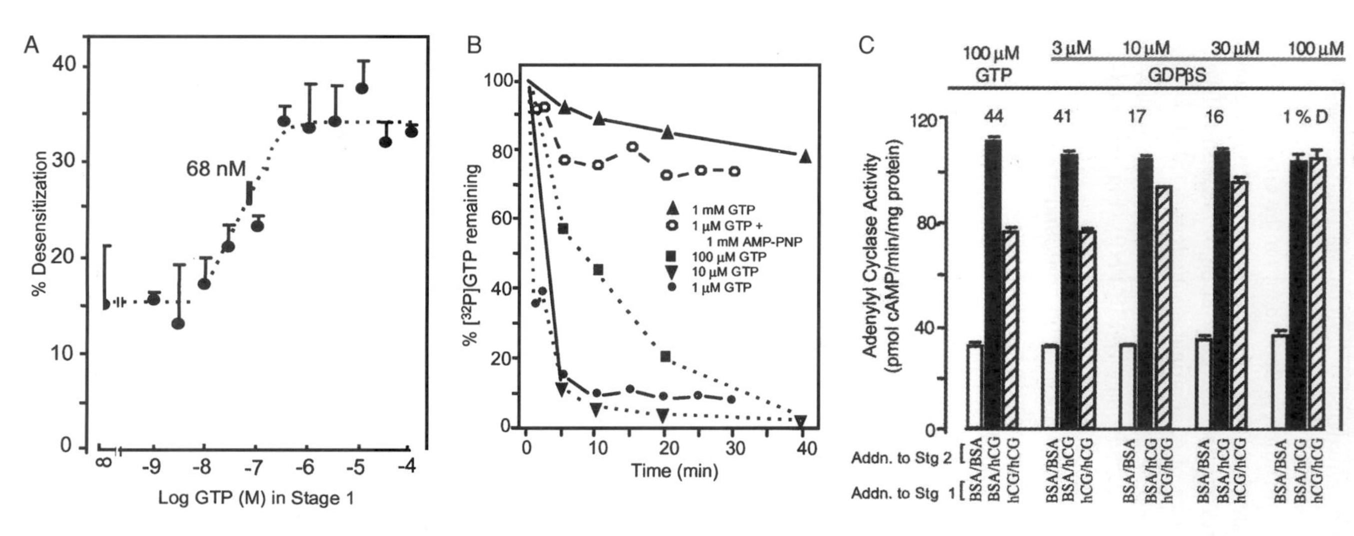

FIG. 3. Dependence of LH/CG receptor desensitization on GTP. (A) Concentrations of GTP are varied in stage 1; GTP in stage 2 is 100 μM. From R. C. Ekstrom and M. Hunzicker-Dunn, *Endocrinology* **124,** 956 (1989). (B) Follicular membranes were incubated with stage 1 reagents, but in the absence of agonist and with indicated concentrations of indicated nucleotides and 1 μCi [α-^{32}P]GTP. Percentage of ^{32}P remaining as GTP was determined by thin-layer chromatography. From R.C. Ekstrom and M. Hunzicker-Dunn, *Endocrinology* **125,** 2470 (1989). (C) Addition of GDPβS to an intermediate 40-min incubation (between stages 1 and 2 shown in Fig. 1A) selectively reverses LH/CG receptor desensitization. From R. C. Ekstrom, E. M. Carney, M. L. G. Lamm, and M. Hunzicker-Dunn, *J. Biol. Chem.* **267,** 22183 (1992).

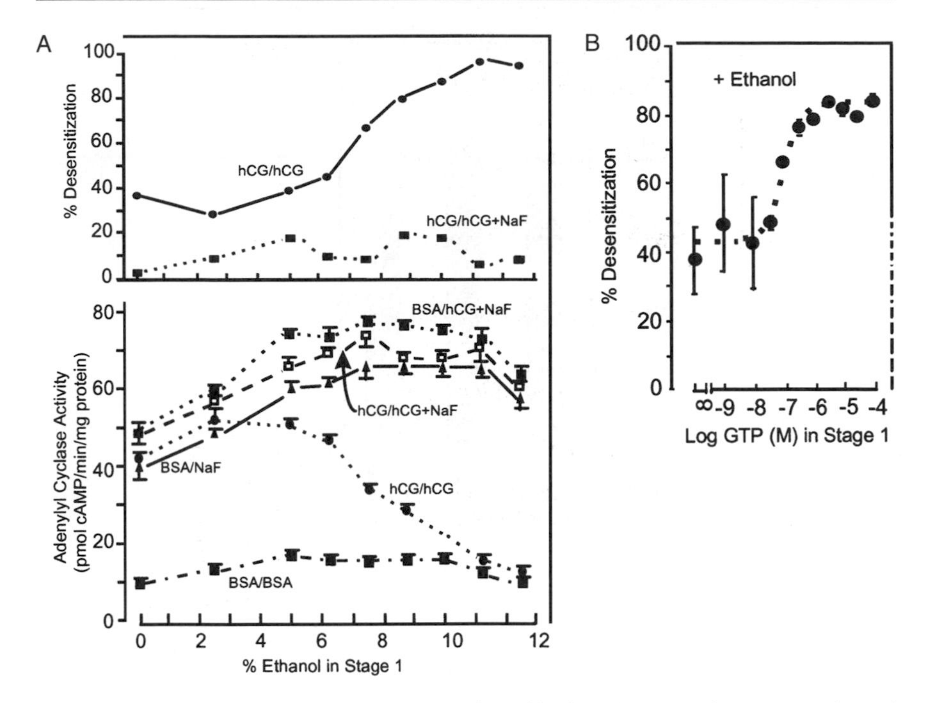

FIG. 4. Effect of ethanol on LH/CG receptor desensitization. (A) Increasing concentrations of ethanol in the stage 1 reaction promote a selective hyperdesensitization of LH/CG receptor-stimulated adenylyl cyclase activity but do not affect basal, full hCG-stimulated, or fluoride-stimulated adenylyl cyclase activities. (B) GTP concentration in stage 1 is varied. From R. C. Ekstrom and M. Hunzicker-Dunn, *Endocrinology* **127,** 2578 (1990).

is that the resulting increased membrane fluidity enhances the extent of LH/CG receptor desensitization in some manner. Desensitization of the gonadotropin-releasing hormone receptor is also reported to be facilitated by ethanol.[30]

This cell-free membrane model is very amenable to the addition of peptides, antibodies, or proteins to elucidate their effects on receptor activity. For example, addition of the $G_s\alpha_{354-372}$ peptide, which has been reported to partially inhibit α_s signaling to adenylyl cyclase,[31] reduced full hCG-stimulated adenylyl cyclase activity by nearly 50% (when basal cyclase activity is subtracted) but did not inhibit LH/CG receptor desensitization (Fig. 5). A second control $G_s\alpha_{384-395}$ peptide reported to be ineffective in modulating α_s signaling to adenylyl cyclase[31] did not

[30] W. C. Gorospe and P. M. Conn, *Mol. Cell. Endocrinol.* **53,** 131 (1987).

[31] M. M. Rasenick, M. Watanabe, M. B. Lazarevic, S. Hatta, and H. E. Hamm, *J. Biol. Chem.* **269,** 21519 (1994).

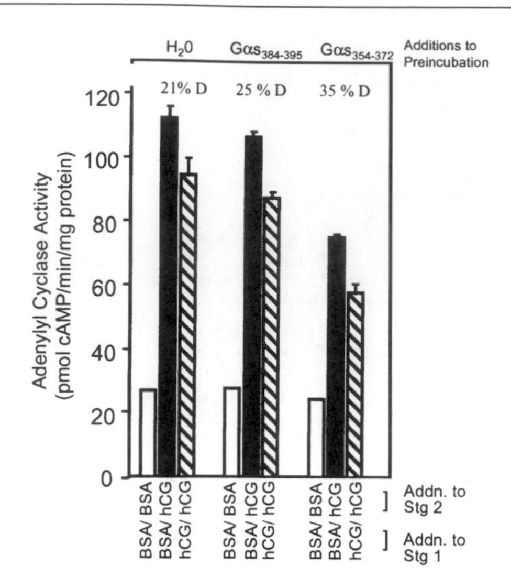

FIG. 5. Effect of addition of synthetic $G_{\alpha s}$ peptides on LH/CG receptor-stimulated adenylyl cyclase activity. Membranes were preincubated for 1 hr at 4° with peptides (final assay concentration of 100 μM) followed by a two-stage reaction. From R. M. Rajagopalan-Gupta, S. Mukherjee, X. Zhu, Y.-K. Ho, H. Hamm, M. Birnbaumer, L. Birnbaumer, and M. Hunzicker-Dunn, *Endocrinology* **140,** 1612 (1999).

affect full hCG-stimulated adenylyl cyclase activity or receptor desensitization (Fig. 5). This result suggests that LH/CG receptor desensitization is independent of G_s activation.

Like desensitization of the β_2-adrenergic,[6] rhodopsin,[8] and muscarinic acetylcholine[32] receptors, LH/CG receptor desensitization is dependent on an arrestin.[14] By adding neutralizing arrestin antibodies to the membranes in a preincubation stage, we have demonstrated that LH/CG receptor desensitization is dependent on an endogenous arrestin present in the membrane preparation.[14] Neutralizing N-terminal-directed A9C6 and C-terminal-directed F4C1 arrestin antibodies added together completely prevented the development of LH/CG receptor desensitization (Fig. 6)

LH/CG receptor desensitization can also be prevented by the addition of a synthetic peptide corresponding to the third intracellular (3i) loop of the LH/CG receptor and not by a peptide containing these amino acids in a scrambled sequence (Fig. 7). Peptides corresponding to the second intracellular loop or N-terminal amino acids in the C-terminal tail of the LH/CG receptor or the 3i loop of the FSH receptor do not prevent LH/CG receptor desensitization.[15] We have shown

[32] S. J. Mundell and J. L. Benovic, *J. Biol. Chem.* **275,** 12900 (2000).

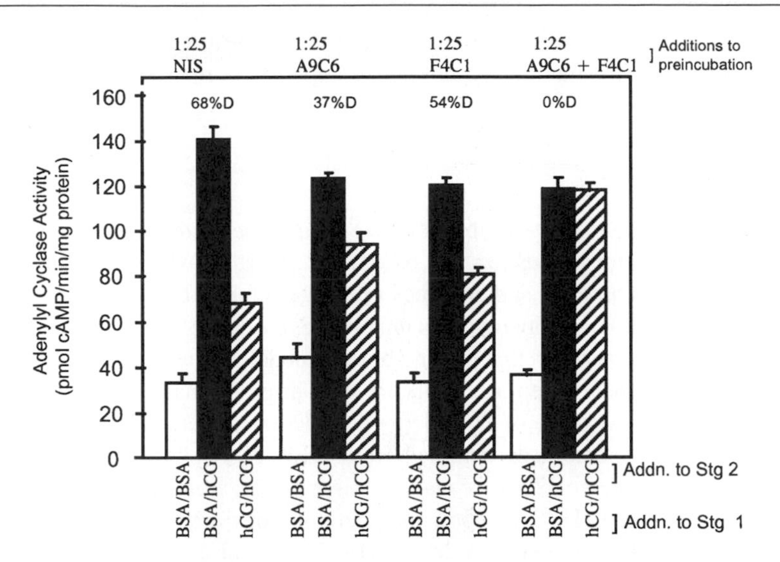

FIG. 6. Effect of addition of neutralizing arrestin antibodies on LH/CG receptor desensitization. Membranes were preincubated with nonimmune serum (NIS) or indicated arrestin antibodies, followed by a two-stage reaction. From S. Mukherjee, K. Palczewski, V. Gurevich, J. L. Benovic, J. P. Banga, and M. Hunzicker-Dunn, *Proc. Natl. Acad. Sci. U.S.A.* **96,** 494 (1999).

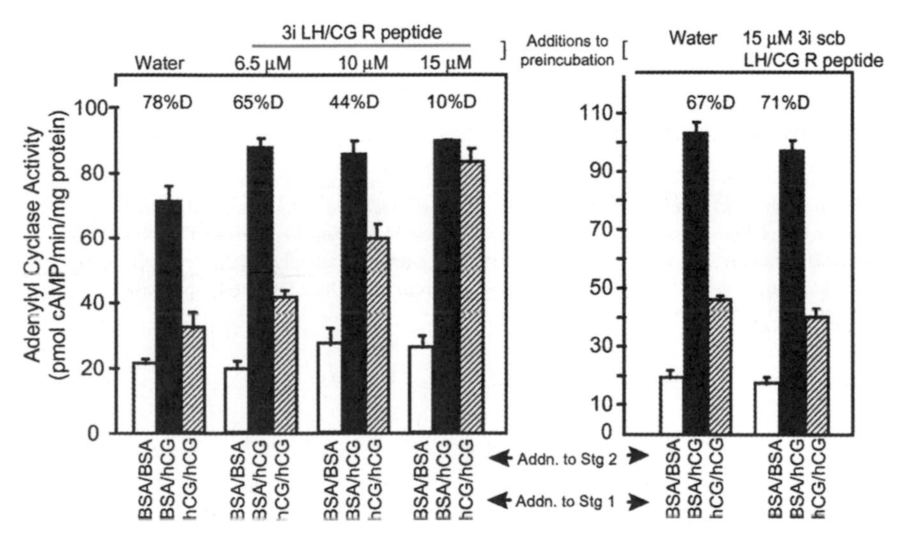

FIG. 7. Effect of addition of synthetic peptide corresponding to the third intracellular (3i) loop of the LH/CG receptor on LH/CG receptor desensitization. Membranes were preincubated with 3i peptide or a peptide containing amino acids of the 3i loop in a scrambled (scb) order, followed by a two-stage reaction. From S. Mukherjee, K. Palczewski, V. V. Gurevich, and M. Hunzicker-Dunn, *J. Biol. Chem.* **274,** 12984 (1999).

that the 3i loop LH/CG receptor peptide blocks receptor desensitization based on its ability to bind arrestin directly and thus compete with the LH/CG receptor for endogenous β-arrestin.[15,16]

Discussion

Widespread applicability of this cell-free membrane model to evaluate G protein-coupled receptor desensitization is untested. Clearly, when receptor desensitization requires proteins and enzymes present in compartments other than the plasma membrane, this assay model is unsuitable. Evidence from the older literature, however, demonstrates that this membrane model can be applied to receptors other than the LH/CG receptor.[20–24] Moreover, in preliminary studies, we have demonstrated the existence of GTP-dependent desensitization of the LH/CG receptor in plasma membranes obtained from HEK 293 cells transfected stably with the LH/CG receptor. That we can demonstrate cell-free, GTP-dependent desensitization of the LH/CG receptor in a heterologous membrane suggests that the proteins required to mediate at least LH/CG receptor desensitization are also present in nonovarian membranes and that these events may not be unique to ovarian follicular membranes and the LH/CG receptor. This preliminary result is consistent with the possibility that a GTP-dependent step is also obligatory for the desensitization of other G protein-coupled receptors. It is therefore likely that there is additional information still to be learned about the cellular mechanism of G protein-coupled receptor desensitization.

Acknowledgments

We thank the many students, postdoctoral fellows, colleagues, and collaborators who contributed ideas, reagents, and experimental expertise to this work. Included in this list are Richard C. Ekstrom, Marilyn L.G. Lamm, Minnie Rajagopalan-Gupta, Sutapa Mukherjee, Lisa Salvador, Evelyn T. Maizels, Mariel Birnbaumer, Krzysztof Palczewski, Vsevolod Gurevich, Richard A. Kahn, and James Casanova. We acknowledge the following grant support to the Hunzicker-Dunn laboratory: NIH P01 HD21921, NIH R01 HD/DK 38060.

[25] Methods to Determine the Constitutive Activity of Histamine H_2 Receptors

By ROB LEURS, MARCEL HOFFMANN, ASTRID E. ALEWIJNSE, MARTINE J. SMIT, and HENK TIMMERMAN

Introduction

Classical models of G protein-coupled receptors (GPCRs) require agonist occupation of receptors to activate signal transduction pathways. It is well documented that GPCRs can be spontaneously active, and this agonist-independent receptor activity is often referred to as constitutive receptor activity.[1–5] Inverse agonists reduce constitutive GPCR activity, whereas neutral antagonists do not affect basal GPCR activity. Constitutive activity is often shown for mutant GPCRs and prove to be the mechanistic basis for several genetic diseases.[6] Moreover, a detailed understanding of constitutive GPCR activity is thought to give better insights in agonist-induced GPCR activation. Because of these important implications, the phenomenon of constitutive GPCR activity and inverse agonism currently receives considerable attention.

Constitutive activity of wild-type receptors is generally low and inverse agonism is therefore often studied using constitutively active mutant (CAM) GPCRs. However, as a result of introduced mutations, the pharmacological behavior of ligands can be seriously affected. For example, as a result of a single point mutation in TM4 of the δ-opioid and μ-opioid receptors, antagonists became surprisingly full agonists.[7] Moreover, the β-blocker propranolol is reported to be an inverse agonist at the wild-type β_2-adrenergic receptor,[8] but a neutral antagonist at a constitutively active β_2-adrenergic receptor mutan.[9] In view of these findings, the use of wild-type receptors in the study of inverse agonism is preferred.

The histamine H_2 receptor belongs to the large family of GPCRs and is positively coupled to adenylate cyclase. We have demonstrated that this receptor stably expressed in Chinese hamster ovary (CHO) cells shows a considerable degree of

[1] T. Costa, Y. Ogino, P. J. Munson, H. O. Onaran, and D. Rodbard, *Mol. Pharmacol.* **41,** 549 (1992).
[2] W. Schütz and M. Freissmuth, *Trends Pharmacol. Sci.* **13,** 376 (1992).
[3] R. L. Lefkowitz, S. Cotecchia, P. Samama, and T. Costa, *Trends Pharmacol. Sci.* **14,** 303 (1993).
[4] G. Milligan, R. A. Bond, and M. Lee, *Trends Pharmacol. Sci.* **16,** 10 (1995).
[5] A. Scheer and S. Cotecchia, *J. Recept. Signal Transduct. Res.* **17,** 57 (1997).
[6] A. M. Spiegel, *Annu. Rev. Physiol.* **58,** 143 (1996).
[7] P. A. Claude, D. R. Wotta, X. H. Zhang, P. L. Prather, T. M. McGinn, L. J. Erickson, H. H. Loh, and P. Y. Law, *Proc. Natl. Acad. Sci. U.S.A.* **93,** 5715 (1996).
[8] E. J. Adie and G. Milligan, *Biochem. J.* **300,** 709 (1994).
[9] D. J. MacEwan and G. Milligan, *FEBS Lett.* **399,** 108 (1996).

METHODS IN ENZYMOLOGY, VOL. 343
0076-6879/02 $35.00

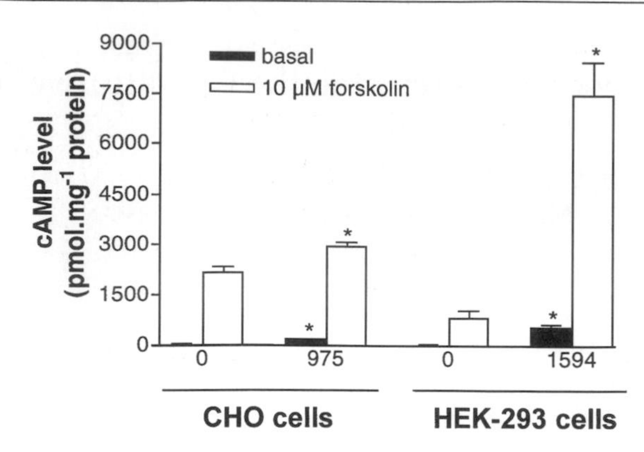

FIG. 1. Modulation of basal- and forskolin-induced cAMP levels by the rat H_2 receptor after stable expression in CHO and HEK-293 cells. Receptor expression levels in cell homogenates were determined by [^{125}I]APT radioligand binding. Cells were also assayed for basal (closed bars)- and forskolin (open bars)-mediated cAMP generation. The inverse agonist ranitidine reduces the increased basal or forskolin levels to the level of cAMP present in mock-transfected cells (not shown). Data represent the mean ± SEM of al least three independent experiments.

constitutive receptor activity (Fig. 1).[10] In addition, we showed that some classical H_2 antagonists, e.g., ranitidine and cimetidine, behave as inverse agonists, whereas burimamide behaves either as a neutral antagonist or as a weak partial agonist.[11,12] This article discusses a variety of experimental approaches that can be used to determine the constitutive activity of the H_2 receptor or other G_s-coupled receptors.

Receptor Overexpression

In the oversimplified two-state model, GPCRs are considered to be in equilibrium between an inactive state R and an active conformation R*. Agonist are thought to show higher affinity for the R* state and, on binding, shifting the equilibrium to the active conformation. In contrast, inverse agonists are considered to bind preferentially to the inactive conformation.[13–15] In most instances the R*

[10] M. J. Smit, R. Leurs, A. E. Alewijnse, J. Blauw, G. P. van Nieuw Amerongen, Y. van de Vrede, E. Roovers, and H. Timmerman, *Proc. Natl. Acad. Sci. U.S.A.* **93,** 6802 (1996).

[11] A. E. Alewijnse, H. Timmerman, E. H. Jacobs, M. J. Smit, E. Roovers, S. Cotecchia, and R. Leurs, *Mol. Pharmacol.* **57,** 890 (2000).

[12] A. E. Alewijnse, M. J. Smit, M. Hoffmann, D. Verzijl, H. Timmerman, and R. Leurs, *J. Neurochem.* **71,** 799 (1998).

[13] R. L. Lefkowitz, S. Cotecchia, P. Samama, and T. Costa, *Trends Pharmacol. Sci.* **14,** 303 (1993).

[14] G. Milligan, R. A. Bond, and M. Lee, *Trends Pharmacol. Sci.* **16,** 10 (1995).

[15] A. Scheer and S. Cotecchia, *J. Recept. Signal Transduct. Res.* **17,** 57 (1997).

state is difficult to detect in the absence of any agonist as the absolute number of receptor protein molecules in the R* state is too low to give rise to detectable levels of second messenger production. However, with the various heterologous expression techniques it is currently possible to attain high GPCR expression levels. Consequently, relatively high absolute numbers of receptors in an active conformation are obtained and an agonist-independent response can be detected. This section describes how modulation of the H_2 receptor number allows the detection of constitutive activity of this receptor.

H_2 Receptor Overexpression Using Conventional Expression Technologies

H_2 receptors can be efficiently overexpressed in a variety of cell lines that are commonly used in heterologous expression experiments. Stable expression of the H_2 receptor in CHO or human embryonic kidney 293 (HEK-293) cells after lipofection-mediated transfection usually results in expression levels of 1 pmol mg^{-1} protein or more. Constitutive activity of the H_2 receptor was originally detected in a CHO clone stably expressing the rat H_2 receptor at 1 pmol mg^{-1} protein.[16] Testing a variety of clones at different expression levels clearly showed a receptor expression-dependent increase of the basal cAMP levels that could be inhibited by some H_2 antagonists (cimetidine, ranitidine) but not by others (burimamide) (Fig. 1). To assure that the agonist-independent response is not due to residual histamine in the culture medium, we determined the actual concentration of histamine in the culture medium, but were unable to measure detectable levels of histamine (<1 n*M*).[16]

Because HEK-293 cells are very sensitive to changes in adenylate cyclase activity, these cells are very useful for the detection of activation of G_s-coupled GPCRs. We therefore also stably expressed the H_2 receptor in HEK-293 cells. As expected, expression of the H_2 receptor (1.5 pmol mg^{-1} protein) resulted in the constitutive activation of adenylate cyclase,[17] which was inhibited by inverse agonists. From Fig. 1 one can deduce that the constitutive activity of the H_2 receptor is more marked in HEK-293 cells compared to CHO cells, rendering HEK-293 cells a very suitable cell system for the detection of constitutive activation of G_s-coupled receptors. Moreover, transient expression of the H_2 receptor in HEK-293 cells using standard calcium phosphate-mediated transfection also results in high levels of receptor expression. Concomitantly, constitutive H_2 receptor activation of wild-type H_2 receptors is detected easily,[8] making HEK-293 cells

[16] M. J. Smit, R. Leurs, A. E. Alewijnse, J. Blauw, G. P. van Nieuw Amerongen, Y. van de Vrede, E. Roovers, and H. Timmerman, *Proc. Natl. Acad. Sci. U.S.A.* **93,** 6802 (1996).

[17] A. E. Alewijnse, M. J. Smit, M. S. Rodriguez Pena, D. Verzijl, H. Timmerman, and R. Leurs, *FEBS Lett.* **419,** 171 (1997).

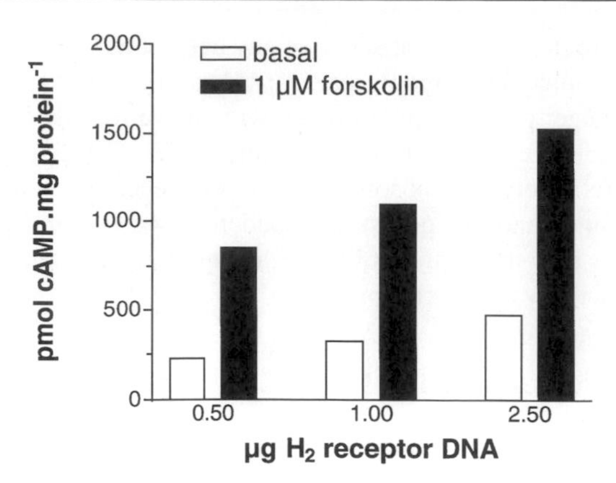

FIG. 2. Expression-dependent modulation of basal- and forskolin-induced cAMP production after H_2 receptor expression in COS-7 cells. COS-7 cells were transfected with different amounts of pcDNA3-rH_2. Cells were allowed to express the receptor for 24 hr and were also assayed for basal (open bars)- and forskolin (closed bars)-mediated cAMP generation. Data represent the mean of three independent experiments.

a suitable model for the evaluation of large numbers of mutant receptors. In our laboratory, COS-7 cells are also often used in the study of mutant GPCRs. Expression of the H_2 receptor using the DEAE-dextran method results in high levels of receptor expression and the subsequent detection of constitutive H_2 receptor activity. As can be seen in Fig. 2, an expression-dependent increase of basal- and forskolin-induced cAMP levels is also observed 24 hr after transfection of COS-7 cells with H_2 receptor cDNA.

H_2 Receptor Overexpression Using the Semliki Forest Virus

Viral expression systems are highly efficient tools for protein expression in higher eukaryotic cells. The Semliki Forest (SFV), a member of the family of Togaviridae, is one of the best-studied alphaviruses and has been adopted successfully for heterologous gene expression.[19] Because of its self-amplifying viral genome, the SFV takes over the host cell translation machinery for high-level expression of recombinant proteins.[20]

SFV can be used to infect a broad range of commonly used mammalian cell lines and can be used effectively to express H_2 receptors. Although H_2 receptor

[18] A. E. Alewijnse, H. Timmerman, E. H. Jacobs, M. J. Smit, E. Roovers, S. Cotecchia, and R. Leurs, *Mol. Pharmacol.* **57,** 890 (2000).

[19] P. Liljestrom and H. Garoff, *Biotechnology* **9,** 1356 (1991).

[20] K. Lundstrom, *J. Recept. Signal Transduct. Res.* **19,** 673 (1999).

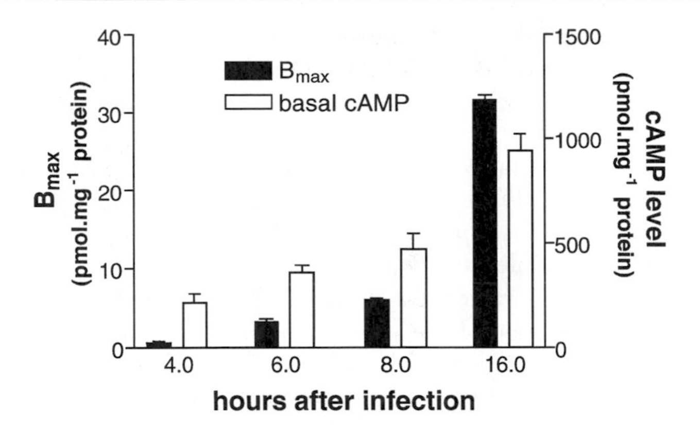

FIG. 3. Expression of rat histamine H_2 receptor and constitutive production of cAMP at various times after SFV-rH_2 infection. COS-7 cells were infected with recombinant SFV-rH_2 as indicated in the text. Cells were allowed to express the receptor for the indicated time and were harvested, and expression levels in cell homogenates were determined by [^{125}I]APT radioligand binding (solid bars). Cells were also assayed for basal cAMP generation (open bars) at various time points after infection. The inverse agonist ranitidine reduces the increased basal levels to the level of cAMP present in mock-infected cells (not shown). Data represent the mean $\pm$ SEM of three independent experiments.

expression varies from one cell line to another in general, the expression levels exceed those obtained from any other mammalian transfection system (unpublished observations). As early as 4 hr postinfection the expression of the H_2 receptor in COS-7 cells is 700 fmol mg^{-1} protein and the expression reaches a maximum at 40 hr postinfection (50 pmol mg^{-1} protein). The high-level expression of the H_2 receptor obtained shortly after recombinant SFV infection also makes the SFV system a very versatile tool in studying the constitutive activity of the wild-type H_2 receptor. From 4 to 16 hr postinfection the basal cAMP levels generated by the H_2 receptor increase almost fivefold. Already at 4 hr postinfection the basal cAMP level is increased compared to mock-infected COS-7 cells (Fig. 3). Incubation with 100 μM ranitidine reduced the basal cAMP levels.

The SFV system also offers an excellent tool to study constitutively active mutant (CAM) receptors that usually express at very low levels using conventional transfection methods, e.g., the $D^{115}N$ mutant of the H_2 receptor has a mutation in the highly conserved DRY motif on the boundary of TM3 and the second intracellular loop. The mutated aspartate is highly conserved within the GPCR family and is suggested to play an important role in receptor activation.[21–23] Mutation

[21] A. E. Alewijnse, H. Timmerman, E. H. Jacobs, M. J. Smit, E. Roovers, S. Cotecchia, and R. Leurs, *Mol. Pharmacol.* **57,** 890 (2000).

[22] S. G. Rasmussen, A. D. Jensen, G. Liapakis, P. Ghanouni, J. A. Javitch, and U. Gether, *Mol. Pharmacol.* **56,** 175 (1999).

[23] A. Scheer, F. Fanelli, T. Costa, P. G. De Benedetti, and S. Cotecchia, *EMBO J.* **15,** 3566 (1996).

of the aspartate in the H_2 receptor results in high levels of constitutive activity and structural instability.[21] Because of this structural instability, pharmacological and biochemical studies of these constitutive active GPCRs are hampered by low expression levels. In HEK-293 cells expressing the $H_2D^{115}N$ mutant after calcium phosphate transfections, the B_{max} is only 246 fmol mg^{-1} protein against 2.5 pmol mg^{-1} protein for the wild-type receptor.[21] Using a recombinant SFV, the $H_2D^{115}N$ mutant transiently expressed in COS-7 cells reaches expression levels of 5 pmol mg^{-1} protein 16 hr postinfection. These expression levels allow an easy pharmacological characterization, whereas the study of constitutive activity of the $H_2D^{115}N$ mutant in COS-7 cells is already possible 6 hr postinfection with recombinant SFV particles. At 6 hr postinfection, the basal level of the $D^{115}N$ mutant is increased 3-fold over that of the wild-type receptor. At 16 hr the basal level of the $D^{115}N$ mutant is still 1.3-fold higher compared to the wild-type receptor. Treatment of the $D^{115}N$ H_2 receptor with the inverse agonist ranitidine reduced cAMP levels to approximately the basal cAMP level of mock-transfected cells.

Methods

Cell Culture and Transfection

HEK-293 cells and COS-7 cells are grown at 37° in a humidified atmosphere with 5% CO_2 in Dulbeco's modified Eagle's medium (DMEM) containing 10% (HEK-293 cells) (v/v) or 5% (COS-7 cells) fetal calf serum supplemented with 2 m*M* L-glutamine, 50 IU/ml penicillin, and 50 μg/ml streptomycin. HEK-293 cells are transiently transfected with 1–10 μg DNA using calcium phosphate precipitation, whereas COS-7 cells are transiently transfected with 1–10 μg DNA using DEAE-dextran.[24] The rat H_2 receptor cDNA is inserted into the commercially available expression vector pcDNA3.

Generation of Recombinant SFV Particles

Plasmid handling is done by standard laboratory procedures.[25] Full-length rat H_2 receptor cDNA is subcloned into the *Xho*I and *Bam*HI sites of the polylinker of pSFV2gen. Recombinant pSFV2gen and pSFV2-helper plasmids are linearized with *Sap*I and purified by phenol extraction prior to *in vitro* transcription. RNA is synthesized *in vitro* driven from the bacterial Sp6 promoter. Linearized pSFV2gen (2.5 μg) is transcribed with 7 U SP6 RNA polymerase (Pharmacia Biotech) in a buffer containing 40 m*M* HEPES (pH 7.4), 6 m*M* MgOAc, 2 m*M* spermidine, 1 m*M* of ATP, CTP, and UTP, 0.5 m*M* GTP, 1 m*M* m7G(5′)ppp(5′)G (CAP, Pharmacia Biotech), 1.5 U RNasin (Roche), and 5 m*M* dithiothreitol (DTT) in

[24] R. H. Brakenhoff, E. M. Knippels, and G. A. van Dongen, *Anal. Biochem.* **218,** 460 (1994).

[25] T. Maniatis, E. F. Fritsch, and J. Sambrook, "Molecular Cloning, a Laboratory Manual." Cold Spring Harbor Laboratory Press, Cold Spring Harbor, NY, 1982.

a final volume of 50 μl for 1 hr at 37°. One semiconfluent 168-cm^2 flask with BHK-21 cells is harvested with Versene/Trypsin solution. Cells are collected by centrifugation for 5 min at 1500 rpm and washed twice with 10 ml phosphate-buffered saline (PBS; 137 m*M* NaCl, 2.7 m*M* KCl, 10.1 m*M* Na_2HPO_4, 1.8 m*M* KH_2PO_4, pH 7.4). Cells are resuspended in 2.4 ml PBS prior to electroporation with *in vitro*-synthesized RNA. Four hundred microliters of BHK-21 cell suspension is transferred to Bio-Rad Genepulser cuvettes (Bio-Rad, 165-2086) together with 50 μl of transcribed pSFV2gen RNA and 25 μl of transcribed pSFV2-helper RNA. Cells are electroporated twice with a Bio-Rad Genepulser (25 μF, 1500 V, $\infty\Omega$). Following electroporation, cells are immediately resuspended in growth medium and seeded in 25-cm^2 flasks. Twenty-four hours after electroporation, the medium from the cells is collected, passed through a 022-μm filter, and used as viral stock. Viral particles are stored at −80°.

H_2 Receptor Expression with the SFV System

Prior to infection the recombinant virus is activated by chymotrypsin treatment. Five hundred micrograms of chymotrypsin (Roche Biochemicals) is added per 10 ml of virus suspension and incubated for 15 min at room temperature. Subsequently, the chymotrypsin is inactivated by the addition of 250 μl aprotinin (10 mg/ml, Sigma). Cells are grown to 80% confluency, the medium is aspirated, and cells are washed once with PBS. Diluted viral suspensions are added in a small volume, just enough to cover the cells. In most experiments, the original viral suspension is diluted 10-fold. This does not result in a significant loss of expression. Cells are incubated with the virus suspension at 37° for 1 hr, after which appropriate culture medium is added and the cells are cultured for 4 to 40 hr. For measurements of cAMP levels, COS-7 cells are seeded in 24-well plates and allowed to attach prior to infection with recombinant SFV.

cAMP Determination Using a Competitive Binding Assay

The day before the experiment, transfected cells are seeded in 24-well plates. Thirty minutes before the experiment, cells are washed with plain DMEM and allowed to equilibrate. Thereafter cells are incubated for 10 min at 37° with the appropriate drugs in DMEM supplemented with 25 m*M* HEPES (pH 7.4 at 37°) and 300 μM of the phosphodiesterase inhibitor isobutylmethylxanthine (IBMX). Aspiration of the medium and addition of 200 μl ice-cold HCl terminate the incubation. Cells are disrupted by sonication (5 secs, 50 W, Labsonic 1510, Braun-Melsungen, Germany), and the homogenate is neutralized with 1 *N* NaOH. The amount of cAMP present in samples is determined using a competitive protein kinase A binding assay according to Nordstedt and Fredholm,[26] with some minor modifications.[27] cAMP data are analyzed using Assayzap and Graphpad Prism

[26] C. Nordstedt and B. B. Fredholm, *Anal. Biochem.* **189,** 231 (1990).

version 2.01 software packages. To compare individual 24-well plates, the cAMP levels are corrected for the actual total protein concentration per well. Protein concentrations are determined according to Bradford using bovine serum albumin as a standard.[28]

Modulation of Forskolin Response in Order to Detect Inverse Agonism

Activation of G_s-coupled receptors and reconstitution of adenylyl cyclase isoenzymes with purified G_{α_s} subunits are known to have a synergistic effect on forskolin-induced adenylyl cyclase activation.[29,30] Consequently, constitutively active G_s-coupled receptors can be anticipated to potentiate forskolin responses by increasing the levels of free G_{α_s} subunits. As mentioned earlier, stable expression of the rat H_2 receptor in CHO cells enhanced the basal cAMP level. However, forskolin-induced cAMP production is also potentiated by overexpression of the H_2 receptor (Fig. 1). In CHO cells the increase in forskolin-induced cAMP production was maximal at a concentration of 1 μM. At higher forskolin concentrations, the relative increase in forskolin-induced cAMP production decreased due to the increased cAMP response in CHOdhfr cells. A dose-dependent inhibition of forskolin-induced cAMP production by inverse agonists was observed in H_2 receptor-expressing cells. The pIC_{50} values of inverse agonists determined from the inhibition of the forskolin responses were not significantly different from values determined from inhibition of the basal cAMP level (Table I).

Modulation of the forskolin response by constitutively active H_2 receptors thus represents a useful tool to study inverse agonism at the H_2 receptor. The method seems quite sensitive, as cimetidine in transfected CHO cells could be clearly classified as a partial inverse agonist at forskolin-stimulated cAMP levels. At the basal cAMP level, differences in negative intrinsic activity among cimetidine, famotidine, and ranitidine could not be detected easily. This is probably due to the fact that the forskolin-induced cAMP level is approximately 20-fold higher than the basal cAMP level and small differences can consequently be detected easier.

The increase of the forskolin response by H_2 receptor expression is not limited to a particular cell type. Compared to CHO cells, the constitutive activity of the H_2 receptor is more pronounced in HEK-293 cells, which are known to be very responsive to G_s-coupled receptors. Stable expression of the H_2 receptor in this cell line resulted in a strongly enhanced basal- and forskolin-induced cAMP production

[27] M. J. Smit, R. Leurs, A. E. Alewijnse, J. Blauw, G. P. van Nieuw Amerongen, Y. van de Vrede, E. Roovers, and H. Timmerman, *Proc. Natl. Acad. Sci. U.S.A.* **93,** 6802 (1996).

[28] M. M. Bradford, *Anal. Biochem.* **72,** 248 (1976).

[29] J. W. Daly, W. Padgett, and K. B. Seamon, *J. Neurochem.* **38,** 532 (1982).

[30] E. M. Sutkowski, W. J. Tang, C. W. Broome, J. D. Robbins, and K. B. Seamon, *Biochemistry* **33,** 12852 (1994).

TABLE I
COMPARISON OF POTENCIES AND INTRINSIC ACTIVITY OF H_2 LIGANDS DETERMINED IN THE cAMP ASSAY AGAINST BASAL OR FORSKOLINE (1 μM)-INDUCED PRODUCTION OF cAMP[a]

	Basal cAMP		Forskoline	
Ligand	pIC_{50}	α	pIC_{50}	α
Cimetidine	6.3	−0.88	6.3	−0.83
Ranitidine	6.8	−1.0	6.7	−1.0
Famotidine	6.7	−0.88	6.7	−0.93

[a] Intrinsic activity (α) and potencies (pEC_{50}) were determined from the dose–response curves of the indicated drugs. The inhibition by ranitidine is defined as −1, and the intrinsic activity of cimetidine and famotidine is expressed relative to ranitidine.

(Fig. 1), which is inhibited by the inverse agonist cimetidine with comparable potency as observed in CHO cells. Moreover, transient expression of the H_2 receptor in HEK-293 or COS-7 cells also leads to increased forskolin responses that can be inhibited by inverse agonists (Fig. 2). Importantly, transient expression of the G_s-coupled TSH receptor, and constitutively active TSH mutant receptors in COS-7 cells also clearly enhanced the forskolin response.[31] In studing inverse agonism at the H_2 receptor, this method may thus be of general use in the study of inverse agonism at G_s-coupled receptors.

In conclusion, modulation of the forskolin response offers a useful and simple method to investigate constitutive GPCR activity in several cell lines and inverse agonism at wild-type G_s-coupled receptors. In line with these findings, constitutively active G_i-coupled receptors also affect the forskolin response.[32,33] As expected, the forskolin response is modulated in a way opposite to that of G_s-coupled receptors; consequently, elevation of the forskolin response by inverse agonists can be regarded as a related assay for inverse agonism at G_i-coupled receptors.

Use of Reporter Gene Assay to Detect Inverse Agonism

Several studies utilizing a reporter gene assay to measure the activation of GPCRs linked to adenylate cyclase have been published.[34–37] Reporter gene assays are quite often more sensitive compared to the direct measurement of second

[31] A. E. Alewijnse, M. J. Smit, M. S. Rodriguez Pena, D. Verzijl, H. Timmerman, and R. Leurs, *FEBS Lett.* **419,** 171 (1997).

[32] T. T. Chiu, L. Y. Yung, and Y. H. Wong, *Mol. Pharmacol.* **50,** 1651 (1996).

[33] N. Griffon, C. Pilon, F. Sautel, J. C. Schwartz, and P. Sokoloff, *J. Neural. Transm.* **103,** 1163 (1996).

[34] A. Himmler, C. Stratowa, and A. P. Czernilofsky, *J. Recept. Res.* **13,** 79 (1993).

[35] M. J. Castanon and W. Spevak, *Biochem. Biophys. Res. Commun.* **198,** 626 (1994).

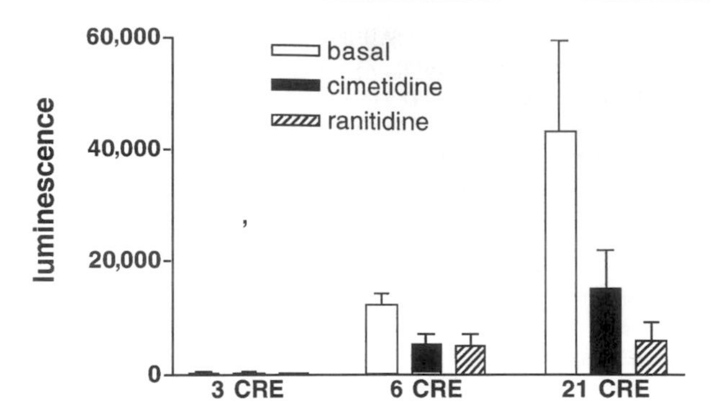

FIG. 4. Luciferase activity in COS-7 cells transiently cotransfected with the H_2 receptor cDNA and luciferase reporter gene constructs with different numbers of CRE elements. Basal luciferase activity (open bars) expressed as luciferase-mediated luminescence after CRE-mediated luciferase transcription from a reporter gene plasmid containing 3, 6, or 21 CRE elements. The inverse agonists cimetidine (100 μM, closed bars) and ranitidine (100 μM, striped bars) inhibit basal luciferase activity to levels obtained in cells transfected without the H_2 receptor cDNA. Data shown represent the mean $\pm$ SEM of at least three independent experiments.

messengers following GPCR (de)activation, presumably because of the signal amplification along the pathway. Consequently, these assays might also facilitate the investigation of constitutive GPCR activity and inverse agonism.

To develop a highly sensitive reporter gene assay for the measurement of inverse agonism, different reporter gene plasmids were compared. The main difference between the tested plasmids is the number of cAMP response elements (CREs) that control the firefly luciferase gene expression. Plasmids with 3, 6, or 21 CREs were compared and coexpressed with the histamine H_2 receptor in COS-7 cells (Fig. 4). The receptor density after coexpression with either one of the three different constructs tested was not significantly different and amounted to 8 pmol mg^{-1} protein on average. Basal luciferase activity in COS-7 cells transiently expressing the H_2 receptor and a luciferase gene is highly dependent on the number of CREs that control the transcription of the luciferase gene. The luciferase construct containing 3 CRE elements is not useful to measure inverse agonism, as the basal luciferase activity is only slightly above the detection limit. Moreover, treatment with the inverse agonist ranitidine does not inhibit basal luciferase expression. Cotransfection of the H_2 receptor with luciferase reporter gene plasmids containing 6 or 21 CRE elements results in a readily detectable increase in basal luciferase expression. Moreover, under these experimental conditions, ranitidine

[36] C. Stratowa, H. Machat, E. Burger, A. Himmler, R. Schafer, W. Spevak, U. Weyer, M. Wiche-Castanon, and A. P. Czernilofsky, *J. Recept. Signal Transduct. Res.* **15,** 617 (1995).

[37] B. Fluhmann, U. Zimmermann, R. Muff, G. Bilbe, J. A. Fischer, and W. Born, *Mol. Cell. Endocrinol.* **139,** 89 (1998).

TABLE II
COMPARISON OF POTENCIES AND INTRINSIC ACTIVITY OF H_2 LIGANDS DETERMINED IN THE LUCIFERASE ASSAY OR THE cAMP ASSAY[a]

	Luciferase assay		cAMP assay	
Ligand	pEC_{50}	α	pEC_{50}	α
Histamine	6.9	1.0	7.1	1.0
Cimetidine	5.7	−0.7	5.9	−0.8
Ranitidine	6.3	−1.0	6.9	−1.0
Famotidine	7.6	−0.9	7.3	−1.0

[a] Intrinsic activity (α) and potencies (pEC_{50}) were determined from the dose–response curves of the indicated drugs. The inhibition by ranitidine is defined as −1, and the intrinsic activity of cimetidine and famotidine is expressed compared to ranitidine.

decreases basal luciferase activity for both constructs. The partial agonistic properties of cimetidine[38] are, however, only evident using the 21 CRE construct, and the use of this construct to investigate inverse agonism is consequently preferred.

To validate the reporter gene assay for the measurement of inverse agonism at the H_2 receptor, the intrinsic activity and the potency of a selected group of H_2 ligands were compared with data obtained in the classical cAMP assay (Table II). COS-7 cells were transiently transfected with the 21-CRE luciferase plasmid and the histamine H_2 receptor plasmid. Histamine dose dependently increases luciferase activity up to 165%, whereas cimetidine, famotidine, and ranitidine, previously identified as (partial) inverse agonists,[38] inhibit basal luciferase activity dose dependently. A similar rank order of potencies and comparable intrinsic activities (α) is found in comparison with cAMP assays in CHO cells stably expressing the H_2 receptor.[39] Moreover, the potency correlates well with the H_2 receptor affinity, as determined by [^{125}I]APT-binding studies.[39]

Methods to Determine Constitutive GPCR Activity via CRE-Mediated Luciferase Transcription

COS-7 cells are transfected transiently by the DEAE-dextran method with the CRE-luciferase reporter plasmid pTLNC121-3[40] and pcDNA3 rat H_2

[38] M. J. Smith, R. Leurs, A. E. Alewijnse, J. Blauw, G. P. van Nieuw Amerongen, Y. van de Vrede, E. Roovers, and H. Timmerman, *Proc. Natl. Acad. Sci. U.S.A.* **93,** 6802 (1996).

[39] M. J. Smit, R. Leurs, A. E. Alewijnse, J. Blauw, G. P. van Nieuw Amerongen, Y. van de Vrede, E. Roovers, and H. Timmerman, *Proc. Natl. Acad. Sci. U.S.A.* **93,** 6802 (1996).

[40] B. Fluhmann, U. Zimmermann, R. Muff, G. Bilbe, J. A. Fischer, and W. Born, *Mol. Cell. Endocrinol.* **139,** 89 (1998).

(1 : 4 ratio, 5 μg DNA/10^6 cells) and seeded at a density of 2×10^4 cells/well in a 96-well black view plate (Corning-Costar). Directly after transfection, 50-μl solutions of the tested drugs are added, and the plates are incubated at 37° in a humidified atmosphere with 5% CO_2 for 48 hr. Thereafter the incubation mixture is discarded, and 25 μl of a luciferase reagent mixture (2.6 m*M* DTT, 39 m*M* Tris–H_3PO_4, 38.7% glycerol, 2.6% Triton X-100, 18.6 m*M* $MgCl_2$, 78 μM NaPPi, 0.8 m*M* ATP, 0.8 m*M* D-luciferin) is added. The plate is mixed for 3 sec, and luminescence is measured after 20 min on a Victor^2, 1420 multilabel counter (Perkin Elmer-Wallac).

Conclusions

With the aid of different heterologous expression techniques, constitutive activity of the H_2 receptor is detected easily in a wide variety of cell lines. Especially at high levels of receptor expression the increase in basal cAMP production can be observed easily (either direct or via a reporter gene assay). For a variety of other GPCRs, similar examples can be found in the literature. Whereas modulation of GPCR expression is an important factor in the detection of constitutive signaling, the level of GPCR expression is also an important determinant for the extent of the constitutive responsiveness. One can easily understand that the expression levels of the various partners in the GPCR signaling pathway also determine the final agonist-independent output. An obvious example is the expression of the G_α subunit. For the G_q-coupled muscarinic m3 receptor,[41] overexpression of G_{α_q} increases the basal activation of signal transduction pathways. However, this strategy highly depends on the actual expression level of the appropriate G_α subunit. If the expression of the G_α subunit is not the limiting factor in the signal transduction cascade, overexpression of the protein is not useful. In the case of the β_2 receptor, the expression level of the G_s protein in NG108-15 cells is not the limiting factor in the signaling pathway of the β_2 receptor.[42] However, in these cells, overexpression of adenylate cyclase II enhances β_2 receptor signaling and has been shown to increase the observed constitutive responsiveness.[43] By experimental manipulation of the stochiometry of various signaling partners, these data show that the sensitivity of the detection of constitutive GPCR activity can be enhanced greatly and offer the potential to engineer suitable cell systems for the detection of constitutive GPCR activity.

[41] E. S. Burstein, T. A. Spalding, H. Braüner-Osborne, and M. R. Brann, *FEBS Lett.* **363,** 261 (1995).
[42] I. Mullaney, I. C. Carr, and G. Milligan, *FEBS Lett.* **397,** 325 (1996).
[43] P. A. Stevens and G. Milligan, *Br. J. Pharmacol.* **123,** 335 (1998).

[26] Expression of G Protein-Coupled Receptors and G Proteins in Sf9 Cells: Analysis of Coupling by Radioligand Binding

By ROLF T. WINDH and DAVID R. MANNING

Introduction

Radioligand-binding assays are a mainstay in the pharmacological analysis of receptor function. With these assays, the location, expression levels, and pharmacological specificity of a large number of G protein-coupled receptors (GPCRs) have been characterized, in many cases for the first time.

Just as receptor occupation by ligands can alter the activity of G proteins, G proteins can influence the affinity of GPCRs for ligands. The extended ternary complex theory, which describes the interactions of ligands, GPCRs, and G proteins, is in part based on observations that the affinity of agonists, but not antagonists, for GPCRs is sensitive to the G protein-coupling status of the receptor.[1] The receptor normally exists in at least two interconvertible affinity states for agonist. Addition of GTP, or nonhydrolyzable analogs such as guanosine 5′-(γ-thio)triphosphate (GTPγS) or guanylylimidodiphosphate (GppNHp), causes a dissociation of G proteins from the receptor and a consequent shift to a lower, single affinity for the agonist.

Although the GTP sensitivity of agonist binding has been used commonly to demonstrate functional coupling of GPCRs to G proteins in tissue preparations and in transfection paradigms, the amount of information that can be obtained about the identity of the G proteins that are coupled to the receptor is limited. Other methods of altering G protein coupling, such as using pertussis toxin (PTX) to uncouple members of the G_i family from receptor or using G_α subunit-specific antibodies to interrupt communication with receptor, can be used in conjunction with radioligand binding to assess G protein identity.

Radioligand binding in insect Sf9 cells is an especially useful paradigm for delineating G protein coupling of GPCRs. Sf9 cells, derived from the fall army worm *Spodoptera frugiperda,* express relatively low levels of G proteins and no GPCRs that have interacted with ligands for mammalian receptors (see Windh *et al.*[2] for review). Thus they can be used essentially as an intact cell reconstitution system: mammalian GPCRs can be introduced with or without G proteins

[1] T. Kenakin, *Pharmacol. Rev.* **48,** 413 (1996).

[2] R. T. Windh, A. J. Barr, and D. R. Manning, *in* "The Pharmacology of Functional, Biochemical and Recombinant Receptor Systems" (T. Kenakin and J. A. Angus, eds.), p. 335. Springer-Verlag, Heidelberg, 2000.

METHODS IN ENZYMOLOGY, VOL. 343
0076-6879/02 $35.00

of defined α, β, and γ subunit composition through infection with recombinant baculoviruses, and receptor binding and function can be assessed in subsequently prepared membranes.[3–6] Proteins expressed in Sf9 cells undergo substantial co- and posttranslational modifications.[7] Fatty acid acylation of G protein subunits and GPCRs occurs in a manner almost identical to that in mammalian cells.[3,8–12] Although heterologously expressed proteins, including GPCRs, are glycosylated, the extent and specific glycosylation events do not always match exactly those in mammalian cell lines.[3,13,14] With few exceptions, the expressed receptors and G proteins function and interact in a manner consistent with those in other expression systems and in tissues.

From the standpoint of receptor binding as a means of assessing G protein communication, the Sf9 cell expression system offers additional advantages. Expression of mammalian GPCRs is often high enough that the endogenous insect G protein complement is stoichiometrically insignificant, and the heterologously expressed GPCR assumes a completely uncoupled phenotype, i.e., the receptor binds agonist only with low affinity (see Windh *et al.*[2] for review). The ability to generate homogeneous populations of uncoupled receptors is an unusal and desirable feature of the Sf9 expression system. G proteins can be added back into the cells through coinfection with recombinant baculoviruses, and the emergence of high-affinity agonist binding can be measured. This is in contrast to mammalian expression systems where it is the loss of high-affinity agonist binding, induced by GTP analogs or PTX, that is usually measured. By measuring the promotion of high-affinity binding with individual G proteins (or subunits), the influence of each G protein or subunit can be assessed individually.

[3] P. Butkerait, Y. Zheng, H. Hallak, T. E. Graham, H. A. Miller, K. D. Burris, P. B. Molinoff, and D. R. Manning, *J. Biol. Chem.* **270,** 18691 (1995).

[4] A. J. Barr, L. F. Brass, and D. R. Manning, *J. Biol. Chem.* **272,** 2223 (1997).

[5] A. J. Barr and D. R. Manning, *J. Biol. Chem.* **272,** 32979 (1997).

[6] R. T. Windh, M.-J. Lee, T. Hla, S. An, A. J. Barr, and D. R. Manning, *J. Biol. Chem.* **274,** 27351 (1999).

[7] D. O'Reilly, L. K. Miller, and V. A. Luckow, "Baculovirus Expression Vectors: A Laboratory Manual." Oxford Univ. Press, Oxford, 1994.

[8] L. K. Miller, *Annu. Rev. Microbiol.* **42,** 177 (1988).

[9] B. Mouillac, M. Caron, H. Bonin, M. Dennis, and M. Bouvier, *J. Biol. Chem.* **267,** 21733 (1992).

[10] M. E. Linder, C. Kleuss, and S. M. Mumby, *Methods Enzymol.* **250,** 314 (1995).

[11] S. Grünewald, W. Haase, H. Reiländer, and H. Michel, *Biochemistry* **35,** 15149 (1996).

[12] M. A. Lindorfer, N. E. Sherman, K. A. Woodfork, J. E. Fletcher, D. F. Hunt, and J. C. Garrison, *J. Biol. Chem.* **271,** 18582 (1996).

[13] H. Reiländer, F. Boege, S. Vasudevan, G. Maul, M. Hekman, C. Dees, W. Hampe, E. J. Helmreich, and H. Michel, *FEBS Lett.* **282,** 441 (1991).

[14] T. Kusui, M. R. Hellmich, L. H. Wang, R. L. Evans, R. V. Benya, J. F. Battey, and R. T. Jensen, *Biochemistry* **34,** 8061 (1995).

Experimental Procedures

Baculovirus Construction

Construction of recombinant baculoviruses was quite time-consuming previously, representing perhaps the greatest obstacle in starting up an Sf9 expression system. Several new commercially available methods significantly reduce the time required to produce recombinant baculoviruses from over a month to as little as 10 days. One such method, marketed as the Bac-to-Bac system by Life Technologies (Gaithersburg, MD), generates recombinant baculoviruses by site-specific transposition in *Escherichia coli* rather than homologous recombination in the insect cells themselves.[15] Other systems, such as the Echo Cloning System marketed by Invitrogen (Carlsbad, CA), make it easier to express the protein of interest in bacterial, yeast, insect, and mammalian expression systems by using a universal donor plasmid.[16] The choice of a transfer vector also determines which method will be used to isolate recombinants.[7] Choosing a recombinant baculovirus construction method therefore depends largely on what spectrum of studies will be performed. In the most widely used baculovirus expression systems, the *Autographa californica* nuclear polyhedrosis virus (AcNPV), the gene of interest replaces the viral gene for polyhedrin, a protein matrix that is produced late in infection to protect newly made virions. Under control of the *polh* promoter, the introduced gene is highly expressed 24–72 hr following infection, potentially representing up to 50% of the total new protein synthesis.[7]

Culture and Infection of Sf9 cells

Sf9 cells [ATCC # CRL-1711 or Invitrogen (Carlsbad, CA)] typically double in number every 18–24 hr and can be grown in monolayer or suspension. In either case, cells are cultured at 27° at ambient O_2 and CO_2. Whereas suspension cultures are most convenient for the regular propagation of cells, infection is performed most easily in cells growing in monolayer. Cells can be cultured in media containing fetal bovine serum (FBS), such as Grace's media supplemented with 3.3 g/liter yeastolate and 3.3 g/liter lactalbumin hydrolysate or TNM-FH or in optimized serum-free media such as Sf900II (Life Technologies, Gaithersburg, MD).[17,18] The presence of ligands for many GPCRs in serum can be problematic in the later stages of infection, as once the ligand binds to the newly expressed receptor, it is

[15] V. A. Luckow, S. C. Lee, G. F. Barry, and P. O. Olins, *J. Virol.* **67,** 4566 (1993).

[16] Q. Liu, M. Z. Li, D. Leibham, D. Cortez, and S. J. Elledge, *Curr. Biol.* **8,** 1300 (1998).

[17] S. Weiss, W. G. Whitford, S. F. Gorfien, and G. P. Godwin, *in* "Baculovirus Expression Protocols" (C. D. Richardson, ed.), p. 65. Humana Press, Totowa, NJ, 1995.

[18] S. Weiss, G. P. Godwin, S. F. Gorfien, and W. G. Whitford, *in* "Baculovirus Expression Protocols" (C. D. Richardson, ed.), p. 79. Humana Press, Totowa, NJ, 1995.

very difficult to remove completely. We have found that serotonin concentrations in FBS can be as high as 30 μM, more than 1000-fold higher than the K_d for the 5-HT_{1A} receptor. Charcoal treatment of the serum reduces 5-HT concentrations by nearly 1000-fold. We thus routinely culture and infect Sf9 cells in charcoal-treated FBS, and also switch to Sf900II when required.

To infect Sf9 cells, subculture cells from suspension to monolayer, using approximately 6 million cells in 10 ml of medium per T75-cm^2 flask. Allow the cells to attach for 30 min, and add virus for receptor and/or G protein subunits, using a multiplicity of infection (MOI, the number of infectious virus particles per subcultured Sf9 cell) of two for receptor and one for G protein subunits. The chosen ratio of G protein-to-receptor MOIs can significantly affect the coupling characteristics of the expressed receptor.[19] If the medium contains appreciable amounts of the ligand for the expressed receptor, it may be necessary to change the medium to an optimized serum-free medium before significant receptor expression occurs, generally 16–20 hr; wash the cells gently once or twice with 5 ml of the serum-free medium and then add 10 ml of fresh serum-free medium. By changing the medium before substantial expression of receptor has occurred, carryover of ligand into the membrane preparation by the receptor is minimized. Note that this abrupt change in medium is likely to reduce the expression of the recombinant proteins.

Preparation of Sf9 Cell Membranes

Harvest cells 48 hr following infection by scraping without changing the medium, as some of the infected cells will have detached following the infection cycle. Transfer the harvested cells to a 50-ml centrifuge tube. Rinse the plates with 5 ml of 0.9% NaCl and add this to the collected cells. Pellet the cells by spinning 3–5 min at 500*g* at room temperature, and resuspend the pellet very gently (Sf9 cells are susceptible to shearing) in 5 ml of 0.9% saline per original plate. Pellet the washed cells as before, wash an additional two times, and place the final pellet on ice. From this point, all procedures should be performed on ice.

To the pelleted cells, add 0.5 ml of ice-cold HE/PI buffer (20 m*M* HEPES, pH 8.0, 2 m*M* EDTA; add protease inhibitors [2 μg/ml aprotinin, 10 μg/ml leupeptin, and 0.1 m*M* phenylmethylsulfonyl fluoride (PMSF) immediately before using]) per original plate of cells, resuspend cells gently with a pipette, and incubate on ice at least 5 min to let the cells swell, as this will make lysis more complete. Break the cells by passing through a 26-gauge needle 15 times. Pellet unbroken cells and nuclei by centrifuging at 110*g* for 5 minutes at 4° and then pellet the membranes by spinning the supernatant for 30 min at 20,000*g* at 4°. Resuspend the pelleted membranes by passing through a 26-gauge needle 15 times in HE/PI buffer such that the final protein concentration is approximately 1.5 μg/μl

[19] S. Grünewald, H. Reiländer, and H. Michel, *Biochemistry* **35,** 15162 (1996).

(usually about 150–200 μl of HE/PI buffer per original plate of cells). Determine the protein concentration, aliquot membranes into convenient lots, quick-freeze in a dry ice-ethanol bath, and store at -70° until use. Membranes prepared in this way can be thawed for use only once and can be stored for at least 2–3 months without harm. A single T75-cm^2 flask of 6 million cells in monolayer will yield approximately 400 μg of membrane protein.

Overview of Radioligand-Binding Assays

We have used three basic types of radioligand-binding assays in characterizing the G protein coupling of the human 5-HT_{1A} receptor: saturation binding to determine receptor expression levels, competition binding to determine the affinities of several 5-HT_{1A} receptor ligands, and single agonist concentration binding to measure the promotion of high-affinity agonist binding upon coexpression of different G protein α, β, and γ subunits. A detailed discussion of the principles and theories of radioligand binding is beyond the scope of this article; the reader is referred to any of a number of good books on this subject.[20–22] In the simplest case, only three conditions are necessary to demonstrate specific binding in a membrane or whole cell preparation. First, the total binding needs to be defined; this is simply the amount of radioligand binding on the filter. Second, nonspecific binding, or the binding of the radioligand to sites other than the receptor, is defined. Specific binding is the difference between total and nonspecific binding. Third, the total amount of radioligand ("counts") added in the assay must be determined. Because assumptions about the concentration of unbound radioligand are important in any calculations made from data, total binding must be less than 10% of the total number of counts added to the reaction. We find that performing all three measurements in triplicate rather than duplicate improves the consistency of the results greatly. These three conditions need to be defined for each concentration of radioligand used.

Binding Protocol

We have found that binding protocols developed for tissue or mammalian expression systems do not need to be changed markedly for use in membranes from Sf9 cells. For all of the 5-HT_{1A} ligands we have used in binding studies ([^{125}I]*p*-MPPI, [^{125}I]8-OH-PIPAT, [^{3}H]8-OH-DPAT, and [^{3}H]spiperone), a standard binding buffer of 50 m*M* Tris, pH 7.4, with 2.5 m*M* $MgCl_2$ has been used. Due to the high specific activity of [^{125}I]*p*-MPPI and [^{125}I]8 OH-PIPAT (NEN

[20] A. Levitzki, "Receptors: A Quantitative Approach." Benjamin/Cummings, Menlo Park, CA, 1984.

[21] L. E. Limbird, "Cell Surface Receptors: A Short Course on Theory and Methods." Kluwer Academic Publishers, Boston, MA, 1996.

[22] T. Kenakin, "Pharmacologic Analysis of Drug-Receptor Interaction." Raven, New York, 1997.

Life Science Products, Boston, MA; 2200 Ci/mmol), only 0.5–2 μg of membrane protein per tube is needed when using these ligands so 0.1% BSA is added to the binding buffer just before use. In this range of membrane protein, total specific binding is linear with protein, and total binding is less than 8% of the total number of counts added to the reaction.

For [^{125}I]*p*-MPPI and [^{125}I]8-OH-PIPAT binding, the total reaction volume is 100 μl: 25 μl each of radioligand and competitor (or ligand for nonspecific binding) and 50 μl of membranes. Thaw frozen membranes at 30° and pellet for 30 min at 20,000*g* at 4°. Resuspend the pellet in binding buffer (with BSA added) by passing through a 26-gauge needle 15 times such that 50 μl contains 2 μg of membrane protein.

Dilute the radioligand in binding buffer. For competition-binding experiments, a concentration near the K_d is typically used: 0.3 n*M* in the case of [^{125}I]*p*-MPPI and [^{125}I]8-OH-PIPAT. For saturation-binding experiments, many concentrations of radioligand will be used, generally ranging from one-tenth to 10-fold the expected K_d. In using a single concentration of radiolabeled agonist to measure increased agonist affinity promoted by G protein subunit coexpression, a sub-K_d concentration of radioligand is used.[3] Taking into consideration the decay of the ^{125}I label, determine the current concentration of the radioligand and dilute to four times the final desired concentration (e.g., 1.2 n*M* in the case of [^{125}I]*p*-MPPI and [^{125}I]8-OH-PIPAT) in binding buffer. Dilute competitor, if using, in binding buffer to four times the desired final concentration.

Add 25 μl of the diluted competitor (or binding buffer) and 25 μl of the radioligand to glass 12 × 75-mm tubes and then start the reaction by adding 50 μl of membranes. Incubate the tubes at 37° for 45 min and terminate the reaction by adding 5 ml/tube of ice-cold wash buffer (50 m*M* Tris, pH 7.4) and filtering through a glass filter (#32; Schleicher and Schuell, Keene, NH) presoaked in 0.3% polethylenimine using a cell harvester. Quickly wash the filter three times with 5 ml each of cold wash buffer. Transfer filters to fresh tubes and count in a γ counter. Using this protocol with [^{125}I]*p*-MPPI or [^{125}I]8-OH-PIPAT, the specific binding at K_d concentrations of the radioligand is approximately 80% of the total binding.

Binding assays using tritiated ligands, e.g., [^{3}H]8-OH-DPAT (Amersham Pharmacia Biotech, Picataway, NJ) and [^{3}H]spiperone (NEN Life Science Products), are performed in essentially the same manner as those using iodinated ligands. Differences include the total assay volume, which is increased from 100 μl for the iodinated ligands to 400 μl for the tritiated ligands, and the amount of protein used per tube, which is increased 10-fold. Due to the lower specific activity of these radioligands (220 and 15–30 Ci/mmol for [^{3}H]8-OH-DPAT and [^{3}H]spiperone, respectively), more membrane protein is needed for robust results. Resuspend the membrane pellet in binding buffer such that 200 μl contains 20 μg of membrane protein. The increased amount of membrane protein means that the binding buffer does not need BSA. Dilute the radioligand and competitor to 4-fold their final concentrations, as 100 μl of each will be added per reaction tube. Add the competitor (or binding buffer) and radioligand to 12 × 75-mm glass tubes, and start the reaction by adding 200 μl of the membranes. Incubate the tubes at 25° for 45 min, and terminate the reaction as described earlier. Transfer the filters to scintillation vials, add scintillation fluid, and count in a liquid scintillation spectrometer. The specific binding at K_d concentrations of radioligands such as $[^3H]$8-OH-DPAT and $[^3H]$spiperone should be approximately 85% of the total binding.

Separating Bound from Free Radioligand

Several methods of separating the bound radioligand from free radioligand have been used for binding assays. The key consideration in choosing a method of separation is the rate of dissociation of the ligand from the receptor under the separation conditions. Because it is rapid and can be performed with cold reagents, filtration is used most commonly. However, centrifugation of membranes (or in some cases whole cells) and dialysis have also been used successfully. When only a few small assays will be performed, filtration apparati that require handling each reaction tube separately can be used effectively. When larger assays are required, for example, a 14 concentration competition experiment, it is almost impossible to obtain reproducible results using these smaller units. For these experiments, cell harvesters are a great help. Connected to a vacuum pump, cell harvesters aspirate from 12 to 48 tubes at once, and they dispense wash buffer to all tubes simultaneously.

Use of Assays and Analysis of Results

Saturation Binding

Saturation binding is used most commonly for determinating receptor expression levels and radioligand affinity. Use of antagonists in Sf9 cells most often reveals a single binding site, regardless of G protein expression. Data can be plotted and analyzed in one of two ways. Rosenberg/Scatchard transformation, which is plotted as bound/free vs bound, provides a linear curve fit for a single binding site.[23,24] The x intercept is the B_{max}, whereas the negative reciprocal of the slope is the K_d. While this method is useful for visualizing results, the X value (bound) is used to calculate the Y value (bound/free), and thus the two variables are not truly independent. A more accurate description of the binding parameters can be made by plotting specific binding (total binding-nonspecific) vs free radioligand (total radioactivity added-total binding). A rectangular hyperbola will fit data for

[23] G. Scatchard, *Ann. N.Y. Acad. Sci.* **51,** 660 (1949).
[24] H. E. Rosenthal, *Anal. Biochem.* **20,** 525 (1967).

a single binding site. Using a variation of this method, two-site fits can also be determined accurately. Computerized pharmacological curve-fit programs such as LIGAND[25] or GraphPad (GraphPad Software Inc., San Diego, CA) that perform nonlinear regression are almost indispensible tools in the analysis of these data.

Saturation binding is also useful for determining whether displacement of one ligand by another is competitive or noncompetitive. Competitive binding by another ligand reduces the apparent affinity of the radioligand without changing the number of binding sites (B_{max}), whereas the opposite results are obtained with ligands that bind noncompetitively. The knowledge that only a single receptor type and G protein type are expressed in infected Sf9 cells makes this a particularly clean pharmacological system for these kinds of determinations. Competitive or noncompetitive inhibition of binding can be examined with respect to the G protein coupling state of the receptor. Ligands that bind coupled (agonists) and uncoupled (inverse agonists) states preferentially inhibit the binding of each other in a noncompetitive manner, whereas each inhibits the binding of a neutral antagonist competitively[26] (Fig. 1). The ability to label a homogeneous population of receptors in either state in Sf9 cells is a real advantage of this system. The uncoupled receptor state can be labeled with a neutral antagonist in membranes that express receptor in the absence of coexpressed G proteins, whereas the coupled state can be labeled by a concentration of agonist at or below its K_d for the coupled form in membranes expressing both receptor and G protein.

Competition Binding

Competition binding studies are used most frequently to determine the affinities of ligands that may not be available in radiolabeled form. Because neutral antagonists bind both coupled and uncoupled forms of the receptor with essentially identical affinity, they are used most commonly as radioligands for this assay type, as agonists will preferentially label only the G protein-coupled state of the receptor (see earlier discussion). The radioligand is used at a concentration near its K_d so that one-half of the total receptor population is labeled. Increasing concentrations of an unlabeled competitor are used to displace the radiolabeled antagonist. When performed in Sf9 membranes expressing receptor without G protein, agonists will often displace the labeled antagonist from a single low-affinity site. In membranes from cells expressing both the receptor and the G protein, however, an additional high-affinity binding site for the agonist is revealed (Fig. 2). The high-affinity binding can be reduced with the addition of GTP or its analogs. Results can be fit with one- or two-site sigmoidal curves, respectively, yielding K_i values of the agonist for each site. Because the concentration of the radiolabeled compound is

[25] P. J. Munson, *Methods Enzymol.* **92,** 543 (1983).

[26] K. A. Wreggett and D. Léan, *Mol. Pharmacol.* **26,** 214 (1984).

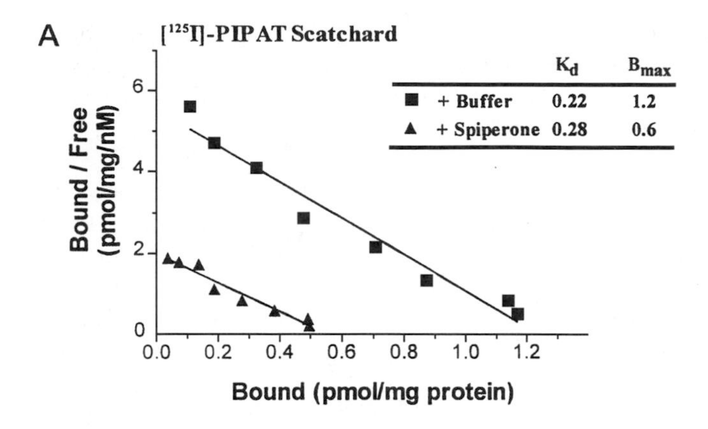

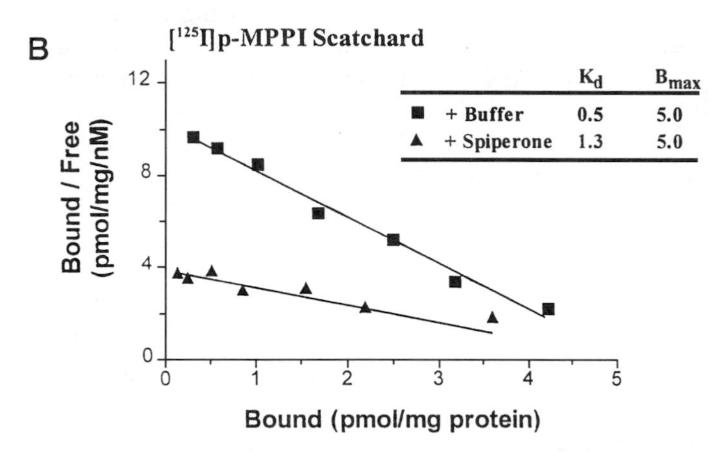

FIG. 1. Linear transformation of [^{125}I]8-OH-PIPAT (A) and [^{125}I]*p*-MPPI (B) saturation binding in the absence and presence of spiperone. Sf9 membranes expressing the human 5-HT_{1A} receptor with G_z (α_z, β_1, γ_2) were incubated in the absence or presence of the inverse agonist spiperone (300 n*M*) and various concentrations of the radiolabeled agonist ([^{125}I]8-OH-PIPAT) or antagonist ([^{125}I]*p*-MPPI) as described in the detailed protocol. Spiperone displaced the antagonist in a competitive manner, but it displaced agonist binding in a noncompetitive manner. Reprinted from Ref. 4, with permission from the publisher.

known, K_i values determined from the curve fits can be converted to K_d values using the equation from Cheng and Prusoff.[27] Concentrations of the competitor need to range from at least one-tenth the K_d for the high-affinity binding site to at least 10-fold the K_d for the low-affinity binding site. We typically use 14 different concentrations of competitor.

[27] Y. Cheng and W. H. Prusoff, *Biochem. Pharmacol.* **22,** 2999 (1973).

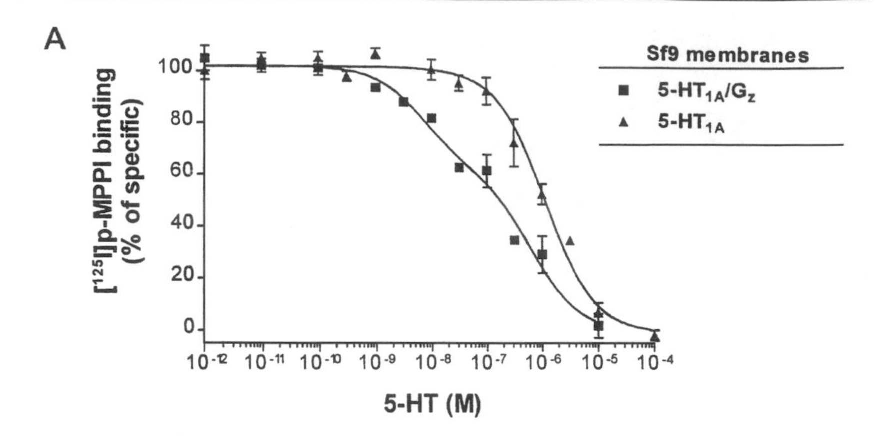

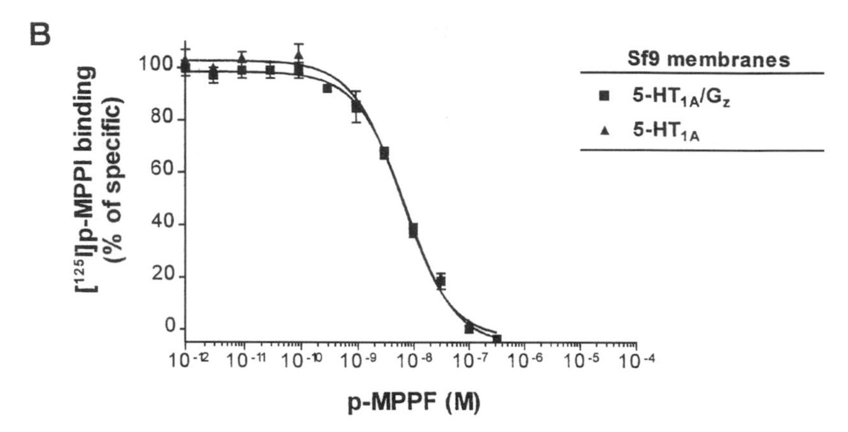

FIG. 2. Displacement of [^{125}I]*p*-MPPI by 5-HT (A) or *p*-MPPF (B) as a function of G protein expression. Sf9 membranes expressing the human 5-HT_{1A} receptor with or without G_z (α_z, β_1, γ_2) were incubated with 0.4 n*M* [^{125}I]*p*-MPPI and various concentrations of the agonist 5-HT or neutral antagonist *p*-MPPF. In the absence of heterologously expressed G protein, 5-HT displaces [^{125}I]*p*-MPPI binding at a single, low-affinity site (K_d 640 n*M*). Coexpression of the G protein promotes the formation of an additional high-affinity site for agonist (K_d 6 n*M*). No change in affinity of the receptor for *p*-MPPF is observed on G protein coexpression. Reprinted from Ref. 4, with permission from the publisher.

Single Agonist Concentration Binding

To determine which G proteins a receptor engages, the increase in the number of receptors exhibiting high affinity for an agonist can be used in a quick radioligand-binding screen. When membranes expressing only receptors are incubated with concentrations of radiolabeled agonist at or below the expected K_d of

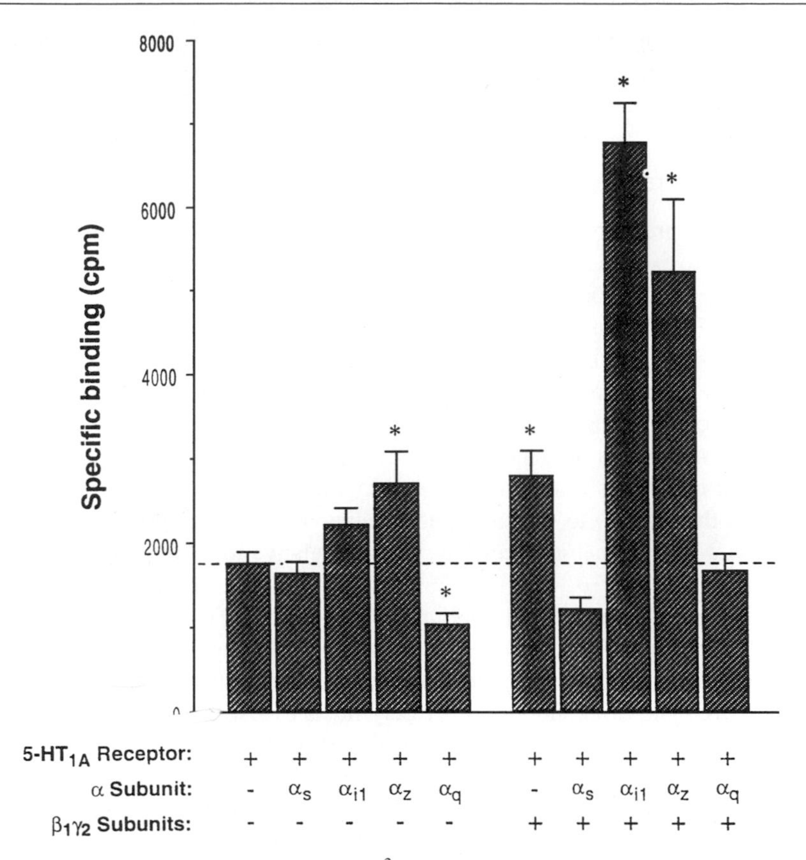

FIG. 3. Effects of G protein expression on [^{3}H]8-OH-DPAT binding in Sf9 cells expressing the human 5-HT_{1A} receptor. Specific binding of 0.5 n*M* [^{3}H]8-OH-DPAT was determined in Sf9 cells expressing the 5-HT_{1A} receptor and G protein α and $\beta_1\gamma_2$ subunits as indicated. Coexpression of heterotrimeric G proteins in the G_i, but not G_s or G_q, family promoted significant increases in [^{3}H]8-OH-DPAT binding. An asterisk indicates significantly different from membranes expressing the 5-HT_{1A} receptor alone, far left column. Reprinted from Ref. 3, with permission from the publisher.

that agonist for the high-affinity site, little specific binding will be apparent, as few of the receptors are coupled to G proteins. Because coexpression of the appropriate G protein will convert some of the receptors to a high-affinity state for agonist, the amount of G protein coupling induced by expression of a given G protein can be determined by comparing the amount of specific binding at that concentration of radiolabeled agonist in membranes expressing receptor with or without G proteins (Fig. 3). Similarly, the effects of different β and γ subunits on the coupling state of the receptor can also be determined. It should be noted that whereas this method is useful for screening purposes, full competition and saturation binding experiments need to be performed to quantitatively assess the changes induced by G protein coexpression.

Technical Considerations

The Sf9 expression system is most useful for binding studies where one wants the receptor to exist in an essentially purely uncoupled form in the absence of coexpressed G proteins or to couple to a single G protein without interference from others. Studies of agonist-induced signaling of GPCRs in Sf9 cells expressing the receptor alone indicate that these GPCRs can indeed interact functionally with endogenous insect G proteins.[28–30] However, because activation assays require a lower ratio of G protein-to-receptor than the stabilization of high-affinity agonist binding in radioligand-binding studies,[31] it is possible to see activation of G proteins under conditions where no high-affinity agonist binding is observed.[28,32,33] It appears that the key to producing homogeneous populations of uncoupled receptors is a stoichiometric excess of receptor relative to the endogenous insect G proteins.

If there is GTP-sensitive, high-affinity agonist binding in the absence of coexpressed G proteins, increasing the level of receptor expression may eliminate it. Using a higher MOI often promotes greater expression, although the relationship between MOI and expression level is rarely linear.[19] Using a different virus construct may also help, as many aspects of the gene structure and its surrounding sequence can affect expression (see O'Reilly *et al.*[7] for detailed information). Not changing the medium to an optimized serum-free medium following infection may also increase expression, as the cells do not have time to adapt to an abrupt change in conditions. We find that the change of medium following infection can lower expression levels of the 5-HT_{1A} receptor by as much as 50%.[34] Finally, harvesting the cells at longer time points following infection may also help increase

[28] E. M. Parker, K. Kameyama, T. Higashijima, and E. M. Ross, *J. Biol. Chem.* **266,** 519 (1991).

[29] S. Vasudevan, L. Premkumar, S. Stowe, P. W. Gage, H. Reiländer, and S. H. Chung, *FEBS Lett.* **311,** 7 (1992).

[30] J. G. Mulheron, S. J. Casañas, J. M. Arthur, M. N. Garnovskaya, T. W. Gettys, and J. R. Raymond, *J. Biol. Chem.* **269,** 12954 (1994).

[31] K. Haga, H. Uchiyama, T. Haga, A. Ichiyama, K. Kanagawa, and H. Matuo, *Mol. Pharmacol.* **35,** 286 (1989).

[32] C. C. Jansson, M. Karp, C. Oker-Blom, J. Nasman, J. M. Savola, and K. E. Akerman, *Eur. J. Pharmacol.* **290,** 75 (1995).

[33] L. A. Obosi, D. G. Schuette, G. N. Europe-Finner, D. J. Beadle, R. Hen, L. A. King, and I. Bermudez, *FEBS Lett.* **381,** 233 (1996).

[34] R. T. Windh and D. R. Manning, unpublished observations.

expression,[14,35] although harvesting at later time points can result in significant proteolysis of the receptor. Once the extent of coupling with endogenous G proteins has been determined, more probing binding studies can be performed.

Because some amount of the radioligand will bind to sites other than the receptor, e.g., to the glass filter, nonspecific binding must be determined in each assay and subtracted from the total amount of binding on the filter. Nonspecific binding is usually determined by measuring the amount of radioligand binding in the presence of a supersaturating concentration of another ligand for the receptor. With all of the receptor sites occupied by this competitor, the only radioligand binding that remains is assumed to be nonspecific. Because the goal is to abolish binding of the radioligand to the receptor, but not other binding sites, it is best to use a competitor that is structurally dissimilar to the radioligand, thereby minimizing the chances that it will displace the radioligand from sites other than the receptor. We usually use 10 μM 5-HT for defining nonspecific binding in 5-HT_{1A} receptor-binding assays. In some cases, using structurally dissimilar ligands to define nonspecific binding is not feasible, and it may be necessary to use the unlabeled form of the radioligand for this purpose. In these cases, keep in mind that the specific binding may be underestimated.

Whereas the advantages of the Sf9 expression system for identifying receptor–G protein coupling, and for rapid screening of ligands, are clear, it is important to note that the promotion of high-affinity agonist binding by G proteins is not necessarily an attribute intrinsic to all G proteins. The ternary complex theory is based largely on results with receptors that couple to members of the G_s and G_i families of G proteins. Promotion of high-affinity agonist binding has been observed less consistently for G_q,[36] and little is known of the manner in which G_{12} members interact with receptor. Therefore, using receptor binding to probe G protein coupling should be just one of several tools used.

Acknowledgment

The authors wish to acknowledge the support of NIH Grant MH48125.

[35] V. A. Boundy, L. Lu, and P. B. Molinoff, *J. Pharmacol. Exp. Ther.* **276,** 784 (1996).

[36] F. G. Szele and D. B. Pritchett, *Mol. Pharmacol.* **43,** 915 (1993).

[27] G Protein-Coupled Receptors and Proliferative Signaling

By MARTINE J. SMIT, REMKO A. BAKKER, and ETHAN S. BURSTEIN

Introduction

The superfamily of G protein-coupled receptors (GPCRs) plays a fundamental role in cellular communication. GPCRs convert diverse extracellular stimuli such as light, odors, hormones, and neurotransmitters into distinct intracellular signals. GPCRs activate heterotrimeric G proteins, which in turn regulate the activities of a wide variety of cellular effectors,[1] among them signal transduction pathways involved in proliferative signaling such as the mitogen-activated protein (MAP) kinase (MAPK) pathways.[2–4] Both α and $\beta\gamma$ subunits of heterotrimeric G proteins have been shown to activate MAPK and Jun kinase (JNK) pathways.[5]

Accumulating evidence indicates that GPCRs harbor oncogenic potential. Many GPCRs, including $5HT_{2C}$ serotonergic, α_{1B}-adrenergic, and m1, m3, and m5 muscarinic receptors, have been found to induce oncogenic transformation in an agonist-dependent manner.[5] Paracrine and autocrine stimulation of GPCRs for neuropeptides and prostaglandins has been implicated in a number of proliferative disorders, such as small lung carcinoma, colon adenomas, and carcinoma and gastric hyperplasia.[5]

Expression of constitutively active GPCRs[6] can also induce oncogenic transformation *in vitro*[6,7] and tumor formation in nude mice.[7] A variety of human diseases, including some proliferative disorders, have been ascribed to mutations that constitutively activate GPCRs.[8] Transforming DNA viruses, including Kaposi sarcoma-associated herpesvirus[9] and cytomegalovirus,[10] contain genes that encode constitutively active GPCRs. The GPCR encoded by Kaposi's sarcoma-associated

[1] H. Bourne, *Curr. Opin. Cell. Biol,* **9,** 134 (1997).
[2] T. van Biesen, L. M. Luttrell, B. E. Hawes, and R. J. Lefkowitz, *Endoc. Rev.* **17,** 698 (1996).
[3] T. Gudermann, R. Grosse, and G. Schultz, *Naunyn Schmiedebergs Arch. Pharmacol.* **361,** 345 (2000).
[4] C. Murga, S. Fukuhara, and J. S. Gutkind, *Trends Endocrinol. Metab.* **10,** 122 (1999).
[5] J. S. Gutkind, *J. Biol. Chem.* **273,** 1839 (1998).
[6] L. F. Allen, R. J. Lefkowitz, M. G. Caron, and S. Cotecchia, *Proc. Natl. Acad. Sci. U.S.A.* **88,** 11354 (1991).
[7] R. S. Westphal and E. Sanders-Bush, *Mol. Pharmacol.* **49,** 474 (1996).
[8] A. M. Spiegel, *Annu. Rev. Physiol.* **58,** 143 (1996).
[9] L. Arvanitakis, E. Geras-Raaka, A. Varma, M. C. Gershengorn, and E. Cesarman, *Nature* **385,** 347 (1997).
[10] P. Casarosa, R. A. Bakker, D. Verzijl, M. Navis, H. Timmerman, R. Leurs, and M. J. Smit, *J. Biol. Chem.* **276,** 1133 (2001).

0076-6879/02 $35.00

herpesvirus (ORF74) has been clearly linked to tumorigenesis, as transgenic mice expressing ORF74 develop symptoms resembling that of Kaposi's sarcoma.[11] In view of the potential therapeutic significance of constitutive activity of GPCRs, the development of inverse agonists, which are able to inhibit constitutive activity of GPCRs has gained widespread interest.

This article outlines different techniques used to assay proliferative signaling by GPCRs. Advantages and disadvantages of these assays with respect to cost and throughput are described. Assays that are suited to pharmacological characterization of ligands such as inverse agonists are indicated.

Conventional, Colorimetric, and Fluorescence-Based Proliferation Assays

Cell Growth

A straightforward way to determine whether GPCRs induce proliferation is by generating cell growth curves. Cells are plated at low densities and switched to serum-free or low serum conditions for 12 to 24 hr to synchronize the cells in G_0 phase and are then stimulated with an agonist. Constitutive activity can be measured by comparing the growth rates of cells expressing the GPCR of interest to mock-transfected cells, either in the presence or in the absence of serum. Cells are counted by a hemocytometer each day. Media (and ligands) are replenished every 3 days. Counting cells using a hemocytometer is simple but labor-intensive. This article lists some alternatives.

Colorimetric Proliferation Assays

There are various colorimetric ways to determine the number of (viable) cells in proliferation assays that are based on the measurement of either enzymatic activity or on the total amount of proteins present in the cells. Mosmann[12] established the MTT [3-(4,5-dimethylthiazol-2-yl)-2,5-diphenyltetrazolium bromide] colorimetric tetrazolium assay and, subsequently, other tetrazolium-based assays [XTT, MTS (Promega), WST-1] were developed. The conversion of tetrazolium compounds into formazan is accomplished by dehydrogenase enzymes found predominantly in metabolically active cells. The quantity of formazan can be measured at 490 nm and is directly proportional to the number of living cells in culture. The use of MTS in combination with phenazine methosulfate (PMS; Promega) is advantageous, as an aqueous soluble formazan product is formed that does not need solubilization (i.e., MTT). In addition, XTT has limited solubility and stability.

[11] T. Y. Yang, S. C. Chen, M. W. Leach, D. Manfra, B. Homey, M. Wiekowski, L. Sullivan, C. H. Jenh, S. K. Narula, S. W. Chensue, and S. A. Lira, *J. Exp. Med.* **191,** 445 (2000).

[12] T. Mosmann, *J. Immunol. Methods* **65,** 55 (1983).

A low-cost alternative is the use of sulforhodamine B (SRB), which binds to basic amino acids, measuring cellular protein content. This assay can be used for cell proliferation and cytotoxicity measurements in both adherent and suspension cultures.[13,14] Cells are fixed with trichloric acid, SRB is added and solubilized with unbuffered Tris-base solution, and the absorbance at 492 nm minus the absorbance at 620 nm is read to determine the response. This protocol is amenable to 96-well plates.

Fluorescence-Based Proliferation Assays

The number of cells can also be quantified by means of fluorescent dyes (e.g., Cyquant Molecular Probes, fluorescent CellTiter 96 Promega). After a freeze-thaw cycle or addition of lysis solution, a fluorescent dye is added to cells. The dye exhibits strong fluorescence enhancement upon binding nucleic acids, providing a simple and accurate indicator of cell proliferation. Fluorescence is measured using a microplate reader to determine excitation at 485 nm and emission at 530 nm. Similar to colorimetric assays, fluorescence-based assays are compatible with 96-well plate formats.

DNA Synthesis

Proliferation of cells is accompanied by the initiation of DNA synthesis. This process can be followed in a quantitative manner by measuring the uptake of [^{3}H]thymidine during the S phase of the cell cycle. In order to measure [^{3}H]thymidine uptake, cells need to be serum starved to enter G_0. Cells whose basal uptake of thymidine is, high in the absence of serum (i.e., COS cells) are not suitable for this assay. The following protocol describes [^{3}H]thymidine incorporation into NIH-3T3 cells.

Materials

Phosphate-buffered saline (PBS)
[^{3}H]Thymidine
10% trichloroacetic acid (TCA)
0.1 *N* NaOH
Scintillation counter

Cells are seeded in normal growth medium into 6- or 24-well plates. The following day, medium is removed and cells are washed once with PBS and switched

[13] P. Skehan, R. Storeng, D. Scudiero, A. Monks, J. McMahon, D. Vistica, J. T. Warren, H. Bokesch, S. Kenney, and M. R. Boyd, *J. Natl. Cancer Inst.* **82,** 1107 (1990).

[14] K. T. Papazisis, G. D. Geromichalos, K. A. Dimitriadis, and A. H. Kortsaris, *Methods* **208,** 151 (1997).

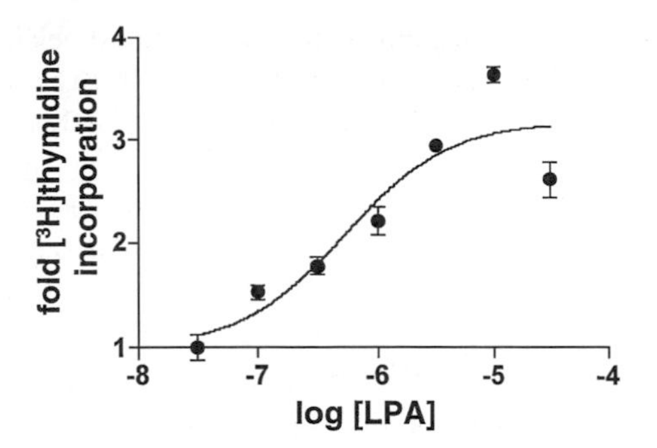

FIG. 1. LPA-induced incorporation of [^{3}H]thymidine in NIH-3T3 cells. Cells are serum starved for 24 hr, stimulated with increasing concentrations of LPA for 18 hr, and labeled for 6 hr with 0.5 μCi/ml [^{3}H]thymidine prior to harvesting of cells.

to serum-free or low-serum (0.1–0.5%bovine serum albumin for NIH-3T3 cells) media. The stimulus is added after 4 to 24 hr (fibroblasts). Use of serum (10%) is recommended as a positive control. Cells are labeled 6 to 24 hr with 0.5–1 μCi/ml [^{3}H]thymidine (NEN) per well prior to harvesting. Upon harvesting, medium is removed and cells are washed twice with cold PBS. One milliliter of 10% TCA is added to the cells and left on ice for 30 min. Cells are washed twice with cold PBS and solubilized by the addition of 1 ml 0.1 *N* NaOH. After 1 hr the samples are pipetted up and down to ensure complete solubilization of cells and are transferred to scintillation vials containing 5 ml scintillation fluid and counted. As shown in Fig. 1, lysophosphatidic acid (LPA) induces a dose-dependent increase in [^{3}H]thymidine incorporation in NIH-3T3 cells. It is currently also possible to measure DNA synthesis in a nonradioactive manner using an immunoassay for bromodeoxyuridine (BrdU), a thymidine analogue (Boehring Mannheim).

MAP Kinase Activity

Activation of the MAPK pathway is crucial for cellular proliferation.[15] On activation, MAP kinases translocate to the nucleus where they can differentially alter the phosphorylation status of transcription factors such as AP-1, Elk-1, c-jun, and ATF-2, regulating the transcription of genes that are involved in cellular proliferation.[4]

Three major classes of MAPK have been identified so far. The first class consists of p42/p44 MAPK, also referred to as ERK1 and ERK2, and is activated by many hormones and growth factors. The other two classes, c-jun kinases (JNK) and p38

[15] D. M. Payne, A. J. Rossomando, P. Martino, A. Erickson, J. H. Her, J. Shabanowitz, D. F. Hunt, M. J. Weber, and T. W. Sturgill, *EMBO J.* **10,** 885 (1991).

kinases, are activated by stress responses such as osmotic shock and irradiation.[16,17] This section focuses on the activation of p44/p42 MAPK, from here on referred to as MAPK. Assays measuring JNK and p38 activity are outlined elsewhere in this series.[18,19]

Activation of MAPK requires phosphorylation by MAPK kinase (MEK) on a threonine and a tyrosine residue that are separated by a glutamic acid residue. This TEY sequence motif is conserved in MAPK homologues from yeast to humans. JNK and p38 are similarly activated, although for JNK a proline and for p38 a glycine residue reside between the phosphorylated residues. Early work with partially purified MAPK and antibodies that recognize phosphotyrosine revealed that activation and phosphorylation of MAPK are closely linked. Synthesis of peptides containing the (non)phosphorylated T-x-Y made it possible to generate antibodies specifically recognizing the (non)phosphorylated form of MAPK.

Determination of MAPK Activity by Means of Gel Electrophoresis and Immunoblotting

Activation of MAP kinases occurs in most cell types in response to a wide variety of stimuli. To determine whether activated GPCRs stimulate MAPK, it is usually necessary to allow cells to enter G_0, as MAPK activity is low and activation of MAPK occurs upon reentry of quiescent cells into S phase. Activation of MAPK is usually rapid and transient. Difficulties detecting changes in MAPK levels can be overcome by using hemagglutinin (HA)-tagged MAPK.

Commercially available antibodies specifically recognizing either proteins phosphorylated on tyrosine residues (anti-PTyr antibodies) or the nonposhophorylated form of MAPK serve as useful tools for characterizing the involvement of MAPK. Protein extracts are denaturated, electrophoresed on SDS–PAGE, and immunoblotted with the appropriate antibodies. Clone 4G10 antibodies (Upstate Biotechnology) are recommended for detecting phosphorylated MAPK; however, these antibodies also detect other tyrosine posphoproteins. Thus, reprobing the blot with antibodies specific for MAPK is recommended. MAPK activity can also be detected based on the gel mobility shift of phosphorylated and nonphosphorylated forms of MAPK. This method gives an estimate of the fraction of MAPK in the active form, however, this method can overestimate the active fraction, as singly phosphorylated ERK2, which is not active, has been detected and shown to exhibit retarded mobility of ERK2.[20] The following protocol details the procedure for the preparation of lysates from stimulated adherent cells such as NIH-3T3 cells.

[16] R. J. Davis, *Trends Biochem. Sci.* **23,** 481 (1998).

[17] A. R. Nebreda and A. Porras, *Trends Biochem. Sci.* **25,** 257 (2000).

[18] M. Cheariello and J. S. Gutkind, *Methods Enzymol.* **344,** [34] 2001.

[19] J. Perry Hall and R. J. Davis, *Methods Enzymol.* **344,** [32] 2001.

[20] N. G. Anderson, *Adv. Prot. Phosphatases* **6,** 119 (1991).

Materials

0.5 *M* HEPES

Cold RIPA: 1× PBS, 1% NP-40, 0.5% sodium deoxycholate, 0.1% SDS with freshly added protease inhibitors, 1 m*M* phenylmethylsulfonyl fluoride (PMSF), 2 μg/ml aprotinin, 2 μg/ml leupeptin, and 1 m*M* orthovanadate

Orthovanadate: Make 1 m*M* solution of orthovanadate in water, adjust pH to 10, boil and readjust pH to 10. Repeat procedure until colorless solution with pH 10 is obtained.

Sonicator or 21-gauge needle

Preparation of Extracts from Cells

Cells are seeded in normal growth medium and switched to serum-free or low serum for 12 to 24 hr. HEPES (25 m*M* final concentration) may be added to maintain physiological pH during subsequent procedures. Cells are stimulated by a 5- to 15-min incubation with agonist. A time course from 2 to 30 min is recommended to determine the optimal exposure to activate MAPK. After stimulation, cells are placed on ice, medium is removed, cells are rinsed twice with cold PBS, and 150 μl cold RIPA lysis buffer is added per 6 well (300 μl/ 10-cm dish). Swirl plates and incubate cells on ice for 10 min. Gently scrape cells with cell scraper. Transfer the cell suspension to an Eppendorf tube and either sonicate cells 3 sec or use a 21-gauge needle to shear DNA. Centrifuge the cell lysate at 15,000*g* for 10 min at 4°. Transfer the supernatant solution (lysate) to a clean Eppendorf tube. Take a sample and determine protein concentration, freeze samples (preferably quick in liquid nitrogen before storage), or use immediately for further analysis. A protein determination assay that is not affected by the presence of SDS and DTT (e.g., BCA from Pierce) is recommended. Preparation of extracts from cells growing in suspension is identical with the exception that after stimulation, cells are sedimented rapidly, medium is removed, and cells are washed once with cold PBS before solubilization in lysis buffer.

Materials

Minigel electrophoresis unit and power supply

Separating gel (10%) per 10 ml: 4.0 ml H_2O, 3.3. ml 30% acrylamide mix (30 : 0.8), 2.5 ml 1.5 *M* Tris (pH 8.8), 0.1 ml 10% SDS, 0.1 ml 10% ammonium persulfate (APS, prepare fresh), 6 μl TEMED

Stacking gel (5%) per ml: 0.68 ml H_2O, 0.17 ml 30% acrylamide mix (30 : 0.8), 0.13 ml 1.0 *M* Tris (pH 6.8), 10 μl 10% SDS, 10 μl APS (prepare fresh), 1 μl TEMED

Running buffer: 25 m*M* Tris, 250 m*M* glycine (electrophoretic grade), pH 8.3, 0.1% SDS

6x loading buffer per 10 ml: 7.0 ml 0.5 *M* Tris (pH 6.8)/SDS, 3 ml glycerol, 1.0 g SDS, 0.93 g dithiothreitol (DIT), 1.2 mg bromphenol blue
Water-saturated butanol
Prestained marker

Gel Electrophoresis

Prepare SDS–PAGE by pouring a separating gel (10%). Immediately overlay with water-saturated butanol or water to ensure a horizontal meniscus. Remove layer and add on the stacking gel (5%) and place multiwell comb. Remove the comb when the gel is solidified, submerge gel in running buffer, and rinse wells. Mix samples 10–50 μg whole cell lysate with 6x loading buffer. Boil samples for 5 min, place on ice for 5 min, and centrifuge for 3 min at 15,000*g*. Load samples and prestained marker or rainbow marker (Amersham) and run gel (mini-Protean gel Bio-rad: 100 V/~30 mA) for ~1.5 hr.

Materials

Blotting apparatus
Transfer buffer: 20% MeOH, 25 m*M* Tris base, 0.2 M glycine, pH 8.3
Nitrocellulose or PVDF (prewash 3 sec in MeOH, rinse 10 sec with water) membrane
Ponceau S: 0.2 % Ponceau S in 0.1% acetic acid
TBS-T: 10 m*M* Tris base, pH 8.0, 150 m*M* NaCl, 0.1% Tween 20
PBS-T: phosphate-buffered saline, 0.1% Tween 20
Phospho MAPK: antibody specifically recognizing the phosphorylated form of MAPK
Secondary antibody
0.2 *N* NaOH
Agents and apparatus for detection of chemiluminescence

Immunoblotting

Transfer gel carefully in transfer buffer and soak gel and nitrocellulose (Hybond C) or PVDF membrane in transfer buffer for 15 min. Proteins are transfered to the membrane in a transfer unit in the cold room or with a cooler unit while stirring for 1 hr at ~100 V/200 mA. The transfer time and setting are different for each electrophoresis unit. Confirm proper transfer and equal protein loading by staining with Ponceau S. Block blot in blocking buffer [TBS-T or PBST with dry milk (3 or 5%)] according to the manufacturer's protocol for 0.5 to 1 hr at room temperature or overnight at 4° on shaker. Incubate blot with antibody specifically recognizing the phosphorylated form of MAPK (e.g., New England Biolabs, Promega) according to the manufacturer's protocol: 1–4 hr at room temperature or overnight at 4° on shaker. Rinse blot three times 5 min with wash buffer (T-TBS or T-PBS)

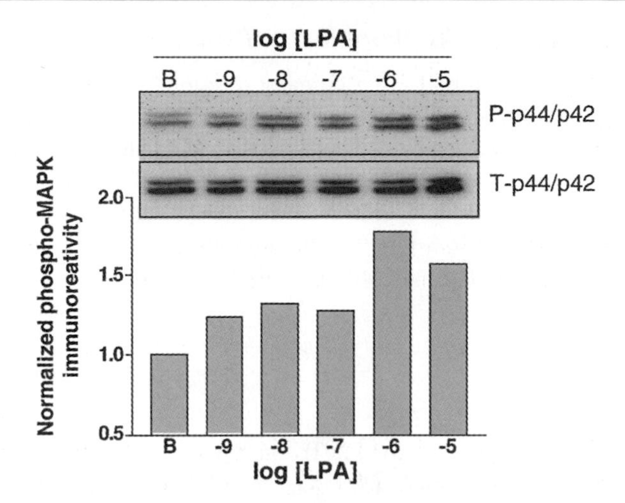

FIG. 2. LPA-induced increase of phosphorylation of p44/p42 MAPK in NIH-3T3 cells as determined by Western blot analysis using specific anti-p44/p42-(phospho) antibodies. Cells are serum starved for 24 hr and stimulated for 5 min with increasing concentrations of LPA. Phosphorylation of p44/p42 MAPK is quantified using chemiluminescence imaging and corrected for total p44/p42 MAPK on stripped blots. Data are expressed as fold basal and are from a representative experiment.

on shaker. Probe blot with horseradish peroxidase-conjugated secondary antibody (1 : 5000–10,000) in blocking buffer at room temperature for 45 min–1.5 hr. Rinse blot three times for 5 min with wash buffer (TBS-T or PBS-T). Apply agents for detection of chemiluminescence [e.g., Rennaissance blot (NEN) or ECL (Amersham)] and expose and develop X-ray film. Band densities may be estimated from the autoradiogram by densitometry. However, due to the nonlinear characteristics of X-ray film, direct chemiluminescence measurements with Imagestation (NEN Kodak) are recommended.

To ensure that the levels of total MAPK are equal in all lanes, blots can be stripped and reused for total MAPK antibody detection. Wash blots 15 min in water while gently rocking, remove water, add 0.2 *N* NaOH for 15 min, and wash blots once more with water for 15 min. Check before further use with Ponceau S for the presence of protein bands. Repeat blocking and other procedures as described earlier. Figure 2 illustrates the activation of MAPK by LPA in NIH-3T3 cells. LPA dose dependently stimulates phosphorylation and activation of p44/p42.

Phospho-specific antibodies can also be used for immunohistochemistry studies, permitting correlation of subcellular localization of MAPK with its activation status. Giovannini in this series illustrates this further.[21]

[21] M. G. Giovannini, *Methods Enzymol.* **343,** [33] 2001.

Phosphospecific Antibody Cell-Based ELISA (PACE)

The availability of antibodies specifically recognizing the active phosphorylated form of MAPK has facilitated research in this field. However, the analysis requires Western blotting, which is labor-intensive and is low throughput. Peppelenbosch *et al.*[22] developed a new phospho-specific antibody cell-based ELISA (PACE) for the determination of MAPK activity using phospho-specific antibodies. PACE and similar ELISA-based techniques allow high throughput analysis. NIH-3T3 cells, macrophages, and primary cultures of neutrophils have been used successfully in this assay.

Materials

4 or 8% formaldehyde in PBS
PBS/Triton: PBS containing 0.1% Triton X-100
0.6 % H_2O_2 in PBS/Triton or 1% H_2O_2 and 0.1% sodium azide in PBS/Triton
Phospho-specific antibody
Peroxidase-conjugated secondary antibody
5% BSA/PBS/Triton
0.4 mg/ml OPD, 11.8 mg/ml $Na_2HPO_4 \cdot 2H_2O$, 7.3 mg/ml citric acid, and 0.015% H_2O_2
1 *M* H_2SO_4
Crystal violet: 0.04% crystal violet in 4% (v/v) ethanol
1% SDS solution
Poly-L-lysine (10 mg/ml)
Shaker, microplate reader

Cell-Based ELISA for Adherent Cells

Cells are seeded in a 96-well plate in normal growth medium. The following day, cells are serum starved for appropriate time periods (4 hr for macrophages, 18–24 hr for fibroblasts) in medium without serum or 0.1–0.5% BSA. Cells are stimulated for 10–15 min. As a positive control, an activator of protein kinase C (e.g., 100 n*M* PMA) can be used. After stimulation, cells are fixed with 100 μl 4% formaldehyde in PBS at 4° for 20 min and washed three times with 200 μl PBS/Triton. To quench endogenous peroxidase activity, cells are treated with 100 μl 0.6 % H_2O_2 in PBS/Triton for 20 min. Cells are washed three times with PBS/Triton, blocked with 10% fetal calf serum in PBS/Triton for 1 hr, and incubated overnight with the primary antibody (phospho-specific) antibody (1 : 100–1 : 500, optimization is required for best signal/noise ratio) in 5% BSA/PBS/Triton.

[22] H. H. Versteeg, E. Nijhuis, G. R. Van Den Brink, M. Evertzen, G. N. Pynaert, S. J. Van Deventer, P. J. Coffer, and M. P. Peppelenbosch, *Biochem. J.* **15,** 717 (2000).

The next day, cells are washed three times with PBS/Triton for 5 min and incubated with the secondary antibody (peroxidase-conjugated, 1 : 100) in PBS/Triton with 5% BSA for 1 hr at room temperature and washed three times with PBS/Triton and twice with PBS. Thereafter, cells are incubated with 50 μl solution containing 0.4 mg/ml OPD, 11.8 mg/ml $Na_2HPO_4 \cdot 2H_2O$, 7.3 mg/ml citric acid, and 0.015% H_2O_2 for 15 min at room temperature in the dark. The reaction is stopped with 25 μl of 1 *M* H_2SO_4, and the OD is measured at 490 nm/650 nm.

Cell-Based ELISA for Nonadherent Cells

This assay can also be used for nonadherent cells by coating the plates with poly-L-lysine (10 mg/ml) for 30 min at 37°. Cells are washed twice with PBS, counted and plated into wells, and incubated for 30 min at 37°. Thereafter, cells are stimulated with the drug and fixed with 8% formaldehyde for 20 min followed by three washes with PBS/Triton. Endogenous peroxidase is quenched with 1% H_2O_2 and 0.1% sodium azide in PBS/Triton for 20 min. Further treatment is as described earlier with the exception of the use of a 1 : 500 dilution of the peroxidase-conjugated secondary antibody.

Crystal Violet Cell Quantification Assay

In order to correct for differences in cell number, a crystal violet-staining procedure may be applied after the PACE procedure. After the peroxidase reaction, cells are washed twice with PBS/Triton and twice with demineralized water. Wells are dried for 5 min, and 100 μl of crystal violet [0.04% crystal violet in 4% (v/v) ethanol] is added for 30 min at room temperature. Cells be then washed at least three times with demineralized water 100 μl 1% SDS solution is added, and cells are incubated for 1 hr at room temperature while shaking. Absorbance is measured at 595 nm with a microplate reader. As shown in Fig. 3, LPA caused a dose-dependent increase in p44/p42 immunoreactivity as determined by the PACE procedure. The dose–response curve obtained for LPA is in accordance with that found using the phospho-specific antibody in the Western blot analysis procedure (Fig. 2).

The procedures just outlined may be sufficient to assess the relative activation of MAPK. Direct measurement of MAPK enzyme activity using specific substrates of MAPK may be desirable. MAP kinase assays are generally performed in partially purified cell extracts or following immunoprecipitation with an antibody that preserves the enzyme in its activated state. The reader is referred to Volume 344 for more details.[18]

Soft Agar Transformation Assay

Soft agar assays are used to determine the ability of cells to proliferate in an anchorage-independent manner, a characteristic indicative of a transformed

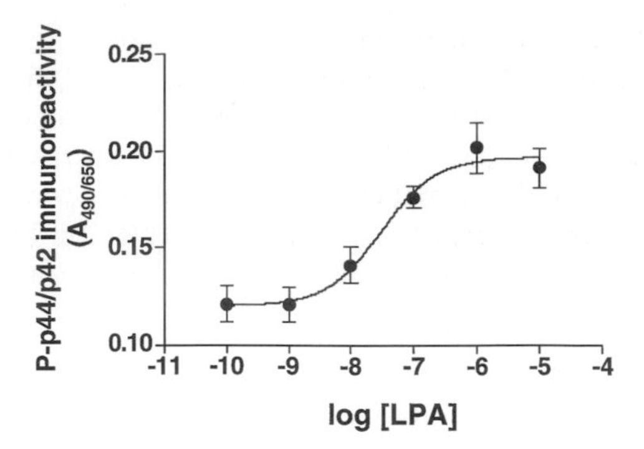

FIG. 3. LPA-induced increase of p44/p42 MAPK in NIH-3T3 cells measured by PACE. Cells are serum starved for 24 hr and stimulated for 5 min with increasing concentrations of LPA. Phosphorylation of p44/p42 MAPK is quantified using the specific anti-p44/p42-phospho antibody at a dilution of 1 : 250.

phenotype. NIH-3T3 cells are commonly used in this assay, as they are transformed easily. When transfected with genes that possess proliferative/oncogenic properties, NIH-3T3 cells will grow in an anchorage-independent manner. It is therefore important to ensure that the control cells themselves do not spontaneously adopt a transformed phenotype, as this will increase the background of the assay. Growth of NIH-3T3 cells in fetal calf serum instead of calf serum will increase the rate of spontaneous transformation and is therefore not recommended. Other cell lines, such as human small cell lung cancer cells (SCLC) and human mammary epithelial cells (MCF-10), can be used in this type of assay. To assess transforming ability, cells expressing a GPCR of interest are suspended in agar, allowed to proliferate in the presence or absence of ligand, and assayed for the number of colonies formed. GPCRs endogenously or stably expressed are preferable, but transiently expressed GPCRs may be used. As a positive control, one can use cells transfected with v-ras.[23]

Materials

60-mm gridded dishes
0.3 and 0.5% Agar Noble
30% *p*-iodonitrotetrazolium

First a layer of agar is poured in 60-mm gridded dishes. This layer consists of DMEM, 10% calf serum, 100 U/ml penicillin, 100 μg/ml streptomycin, 500 ng/ml fungizone, and 0.5% Agar Noble (Difco, 3 ml per well). Cells (10^3–10^4) are seeded

[23] M. Barbacid, *Annu. Rev. Biochem.* **56,** 779 (1987).

in DMEM supplemented with 10% calf serum containing 100 U/ml penicillin, 100 μg/ml streptomycin, 500 ng/ml fungizone, and 0.3% Agar Noble (3 ml per well) with or without agonist (if necessary). The cells are fed twice a week with 0.5 ml DMEM, replenishing ligand at each feeding. Numbers of colonies larger than 0.15 mm 6 weeks after the cells were seeded are counted. To visualize the colonies, stain the plates with 0.5 ml 30% solution *p*-iodonitrotetrazolium (Sigma) for 2 days at 37°.

Reporter Gene Assays

Activation of proliferative pathways is accompanied by the onset of gene transcription of genes involved in cellular proliferation. Gene transcription in transfected eukaryotic cells is generally studied by linking a promoter sequence to an easily detectable "reporter gene." Examples of commonly used reporter genes include firefly luciferase, chloramphenicol acetyltransferase (CAT), and β-galactosidase.

Materials

Reporter gene construct, e.g., pTLNSRE3-Luc, pNF-κB-Luc
D-Luciferin
Luciferase assay reagent: 0.83 m*M* ATP, 0.83 m*M* D-luciferin, 18.7 m*M* $MgCl_2$, 0.78 μ*M* $Na_2H_2P_2O_7$, 38.9 m*M* Tris (pH 7.8), 0.39% (v/v) glycerol, 0.03% (v/v) Triton X-100, and 2.6 μ*M* DTT
TopCount (Packard) or a Victor^2 (Wallac)

Firefly-Luciferase Reporter Gene Assay

Cells (e.g., COS-7, NIH-3T3, or HeLa) are cotransfected with the GPCR of interest together with a suitable reportergene such as a reporter plasmid containing a firefly luciferase gene[24,25] under the transcriptional control of three serum responsive elements (pTLNSRE3-Luc)[26] or five NF-κB (pNF-κB-Luc, Stratagene) enhancer elements, seeded into 96 (or 384)-well plates in either serum-free or low serum medium, and incubated with drugs. Transparent plates should be avoided because these plates will transduce the emitted light that is generated upon addition of D-luciferin, the substrate for firefly luciferase. After 24 to 48 hr, cells

[24] S. J. Gould and S. Subramani, *Anal. Biochem.* **15,** 5 (1998).
[25] L. H. Naylor, *Biochem. Pharmacol.* **58,** 749 (1999).
[26] B. Fluhmann, U. Zimmerman, R. Muff, G. Bilbe, J. A. Fischer, and W. Bom, *Mol. Cell. Endorinol.* **139,** 89 (1998).

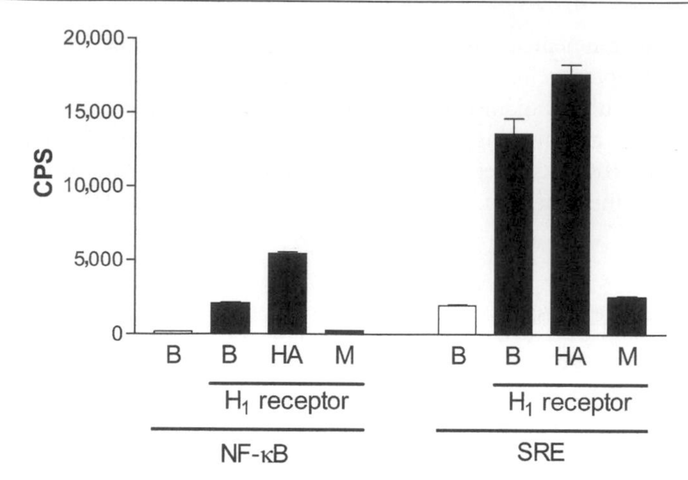

FIG. 4. Histamine H_1 receptor-mediated activation of SRE and NF-κB. COS-7 cells were transiently cotransfected with cDNA encoding the human histamine H_1 receptor together with a reporter plasmid containing a firefly luciferase gene under the transcriptional control of either five NF-κB-(pNF-κB-Luc) or three SRE (pTLNSRE3-Luc) enhancer elements and incubated in the absence (basal, B) or presence of either 10 μM histamine (HA) or 10 μM mepyramine (M) for 48 hr after which luminescence was measured.

are assayed for luminescence by aspiration of the medium and the addition of 25 μl/well luciferase assay reagent. Luminescence is measured 30 min after addition of the luciferase assay reagent for 3 sec/well in a TopCount (Packard) or a Victor^2 (Wallac).

As shown in Fig. 4, the histamine H_1 receptor can induce expression of genes that are under the transcriptional control of NF-κB or SRF transcription factors. The extent of gene expression differs depending on the transcriptional elements used. Although both reporter genes can be used to measure H_1 receptor-mediated agonist responses, the fold activation is greater using the NF-κB reporter gene. The constitutive activity of the H_1 receptor, however, is detected more readily using the SRE reporter gene (note the inhibition of basal signaling by the inverse agonist mepyramine). Therefore, the choice of reporter genes used may differentially affect sensitivity to constitutive and agonist-stimulated responses.

R-SAT and Focus-Forming Assays

R-SAT (Receptor Selection and Amplification Technology) is a high-throughput functional assay for GPCRs that is simple to perform, applicable to most GPCRs, and very sensitive. Due to the simplicity of the assay, R-SAT is well suited to high-throughput applications and in fact was originally developed to screen libraries of randomly mutated GPCRs to study the structure–function relationships

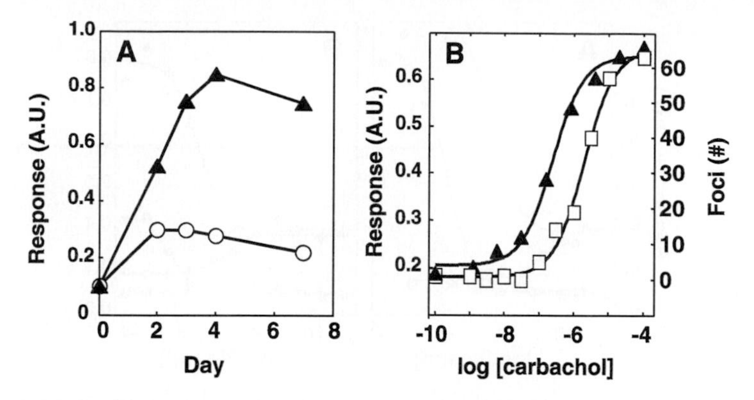

FIG. 5. Receptor Selection and Amplification Technology (R-SAT) assay. (A) Time course. m5 and β-galactosidase cDNAs were cotransfected into NIH-3T3 cells and induced β-galactosidase activity assayed at the indicated times as described under Materials and Methods. Responses were calculated from the baseline, and maximum responses were derived from least-square fits of full dose/responses at each time point (▲, maximal response; ○, basal response). (B) Carbachol dose response of β-galactosidase amplification on day 5 and focus formation on day 14. Plotted are absorbance (R-SAT, ▲) and number of macroscopic foci per 10-cm^2 dish (focus assay, □) versus carbachol concentration. Assays and curve fits were performed as described under Materials and Methods. Figure taken from Burstein *et al.*[27]

of receptor activation.[27–29] R-SAT is based on the observation that oncogenes and many receptors induce proliferation or transformation responses in NIH-3T3 cells. On forming a monolayer, NIH-3T3 cells normally stop growing due to contact inhibition. However, NIH-3T3 cells transfected with genes that promote cell growth overcome contact inhibition and proliferate. Thus for GPCRs, ligands select and amplify cells that express functional receptors. In focus-forming assays, activity is measured by counting the number of macroscopic foci formed, a process that takes 14–20 days. These stimulatory effects can be quantified using a marker gene,[27,30,31] which allows graded responses to be measured, permitting precise determinations of ligand potency and efficacy, and reducing the length of the assay significantly. The marker gene is used to quantify the proliferative response and is not under direct transcriptional control as is the case in reporter gene assays (e.g., using cyclic AMP response elements to report production of cyclic AMP). As shown in Fig. 5, there is a time- and dose-dependent amplification of cells transfected with

[27] E. S. Burstein, T. A. Spalding, D. Hill-Eubanks, and M. R. Brann, *J. Biol. Chem.* **270,** 3141 (1995).

[28] E. S. Burstein, T. A. Spalding, and M. R. Brann, *J. Biol. Chem.* **273,** 24322 (1998).

[29] T. A. Spalding, E. S. Burstein, S. C. Henderson, K. R. Ducot, and M. R. Brann, *J. Biol. Chem.* **273,** 21563 (1998).

[30] T. L. Messier, C. M. Dorman, H. Bräuner-Osborne, D. Eubanks, and M. R. Brann, *Pharmacol. Toxicol.* **76,** 308 (1995).

[31] H. Bräuner-Osborne and M. R. Brann, *Eur. J. Pharmacol.* **295,** 93 (1996).

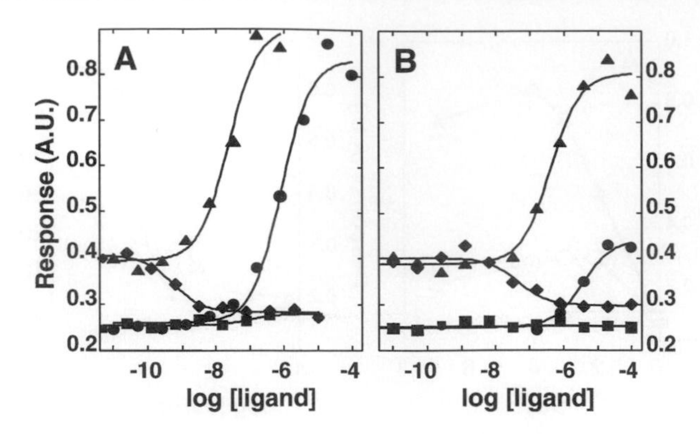

FIG. 6. Constitutive activation of the m3 muscarinic receptor by G_{α_q}. m3 and β-galactosidase were transfected into NIH-3T3 cells either with or without G_{α_q} and cultured in the presence of the indicated concentrations of ligands. Ligands are represented by the same symbols in A and B: (A) carbachol (▲, m3 + G_q; ●, m3); atropine (◆, m3 + G_q; ■, m3) and (B) McN-A-343 (▲, m3 + G_q: ●, m3); pirenzepine (◆, m3 + G_q; ■, m3). Plotted are absorbance of the β-galactosidase substrate ONPG at 420 nm versus ligand concentration. Each data point represents the average of two determinations. Figure taken from Burstein *et al.*[32]

the human m5 muscarinic acetylcholine receptor and β-galactosidase as measured by the absorbance of a β-galactosidase substrate. Similar results were obtained in the focus-forming assays, but the assay took much longer, it was 10-fold less sensitive to carbachol, and 10-cm^3 plates were needed to obtain sufficient numbers of foci for statistically meaningful results. In contrast, R-SAT is compatible with 96- and 384-well plates. R-SAT is particularly well suited to discriminating ligand efficacy and for detecting the constitutive activity of receptors[28,29,32] (see Fig. 6). Extensive work has demonstrated that the pharmacology of agonists, antagonists, and inverse agonists at receptors evaluated using R-SAT accurately reflects results obtained with other commonly used functional assays.[31,32]

The abilities of genes to transform NIH-3T3 cells and to induce responses in R-SAT are strongly correlated. GPCRs coupled to G_{α_q} (which induce focus formation in NIH-3T3 cells[33]), as well as many growth factor receptors and nonreceptor protooncogenes, such as G proteins and MAP kinases, induce strong responses in R-SAT.[30,34–36] GPCRs coupled to the G_{α_i} and G_{α_s} families of heterotrimeric

[32] E. S. Burstein, T. A. Spalding, and M. R. Brann, *Mol. Pharmacol.* **51,** 312 (1997).

[33] J. S. Gutkind, E. A. Novotny, M. R. Brann, and K. C. Robbins, *Proc. Natl. Acad. Sci. U.S.A.* **88,** 4703 (1991).

[34] E. S. Burstein, D. J. Hesterberg, J. S. Gutkind, M. R. Brann, E. A. Currier, and T. L. Messier, *Oncogene* **17,** 1617 (1998).

[35] E. S. Burstein, H. Brauner-Osborne, T. A. Spalding, B. R. Conklin, and M. R. Brann, *J. Neurochem.* **68,** 525 (1997).

[36] E. S. Burstein, unpublished observations.

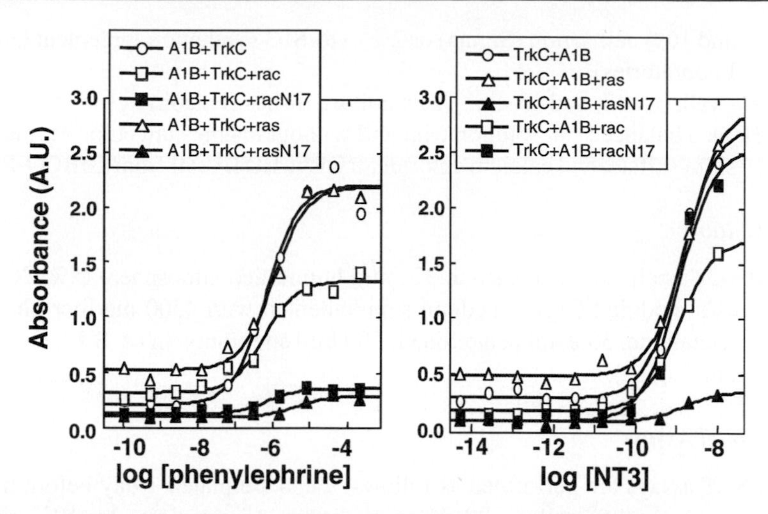

FIG. 7. Multiplexed assay of the effects of ras and rac on proliferative responses to the α_{1B}-adrenergic receptor and the TrkC neurotrophin receptor. NIH-3T3 cells were cotransfected with 40 ng of each receptor and the indicated G proteins into a 6-well dish and assayed as described. Cells were trypsinized, divided, and aliquoted into the wells of a 96-well rack as described under Methods and cultured in the presence of the indicated concentrations of either phenylephrine or neurotrophin-3 (NT-3). Control experiments have verified that there is no cross-reactivity of either ligand with either receptor (data not shown). racN17 and rasN17 are dominant-negative mutants of rac and ras, respectively. Figure taken from Burstein *et al.*[34]

G proteins do not transform NIH-3T3 cells and do not by themselves induce responses in R-SAT. We have shown that cells can be engineered to respond robustly in R-SAT to G_i and G_s-coupled receptors (e.g., cotransfection of chimeric G proteins), although such receptors do not also gain the ability to induce focus formation in NIH-3T3 cells.[35,36] Thus, either only subsets of signals needed to transform cells are needed to generate a response in R-SAT or R-SAT detects signals other than transformation. Previously, we have shown that GPCRs induce responses in R-SAT through a Rac-dependent pathway that is not utilized by tyrosine kinase or JAK/STAT-linked receptors, indicating that R-SAT can discriminate and detect multiple signal transduction inputs[34] (Fig. 7). In addition, the receptors were assayed in a multiplexed manner, suggesting that the effects of the dominant-negative mutants of Ras and Rac were selective.

Materials

NIH-3T3 cells (ATCC CRL 1658)

o-Nitrophenyl-β-D-galactopyranoside and Nonidet P-40 (Sigma)

Dulbecco's modified Eagles medium (A.B.I.) supplemented with 4500 mg/liter glucose, 4 n*M* L-glutamine, 50 U/ml penicillin G, 50 U/ml streptomycin, and 10% calf serum (Sigma) or 2% cyto-SF3 synthetic supplement (Kemp Laboratories)

96-well, 6-well, and 15-cm^2 tissue culture dishes (Falcon)

Hank's balanced salt solution with and without magnesium chloride, magnesium sulfate, and calcium chloride, trypsin-EDTA (all from GIBCO-BRL)

Cell Culture

NIH-3T3 cells are incubated at 37° in a humidified atmosphere (5% CO_2) in Dulbecco's modified Eagles medium supplemented with 4500 mg/liter glucose, 4 n*M* L-glutamine, 50 U/ml penicillin G, 50 U/ml streptomycin (A.B.I.), and 10% calf serum (Sigma).

Functional Assays

R-SAT assays are performed as follows: Cells are plated 1 day before transfection using 2×10^6 cells in 20 ml of media per 15-cm plate, 2×10^5 cells in 2 ml of media per 6-well plate, or 7.5×10^3 cells in 0.1 ml of media per well of a 96-well plate. Cells are transfected by calcium precipitation as described by Wigler *et al.*[37] or using Lipofectamine (GIBCO) or Superfect (Qiagen) according to the manufacturer's instructions using 0.5–10 μg of DNA and 5 μg pSl-β-galactosidase (Promega, Madison, Wl) per 15-cm^2 plate and proportionately less for smaller dishes. One day after transfection, medium is changed and cells are combined with ligands in DMEM supplemented with 2% cyto-SF3 synthetic supplement (Kemp Laboratories) instead of calf serum to a final volume of 200 μl/well. For 15-cm plates, cells are trypsinized and aliquoted into the wells of a 96-well plate (100 μl/well). One 15-cm^2 plate yields enough cells for two 96-well plates. After 5 days in culture, β-galactosidase levels are measured essentially as described.[38] The medium is aspirated from the wells, and the cells are rinsed with PBS, pH 7.4. Two hundred microliters of PBS with 3.5 m*M* *o*-nitrophenyl-β-D-galactopyranoside and 0.5% Nonidet P-40 (both Sigma, St. Louis, MO) is added to each well and the 96-well plates are incubated at room temperature. After 3 hr the plates are read at 420 nm on a plate reader (Bio-Tek EL 310 or Molecular Devices). Dose–response data from R-SAT assays are fit to the equation:

$$R = D + (A - D)/[1 + (x/c)]$$

where A is the minimum response, D is the maximum response, and $c = EC_{50}$ (R is the response and x is the concentration of ligand).

For focus-forming assays, 2×10^5 cells are plated into 10-cm dishes 1 day prior to transfection. The m5 muscarinic receptor (1 μg) is transfected as described

[37] M. Wigler, S. Silverstein, L.-S. Lee, A. Pellicer, Y.-C. Cheng, and R. Axel, *Cell* **11,** 223 (1977).

[38] K. Lim and C.-B. Chae, *BioTechniques* **7,** 576 (1989).

earlier. Cells are cultured for 2 weeks in the presence of the indicated concentrations of carbachol until macroscopic foci appear. Foci are visualized by Giemsa staining. Plates are washed with PBS, fixed for 30 min with 70% ethanol, stained with filtered Giemsa for 30 min, washed with PBS, and allowed to air dry. Foci are scored visually.

Other Considerations

The R-SAT assay is based on the concept of using genetic selection as a functional assay. Because the assay is currently configured, selection is based on acquiring a proliferative advantage that allows transfected cells to overcome contact inhibition. NIH-3T3 cells are well contact inhibited, but are transformed easily by transfected genes, thus they are the cell line of choice. NIH-3T3 cells at low passage numbers work best. We have observed that other contact-inhibited cell lines, such as RAT1 cells and C3H10T1/2 cells, also work. In its current configuration, R-SAT will not work for nonadherent cell lines. Calf serum is required for propagation of the cells, recovery of the cells from transfection, and for inactivation of trypsin. A stable ligand is a prerequisite. Most commercially available ligands, including peptide ligands, give satisfactory results. The assay works with calf serum, but the use of defined supplements during incubation of cells with ligands is recommended, as the likelihood of ligand degradation is reduced and the possibility that endogenous ligands are introduced is reduced.

Overall, a number of strategies for measuring proliferative responses to GPCRs have been described. Assays for DNA synthesis, MAPK activity, and tranformation are useful for examining specific questions about signal transduction, but are not high throughput and not useful for pharmacological experiments. For the pharmacological characterization of ligands, the PACE assay and in particular the reporter gene and R-SAT assays are recommended as they allow high-throughput analysis.

Acknowledgment

M.J.S. is supported by the Royal Netherlands Academy of Arts and Sciences.

[28] Genetic Analysis of G Protein-Coupled Receptor Genes

By DENNIS P. HEALY

Introduction

G protein-coupled receptors (GPCRs) are a family of genes composed of anywhere from 300 to 500 distinct members and subdivided into more than 100 subfamilies.[1] The diversity among the GPCR gene family is consistent with the wide variety of extracellular signals mediated by GPCRs, including sensitivity to photons of light, odorants, and molecules ranging from single amino acids and biogenic amines to peptides and large glycopeptide hormones.

GPCRs share a common structural topology characterized by seven hydrophobic domains, each 20–27 amino acids in length and presumed to correspond to transmembrane-spanning segments. N termini are extracellular, C termini are intracellular, and the intervening transmembrane segments are connected by three extracellular and three intracellular loops. N termini, C termini, and the third cytoplasmic loops vary greatly in size. Depending on the receptor, endogenous ligands bind by associating with residues in transmembrane domains, extracellular loops, or the N terminus. Agonist binding induces a conformational change in the receptor and subsequent coupling and activation of specific G proteins.[2] G proteins couple to the intracellular loops of GPCRs, particularly the third cytoplasmic loop, and the membrane proximal portion of the cytoplasmic tail.

GPCRs can be broadly classified into three distinct classes. Class I GPCRs are rhodopsin-like receptors. This is the largest class of GPCRs, with rhodopsin and the β_2-adrenergic receptor being the mostly extensively characterized. While some specific residues tend to be conserved within the entire class, distinct residues are generally conserved within specific subclasses. The second class of GPCRs are glycoprotein hormone receptors. This class includes receptors for secretin, glucagon, VIP, calcitonin, and PTH. This class has the seven transmembrane topology but lacks the structural signature sequences of class I receptors. This class is characterized by large N-terminal extracellular domains (~350 amino acids) that contain ligand-binding sites. The liganded N-terminal segment then interacts with the extracellular loops connecting the transmembrane domains. Class III GPCRs are metabotropic receptors. This class includes receptors for Ca^{2+}, glutamate, and GABA and is characterized by large N-terminal extracellular domains

[1] T. H. Ji, M. Grossmann, and I. Ji, *J. Biol. Chem.* **273,** 17299 (1998).

[2] U. Gether and B. K. Kobilka, *J. Biol. Chem.* **273,** 17979 (1998).

0076-6879/02 $35.00

(~650 amino acids). Again, the ligands bind to the N termini of these receptors, which then interact with the transmembrane domains to activate signaling.

Based on similarities with bacteriorhodopsin, a light-sensitive proton pump from *Halobacterium halobium,* the hydrophobic amino acid sequences of GPCRs are believed to form an α-helical secondary structure that spans the membrane.[1] Within particular subfamilies, the amino acid sequence of GPCRs tend to be relatively conserved, with the highest degree of conservation being those residues that are critical for ligand binding. For example, for catecholamine receptors, the amine portion of the catecholamine is known to pair with the carboxyl group of a conserved Asp residue deep within transmembrane 3. The catechol ring appears to form hydrogen bonds with two conserved Ser residues in transmembrane 5. In the case of histamine receptors, these Ser residues have been replaced with conserved Asp and Thr residues in transmembrane 5. These residues are believed to associate with the imidazole ring of histamine. Thus, conservation of functionally significant residues aids in the classification of newly described gene products to a particular family.

Whereas some residues are conserved within class I GPCR subfamilies, a number of residues tend to be conserved across GPCR subfamilies. These include, for example, a highly conserved DRY motif at the intracellular surface of transmembrane 3, conserved Cys residues in extracellular loops 1 and 2 that are thought to form disulfide bonds, a CWxP motif in transmembrane 6, and a NPxxY sequence or analogous motif present at the C-terminal region of transmembrane 7. The highly conserved nature of these residues suggests that they provide some structural information that is critical to GPCR function.

Conventional Cloning

Receptors were first characterized by conventional purification approaches based on photoaffinity cross-linking or affinity chromatography. The speed at which receptors were characterized accelerated with the advent of molecular cloning techniques. Here too, however, early approaches utilized a combination of the two approaches. For example, the β_2-adrenergic receptor was purified to homogeneity from hamster lung by affinity chromatography.[3] Peptide fragments from the purified receptor were generated by CNBr cleavage and the N-terminal peptide sequence was obtained. Oligonucleotides complementary to one such peptide were used as a probe for hybridization screening of a hamster genomic library. Sequence analysis of positive clones confirmed that the peptide sequences were contained within the cDNA open reading frame.

As has often been the case in the history of characterizing GPCR genes, the elucidation of one GPCR gene serves as the catalyst for the rapid identification

[3] R. A. Dixon, B. K. Kobilka, D. J. Strader *et al., Nature* **321,** 75 (1986).

of additional genes. Again using the β_2-adrenergic receptor as an example, the elucidation of the hamster β_2-adrenergic gene led to its use as a probe to screen cDNA libraries with low stringency. One such cDNA clone that was isolated using this approach was the dopamine D2 receptor.[4] The cloning of the D2 receptor cDNA, in turn, led to the cloning of the closely related dopamine D3[5] and D4[6] receptors and the more distantly related dopamine D1 receptor, either by hybridization screening with low stringency[7] or by polymerase chain reaction (PCR)-based approaches.[8,9] The D1 receptor was then used in turn to isolate the dopamine D5 receptor gene from a human genomic library.[10] Thus the identification of one GPCR gene or cDNA provided the reagents that were then used to isolate and characterize additional GPCR genes.

The isolation of GPCR genes or cDNAs using hybridization-based methods was then followed up by functional analysis of the cloned and expressed proteins. Based on binding profiles and secondary messenger systems activated by appropriate ligands, the functional identity of clones encoding putative GPCRs could be verified or newly ascribed.

Expression Cloning

Expression cloning has been utilized as an alternative approach to isolate GPCR cDNAs in the absence of available protein or nucleotide sequence information. This approach takes advantage of the inherent binding properties of GPCRs and the availability of selective radioligands. cDNA libraries are constructed from mRNA isolated from tissues believed to express the receptor of interest using plasmid expression vectors. The plasmid libraries are fractionated into smaller pools and transfected into cells. The proteins are expressed and the pools screened for specific radioligand binding of cell surface receptors using autoradiography. Factors affecting the likelihood of detecting a positive signal on the first round of screening include the degree to which the library was diluted, the quality of the mRNA used to construct the library, the efficiency of transfection and expression, the specific activity of the radioligand, and the signal : noise ratio for specific binding. For example, if the library contains 10^6 copies, let us assume that this indicates 10^5 distinct cDNAs. Of those, only one-third will have been in the correct reading frame, leaving a total of 33,333 individual cDNAs. If the library were fractionated into

[4] J. R. Bunzow, H. H. Van Tol, D. K. Grandy *et al.*, *Nature* **336,** 783 (1988).
[5] P. Sokoloff, B. Giros, M. P. Martres, M. L. Bouthenet, and J. C. Schwartz, *Nature* **347,** 146 (1990).
[6] H. H. Van Tol, J. R. Bunzow, H. C. Guan *et al.*, *Nature* **350,** 610 (1991).
[7] J. A. Gingrich, A. Dearry, P. Falardeau *et al.*, *Neurochem. Int.* **20** (Suppl), 9s (1992).
[8] F. J. Monsma, Jr., L. C. Mahan, L. D. McVittie, C. R. Gerfen, and D. R. Sibley, *Proc. Natl. Acad. Sci. U.S.A.* **87,** 6723 (1990).
[9] Q. Y. Zhou, D. K. Grandy, L. Thambi *et al.*, *Nature* **347,** 76 (1990).
[10] R. K. Sunahara, H. C. Guan, B. F. O'Dowd *et al.*, *Nature* **350,** 614 (1991).

10 pools, that would leave transfection and expression of 3333 individual cDNAs. Would there be sufficient amplification and expression of an individual cDNA to be detectable by radioligand binding? The answer to this question is determined empirically. Clearly, division of the library into a larger number of pools, e.g., 100, decreases the number of individual cDNAs that are transfected into cells and increases the relative expression of each clone. The trade-off is the increased labor associated with a larger number of pools to screen. One therefore has to take into account all the factors listed earlier in deciding the number of pools of cDNA that will be initially screened. Assuming a positive signal is detected, the pool(s) of plasmids that give rise to the signal is subdivided again and the procedure repeated. With each round of screening the proportion of plasmids expressing the receptor increases, intensifying the signal detected by radioligand binding until individual plasmid clones are recovered and sequenced.

A variety of mammalian cells have been used for expression cloning, such as COS-7 or CHO cells.[11,12] An alternative expression cloning approach has utilized the large size and electrophysiological properties of *Xenopus laevis* oocytes.[13] cDNA libraries constructed with RNA expression vectors are fractionated and injected into *Xenopus* oocytes. Receptor expression is monitored by electrophysiological recording of cells sensitive to agonist. Again, positive fractions are further diluted and rescreened until single cDNA clones are isolated. *Xenopus* expression cloning has been instrumental in isolation of a large number of GPCR clones.[14–16]

Virtual Cloning

It is not an exaggeration to declare that bioinformatics has revolutionalized the way in which receptor cloning is approached. The creation and accessibility to databases of primary sequence information have transformed the way in which cloning experiments are conducted. Initially, databases such as GenBank served as the repository of published sequence information for GPCR genes. However, with the expanded efforts of the Human Genome Project from both federally and privately supported sources, more and more primary sequence information was submitted independent of functional identification. Much of the most useful genetic material came from so-called expressed sequence tags (ESTs). This approach entailed large-scale partial sequencing of human or mouse cDNAs. The

[11] M. Mukoyama, M. Nakajima, M. Horiuchi *et al., J. Biol. Chem.* **268,** 24539 (1993).

[12] T. J. Murphy, R. W. Alexander, K. K. Griendling, M. S. Runge, and K. E. Bernstein, *Nature* **351,** 233 (1991).

[13] Y. Masu, K. Nakayama, H. Tamaki *et al., Nature* **329,** 836 (1987).

[14] K. Tanaka, M. Masu, and S. Nakanishi, *Neuron* **4,** 847 (1990).

[15] Y. Yokota, Y. Sasai, K. Tanaka *et al., J. Biol. Chem.* **264,** 17649 (1989).

[16] D. Julius, A. B. MacDermott, R. Axel, and T. M. Jessell, *Science* **241,** 558 (1988).

partial sequences, containing anywhere from 100 to 800 bp of nucleotide sequence obtained from either the 5′ or 3′ direction, would be initially compared to existing identified GenBank sequences. If the sequence were similar to an existing GenBank sequence, then it would be annotated as such. The major advantage of the EST database, however, comes from the fact that these clones are made available to investigators through various consortiums (e.g., I.M.A.G.E. Consortium at http://image.llnl.gov/). Human and mouse genomes were the first to become available, but plans are in place for rat, zebrafish, and *Xenopus*, and possibly other species, to be added over time.

Sequence databases and accessibility to cDNA clones to investigators have proved to be invaluable resources. Previously, the classic approach was to purify a particular receptor protein, obtain peptide sequence information, synthesize degenerate complementary oligonucleotides, and screen a cDNA or genomic library. Likewise, PCR amplification of cDNA libraries using degenerate primers has been widely successful at generating partial cDNA clones for novel GPCR genes, which they would then use to screen a cDNA or genomic library. However, with the availability of the sequence databases, experimenters with peptide or primary sequence data can now go directly to the databases and "screen" them, especially the EST databases, for clones that have been only partially characterized but that may nonetheless represent full-length but as yet unidentified clones. Likewise, even if the detected EST clone were not full-length, the nucleotide sequence of the partial EST clones could be used in turn to do a nucleotide screening of the database. Thus, systematic screening of the database could eventually yield full-length cDNA clones that could be purchased and subsequently characterized in greater detail.

Proteomics is another area of investigation that has grown as a result of the human genome project. Proteomics complements genomics by providing ways to rapidly identify the products of predicted genes. A wide variety of functional motifs have been reported for proteins. These motifs have been duly recorded and represented in databases. Protein sequences are checked routinely against these databases for motifs that might provide some clue as to protein function. Alternatively, proteomics can be used to assign possible functions to novel proteins that are predicted based on primary sequence data obtained from human genome or EST databases. Indeed, as noted earlier, GPCRs have certain structural motifs that can be used to assign novel clones to belonging to the GPCR gene family. For example, class I genes have approximately 20 amino acids that are highly conserved across all members of the subfamily. A putative protein with a series of seven hydrophobic domains and conserved residues in these critical locations would have a high probability of being novel members of the GPCR gene family. Indeed, using this approach, 80+ novel GPCR-like genes have been identified. These receptors have been termed orphan receptors. In some cases, the sequences are similar so as to assign the orphan to a particular subfamily. By comparing the

TABLE I
BLAST PROGRAMS

blastp	Compares an amino acid query sequence against a protein sequence database
blastn	Compares a nucleotide query sequence against a nucleotide sequence database
blastx	Compares a nucleotide query sequence translated in all reading frames against a protein sequence database
tblastn	Compares a protein query sequence against a nucleotide sequence database translated dynamically in all reading frames
tblastx	Compares six-frame translations of a nucleotide query sequence against six-frame translations of a nucleotide sequence database

functional properties of these orphan receptors to other members of the family, one might be able to ascertain whether the receptor is likely to represent a new subtype sensitive to a known endogenous agonist or a novel subtype that is likely to be the target for a novel endogenous ligand.

The existence of G protein-coupled orphan receptors has led to speculation that orphan receptors may be valid targets for drug development. Pursuant to that goal, investigators have taken what has been termed a "reverse pharmacology" approach to identify the presumptive endogenous agonist for each receptor. Once the endogenous ligand is identified and the physiology of the system is better understood, drugs that are either agonists or antagonists can be developed rationally. Thus we now have the situation whereby the screening process has been reversed, i.e., instead of screening for receptors using known endogenous ligands, many are using orphan receptors to scan for novel endogenous ligands. There have been some notable successes using the reverse pharmacology approach. For example, an orphan receptor, GPR 14, was believed to be a new member of the somatostain receptor family. This receptor was found to be highly sensitive to the vasoactive fish peptide urotensin. This led to the cloning of human urotensin, the most potent vasoactive peptide constrictor substance ever reported.[17] Because most known GPCRs have been the basis for the design of therapeutic agents, the expectation is that these orphan receptors are also targets for drug discovery.[18]

The search tool that was created by scientists at the National Center for Biotechnology Information and used to scan sequence databases is termed BLAST, which stands for basic local alignment search tool.[19] The program accommodates searches using either protein or DNA (Table I) and a variety of sequence database repositories (Table II). BLAST uses a heuristic algorithm for local alignment and assigns a maximal segment pair score that can be used to distinguish statistically

[17] R. S. Ames, H. M. Sarau, J. K. Chambers *et al., Nature* **401,** 282 (1999).

[18] S. Wilson, D. J. Bergsma, J. K. Chambers *et al., Br. J. Pharmacol.* **125,** 1387 (1998).

[19] S. F. Altschul, W. Gish, W. Miller, E. W. Myers, and D. J. Lipman, *J. Mol. Biol.* **215,** 403 (1990).

TABLE II
PEPTIDE AND NUCLEOTIDE SEQUENCE DATABASES

Peptide sequence databases	
nr	All nonredundant GenBank CDS translations+PDB+SwissProt + PIR+PRF
month	All new or revised GenBank CDS translation+PDB+SwissProt + PIR+PRF released in the last 30 days
swissprot	Last major release of the SWISS-PROT protein sequence database
Drosophila genome	*Drosophila* genome proteins provided by Celera and Berkeley Drosophila Genome Project (BDGP)
yeast	Yeast (*Saccharomyces cerevisiae*) genomic CDS translations
E coli	*Escherichia coli* genomic CDS translations
pdb	Sequences derived from the three-dimensional structure from Brookhaven Protein Data Bank
kabat [kabatpro]	Kabat's database of sequences of immunological interest
alu	Translations of select Alu repeats from REPBASE, suitable for masking Alu repeats from query sequences
Nucleotide sequence databases	
nr	All GenBank+EMBL+DDBJ+PDB sequences (but no EST, STS, GSS, or phase 0, 1, or 2 HTGS sequences)
month	All new or revised GenBank+EMBL+DDBJ+PDB sequences released in the last 30 days
Drosophila genome	*Drosophila* genome provided by Celera and Berkeley Drosophila Genome Project (BDGP)
dbest	Database of GenBank+EMBL+DDBJ sequences from EST divisions
dbsts	Database of GenBank+EMBL+DDBJ sequences from STS divisions
htgs	Unfinished high throughput genomic sequences: phases 0, 1, and 2 (finished, phase 3 HTG sequences are in nr)
gss	Genome survey sequence, includes single-pass genomic data, exon-trapped sequences, and Alu PCR sequences
yeast	Yeast (*Saccharomyces cerevisiae*) genomic nucleotide sequences
E. coli	*Escherichia coli* genomic nucleotide sequences
pdb	Sequences derived from the three-dimensional structure from Brookhaven Protein Data Bank
kabat [kabatnuc]	Kabat's database of sequences of immunological interest
vector	Vector subset of GenBank(R), NCBI, in ftp://ncbi.nlm.nih.gov/blast/db/
mito	Database of mitochondrial sequences
alu	Select Alu repeats from REPBASE, suitable for masking Alu repeats from query sequences
epd	Eukaryotic promotor database found on the web at http://www.genome.ad.jp/dbget-bin/www_bfind?epd

between real matches and random background hits. The BLAST site can be reached at http://www.ncbi.nlm.nih.gov:80/BLAST/. A BLAST tutorial is available that provides information as how to best set the criteria for searching.

As more and more human genomic information becomes deposited and/or finalized within databases, the existence of novel GPCR clones or new members of receptor subfamilies can be ascertained by periodic screening of the databases with sequences of interest. There are services, for example, that will conduct the screening routinely and update the subscriber when new sequences become submitted. One such service is available from Doubletwist (http://www.doubletwist.com/).

To illustrate the type of approach that can be used to detect novel GPCRs gleaned from databases and based on searches using known protein sequences, it might be helpful to follow a simple tutorial. This example also highlights possible limitations with this approach. There are two known D1-like receptors, namely, D1 and D5 (or D1A and D1B). However, there has been speculation that additional D1-like receptors may exist.[20] If one suspected that additional D1-like receptors existed, then one might further expect that they would eventually be submitted into one of the databases as part of the Human Genome Project. To identify such sequences, one approach would be to screen the database with either the cDNA or the protein sequence for the D1 receptor and sort out whether the hits are from previously identified or novel clones.

We have previously cloned the porcine D1 (or D1A) receptor.[21] We used a portion of the porcine D1 receptor protein sequence to conduct a search of the database. We screened using amino acids 190 to 298 of the D1 receptor (Fig. 1), sequence which includes transmembrane 5, the third cytoplasmic loop, and transmembrane 6. Contained within this region would be two Ser residues in transmembrane 5 that are conserved among catecholamine receptors and the CWxP motif within transmembrane 6, a highly conserved sequence among class I receptors. The initial screen used the BLAST program blastp and the nr database limited to *Homo sapiens* clones.

As expected, human D1 receptor clones submitted previously led the list of hits from this screen. These were followed by D5, β_2-adrenergic, β_1-adrenergic, etc., receptors. These receptors are closely related and all couple to the stimulatory G protein, G_s. Figure 1 lists the first five hits whose annotations were not of a previously known receptor. Listed are the accession numbers, the percentage identity over certain amino acids, and the degree to which the porcine/human/rat D1 receptor sequence CWLPFFI was conserved. The annotations were 131134A, G protein-coupled receptor; NP_003958, putative neurotransmitter receptor; NP_055442, G protein-coupled receptor 57; CAB55871, novel 7 transmembrane receptor protein; and NP_055441, G protein-coupled receptor 58. Each accession was then accessed and used to again screen the database using blastp (nr, *Homo sapiens*) to confirm their identities and to see which proteins they were most closely related to. The clone designated 1311340A was found to be identical with the human 5HT1A receptor in the database. The original annotation of this clone as "G protein-coupled receptor" was thus not fully accurate. The next four clones each had low homology with the D1 receptor (23–28%). Three of the four appeared to be related proteins[22] and the fourth (CAB55871) was related to the LPA receptor homologue. Thus, as

[20] J. J. Clifford, O. Tighe, D. T. Croke *et al., Neuroscience* **93,** 1483 (1999).

[21] A. C. Grenader, D. A. O'Rourke, and D. P. Healy, *Am. J. Physiol.* **268,** F423 (1995).

[22] D. K. Lee, K. R. Lynch, T. Nguyen *et al., Biochim. Biophys. Acta.* **1490,** 311 (2000).

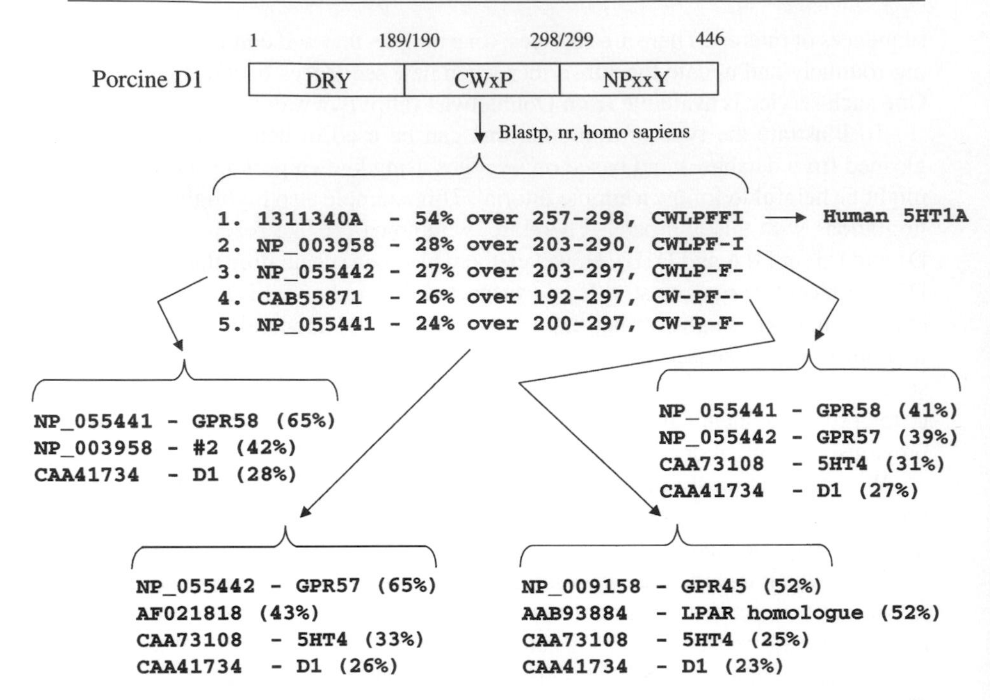

FIG. 1. Flow chart of results obtained from a BLAST search of *Homo sapiens* protein sequences using a peptide sequence from the porcine D1 receptor (shown schematically at the top). The five hits from the database either generically annotated or orphan receptors are numbered 1–5. The percentage identity over the specified amino acids is listed, as is the similarity surrounding the CWxP sequence from transmembrane 6. Each clone in turn was used to screen the protein database and the hits with highest scores listed (excluding self). The degree of identity with the human D1 receptor is also listed for each.

a result of the protein database search, there did not appear to be any novel human proteins present within the database that were closely related to the dopamine D1 receptor but not yet characterized.

The "nr" database does not contain EST fragments. Therefore, the initial blastp search would have failed to pick up any partial D1-like receptor clones. We next searched the human EST database using the tblastn search program with the number of expected hits set high so as to include hits of even small fragments. The tblastn search program compares a protein query sequence against a nucleotide sequence database, in this case human ESTs, translated dynamically in all reading frames. Again, a large number of partially characterized EST clones were picked up by this screen (Fig. 2). In some cases, the annotations were simply identifying numbers, whereas in other cases, they went on to say ". . . similar to . . ." and list the accession number and the specific receptor identifier. We then proceeded systematically to

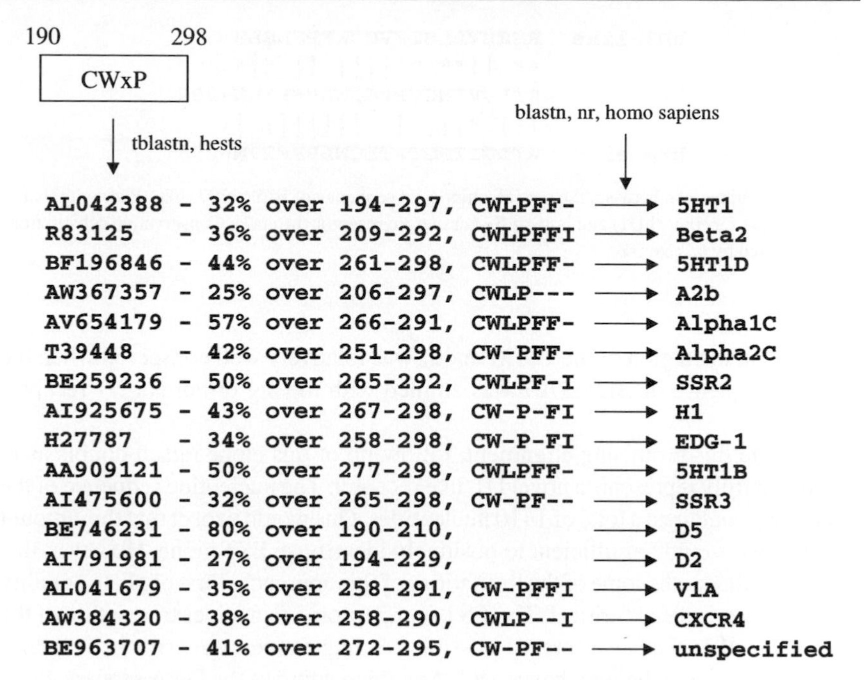

FIG. 2. Flow chart of results obtained from a BLAST search of *Homo sapiens* EST sequences using a peptide sequence from the porcine D1 receptor (shown schematically at the top). Accession numbers for the clones with the highest scores are listed, as are their percentage identity over a specified number of amino acids and the degree of conservation surrounding the CWxP sequence. The nucleotide sequence from each clone was used in turn to screen the nucleotide database. The (high probability) identities of each EST are also listed.

examine each clone via its accession number and screen the nucleotide database using the EST nucleotide sequence and the blastn search program so as to confirm each clones identity as best we could. Clones that had identity in the range of 95% and above were considered identical (small sequencing errors, especially in sequence >400 nucleotides in length and resulting in mismatches, were not uncommon). As can be seen in Fig. 2, with the exception of the D5-like receptor EST clone (BE746231), most clones did not have high identity with the pig D1 and were indeed found to be identical to previously characterized human GPCRs. The one exception was clone BE963707, which aligned over a short region with the D1 receptor with an identity of 41% over 24 amino acids (Fig. 2). This clone was not annotated as being similar to a known receptor. We then examined this clone further.

Screening of the protein database (blastp, nr) with the partial amino acid sequence of BE963707 surprisingly yielded hits that had the highest homology with D1-like receptor proteins, from both vertebrate and mammalian species. The

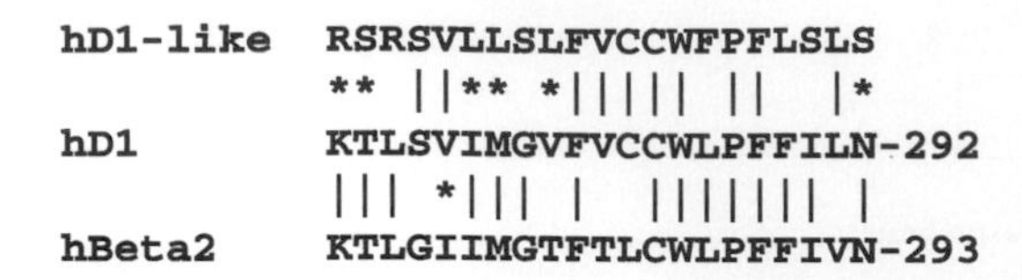

FIG. 3. Comparison between the partial amino acid sequence of BE963707 (hD1-like) with that of the human D1 receptor (hD1) and human β_2-adrenergic receptor (hBeta2). Conservative substitutions are designated by an asterisk.

sequence in the region of the CWxP motif was relatively well conserved when the putative sequence of BE963707 was aligned with the pig or human D1 receptor (Fig. 3).

Despite this promising alignment, follow-up of this clone raised doubts as to whether it truly represents a novel D1-like receptor. The nucleotide sequence of the EST clone contained a total of 1410 nucleotides. One would expect that this amount of sequence would be sufficient to positively identify an EST clone. However, this clone also illustrates some of the limitations of this approach. First of all, the quality of the sequence data is critical. The EST search protocol involves translation of the EST nucleotide sequence into all possible reading frames. An error in sequence data will confound the search protocol. Any misreading of the DNA sequence will result in frame shifts, premature stops, and so on, thus leading to spurious protein sequences and negatively impacting the screening. For example, we downloaded the EST sequence (3′) and translated it in all six reading frames to identify the major open reading frame. Interestingly, the translated sequence that led us to originally select this clone was not even in the largest open reading frame. One way around this dilemma at this point would be to do a nucleotide-based screen, which would be more forgiving of single nucleotide errors. However, in this case, a blastn search (nr) did not result in any significant sequence homology beyond short fragments.

Further comparison (Fig. 3) of the partial amino acid D1-like sequence of BE963707 to human D1 and the closely related human β_2-adrenergic receptor indicated that it may be overly presumptive to conclude that the BE963707-encoded protein is D1-like. Indeed, human D1 appears to be more closely related to the human β_2 receptor across this region than it is to the D1-like peptide fragment. Thus, on the basis of the inspection of the BE963707 EST clone, one would have serious doubts that it (a) encodes a GPCR and (b) represents a novel D1-like receptor clone. Further experiments, led off by more complete sequence analysis, would be necessary before this clone could be identified unequivocally.

Thus, the genetic analysis of GPCR begins with the molecular cloning of the individual genes. Subsequent to characterization of the GPCR genes, studies relating polymorphisms for specific receptors, both within coding and noncoding regions, with various diseases have been employed.[23–27] As discussed here, the molecular cloning of receptor genes and cDNAs appears to have been transformed from a molecular biology-based method to a bioinformatics-based method. Although the latter approach is more passive, i.e., waiting for submission of primary sequence information and database searches, as opposed to the conventional proactive approach of purifying, protein sequencing, and library screening, the bioinformatics approach is potentially much more expeditious. Likewise, as time goes by and a more complete human genome sequence is assembled in the not too distant future, molecular biology-based library screening will only be necessary for species not included in the consortium genomic databases.

[23] K. Blum, E. P. Noble, P. J. Sheridan *et al., JAMA* **263,** 2055 (1990).
[24] S. Cichon, M. M. Nothen, J. Erdmann, and P. Propping, *Hum. Mol. Genet.* **3,** 209 (1994).
[25] Q. Liu, J. L. Sobell, L. L. Heston, and S. S. Sommer, *Am. J. Med. Genet.* **60,** 165 (1995).
[26] D. W. McGraw, S. L. Forbes, L. A. Kramer, and S. B. Liggett, *J. Clin. Invest.* **102,** 1927 (1998).
[27] M. Sato, M. Soma, T. Nakayama, and K. Kanmatsuse, *Hypertension* **36,** 183 (2000).

[29] Identification of Adrenergic Receptor Polymorphisms

By KERSTEN M. SMALL, DEBORAH A. RATHZ, and STEPHEN B. LIGGETT

Adrenergic receptors (AR) are cell surface G protein-coupled receptors that transduce signals due to binding of the endogenous catecholamines epinephrine and norepinephrine. Adrenergic receptors are the receptor component of the sympathetic nervous system, which regulates a host of homeostatic functions, including those of the cardiovascular, pulmonary, endocrine, and central nervous systems. There are nine cloned human adrenergic receptor subtypes: α_{1A}, α_{1B}, α_{1D}, α_{2A}, α_{2B}, α_{2C}, β_1, β_2, and β_3. Each is encoded by a distinct gene. While recent evidence shows previously unrecognized coupling pathways for some subtypes, in general the α_1ARs are considered to primarily couple to $G_{q/11}$, the α_2ARs to G_i and G_o, and the βARs to G_s.

Adrenergic receptors are also targets for a number of pharmacologic agents acting as agonists or antagonists. Interestingly, the activity of the sympathetic nervous system and the therapeutic responsiveness to agonists and antagonists acting at these receptors are known to be highly variable in the population. This has led us to investigate whether such variation is due to polymorphisms of these receptors. Our approach has been to ascertain the sequences of the coding region and, in some cases, the 5′-untranslated regions of these genes in ~50 normal individuals, which

0076-6879/02 $35.00

make up a reference population. To date, all the polymorphisms we have found in the adrenergic receptors are present not only in certain disease populations, but also in the normal population as well.[1-16] Thus while there may be a clustering of a polymorphism in a certain disease, it is still present in healthy individuals at some frequency. This has given rise to the notion that on an individual basis these polymorphisms may represent low-level risk factors for a disease (particularly complex diseases with environmental interactions) and that certain combinations of variants of a number of genes are necessary for causation.[17-19] Nevertheless, an adrenergic receptor polymorphism may alter disease characteristics.[8,9,12,14,16,20-23] As discussed earlier, polymorphisms of these receptors may represent important determinants of drug response. After initial polymorphism discovery, our bias has been to mimic the polymorphism by site-directed mutagenesis or other techniques and express the wild-type and polymorphic receptor in cells in order to determine

[1] K. M. Small, S. L. Forbes, K. M. Brown, and S. B. Liggett, *J. Biol. Chem.* **275,** 38518 (2000).

[2] K. M. Small, K. M. Brown, S. L. Forbes, and S. B. Liggett, *J. Biol. Chem.,* in press.

[3] K. M. Small, S. L. Forbes, F. F. Rahman, K. M. Bridges, and S. B. Liggett, *J. Biol. Chem.* **275,** 23059 (2000).

[4] D. A. Mason, J. D. Moore, S. A. Green, and S. B. Liggett, *J. Biol. Chem.* **274,** 12670 (1999).

[5] J. D. Moore, D. A. Mason, S. A. Green, J. Hsu, and S. B. Liggett, *Hum. Mutat.* **14,** 271 (1999).

[6] S. Green, J. Turki, M. Innis, and S. B. Liggett, *Biochemistry* **33,** 9414 (1994).

[7] S. A. Green, G. Cole, M. Jacinto, M. Innis, and S. B. Liggett, *J. Biol. Chem.* **268,** 23116 (1993).

[8] J. Turki, J. Pak, S. Green, R. Martin, and S. B. Liggett, *J. Clin. Invest.* **95,** 1635 (1995).

[9] T. D. Weir, N. Mallek, A. J. Sandford, T. R. Bai, N. Awadh, J. M. FitzGerald, D. Cockcroft, A. James, S. B. Liggett, and P. D. Pare, *Am. Respir. Crit. Care Med.* **158,** 787 (1998).

[10] S. B. Liggett, L. E. Wagoner, L. L. Craft, R. W. Hornung, B. D. Hoit, T. C. McIntosh, and R. A. Walsh, *J. Clin. Invest.* **102,** 1534 (1998).

[11] D. W. McGraw, S. L. Forbes, L. A. Kramer, and S. B. Liggett, *J. Clin. Invest.* **102,** 1927 (1998).

[12] I. P. Hall, A. Wheatley, P. Wilding, and S. B. Liggett, *Lancet* **345,** 1213 (1995).

[13] E. Reihsaus, M. Innis, N. MacIntyre, and S. B. Liggett, *Am. J. Resp. Cell. Mol. Biol.* **8,** 334 (1993).

[14] J. C. Dewar, J. Wilkinson, A. Wheatley, N. S. Thomas, I. Doull, N. Morton, P. Lio, J. Harvey, S. B. Liggett, I. S. Holgate, and I. P. Hall, *J. Allergy Clin. Immunol.* **100,** 261 (1997).

[15] B. D. Hoit, D. P. Suresh, L. Craft, R. A. Walsh, and S. B. Liggett, *Am. Heart J.* **139,** 537 (2000).

[16] L. E. Wagoner, L. L. Craft, B. Singh, D. P. Suresh, P. W. Zengel, N. McGuire, W. T. Abraham, T. C. Chenier, G. W. Dorn II, and S. B. Liggett, *Circ. Res.* **86,** 834 (2000).

[17] D. W. McGraw and S. B. Liggett, *Clin. Exp. Allergy* **29,** 43 (1999).

[18] S. B. Liggett, *J. Allergy Clin. Immunol.* **103,** S42 (1999).

[19] S. B. Ligget, *in* "The Genetics of Asthma" (S. B. Liggett *et al.,* eds.), p. 455–478. Dekker, New York, 1996.

[20] S. B. Liggett, *Eur. J. Pharmacol.* **163,** 171 (1989).

[21] S. Tan, I. P. Hall, J. Dewar, E. Dow, and B. Lipworth, *Lancet* **350,** 995 (1997).

[22] F. D. Martinez, P. E. Graves, M. Baldini, S. Solomon, and R. Erickson, *J. Clin. Invest.* **100,** 3184 (1997).

[23] E. Israel, J. M. Drazen, S. B. Liggett, H. A. Boushey, R. M. Cherniack, V. M. Chinchilli, D. M. Cooper, J. V. Fahy, J. E. Fish, J. G. Ford, M. Kraft, S. Kunselman, S. C. Lazarus, R. F. Lemanske, R. J. Martin, D. E. McLean, S. P. Peters, E. K. Silverman, C. A. Sorkness, S. J. Szefler, S. T. Weiss, and C. N. Yandava, *Am. J. Resp. Crit. Care. Med.* **162,** 75 (2000).

```
2A GGPQPAEPRCEINDQKWYVISSCIGSFFAPCLIMILVYVRIYQIAKRRTRVPPSRRGPDA
2B   PR-GRPQCKLNQEAWYILASSIGSFFAPCLIMILVYLRIYLIAKRSNRRGPRAKG---
2C     AAYPQCGLNDETWYILSSCIGSFFAPCLIMGLVYARIYRVAKRRTRTLSEKR----

2A VAAPPGGTER--RPNGLGPERSAGPGGAEAEPLPTQLNGAPGEPAPAGPRDTDALDLEES
2B --GPGQGESKQPRPDHGGALASAKLPALASVASAREVNGHSKSTGEKEEGETPEDTGTRA
2C --APVG-------PDGASPTTENGLGAAAGEARTGTARPRP----PTWSR-TRAAQRPRG

2A SSSDHAERPPGPRRPERGPRGKGKAR------------ASQVKPGDSLP-----RRGPG--
2B LPPSWAALPNSGQGQKEGVCGASPEDEAEEEEEEEEEEEEECEPQAVPVSPASACSPPLQQ
2C GAPGPLRR--GGRRRAGAEGGAGGAD------------GQGAGPGAAESGALTASRSPG--

2A ATGIGTPAAGPGEE------RVGAAKASRWRGRQNREKRFTF
2B PQGSRVLATLRGQVLLGRGVGAIGGQWWRRRAHVTREKRFTF
2C PGGRLSRASSRSVEFFLSRRRRARSSVCRRKVAQAREKRFTF
```

FIG. 1. Amino acid sequence of the third intracellular loop regions of α_{2A}-, α_{2B}-, and α_{2C}ARs. The amino acid alignment of the third intracellular loop regions of each of the three α_2AR subtypes is shown. Shaded regions denote locations of the various α_2AR polymorphisms.

whether the variation has a meaningful biologic consequence. Indeed, knowing the phenotype of the polymorphic receptor can be helpful in designing clinical trials to determine relevance in human disease.

This article focuses on the specific methods used to determine sequence variants in these receptors. In addition, techniques for detecting specific polymorphisms are provided so that investigators can genotype large populations rapidly. Finally, we briefly review the consequences of these polymorphisms on signal transduction.

Identification of α_2-Adrenergic Receptor Polymorphisms

To identify sequence variation of the α_2ARs, as well as the other adrenergic receptors discussed here, automated sequence analysis of polymerase chain reaction (PCR) fragments spanning the coding region of these genes was performed using an ABI 377 or 373A sequencer. In doing so, careful consideration was given to the generation of each PCR product as well as to the sequencing methods themselves. While these methods are specifically addressed with α_2ARs, they are also applicable to other adrenergic receptor genes.

For each of the α_2AR subtypes, we chose to analyze the third intracellular loop region for sequence variation, a region known to be important for G protein coupling (Fig. 1).[24] Because each of the α_2AR genes is intronless, it was possible to amplify this region of interest from genomic DNA in one or two overlapping PCR fragments. For α_{2A}- and α_{2B}AR genes, two sets of forward and reverse primers were chosen to amplify overlapping fragments (~500 bp) spanning the region encoding the third intracellular loop (Table I). For the α_{2C}AR gene, primers that amplified

[24] M. G. Eason and S. B. Liggett, *J. Biol. Chem.* **271,** 12826 (1996).

TABLE I
PCR PRIMERS AND CONDITIONS USED FOR AMPLIFICATION OF THIRD INTRACELLULAR LOOP REGION OF α_2AR GENES[a]

Receptor	Primer sets	Buffer	Annealing temperature (°C)	Product size (bp)
α_{2A}	5′-GCCATCATCATCACCGTGTGGGTC 5′-GGCTCGCTCGGGCCTTGCCTTTG	A	56	556
	5′-GACCTGGAGGAGAGCTCGTCTT 5′-TGACCGGGTTCAACGAGCTGTTG	A	58	436
α_{2B}	5′-GCTCATCATCCCTTTCTCGCT 5′-AAAGCCCCACCATGGTCGGGT	J	58	534
	5′-CTGATCGCCAAACGAGCAAC 5′-AAAAACGCCAATGACCACAG	J	60	588
α_{2C}	5′-CCACCATCGTCGCCGTGTGGCTCATCT 5′-AGGCCTCGCGGCAGATGCCGTACA	E	65	723

[a]The 5′ end of each sense (top) and antisense (bottom) primer for α_{2A}- and α_{2B}AR genes also contained sequences corresponding to the M13 forward (5′-TGTAAAACGACGGCCAGT) and M13 reverse (5′-CAGGAAACAGCTATGACC) universal sequencing primers, respectively. PCR reactions consisted of ~100 ng genomic DNA, 20 μl 5× buffer (Invitrogen), 5 pmol of each primer, 0.8 n*M* dNTPs, 10% DMSO, and 2.5 units Platinum *Taq* DNA polymerase (GIBCO/BRL) in a 100-μl reaction volume. PCR cycling started at 94° for 4 min, followed by 94° for 30 sec, the annealing temperature indicated above for 30 sec, and 72° for 30 sec for 35 cycles, and a final extension of 72° for 7 min.

a single 723 bp product encompassing the entire third intracellular loop region were selected. MacVector (Oxford Molecular) as well as Amplify 1.2 software facilitated the selection of these primers. Sequences corresponding to the M13 Forward (−20) and M13 Reverse universal sequencing primers were also added to the 5′ end of each forward and reverse primer, respectively, so that dye primer sequencing methods could be utilized (discussed later). In order to obtain high-quality sequencing data, it was very important that the PCR conditions be as specific and robust as possible. Therefore, genomic DNA was extracted from blood using the Genomic Prep blood DNA isolation kit (Pharmacia), and PCR reactions for each set of primers were optimized using the PCR optimizer kit (Invitrogen). Because of the GC-rich nature of the α_2AR genes, we found that including 10% dimethyl/sulfoxide (DMSO) in each PCR reaction, as well as Platinum *Taq* DNA polymerase (GIBCO/BRL), significantly reduced the level of nonspecific products generated for all three of the α_2AR PCR products. Finally, PCR products were purified using the Qiaquick PCR purification system (Qiagen).

Because only one chromosome copy of a particular α_2AR gene may contain a changed base at a particular location, heterozygous sequencing using dye primer methods was performed using the BigDye primer cycle sequencing ready reaction

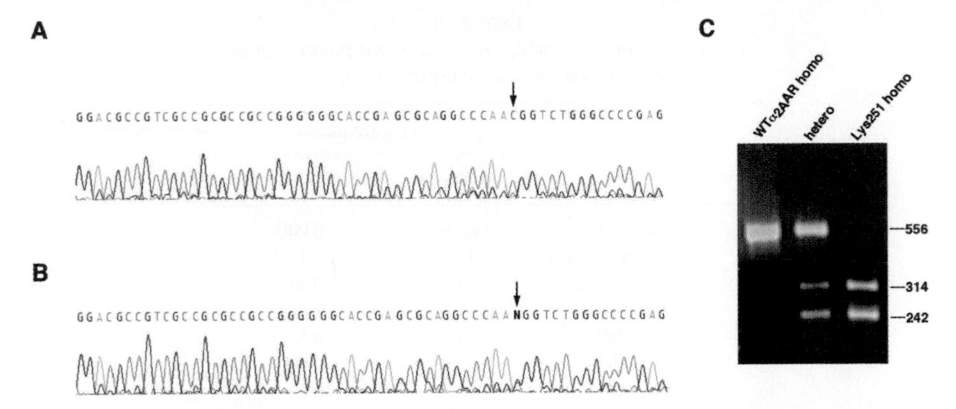

FIG. 2. Sequence variation of the human α_{2A}AR at nucleotide 753. Automated sequencing electropherograms (sense strand) from individuals homozygous for the wild-type α_{2A}AR (A) and heterozygous for Lys-251 polymorphism (B) are shown. (C) Agarose gel of PCR products (see Table I for methods) from wild-type homozygous, heterozygous, and Lys-251 homozygous individuals digested with *Sty*I. The C-to-G transversion at nucleotide 753 creates a unique *Sty*I site that results in partial and complete digestion of the 556-bp PCR fragment amplified from Lys-251 heterozygous and homozygous individuals, respectively.

kit with Amplitaq DNA polymerase, FS (ABI Prism). As stated previously, the α_{2A}- and α_{2B}AR PCR products were generated using primers "tailed" with M13 forward and reverse primer sequences. In doing so, both strands of the PCR product could be sequenced using standard M13 primers labeled with high-sensitivity dyes, a chemistry that produces data with relatively even peak heights for easy and reliable detection of heterozygotes, visualized as two electropherogram peaks at the same position. In this manner, a single nucleotide polymorphism (SNP) was identified within the third intracellular loop region of the α_{2A}AR gene.[1] Figure 2b shows a representative electropherogram of a heterozygous α_{2A}AR sequence in which both a cytosine (C) and a guanine (G) appear at nucleotide 753. Subsequent translation of these α_{2A}AR polymorphic sequences resulted in either an asparagine (Asn) or a lysine (Lys) at amino acid 251 (Fig. 1). Rapid analysis of multiple samples was achieved using the sequence analysis program MacVector (Oxford Molecular), which aligned test sequence data with a wild-type α_2AR reference sequence. In doing so, sequence discrepancies were detected easily and variation in protein translation was determined. Sequencing both strands of the test DNA also helped rule out equivocal sequence as opposed to legitimate heterozygotes. Verification of the Lsy-251 polymorphism was performed by restriction enzyme digestion of the pertinent PCR product with *Sty*I. The presence of a G at nucleotide 753 created a unique *Sty*I restriction enzyme site in PCR amplicons, whose presence or absence readily distinguished each of the various genotypes (Fig. 2c). Furthermore, this restriction enzyme digest was also used to rapidly screen a large number of

TABLE II
ALLELE FREQUENCIES OF α_2-, β_1-, AND β_2AR POLYMORPHISMS IN CAUCASIANS AND AFRICAN-AMERICANS

Gene	Allele	Allele frequency	
		Caucasians	African-Americans
α_{2A}	Lys-251	0.0040	0.040
α_{2B}	Del301-303	0.31	0.12
α_{2C}	Del322-325	0.04	0.43
β_1	Ser-49	0.85	0.87
	Arg-389	0.73	0.58
β_2	Gly-16	0.61	0.50
	Glu-27	0.43	0.27
	Ile-164	0.05[a]	Unknown
BUP	Cys-19	0.63	Unknown

[a] Heterozygous frequency.

samples to determine the frequencies of each allele in the population. We found that Lys-251 α_{2A}AR is relatively rare, with allele frequencies of 0.04 and 0.004 in African-Americans and Caucasians, respectively (Table II).[1]

Similar methods were used to detect and verify polymorphisms in α_{2B}- and α_{2C}AR genes, with some notable differences. For the α_{2B}AR gene (Fig. 3), an in-frame 9-bp deletion beginning at nucleotide 901 was detected using dye primer sequencing methods.[2] This deletion resulted in loss of three glutamic acid (Glu) residues (amino acids 301–303) from an acidic motif within the third intracellular loop known to be important in agonist-promoted phosphorylation and desensitization (Fig. 1).[25] While homozygous DNAs (for either allele) were clearly identified by sequence analysis (Figs. 3A and 3B), sequence data from heterozygous individuals, which contained a mixture of fragments of two different lengths, routinely gave inconclusive sequence data (Fig. 3C). Furthermore, this deletion did not result in the alteration of any restriction enzyme sites. Therefore, to verify the existence of this deletion polymorphism, as well as to develop a rapid screen that would clearly identify all three genotypes, PCR primers were designed to generate fragments that were clearly distinguishable by size when run on a 4% Nusieve agarose gel (Fig. 3D). We found that this deletion polymorphism was more common in Caucasians than African-Americans, with allele frequencies of 0.31 and 0.12, respectively (Table II).[2]

A slightly different strategy was taken in screening the α_{2C}AR gene for polymorphisms. Numerous attempts were made to optimize sequencing of a double-stranded PCR fragment encompassing the third intracellular loop region of this receptor using dye primer methods with limited success. Therefore, this PCR

[25] E. A. Jewell-Motz and S. B. Liggett, *Biochemistry* **34,** 11946 (1995).

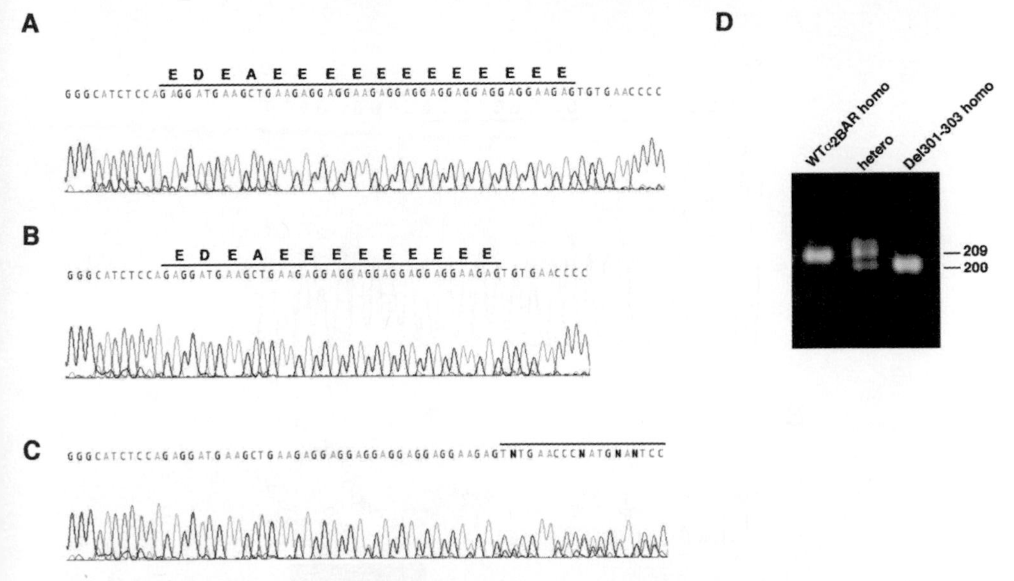

FIG. 3. Sequence variation of the human α_{2B}AR at nucleotides 901–909. Automated sequencing electropherograms (sense strand) of the α_{2B}AR gene from wild-type homozygous (A), Del301-303 homozygous (B), and heterozygous individuals (C), are shown. The line in A highlights nucleotides encoding a glutamic acid-rich region of the α_{2B}AR. Nine bases encoding three glutamic acid residues were found deleted in B. (C) Sequence analysis of PCR products amplified from heterozygous DNA gives inconclusive data (indicated by line). (D) PCR products encompassing nucleotides 901–909 from wild-type (209 bp), Del301-303 (200 bp), and heterozygous individuals run on a 4% Nusieve agarose gel. Primers used to amplify these products were 5′-AGAAGGAGGGTGTTTGTGGGG (sense) and 5′-ACCTATAGCACCCACGCCCCT (antisense), with conditions the same as those described in Table I, except that buffer F was used.

fragment, generated using Platinum *Taq* DNA polymerase high fidelity (GIBCO/BRL) to eliminate PCR errors, was TA subcloned into the plasmid PCR2.1-TOPO (Invitrogen), and multiple subclones of each sample were sequenced using dye terminator methods (ABI Prism). In doing so, high-quality sequence data for α_{2C}AR were obtained. Although this method was more costly and cumbersome due to the increased number of clones analyzed, we found it necessary in order to identify sequence variation in the coding region of the α_{2C}AR. Within the α_{2C}AR gene, a 12 bp in frame deletion polymorphism beginning at nucleotide 964 that resulted in the deletion of amino acids 322–325 (glycine-alanine-glycine-proline) (Figs. 1 and 4) was found.[3] This polymorphism altered the *Nci*I restriction enzyme pattern of a 384-bp PCR fragment, with loss of one of six of these sites (Fig. 4). Again, this method was used to determine the allele frequencies in Caucasians and African-Americans. In this case, we found that the α_{2C}AR deletion polymorphism was ~10-fold more common in African-Americans than in Caucasians, with allele frequencies of 0.43 and 0.04, respectively.[3]

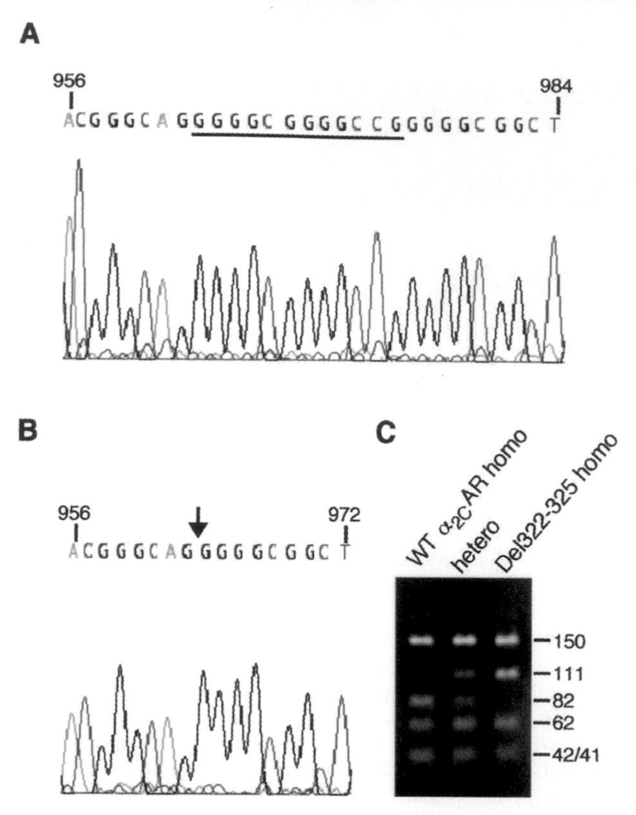

FIG. 4. Sequence variation of the human α_{2C}AR at nucleotides 964–975. Automated sequencing electropherograms (sense strand) for individuals homozygous for the wild-type α_{2C}AR (A) and Del322-325 polymorphism (B) are shown. The underlined bases in A represent nucleotides that were deleted in the polymorphic sequence (see arrow in B). (C) Agarose gel of PCR products (see Table I for methods) from wild-type homozygous (384 bp), Del322-325 homozygous (372 bp), and heterozygous individuals digested with *Nci*I. The wild-type receptor provides for the bands at the indicated molecular sizes (two products of 6 and 1 bp are not shown). The loss of one of the six *Nci*I sites due to polymorphism results in a unique product of 111 bp and loss of the 82- and 41-bp products. Heterozygotes have all six fragments. PCR conditions for the digested fragments in C are the same as those described in Table I, except that the PCR sense primer was 5′-AGCCCGACGAGAGCAGCGCA.

Consequences of α_2AR Polymorphisms

The effects of each polymorphism on signal transduction were examined using cell culture-based structure/function studies. For each case, permanent mammalian cell lines were generated that expressed either wild-type or polymorphic α_2AR, and to date, we have observed alterations in function for all three polymorphic α_2AR. Our studies show that the Lys-251 α_{2A}AR confers enhanced signaling to inhibition of adenylyl cyclase (Fig. 5A), as well as activation of the MAP kinase mitogenic signaling pathway, with no differences seen in the α_2 agonist-stimulated

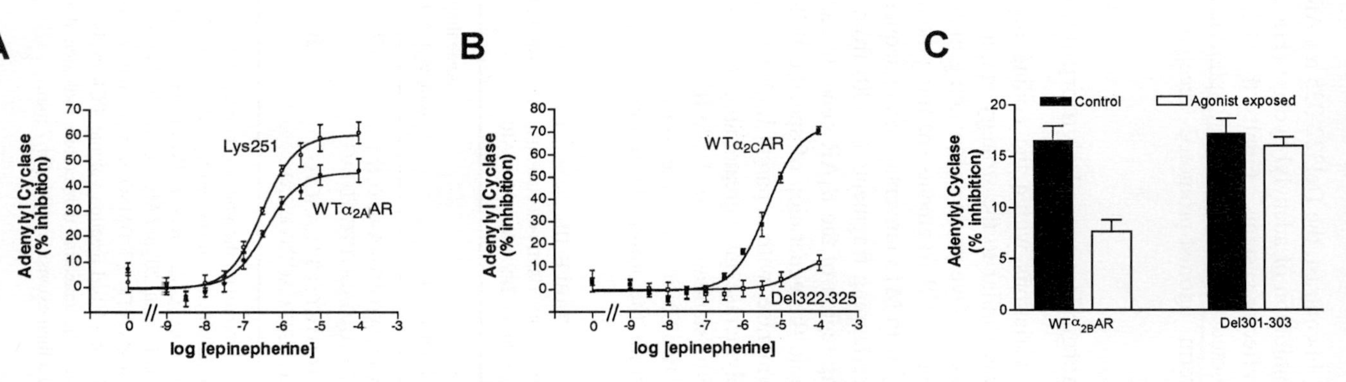

FIG. 5. Phenotypes of the α_2AR polymorphisms. Membranes from Chinese hamster ovary cells were prepared, and adenylyl cyclase activities were determined in the presence of 5.0 μM forskolin as described previously.[28] Results are shown as the percentage inhibition of forskolin-stimulated activities. (A) Results from two cell lines expressing the wild-type α_{2A}AR and Lys-251 receptor at ~2500 fmol/mg. (B) Results from cell lines expressing the wild-type α_{2C}AR and Del322-325 receptor at ~550 fmol/mg. (C) Results are from cell lines expressing wild-type α_{2B}AR and Del301-303 receptors at ~600 fmol/mg following a 30-min pretreatment with media alone or media containing 10 μM norepinephrine. Results shown are from four or five independent experiments.

activation of phopholipase C.[1] In contrast, the Del322-325 α_{2C}AR causes significant loss of signaling to the inhibition of adenylyl cyclase (Fig. 5B), as well as the other two aforementioned effector systems.[3] Finally, the Del301-303 α_{2B}AR shows decreased agonist-promoted GRK-mediated phosphorylation, as well as a complete loss of short-term agonist-promoted receptor desensitization (Fig. 5C).[2]

Identification of β_1-Adrenergic Receptor Polymorphisms

The approach to identify variations within the coding region of the β_1-adrenergic receptor (β_1AR) was similar to the strategy employed for the identification of α_2AR polymorphisms. Dye primer sequencing methods (discussed earlier) were also used to sequence PCR fragments spanning the intronless β_1AR gene. Five primer sets, "tailed" with M13 universal primer sequences, were generated and used to amplify overlapping fragments (~500) from genomic DNA encompassing the entire coding region of the β_1AR gene. In each reaction, 5% DMSO was included to promote the generation of specific PCR products (Table III). Here, *Pwo* DNA polymerase, with apparent higher fidelity and greater heat stability than *Taq* DNA polymerase, gave greater amplification of β_1AR PCR products than was observed with "standard" *Taq* (Table III).

Sequence analysis revealed two SNPs within the coding region of the β_1AR.[4,5] The first variant occurred within the amino terminus of the β_1AR gene, at

TABLE III
PCR PRIMERS AND CONDITIONS USED FOR AMPLIFICATION OF AMINO AND CARBOXY TERMINUS OF β_1AR[a]

Location	Primer sets	Annealing temperature (°C)	Product size (bp)
Amino terminus	5′-CGCTCAGAAACATGCTGAAGTCC 5′-GGACATGATGAAGAGGTTGGTGAG	67	602
Carboxy terminus	5′-CCGCCTCTTCGTCTTCTTCAACTG 5′-TGGGCTTCGAGTTCACCTGCTATC	60	488

[a] The 5′ end of the antisense primer (bottom) for the amino-terminal polymorphism and the sense primer (top) for the carboxy-terminal polymorphism were tagged with the M13 reverse universal primer (5′- CAGGAAACAGCTATGACCACTGGAG). PCR reaction conditions consisted of ~300 ng DNA, 5 μl 10X Pwo buffer with 20 m*M* $MgSO_4$ (Roche Molecular Biochemicals), 100 pmol of each primer, 0.25 m*M* dNTP, 5% DMSO, and 1.5 units *Pwo* DNA polymerase (Roche Molecular Biochemicals) in a 50-μl reaction volume. PCR cycling started at 98° for 5 min, followed by 98° for 45 sec, the annealing temperature indicated above for 1 min, and 72° for 1 min for 35 cycles, and a final extension at 72° for 7 min.

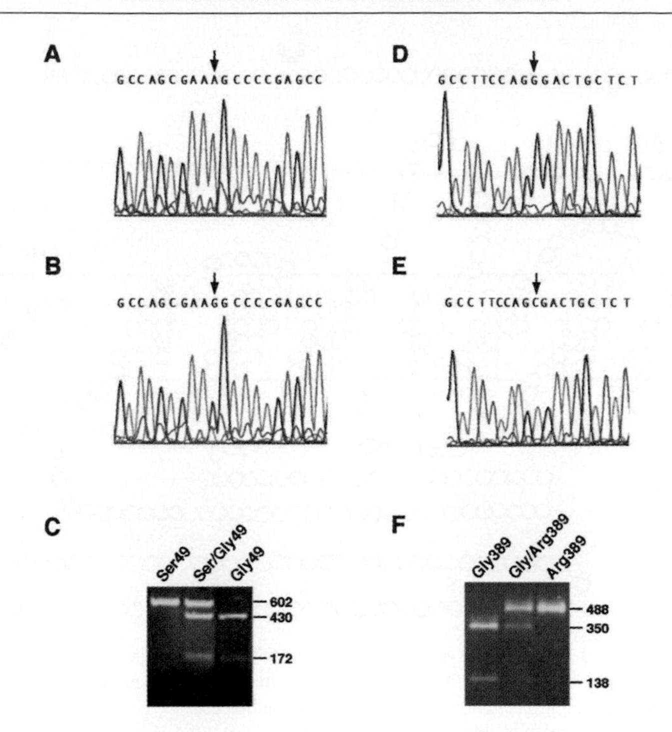

FIG. 6. Sequence variation of human β_1AR at nucleotides 145 and 1165. Automated sequencing electropherograms (sense strand) from individuals homozygous for the wild-type β_1AR (A and D) and Gly-49 (B) and Arg-389 (E) polymorphisms are shown. (C and F) Agarose gels of PCR products (see Table II for methods) from wild-type homozygous, heterozygous, and polymorphism homozygous individuals digested with *Eco*0109I and *Bsm*FI, respectively.

nucleotide 145, where guanine (G) replaced adenine (A), resulting in a serine (Ser) to glycine (Gly) change at amino acid 49. The second variation occurred at nucleotide 1165, within the proximal portion of the carboxy terminus. At this location, cytosine (C) replaced guanine (G) resulting in a change from glycine (Gly) to arginine (Arg) at amino acid 389 (Figs. 6 and 7). The A-to-G transition at nucleotide 145 introduced a unique *Eco*0109I restriction enzyme site, and the G-to-C transversion at nucleotide 1165 resulted in loss of a *Bsm*FI site. Restriction enzyme digests of the pertinent PCR fragments (Table II) were analyzed by standard agarose gel electrophoresis (2% gel, Fig. 6). In this case, 10% glycerol was added to each digest prior to electrophoresis to eliminate a smeared appearance of the fragments. This rapid detection method was used to determine the allele frequencies of each polymorphism, which were 0.15 and 0.13 for the Gly-49 polymorphism and 0.73 and 0.58 for the Arg-389 polymorphism in Caucasian and

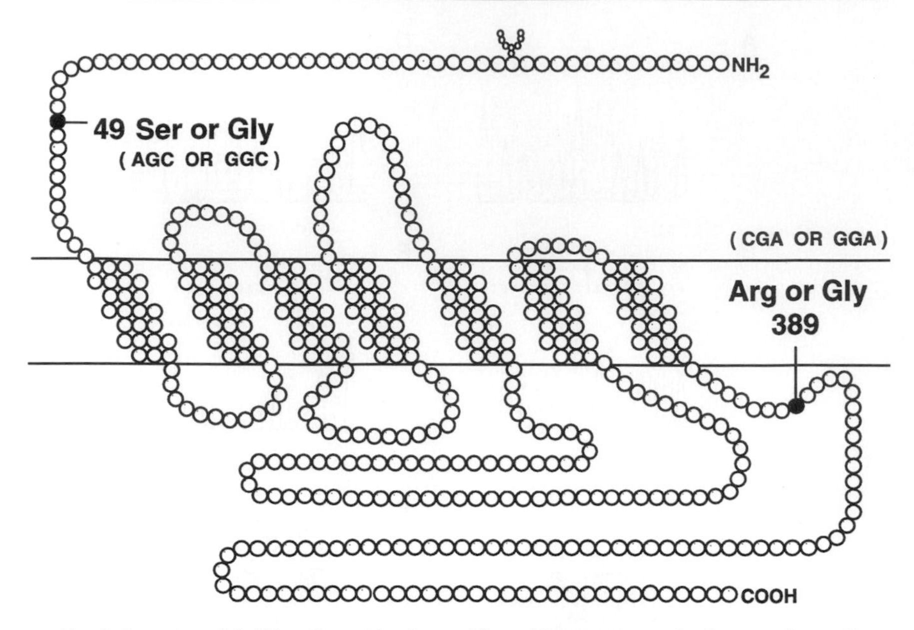

FIG. 7. Location of β_1AR amino acid polymorphisms. The structure and primary amino acid sequence of the β_1AR are shown. The two possible residues in the amino terminus (amino acid 49) and the carboxy terminus (amino acid 389) of the receptor are indicated.

African-Americans, respectively (Table II). In addition, because of the importance of β_1AR function in both healthy and diseased myocardium, we investigated whether either β_1AR polymorphism was associated with heart failure. These studies, however, showed no significant differences in the presence of either polymorphic allele in Caucasian or African-American heart failure patients as compared with controls.

Consequences of β_1-Adrenergic Receptor Polymorphisms

To date, the molecular consequences of the Gly-389 β_1AR polymorphism have been investigated.[4] As described previously, functional differences between wild-type (Gly-389) and polymorphic (Arg-389) were analyzed in studies of mammalian cells, Chinese hamster fibroblasts (CHW-1102), permanently expressing each receptor. These studies showed that the Arg-389 receptor displayed markedly enhanced agonist-promoted stimulation of adenylyl cyclase activity as compared with the wild-type Gly389 receptor, indicating enhanced efficiency of the Arg-389 receptor to couple to G_s (Fig. 8). Agonist-promoted [^{35}S]guanosine 5′-*0*-(thiotriphosphate) binding was also increased with the Arg-389 receptor, and

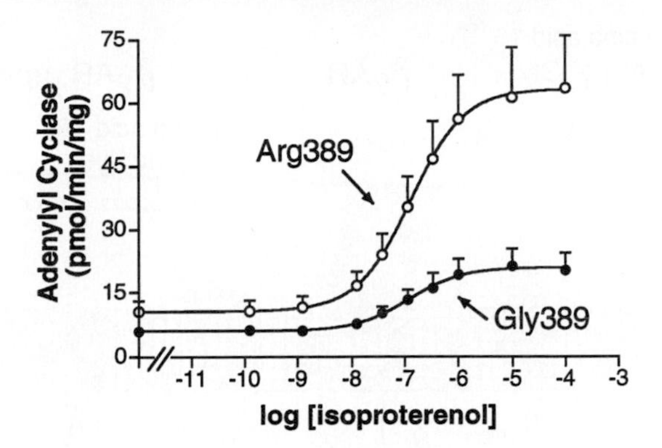

FIG. 8. Coupling of the wild-type Gly-389 and polymorphic Arg-389 β_1AR to stimulation of adenylyl cyclase. Membranes from Chinese hamster fibroblasts (CHW cells) were prepared, and adenylyl cyclase activities were determined as described previously.[4] Results from two cell lines expressing the wild-type and polymorphic receptor at ~150 fmol/mg each are shown.

in the absence of GTP, agonist competition-binding assays showed high-affinity binding for the Arg-389 receptor only, results also consistent with enhanced coupling of the Arg-389 receptor to G_s.

Identification of β_2-Adrenergic Receptor Polymorphisms

The human β_2AR open reading frame (ORF) has four nonsynonymous SNPs at nucleotides 46, 79, 100, and 491.[13] These correspond to changes in amino acid residues at 16, 27, 34, and 164 (Fig. 9). Allele frequencies for these are shown in Table II. As shown, the amino-terminal polymorphisms at amino acid positions 16 and 27 are common, and indeed Arg at position 16 was considered the wild-type receptor prior to our identification of polymorphisms. Initial SNP discovery was carried out by dideoxy sequencing of overlapping PCR products.[13] Subsequently, we have utilized temperature gradient gel electrophoresis,[13] allele-specific PCR,[8] differential hybridization,[12] single base primer extension,[10] and differential enzyme digestion.[22] Because of the varying quality of DNA samples, we have found that the former three techniques provide equivocable results depending on the template. Thus, we now utilize single base primer extension via an automated system[10] or differential restriction nuclease digestion of PCR products to identify the SNPs at codons 16, 27, and 164 (Fig. 10 and Table IV). Similar to the methods described earlier for α_2AR and β_1AR, the restriction nuclease methods are based on the loss or gain of a restriction site. A common set of primers, which results in a 168-bp PCR product, are used to identify the amino acid at positions 16 and 27 (nucleotides 46 and 79, respectively). At nucleotide 46, the A-to-G transition

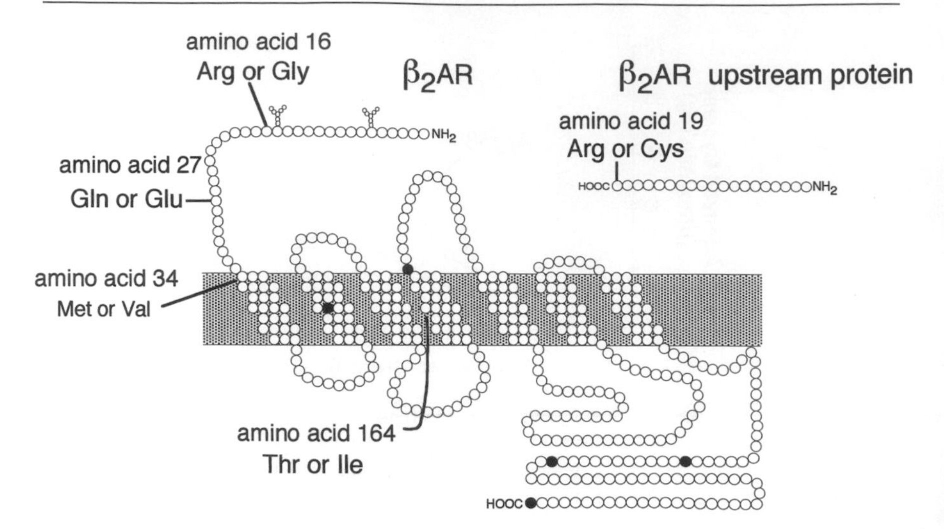

FIG. 9. Localization of polymorphisms of human β_2AR and β_2AR upstream protein genes. Two possible residues at the indicated amino acid positions due to nonsynonymous SNPs in the respective open reading frames are shown. Darkened positions without amino acid changes represent codons where synonymous SNPs are found.

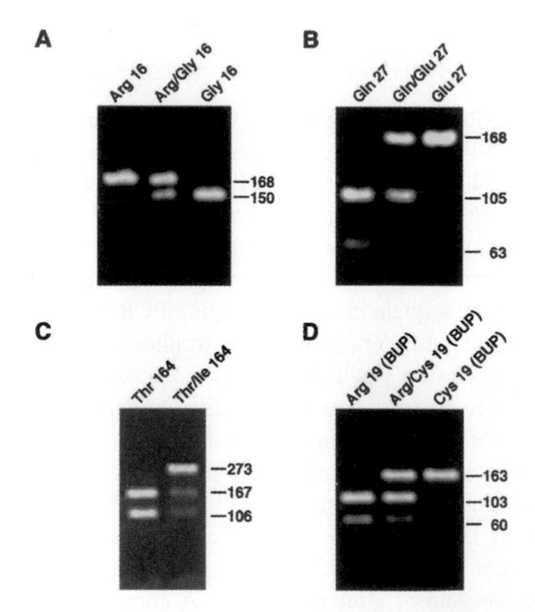

FIG. 10. Identification of β_2AR polymorphisms by restriction enzyme analysis. *Nco*I (A), *Bbv*I (B), and *Mnl*I (C) digests of PCR products amplified from wild-type homozygous, heterozygous, and polymorphic homozygous individuals identifying the amino acids at positions 16, 27, and 164 of the β_2AR (see Table IV and text for details). (D) *Msp*AII digest of the 163-bp PCR fragment reveals the identity of amino acid 19 of BUP.

TABLE IV
PCR PRIMERS AND CONDITIONS USED FOR AMPLIFICATION AND IDENTIFICATION OF β_2AR POLYMORPHISMS[a]

Polymorphic amino acid	Primer sets	Annealing temperature (°C)	Product size (bp)
16 and 27	5′-GCCTTCTTGCTGGCACCCCAT 5′-CAGACGCTCGAACTTGGCAATG	64	168
164	5′-ATTGATGTGCTGTGCGTCACGG 5′-AGGCAATGGCATAGGCTTGG	58	273
19 (5′LC)	5′-CCACCATCGTCGCCGTGTGGCTCATCT 5′-AGGCCTCGCGGCAGATGCCGTACA	56	163

[a] PCR reactions consisted of ~100 ng genomic DNA, 2.5 μl 10X Platinum *Taq* buffer (GIBCO/BRL), 1.5 m*M* $MgCl_2$, 25 ng each primer, 0.8 m*M* dNTPs, and 0.2 units Platinum *Taq* DNA polymerase (GIBCO/BRL) in a 25-μl reaction volume. PCR cycling started at 94° for 3 min, followed by 94° for 30 sec, the annealing temperature indicated above for 30 sec, and 72° for 30 sec for 35 cycles, and a final extension of 72° for 7 min. For the 5′LC polymorphism, 5 ml of buffer C (Invitrogen) that included 2.5 m*M* $MgCl_2$ was used.

does not result in such a change with known restriction endonucleases. However, a 5′ primer-induced mutation (A to C at nucleotide 76) results in a product that is digested by *Nco*I due to the Gly-16 polymorphism. The Arg-16 allele is not digested. This primer-induced restriction digest assay using the conditions shown in Table IV results in loss of an 18-bp fragment with the Gly-16 allele, which is visualized after digestion as a decrease in the PCR fragment from 168 to 150 bp on a 3% Nusieve gel. The Gln-27 allele is observed on digestion of the PCR product with *Bbv*I, whereas the Glu-27 variant fails to cut. For the amino acid 164 variants, PCR primers are utilized to generate a 273-bp product, which is digested with *Mnl*I to produce 167 and 106-bp fragments with the Thr-164 allele. The homozygous Ile-164 product (which we have yet to observe) fails to cut, whereas the heterozygous product appears as three bands as shown after *Mnl*I digestion. In the region 5′ upstream of the β_2AR-coding sequence, a small ORF encodes a 19 amino acid peptide. This ORF has been denoted as the β_2AR 5′ leader cistron and the peptide as the β_2AR upstream peptide (BUP). The initiation of transcription of both the BUP and the β_2AR occurs at a common site 5′ of the former. The BUP has been shown to modulate β_2AR transcription.[11,26] We have delineated a SNP in the most 3′ codon such that Arg or Cys can be encoded.[11] As indicated in Fig. 10 and Table IV, this variation can be identified by an *Msp*A1I digestion, which uniquely cuts the Arg allele.

[26] A. L. Parola and B. K. Kobilka, *J. Biol. Chem.* **269,** 4497 (1994).

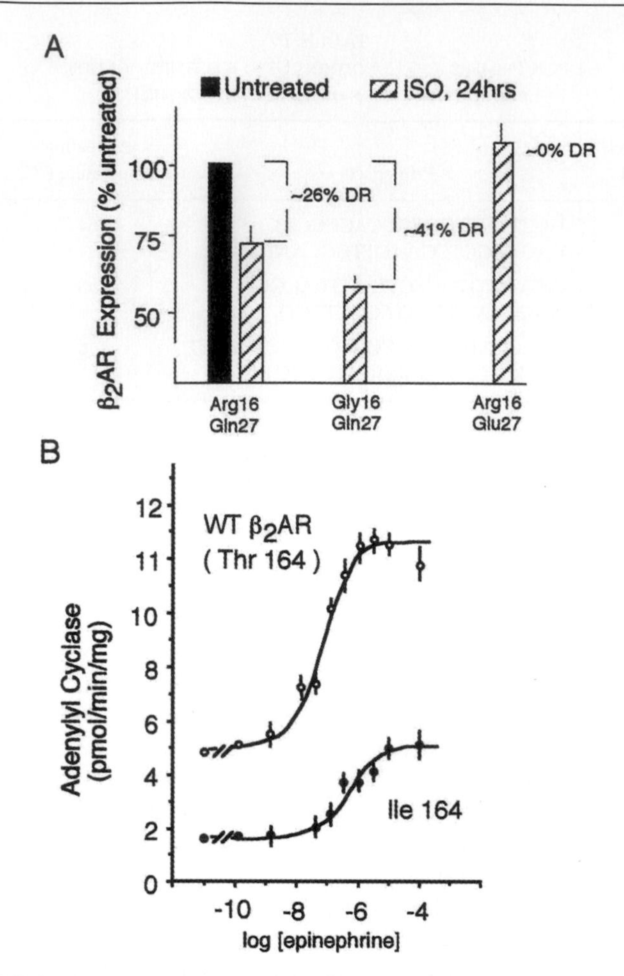

FIG. 11. Cellular phenotypes of β_2AR polymorphisms. (A) Differential agonist-promoted down-regulation of position 16 and 27 variants in permanently transfected CHW cells. (B) Adenylyl cyclase activities of the Thr-164 and Ile-164 receptors expressed in CHW cells.

Consequences of β_2AR Polymorphisms

Polymorphisms of the amino terminus at amino acid positions 16 and 27 alter receptor trafficking induced by prolonged agonist exposure (receptor down-regulation).[6,27] Agonist- and antagonist-binding affinities and G_s coupling are

[27] S. A. Green, J. Turki, P. Bejarano, I. P. Hall, and S. B. Liggett, *Am. J. Resp. Cell. Mol. Biol.* **13,** 25 (1995).

[28] W. E. Kraus, J. P. Longabaugh, and S. B. Liggett, *Am. J. Physiol.* **263,** E226 (1992).

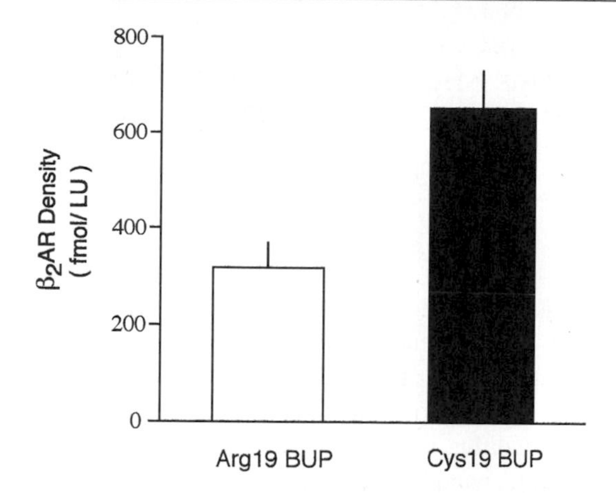

FIG. 12. Cellular phenotype of β_2AR upstream protein (BUP) polymorphism. The polymorphism at amino acid position 19 of BUP causes differential expression of β_2AR due to the altered modulation of receptor translation. Studies were performed in transiently transfected COS-7 cells.

unaffected. In transfected CHW cells, the "wild-type" receptor (Arg-16/Gln-27) displays ~26% loss of receptor expression after 24 hr of exposure of the cells to isoproterenol (Fig. 11A). In contrast, cells expressing the Gly-16/Gln-27 receptor underwent ~41% downregulation, and the Arg-16/Gln-27 receptor displayed no downregulation.[6] The Ile-164 receptor is significantly uncoupled from G_s, with decreased basal and agonist-stimulated adenylyl cyclase activities (Fig. 11B).[7] The 5′ leader cistron polymorphism alters receptor expression independent of agonist exposure. The Cys-19 variant results in a greater expression than the Arg-19 variant (Fig. 12).[11]

Conclusions

Genetic variation of G protein-coupled receptors appears to be the basis for individual differences in physiologic responses or the efficacy of pharmacologic agents targeting this large class of proteins. Herein we have detailed methods for the rapid detection of known polymorphisms of selected adrenergic receptors, which should facilitate additional physiologic and pharmacogenetic studies.

Acknowledgment

This work was supported in part by the National Institutes of Health Grants HL45967, ES06096, and HL53218.

[30] Strategies and Requirements for the Detection of RNA Editing in G Protein Coupled-Receptor RNA

By COLLEEN M. NISWENDER

Introduction

RNA editing is broadly defined as any RNA processing event (excluding RNA splicing) that generates an RNA message with a nucleotide sequence that differs from the corresponding genomic DNA. Mammalian RNA editing generally occurs by either cytidine-to-uridine (C-to-U) or adenosine-to-inosine (A-to-I) deamination. C-to-U editing is best represented by the conversions found within apolipoprotein B RNA[1,2] and is mediated by a cytidine deaminase in conjunction with additional cellular machinery. A-to-I conversions, regulated by a growing family of adenosine deaminases, have been found within a steadily increasing number of mammalian RNAs, including those encoding inotropic glutamate receptor subunits,[3–5] the G protein-coupled serotonin 2C receptor,[6] and the antigenome of the hepatitis delta virus.[7,8] This article describes strategies aimed at the identification of new A-to-I RNA editing events within mammalian RNA substrates, with a focus on RNA molecules encoding G protein-coupled receptors.

Identification of Potential RNA Editing Events within G Protein-Coupled Receptor RNA

Prior to the initiation of studies aimed at quantitating and determining the mechanism of RNA editing within a given RNA substrate, a few simple analyses can be performed to further substantiate that editing might be operating to posttranscriptionally regulate a specific RNA molecule. These analyses will be addressed in general terms later; if an RNA substrate fits these criteria, more detailed biochemical and molecular biological experiments can be performed and are described in the following sections.

[1] S. H. Chen, G. Habib, C. Y. Yang, Z. W. Gu, B. R. Lee, S. A. Weng, S. R. Silberman, S. J. Cai, J. P. Deslypere, M. Rosseneu, A. M. J. R. Gotto, W. H. Li, and L. Chan, *Science* **238,** 363 (1987).

[2] L. M. Powell, S. C. Wallis, R. J. Pease, Y. H. Edwards, T. J. Knott, and J. Scott, *Cell* **50,** 831 (1987).

[3] B. Sommer, M. Kohler, R. Sprengel, and P. H. Seeburg, *Cell* **67,** 11 (1991).

[4] M. Kohler, N. Burnashev, B. Sakmann, and P. H. Seeburg, *Neuron* **10,** 491 (1993).

[5] H. Lomeli, J. Mosbacher, T. Melcher, T. Hoger, J. R. Geiger, H. Kuner, H. Monyer, A. Bach, and P. H. Seeburg, *Science* **266,** 1709 (1994).

[6] C. M. Burns, H. Chu, S. M. Rueter, L. K. Hutchinson, H. Canton, E. Sanders-Bush, and R. B. Emeson, *Nature* **387,** 303 (1997).

[7] J. L. Casey and J. L. Gerin, *J. Virol.* **69,** 7593 (1995).

[8] A. G. Polson, B. L. Bass, and J. L. Casey, *Nature* **380,** 454 (1996).

0076-6879/02 $35.00

Comparison of Genomic and cDNA Sequences

Adenosine-to-guanosine discrepancies between genomic and cDNA sequences provided the initial clues that RNA editing might be operating to posttranscriptionally modify serotonin 2C receptor (5-HT_{2C}R) RNA.[6] Because inosine is read as guanosine by both cellular translational machinery and reverse transcriptase, edited nucleotides appear as guanosines within cDNA sequences after subsequent amplification by the polymerase chain reaction (PCR). Therefore, adenosine-to-inosine RNA editing can be suspected as a regulatory mechanism when guanosines are noted in the cDNA sequence at sites corresponding to adenosines within the gene.

Several mechanisms might account for the presence of nucleotides within cDNA sequences that do not correspond to the bases encoded by the gene; these include multiple genes, polymorphisms, alternative splicing, and RNA editing. In addition to studies aimed at elucidating whether RNA editing is responsible for a particular modification, experiments should be designed to rule out these alternate possibilities using genomic Southern analyses and accurate sequencing of genomic DNA.

It should be noted that when nucleotide differences are observed between genomic and cDNA sequences, it is also critical to eliminate the possibility of cloning artifacts, particularly within the cDNA sequence. Due to the potential for error introduced by reverse transcriptase or polymerases used during PCR, adenosine-to-guanosine discrepancies should be verified by the individual cloning and sequencing of a statistically significant number of cDNA isolates (described in detail later).

Examination of Introns

In mammalian A-to-I RNA editing events identified to date, editing relies on the presence of a stem–loop RNA secondary structure generated by the base pairing of exonic and intronic nucleotides.[5,6,8–11] This RNA duplex directs the activity of a family of adenosine deaminases that require double-stranded RNA as a catalytic substrate to convert adenosine to inosine. In the case of the GluR-B subunit of 3-amino-5-methylisoxazole-4-propionic acid (AMPA)-subtype glutamate receptors and the 5-HT_{2C}R, the RNA duplex structure necessary for editing is located immediately 3′ of the exon containing the edited nucleotides.[9,10] Therefore, sequencing of the gene of interest immediately downstream of the potential editing targets may yield a sequence that models a duplexed secondary structure when entered into an RNA folding program (for a comparison of various structures

[9] M. Higuchi, F. N. Single, M. Kohler, B. Sommer, R. Sprengel, and P. H. Seeburg, *Cell* **75,** 1361 (1993).

[10] J. Egebjerg, V. Kukekov, and S. F. Heinemann, *Proc. Natl. Acad. Sci. U.S.A.* **91,** 10270 (1994).

[11] A. Herb, M. Higuchi, R. Sprengel, and P. H. Seeburg, *Proc. Natl. Acad. Sci. U.S.A.* **93,** 1875 (1996).

identified to date, see Rueter and Emeson[12]). Occasionally, however, critical *cis*-active elements reside some distance downstream from the putative editing sites[11] or in an upstream intron,[13] preventing obvious identification of potential editing regulatory elements.

Species Conservation of Editing Events and Secondary Structures

Editing events within several A-to-I substrates, represented here by the 5-HT_{2C}R, have been observed to be conserved across species.[14] For example, edited nucleotides have been observed at the same sites in templates isolated from rat, mouse, gerbil, and human 5-HT_{2C}R RNA.[14] Sequencing of the introns downstream of the 5-HT_{2C}R editing sites has also revealed a striking nucleotide conservation between species.[14] High sequence homology within introns from different species is rare and points to a conserved role.[15] Conservation of nucleotide discrepancies, as well as identification of intronic sequence similarity from a variety of species, is further suggestive that RNA editing is a potential mechanism for observed nucleotide differences.

Biochemical Mechanisms Mediating Substitutional RNA Editing

Once RNA editing has been determined to be a potential posttranscriptional modification, biochemical assays can be employed to assess the possibility that adenosine-to-inosine conversion is occurring within the RNA substrate. To date, all editing events represented by the appearance of adenosines in the genome and guanosines within the cDNA sequence are due to the conversion of adenosine to inosine. It remains possible, however, that direct conversion of adenosine to guanosine or another "guanosine interpreted" nucleotide could occur. The appearance of inosine within RNA can be assessed and confirmed using several biochemical techniques developed for the measurement of substitutional RNA editing events.

In theory, substitutional editing can be mediated by one of three mechanisms: nucleotide removal and replacement, base transglycosylation, and nucleotide modification (Fig. 1). In nucleotide replacement, the phosphodiester backbone is cleaved on each side of the target nucleotide and a new base, presumably donated from a nucleotide triphosphate, is ligated into the backbone. In transglycosylation, only the ring of the target nucleotide is removed from the sugar and then replaced, maintaining the integrity of the phosphodiester bonds. Direct

[12] S. Rueter and R. Emeson, *in* "Modification and Editing of RNA" (R. Benne, ed.), p. 343. ASM, Washington, DC, (1998).

[13] S. M. Rueter, T. R. Dawson, and R. B. Emeson, *Nature* **399,** 75 (1999).

[14] C. M. Niswender, E. Sanders-Bush, and R. B. Emeson, *Ann. N. Y. Acad. Sci.* **861,** 38 (1998).

[15] B. Lewin, *in* "Genes IV" (B. Lewin, ed.), p. 488. Oxford Univ. and Cell Press, 1991.

A. 1. Base replacement

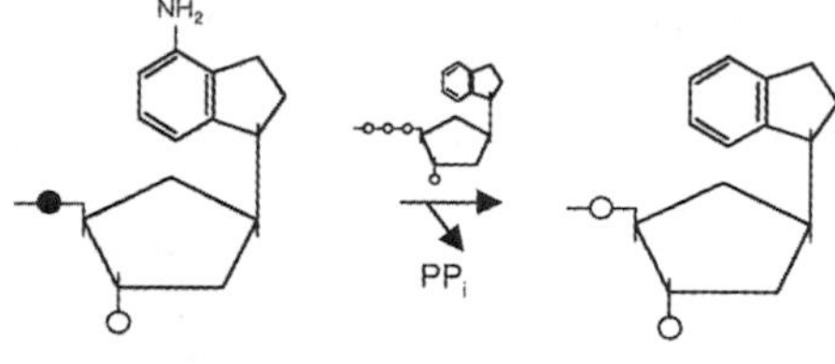

2. Transglycosylation

3. Base modification

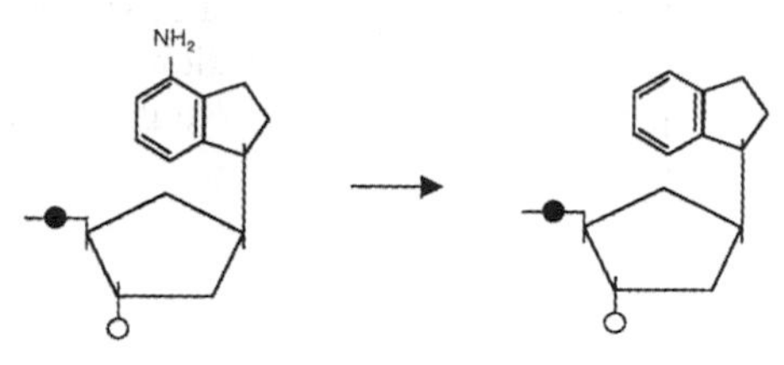

B. 2. Transglycosylation

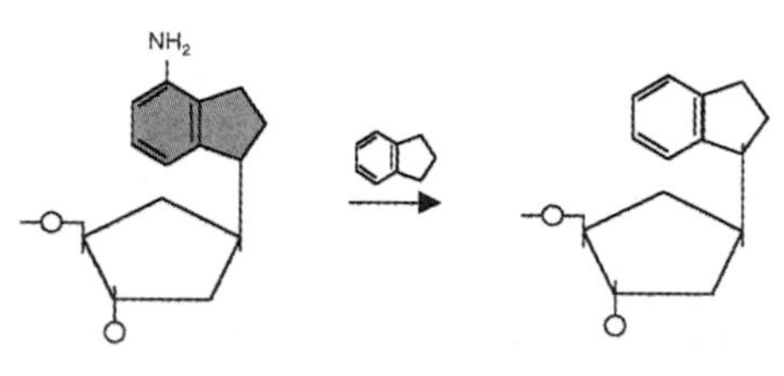

3. Base modification

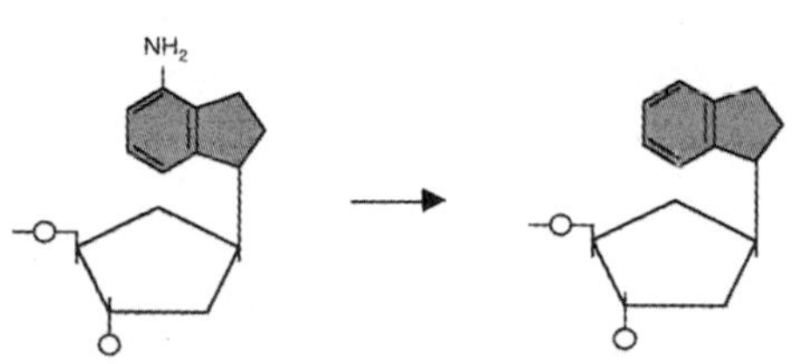

FIG. 1. Comparison of biochemical mechanisms mediating substitutional RNA editing. (A) RNA transcribed *in vitro* in the presence of [α-^{32}P]ATP is labeled within the phosphodiester backbone of the RNA at adenosines (black circles). In the case of nucleotide replacement (1), an unlabeled nucleotide is exchanged into the backbone, resulting in the loss of the radiolabel (white circle). In transglycosylation and nucleotide modification (2 and 3), the backbone remains intact after modification and the radiolabel is retained. (B) RNA transcribed in the presence of the radiolabeled ring constituent [2,8-^{3}H]ATP (gray base) can distinguish transglycosylation from base modification.

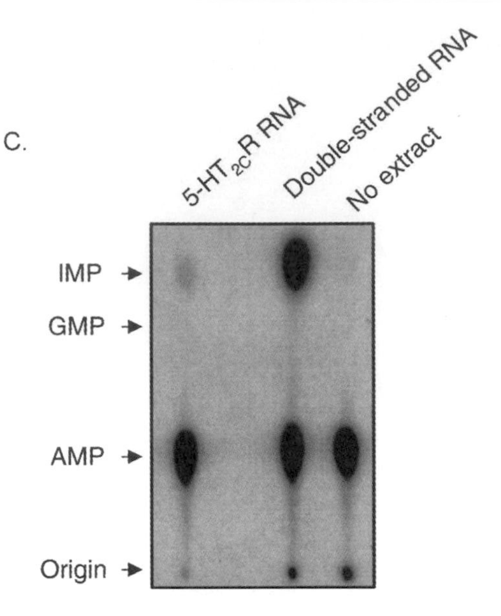

FIG. 1. (*Continued*)

nucleotide modification involves an alteration of only a portion of the ring. In cases of adenosine-to-inosine editing, enzymatic conversion of the adenosine C-6 amino group by an adenosine deaminase results in the ketone group characteristic of inosine.

These mechanisms can be distinguished using a radiolabeled RNA substrate and an *in vitro* editing system. Nuclear extracts prepared from HeLa cells[16] have been shown to serve as a competent source of adenosine-to-inosine deaminase activity.[17–19] An editing-competent RNA message is incubated with these extracts to allow for potential A-to-I conversion. After an extraction procedure, the RNA is incubated with nuclease P1, resulting in cleavage of the RNA to free 5′-nucleoside

Radiolabel will remain with the parent nucleotide only if base modification is the correct mechanism (represented by retention of the gray base). The identity of the modified base is left ambiguous, reflecting the presence of either inosine or guanosine. (C) Thin-layer chromatographic analysis revealing that editing of 5-HT$_{2C}$R RNA occurs by the conversion of adenosine to inosine rather than guanosine. Use of the proper solvent system will distinguish inosine from guanosine due to their differential mobilities (the relative positions of CMP and UMP, which migrate more quickly than IMP in the solvent described in the text, are not shown). The relative conversion of 5-HT$_{2C}$R RNA versus a generic double-stranded RNA is compared after incubation of each RNA with extracts prepared from rat brain. Due to the specificity of editing at only five positions within 5-HT$_{2C}$R RNA, the intensity of the radioactive inosine spot is much less than for a completely double-stranded RNA where up to 50% of the adenosines may be modified.

monophosphates. These free nucleosides can then be separated and identified using an appropriate thin-layer chromatographic system.

1. Prepare nuclear extracts from a known editing competent source, such as HeLa cells, according to the method of Dignam *et al.*[16] This method is described in detail elsewhere.[20]

2. Linearize the DNA construct containing the template of interest with a unique restriction enzyme such that the sense orientation of the RNA will be transcribed. Transcribe RNA in the presence of [α-^{32}P]ATP (800 Ci/mmol) with the appropriate RNA polymerase (T3, T7, or SP6; Promega).

For a 20-μl reaction:
1 μg of linearized DNA template
40 m*M* Tris–HCl (pH 7.9)
6 m*M* $MgCl_2$
2 m*M* spermidine
10 m*M* NaCl
5 m*M* dithiothreitol
0.5 m*M* each ATP, CTP, GTP, and UTP
20 units of RNAsin (Promega)
10–20 units of the appropriate RNA polymerase (Promega)
10 μCi of [α-^{32}P]ATP (800 Ci/mmol)
DEPC-treated water to 20 μl

3. Treat the RNA with 10 units of ribonuclease-free deoxyribonuclease at 37° for 40 min and then bring the total volume to 100 μl with DEPC-treated water.

4. Add an equal volume of phenol–chloroform (1:1), vortex, and spin at 10,000*g* for 3 min.

5. Remove the top (aqueous) layer and add 20 μl of 3 *M* ammonium acetate and 2.5 volumes of 95% ethanol.

6. Place the tube on dry ice for 10 min, and then spin at 10,000*g* for 15 min at 4°.

7. Remove the supernatant and rinse the RNA with 100 μl of 70% ethanol to remove excess salt; spin at 10,000*g* for 5 min.

8. Remove the wash and resuspend the RNA in 30 μl of DEPC-treated water. The desired specific activity of labeling is 2.2×10^8. One microliter of the reaction should be counted and corrected for the decay date of the radioactive ATP.

[16] J. D. Dignam, R. M. Lebovitz, and R. G. Roeder, *Nucleic Acids Res.* **11,** 1475 (1983).
[17] S. M. Rueter, C. M. Burns, S. A. Coode, P. Mookherjee, and R. B. Emeson, *Science* **267,** 1491 (1995).
[18] T. Melcher, S. Maas, M. Higuchi, W. Keller, and P. H. Seeburg, *J. Biol. Chem.* **270,** 8566 (1995).
[19] J. H. Yang, P. Sklar, R. Axel, and T. Maniatis, *Nature* **374,** 77 (1995).
[20] "Current Protocols in Molecular Biology." Wiley, New York, 1989.

In addition, the number of adenosines within the transcript should be known; this is necessary to calculate accurately the amount of RNA that should be incubated with extract to remain within the linear range for enzymatic activity.

9. Incubate the RNA with editing competent extracts for 3 hr at 30°.

For a 50-μl reaction:
50 fmol of RNA substrate
50 μg of HeLa extract
10 m*M* HEPES (pH 7.9)
10% glycerol
20 units of RNAsin
0.25 m*M* dithiothreitol
KCl or NaCl appropriate for the experimental conditions (usually 90–120 m*M*, but this may need to be determined empirically)

10. Stop the reaction with the following: 50 m*M* Tris–HCl (pH 7.9), 5 m*M* EDTA, 100 m*M* NaCl, 0.2% SDS, and 100 μg proteinase K (GIBCO BRL). Allow the reaction to incubate for 15 min at 30°.

11. Extract the RNA with phenol/chloroform (1:1) and precipitate with ethanol as described earlier. Resuspend the RNA in 5 μl of DEPC-treated water.

12. Add 2 μl of nuclease P1, diluted 1:10 in cold sodium acetate (50 m*M* final), and digest for 1 hr at 37°.

13. Spot the samples onto a cellulose TLC plate (Sigmacell Type 100) and allow chromatography to proceed for 4 hr in 0.1 *M* NaH_2PO_4 (pH 6.8), ammonium sulfate, and *n*-propanol (100 : 60 : 2, v/w/v). Because inosine and guanosine migrate at distinct positions within this solvent, use of this TLC system can discriminate between the two bases. One microliter of 100 m*M* of cold AMP, IMP, GMP, CMP, and UMP standards should be run on the same plate to compare the migration positions of the unknown radioactive spots; these standards can be seen on the plate using UV light. Dry the plates and expose to film.

As a control for adenosine-to-inosine editing activity within a particular extract, conversion within a completely double-stranded RNA substrate can be monitored using the labeling techniques just described. This double-stranded RNA control can be produced by performing the following steps.

A. Transcribe two complementary RNA molecules as described previously. These complementary templates can be generated from the same plasmid; general vector sequences derived from any plasmid can work well for this technique provided that they will produce a double-stranded RNA template in the range of 200–500 bp. Proceed with steps 2–7 as described previously.

B. Anneal equimolar amounts of the complementary RNA molecules in the following: 80% formamide, 40 m*M* PIPES (pH 6.4), 400 m*M* NaCl, and 1.25 m*M* EDTA. The volume of this reaction is not critical as the annealed RNA will be subsequently precipitated and resuspended in an appropriate volume. Slowly cool the reaction from 85° to 25°; allow the annealing reaction to remain at 25° overnight.

C. After annealing, extract the double-stranded RNA with phenol–chloroform (1:1), precipitate with ethanol as described earlier, and proceed to step 8.

Utilizing these steps can distinguish base excision/replacement from the two remaining mechanisms, as only base excision will remove the radiolabeled phosphate from its original base (Fig. 1A). It is possible to further discriminate between transglycosylation and base modification by transcribing the RNA in the presence of a radiolabeled ring compound, such as [2,8-^{3}H]ATP (30 Ci/mmol; RNA specific activity 9.6×10^9 cpm/mmol), and performing the steps. In this situation, the radiolabel will remain with the original nucleoside after nuclease P1 treatment only if base modification is responsible for editing (Fig. 1B).[17,21] Using combinations of these methods, base modification has been shown to be definitively responsible for mediating many known C-to-U and A-to-I conversions.[6,17–19,21–23] Once a mechanism for editing has been identified, it is necessary to develop assays for the sensitive and quantitative measurement of levels of editing at particular sites and to determine editing patterns within individual RNA molecules.

Sensitive and Quantitative Assays for Detection of A-to-I Editing

Quantitation of Editing Based on the Introduction of Restriction Sites Specific for Edited or Nonedited RT-PCR Products

The ability to quantitate editing of a single nucleotide within a large RNA template relies not only on detection but on sensitivity as well. A simple assay useful for editing measurement relies on the potential introduction of restriction sites as a result of editing. If restriction site alterations are present, this method provides the potential to examine the editing level at a particular site within a large RNA population after reverse transcriptase–polymerase chain reaction (RT-PCR). In the case of 5-HT_{2C}R, genomic DNA and nonedited RNA encode a *Sna*Bl restriction site. Editing at the B or E positions of the RNA results in the loss of this *Sna*Bl site (Fig. 2). In the case of 5-HT_{2C}R RNA, however, restriction site analysis will not determine the editing status of the A, C, and D positions.

In this assay, total RNA is isolated from tissues or from tissue culture cells using the guanidinium isothiocyanate method of Chirgwin *et al.*[24] or by employing

[21] D. F. Johnson, K. S. Poksay, and T. L. Innerarity, *Biochem. Biophys. Res. Commun.* **195,** 1204 (1993).

[22] K. Bostrom, Z. Garcia, K. S. Poksay, D. F. Johnson, A. J. Lusis, and T. L. Innerarity, *J. Biol. Chem.* **265,** 22446 (1990).

[23] P. E. Hodges, N. Navaratnam, J. C. Greeve, and J. Scott, *Nucleic Acids Res.* **19,** 1197 (1991).

[24] J. M. Chirgwin, A. E. Przybyla, R. J. MacDonald, and W. J. Rutter, *Biochemistry* **18,** 5294 (1979).

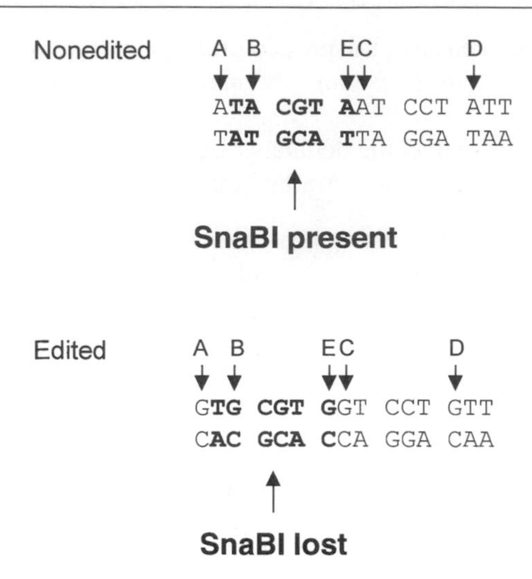

FIG. 2. Use of a *Sna*Bl restriction site in 5-HT_{2C}R RT-PCR products to assess editing. Positions of the five editing sites within 5-HT_{2C}R RNA are shown. An *Sna*Bl restriction site is present within PCR products amplified from nonedited 5-HT_{2C}R RNA. Editing at B or E sites results in the loss of this cleavage site.

a commercially available kit such as Tri-Reagent or Trizol. Messenger RNA is then reverse transcribed to produce first-strand complementary DNA (cDNA) and PCR amplification is employed. Quantitation of the relative amounts of edited and nonedited products is accomplished by separation of the restriction enzyme products on an agarose or acrylamide gel followed by detection with ethidium bromide or by autoradiography.

1. RT-PCR can be performed reliably using any number of commerically available kits (Qiagen, Promega, Clontech, etc.); therefore, detailed protocols for these methods will not be addressed here. However, it should be noted that using a primer specific to the RNA of interest during the cDNA synthesis step (versus oligo (dT) or random hexamer priming) can result in enhanced specificity for the RNA template of interest. In this case, 1 μg of specific primer should be used.
2. After amplification, bring the volume to 100 μl and extract with phenol/chloroform (1:1) as described earlier.
3. Remove the aqueous phase and precipitate as described previously with the exception of adding 5 μl of 10 *M* LiCl in substitution for ammonium acetate.
4. To analyze by ethidium staining on an agarose gel, resuspend the pellets in 20 μl of water. Digest 10 μl of the reaction with 2 μl of the appropriate restriction enzyme and 2 μl of 10$\times$ reaction buffer in a total volume of 20 μl for 1.5 hr at 37°.

Ten microliters of undigested PCR product should be processed in the absence of enzymatic digestion for size comparison.

5. Separate the samples using agarose gel electrophoresis and determine the relative amounts of edited and nonedited products by ethidium bromide staining and subsequent densitometry.

6. More accurate quantitation of the editing status after restriction endonuclease digestion can be obtained by including 0.5 μl [α-^{32}P]dATP in the PCR reaction. For detection of radiolabeled products, pellets are resuspended in 100 μl of water, and 4 μl is digested with 2 μl of the appropriate enzyme and 2 μl of 10X reaction buffer in a total volume of 20 μl for 1.5 hr at 37°.

7. Five microliters of these reactions is added to 5 μl of formamide dye (92% formamide, 0.25% bromphenol blue in 50% ethanol, 0.25% xylene cyanole in water, 0.5X Tris–HCl/boric acid/EDTA) and heated for 10 min at 85° before loading onto a polyacrylamide/7 *M* urea gel. One microliter of undigested sample is added to 9 μl of formamide dye and run in parallel. Positive control DNAs, which should unequivocally digest to completion, should be used to assess enzymatic efficiency. Densitometry or phosphorimager analysis can be used to quantitate the relative levels of edited to nonedited products.

A difficulty with this method is that the restriction enzymes may not completely digest the template DNAs; this can often be seen in the percentage editing of control plasmids and introduces error when the percentage editing of unknown samples is to be measured. Therefore, it is often difficult to pinpoint the exact level of editing within a given sample and it is only possible to predict an editing "window." Optimization of restriction digest conditions, including the amount of template and enzyme, may improve digestion conditions and allow for sufficient measurement of editing levels. Often, however, digestion of the control plasmids does not reach 100%, limiting the quantitative abilities of this technique. In addition, many putative editing sites do not result in a restriction endonuclease site change, necessitating use of an alternate method for editing measurement.

Quantitation of RNA Editing-Based on Primer Extension Analysis of RT-PCR Products

Editing studies for apolipoprotein B C-to-U editing have commonly employed an assay termed RNA primer extension.[22,25–27] This method involves direct detection of editing within the RNA itself by annealing a primer and extending it with reverse transcriptase and a deoxy/dideoxy nucleotide mixture, differential

[25] D. M. Driscoll, J. K. Wynne, S. C. Wallis, and J. Scott, *Cell* **58,** 519 (1989).

[26] M. S. Davies, S. C. Wallis, D. M. Driscoll, J. K. Wynne, G. W. Williams, L. M. Powell, and J. Scott, *J. Biol. Chem.* **264,** 13395 (1989).

[27] J. H. Wu, C. F. Semenkovich, S. H. Chen, W. H. Li, and L. Chan, *J. Biol. Chem.* **265,** 12312 (1990).

termination occurs depending on the editing status at the relevant position. Due to exonic nucleotide similarities among the 5-HT_{2A}, 5-HT_{2B}, and 5-HT_{2C} receptors, however, it is impossible to directly measure editing of 5-HT_{2C}R RNA without contamination by other receptor RNAs. Therefore, it became necessary to selectively amplify 5-HT_{2C}R RNA by RT-PCR before performing a primer extension assay based on the dideoxynucleotide sequencing method of Sanger *et al.*[28] to detect the percentage of editing at the relevant position. As an example, the assay used to determine editing at the A position of the 5-HT_{2C}R will be described; the primers and concentrations of deoxy and dideoxy mixes will need to be altered depending on the substrate and editing sites to be measured.

1. Total RNA isolation, cDNA production, and RT-PCR steps are performed as described previously for the restriction enzyme-based editing assay.
2. Resolve the PCR fragments on an appropriate percentage agarose gel and isolate the fragments from the gel to remove free nucleotides and excess primer. A variety of commercially available purification kits are suitable for this purpose (Gelase, GeneClean).
3. The primer used for this assay can be designed to extend in either the sense or the antisense direction (extension in the sense direction is shown in this example). It is important to design a primer that stops at least two nucleotides away from the first possible termination site, as excess labeled primer can obscure quantitation if it migrates too closely to the bands of interest. Phosphorylate the primer as follows: 1 pmol of primer, 30 μCi [γ-^{32}P]ATP (3000 Ci/mmol), 70 m*M* Tris–HCl, pH 7.6, 10 m*M* $MgCl_2$, 5 m*M* dithiothreitol, and 10 units of T4 polynucleotide kinase (New England Biolabs). Incubate for 30 min at 37°. Heat the reaction to 99° to inactivate the enzyme.
4. Combine the following reagents in a 10-μl volume to anneal the primer to the template: 2 pmol PCR product, 0.1 pmol [γ-^{32}P]ATP-labeled primer, 40 m*M* Tris–HCl (pH 7.5), 50 m*M* NaCl, and 20 m*M* $MgCl_2$. Anneal by cooling from 85° to 25°. This can be done efficiently in a PCR machine with a ramping function.
5. For editing at the A site of the 5-HT_{2C}R, extend the primer as follows: 10 μl annealing reaction, 6.25 m*M* $MnCl_2$, 6.25 m*M* dithiothreitol, 3.2 units Sequenase (United States Biochemical), and H_2O to 16 μl. Mix the reagents and remove 8 μl to a new tube containing the following final concentrations of nucleotides in a total volume of 14 μl : 90 μM deoxyadenosine 5′-triphosphate, 90 μM deoxycytosine 5′-triphosphate, 360 μM of dideoxyguanosine 5′-triphosphate, and 360 μM of dideoxythymidine 5′-triphosphate. Incubate for 10 min at 37°.
6. After extension, add 8 μl formamide-loading dye and heat the samples for 5 min at 99°. Separate the products on a 20% polyacrylamide/7 *M* urea gel; the ratio of edited and nonedited bands can be quantitated with a phosphorimager or using densitometry.

[28] F. Sanger, S. Nicklen, and A. R. Coulson, *Proc. Natl. Acad. Sci. U.S.A.* **74,** 5463 (1977).

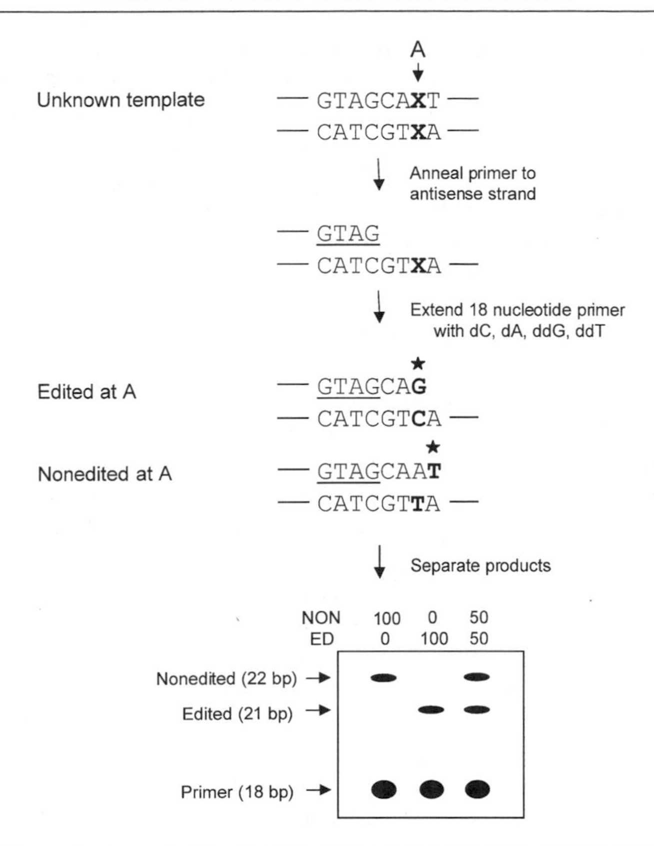

FIG. 3. Primer extension analysis to determine editing levels at the A site of 5-HT$_{2C}$R PCR products. An 18 nucleotide primer is annealed to 5-HT$_{2C}$R RT-PCR products 5′ of the A editing site and extended. If editing at the A site has occurred, extension ceases at position 21; nonedited products terminate at position 22. Shown at the bottom is an example of control separated products. The ratio of the bands at positions 21 and 22 reveals the level of editing at the A site.

In this example, inclusion of the A and C deoxynucleotides extends the sequence to the 5-HT$_{2C}$R A editing site. If editing has occurred, a cytosine will be present in the complementary strand (reflecting the edited guanosine) and the polymerase will incorporate a dideoxyguanosine into the chain, resulting in strand termination at this position (Fig. 3). If editing has not occurred, a thymidine is present at this site. In this case, a deoxyadenosine is incorporated at this position and extension continues to the next complementary nucleotide, an adenosine. At this point a dideoxythymidine is added and all strands are terminated. Manganese chloride is also included in the extension reaction to enhance the visualization of termination products close to the primer (USB Sequenase protocol). In the case of the A site of the 5-HT$_{2C}$R, products will terminate at nucleotides 21 (edited) and 22 (nonedited), respectively.

This method has become widely employed in editing studies due to its technical ease, reproducibility, and sensitivity. Unfortunately, several of the edited adenosines within 5-HT_{2C}R RNA are difficult to analyze using this method due to their spatial position. Because the five editing sites span only 12 nucleotides, some primers must anneal to one editing position to analyze another. The presence of either an A or a G at a given editing site often prevents annealing of a primer that is complementary to only one of these nucleotides. Attempts to utilize primers that are degenerate at known editing positions often result in background problems; therefore, the absolute editing levels at some sites may be difficult to quantitate.

While primer extension is a useful method to determine editing levels, the potentially close proximity of editing sites can limit its usefulness. Therefore, direct sequencing of PCR fragments, either as a whole population or after subcloning into individual expression vectors, provides an alternate mechanism for assessment of editing levels.

Quantitation of Editing Based on Direct Sequencing of RT-PCR Products

Another assay described previously for apolipoprotein B involves sequencing a population of PCR products within one reaction and quantitating the ratio of edited to nonedited bands within the sequencing ladder.[27] The method described here has been modified slightly to employ a phosphorylated primer within the sequencing reaction rather than directly incorporating radioactively labeled ^{35}S-dATP during the extension phase of the reaction.

1. After PCR product purification, incubate the following in a 16-μl reaction: 100–200 ng DNA, 5 μl of 5X fmol sequencing buffer (Promega; 250 m*M* Tris–HCl, pH 9.0, 10 m*M* $MgCl_2$), 5 units of sequencing grade *Taq* polymerase (Promega), and 1.5 pmol of specific kinased primer. In this assay, the use of ^{33}P-ATP rather than ^{32}P-ATP to phosphorylate the primer often results in better resolution of bands after gel separation.
2. Add 4 μl of this reaction to 2 μl of each deoxy/dideoxy mixture (these mixtures, in the initial concentrations given, can be purchased directly from USB):

   dATGC:ddATP 20 μM : 350 μM
   dATGC:ddTTP, 20 μM : 600 μM
   dATGC:ddCTP, 20 μM : 200 μM
   dATGC:ddGTP, 20 μM : 30 μM

3. Cycle in a thermal cycler at 95° for 1 min 15 sec, 42° for 1 min 15 sec, and 72° for 1 min 30 sec for 30 cycles.
4. Terminate the reaction by adding 4 μl of formamide dye and heating for 10 min at 85° before loading onto a 6% polyacrylamide/7 *M* urea gel. Run, dry, and expose the gel to autoradiography film.

A

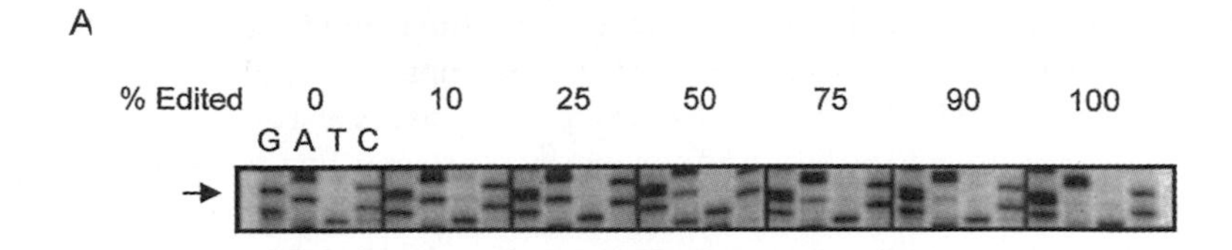

B

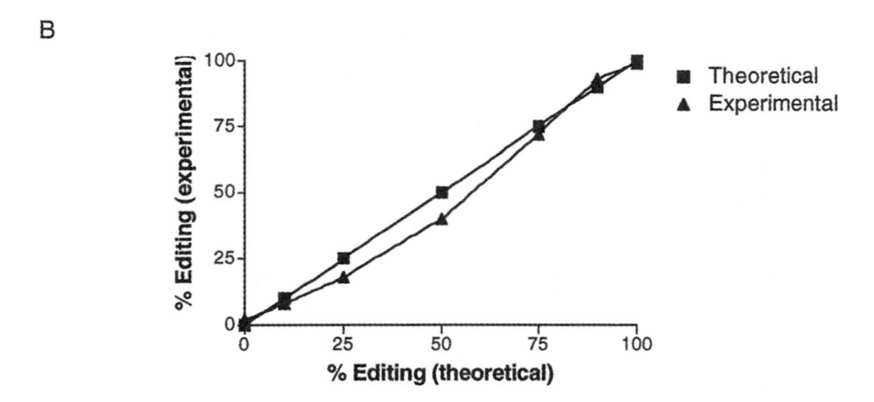

FIG. 4. Sequencing of pooled RT-PCR products. Increasing ratios of edited to nonedited GluR-B RT-PCR products were sequenced and separated on a 6% acrylamide gel. (A) The level of editing at a given position is obtained by comparing the ratio of the radioactivity in A versus G lanes at the editing site. To normalize for relative band intensities, four bands representing nucleotides that are not editing targets are chosen in the G and A lanes. These bands are used as a reference for the potential differences in the efficiency of extension of the two deoxy/dideoxy nucleotide mixtures. The area of these bands is quantitated, and the amount of signal present within these reference bands in the G lane is divided by the area of the reference bands in the A lane. The resulting number generates a "correction factor" that is multiplied by the area of each edited band within the A lane; the final percentage of editing is then determined. (B) Graph of the percentage editing theoretical versus percentage editing experimental for raw data presnted in this example.

Autoradiographs can then be scanned and analyzed using a quantitation program such as Molecular Dynamics ImageQuant. A standard curve obtained using this assay is shown in Fig. 4. For simplicity of this example, a GluR-B template was chosen and a single editing site is shown. In Fig. 4A, increasing ratios of edited to nonedited GluR-B template were added to the sequencing reaction and the percentage of editing was quantitated. A graph of percentage editing theoretical (the amount of editing predicted by the ratio of the input templates) versus percentage editing experimental (the actual amount of editing detected), showing good correlation, is shown in Fig. 4B. Editing at this position within GluR-B RNA can also be examined accurately using primer extension; values obtained using the two assays in parallel are in good agreement (data not shown). While this assay does not exhibit the sensitivity of primer extension at high and low values of editing, the advantages of the method are many. For example, editing levels at the B and E sites of the 5-HT$_{2C}$R are difficult to examine by primer extension due to the primer annealing difficulties discussed previously; the sequencing assay allows for an assessment of editing at these positions. Also, this assay allows simultaneous detection of multiple editing sites within one pool of RNA, whereas the primer extension assay can often only examine one position within a given reaction.

Determination of Editing Pattern within an Individual RNA Transcript

The methods described determine the level of editing at a given position but do not reveal the editing pattern, and therefore the encoded amino acids, within an individual RNA molecule. If editing levels at all sites can be determined accurately and if editing events are independent of each other, then multiplication of the editing levels at each site should reflect the editing pattern and the corresponding amino acids. In the case of 5-HT$_{2C}$R, however, editing at some positions may be linked,[29] preventing the use of editing levels to predict the expression of specific protein isoforms. In this case, it is necessary to individually subclone and sequence RT-PCR products derived from the RNA population of interest. This method involves ligation of the PCR products into individual expression vectors. These ligations are then transformed into bacteria and prepped using standard molecular biological techniques. It should be noted that to obtain a relevant number of independent isolates, the transformation procedure should not be allowed to proceed for more than one doubling period. This prevents the examination of replicate clones and the potential skewing of the analyzed population. After DNA isolates have been prepared and confirmed for the presence of the correct insert, they should be sequenced by either the Sanger method or by automated sequencing. A statistically relevant number of isolates should be examined to reliably predict the editing pattern. The number of editing sites will determine the number of clones that must be sequenced to provide statistical relevance. While time-consuming, this method provides the most accurate assessment of editing patterns and is the best predictor of potential protein changes induced by editing.

Strategies for Identifying New A-to-I Editing Events within G Protein-Coupled Receptor RNA

Although most A-to-I editing events have been identified serendipitously, techniques are becoming available to attempt a more focused approach at identifying editing sites. Obviously, once an editing event has been identified within one member of a G protein-coupled receptor family, a concerted search of related family

[29] Y. Liu, R. B. Emeson, and C. E. Samuel, *J. Biol. Chem.* **274,** 18351 (1999).

members should be performed. For example, 5-HT_{2A}, 5-HT_{2B}, and 5-HT_{2C} receptor RNA share high exonic nucleotide homology and similar exon–intron boundary locations, suggesting that all three might be subject to a similar editing mechanism. An examination of the homologous sites within 5-HT_{2A} and 5-HT_{2B} receptor RNA revealed no editing; it was further shown that 5-HT_{2A} and 5-HT_{2B} receptor RNA lack the intronic regulatory elements necessary for forming the critical RNA secondary structure.[14] The unique presence of intronic editing sequences within one member of the 5-HT_2 receptor family reiterates the importance of introns in A-to-I editing and suggests a mechanism by which highly similar premessenger RNAs may be modified in a transcript-specific manner. Introns are also an attractive source of editing regulatory information, as they allow for variation in the placement and nucleotide sequence of the inverted repeats required for base pairing and duplex formation. In addition, introns effectively accommodate base alterations without affecting protein function, helping to conserve editing during evolutionary pressures that might otherwise result in deleterious amino acid mutations.

Due to the observation that editing events identified to date rely on direction by intronic information, it might be hypothesized that RNAs encoding many of the intronless G protein-coupled receptors would not be candidates for editing. This is not necessarily the case; it might be hypothesized that highly conserved regions of the genome might employ exonic information to direct A-to-I editing if these areas were sufficiently protected from evolutionary stress. In addition, 5′ and 3′ exonic untranslated regions could also serve as important editing regulatory elements, making all G protein-coupled receptor RNAs potential editing targets.

Studies have described the use of enzymes to identify RNA molecules encoding inosine.[30] Inosine and guanosine are both substrates for ribonuclease T1, which cleaves on the 3′ side of these residues. An initial treatment of the RNA with a mixture of glyoxal and borate, however, renders the guanosine residues insensitive to cleavage and promotes cleavage of the RNA only at inosines. Ligation of a known primer to the cleaved end of the RNA can then be used for amplification with a sense primer specific to an RNA substrate of interest or nonspecific primers can be employed to search for unknown editing targets. This technique has been used successfully to identify novel A-to-I editing targets within *C. elegans*.[31] While this technique may be difficult for general examination of editing within G protein-coupled receptor RNAs due to their primary sequence dissimilarities, it may prove useful for the analysis of members of a specific G protein-coupled receptor subfamily.

Due to the lack of a conserved sequnce in A-to-I editing and the reliance on base pairing, searching databases for new A-to-I editing substrates has been difficult. Genomic sequences are rarely presented in current databases in their entirety; often

[30] D. P. Morse and B. L. Bass, *Biochemistry* **36,** 8429 (1997).

[31] D. P. Morse and B. L. Bass, *Proc. Natl. Acad. Sci. U.S.A.* **96,** 6048 (1999).

the gene exon–intron boundaries are the only published information that can be compared to a cDNA sequence. The rapid progress in genome sequencing should prove invaluable to the identification of editing regulatory elements and novel editing targets.

Conclusions

It has been estimated that one IMP molecule is present for every 17,000 ribonucleotides in the brain; in addition, IMP has been detected in poly(A^+) RNA from a variety of tissues such as lung, thymus, and heart.[32] Coupled with the widespread expression of adenosine-to-inosine editing enyzme family members,[33–35] these results suggest that a potentially large number of RNAs are targets for substitutional A-to-I editing. It is anticipated that the number of edited substrates will continue to grow, further substantiating the role of RNA editing in the modulation of cellular diversity.

Acknowledgments

The author thanks Dr. Elaine Sanders-Bush, Dr. Margaret Allen, and Linda Hutchinson for critical reading of the manuscript and Dr. Susan Rueter for protocol assistance. This work was supported by a neurobiology and behavior postdoctoral training grant from the NIH.

[32] M. S. Paul and B. L. Bass, *EMBO J.* **17,** 1120 (1998).
[33] U. Kim, Y. Wang, T. Sanford, Y. Zeng, and K. Nishikura, *Proc. Natl. Acad. Sci. U.S.A.* **91,** 11457 (1994).
[34] M. A. O'Connell, S. Krause, M. Higuchi, J. J. Hsuan, N. F. Totty, A. Jenny, and W. Keller, *Mol. Cell. Biol.* **15,** 1389 (1995).
[35] T. Melcher, S. Maas, A. Herb, R. Sprengel, P. H. Seeburg, and M. Higuchi, *Nature* **379,** 460 (1996).

[31] Fluorescence Microscopy Techniques for the Study of G Protein-Coupled Receptor Trafficking

By LORENA KALLAL and JEFFREY L. BENOVIC

Introduction

This article outlines various methods that have been used to examine G protein-coupled receptors (GPCRs) by fluorescence microscopy. Because many GPCRs exhibit rapid agonist-dependent trafficking, a better understanding of GPCR biology must include knowledge of subcellular distribution patterns and how these relate to regulatory processes such as desensitization and resensitization. Receptor

0076-6879/02 $35.00

localization in the absence of ligand is also an important aspect of GPCR characterization. Finally, simple microscopy techniques are invaluable in troubleshooting efforts to express cloned GPCRs in model cell culture systems. This article describes methods that can be used to examine GPCR localization and trafficking by fluorescence microscopy both in living cells and in fixed cells.

Epitope Tagging GPCRs

Although GPCR immunofluorescence may be performed with antibodies raised against a particular receptor or receptor peptide, antibodies of sufficient quality are often difficult to obtain. We thus focus on the use of two different types of epitope tags used for GPCR fluorescence microscopy: small peptide sequence tags against which there are commercially available high-affinity antibodies and green fluorescent protein (GFP) tags. When small epitope tags are used, a primary or secondary antibody that is conjugated to a fluorescent molecule such as fluorescein isothiocyanate (FITC) or tetrarhodamine isothiocyanate (TRITC) is used to visualize the receptor. With GFP, the protein tag itself is fluorescent and therefore the protocols do not involve the use of antibodies. In both cases, the encoding DNA sequence of either the small epitope or the GFP tag is cloned in frame to the GPCR DNA sequence to generate a protein fusion or chimera. We describe tagging GPCRs with small epitope tags at the N terminus and with GFP at the C terminus. Small epitope tags have also been used successfully at receptor C termini, but we are not aware of reports on functional GFP–GPCR chimeras with GFP appended at the N terminus.

This article describes how GPCR chimeras are constructed, expressed, and subsequently observed by fluorescence microscopy. Many tagged GPCR chimeras have been shown to retain their pharmacological characteristics, and specific examples of epitope-tagged GPCRs and GPCR–GFP chimeras are prevalent in the literature.[1,2]

Tagging Receptors with FLAG or HA Epitopes at the N Terminus

Use of an N-Terminal-Cleaved Signal Sequence

With the addition of an N-terminal epitope tag such as FLAG or hemagglutinin (HA), one must be concerned not only with the possibility of disturbing native ligand binding and signaling properties of a particular GPCR (as with any sequence modification), but also with the potential alteration of receptor localization to cell membranes. The correct insertion of a GPCR into lipid bilayers in some cases is directed by a cleaved signal sequence, but in other cases is directed by the

[1] G. Milligan, *Br. J. Pharmacol.* **128,** 501 (1999).
[2] L. Kallal and J. L. Benovic, *Trends Pharmacol. Sci.* **21,** 175 (2000).

hydrophobic protein sequence of the first transmembrane domain, which functions as a type of noncleaved signal sequence. A cleavable signal sequence may be added along with the N-terminal epitope tag to improve the membrane localization of the receptor. This is particularly useful if localization is disrupted by the N-terminal sequence modification. Many GPCRs have been tagged using this approach with excellent results.

This article describes tagging a human somatostatin-2 receptor (SST2R) with a cleavable influenza virus hemagglutinin protein signal sequence and the FLAG epitope at the N terminus. The DNA and protein sequences of the commonly used T8 signal sequence and HA epitope tag are subsequently listed and can be added at receptor N termini using the same strategies that we describe in detail for the FLAG epitope. Note that if there is a cleaved signal sequence in the native receptor, a new signal sequence and epitope tag can be fused to the first amino acid after the original cleavage site to retain the epitope tag in the translated protein (e.g., see construction of HA-tagged FSHR[3]). In addition, it is important to note that both HA and FLAG epitope tags are often used without an added signal sequence and may also be placed at receptor C termini. The use of a particular tag or signal sequence depends on the specific receptor being tagged and on any experimental applications planned in addition to fluorescence microscopy.

Construction of a Signal-FLAG-SST2R Plasmid

In this case, a plasmid vector containing a signal sequence and FLAG tag is first prepared. The SST2R is then amplified by PCR, with an N-terminal primer designed to produce an in-frame fusion of the receptor to the signal sequence and FLAG tag.

Signal-FLAG Plasmid

Two complementary oligonucleotides (oligos) were designed to anneal and create *Hin*dIII and *Eco*RI overhangs. Once annealed, they were ligated into pCDNA3.1, which had been digested with the same enzymes (see Ref. 4 for details). The following oligo sequences are shown with the restriction site overhangs underlined and the signal sequence initiator methionine codon in bold. The amino acids are shown in single letter code (signal sequence in lowercase, FLAG epitope in uppercase) below the corresponding DNA sequence of the encoding sense oligo. The final amino acid of the FLAG epitope shown in parentheses has been widely varied, and anti-FLAG antibodies seem to bind very well whether or not this is

[3] K. J. Koller, E. A. Whitehorn, E. Tate, T. Ries, B. Aguilar, T. Chernov-Rogan, A. M. Davis, A. Dobbs, M. Yen, R. W. Barrett, *Anal. Biochem.* **250,** 51 (1997).

[4] J. L. Parent, P. Labrecque, M. J. Orsini, and J. L. Benovic, *J. Biol. Chem.* **274,** 8941 (1999).

a lysine (which is the amino acid listed by commercial suppliers of anti-FLAG antibodies).

sense oligo:

5′-AGCTTGGGCACC **ATG** AAG ACG ATC ATC GCC CTG AGC TAC ATC

m k t i i a l s y i

TTC TGC CTG GTG TTC GCC GAC TAC AAG GAC GAT GAT GAC ACC G-3′

f c l v f a D Y K D D D D (T)

complementary oligo:

5′-AATTC GGT GTC ATC ATC GTC CTT GTA GTC GGC GAA CAC CAG

GCA GAA GAT GTA GCT CAG GGC GAT GAT CGT CTT CAT GGT GCC CA-3′

Once ligated into pCDNA3.1, the resulting pCDNA3-signal-FLAG plasmid may be used as a recipient plasmid vector for various receptor clones. Receptor cDNAs are amplified by polymerase chain reactions (PCR) with oligos designed to insert them in frame with the signal-FLAG sequence. Other restriction sites may be used to accommodate receptor genes that contain internal *Eco*RI sites. Note that in this example the signal sequence and epitope tag are shown directly in tandem, however, they can also be separated by a restriction site. The amino acids encoded by the intervening restriction site would then be appended at the N terminus of the receptor in addition to the epitope tag. If a receptor is tagged in this manner, particular care should be taken in purchasing anti-FLAG antibodies, as some commercially available preparations only recognize the epitope at the free N terminus of a protein.

5′ oligo for SST2R amplification: The restriction site is underlined and the receptor ATG is shown in bold.

5′-GGAATTC**ATG**GACATGGCGGATGAG-3′

*Eco*RI

3′ oligo for SST2R amplification: The restriction site is underlined and the receptor stop codon is shown in bold.

5′-GCCCTCGAGAAGCTT**TCA**GATACTGGTTTGGAG-3′

*Xho*I stop

The receptor cDNA is amplified by PCR (see later) and the resulting product is cut with *Eco*RI and *Xho*I, gel purified, and ligated into the pCDNA3-signal-FLAG plasmid described earlier. The two amino acids (E and F) encoded by the *Eco*RI site become part of the final chimera. DNA inserts generated by PCR should be sequenced entirely.

PCR Conditions

General conditions for use with Expand Hi Fidelity polymerase (Roche Molecular Biochemicals) using the manufacturer's buffers and recommended conditions are oulined.

1 μl template DNA (2–50 ng of plasmid containing the receptor of interest)
5 μl 10× PCR buffer with magnesium
5′ primer at 250–500 n*M* final concentration
3′ primer at 250–500 n*M* final concentration
200–500 μ*M* dNTPs
0.5–1.0 μl Expand Hi Fidelity polymerase
Sterile distilled H_2O to final volume of 50 μl

Typical thermocycling parameters:

1 cycle:	92°, 2 min (initial denaturation)
25–35 cycles:	30–60 sec, 92° (denaturation)
	30–60 sec, 40–65° (annealing, temperature depends on primers)
	45 sec–2.5 min, 68° (elongation)
1 cycle:	2–5 min, 68° (final elongation step)

Notes on PCR

Troubleshooting can involve increasing the concentration of magnesium up to 5 m*M* (from 1.7 m*M*) and varying the concentration of primer or template DNA within the ranges shown previously. The addition of up to 5% dimethyl sulfoxide (DMSO) and/or 5% glycerol can also improve PCR product formation in some cases.

PCR products may be purified and cut with restriction enzymes using general molecular biology techniques.[5,6] We generally use molecular biology enzymes from Roche and Promega and DNA purification products from Qiagen.

Alternative Templates for PCR

RNA Template. If receptor clones are not available to use as a DNA template in PCR, one can isolate RNA from tissues or cultured cells and synthesize DNA using reverse transcriptase (RT). RNA may be prepared with TRI-Reagent (Sigma) following the manufacturer's instructions. cDNA may be synthesized using total RNA

[5] F. M. Ausubel, R. Brent, R. E. Kingston, D. D. Moore, J. G. Seidman, J. A. Smith, and K. Struhl (eds.), "Current Protocols in Molecular Biology." Wiley, New York, 2000.

[6] J. Sambrook, E. F. Fritsch, and T. Maniatis (eds.), "Molecular Cloning, a Laboratory Manual," 2nd Ed. Cold Spring Harbor Laboratory Press, Cold Spring Harbor, NY, 1989.

as template using Superscript [derivatives of Moloney Murine Leukemia Virus (MMLV) reverse transcriptase] or avian myeloblastis virus (AMV) reverse transcriptase enzymes, following the manufacturer's instructions. Random primers, oligo(dT) primers, or 3′ receptor-specific primers may be used to prime cDNA synthesis in these reactions. The resulting cDNA is used as the template for amplification of a particular receptor in a subsequent PCR. There are also single-tube RT-PCR kits available where reverse transcription and PCR are performed in a single reaction tube.

Genomic DNA Template. Genomic DNA should be used as template only if the receptor gene is known to be intronless. Because this is very common for genes encoding GPCRs, one can avoid RNA preparation and subsequent cDNA synthesis in these cases. Human genomic DNA may be easily prepared from 0.2 ml blood using the QIAamp blood kit from Qiagen.

Specific conditions for PCR of the human somatostatin 2 receptor:

100 ng human genomic DNA
5 μl 10× buffer with magnesium (1.7 m*M*)
5′ primer at 500 n*M* final concentration
3′ primer at 500 n*M* final concentration
200 μM dNTPs
0.5 μl Expand Hi Fidelity polymerase
Sterile distilled H_2O to 50 μl volume

Cycling parameters: 1 cycle, 2 min, 92°; 30 cycles (92°, 45 sec; 50°, 45 sec; 68°, 2 min), and 1 cycle 68°, 3 min.

Use of a T8 Signal Sequence and HA Tag

The DNA sequence encoding the cleaved signal sequence of the human T-cell surface protein T8 (see Ref. 7) followed by the sequence of the HA epitope tag is shown with the amino acids in single letter code below the DNA sequences (signal sequence in lowercase, HA epitope in uppercase). Restriction sites for creating a receptor fusion are not shown, but should be added so that this signal/tag combination can be ligated into the vector of choice (as per the signal-FLAG construction).

5′-ATG GCC TTA CCA GTG ACC GCC TTG CTC CTG CCG CTG GCC TTG CTG
m a f p v t a l l l p l a l l
CTC CAC GCC GCC AGG CCG TAC CCA TAC GAT GTT CCA GAT TAC GCT
l h a a r p Y P Y D V P D Y A

[7] D. R. Littman, Y. Thomas, P. J. Maddon, L. Chess, and R. Axel, *Cell* **40,** 237 (1985).

G CTA GCG CTA CCG GAC TCA GAT CTC GAG CTC AAG CTT CGA ATT CTG CAG TCG AC

Nhe I *Eco*47 III *Bgl* II *Xho* I *Sac* I *Hin*dIII *Eco*RI *Pst* I *Sal* I

G GTA CCG CGG GCC CGG GAT CCA CCG GTC GCC ACC **ATG GTG.......** ⟶ **(EGFP)**

Kpn I *Sac* II *Apa* I *Xma* I *Bam* HI *Age* I

FIG. 1. Multiple cloning site of the ClonTech pEGFP-N1 vector. The multiple cloning site of the vector is shown with the first two codons of EGFP in bold. More information about pEGFP and other ClonTech vectors can be obtained at the ClonTech website http://www.clontech.com. Note that not all of the restriction enzymes that cut at the sites underlined are shown.

Construction of a GPCR–GFP Chimera

The following example describes the cloning of a FLAG-tagged β_2-adrenergic receptor (β_2AR) into the ClonTech pEGFP-N1 vector creating a FLAG-β_2AR-GFP chimera. Figure 1 shows the sequence of the multiple cloning site in pEGFP-N1.

To insert the β_2AR into pEGFP-N1, a β_2AR cDNA was initially amplified by PCR. The source plasmid for the receptor was a pCDNA3.1 clone that already contained an N-terminal signal sequence and FLAG tag (see Ref. 8). Because pCDNA3.1 contains a generic T7 sequence 5′ to the multiple cloning site, the T7 primer could be used for PCR amplification. The 3′ primer was a β_2AR-specific primer that was designed to delete the stop codon and add a *Sac*II restriction site. The sequences for these primers are as follows with the *Sac*II site underlined:

5′primer(T7): 5′TAA TAC GAC TCA CTA TA 3′

3′primer: 5′GTC CCC GCG GCA GCA GTG AGT CAT TTG TAC 3′

The general PCR conditions that can be used to amplify the β_2AR were described earlier. PCR products are digested with *Hin*dIII (site in the pCDNA3.1 polylinker 3′ to the receptor insert) and *Sac*II and ligated into pEGFP-N1, creating a β_2AR–GFP chimera. Because the template plasmid includes an N-terminal signal sequence and FLAG tag, the final construct includes both N-terminal FLAG and C-terminal GFP tags. Amino acids encoded by extra restriction sites in the multiple cloning site remain in the chimera, with the number being dependent on the particular sites used.

Expression of Receptor Constructs in Mammalian Cells

The following protocols have been applied to adherent cell lines. Cells transiently expressing tagged receptors are generally observed 24–48 hr

[8] L. Kallal, A. W. Gagnon, R. B. Penn, and J. L. Benovic, *J. Biol. Chem.* **273,** 322 (1998).

after transfection. The specific steps for transfection and observation are described.

Day 1: seed cells into 6-well plates or 35-mm dishes
Day 2: transfect (see later)
Day 3: split transfected cells and seed onto coverslips
Day 4: treat cells and prepare slides for microscopy
(For observation of cells 24 hr after transfection, cells are transfected on the morning of day 2 and split onto coverslips later on day 2. Microscopy is then carried out on day 3).

Transfection

Any commercially available transfection reagent can be used with cell lines that are transfected easily, such as HEK293 and COS. We routinely use Lipofect-AMINE (Life Technologies) or FuGene 6 (Roche) for both transient and stable expression of tagged receptors.

Plating Cells on Coverslips for Microscopy

Adherent cells are plated the day before the experiment is to be performed. They should be seeded to achieve a confluency of ~50% on the following day. For very adherent cell lines such as COS-1, it is generally not necessary to treat glass slides or coverslips. For less adherent cell lines such as HeLa or HEK293, one can prevent loss of cells from the coverslips during the fixing, antibody incubations, and washing steps by treating the slides or coverslips with poly-L-lysine before use. Other agents such as gelatin or collagen may also be used depending on the cell line.

Any sterile glass coverslip or slide can be coated with poly-L-lysine using this protocol. To render coverslips "sterile," soak them in 95% ethanol and dry in a tissue culture hood before use. We typically coat 12-mm circular coverslips for use in 24-well plates or 18 × 18 or 22 × 22-mm^2 coverslips for 6-well plates. It is suggested that No. 1.5 thickness coverslips be used for microscopy, but we have obtained similar results with No. 1 coverslips.

1. Place clean dry coverslips in wells of appropriate culture plates in a tissue culture hood.
2. Add enough sterile 0.1 mg/ml poly-L-lysine to each well to submerge the coverslip or glass surface (poly-L-lysine is stored frozen at −20° and can be reused two to three times if desired).
3. Aspirate the poly-L-lysine solution after ~5 min.
4. Wash once with sterile water (optional) and aspirate.
5. Allow to air dry.

Immunofluorescence with Anti-FLAG and Anti-HA Antibodies

The use of antibodies to detect GPCRs allows flexibility in methods, as one can choose to observe total cellular receptors or only cell surface receptors if the epitope tag is at the N terminus (i.e., extracellular). This is in contrast to the use of GFP chimeras, where all of the receptors in the cell are viewed during microscopy.

Detection of Total Cell Receptors with Antibodies

Cells plated on coverslips are treated as desired (e.g., incubation with agonist for 20 min at 37°) along with untreated controls for comparison. Conditions and media additives that are particular to each agonist should be used according to established protocols. After treatment, plates are removed from the incubator and subsequent work is done at the bench under nonsterile conditions.

In 24-well plates (using 12-mm circular coverslips), wash volumes should be approximately 0.5 ml/wash and antibodies prepared to allow 0.25–0.3 ml/well. To perform experiments in 6-well plates (18×18 or 22×22-mm^2 coverslips), volumes required are approximately 10-fold larger (2 ml for antibodies and 4–5 ml for washes). Cells should not be allowed to dry out during the procedure. All buffers should be prepared fresh except for phosphate-buffered saline (PBS), which can be stored for extended periods.

1. Aspirate media, rinse cells in PBS, and aspirate.
2. Add PBS/3.7% formaldehyde (fixing buffer).
3. Incubate at room temperature for 10–20 min.
4. Aspirate and wash three times in PBS/ 0.01–0.05% Triton X-100 (permeabilization buffer).
5. Add permeabilization buffer +1% BSA (blocking buffer) (note that 5% nonfat milk can be substituted for BSA) at 37° for 30–60 min.
6. Aspirate blocking buffer.
7. Add primary antibody to cover cells, antibody is diluted 1 : 150–1 : 1000 in blocking buffer. Antibodies are prepared immediately prior to use and prespun for 10 min at maximum speed in a microcentrifuge at room temperature (or larger volumes spun for 10 min, 8000–12,000 rpm in 10- to 40-ml tubes). The cleared antibody supernatant is transferred to a new tube and distributed into the desired experimental wells.
8. Incubate cells with antibody at 37° for 30–60 min (carried out in a humidified incubator to prevent sample drying).
9. Wash five times with permeabilization buffer, leaving the fourth wash on for 30 min at 37°.
10. Add blocking buffer, incubate for 10–15 min at 37°, and aspirate.
11. Add secondary antibody (prepared as described earlier), incubate for 30–60 min at 37°, and aspirate.

12. Wash five times with permeabilization buffer, leaving the fourth wash on for 30 min at 37°.

13. Add fixing buffer, incubate for 10–20 min at room temperature, and aspirate.

14. Wash twice with PBS and add a third aliquot of PBS to keep samples wet.

15. Mount slides or coverslips using the Slow-Fade mounting kit (Molecular Probes) following the manufacturer's instructions. Briefly, the kit provides an equilibration buffer and mounting buffers that contain agents that help preserve fluorophores and prevent photobleaching. Coverslips are removed from plates carefully with forceps, and the noncell side is gently wiped with a Kimwipe. Equilibration buffer is added dropwise to the cell side, excess is drained, and a drop of mounting buffer is added. The coverslip is turned over onto a glass slide, and excess buffer at the edges is removed by wicking with absorbant paper or with a 200-μl pipette tip attached to an aspirator. The edges of the coverslip are then sealed with nail polish, and the slides are observed after the nail polish has dried. Preferably, slides are observed the same day of preparation but can also be stored for several days at 4°. To preserve the three-dimensional character of cells, coverslips can be mounted on glass slides that have plastic rings or small drops of dried nail polish. This raises the coverslip and helps preserve cellular morphology.

Labeling of Cell Surface Receptors

In this protocol, receptors on the surface of live cells are labeled with primary antibody at 4° (a temperature that blocks receptor trafficking). Excess antibody is washed away, and cells are then warmed to 37° and treated. If the treatment results in receptor internalization, bound antibodies are internalized along with the receptor. After cells are fixed and permeabilized, one can visualize internalized receptors with secondary antibodies. This protocol offers the benefit of avoiding interference from receptors that were intracellular at the onset of the experiment.

1. Place cells on a bed of wet ice.
2. Rinse cells in ice-cold media containing 1% BSA.
3. Add ice-cold primary antibody solution prepared in media/1% BSA.
4. Incubate on ice or at 4° for 1 hr.
5. Rinse twice with cold media/1% BSA.
6. Add agonists/treatments in media/1% BSA (BSA optional).
7. Incubate at desired temperature and time (e.g., 20 min at 37°).
8. Rinse cells quickly twice in PBS at room temperature.
9. Fix in fixing buffer (PBS/3.7% formaldehyde) for 10–20 min at room temperature and aspirate.
10. Wash three times in permeabilization buffer.

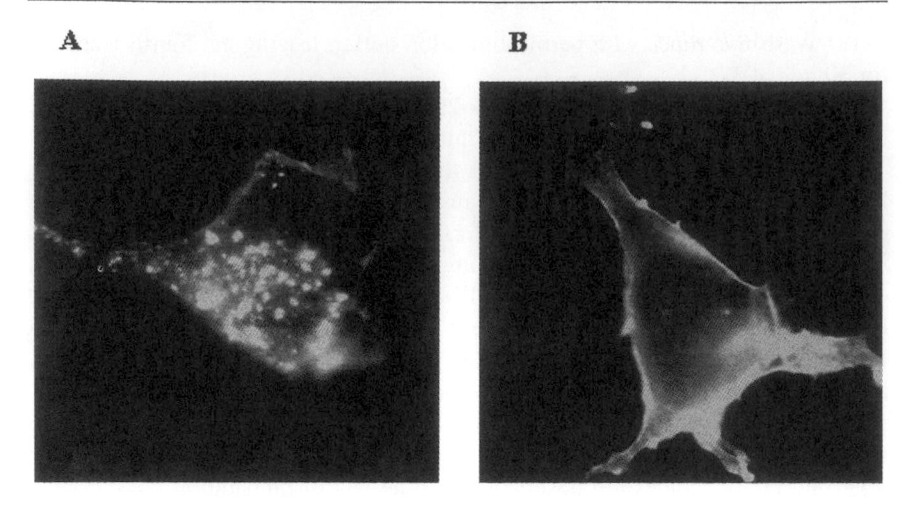

FIG. 2. HEK293 cells transiently transfected with signal-FLAG-SST2R were labeled using the cell surface labeling protocol. (A) Cells were treated with 1 μM somatostatin-14 peptide at 37° for 15 min, after surface labeling with anti-FLAG antibody at 4°. (B) Cells were left untreated after surface labeling. The M1 mouse monoclonal anti-FLAG antibody (5 mg/ml stock from Sigma) was used at 1 : 500 dilution, and the FITC-conugated goat antimouse antibody (2 mg/ml from Molecular Probes) was used at 1 : 200 dilution, as per the protocols described.

11. Incubate in blocking buffer for 30 min at 37° and aspirate.
12. Add secondary antibody, incubate for 30–60 min at 37°, and aspirate.
13. Wash five times in permeabilization buffer, leaving the fourth wash on for 30 min at 37°.
14. Incubate in fixing buffer for 10–20 min at room temperature and aspirate.
15. Wash three times in PBS and mount with Slow-Fade kit.

For untreated controls, some samples should be processed by skipping steps 6 through 8 so that only cell surface receptors are labeled. Additional controls should be included such as incubations at 37° in the absence of agonist and a negative control in which the primary antibody is not added. An example of an internalized (A) and cell surface (B) FLAG-tagged SST2 receptor is shown in Fig. 2.

Anti-FLAG and Anti-HA Primary Antibodies

Mouse monoclonal and rabbit polyclonal antibodies against both FLAG and HA epitopes are available from various companies. M1 and M2 monoclonal antibodies for FLAG tags (Sigma) and 12CA5 (Roche) and 101R (Babco) antibodies for HA are routinely used. We typically dilute M1 and M2 antibodies 1 : 500 for immunofluorescence.

Secondary Antibodies

We generally use FITC-, TRITC-, or ALEXA-fluor488- conjugated secondary antibodies (rabbit or goat antimouse or goat antirabbit) from Molecular Probes at dilutions of 1 : 150–1 : 500.

Fluorescence Microscopy with GPCR–GFP Chimeras

Fixing Cells to Observe GPCR–GFP Chimeras

Although a major advantage of using GFP is that one can observe receptors in living cells, for convenience the samples may also be fixed. In either case, the use of GFP still greatly simplifies the experiment, because there is no need for antibody labeling. Specimens may be observed quickly or mounted using mounting medium and sealed with nail polish to prolong viewing time. A simple protocol is as follows.

1. Treat cells previously seeded on coverslips with or without agonist as desired.
2. Aspirate media and rinse with PBS.
3. Fix cells in PBS/3.7% formaldehyde for 10–20 min at room temperature.
4. Wash three times with PBS.
5. Remove coverslip with forceps and gently dry bottom (side without cells) with a Kimwipe.
6. Mount coverslips with Slow-Fade kit.

To observe cells quickly, e.g., when screening clones during the selection of stable cell lines expressing GFP chimeras, one can skip fixation. Coverslips are simply rinsed in PBS, turned onto a glass slide, and observed directly under the microscope.

Live Cell GPCR-GFP Time Courses

Cells are transfected and prepared as per other protocols, growing the cells onto coverslips or slides. It is desirable to use an imaging chamber if real-time analysis of GPCR-GFP movement is being studied. We have employed an imaging chamber from Warner Instrument Corp, which is assembled using two 22 × 22-mm coverslips (one coverslip contains the cells and the other is left blank). The chamber is designed such that the coverslip with the cells can be placed on either the top or the bottom of the chamber to accommodate either an upright or an inverted microscope.

Imaging chambers generally allow treatments to be injected into the chamber either from a source reservoir using a pump or through a simple injection port on the chamber using a pipette. Injected media should be warmed to the desired temperature, and the entire stage and chamber should be temperature controlled during the course of the experiment. This can be accomplished by purchasing available temperature control devices for microscope stages and imaging chambers (e.g., Warner Instruments sells controllers for their imaging chambers). Alternatively, one can use a warm air flow over the microscope and stage, although this does not provide precise control of the temperature. The basic steps are as follows.

1. To reduce background/autofluorescence, prepare media and media/treatments without phenol red and reduce serum concentrations to 1% or less.
2. Mount the coverslips containing the cells into the imaging chamber, fill with warm media, and mount the chamber onto the microscope stage.
3. Focus the objective and locate a cell or group of cells to image.
4. Take a pretreatment image and start the time course by adding the agonist or treatment.
5. Take images at desired time points, closing shutters between images to minimize photobleaching.

Colocalization Studies

Whether a GPCR has been tagged with GFP or with epitopes, comparing its intracellular localization to that of other molecules of known localization is invaluable in deciphering GPCR trafficking patterns. One can use antibodies against proteins known to be localized in specific organelles or compartments or fluorescent compounds that accumulate in known subcellular locations. If two antibodies are used, one for the receptor tag and one for another cellular protein, it is important to use noncross-reacting systems. This requires the use of primary antibodies derived from different animal species, e.g., mouse and rabbit primary antibodies, and fluor-conjugated secondary antibodies that contain different fluorophores and do not cross-react with one another. When dual antibody systems are used, the primary and secondary antibodies may be mixed and centrifuged together following the protocols described.

With GFP-tagged GPCRs, the second fluor used in colocalization studies should be a red fluorescent compound. It can either be an antibody or a compound such as dextran or transferrin. For fluorescent compounds such as rhodamine-conjugated dextran or transferrin, labeling procedures are specific for each compound and should be followed according to the manufacturer's instructions. TRITC–transferrin is used commonly to label early endosomes and endocytic recycling compartments, whereas dextran is a good marker for late endosomes and lysosomes. A protocol for examining GPCR–GFP colocalization with TRITC–transferrin follows (see Ref. 8 for more details).

β_2AR-GFP Colocalization with TRITC–Transferrin in Endosomes

Cells are transfected and plated on coverslips as described previously.

1. Replace media with media lacking serum and incubate for 30 min at 37°.
2. Add 1 μM (-)isoproterenol (β_2AR agonist) and 20–200 μg/ml rhodamine–transferrin.
3. Incubate for 20 min* at 37° in a tissue culture incubator.
4. Rinse cells briefly three times with PBS at room temperature.
5. Fix in PBS/3.7% formaldehyde.
6. Rinse three times in PBS and mount with Slow-Fade mounting kit.

Microscope Considerations

The combined cell expression conditions and microscope system should be arranged to minimize exposure times because long exposures bleach the samples and might result in blurred images due to movement of the equipment and/or cells during the exposure. Short exposures are obviously more critical for live cell imaging than for fixed cell imaging. For live cell studies, exposures in the 50- to 100-msec range are achieved easily with available charged-coupled device (CCD) digital cameras. Paired with imaging software such as QED (QED Imaging Inc.), time course parameters can be specified and shutters opened and closed automatically during the time course. The time between each acquired image is dependent on how rapidly the receptor moves and how many total frames are to be collected. Only 5–10 consecutive images need to be collected during the treatment course if the goal is to display a short series of still shots. A short animation, however, contains at least 25–50 image frames, and more comprehensive biological "movies" contain several hundred frames. Because each frame generally requires 1–2 M-bytes of memory, computer systems attached to digital cameras must be equipped with sufficient data storage capabilities.

For fixed cell samples, it is acceptable to expose the sample for up to 10 sec or more to acquire an image, providing the equipment is in a relatively vibration-free environment. Acquiring images with traditional print or slide film generally requires longer exposures, whereas short exposures are achieved more easily with

*Incubation times for both the agonist treatment and the transferrin can be varied. When rhodamine–transferrin is incubated with cells for longer than 5 min, the endocytic recycling compartments accumulate the compound and become strongly visible. If only early endosomes are of interest, incubation time should be limited; the range of time points used most commonly for transferrin labeling falls between 5 and 30 min. Agonists and rhodamine–transferrin can be added separately so that the desired incubation end points coincide.

digital cameras. Digitally acquired images have the added advantage of being imported and manipulated easily in graphics software such as Photoshop.

For most mammalian cells, 40, 60, and 100X objectives will cover all applications for observing GPCRs by fluorescence microscopy. Oil-immersion or water-immersion objectives provide added light-gathering characteristics compared with nonimmersed objectives. There are numerous suppliers of microscope equipment and specific details on each system can be obtained from the vendors.

Satisfactory results for most studies may be obtained with regular fluorescence microscopy, more specifically defined as epifluorescence. However, confocal microscopy eliminates some of the interference from light outside the focal plane, generally producing a higher contrast image. Moreover, because focal planes are narrower in confocal microscopy, serial cross-section images of a cell can be acquired (a Z series). With multiple images at defined intervals, a crude three-dimensional image of a cell can be reconstructed. Confocal techniques also generally allow more detailed analyses of intracellular structures than classic fluorescence microscopy. The disadvantages of confocal microscopy are that the systems are more expensive, require more space, and require more training to operate.

[32] Measurement of Receptor Desensitization and Internalization in Intact Cells

By RICHARD B. CLARK and BRIAN J. KNOLL

Introduction

The goal of studying agonist-induced and heterologous receptor-level desensitization is to determine the extent to which the coupling efficiency of the receptor has been compromised and how uncoupling correlates with phosphorylation and internalization. A number of successful intact cell and cell-free approaches and methodologies are discussed. The primary emphasis in this article is on the β_2-adrenergic receptor (β_2AR) because of our familiarity with this system and because historically the β_2AR has been the archetype for studies of G protein-coupled receptor (GPCR) regulation. Discussion focuses on those methodologies used to elucidate mechanisms of desensitization that are clearly receptor level. Effects on downstream components are also considered because they must be evaluated if they confound interpretation of receptor-level events.

Intact cell approaches include the measurement of (i) the turnover of second messengers in intact cells, (ii) the change in receptor affinity for agonists by intact cell binding, and (iii) receptor internalization (endocytosis and recycling) by ligand

0076-6879/02 $35.00

binding to surface receptors and by related methods such as immunofluorescence. The intact cell approach can be used for the manipulation of protein kinase levels and other components that alter desensitization such as dominant-negative or wild-type arrestins and dynamin. In some instances, protein kinase or phosphatase inhibitors have been employed, and although use of inhibitors suffers from lack of specificity, it is the approach that may be of greatest use in the manipulation of desensitization in the treatment of disease. Selective use of antagonists and partial or negative agonists to alter desensitization has also been widely used.

The most useful and widely used techniques for the study of receptor-level desensitization of G_s- and G_i-coupled receptors are based on the assays of receptor stimulation or inhibition of adenylyl cyclase (AC) in cell-free preparations. Receptor movement to endocytic vesicles has also been examined using gradient separation of plasma membrane from endocytic vesicles. The study of receptor phosphorylation, while an indirect measure of desensitization, has received intense scrutiny, as protein kinase-mediated desensitization plays the major role in driving β_2AR desensitization. These approaches, as well as commentary on the use of stable and transient transfection, are considered.

Intact Cell Measurements of Receptor-Level Desensitization

Measurement of receptor desensitization by monitoring second messenger accumulation is ideal because it is performed in intact cells. For β_2AR stimulation of cAMP levels, any change in cAMP accumulation over a given time period is equal to the amount synthesized minus the amount lost by hydrolysis and egress. In equation form, this can be written as

$$d[\text{cAMP}]/dT = V_s - k_e[\text{cAMP}] \tag{1}$$

where the rate of loss of cAMP, k_e, is a first order function of cAMP concentration and the rate of synthesis is given by V_s. In theory the decrease in V_s with time of agonist treatment can be estimated from the time course of cAMP accumulation in conjunction with estimates of k_e.[1,2] However, there are several problems with this approach for estimating receptor desensitization. As we have shown, β_2AR activation of AC occurs by a collision-coupling mechanism (see discussion later and Fig. 1). In many cell types, agonist-induced desensitization causes an increase in the EC_{50} with little or no decrease in V_{max}. As a consequence, even a massive desensitization induced by a high concentration of a full agonist may cause no change in V_{max}.[2] k_e can be evaluated by addition of an antagonist to block synthesis with subsequent measure of cAMP decay. Confounding this method is the fact that

[1] R. Barber and T. J. Goka, *J. Cyclic Nucleotide Res.* **7,** 353 (1981).

[2] R. Barber, *Mol. Cell. Endocrinol.* **46,** 263 (1986).

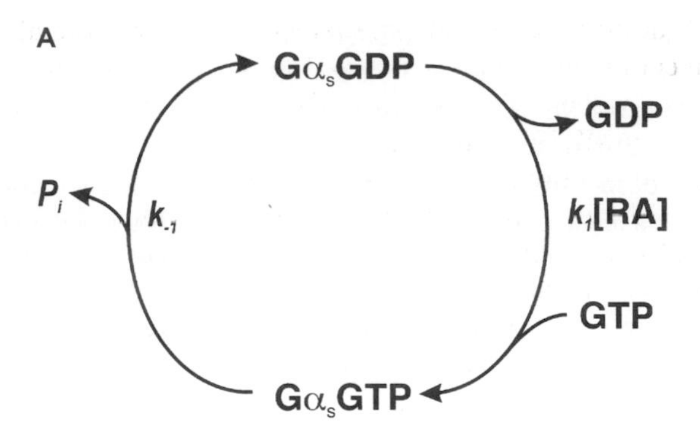

B

$$\frac{V_{\max}}{V_{100}} = \frac{\dfrac{k_1 r}{k_1 r + k_{-1}}[\mathrm{A}]}{\dfrac{k_{-1}}{k_1 r + k_{-1}} K_\mathrm{d} + [\mathrm{A}]}$$

FIG. 1. (A) Cassel–Selinger scheme for receptor activation of G_s protein, where k_1 is the rate constant for agonist receptor [AR] catalysis of GTP for GDP exchange and k_{-1} is the rate constant for the inactivation of G_s protein. (B) Equation describing this scheme from Whaley *et al.*,[15] where r is the total receptor present, [A] is the agonist concentration, K_d is the low-affinity agonist-binding constant (i.e., measured + GTP), V_{max} is the AC activity at any given agonist concentration, and V_{100} is the theoretical maximal activity if k_1 were infinitely large.

agonist treatment may activate phosphodiesterases (PDEs),[3] thus addition of an antagonist may eliminate PDE regulation. The measurement of the contribution of PDE regulation to overall desensitization has not received much attention; however, it may be a major determinant of desensitization in intact cells, greatly complicating the evaluation of receptor-level desensitization.

One approach to circumvent these problems in measuring receptor desensitization is to inhibit turnover of cAMP or IP_3 by blocking the relevant PDEs and phosphatases, respectively. The difficulty is that blockade of PDEs by inhibitors is rarely complete, and, unless inhibition approaches 100% of turnover, estimates of desensitization may be inaccurate. This problem is accentuated in cells that have a positively cooperative PDE such as mouse L cells[4] or in cells in which IP_3 turnover is very rapid.

[3] J. Seybold, R. Newton, L. Wright, P. A. Finney, N. Suttorp, P. J. Barnes, I. M. Adcock, and M. A. Giembycz, *J. Biol. Chem.* **273,** 20575 (1998).

[4] R. Barber, T. J. Goka, and R. W. Butcher, *Second Messengers Phosphoproteins* **14,** 77 (1992).

A second method used to estimate receptor-level desensitiztion is through the measurement of *initial rates* of second messenger production after agonist pretreatment and washout. The challenge of this method is that second messenger production (be it cAMP, IP_3, DAGs, or Ca^{2+}) from the desensitizing treatment must be distinguished from the test stimulus. If the washout phase requires extended time for removal of high-affinity agonist and decay of second messenger, receptors may resensitize significantly. Further, for many receptors, agonist-induced desensitization is so rapid that the measurement of initial rates must be accomplished within seconds after the addition of agonist. Application of this technique to the study of cholecystokinin receptor desensitization of IP_3 production has been demonstrated in pancreatic acinar cells.[5] As noted earlier, blocking agonist activation to allow decay of intracellular second messengers may prevent agonist regulation of turnover.

Another problem related to estimates of *receptor-level desensitization* based on intact cell second messenger production is that they do not distinguish effects of pretreatments on the downstream regulation of G proteins (e.g., by RGS protein stimulation of G protein off rates) and effectors (e.g., PKA and PKC regulation of AC and PLC). For these reasons, it is often necessary to employ cell-free assays to obtain an accurate measure of the relative contributions of downstream effects relative to receptor-related effects to overall desensitization.

Cell-Free Measurement of Receptor Desensitization after Intact Cell Treatment

Measurement of the desensitization of AC responses to G_s- or G_i-coupled receptors has generally been an accurate and feasible approach, with the limitation that there are few instances where cell-free measurements were shown to be in agreement with intact cell estimates. A problem with cell-free preparations is that if the homogenization procedure is too harsh or coupling too labile, receptor stimulation may be grossly uncoupled and the EC_{50} far right-shifted artifactually. There are many tissues, such as brain, heart, and smooth muscle, that exhibit enormous discrepancies between intact cell and cell-free EC_{50} values for agonist stimulation of AC. With the exception of a study by Martin and Harden,[6] cell-free assays of receptors coupled to $G_{q/11}$ have been consistently unsuccessful.

Cell Lysis

For receptors coupled to G_s or G_i, the first critical step is selecting a reproducible and gentle method for breaking cells without denaturing the system. For cultured

[5] R. V. Rao, B. F. Roettger, E. M. Hadac, and L. J. Miller, *Mol. Pharmacol.* **51,** 185 (1997).
[6] M. W. Martin and T. K. Harden, *J. Biol. Chem.* **264,** 19535 (1989).

cells this is not usually a problem, and cell lysis with a Dounce homogenizer in a hypotonic buffer (0–4°) containing 20 m*M* HEPES, pH 7.4, 1 m*M* EDTA (HE) is most reproducible. For the study of desensitization, we include a cocktail of both protease and phosphatase inhibitors in the lysis buffer.[7] It is important to assess the degree of cell lysis and the stability of nuclei. For the latter, inclusion of Mg^{2+} is often essential. Following cell lysis, the most active preparations are achieved if the homogenate is immediately placed on either a continuous sucrose gradient in HE buffer[8] or a simple step gradient (23%/43% sucrose in HE buffer) followed by ultracentrifugation at 45,000*g*.[7] This manipulation purifies membranes and quickly separates the cytosol from the crude plasma membrane fractions, and the use of 23% sucrose allows light vesicles (endosomes) containing internalized receptor to be included with the plasma membranes. Membranes at the 23/43% sucrose interface are removed and either further washed in HE buffer or flash frozen in liquid N_2 without removing sucrose. It is important to realize that agonist stimulation of AC activity in the cell lysate is very unstable, with a half-life of about 30 min, and that controls must be included at the beginning and end of pretreatment protocols.

Adenylyl Cyclase Assay

AC activity is assayed using a modification of a published method.[9] Membranes are thawed and diluted to a final protein concentration of 0.1–0.2 mg/ml to achieve 5–10 μg per reaction. Thawed membrane fractions must be used immediately and cannot be refrozen. Incubations are carried out at 30° for 10 min with 40 m*M* HEPES, pH 7.7, 6 m*M* $MgCl_2$, 1 m*M* EDTA, pH 7.0, 100 μM ATP, 1 μM GTP, 0.1 m*M* 1-methyl-3-isobutylxanthine, 8 m*M* creatine phosphate, 16 units/ml creatine phosphokinase, and 2 μCi of [α-^{32}P]ATP (30 Ci/mmol) in a total volume of 100 μl. ATP and GTP solutions are unstable and should be brought to pH 7.4 and 8.0, respectively, and stored frozen. EDTA should be neutralized prior to addition to incubation solutions. Each point is assayed in triplicate. For dose–response data, 6–10 concentrations of hormone bracketing the EC_{50} are used. The reaction is stopped by the addition of 0.5 ml of 5% TCA containing 1 m*M* cAMP ($\sim$10,000 cpm [^{3}H]cAMP/assay for estimating recoveries). The cAMP is purified using Dowex and Alumina columns as described previously,[10] and samples are counted in a dual-label scintillation counter. The EC_{50} and the V_{max} for hormonal activation are estimated using the algorithms in Prism 3 (GraphPad Software, Inc).

Most AC assays are not linear for greater than 8–10 min or at protein concentrations greater than 20 μg/100 μl. Methylxanthines are utilized to inhibit both

[7] B. January, A. Seibold, B. S. Whaley, R. W. Hipkin, D. Lin, A. Schonbrunn, R. Barber, and R. B. Clark, *J. Biol. Chem.* **272,** 23871 (1997).

[8] E. M. Ross, A. C. Howlett, K. M. Ferguson, and A. G. Gilman, *J. Biol. Chem.* **253,** 6401 (1978).

[9] Y. Salomon, C. Londos, and M. Rodbell, *Anal. Biochem.* **58,** 541 (1974).

[10] R. B. Clark, M. W. Kunkel, J. Friedman, T. J. Goka, and J. A. Johnson, *Proc. Natl. Acad. Sci. U.S.A.* **85,** 1442 (1988).

PDE activity and stimulation or inhibition by adenosine that is a contaminant of most ATP preparations and membrane vesicles. Creatine phosphate and creatine phosphokinase are used to regenerate ATP that is otherwise hydrolyzed rapidly by ATPase activity. For some cell lines, relatively low levels of Mg^{2+} ($\sim$0.3 mM) are necessary for the measurement of desensitization[11,12] and for hormonal inhibition of AC.[13,14] The free level of Mg^{2+} can be set by inclusion of appropriate levels of EDTA with the free level of Mg^{2+} calculated as described previously.[10]

Methods for the Quantitation of Desensitization/Coupling Efficiency

To quantitate desensitization accurately, it is imperative that the coupling efficiency of the desensitized receptor be compared with that of controls, and we have devised a straightforward analysis that accomplishes this. Oftentimes, adequate analysis of data is not performed, and many groups assume that measuring the decrease in efficacy (V_{max}) of hormonal stimulation alone is sufficient.

Formulations that accurately describe agonist/receptor stimulation of AC[15] for the majority of GPCRs and provide for the quantitation of desensitization are described later and in Fig. 1. In the model shown in Fig. 1, the rate of activation of AC is given by $k_1 r$, which we termed the coupling capacity, and the formulations shown later allow the comparison of this term for control and desensitized receptors. All that is required for this calculation is the measurement of the changes in the EC_{50} and the V_{max} for hormonal stimulation of AC after agonist treatment of cells. In addition to methods for calculating desensitization by comparing the ratio of coupling capacities, we also discuss the equations that predict the change in the EC_{50}/K_d ratio and V_{max} as a function of changes in $k_1 r$. A more detailed discussion of the model, the experimental evidence for it, and its relationship to the Furchgott model are covered in another review,[16] as well as in Whaley *et al.*[15]

Formulations That Allow Predictions of the Effects of Receptor Number

The equation in Fig. 1 describes a hyperbolic relationship between agonist concentration (A) and the fractional activation of AC. From this relationship, it can be shown that the V_{max} and the EC_{50} for hormonal activation as a function of

[11] M. W. Kunkel, J. Friedman, S. Shenolikar, and R. B. Clark, *FASEB J.* **3,** 2067 (1989).

[12] R. B. Clark, J. Friedman, J. A. Johnson, and M. W. Kunkel, *FASEB J.* **1,** 289 (1987).

[13] R. P. Xiao, X. Ji, and E. G. Lakatta, *Mol. Pharmacol.* **47,** 322 (1995).

[14] R. W. Hipkin, J. Friedman, R. B. Clark, C. M. Eppler, and A. Schonbrunn, *J. Biol. Chem.* **272,** 13869 (1997).

[15] B. S. Whaley, N. Yuan, L. Birnbaumer, R. B. Clark, and R. Barber, *Mol. Pharmacol.* **45,** 481 (1994).

[16] R. B. Clark, B. J. Knoll, and R. Barber, *Trends Pharmacol. Sci.* **20,** 279 (1999).

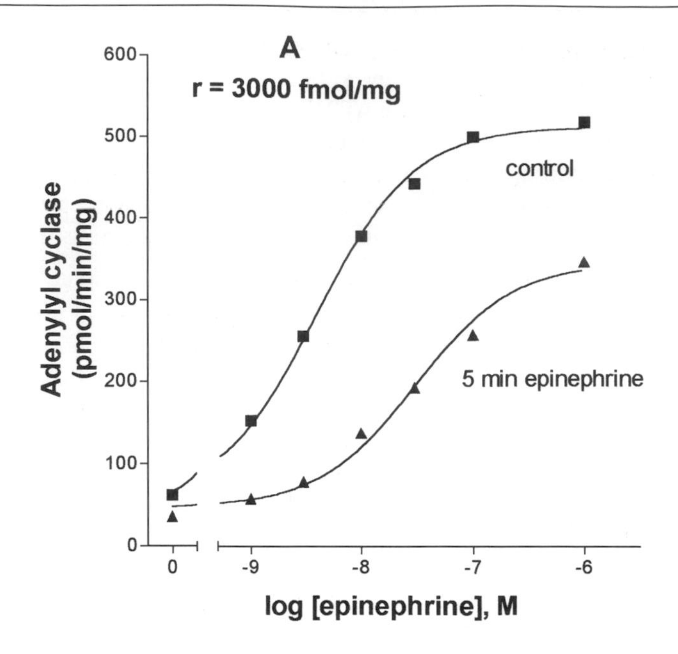
A
r = 3000 fmol/mg
Adenylyl cyclase (pmol/min/mg)
600
500
400
300
200
100
0
control
5 min epinephrine
0
-9
-8
-7
-6
log [epinephrine], M

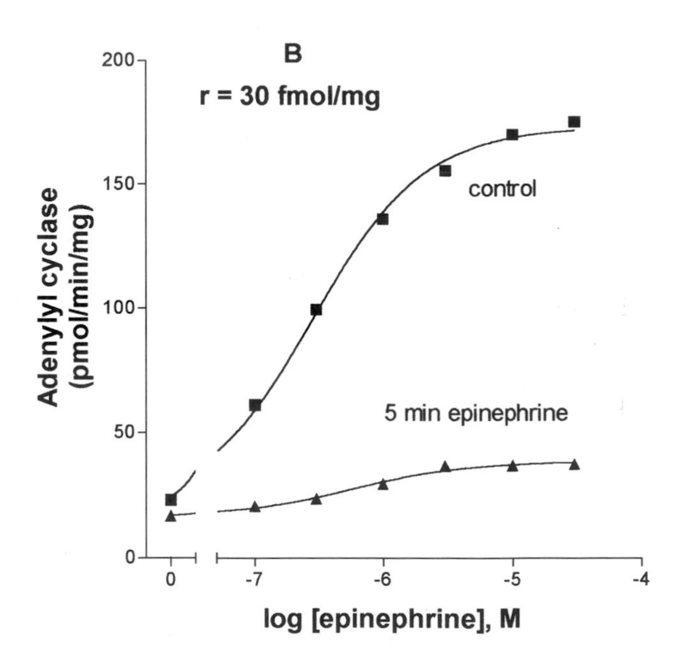
B
r = 30 fmol/mg
Adenylyl cyclase (pmol/min/mg)
200
150
100
50
0
control
5 min epinephrine
0
-7
-6
-5
-4
log [epinephrine], M

receptor level, r, are given by Eqs. (2) and (3):

$$V_{\max} = V_{100}(k_1 r / k_1 r + k_{-1}) \tag{2}$$

$$\mathrm{EC}_{50} = K_d(k_{-1} / k_1 r + k_{-1}) \tag{3}$$

where k_1 is the rate constant (coupling efficiency) for receptor activation of GTP for GDP exchange on the G protein, and k_{-1} is the first-order rate constant for inactivation. $V_{\max}$ is AC activity at any given concentration of agonist (A) and V_{100} is the theoretical maximum activity of AC if k_1 is infinite (in practice, this is typically the $V_{\max}$ for a full agonist when the $\mathrm{EC}_{50} \ll K_\mathrm{d}$). Equations (2) and (3) can be used to predict the changes in EC_{50} and $V_{\max}$ that occur with decreases in $k_1 r$ as a result of desensitization or experimental manipulation of r levels. Inspection of Eq. (2) and Eq. (3) shows that when the $\mathrm{EC}_{50} \ll K_\mathrm{d}$, as in a well-coupled system, a decrease in $k_1 r$ results in a large change in EC_{50} and almost no decrease in $V_{\max}$ [because the r term is only in the denominator of Eq. (3), but in both the denominator and the numerator in Eq. (2)]. Note also that Eq. (3) shows that as receptor levels approach zero, the EC_{50} approaches the K_d. When r is zero, the EC_{50} equals the K_d and of course there is no activation. It is important to realize that in well-coupled systems, receptor-level desensitization will be represented primarily through EC_{50} shifts initially, but that as desensitization proceeds, decreases in $V_{\max}$ will be significant. Decreases in $V_{\max}$ due to receptor-level desensitization are likely to be added to by downstream inhibition of G protein or effector function. For poorly coupled receptors or those expressed at very low levels, such as the endogenous level of β_2AR in HEK 293 or BEAS-2B cells (~30 fmol/mg), desensitization will be reflected almost entirely by changes in the $V_{\max}$.[17] Data illustrating these points are shown in Fig. 2. Figure 2A shows the shifts in EC_{50} and the $V_{\max}$ after 10 μM epinephrine-induced desensitization (5 min) of HEK 293 cells stably expressing high β_2AR levels (~3000 fmol/mg), and Fig. 2B shows similar data for HEK 293 cells expressing low levels of the endogenous β_2AR. Note that in the overexpressing cell line the EC_{50} is about 3 nM, whereas for the endogenous receptor the EC_{50} (~200 nM) approaches the K_d for epinephrine (~500 nM).

[17] B. January, A. Seibold, C. Allal, B. S. Whaley, B. J. Knoll, R. H. Moore, B. F. Dickey, R. Barber, and R. B. Clark, *Br. J. Pharmacol.* **123,** 701 (1998).

FIG. 2. Comparison of epinephrine-induced desensitization of HEK 293 cells expressing either very high or low levels of the β_2AR. (A) HEK 293 cells expressing either very high levels (3000 fmol/mg) of a double epitope-tagged wild-type β_2AR (HA-β_2AR-His$_6$)[7] or (B) very low levels (30 fmol/mg) of the endogenous wild-type β_2AR pretreated with carrier (■) or with 10 μM epinephrine for 5 min (▲). Membranes were prepared and assayed for AC activity in triplicate with the indicated epinephrine concentrations. EC_{50} values for epinephrine in A were 4 nM for control and 31 nM for epinephrine treated. For B, EC_{50} values were 300 nM for control and ~580 nM for the epinephrine treated. Decreases in $V_{\max}$ values for A and for B were 36 and 85%, respectively.

It should be emphasized that some receptors do not appear to follow the collision-coupling mechanism and neither desensitization nor decreasing receptor levels alter the EC_{50} for activation. It is possible that these receptors are either locked into their G proteins by virtue of a very high-affinity interaction or immobilized in close proximity to G proteins such that the classic mobility of the receptor in the membrane is not observed, a state referred to as private receptors. In these instances, receptors may exhibit a desensitization characterized by a decrease in V_{max} only with no change in the EC_{50}. In this circumstance the EC_{50} is likely to be identical to the K_d. Also, in some cases the EC_{50} for activation of effector is actually greater than agonist K_d, and this is not possible with the collision-coupling model in Fig. 1. One explanation for this is that there is high-affinity binding of agonist with the majority of receptors, but they are uncoupled from effector stimulation, which occurs with a relatively low affinity.

Formulations for Quantitation of Desensitization: Coupling Efficiency, Coupling Capacity, and Fraction Activity Remaining

In the model shown in Fig. 1 the coupling capacity for receptor activation of AC is defined by $k_1 r$. For the quantitation of desensitization, we have developed an equation that allows the calculation of the ratio of coupling efficiency of control and desensitized receptors [Eq. (5)]. This equation requires only the values for the EC_{50} and V_{max} obtained from the dose response for hormonal stimulation of AC in the desensitized (D) and naive (N) states. The derivation of this equation is as follows.

Dividing Eq. (2) by Eq. (3) and rearranging gives the following expression:

$$(k_1/k_{-1})r = (V_{max})(K_d)/(V_{100})(EC_{50}) \tag{4}$$

The term $k_1 r$ describes the receptor *coupling capacity*. The extent of desensitization is quantitated as the *ratio* of the coupling capacity in the desensitized state relative to the naïve state. Because the values for k_{-1}, K_d, and V_{100} do not change with desensitization of the β_2AR, this ratio is described as

$$(k_1 r)_D/(k_1 r)_N = (V_{max})_D(EC_{50})_N/(V_{max})_N(EC_{50})_D \tag{5}$$

The term $(k_1 r)_D/(k_1 r)_N$, the ratio of the coupling capacity of desensitized receptor to naïve, actually gives the fraction of receptor activity remaining relative to control. Typical data for the time course of epinephrine-induced desensitization of the overexpressed wild-type β_2AR in HEK 293 cells using this method of analysis are shown in Fig. 3. Note that when the fraction activity remaining is 0.1, the desensitization expressed as a percentage is 90%. A difference in fraction activity remaining from 0.1 to 0.03 means that there is actually a large difference in the extent of desensitization, i.e., three times more active receptor with the 0.1 versus the 0.03 levels. Use of percentage desensitization to express this result (90 versus 97% desensitization) would indicate wrongly that there is little difference in the extent of desensitization (i.e., a 7% difference). Use of Eq. (5) to calculate fraction

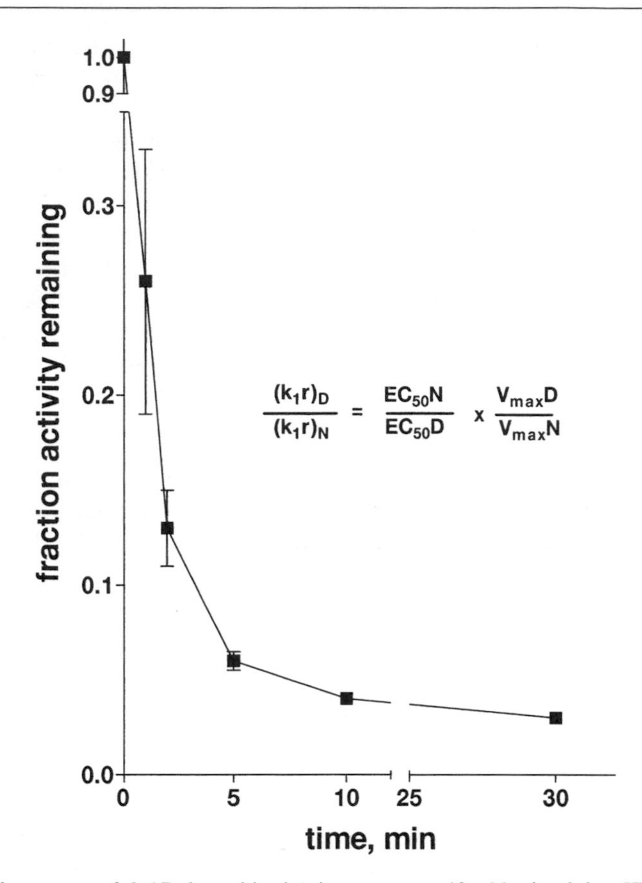

FIG. 3. Time course of β_2AR desensitization in response to 10 μM epinephrine. HEK 293 cells stably expressing the double epitope-modified β_2AR (HA-βAR-His_6) were pretreated with 10 μM epinephrine for various times from 1 to 30 min. Membranes were prepared and assayed for epinephrine-stimulated AC activity as described.[7] Values for fraction activity remaining were calculated according to Eq. (5). Each point was calculated using data from at least three experiments, each of which included separate AC dose–response curves for epinephrine-treated and carrier-pretreated (control) samples to determine the EC_{50} and V_{max}.

activity remaining for the experiment in Fig. 2 gives a value of 0.08 for both high and low receptor-expressing lines.

Calculation of Coupling Efficiency Ratios for Mutated Receptors

The generation of mutant receptors for desensitization studies may alter the coupling efficiency of the receptor. A method to quantitate any alterations in receptor coupling efficiency is extremely important. A useful formulation we have derived from the expressions given earlier is the coupling efficiency ratio as shown:

$$\text{coupling efficiency ratio} = k_1/k_{-1} = [K_d - EC_{50}]/EC_{50}r$$

From this formulation the coupling efficiency ratio can be calculated from the experimentally derived values for the K_d (+GTP) for agonist, the EC_{50} for its stimulation of AC, and the B_{max} for the expression level of the receptor, r, in the plasma membrane.[7,18,19] Comparing the values for the k_1/k_{-1} ratios for mutant versus wild-type receptors gives the ratios of their coupling efficiencies (with the assumption that k_{-1} is not altered by receptor mutations). We have shown that similar coupling efficiencies are obtained over a hundredfold range of receptor levels in mouse L cells,[15] and this is illustrated as well for HEK 293 cells in Fig. 2.

Methods for Evaluation of Downstream Effects on G_s/AC

It should be emphasized that the calculation of coupling efficiency following desensitization does not distinguish between receptor-level and downstream changes that are provoked by agonist treatment. For G_s/G_i-coupled receptors, estimates of receptor-level desensitization can, with some assumptions, be distinguished from the downstream regulation of G_s, G_i or effectors by measuring GTPγS, $(AlF_4)^{-1}$, and forskolin stimulation or heterologous activation of G_s by other receptors. For G_s-coupled receptors, G_i effects can be eliminated by pretreating intact cells with pertussis toxin. Regulation of G_i can be monitored by assaying the hormonal inhibition of AC or, in some instances, by using forskolin plus GTPγS (or GppNHp).[20]

A general problem in the use of nonhydrolyzable GTP analogs is that all G proteins will be activated, which can complicate interpretation. Also, forskolin synergizes with any agonist that is not removed from the pretreatment of cells and it activates all AC subtypes with the exception of type IX. We have found that synergy between forskolin and agonist is particularly acute when receptor expression is high and EC_{50} values are low, and even trace levels of a full agonist carried through the membrane preparation into the assay will trigger some synergism. Carryover of agonist may be caused by its penetration of cells during pretreatment.

There is now considerable evidence that β_2AR activates G_i as well as G_s in cardiac myocytes,[13] and this introduces still more uncertainty in the use of nonspecific interventions. In myocytes, treatment with β_2AR agonists may not only desensitize the receptor stimulation of G_s, but also promote desensitization by activating G_i. However, there is no evidence for involvement of G_i in agonist-induced desensitization of HEK 293 cells or of S49 lymphoma cells.[20]

[18] A. Seibold, B. G. January, J. Friedman, R. W. Hipkin, and R. B. Clark, *J. Biol. Chem.* **273,** 7637 (1998).

[19] B. S. Whaley, N. Yuan, R. Barber, and R. B. Clark, *Pharmacol. Comm.* **6,** 203 (1995).

[20] R. B. Clark, T. J. Goka, M. A. Proll, and J. Friedman, *Biochem. J.* **235,** 399 (1986).

Receptor Phosphorylation

Phosphorylation by protein kinases is thought to be the major driving force of desensitization for most receptors. Correlation of phosphorylation with measurements of desensitization and internalization offers important insights into the mechanisms involved. Although it is often assumed that the measurement of phosphorylation correlates with desensitization, this is clearly not the case for many receptors. This reflects the fact that phosphorylation is a complex dynamic process. For example, the β_2AR once internalized is likely dephosphorylated rapidly and yet this receptor remains uncoupled. Also, phosphorylation of receptors shows a complex dependence on agonist coupling efficiency and concentration.[7] For the β_2AR, phosphorylation and desensitization correlate best at very early times prior to significant internalization and subsequent turnover of the [^{32}P]β_2AR.

General Protocols for Measuring Phosphorylation

Early methods were geared toward the purification of endogenous β_2ARs from lung tissue and cell lines.[21,22] A major limitation was the low level of receptor in these tissues, requiring heroic measures and multistep purifications using an antagonist (alprenolol) affinity column[21,23] from which recovery was quite low. This problem has been circumvented by the use of either stable or transient overexpression of receptors in various cell lines. Also, epitope tagging of receptors is widely used because it allows use of commercially available monoclonal antibodies. With either overexpression or epitope tagging, extensive controls must be run to determine that they do not significantly alter either coupling efficiency or kinetics of phosphorylation and desensitization. We have found that HEK 293 cells demonstrate the same time course of epinephrine-induced desensitization over a 100-fold level of β_2AR expression. However, there are many cell lines, such as COS, where this is likely not the case because either GRKs or arrestin may be limiting.[24] In our hands, stable transfection of appropriate cell lines has been far superior to transient overexpression. Transient overexpression gives a high receptor number, but these receptors are, for the most part, uncoupled, perhaps a reflection of there being a small population of cells that are expressing a huge level of receptors. The use of transient transfection for studies of phosphorylation and internalization should always include controls to establish that coupling to stimulation of AC and subsequent desensitization kinetics are normal.

[21] R. H. Strasser, G. L. Stiles, and R. J. Lefkowitz, *Endocrinology* **115,** 1392 (1984).

[22] J. L. Benovic, L. J. Pike, R. A. Cerione, C. Staniszewski, T. Yoshimasa, J. Codina, M. G. Caron, and R. J. Lefkowitz, *J. Biol. Chem.* **260,** 7094 (1985).

[23] J. L. Benovic, R. G. Shorr, M. G. Caron, and R. J. Lefkowitz, *Biochemistry* **23,** 4510 (1984).

[24] L. Ménard, S. S. G. Ferguson, J. Zhang, F.-T. Lin, R. J. Lefkowitz, M. G. Caron, and L. S. Barak, *Mol. Pharmacol.* **51,** 800 (1997).

Another limiting factor in the measure of phosphorylation and its correlation to desensitization is that phosphorylation of most receptors involves multiple residues and protein kinases. Each phosphorylation may exhibit hierarchy and differential kinetics of phosphorylation and dephosphorylation, as well as differential effects on desensitization and other processes contributing to desensitization, such as arrestin binding and internalization. Hierarchical effects have been demonstrated convincingly for the CCK receptor.[5]

With the long-term goal of developing a system in which these important questions can be addressed, we constructed an epitope-tagged β_2AR [C-tail His_6 and N-terminal hemmagglutinin (HA) tag] and overexpressed it in HEK 293 cells. Several purification protocols have been developed to measure phosphorylation of the β_2AR, making use of various combinations of immunoprecipitation, lectin affinity columns, and Ni^{2+}-NTA and Co^{2+}-Talon resins. The use of epitope tags increases the number of possible purification strategies greatly and also allows the use of commercial antibodies and resins. Epitope tags on the N- and C-terminal tails do not alter the desensitization parameters relative to wild-type β_2ARs. However, we have found that the C-tail His_6 epitope blocks interactions with PDZ domain-containing proteins (Ruoho *et al.*, unpublished data).

Two types of procedures have been developed in our laboratory: one for purification of intact glycosylated β_2AR receptor as published in detail previously[18] and a second for the purification of deglycosylated β_2AR. Introducing a deglycosylation step into the purification scheme has considerably improved quantitation of phosphorylation and results in one tight band after SDS–PAGE at a molecular weight of 48,000-50,000 instead of the typical 20,000 spread of the glycosylated β_2AR on SDS gels from 60,000 to 80,000. It has also made it possible to obtain preliminary results with matrix-assisted laser desorption ionization/time of flight (MALDI-TOF) analysis. Each of these procedures for purification of the β_2AR incorporates two affinity steps for purification. The method employed most often for measuring phosphorylation of receptors is through a one-step immunoprecipitation with an antireceptor antibody; however, we have found that to be inadequate for both the somatostatin receptor and the β_2AR.[7,14] Basal activities are high and fold stimulations low if immunoprecipitation alone is employed.

An outline of the deglycosylation procedure currently in use is as follows (see details later). His_6-tagged β_2ARs are solubilized from intact cells with 2% *n*-dodecyl-β-D-maltoside (DβM) and purified by use of anti-C-tail antibody to the β_2AR (Santa Cruz) linked to agarose. β_2AR is eluted by a pH 2.0 glycine buffer containing 0.05% DβM. We use DβM for solubilization of the β_2AR in preference to digitonin because DβM is highly purified and shows no variation between lots and because it has a high CMC, making its removal easier. After neutralization the receptor is treated for 2 hr with *N*-glycosidase F and is then passed over either a Talon or a Ni-NTA resin. The β_2AR is eluted batch style with 100 m*M* imidazole, concentrated, and run on SDS–PAGE. Inclusion of the *N*-glycosidase

F treatment to strip glycosylated residues greatly improves quantitation of receptor phosphorylation and immunoblots.

Specific Protocols for Determination of β_2AR Phosphorylation

The methods used are modifications of those described previously.[7] Cells are grown to confluence in 100-mm dishes precoated with poly-L-lysine to improve adherence. Cells are rinsed once with phosphate-free Dulbecco's modified Eagle medium (DMEM) and then incubated with 0.5 mCi [^{32}P]orthophosphate in phosphate-free DMEM containing 1% fetal bovine serum (FBS) for 3 hr at 37° in 5% CO_2. The labeling medium is removed and replaced with 5 ml of bicarbonate- and phosphate-free DMEM containing 10% FBS. After 30 min of equilibration, cells are treated with 10 μM epinephrine or ascorbate/thiourea (AT carrier) for varying times. The medium is removed, and cells are rinsed with 5 ml ice-cold phosphate-bufferd saline (PBS) and then scraped into 3 ml of ice-cold PBS with 10 μg/ml leupeptin and 100 n*M* okadaic acid. Cells are collected by centrifugation at 600*g* for 5 min at 4°. The cell pellet is solubilized by vortexing in 100 ml of buffer containing 20 m*M* HEPES, pH 7.4, 300 m*M* NaCl, 0.8% DβM, 5 m*M* EDTA, 3 m*M* EGTA, 20 m*M* sodium pyrophosphate, 10 m*M* sodium fluoride, 25 m*M* imidazole, 10 μg/ml benzamidine, 10 μg/ml trypsin inhibitor, 100 n*M* okadaic acid, 10 μg/ml leupeptin, and 14 m*M* 2-mercaptoethanol. After 30 min rocking at 4°, the solubilized cells are centrifuged for 30 min at 45,000 rpm in a Beckman 50 Ti rotor.

Equal amounts of solubilized β_2AR from control and epinephrine-treated cells are applied to a 100-μl packed volume of antibody resin in a column (#SC-569 anti-β_2AR C-tail antibody from Santa Cruz coupled to agarose) that has been prewashed with PBS, pH 7.0. β_2AR levels in the extracts are determined using a new procedure we developed as discussed later. After recycling the extract three times through the column, it is washed once with 3 ml of 10 m*M* phosphate buffer, pH 6.8, containing 0.05% DβM. The β_2AR is eluted with 1 ml of 100 m*M* glycine buffer, pH 2.5, plus 0.05% DβM, and the eluate is collected in 0.3 ml 1 *M* phosphate buffer, pH 8.0, for neutralization. The eluate is digested with 1500 units of *N*-glycosidase F (New England Biolabs) for 2 hr at 37° and is applied to a Clontech Talon Co^{2+}-carboxymethylaspartate-agarose column (0.5 ml packed resin) that has been prewashed with Talon buffer (0.05% DβM, 20 m*M* HEPES, pH 7.4, 150 m*M* NaCl) and recycled through the resin twice. The column is washed twice with Talon buffer, once with 4 ml of 10 m*M* imidazole, and finally with 0.25 ml of 20 m*M* imidazole, all in the same buffer. The β_2AR is eluted with 0.75 ml of Talon buffer containing 100 m*M* imidazole and concentrated to 50 μl in a Centricon (Amicon, 30-kDa cutoff). 2X-SDS sample buffer is added to the 50 μl of eluate and heated at 60° for 15 min (there is extensive cross-linking of the receptor if samples are heated to 100°). Samples are run on SDS–PAGE (12% gel) and transferred to nitrocellulose.

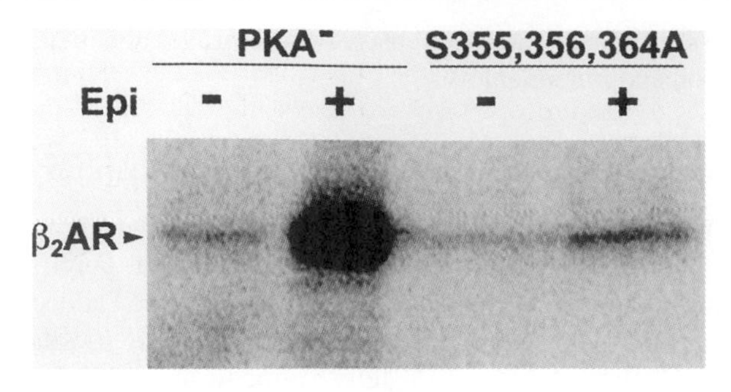

FIG. 4. Phosphorylation of PKA and S355,356,364A mutants in response to a 2-min pretreatment with 10 μM epinephrine. Cells expressing the PKA (a mutant of the β_2AR with both PKA consensus site serines substituted with alanines) or the S355,356,364A (the PKA mutant with additional alanine substitutions of serines 355, 356, and 364) were labeled with [^{32}P]orthophosphate for 3 hr and then exposed to either 10 μM epinephrine (+) or carrier (−) for 2 min. The cells were solubilized, and the receptor protein was purified using antibody, deglycosylation with PNGase F, Talon chromatography, and SDS–PAGE as described in the text. The β_2AR was transferred to nitrocellulose membranes, and phosphorimager analysis was performed as shown. Immunoblots performed on the same membranes using antibody directed against the HA epitope in the β_2AR amino terminus (2) showed that equal levels of receptor were recovered for control and epinephrine-treated samples (data not shown).

Phosphorimager analysis is performed on the PVDF or nitrocellulose membranes using a Molecular Dynamics Storm Phosphorimager Model 860 and ImageQuant software. Immunoblotting is performed using an anti-HA antibody (12CA5 or mHA.11) or the anti-β_2AR C-terminal antibody (Santa Cruz) as described previously.[18] After incubation with the secondary antibody (goat antirabbit-HRP), enhanced chemiluminescence is performed. A typical phosphorylation result is given in Fig. 4, which shows the effect of a 2-min treatment with 10 μM epinephrine of cells expressing mutant β_2ARs.

The procedure can be scaled up to obtain material pure enough for mass spectrometry. Briefly, ten 150-mm dishes of HEK 293 cells expressing 5–10 pmol β_2AR/mg membrane protein are extracted and purified through the two-step procedure outlined earlier. The receptor is identified on SDS gels by Coomassie stain and by immunoblots on parallel gels, and the protein is estimated by comparison of the staining with known amounts of standard (a tyical yield is 200–400 ng of β_2AR). The method for proteolysis of the receptor and preparation for MALDI-TOF and electrospray are exactly as published previously by Zhang *et al.*[25] In this procedure, the β_2AR band is excised and treated with trypsin to generate fragments, which are then extracted from the gel. An aliquot of the peptide extract is treated further with phosphatase to locate phosphorylated peptides by band shifts of 80 kDa, or multiples thereof.

[25] X. Zhang, C. J. Herring, P. R. Romano, J. Szczepanowska, H. Brzeska, A. G. Hinnebusch, and J. Qin, *Anal. Chem.* **70,** 2050 (1998).

We have also devised a procedure for measuring receptor levels after solubilization. This procedure makes use of either receptor glycosylation or the HIS_6 epitope tag. Solubilized receptor is incubated with [^{125}I]ICYP for 30 min in the presence of Ni^{2+}-NTA resin (and/or lectin affinity resin for glycosylated receptor) in microfuge tubes in Tris/Mg^{2+} buffer (as given later), pelleted, and washed three times with the same buffer in the cold. The bottoms of the tubes with pelleted resin are then cut off and γ counted.

Measurement of Receptor Levels in Intact Cells and Membranes

Cells are cultured in 12-well dishes for measurement of intact cell receptor number by [^{125}I]iodocyanopindolol ([^{125}I]ICYP) binding. The radioligand is prepared as described previously.[26,27] The cells are rinsed in serum-free DMEM (SF-DMEM) and then removed from the plates by pipeting up and down with 0.5 ml of SF-DMEM. Aliquots (25–50 μl containing 0.1–1 μg of protein) of the resuspended cells are used in triplicate-binding reactions, each containing about 200 p*M* [^{125}I]ICYP. Nonspecific binding is measured in triplicate reactions containing 1 μM alprenolol. Binding is performed on ice for 50 min and terminated by the addition of 2.5 ml ice-cold 50 m*M* Tris–HCI, pH 7.5, 10 m*M* $MgCl_2$. The [^{125}I]ICYP-bound β_2AR is collected by filtration through Whatman GF/C filters. The filters are rinsed three times with 2.5 ml of cold binding buffer and then γ counted. Receptor levels in cell membranes are measured in binding reactions using 2–5 μg membrane protein, 0.1 m*M* phentolamine, 40 m*M* HEPES, pH 7.2, 2 m*M* EDTA, 0.2 m*M* ascorbate, 2 m*M* thiourea, and about 200 p*M* ^{125}ICYP. Nonspecific binding is determined in the presence of 1 μM alprenolol. Reactions are performed at 30° for 50 min and terminated as described for the intact cell-binding assay.

Measurement of Equilibrium-Binding Constants for ICYP and Epinephrine

K_d values for [^{125}I]ICYP and epinephrine are determined for each of the mutant β_2ARs as described previously,[7,17] with a range of 1–150 p*M* [^{125}I]ICYP concentrations. Reactions are performed in triplicate with 1 μg of membrane protein (for β_2AR-overexpressing cell lines), and nonspecific binding is measured by the inclusion of alprenolol at 1 μM. The K_d is estimated by fit of data to a rectangular hyperbola. Assays to measure epinephrine K_d included 40–50 p*M* [^{125}I]ICYP, 10 μM GTPγS, and concentrations of epinephrine ranging from 0.1 to 100 μM. Reactions are performed with triplicate points using 1 μg of membrane protein. The algorithms in Prism (GraphPad Software, Inc.) are used to fit data to a one component sigmoidal curve with a Hill coefficient of -1. The K_d values are calculated using the Cheng–Prusoff correction.

[26] K. Barovsky and G. Brooker, *J. Cyc. Nucleic Res.* **6,** 297 (1980).

[27] D. Hoyer, E. E. Reynolds, and P. B. Molinoff, *Mol. Pharmacol.* **25,** 209 (1984).

Receptor Internalization

Internalization is a measure of the fraction of total receptors inside the cell, which is a balance between the rates of endocytosis and recycling (reinsertion of the receptor in the plasma membrane). The contribution of internalization to receptor desensitization is dependent on whether the internalized receptor is uncoupled and the rate and extent of rapid recycling. The degree of desensitization due to internalization is a factor of the relative rates of uncoupling and recycling, and the modeling of these processes has been discussed in several recent reviews.[16,28] For the β_2AR, the rates of endocytosis and recycling are quite similar; therefore, internalization is a major contributor to desensitization. Further, internalization is dependent on agonist concentration and on the time of exposure to agonist. For the β_2AR there is a pronounced lag in internalization relative to desensitization and phosphorylation. As the fraction of receptors internalized increases with time of agonist treatment, the contribution of internalization to overall desensitization increases substantially. Internalization is affected dramatically by the efficacy of the agonist used, i.e., partial agonists cause less internalization than strong agonists.[7,29] In contrast, receptor recycling appears to be unaffected by agonist efficacy.[29,30] Further, at low concentrations of full agonists, there is desensitization and, with extended treatment times, down regulation without detectable internalization,[31,32] as the EC_{50} for internalization (e.g., for epinephrine) is quite high relative to activation of AC[16] and down regulation.

Protocols for Measuring Receptor Internalization

Several techniques can be used, some of which are more quantitative than others and some more revealing of where receptors are in terms of their subcellular trafficking. For the β_2AR, the first GPCR to be studied in this regard, there are at least five methods used to measure internalization. Historically, the first technique used was isolation of light vesicles in density gradients that presumably reflected the movement of receptors into endosomes.[33] Another technique that remains useful is the measurement of the loss of agonist affinity in intact cells,[34] which seems to correlate well with other measurements of internalization. It may also be useful for distinguishing different stages of internalization, i.e., sequestration

[28] J. A. Koenig and J. M. Edwardson, *Trends Pharmacol. Sci.* **18,** 276 (1997).

[29] K. J. Morrison, R. H. Moore, N. D. V. Carsrud, E. E. Millman, J. Trial, R. B. Clark, R. Barber, M. Tuvim, B. F. Dickey, and B. J. Knoll, *Mol. Pharmacol.* **50,** 692 (1996).

[30] P. G. Szekeres, J. A. Koenig, and J. M. Edwardson, *Mol. Pharmacol.* **53,** 759 (1998).

[31] M. A. Proll, R. B. Clark, T. J. Goka, R. Barber, and R. W. Butcher, *Mol. Pharmacol.* **42,** 116 (1992).

[32] B. R. Williams and R. B. Clark, *Mol. Pharmacol.,* in press.

[33] T. K. Harden, C. U. Cotton, G. L. Waldo, J. K. Lutton, and J. P. Perkins, *Science* **210,** 441 (1980).

[34] M. L. Toews, G. L. Waldo, T. K. Harden, and J. P. Parkins, *J. Biol. Chem.* **259,** 11844 (1984).

of receptors from radioligand at the plasma membrane as opposed to actual movement of receptors into endosomes.[35]

Measurement of β_2AR Internalization by Hydrophilic Radioligand Binding

The most widely used technique is to measure the loss of surface receptors using a relatively nonpermeable radioligand such as [^{3}H]CGP 12177 for the β_2AR.[36] If binding is performed at 0–4°, little of the radioligand penetrates cells. It is important to use a sufficiently high concentration such that binding is saturated ($K_d \sim 0.5$ n*M*). At 5–10 n*M* it takes less than 1 hr to reach equilibrium, whereas longer incubation (e.g., overnight) tends to allow ligand penetration of cells and diminishes artifactually the number of internal receptors. For high-affinity radiolabeled peptide agonists, such as somatostatin or vasopressin,[14,37] one can follow the movement of the ligand from the surface by acid-washing techniques that distinguish surface and internal ligand if proper controls are included for destruction of the ligand both at the surface and once internalized.

Cells are grown in 12-well cluster dishes that can be precoated with poly-L-lysine to aid cell adhesion (this may not be necessary for cells that adhere tightly to plastic). The cells are treated with either carrier or carrier plus agonist from 100 × stocks. The final concentration of the carrier components is 0.1 m*M* ascorbate and 1 m*M* thiourea (AT), pH 7.0. After various times at 37° in 5% CO_2, media are removed and cells are washed six times with ice-cold SF-DMEM. To each well, 1 ml of cold SF-DMEM containing 10 n*M* [^{3}H]CGP-12177 is added and incubations are performed on ice for 1 hr. The assays include triplicate wells for each time point, and nonspecific binding is determined by the inclusion of 1 μM alprenolol. Assays may include 0.2% digitonin during the radioligand incubation to measure total receptor number (internal plus surface). After incubation, the radioligand is removed and the wells are washed twice with ice-cold PBS. The cells are scraped into 0.5 ml of trypsin, and liquid scintillation counting is performed. Internalization data are plotted as the fraction of surface receptor number measured in carrier-treated samples as a function of time after agonist addition.

Measurement of β_2AR Internalization by Antibody Binding

If an antibody is available that binds a receptor extracellular epitope, quantitative analysis of antibody binding can be used to measure the loss of surface

[35] J. F. Wang, J. L. Zheng, J. L. Anderson, and M. L. Toews, *Mol. Pharmacol.* **52,** 306 (1997).

[36] M. Stachelin and C. Hertel, *J. Receptor Res.* **3,** 35 (1992).

[37] G. Innamorati, H. Sadeghi, N. T. Tram, and M. Bimbaumer, *Proc. Natl. Acad. Sci. U.S.A.* **95,** 2222 (1988).

receptors. Quantification of surface antibody binding is commonly done by flow cytometry. In the basic protocol using N-terminal HA-tagged β_2ARs expressed in HEK293 cells,[29] cells treated with agonist are chilled and suspended in ice-cold PBS at a concentration of 10^6 cells/ml with 15 μg/ml of monoclonal antibody (12CA5 or mHA.11) or a nonspecific mouse IgG to define background fluorescence. After 60 min on ice, cells are centrifuged at 500*g* for 5 min and washed twice by resuspending in ice-cold PBS and centrifuging at 500*g*. The cells are then incubated with 15 μg/ml of goat or rabbit FITC antimouse IgG for 60 min on ice, washed, and fixed with 1% paraformaldehyde (Electron Microscopy Sciences, Ft. Washington, PA) in PBS. The quality of the paraformaldehyde is critical for reproducibility of the measurements. Surface fluorescence is quantified by flow cytometry using an argon laser tuned to 488 nm. Green fluorescent events are accumulated on a 4 decade logarithmic scale. Background fluorescence with nonspecific IgG is usually less than 10% the fluorescence level obtained with 12CA5 and is subtracted to obtain a specific fluorescence level. In some protocols, antibody is prebound to receptors prior to agonist treatment[29]; however, caution must be taken that the antibody itself does not trigger endocytosis.[38] In addition, receptors bound with antibody may be sorted rapidly to lysosomes following endocytosis.[39] This procedure works well with HEK 293 cells, which can be removed from the monolayer without the use of trypsin, which might digest the epitope.

One potential difficulty with the use of an N-terminal epitope tag is the possibility that it alters trafficking, as early studies of the β_2AR demonstrated that modification of the N terminus (by glycosylation defect) can decrease surface expression of the receptor.[40] Therefore, it is always possible that trafficking might be altered, although we and others have not found that to be a problem with the β_2AR. This technique yields results comparable with those obtained using [^{3}H]CGP-12177 binding.[29,41] Antibody-binding methods are especially useful for the quantification of receptors when internalization is triggered by a high-affinity ligand that is difficult to wash from cells.

Detecting β_2AR Internalization by Immunofluorescence Microscopy

Internalized receptors are detectable using specific antireceptor antibodies or by the use of epitope-tagged receptors. Cells are plated on No. 1 glass coverslips

[38] C. P. Petrou, L. Chen, and A. H. Tashjian, Jr., *J. Biol. Chem.* **272,** 2326 (1997).

[39] G. Raposo, I. Dunia, C. Delavier-Klutchko, S. Kaveri, A. D. Strosberg, and E. L. Benedetti, *Eur. J. Cell Biol.* **50,** 340 (1992).

[40] E. Rands, M. R. Candelore, A. H. Cheung, W. S. Hill, C. D. Strader, and R. A. F. Dixon, *J. Biol. Chem.* **265,** 10759 (1990).

[41] R. H. Moore, H. S. Hall, J. L. Rosenfeld, W. Dai, and B. J. Knoll, *Eur. J. Pharmacol.* **369,** 113 (1999).

(coated if necessary to ensure adhesion) in 35-mm culture dishes and grown to 50–80% confluency prior to agonist treatment. Following a wash with PBS containing 0.12% sucrose (PBSS), the cells are fixed at 4° for 10 min with PBSS containing 4% paraformaldehyde (cells may be left in fixative overnight at 4°). The following steps are done at room temperature. The fixed cells are incubated in 0.34% L-lysine, 0.05% sodium-*m*-periodate in PBSS for 20 min, washed and permeabilized with 0.2% Triton X-100 for 5 min, and then blocked for 15 min with 10% heat-inactivated goat serum (HIGS). (Lysine and periodate may contribute to preserving antigen and cell structure[42]; however, they may be left out.) Primary antibody is diluted to an appropriate concentration in PBSS with 0.2% HIGS and 0.05% Triton X-100, added to the coverslip, and left for 1–12 hr. The cells are washed four times with PBSS, treated with fluorescent secondary antibodies, diluted to 5 μg/ml in PBSS with 0.2% HIGS and 0.05% Triton X-100, and left overnight in the dark. Double labeling (e.g., to localize receptors with other intracellular proteins) is possible if the two antibodies are from different species (e.g., rabbit and mouse) and the secondary antibodies are affinity purified and highly species specific. Primary antibodies labeled directly with fluorescent tags can also be used, although the fluorescent signal can be very weak. A useful adjunct technique is antibody uptake, where antibody to an extracellular receptor epitope is added to cells during agonist treatment at 1–5 μg/ml. The antibody can be tagged directly or detected with a secondary antibody after fixation and cell permeabilization. As mentioned previosly, care must be taken that the bound antibody does not trigger endocytosis or alter intracellular trafficking.

To control for nonspecific binding, cells that are not expressing the receptor can be used. A further control is to leave out the primary antibody to check for noise caused by the fluorescent secondary antibody. The concentration of primary and secondary antibody to be used must be determined empirically to achieve an optimal signal-to-noise ratio. An important control when doing double labeling is to assess "bleed-through," the artifactual detection of one fluorophore using a filter specific for another. This can be a problem if the signal from one protein is much greater than the signal from the other, whether due to overexpression or to higher antibody affinity. To control for bleed-through, cells are single labeled, e.g., with FITC, and then imaged with the Texas Red filter.

Imaging of the cells is done most effectively using confocal or reconstruction microscopy. Confocal instruments, which use a pinhole aperture to remove out-of-focus light, are now widely available and can provide optical sections as small as 500 nm. Reconstruction instruments use computational methods to reconstruct images using out-of-focus light, have increased sensitivity over confocal instruments, and can resolve optical sections to 150 nm.[43] By either of these approaches, a focal plane through the cell center will show surface receptors in a tight

[42] I. W. McLean and P. K. Nakane, *J. Histochem. Cytochem.* **22,** 1077 (1974).
[43] D. A. Agard and J. W. Sedat, *Nature* **302,** 676 (1983).

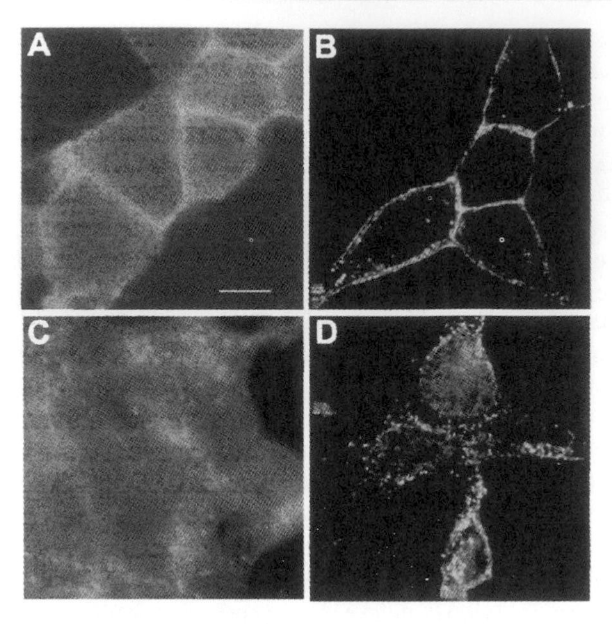

FIG. 5. HEK 293 cells stably expressing human β_2ARs were fixed, permeabilized, and labeled with a rabbit polyclonal antibody against the C-terminal 15 amino acids of the receptor (Santa Cruz). (A and B) Control cells. (C and D) Cells treated with 5 μM isoproterenol for 15 min prior to fixation. (A and C) Images acquired by conventional immunofluorescence microscopy. (B and D) Images acquired by deconvolutional immunofluorescence microscopy (Delta Vision, Issaquah, WA) with approximately 1 μm sections. Bar: 10 μm.

peripheral pattern, limited to the plasma membrane, whereas in agonist-treated cells, receptors will show a pronounced punctate appearance. Using conventional immunofluorescence, the surface receptor pattern will be more diffused, but in agonist-treated cells, a punctate distribution will be noticeable (Fig. 5). A problem when using conventional immunofluorescence is that a punctate appearance of labeling may be caused by receptor clustering on the cell surface rather than receptor internalization[44]; therefore, appropriate controls must be done to distinguish these. For instance, internal receptors will not be labeled efficiently if an antibody against an intracellular epitope is used and the cells are not permeabilized before the addition of antibody.

Recycling

Receptors may be resensitized while trafficking through endocytic compartments en route to the plasma membrane, thus knowing the recycling rate

[44] M. von Zastrow and B. K. Kobilka, *J. Biol. Chem.* **269,** 18448 (1994).

constant is important in the study of resensitization. Also, receptor internalization is a function of both receptor endocytosis and recycling, and both rates must be determined. For example, what appears to be very rapid endocytosis of a receptor may actually be a reflection of very slow recycling. Several GPCRs, such vasopressin and thrombin, are known to undergo very limited recycling.[37,45]

Protocols for Measuring Recycling

To measure the rate at which internalized β_2ARs return to the cell surface, agonist-treated cells are washed thoroughly and then allowed to incubate at 37° for varying times before chilling for radioligand binding with [^{3}H]CGP-12177. In one protocol,[29,46] cells growing in monolayers are treated with agonist until there is steady-state internalization, chilled and washed three times with SF-DMEM, and finally suspended into chilled SF-DMEM. To initiate recycling, the cells are rapidly pipetted into 37° medium. Samples are then withdrawn at intervals, pipetted into tubes containing ice-cold SF-DMEM with 10 n*M* [^{3}H]CGP-12177, and kept on ice for 60 min. Bound ligand is measured by filtration through glass fiber filters (GE/C) followed by scintillation counting. Nonspecific binding of radioligand is determined by additional incubations in the presence of 1 μM alprenolol. In a second protocol, recycling is measured using cells growing in cluster wells, a method that is preferable because there is less manipulation and no chilling of the cells. Chilling disrupts microtubules, which may play a role in the intracellular trafficking of GPCRs.[47] Cells growing in 12-well clusters are treated with agonist, washed three times rapidly with 2 ml of warm medium containing 10% FBS, and incubated in the same medium for varying times between 0 and 60 min to allow recycling. Recycling is stopped by removal of medium and two washes with ice-cold PBS, followed by [^{3}H]CGP-12177 binding as described earlier.

It is important to start the recycling with cells that have receptors internalized to steady state; thus internalization time courses should be performed first to determine the optimal time of agonist exposure. To ensure removal of agonist after internalization, lower concentrations may be employed, as long as steady-state internalization is achieved. Early literature suggests performing β_2AR recycling at 37° in the presence of [^{3}H]CGP-12177[36]; however, increased binding at longer times of incubation at 37° can occur even in the absence of recycling.

[45] J. A. Hoxie, M. Ahuja, E. Belmonte, S. Pizarro, R. Parton, and L. F. Brass, *J. Biol. Chem.* **268,** 13756 (1993).

[46] C. Hertel and M. Staehelin, *J. Cell Biol.* **97,** 1538 (1983).

[47] T. Drmota, G. W. Gould, and G. Milligan, *J. Biol. Chem.* **273,** 24000 (1998).

Derivation of Rate Constants

k_e

For the β_2AR, steady-state internalization is reached within 10 min after agonist addition. The steady state is dynamic, reflecting both endocytosis and recycling. The rate at which the steady state is attained depends on rate constants for endocytosis (k_e) and recycling (k_r). Internalization curves can be modeled to obtain estimates for both these parameters, and if necessary, recycling can be measured separately to obtain an independent estimate for k_r. β_2AR internalization curves fit an equation that assumes two pools of receptors, one internal and one external,[29] shown in Fig. 6. Nonlinear regression using this equation yields unique values for

A

$$\frac{R_s}{R_{s0}} = \frac{k_r(1 - e^{-(k_e+k_r)t})}{k_e + k_r} + e^{-(k_e+k_r)t}$$

B

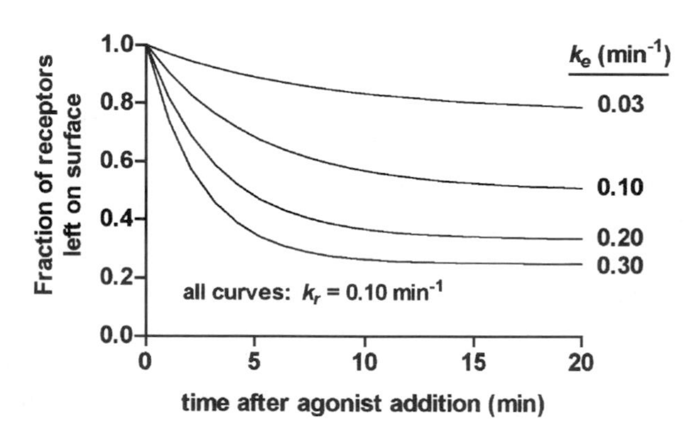

FIG. 6. (A) Equation describing the approach to steady-state internalized receptors as a function of time after agonist addition, where R_{so} is the surface binding of receptors at $t = 0$, R_s is the surface level at each time t of sampling, k_e is the first-order rate constant for receptor endocytosis, and k_r is the first-order rate constant for recycling. (B). Modeling of the equation under conditions where k_r is held constant at 0.1 min^{-1} and k_e is varied.

both k_e and k_r. For the β_2AR in response to a full agonist, k_e is 0.2–0.3 min^{-1}, whereas k_r is ~0.1 min^{-1}.[29,41] Modeling this equation shows that when k_r is kept constant at 0.1 min^{-1}, decreasing k_e will cause less steady-state receptor internalization. It should be noted that with weak agonists, internalization may be nearly undetectable due to the relatively rapid rate of recycling.

k_r

Receptor recycling curves are modeled using a one-phase exponential equation that describes the loss of internal receptors. Internal receptors (R_i) as a function of time after agonist removal are given by $R_i = (R_{s\infty} - R_{st})e^{-kt}$, where $R_{s\infty}$ is the concentration of receptors at the surface at long times after agonist withdrawal and R_{st} is the concentration of surface receptors at time t. This relationship is useful in those instances where a significant fraction of internal receptors fails to recycle due to downregulation or to intracellular trapping of receptors.[41] In practice, k_r obtained this way is in close agreement with that obtained by modeling internalization curves.[29]

These discussions assume that there is very little change of total receptor number during the time course of internalization measurements. This is true for most GPCRs, such as the β_2AR, where steady-state internalization is reached by 10 min after the addition of agonist, whereas receptor degradation is insignificant until after many hours. For other receptors, degradation is quite rapid and is significant by 20–30 min after the addition of agonist, approximately the time needed for transport from the plasma membrane to lysosomes.[45,48] In these circumstances, k_e can be estimated by measuring the loss of surface receptors during the first 5 min after the addition of agonist, before there is significant receptor degradation or recycling. A logarithmic plot of surface fraction remaining versus time of agonist exposure provides estimates of k_e (the slope of the straight line) similar to those obtained by curve fitting, at least for the β_2AR.[29] For a general model that incorporates GPCR degradation, synthesis, recycling, and endocytosis, see Koenig and Edwardson.[49]

Acknowledgments

We thank J. Friedman and J. L. Rosenfeld for help with the figures.

[48] J. Trejo, S. R. Hammes, and S. R. Coughlin, *Proc. Natl. Acad. Sci. U.S.A.* **95,** 13698 (1998).
[49] J.A. Koenig and J. M. Edwardson, *J. Biol. Chem.* **269,** 17174 (1994).

[33] Morphological and Biochemical Strategies for Monitoring Trafficking of Epitope-Tagged G Protein-Coupled Receptors in Agonist-Naive and Agonist-Occupied States

By MAGDALENA WOZNIAK, CHRISTINE SAUNDERS, NICOLE SCHRAMM, JEFFREY R. KEEFER, and LEE E. LIMBIRD

I. Introduction

The goal of this article is to provide protocols that permit using epitope tagging as a helpful tool in the monitoring of receptor protein expression on the cell surface, as well as in localizing receptor proteins as they move through subcellular compartments. We will focus on our application of epitope tagging to the study of trafficking of α_2-adrenergic receptor (α_2-AR) subtypes in Madin–Darby canine kidney (MDCKII) cells as a model system for evaluating the trafficking of G protein-coupled receptors (GPCR) during (1) receptor synthesis, (2) polarization to apical versus basolateral surfaces of polarized epithelia, and (3) receptor turnover at polarized surfaces. We also provide protocols for the use of reversible biotinylation regents to monitor agonist-elicited receptor sequestration and internalization, which we have studied in nonpolarized cells.

Epitope tagging, a recombinant DNA method involving the insertion of a short DNA sequence encoding an epitope tag into a gene of interest, represents a way of utilizing a known antibody to study any receptor protein.[1] The major advantage of epitope tagging of a receptor protein is that it allows use of a commercially available and typically well-characterized antibody to study the protein of interest. Epitope tagging eliminates concerns with endogenous protein-specific antibodies that may manifest cross-reactivity with other closely related endogenous proteins. Epitope tagging also allows investigators to distinguish between the properties of heterologously expressed and endogenous proteins. Potential risks of epitope tagging include alterations of structure and biological activity of the tagged protein. To minimize this risk, a typical epitope tag consists of a small number of amino acids (6 to 30), and preliminary studies must first document that introduction of the epitope does not alter receptor localization or function before embarking on further characterization of the epitope-tagged protein (see Section II,B).

There is a wide variety of commercially available antibodies directed against epitope tags (Table I). Typically, the tag is inserted at or near the amino or carboxy

[1] S. Monro and H. R. B. Pelham, *EMBO J.* **3,** 3087 (1984).

0076-6879/02 $35.00

TABLE I
COMMONLY INTRODUCED EPITOPES INTO HETEROLOGOUSLY EXPRESSED MAMMALIAN PROTEINS AND SOURCES OF COMMERCIAL ANTIBODIES

Epitope tag	Tag sequence	Anti-tag antibody (species)	Antibody source
AU1	DTYRYI	Anti-AU1 (mouse)	BAbCO, Research Diagnostics
AU5	TDFYLK	Anti-AU5 (mouse)	BAbCO, Research Diagnostics
c-myc	EQKLISEEDL	Anti-myc, clone 9E10 (mouse, rabbit, chicken)	Boehringer-Mannheim, Sigma, Chemicon, Invitrogen, Aves Labs
FLAG	DYKDDDDK	M1, M2, M5 (mouse, chicken)	Chemicon, Sigma, BAbCO, Aves Labs
Glu-Glu	EEEEYMPME	GLU-GLU (mouse)	BabCO
HA	YPYDVPDYA	12CA5 (mouse)	Boehringer-Mannheim
		3F10 (rat)	
		HA11 abm (mouse)	Research Diagnostics
		HA11 abr (rabbit)	
		Anti HA (rabbit)	Clontech
		Anti HA (chicken)	Aves Labs
6-His	HHHHHH	Anti-His (C-term) (mouse)	Invitrogen, BAbCO
HSV	QPELAPEDPED	HSV Tag (mouse, rabbit)	Novagen, Clontech
HTTPHH	HTTPHH	PHHTT (mouse)	Universal
IRS	RYIRS	IRS1 (mouse)	BabCO
KT3	PPEPET	KT3 (mouse)	BabCO
Protein C	EDQVDPRLIGK	HPC4 (mouse)	Roche Molecular Biochemicals
T7	MASMTGGQQMG	T7 Tag (mouse)	Novagen
VSV-G	MNRLGK	P5D4 (rabbit)	Boehringer-Mannheim, Clontech, MBL Co.

terminus of the receptor protein in an attempt to minimize any potential effect of the epitope tag on the structure and function of the protein.[2]

II. Use of Epitope-Tagged Receptors to Study GPCR Localization in Polarized Cells

A. *Antibody Purification*

1. Purification of Antibodies Directed against Heterologously Introduced Epitopes on Activated Affi-Gel 10 and Affi-Gel 15. Antibody can be purified against the peptide epitope using standard methodologies for the purification of immunoglobulins. However, to eliminate nonepitope-dependent (or nonspecific) detection of epitopes in target cells incubated with antiepitope antibody, the antibody preparation can be depleted of background-interacting contaminants by adsorbing total

[2] J. W. Jarvik and C. A. Telmer, *Annu. Rev. Genet.* **32,** 601 (1998).

lysates prepared from the heterologous target cells in which the epitope will be expressed to Affi-Gel 10 and Affi-Gel 15 resins and then use these resins to deplete the antibody preparation of nonspecific adsorbents.[3,4] A detailed protocol is provided for adsorbing lysates of MDCK II cells to Affi-Gel 10 and Affi-Gel 15 resins for purification of the 12CA5 mouse monoclonal antibody against the HA epitope.

i. A sufficient number of culture dishes are grown to confluence to yield 50 mg of MDCK II lysate protein; the lysate is *not* fractionated into particulate *versus* cytosolic fractions before adsorption to Affi-Gel 10 or 15.

ii. Before harvesting, the culture medium is removed by aspiration and the cultured cells are washed in Affi-Gel buffers appropriate for subsequent coupling to Affi-Gels. For Affi-Gel 10, a buffer combining 0.1 *M* MOPS, pH 7.5, and 80 m*M* $CaCl_2$ is used, whereas 0.1 *M* MOPS, pH 7.5, 0.3 *M* NaCl is used for coupling to Affi-Gel 15.

iii. Cells are harvested in the smallest possible volume of ice-cold buffer (such that the total cell lysate volume did not exceed 1 ml/100-mm dish) using a rubber policeman, collected in 1.5-ml Eppendorf tubes, and homogenized by 20 up-and-down strokes using a 1-ml syringe with a 20-gauge needle.

iv. Affi-Gel 10 and Affi-Gel 15 (1 ml of each resin) are prewashed three times in separate Eppendorf tubes with ice-cold 10 m*M* sodium acetate, pH 4.5, after which the whole cell lysate preparations are added onto the appropriate Affi-Gel resin.

v. The cell lysate–Affi-Gel coupling reactions are rotated for 4 hr at 4°, followed by multiple washes with the appropriate coupling buffer on a column. Affi-Gels are determined to be free of reactants by monitoring protein at OD_{280}. Affi-Gel 10 and 15 resins free of unbound lysate particles are collected in clean 1.5-ml Eppendorf tubes and incubated in the presence of 1 *M* ethanolamine–HCl, pH 8.0, for 1 hr at 4° in order to block any reactive esters on the resins.

vi. Each Affi-Gel resin is then washed again with the appropriate coupling buffer (see earlier discussion). Following the third wash, an aliquot of 12CA5 antibody is added *sequentially* to Affi-Gel 10 and Affi-Gel 15 and allowed to rotate with each of the Affi-Gels for 4 hr at 4°. Typically, 1 ml of undiluted 12CA5 antibody (10 mg/ml) is added *sequentially* to 500 μl of Affi-Gel 10 and then to 500 μl of Affi-Gel 15.

vii. The final supernatant, consisting of operationlly defined "purified" antiepitope antibody, is aliquoted and stored at −80°. Multiple independent purifications of the 12CA5 antibody against the HA epitope or of the 2E10 monoclonal antibody against the c-myc epitope have demonstrated that there is no loss of the antiepitope

[3] A. E. Thigpen, K. M. Cala, and D. Russell, *J. Biol. Chem.* **268,** 17404 (1993).

[4] M. Wozniak and L. E. Limbird, *J. Biol. Chem.* **271,** 5017 (1996).

antibody using this procedure with Affi-Gel 10/15 coupling to target cell lysates, whereas there is a dramatic decrease in the amount of nonspecific, background staining in various heterologous cell backgrounds.

B. *Documentation That Epitope Insertion Does Not Alter Properties of Receptor under Study*

It is important to establish, *via* as many criteria as possible, that introduction of an exogenous epitope into the molecule of interest does not perturb the functional properties or morphological distribution of the protein of interest. In our studies of α_2-adrenergic receptors, we established that introduction of an amino- or carboxy-terminal HA epitope did not alter receptor-binding capacity or specificity or the ability of agonist-occupied receptors to activate G proteins. Furthermore, we demonstrated that the distribution of all three α_2-adrenergic receptor subtypes (α_2-AR) was not altered by introduction of the HA epitope.[4,5] Figure 1 provides this documentation for the WT α_{2A}-AR and epitope-tagged HA α_{2A}-AR in polarized MDCK II cells.

C. *Growing and Documenting Polarization of MDCK II Cells*

1. Growth of MDCK II Cells on Transwell Culture Plates. MDCK II cells are maintained in Dulbecco's modified Eagles's medium (DMEM) supplemented with 10% fetal calf serum (Sigma), 100 units/ml penicillin, and 100 μg/ml streptomycin (complete DMEM) at 37°, 5% CO_2. MDCK II cells are seeded at a density of 1×10^6 cells/24.5-mm, polycarbonate membrane filter (Transwell chambers, 0.4-μm pore size, Costar) and cultured for 5–8 days, with medium changes at least every 2 days.

2. Monolayer Integrity. To exploit surface biotinylation as a strategy to covalently label receptors at the basolateral *versus* apical surface (described later), it is essential that the MDCK II cell monolayer is nonleaky so that biotinylating reagents added to the basolateral compartment cannot diffuse into the apical compartment and *vice versa*. The integrity of the monolayer can be assessed by adding [^{3}H]methoxyinulin (NEN) to the apical medium and monitoring the "leak" from the apical to the basolateral compartment by sampling and counting the basolateral medium in a β counter (Packard Tricarb) after a 1 hr incubation at 37°. Chambers with greater than 2% leak/hr would not be recommended for use in subsequent cell surface biotinylation experiments.[5]

3. Endogenous Proteins Serve as "Markers" for Polarized Surfaces. Cells are immunostained with antibodies directed against well-characterized polarized epithelial cell surface proteins. The EGF receptor or cadherin serves as a marker for basolateral proteins (enriched on the lateral subdomain of MDCK II cells), ZO1

[5] J. R. Keefer and L. E. Limbird, *J. Biol. Chem.* **268,** 1134 (1993).

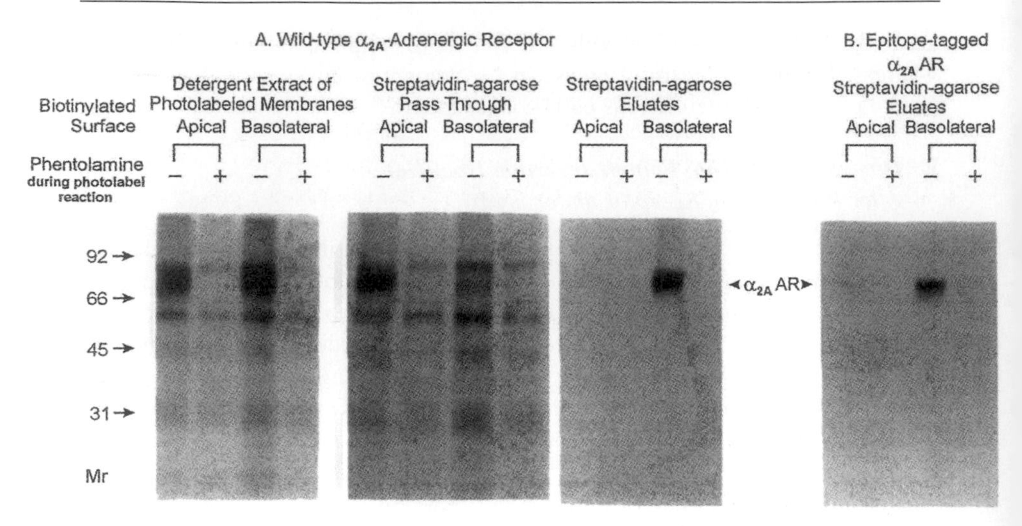

FIG. 1. The α_{2A}-AR is polarized to the basolateral membrane of MDCK II cells at steady state. (A) The Apical versus basolateral surface of a Transwell-cultured MDCK II clonal cell line was treated with biotin-LC-hydrazide in order to selectively label cell surface glycoproteins on one membrane domain or the other. The α_{2A}-AR was subsequently labeled in crude particulate preparations from these cells with the α_2-AR-selective photoaffinity label ^{125}I-Rau-AzPEC in the absence (total photoaffinity labeling) or presence (nonspecific labeling) of the α-adrenergic antagonist phentolamine. Photoaffinity-labeled particulate preparations were extracted in detergent, and one-fifth of the solubilized material was subject to SDS–PAGE and autoradiography. (Left) A representative autoradiogram of these total detergent extracts. The remaining four-fifths of the detergent extracts were adsorbed to and eluted from streptavidin–agarose resin. The entire eluate fraction (right) and one-third of the nonadsorbed material (pass-through) (middle) from the streptavidin–agarose resin were then analyzed by SDS–PAGE and autoradiography (B) An MDCK clonal cell line expressing an epitope-tagged version of the α_{2A}-AR was subjected to the steady-state biotin-LC-hydrazide surface-labeling protocol. An autoradiogram of an SDS–PAGE gel of the streptavidin–agarose eluates from this experiment is shown. Data presented indicate that introduction of the HA epitope into the amino terminus of the α_{2A}-AR does not alter its polarization to the basolateral surface of the MDCK II cells. Data are from Keefer and Limbird.[5]

as a marker for tight junctions, and gp135 for the apical domain.[6] We have used the following concentration of primary antibody to detect these proteins: a 1 : 50 dilution of 12CA5 primary antibody, purified as described earlier, for the localization of hemagglutinin epitope-tagged GPCR (1); 15 μg/ml of a mouse monoclonal antibody against the basolaterally localized EGF receptor[7]; a 1 : 10 dilution of a mouse monoclonal antibody against an apical surface glycoprotein gp135[8]; and a 1 : 8 dilution of a mouse monoclonal antibody against E-cadherin (RR1,

[6] C. Saunders and L. E. Limbird, *J. Biol. Chem.* **272,** 19035 (1997).

[7] P. J. Dempsey and R. J. Coffey, *J. Biol. Chem.* **269,** 16878 (1994).

[8] A. Lee, J. L. Steiner, G. A. Ferrel, J. C. Shaker, and R. N. Sifers, *J. Biol. Chem.* **269,** 7514 (1994).

Developmental Studies Hybridoma Bank), which demarks the lateral subdomain of polarized MDCK II cells.

Routinely, the antibody-containing and antibody wash buffers contain 0.2% Triton X-100 (TX-100) to permit detection of epitope either on the cell surface or in the cell interior. Interestingly, detection of localization of the α_2-AR subtypes at the cell surface is enhanced in TX-100-containing solutions, presumably because the HA epitope is rendered more accessible under these conditions. Adding TX-100 also allows detection of intracellular pools of the α_{2C}-AR[4] and of agonist-internalized α_{2B}-AR.[6] Blocking of nonspecific interactions can be achieved by adding bovine serum albumin, Carnation instant nonfat milk, or nonimmune serum from the species of the secondary antibody to these buffers. Antibodies against endogenous proteins that serve as markers for *intracellular* organelles are described as part of the next section.

D. *Steady-State Localization of HA-Tagged α_2-Adrenergic Receptors in Polarized MDCK II Cells Assessed Using Morphological Strategies*

Clonal cell lines expressing HA-tagged α_2-AR subtypes grown on polycarbonate filters with *daily* media changes (a manipulation that appears to increase cell height) are excised from the Transwell support, rinsed with phosphate-buffered saline (PBS), and fixed for 30 min at 22° (room temperature) with 4% (v/v) paraformaldehyde in PBS (for detecting α_{2A}-AR and α_{2B}-AR subtypes) or 15 min with 2% (v/v) paraformaldehyde in PBS (for detecting α_{2C}-AR subtypes).[4]

Cells are rinsed twice with PBS, incubated in 50 m*M* NH_4Cl solution in PBS for 15 min to quench the fixative, and then permeabilized with 0.2% Triton X-100 in PBS for 15 min at room temperature. Potential sites for nonspecific antibody binding are blocked by a 15-min incubation with 1% BSA, 0.2% Triton X-100, 0.04% sodium azide in PBS at room temperature. As an example, when an HA-tagged receptor is evaluated, the 12CA5 mouse monoclonal primary antibody directed against the hemagglutinin epitope, purified against parental MDCK cell lysates (as described earlier), is diluted 1 : 25–1 : 200 (depending on HA–α_2-AR density) in 1% BSA, 0.2% Triton X-100, 0.04% sodium azide in PBS and incubated with the sample for 1 hr at room temperature. This is followed by four 15-min washes with 0.2% Triton X-100 in PBS and a 1-hr incubation with a secondary Cy3-conjugated donkey antimouse IgG (1 : 100 dilution in 1% BSA, 0.2% Triton X-100, 0.04% sodium azide in PBS). The cells are washed as before with PBS containing 0.1% Triton X-100 and mounted on glass microscope slides with Aqua Polymount (Polysciences, Inc.). The results can be analyzed either on a standard fluorescent microscope (we use a Leitz Fluorescent Microscope using a 40X oil immersion lens) or on a laser-scanning confocal microscope (we use a Zeiss instrument).

Immunological *colocalization* studies are done essentially as described earlier, although expanded by the additional application of another primary and secondary antibody. Dilutions for antibodies against cell surface marker proteins were given in Section II,C,3.

For identifying *intracellular organelles,* polarized MDCK II cells are incubated with a 1:400 dilution of an anticalnexin rabbit polyclonal antibody (used as a marker for the endoplasmic reticulum[8]) or a 1:1000 dilution of antimannosidase II rabbit polyclonal antibody (used to detect *trans*-Golgi network[9]) in 1% BSA, 0.2% Triton X-100, 0.04% sodium azide following the initial detection of the α_{2C}-AR with 12CA5 antibody and FITC (1:60 dilution) or Cy3 (1:100 dilution)-conjugated donkey antimouse IgG. The anticalnexin or antimannosidase II primary antibodies are incubated with the sample for 1 hr at room temperature, followed by four 15-min washes with 0.2% Triton X-100 in PBS, and a 1-hr incubation with a secondary donkey antirabbit FITC (1:60 dilution) or Cy3 (1:200 dilution)-conjugated IgG in 1% BSA, 0.2% Triton X-100, 0.04% sodium azide in PBS. The cells are then washed four times for 15 min with PBS, 0.2% Triton X-100, mounted, and analyzed as described earlier.

E. *Localization of HA-Tagged α_2-Adrenergic Receptors in Polarized MDCK II Cells Assessed Using Surface Biotinylation*

Surface biotinylation provides a quantitative biochemical approach to document surface localization and thus complements the qualitative, morphological approaches outlined in Section II,D. Surface biotinylation strategies can be used to assess receptor localization at steady state (Section II,E,1) or as a function of receptor synthesis and surface delivery (Section II,E,2) or to assess receptor turnover at a biotinylated surface (Section II,E,3).

1. *Localization at Steady State Using Surface Biotinylation*

a. METABOLIC LABELING. Cells grown in Transwell culture are washed twice in PBS containing 0.5 m*M* $CaCl_2$ and 1 m*M* $MgCl_2$ (termed PBS/CM buffer) and incubated for 45 min in methionine/cysteine-free medium (ICN) supplemented with 5% fetal calf serum. Following removal of this medium, the Transwell inserts are inverted (cell side down), and 150 μl of methionine/cysteine-free medium, 5% fetal calf serum containing 1 μCi/μl ^{35}S-Express protein labeling mixture (Dupont-New England Nuclear) is added to the filter surface of the inverted filters. The cells are incubated at 37° with 5% CO_2 in a humidified chamber for 90–120 min for studies examining the steady-state distribution of α_2-AR.[5,10]

For studies of receptor delivery to a particular surface (apical *versus* basolateral) or for assessment of surface half-life, pulse-chase strategies are employed, as described in more detail in Sections II,E,2 and 3.

[9] K. W. Moreman, O. Touster, and P. W. Robbins, *J. Biol. Chem.* **266,** 16876 (1991).
[10] J. R. Keefer, M. E. Kennedy, and L. E. Limbird, *J. Biol. Chem.* **269,** 16425 (1994).

b. BIOTINYLATION WITH SULFO-NHS-BIOTIN. The primary amino groups of cell surface proteins can be modified covalently by sulfo-NHS-biotin as a means to irreversibly "mark" proteins expressed at a particular surface (basolateral *versus* apical) of polarized epithelial cells.[4,5]

i. MDCK cells grown in Transwell culture are washed three times for 15 min in PBS/CM buffer and then rinsed one time in TEA buffer (250 m*M* sucrose, 10 m*M* triethanolamine, 2 m*M* $CaCl_2$, 2 m*M* Mg Cl_2, pH 9.0).

ii. Sulfo-NHS biotin at 1 mg/ml in TEA buffer is then applied to the apical or basolateral surface and incubated for 20 min at 4° in order to covalently label cell surface proteins.

iii. The biotin solution is removed, replaced with fresh Sulfo-NHS-biotin in TEA buffer, and incubated for 20 min at 4°.

iv. Following this labeling, cells are washed three times for 15 min with PBS/CM, rinsed two times with PBS containing 1 m*M* EDTA, and harvested into PBS containing 1 m*M* EDTA and 0.1 m*M* phenylmethylsulfonyl fluoride.

c. BIOTINYLATION WITH BIOTIN-LC-HYDRAZIDE FOR CELL SURFACE PROTEINS THAT ARE GLYCOSYLATED. The apical versus basolateral surface of α_{2A}-AR-expressing MDCK cell clones grown in Transwell culture can be labeled selectively with biotin hydrazide essentially as described by Lisanti and co-workers.[11]

i. Cells are washed three times (3 ml per compartment; 5 min each wash) with ice-cold PBS/CM.

ii. Ice-cold 10 m*M* $NaIO_4$ in PBS/CM is then added to either the apical or the basolateral compartment of the Transwell, and the cells are agitated gently for 15–30 min at 4° in the dark.

iii. The cells are rinsed three times with 3 ml per compartment of PBS/CM and once with 3 ml of 100 m*M* sodium acetate, pH 5.5, containing 1 m*M* $CaCl_2$ and 1 m*M* $MgCl_2$ (referred to as NA/CM buffer).

iv. The apical or basolateral cell surface is then labeled with 1.5–2 ml of 0.8 mg/ml biotin hydrazide in NA/CM for 15–30 min at 4°.

v. Finally, cells are washed three times with PBS/CM and twice with PBS. Both $NaIO_4$ and biotin hydrazide are made fresh from powder for each experiment.

Technical note: Proteins expressed in cells growing up the sides of the plastic insert that holds the filter in place (instead of on the filter itself) can contribute considerably to the presumed apical signal in biotin-based localization experiments, thus leading to an overestimate of the true amount of epitope-tagged protein

[11] M. P. Lisanti, A. Le Bivic, M. Sergiacomo, and E. Rodriguez-Boulan, *J. Cell Biol.* **109,** 2117 (1989).

expressed on the apical surface.[11] A simple solution to this potential problem is to excise the filter from the insert before solubilizing or harvesting cells.[10]

d. SEQUENTIAL DETERGENT EXTRACTION, IMMUNOISOLATION, AND STREPTAVIDIN–AGAROSE CHROMATOGRAPHY TO QUANTIFY THE FRACTION OF THE TOTAL RECEPTOR POPULATION ON BIOTINYLATED BASOLATERAL VERSUS BIOTINYLATED APICAL SURFACES: DETERGENT EXTRACTION. The epitope-tagged receptor is first isolated via detergent extraction, binding to antiepitope antibody, and adsorption to protein A-agarose. Cells are harvested and extracted into RIPA buffer [1% (w/w) NP40, 1% (w/v) sodium deoxycholate, 0.1% (w/v) SDS, 0.15 *M* NaCl, 0.01 *M* sodium phosphate buffer, pH 7.2, 2 m*M* EDTA, and 50 m*M* NaF (as phosphatase inhibitor)] and protease inhibitors, as desired.

e. IMMUNOISOLATION. The precleared RIPA detergent extract is incubated with antiepitope antibody overnight at 4°. For HA-tagged α_2-AR, the detergent extract is incubated with 1 : 25 dilution of 12CA5 antibody. Next, 50 μl of a 1 : 1 slurry of protein A–agarose that had been washed for 1 hr in RIPA containing 2.5 mg/ml BSA and rinsed twice in RIPA alone is added to the samples and incubated at 1–2 hr at 4°. The protein A–agarose is pelleted by centrifugation and washed with RIPA buffer 4 $\times$ 1 min, 1 $\times$ 5 min, and 1 $\times$ 1 min. Following the last wash, proteins are eluted from the protein A–agarose beads in 100 μl of SDS sample buffer at $\geq$75° for 10 min. Boiling of the sample is not indicated, as profound protein denaturation diminishes the yield of biotinylated receptor adsorption to streptavidin–agarose. A second elution is performed in 80 μl of SDS sample buffer, and the eluates are combined and diluted to 1.6 ml in RIPA buffer made without SDS prior to streptavidin–agarose chromatography.

f. STREPTAVIDIN–AGAROSE FRACTIONATION OF IMMUNOISOLATED α_2-AR INTO BIOTINYLATED AND NONBIOTINYLATED FRACTIONS. The eluates of immunoaffinity isolation are incubated with streptavidin–agarose to identify those α_2-AR that are modified by surface biotinylation. A 25-μl aliquot of a 1 : 1 slurry of streptavidin–agarose that is prewashed 1 $\times$ 1 hr with RIPA buffer containing 2.5 mg/ml BSA and 2 $\times$ 30 min with RIPA buffer alone is added to each 500-μl aliquot of the RIPA-solubilized preparation and rotated overnight at 4°. The samples are centrifuged briefly in an Eppendorf centrifuge at 4°, and the supernatant is saved and designated as the streptavidin–agarose "pass through," or nonbiotinylated fraction.[5] Receptor in this fraction is interpreted to represent receptor not existing on the biotinylated surface of interest. The streptavidin–agarose beads are then washed 3 $\times$ 5 min at 4° in 500 μl of RIPA buffer. After the final wash, proteins are eluted from the beads by incubation in 100 μl of SDS sample buffer (1.6% SDS, 8.3% glycerol, 167 m*M* Tris, pH 6.8) containing 5 m*M* dithiothreitol at 90° for 20 min, vortexing every 5 min during the incubation. The elution is repeated, and the eluates are combined and subjected to SDS–PAGE and autoradiography.

g. SDS–PAGE AND AUTORADIOGRAPHY TO DETECT α_2-AR DELIVERED TO BIOTINYLATED BASOLATERAL SURFACE. SDS–polyacrylamide gels are prepared using 7.5 to 20% acrylamide gradients for the best resolution of α_{2A}, α_{2C} ($M_r \sim 65,000$), or α_{2B} ($M_r \sim 45,000$) adrenergic receptor subtypes.

To further increase the resolution of glycosylated GPCR on SDS–PAGE, we routinely incubate DTT-treated eluates with *N*-ethylmaleimide to covalently modify–SH groups and thus minimize broadening of the receptor band due to multiple reduced/oxidized conformations of the receptor. The streptavidin–agarose eluates intended for SDS–PAGE are treated with 5 m*M* DTT for 45 min at 50°. *N*-Ethylmaleimide is then added to the samples to a final concentration of 15 m*M* and the samples are incubated at 50° for 45 min longer.[12] This dithiothreitol/*N*-ethylmaleimide incubation protocol has been used by other investigators to create a more homogeneous population of receptor molecules, which migrate as a more focused band on SDS–PAGE.[13] Following dithiothreitol/*N*-ethylmaleimide treatment, the samples are subjected to SDS–PAGE, fluorography (Entensify, Dupont-New), and autoradiography.

2. *Metabolic Labeling/Biotinylation Strategy for Determining Surface Delivery of the α_{2A}-AR*

a. METABOLIC LABELING. Cells grown in Transwell culture are treated as in Section II,E,1 to deplete endogenous methionine and cysteine before incubation of inserts with 150 μl of methionine/cysteine-free medium containing 1 μCi/μl ^{35}S-Express protein labeling mix (at 37° with 5% CO_2) for various periods of time. We have successfully tracked direct delivery of the α_{2A}-AR to the basolateral surface and random delivery of the α_{2B}-AR to the apical and basolateral surfaces by biotinylating either the apical or the basolateral surface following [^{35}S]cysteine/methionine labeling for 20, 40, 60, 90, and 120 min.

Biotinylation of the apical *versus* basolateral membrane surface is achieved with Sulfo-NHS-biotin or biotin-LC-hydrazide as described earlier. Subsequent sequential immunoisolation of epitope-tagged α_2-AR subtypes and fractionation *via* streptavidin–agarose to capture α_2-AR existing on a particular biotinylated surface is then achieved using the procedures detailed in Section II,D for steady-state localization of epitope-tagged GPCR.

3. *Assessing the Half-Life of Epitope-Tagged GPCR on Biotinylated Surfaces.* The overall procedure for determining GPCR retention time (half-life) on a biotinylated (apical *versus* basolateral) surface is to metabolically label MDCK cells expressing epitope-tagged receptor ("pulse") and then incubate those cells for various periods of time ("chase") before isolating the epitope-tagged receptor using sequential biotinylation, extraction, immunoisolation, and streptavidin–agarose

[12] E. M. Ross, S. K. Wong, R. C. Rubenstein, and T. Higashijima, *Cold Spring Harbor Symp. Quant. Biol.* **53,** 499 (1988).

[13] C. Saunders, J. R. Keefer, A. Pong-Kennedy, J. N. Wells, and L. E. Limbird, *J. Biol. Chem.* **271,** 995 (1996).

chromatography, as described earlier. For examining the surface half-life of epitope-tagged α_2-AR subtypes, the following detailed protocol has been used successfully:

i. MDCK cells in Transwell culture are incubated for 1 to 2 hrs in cysteine/methionine-free DMEM to deplete cellular cysteine/methionine stores.

ii. MDCK cells grown in Transwell culture are labeled metabolically with 1 μCi/μl of ^{35}S-Express protein labeling mix for 30–40 min in cysteine/methionine-free DMEM.[4–6]

iii. Following labeling, the cells are washed twice with PBS and once with complete DMEM medium prior to incubation for various periods of time (generally 2, 8, 14, or 26 hr) at 37°, 5% CO_2 in chase medium (complete DMEM supplemented with 1 m*M* cysteine and 1 m*M* methionine).

iv. At the conclusion of each chase period, HA-tagged α_2-AR residing on the basolateral surface of these cells is isolated by biotinylating the basolateral cell surface with biotin hydrazide or sulfo-NHS-biotin and subjecting detergent extracts of membranes prepared from these metabolically labeled, biotinylated cells to immunoisolation using the 12CA5 anti-HA epitope antibody and adsorption to protein A–agarose followed by streptavidin–agarose chromatography (Section II,E,1). For studies of the apically localized A_1 adenosine receptor[13] or the randomly targeted α_{2B}AR,[4] the apical surface is also biotinylated in separate experiments to assess turnover on that surface.

v. The amount of radiolabeled α_2-AR remaining on the surface of interest at the various durations of chase is determined by subjecting the streptavidin–agarose eluates to SDS–PAGE, performing autoradiography, cutting out the gel area corresponding to the radiolabeled α_2-AR, and counting this gel slice (in 10 ml of scintillation fluor) in a Packard β counter. Similar sized gel slices that do not correspond to any radiolabeled protein band are excised and counted to determine the nonspecific ^{35}S signal in each lane. These background counts are subtracted from the counts in the α_{2A}-AR slices to determine the specific α_{2A}-AR counts/min at each time point. Semiquantitation can also be performed by phosphorimager analysis and quantification by commercially available software programs such as Image Quant. Phosphorimager findings are comparable with those of cutting and counting of radiolabeled bands if imager exposure occurs under conditions where signal intensity correlates linearly with the quantity of radioactivity, confirmed by a "standard curve" included with the sample.

Figure 2 demonstrates the rapid (5–15 min) turnover of the randomly delivered α_{2B}-AR subtype on the apical surface of polarized MDCK II cells in contrast to the long-lived surface retention of the α_{2B}-AR on the basolateral ($t_{1/2}$ = 10–12 hr) surface, comparable to the 10- to 12-hr $t_{1/2}$ of the directly delivered α_{2A}- and α_{2C}-AR subtypes on the basolateral surface.[4]

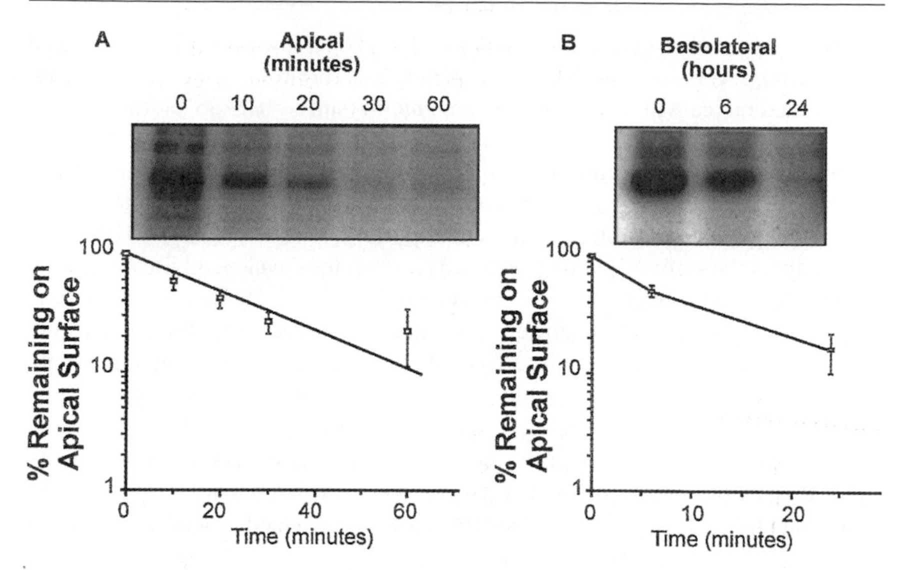

FIG. 2. Differential retention of the α_{2B}-AR subtype on the apical *versus* the basolateral surface of MDCK II cells. MDCK II cells permanently expressing the α_{2B}-TAG-AR were labeled metabolically with 150 μCi of [^{35}S]cysteine/methionine in 150 μl of medium at 37°, 5% CO_2 for 30 (A) or 60 (B) min, and chased in medium supplemented with 1 m*M* methionine and 1 m*M* cysteine for the indicated time periods. After completion of the chase period, cells were biotinylated on either the apical (A) or the basolateral (B) resolved on SDS–PAGE. (Top) Representative autoradiograms. (Bottom) The mean ± SE of radioactivity present in α_{2B}-AR corresponding gel bands from seven (A) and three (B) separate experiments are shown. The calculated half-life of the α_{2B}-AR on the apical surface was 15–30 min, whereas the calculated half-life of the α_{2B}-AR on the basolateral surface was 8–10 hr. Data are from Wozniak and Limbird.[4]

III. Use of Epitope-Tagged Receptors to Study Agonist-Elicited Receptor Redistribution *via* Reversible Biotinylation Strategies

It is possible to exploit reversible biotinylating agents to monitor dynamic receptor localization events, such as agonist-elicited redistribution. Although we have not exploited these strategies in polarized cells, we include here our protocols for examining HEK 293 cells permanently expressing each of the α_2-AR subtypes, as we have found reversible biotinylation a useful tool for quantifying agonist-evoked redistribution of the α_{2B}-AR subtype.[14] The reason for focus on the α_{2B}AR subtype in these studies is threefold: (1) the α_{2B}-AR undergoes rapid agonist-elicited redistribution, (2) the α_{2A}-AR subtype does not undergo rapid agonist-elicited redistribution, and (3) the preexisting distribution of α_{2C}-AR both on the surface and in an intracellular compartment precludes facile evaluation of agonist-elicited changes in α_{2C}-AR redistribution.

[14] N. L. Schramm and L. E. Limbird, *J. Biol. Chem.* **274,** 24935 (1999).

The overall strategy for monitoring cell surface removal of epitope-tagged GPCR using reversible biotinylating strategies is as follows. Surface receptors are biotinylated on ice with disulfide-cleavable biotin (sulfo-NHS-SS-biotin; Pierce). Cells are then warmed to 37° to monitor the trafficking of the receptor, if it occurs, to an intracellular compartment. At the termination of the sulfo-NHS-SS-biotin incubation, thc culture wells are transferred to ice and the cells are treated with a hydrophilic reducing agent, mercaptoethanesulfonic acid (MESNA). MESNA or DTT reduces the sulfo-NHS-SS-biotin and removes the covalent additive from any biotinylated molecules it comes in contact with. Over relatively short incubations at 4°, neither MESNA nor DTT can penetrate the cell and thus release biotin only from molecules still existing at the cell surface. In control wells, the cells are treated with MESNA immediately following biotinylation at 4° to provide an estimate of cell surface molecules accessible to MESNA (or DTT). Receptors inside the cell at the time of MESNA treatment are protected from reduction and therefore subsequently isolated from the detergent extract by adsorption to streptavidin–agarose. The following protocol was adapted from a previously published procedure.[15]

i. For each experiment, HEK 293 clonal cell lines permanently expressing the α_{2A}-, α_{2B}-, or α_{2C}-AR subtype are plated on 60-mm dishes coated with poly-D-lysine and allowed to grow to 70–80% confluence. Cells are serum starved for 16–18 hr prior to each experiment.

ii. On the day of the experiment, cells are washed twice with ice-cold PBS/CM and then incubated with 100 μg/ml sulfo-NHS-SS-biotin in PBS/CM buffer for 30 min at 4°. The cells are washed twice with PBS/CM and once with serum-free DMEM at 4°.

iii. Dishes are incubated at 37° in DMEM and supplemented with 20 m*M* HEPES, pH 7.4, by placement on a rack in a water bath. For agonist activation, the medium is replaced with 37° serum-free DMEM containing 100 μM epinephrine for 5 min at 37° (control cells have their medium replaced with warm medium without agonist).

iv. Incubation is terminated by replacement of the 37° medium with ice-cold DMEM. The culture dishes are returned to 4°, washed twice with ice-cold PBS/CM, and then incubated with 250 μM MESNA in PBS/CM for two 20-min incubations to release the surface-accessible biotinylating reagent via disulfide exchange. The cells are then washed twice with ice-cold serum-free DMEM, and the reducing effect of any residual MESNA is then quenched by incubation with 5 mg/ml iodoacetamide in PBS/CM for 20 min at 4°. As indicated earlier, biotinylated receptors resistant to MESNA reversal of biotinylation are defined as "inaccessible." To define total MESNA-accessible receptors on these cell lines, one 60-mm dish per experiment is treated with MESNA immediately following biotinylation at 4°

[15] J. L. Whistler and M. Von Zastrow, *Proc. Natl. Acad. Sci. U.S.A.* **95,** 9914 (1998).

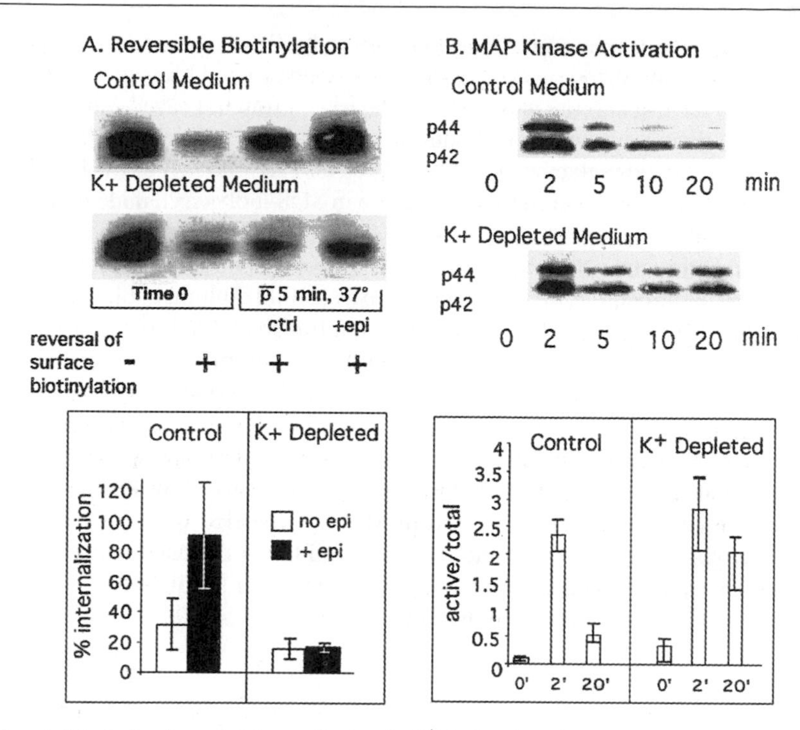

FIG. 3. Blockade of α_{2B}-AR internalization in K^+-depleted medium (documented by reversible biotinylation strategies) does not block epinephrine stimulation of MAP kinase. HEK 293 cells permanently expressing the α_{2B}-AR were incubated in K^+-depleted medium prior to and during stimulation by epinephrine. (A) Receptor redistribution to a MESNA-inaccessible compartment was measured in control cells (top) or cells exposed to K^+-depleted medium (middle) following surface biotinylation with the reversible reagent, sulfo-NHS-SS-biotin. Reversal of biotinylation was performed immediately after biotinylation or after incubation for 5 min at 37° with or without the agonist epinephrine. (Bottom) Quantitation of internalization, mean ± SE for three independent experiments. (B) MAP kinase activation by epinephrine in the absence (top) or presence (middle) of K^+-depleted medium. Stimulation was performed for the indicated times, in minutes. (Bottom) The normalized level of activated MAP kinase (active/total) was calculated as described previously and is plotted for maximal stimulation (2-min time point) and for the point at which desensitization has usually occurred (20-min time point) (mean ± SE for three independent experiments). Data are from Schramm and Limbird.[14]

to reveal the quantity of surface receptor biotinylation that MESNA can reverse efficiently, as indicated earlier.

v. To isolate biotinylated α_2-AR, cells are lysed in solubilization buffer [4 mg/ml dodecyl-β-D-maltoside (Calbiochem), 0.8 mg/ml cholesteryl hemisuccinate (Sigma), 1 mg/ml iodoacetamide (Sigma), 20% glycerol, 25 m*M* glycylglycine, 5 m*M* EDTA, 20 m*M* HEPES, pH 8.0, 0.1 *M* NaCl] and homogenized by five up/down passages through a 25-gauge needle. Cellular debris is removed by centrifugation in a microcentrifuge at 4° for 1 hr. The supernatant (defined as the detergent-solubilized extract) is incubated with streptavidin–agarose (80 μl, 1 : 1 slurry) for 1 hr at room temperature. The pass-through is saved, and the beads are washed three times with a 1 : 8 dilution of the solubilization buffer in "binding buffer" (25 m*M* glycylglycine; 5 m*M* EDTA; 20 m*M* HEPES, pH 8.0; 0.1 *M* NaCl). Proteins are eluted from streptavidin–agarose in SDS–polyacrylamide buffer containing 5 m*M* dithiothreitol.

vi. The entire eluate is separated by SDS–polyacrylamide gel electrophoresis, transferred to nitrocellulose, and probed with a 1 : 1000 dilution in blocking buffer (see earlier discussion) of HA.11 antibody from Babco (Table I) to the HA-tagged α_2-ARs. Horseradish peroxidase-conjugated sheep antimouse secondary antibody (Amersham Pharmacia Biotech) is used at a 1/3000 dilution in blocking buffer. Reactive proteins are visualized by ECL (Amersham Pharmacia Biotech). Bands are quantified by scanning and NIH Image software as described previously.

vii. To calculate percentage internalization, the intensity of the band in the lane where cells are immediately exposed to MESNA (labeled 0', 4°) is subtracted as background from all other band intensities, and then the band intensity in the treated lanes ("no epinephrine" or "+epinephrine" after 5 min at 37°) is divided by the "total" band intensity and the result is multiplied by 100.

Figure 3 demonstrates the rapid agonist-elicited internalization of the α_{2B}-AR and its elimination by treatment of cells with physiological buffers containing low K^+, a manipulation that destabilizes clathrin-coated pH interactions and thus disrupts internalization events that rely on these endofacial structures.

IV. Summary

Epitope tagged α_2-AR subtypes have been used to address a variety of cell biological questions, and the strategies used are readily applicable to all GPCR as well as other cell surface proteins. We have provided detailed protocols for successful utilization of the epitope-tagged receptor in the studies of protein localization and trafficking in epithelial cells, and the mechanisms by which this is achieved. We have also described reversible biotinytion strategies to examine agonist-dependent (and independent) receptor turnover at the cell surface.

Section II

Regulators of GPCR Function

A. G Protein-Coupled Receptor Kinases (GRKs)
Articles 34 through 36

B. Arrestins and Novel Proteins
Articles 37 and 38

[34] Characterization of G Protein-Coupled Receptor Kinases

By ALEXEY N. PRONIN, ROBERT P. LOUDON, and JEFFREY L. BENOVIC

Introduction

A basic feature of most cells is the ability to dynamically regulate their responsiveness to extracellular stimuli. Numerous stimuli transmit their signals via interaction with cell surface G protein-coupled receptors (GPCRs). GPCRs are subject to three principal modes of regulation: (i) desensitization, the process by which a receptor becomes refractory to continued stimuli; (ii) internalization, whereby receptors are physically removed from the cell surface by endocytosis; and (iii) downregulation, where total cellular receptor levels are decreased.[1]

GPCR desensitization is primarily mediated by second messenger responsive kinases, such as protein kinase A (PKA) and protein kinase C (PKC), and by G protein-coupled receptor kinases (GRKs). GRKs specifically phosphorylate agonist-occupied GPCRs and initiate the recruitment of additional proteins, termed arrestins, that further receptor desensitization and internalization. The seven mammalian GRKs identified can be divided into three subfamilies based on their overall structural organization and homology: GRK1 (rhodopsin kinase) and GRK7; GRK2 (βARK1) and GRK3 (βARK2); and GRK4, GRK5, and GRK6. Common features shared by the GRKs include a centrally localized catalytic domain of ~270 amino acids, an N-terminal domain of ~184 amino acids that has been implicated in GPCR interaction and GRK regulation, and a variable length C-terminal domain of 105–233 amino acids involved in phospholipid association (Fig. 1). The ability of GRKs to phosphorylate and regulate the activity of GPCRs has been studied extensively and has been the subject of recent reviews.[2–5]

This article describes some of the current methodologies for analyzing endogenous and expressed GRKs in mammalian cells and strategies for *in vitro* analysis of GRK phosphorylation of GPCRs.

[1] R. B. Penn and J. L. Benovic, *in* "Handbook of Physiology" (P. M. Conn, ed.), Vol. 1, p. 125. Oxford Univ. Press, 1998.

[2] J. A. Pitcher, N. J. Freedman, and R. J. Lefkowitz, *Annu. Rev. Biochem.* **67,** 653 (1998).

[3] J. G. Krupnick and J. L. Benovic, *Annu. Rev. Pharmacol. Toxicol.* **38,** 289 (1998).

[4] C. V. Carman and J. L. Benovic, *Curr. Opin. Neurobiol.* **8,** 335 (1998).

[5] L. Iacovelli, M. Sallese, S. Mariggio, and A. de Blasi, *FASEB J.* **13,** 1 (1999).

0076-6879/02 $35.00

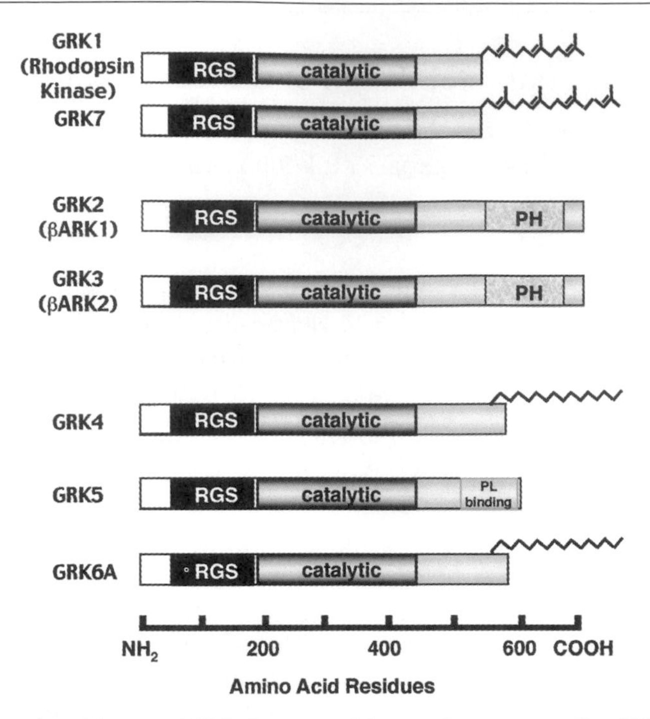

FIG. 1. Domain architecture of GRKs. Sequences of the seven known mammalian GRKs are represented schematically. The putative sites of farnesylation of GRK1, geranylgeranylation of GRK7, and palmitoylation of GRK4 and 6 are shown. RGS, regulators of G protein-signaling homology domain; PH, pleckstrin homology domain; PL, phospholipid.

GRK Expression in Mammalian Cells

Immunoblotting of GRKs

Analysis of endogenous GRK activity using a receptor substrate such as rhodopsin has proved useful. However, it is generally not very sensitive and also does not enable discrimination among the various GRK subtypes. Another method of measuring GRK expression in mammalian cells utilizes GRK specific antibodies to analyze cell or tissue extracts for GRK levels by immunoblot analysis. This method often provides sufficient specificity and sensitivity to enable detection of multiple GRKs in crude cellular extracts. There are three major considerations when attempting to analyze GRK levels by immunoblotting. The first is that GRKs contain phospholipid-binding domains that often result in the kinases being present in both cytosolic and membrane associated fractions following cell lysis. While such fractions can be generated and analyzed individually, it is often

TABLE I
GRK ANTIBODIES

Company/antibody name	Specificity	Type	Epitope
Upstate Biotechnology			
Anti-GRK 2/3 (clone 5/1.1)	GRK2~GRK3	Mouse monoclonal	C-terminal epitope
Anti-GRK 4–6 (A16/17)	GRK4~GRK5~GRK6	Mouse monoclonal	C-terminal epitope
Santa Cruz Biotechnology			
GRK1 (C-20)	GRK1 specific	Rabbit polyclonal	C-terminal epitope
GRK1 (C-8)	GRK1 specific	Mouse monoclonal	C-terminal epitope
GRK2 (C-15)	GRK2 specific	Rabbit polyclonal	C-terminal epitope
GRK2 (H-222)	GRK2/3 specific	Rabbit polyclonal	C-terminal epitope
GRK3 (C-14)	GRK3 specific	Rabbit polyclonal	C-terminal epitope
GRK3 (E-15)	GRK3 specific	Goat polyclonal	C-terminal epitope
GRK4 (I-20)	GRK4 (γ,δ) specific	Rabbit polyclonal	C-terminal epitope
GRK4 (K-20)	GRK4 (α,β) specific	Rabbit polyclonal	N-terminal epitope
GRK4 (K-20)	GRK4 (α,β) specific	Rabbit polyclonal	N-terminal epitope
GRK4 (A-5)	GRK4 specific	Mouse monoclonal	C-terminal epitope
GRK5 (C-20)	GRK5 specific	Rabbit polyclonal	C-terminal epitope
GRK6 (C-20)	GRK6 specific	Rabbit polyclonal	C-terminal epitope

advantageous to homogenize cells or tissues using a buffer that results in the GRKs being primarily in the soluble fraction. A second consideration is the availability and quality of GRK antibodies. Several useful GRK specific antibodies are available commercially (see Table I). A third consideration is the level of a particular GRK in a given cell or tissue. Although endogenous GRK levels are generally low in most cells, GRK2 appears to be the most abundant and can be detected readily by immunoblotting in most mammalian cells. By comparison, GRK3-6 are typically found at lower levels and are thus detected less readily whereas GRK1 and 7 are primarily found in the retina. A partial purification using SP Sepharose enables the generation of samples enriched in GRKs, making their detection easier.

Preparation of Cellular Lysate

The general lysis procedure we use helps to extract both cytosolic and membrane associated proteins, as most GRKs can bind to phospholipids. However, while these extracts are optimal for assessing endogenous (or overexpressed) GRK levels, these preparations should not be used for directly assessing activity because the detergent present in the lysis buffer potently inhibits GRKs. However, a subsequent partial purification on SP Sepharose results in preparations that can be assayed readily.

1. Prepare 5 ml of lysis buffer containg 20 mM Na HEPES, pH 7.5, 1% Triton X-100, 150 mM NaCl, 10 mM EDTA, 0.5 mM phenylmethylsulfonyl fluoride (PMSF), 20 μg/ml leupeptin, and 100 μg/ml benzamidine. The volume of lysis buffer can be scaled up or down as needed but should be made fresh and chilled on ice before use.

2. Cells are rinsed several times with phosphate-buffered saline (PBS). Ice-cold lysis buffer is then added (0.5–1 ml of buffer per 100 mm dish of cells), the cells are scraped off using a rubber policeman, transferred to a centrifuge tube, and then vortexed several times. Tissues can be homogenized with an ~10 : 1 (volume : wet weight) ratio of buffer to tissue (i.e., 1 ml of buffer per 100 mg of minced tissue) using a Polytron homogenizer (two bursts for 20 sec at maximum speed).

3. Lysates are centrifuged (40,000g for 10 min in an SS34 rotor or equivalent) to remove particulate matter, and the supernatant is transferred to an Eppendorf tube and assayed for protein. Cell/tissue supernatants can either be further analyzed immediately or aliquoted, frozen in liquid nitrogen, and stored at −80°C for later analysis. It is important to keep everything on ice and to perform the procedures as rapidly as possible to prevent proteolysis.

Electrophoresis and Immunoblotting Analysis

1. Incubate 20–40 μg of protein from the cell lysate or SP Sepharose-purified sample prepared (described later) with SDS sample buffer for 10 min at room temperature. It is advantageous to also include samples containing 1–5 ng of the purified GRK being assessed as well as a prestained protein standard.

2. Electrophorese the samples on a 10% SDS–polyacrylamide gel at a constant voltage of 120–150 V until the dye front is within a few millimeters of the bottom (~1 hr).

3. Carefully remove the stacking gel and then set up the running gel for transfer to nitrocellulose. This is accomplished by layering a transfer sponge, one piece of Whatman 3 MM paper, the running gel, one piece of nitrocellulose (with one corner cut on the protein standard side), one piece of Whatman 3 MM paper, and another sponge. Everything should initially be wetted in transfer buffer (25 mM Tris base, 192 mM glycine, 20% methanol, *do not adjust pH*) before layering. It is also important to make sure that all air bubbles are removed as each layer is added (a 13 × 100-mm plastic test tube can be rolled over the layer at each step to accomplish this). Transfer the protein samples to a nitrocellulose membrane for 1 hr at 100 V (constant voltage) using cold transfer buffer.

4. After transfer, sample loading and transfer efficiency can be checked by staining the nitrocellulose with 0.2% Ponceau S (prepared in deionized water) for ~1 min with the transfer side up. Transferred proteins can then be visualized by rinsing the nitrocellulose with deionized water to remove excess stain.

5. Rinse the nitrocellulose several times in Tris-buffered saline (20 m*M* Tris–Cl, pH 7.5, 150 m*M* NaCl) containing 0.05% Tween 20 (TTBS). The membrane is then blocked for 1 hr in 10 ml of TTBS containing 5% (w/v) nonfat dry milk or bovine serum albumin (BSA).

6. The nitrocellulose is next incubated with 5 ml of a diluted GRK-specific antibody (see Table I) for ≥1 hr at room temperature. The antibody dilution will vary dependent on the particular antibody being used and the tissue being analyzed.

7. Wash the nitrocellulose three to five times for 10 min each in TTBS.

8. Incubate the nitrocellulose for 1 hr with an ~1 : 3000 dilution of affinity-purified goat antimouse or antirabbit IgG conjugated to horseradish peroxidase (Bio-Rad) in TTBS-5% milk (or 1% BSA).

9. Wash the membrane three to five times for 10 min each in TTBS.

10. Overlay the nitrocellulose with 1–2 ml of SuperSignal West Pico Chemiluminescent Substrate (Pierce) reagent for ~1 min, allow the blot to drip dry, wrap in Saran wrap, and visualize on X-ray film.

Transient Overexpression of GRKs

Overexpression of GRKs in mammalian cells enables functional assessment of GRK/GPCR interaction, as well as the recovery of sufficient quantities of GRKs for biochemical analysis. Transient overexpression of GRKs in mammalian cells allows for a quick assessment of basic *in vitro* activities of newly generated GRK mutants. In addition, intact cell assessment of the effect of wild-type or dominant-negative GRK on the function of a particular GPCR can also be tested by analyzing such processes as GPCR phosphorylation, GPCR/G protein coupling, and receptor internalization. While a number of transfection strategies exist, transient transfection of GRK cDNAs using polycationic lipid reagents, such as LipofectAMINE (Life Technologies, Inc.) and FuGENE (Roche), is a relatively easy and reproducible method of overexpressing GRKs. It should be noted that careful optimization of the transfection protocol should be done to ensure efficient transfection of the cell line of interest. Attention should be paid to factors such as cell type, cell density, DNA and transfection reagent concentrations, and duration of transfection.

Typically, we use an adherent cell line such as the African green monkey kidney COS-1 or the human embryonic kidney cell line HEK293 to perform our transfections. The cells should be seeded 24 hr prior to transfection such that they are 50–90% confluent at the time of transfection. The procedure provided is for the use of FuGENE as the transfection reagent, although LipofectAMINE also works well.

1. Aliquot the plasmid DNA (~3–5 μg per 60-mm dish) into a 1.5-ml Eppendorf tube. In a separate Eppendorf, add 0.3 ml of warmed DMEM followed by 10 μl of FuGENE added dropwise with swirling. Do not allow the FuGENE to directly

contact the sides of the tube. Allow the FuGENE:DMEM mixture to incubate for 5 min at room temperature. We generally maintain a 3 : 1 ratio of FuGENE:DNA (e.g., 12 μl FuGENE to 4 μg DNA). Additional guidelines for optimizing transfection efficiency, depending on the cell type and plasmid, are provided in the manufacturers' directions.

2. Add the DMEM/FuGENE mixture to the DNA dropwise, mix by gentle inversion of the tube, and leave at room temperature for 15 min. The FuGENE:DNA mixture is then added evenly to the cell monolayer dropwise while swirling gently. There is no need to change the existing medium.

3. The transfection mixture can be left on the cells until the time of harvest. The cells are typically harvested 24 to 48 hr posttransfection for immunoblot or biochemical analysis.

Partial Purification of GRKs by SP Sepharose Chromatography

In order to perform biochemical studies of overexpressed or endogenous GRKs, we have found it useful to partially purify these proteins, as various factors within crude lysates inhibit GRK activity. Stepwise fractionation on the cation-exchange resin SP Sepharose permits adequate resolution of GRKs from the majority of lysate proteins and, in addition, results in partial separation of specific GRK isoforms. For example, whereas GRK2, 3, 5, and 6 all bind to SP Sepharose at low ionic strength (e.g., 50 m*M* NaCl), GRK2 and 3 can be eluted with 150 m*M* NaCl, GRK6 with 400 m*M* NaCl, and GRK5 with 600 m*M* NaCl. The following procedure is used for the batchwise purification of GRKs from mammalian cell lysates.

SP Sepharose Purification

1. Prepare a cell or tissue lysate as outlined earlier. Dilute the lysate 1 : 4 with buffer A [20 m*M* HEPES, pH 7.5, 10 m*M* EDTA, 0.02% Triton X-100, 1 m*M* dithiothreitol (DTT), 0.5 m*M* PMSF, 20 μg/ml leupeptin, 200 μg/ml benzamidine] to reduce the concentration of salt (this is particularly important for the purification of GRK2/3).

2. Add the diluted lysate to SP Sepharose resin ($\sim$1 mg total protein/30 μl resin) prewashed in buffer A containing 50 m*M* NaCl and incubate for 1 hr on a rotator at 4°. Pellet the resin by centrifugation ($\sim$10 sec in a microcentrifuge is adequate) and discard the supernatant.

3. Wash the resin two to three times with $\sim$1 ml of cold buffer A containing 50 m*M* NaCl to remove unbound proteins.

4. GRK2 (and GRK3) is eluted with buffer A containing 150–200 m*M* NaCl by incubating the resin for 10 min with elution buffer (100 μl buffer/30 μl resin) on a rotator at 4°. Briefly centrifuge ($\sim$10 sec) the resin, remove the eluate, and repeat the procedure. The eluates are then pooled together.

5. The resin can then be eluted with cold buffer A containing 400 m*M* NaCl to obtain a GRK6-enriched fraction, following the same procedure as used in step 4, whereas GRK5 can subsequently be eluted with buffer A containing 600 m*M* NaCl. It is important to note, however, that some GRK5 will likely elute in the 400 m*M* NaCl fraction. Thus, if complete separation of GRK5 and 6 is needed, it is recommended that the sample be fractionated on a Mono S FPLC column using a linear NaCl gradient.

It is best to aliquot and store the samples at -80° and use the preparations within several weeks of purification. Expression and recovery of GRKs can be evaluated by immunoblotting and rhodopsin phosphorylation assays (see later). If the GRKs are overexpressed, typical yields are 1–10 μg of partially purified GRK per T75 flask of COS-1 cells. Recovery of endogenous GRKs using this procedure will vary depending on the cell type or tissue being analyzed.

Analysis of GRK Activity from Mammalian Cells

Many protein kinases can be assayed using readily available substrates such as histones, phosvitin, casein, or specific synthetic peptides. Unfortunately, these general substrates are relatively poor substrates for GRKs and thus cannot be used to assay GRK activity in crude cell or tissue homogenates. However, GRKs do have the ability to specifically phosphorylate activated GPCRs. Thus, one very useful assay utilizes the ability of GRKs to phosphorylate rhodopsin in a light-dependent fashion. This assay uses urea-treated bovine rod outer segments (ROS) as the substrate. Urea-treated ROS are prepared as described in detail[6] using the procedures of Shichi and Somers[7] and Wilden and Kuhn.[8] The urea-treated ROS are aliquoted (20–30 μl is convenient), wrapped in aluminum foil, frozen in liquid nitrogen, and stored at -80° until needed. The typical yield is 5–10 mg of rhodopsin from a 50 retina preparation, which is enough for several thousand assays.

Phosphorylation of Rhodopsin by GRKs

The phosphorylation of urea-treated ROS by crude kinase preparations requires the separation of the phosphorylated rhodopsin from other endogenous phosphorylated proteins. The separation of rhodopsin is readily accomplished by electrophoresis on a 10% polyacrylamide gel in the presence of sodium dodecyl sulfate (SDS)[9] and autoradiography. The resultant bands can then be cut and counted if quantitative results are needed. It is important to note that GRKs are very

[6] J. L. Benovic, *Methods Enzymol.* **200,** 351 (1991).
[7] H. Shichi and R. L. Somers, *J. Biol. Chem.* **253,** 7040 (1978).
[8] U. Wilden and H. Kuhn, *Biochemistry* **21,** 3014 (1982).
[9] U. K. Laemmli, *Nature* **227,** 680 (1970).

sensitive to inhibition by salt (e.g., 0.1 *M* NaCl inhibits GRK2 activity by ~90%) and detergents (e.g., 0.02% Triton X-100 inhibits ~50% rhodopsin phosphorylation activity of GRK5). Thus, the ionic strength and detergent concentration in the reaction should be kept as low as possible. The procedure for assaying GRK activity using urea-treated ROS is detailed.

1. Prepare an assay mixture containing (per 10 assays): 160 μl 20/2 buffer (20 m*M* Tris–HCl, pH 7.5, 2 m*M* EDTA), 2 μl 0.5 *M* $MgCl_2$, 2 μl 10 m*M* ATP, ~10 μCi [γ-^{32}P]ATP (NEG-002A from NEN Dupont), and 20 μl urea-treated ROS (~1 mg/ml rhodopsin, final concentration ~2.5 μM) (add just before use).

2. Assays are set up containing 18 μl of assay mixture and ~2 μl of the kinase preparation (GRK concentration in the reaction should be $\leq$50 n*M* to prevent complete receptor phosphorylation). The samples are incubated for 5–20 min at 30° in room light (control incubations should be kept in the dark). It is important that the ROS be kept in the dark until just before use or else the amount of phosphate incorporated into rhodopsin will be reduced significantly.

3. Following the incubation period, reactions can be quenched several different ways. If the kinase preparation is a crude homogenate, the reactions can be stopped by the addition of 0.2 ml of cold 100 m*M* sodium phosphate, pH 7, 5 m*M* EDTA buffer followed by centrifugation in a tabletop ultracentrifuge (15 min at 50,000*g*). The pellets can be resuspended in 20 μl of SDS sample buffer (sonication and/or vortexing helps with the resuspension) followed by electrophoresis on a homogeneous 10% polyacrylamide gel. The gel is dried and exposed to X-ray film. The centrifugation step is not absolutely necessary; however, it does improve the signal-to-noise ratio on the autoradiogram significantly. With more purified kinase preparations, the reactions can be directly quenched with SDS sample buffer followed by electrophoresis. For both methods, it is important that the samples be incubated at room temperature for ~30 min following the addition of SDS sample buffer. This is needed to obtain adequate denaturing of the rhodopsin. It is also important to note that the sample should not be boiled, as this will cause aggregation of the receptor.

Expression and Purification of GRKs Using Sf9 Insect Cells

Overexpression and purification of GRKs have provided useful tools for detailed analysis of enzyme kinetics, assessment of *in vitro* substrate specificity, and use as a protein standard for quantitative immunoblotting. While all of the GRKs have been successfully expressed in mammalian cells, expression in insect cells results in a high level of expression and enables purification of milligram quantities of protein at a low cost. Attempts at expressing functional GRKs in bacteria have so far proved unsuccessful. While most of the GRKs have been successfully expressed in insect cells, we will mainly describe the procedures for overexpressing and purifying GRK2[10] and GRK5.[11] The reader should refer to previous publications for specific details of GRK1,[12] GRK3,[10] GRK4,[13] and GRK6[14] overexpression and purification.

Sf9 Cell Culture

Wild-type *Spodoptera frugiperda* (Sf9) cells can be cultured in a 27° incubator either on a monolayer or in suspension (spinner flask, 70 rpm) using Sf-900 II serum-free medium (GIBCO BRL) containing 10 mg/liter gentamicin (GIBCO BRL). Supplementing this medium with 10% of fetal bovine serum (Sigma or Life Technologies, Inc.) will increase cell growth rate and useful cell density by ~1.5- to 2-fold. For large-scale cultures (≥1 liter), Sf9 cells are grown in a shaking incubator (120 rpm) at 27° in complete medium. Cells in monolayer are typically used during the early stages of baculovirus isolation and amplification, whereas suspension cells are used for larger scale baculovirus amplification and protein expression. In general, although Sf9 cells are easy to grow, they need to be split every 2–3 days and kept at a cell density between 0.5 and 2.5×10^6 cells/ml.

Expression of GRKs in Sf9 Cells

1. Several different systems are currently available to prepare recombinant baculovirus containing cDNA of interest. All of the systems at some point involve a step, which requires a recombination of a vector containing the cDNA of interest and linearized viral DNA. The culture supernatant obtained at this step contains recombinant baculovirus and is used to reinfect Sf9 cell for further amplification. The reader should refer to the specific manufacturers' instructions to generate recombinant GRK-containing baculovirus.

2. Culture 1 liter of Sf9 cells at 27° in a 2.8-liter Fernbach flask to a density of $\sim 2.5 \times 10^6$ cells/ml (up to 4×10^6 cells/ml if grown in the presence of serum). This is best accomplished in a shaking incubator, but the cells can also be cultured in a spinner flask to a density of $\sim 2 \times 10^6$ cells/ml.

3. The cells are removed from the incubator and then infected with the recombinant baculovirus [we typically use ~25 ml of amplified virus per liter of cells, resulting in a multiplicity of infection (MOI) of 5–10]. The cells are incubated for 1 hr at room temperature and then returned to the incubator.

[10] C. M. Kim, S. B. Dion, J. J. Onorato, and J. L. Benovic, *Receptor* **3,** 39 (1993).

[11] P. Kunapuli, J. J. Onorato, M. M. Hosey, and J. L. Benovic, *J. Biol. Chem.* **269,** 1099 (1994).

[12] K. Cha, C. Bruel, J. Inglese, and H. G. Khorana, *Proc. Natl. Acad. Sci. U.S.A.* **94,** 10577 (1997).

[13] R. T. Premont, A. D. Macrae, R. H. Stoffel, N. Chung, J. A. Pitcher, C. Ambrose, J. Inglese, M. E. MacDonald, and R. J. Lefkowitz, *J. Biol. Chem.* **271,** 6403 (1996).

[14] R. P. Loudon and J. L. Benovic, *J. Biol. Chem.* **269,** 22691 (1994).

Purification of GRK2 and GRK5 from Sf9 Cells

1. Baculovirus-infected cells are normally harvested at ~40–48 hr postinfection. The cells should be monitored carefully during this period, as longer infection times often result in cell lysis and loss of cytosolic proteins. The cells are centrifuged at 1000*g* for 15 min in a Beckman J-6B centrifuge or equivalent. The supernatant is discarded and the pellet is resuspended in ice-cold PBS and centrifuged again at 1000*g* for 15 min.

2. The washed pellet is resuspended in 150 ml of ice-cold homogenization buffer (20 m*M* sodium HEPES, pH 7.2, 250 m*M* NaCl, 0.02% Triton X-100, 5 m*M* EDTA, 1 m*M* DTT, 5 μM leupeptin, 1 m*M* PMSF, 3 m*M* benzamidine) and homogenized using a Brinkman Polytron (2 × 30 sec at 25,000 rpm). Homogenization and all subsequent steps are performed at 4°. The homogenate is centrifuged at 45,000*g* to remove unbroken cells and nuclei (SS34 rotor or equivalent), and the supernatant is then recentrifuged at 300,000*g* for 60 min (Ti60 rotor or equivalent).

3. Dilute the high-speed supernatant to ~600 ml for GRK2 or ~270 ml for GRK5 with ice-cold buffer C (20 m*M* sodium HEPES, pH 7.2, 2 m*M* EDTA, 0.02% Triton X-100, 1 m*M* DTT) and load the sample on an 18 (GRK2) or 6 (GRK5)-ml SP Sepharose (Pharmacia) column equilibrated with buffer C. The column is washed with 50 m*M* NaCl in buffer C and then eluted with a 120-ml linear gradient from 50 to 300 m*M* NaCl in buffer C at a flow rate of 1 ml/min for GRK2 or a 70-ml linear gradient from 200 to 500 m*M* NaCl in buffer C for GRK5. GRK2 elutes at ~150 m*M* NaCl, whereas GRK5 elutes at ~400 m*M* NaCl. Fractions from the SP Sepharose column are monitored for GRK activity by rhodopsin phosphorylation as described earlier. In addition, protein purity is monitored by SDS–PAGE and Coomassie blue staining (see Fig. 2).

4. The peak GRK2 fractions are pooled (~50 ml), diluted threefold with buffer C, and loaded on a 7-ml heparin-Sepharose column (1.5 × 4 cm). The column is washed with 100 m*M* NaCl in buffer C, and GRK2 activity is then eluted with a 100-ml linear gradient from 100 to 600 m*M* NaCl in buffer C at a flow rate of 1 ml/min. The fractions are analyzed for GRK2 purity, activity, and protein concentration.

5. The peak GRK5 fractions (e.g., 13–16 in Fig. 2A) are pooled (~12 ml), diluted two-fold with buffer C, and loaded on a 1-ml Mono S FPLC column. In order to prevent overloading of the column, the GRK sample can be split and loaded and eluted in two separate runs. The column is washed with 10 ml of 200 m*M* of NaCl in buffer C, and proteins are eluted with a 20-ml linear gradient from 300 to 500 m*M* NaCl in buffer C at a flow rate of 1 ml/min. GRK5 elutes at ~420 m*M* NaCl. The fractions are analyzed for GRK5 purity (Fig. 2B), activity, and protein concentration.

6. The appropriate fractions are then pooled, concentrated (typically to ~1–1.5 mg/ml), stabilized by addition of ~25% glycerol, aliquoted, and stored at either −20° (for short-term storage) or −80° (for long-term storage).

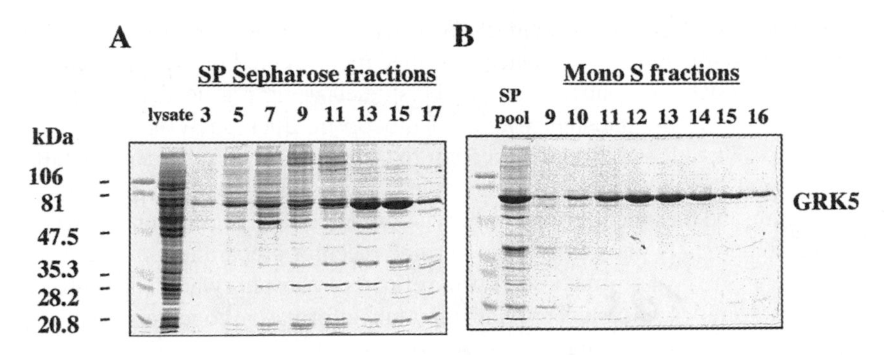

FIG. 2. Purification of GRK5 overexpressed in Sf9 cells. Analysis of proteins eluting in fractions spanning the peak of GRK5 activity from SP Sepharose (A) and Mono S FPLC (B) columns. Ten-microliter aliquots of various fractions were separated on a 10% SDS–PAGE and gels were stained with Coomassie blue. Migration points of marker proteins are indicated at left.

GPCR Phosphorylation by GRKs

While the ability of GRKs to phosphorylate rhodopsin has proved useful for assessing GRK activity and regulation, it is also useful to be able to characterize the phosphorylation of other GPCRs with GRKs. This section describes some of the *in vitro* strategies for characterizing GRK phosphorylation of GPCRs. Successful strategies for studying GPCR phosphorylation by GRKs have utilized purified GPCRs that are present in either phospholipid vesicles or mixed micelles. In addition, strategies for phosphorylating membrane preparations containing overexpressed GPCRs have also been developed.

Phosphorylation of Purified GPCRs

Although a relatively small number of GPCRs have been purified, GRK phosphorylation of several purified receptors (e.g., β_2AR, m2AChR) has provided useful kinetic information.[10,11,14] Such studies require a relatively highly purified receptor that is then reinserted into a phospholipid bilayer (GPCR purification and reconstitution techniques are discussed by Ptasienski and Hosey[15]). Alternatively, the β_2AR has also been used successfully as a GRK substrate in mixed micelle preparations[16] While such purified preparations can be phosphorylated using either highly purified or even relatively crude GRK preparations, a key feature of these assays is the lipid requirement. Reconstitution of the receptor into phospholipid vesicles is often necessary to generate a functionally active receptor that can interact with GRKs in an agonist-dependent manner. In addition, GRKs appear to

[15] J. A. Ptasienski and M. M. Hosey, *in* "Regulation of G Protein-Coupled Receptor Function and Expression (J. L. Benovic, ed.), p. 55. Wiley-Liss, New York, 2000.

[16] J. J. Onorato, M. E. Gillis, Y. Liu, J. L. Benovic, and A. E. Ruoho, *J. Biol. Chem.* **270,** 21346 (1995).

be phospholipid-dependent enzymes whose activity is largely controlled by association with acidic lipids.[2,16,17] Once a purified reconstituted receptor preparation is available, GRK phosphorylation can be studied using a procedure similar to that described for the phosphorylation of rhodopsin. Such analysis can provide information about the ability of a receptor to serve as a GRK substrate, the GRK specificity of the phosphorylation, the stoichiometry and potential sites of phosphorylation, and the kinetics of phosphorylation (K_m, V_{max}). In addition, the ability to stoichiometrically phosphorylate a purified receptor preparation enables additional analysis to be performed, such as the effects of phosphorylation on G protein coupling. Overall, such results provide important information concerning potential mechanisms of GPCR regulation.

Phosphorylation of GPCRs in Membrane Preparations

Because it is not always possible to generate a highly purified receptor preparation, strategies for phosphorylating GPCRs enriched in membrane preparations have also been developed. An important feature of such strategies is an adequate overexpression system for preparing an enriched receptor preparation. The most successful assays for analyzing GRK phosphorylation in membrane preparations have involved the use of GPCRs expressed in Sf9 insect cells. This is partly due to the somewhat higher level of expression that one can achieve in insect cells compared to mammalian cells, as well as the more homogenous glycosylation that occurs in insect cells (the heterogeneity of receptor glycosylation observed in mammalian cells often reduces the signal-to-noise ratio of phosphorylation reactions). Once cells that express a high level of a particular GPCR are generated (ideally $\geq$10 pmol/mg membrane protein), the membranes need to be treated in order to generate a usable substrate for GRK phosphorylation. One way of treating such preparations is essentially identical to the preparation of rhodopsin (i.e., the membranes are treated with urea). This treatment appears to effectively dissociate/denature many of the peripherally associated proteins and protein kinases without harming the GPCR. While such a strategy has been used successfully with the m2AChR,[18] one needs to determine empirically that such a treatment is not harmful to the receptor of interest. If the receptor appears to be denatured by this treatment, antagonist binding to the receptor usually helps stabilize the receptor during urea treatment. Alternatively, other protein denaturants can also be tested.

A second strategy that has been used to generate a GPCR-containing membrane preparation that could serve as an effective GRK substrate involved differential sucrose gradient fractionation.[19] This procedure makes use of the finding that some

[17] S. K. DebBurman, J. Ptasienski, J. L. Benovic, and M. M. Hosey, *J. Biol. Chem.* **270,** 5742 (1995).

[18] S. K. DebBurman, P. Kunapuli, J. L. Benovic, and M. M. Hosey, *Mol. Pharm.* **47,** 224 (1995).

[19] G. Pei, M. Tiberi, M. G. Caron, and R. J. Lefkowitz, *Proc. Natl. Acad. Sci. U.S.A.* **91,** 3633 (1994).

GPCRs are found in a light vesicle fraction that can be separated readily from plasma membranes. Such preparations yield a receptor that is highly enriched (100–300 pmol/mg) and can serve as an effective GRK substrate. As discussed earlier, the use of a purified GRK has proven most effective for looking at GPCR phosphorylation in membranes; however, overexpressed and/or partially purified GRKs should also work in such assays.

[35] Regulation of G Protein-Coupled Receptor Kinase 2

By TATSUYA HAGA, KAZUKO HAGA, KIMIHIKO KAMEYAMA, HIROFUMI TSUGA, and NORIHIRO YOSHIDA

Introduction

G protein-coupled receptor kinase 2 (GRK2) was originally termed a β-adrenergic receptor kinase or βARK because it was purified as a kinase of the adrenergic β_2 receptor,[1] and the adrenergic β_2 receptor was the only known substrate of GRK2. GRK2 is now known to phosphorylate different kinds of G protein-coupled receptors (GPCRs), including rhodopsin, muscarinic acetylcholine receptors, and adrenergic α_2 receptors when they are stimulated by light absorption or agonist binding (see reviews 2–4). There is no appreciable homology among amino acid sequences around phosphorylation sites in these GPCRs, except that they are flanked by acidic amino acid residues.[5] The substrate specificity of GRK2, however, is strict in a sense that only a few proteins are known to be substrates of GRK2 besides agonist-bound GPCRs. One of the reasons why agonist-bound GPCRs are phosphorylated by GRK2 is that GRK2 is activated by agonist-bound receptors. GRK2 is also activated by G protein $\beta\gamma$ subunits[6,7] and is activated synergistically by G protein $\beta\gamma$ subunits and agonist-bound receptors.[8] Thus agonist-bound receptors serve

[1] J. L. Benovic, R. H. Strasser, M. G. Caron, and R. J. Lefkowitz, *Proc. Natl. Acad. Sci. U.S.A.* **83,** 2797 (1986).
[2] T. Haga, K. Haga, and K. Kameyama, *J. Neurochem.* **63,** 400 (1994).
[3] J. A. Pitcher, N. J. Freedman, and R. J. Lefkowitz, *Annu. Rev. Biochem.* **67,** 653 (1998).
[4] J. G. Krupnick and J. L. Benovic, *Annu. Rev. Pharmacol. Toxicol.* **38,** 289 (1998).
[5] J. J. Onorato, K. Palczewski, J. W. Regan, M. G. Caron, R. J. Lefkowitz, and J. L. Benovic, *Biochemistry* **30,** 5118 (1991).
[6] K. Haga and T. Haga, *FEBS Lett.* **268,** 43 (1990).
[7] K. Kameyama, K. Haga, T. Haga, K. Kotani, T. Katada, and Y. Fukada, *J. Biol. Chem.* **268,** 7753 (1993).
[8] K. Haga, K. Kameyama, and T. Haga, *J. Biol. Chem.* **269,** 12594 (1994).

METHODS IN ENZYMOLOGY, VOL. 343
0076-6879/02 $35.00

as both substrates and activators of GRK2. The activation of GRK2 by agonist-bound receptors and $\beta\gamma$ subunits is interfered with by Ca^{2+}-calmodulin[9,10] or Ca^{2+}-S100 protein. Very recently, tubulin,[11–13] synuclein,[14] and phosducin[15] have been reported to be phosphorylated by GRK2, but it is not known whether they serve as activators of GRK2 and why they are substrates of GRK2 unless they are activators.

General Comments on Expression and Purification of Proteins

Many proteins, including GRK2, muscarinic receptors, and G protein $\beta\gamma$ subunits, are expressed in insect cells [*Spodoptera frugiperda* (Sf9) cells] using baculovirus. Constructs of vectors and preparations of viruses are not described here. They will be found in original papers cited here or in the manufacturer's protocol. All the purification procedures of proteins given in the following sections are carried out at 4° unless expressed otherwise. Media used for extraction of proteins or equilibration of columns generally include a "cocktail" of protease inhibitors, which include 0.5 (0.25–1) m*M* phenylmethylsulfonyl fluoride, 0.5 (0.1–1) m*M* benzamidine, 2.5 (0.25–10) μg/ml pepstatin, 2.5 (0.25–10) μg/ml leupeptin, and 0.5 μg/ml aprotinin. The "cocktail" is described as protease inhibitors in the following sections.

Preparation of Enzymes

Purification of Muscarinic Receptor Kinase (mAChR Kinase) from Porcine Brain

A kinase, which phosphorylates muscarinic receptors in an agonist-dependent manner, was purified from porcine brain and termed mAChR kinase.[16,17] The purification procedure is a modification of the one used by Benovic *et al.*[1] for the purification of GRK2. Porcine cerebrum (200 g) is homogenized in a buffer solution [A: 20 m*M* Tris–HCl (pH 7.5), 5 m*M* EDTA, 5 m*M* EGTA, protease

[9] K. Haga, H. Tsuga, and T. Haga, *Biochemistry* **36,** 1315 (1997).

[10] A. N. Pronin, D. K. Satpaev, V. Z. Slepak, and J. L. Benovic, *J. Biol. Chem.* **272,** 18273 (1997).

[11] K. Haga, H. Ogawa, T. Haga, and H. Murofushi, *Eur. J. Biochem.* **255,** 363 (1998).

[12] J. A. Pitcher, R. A. Hall, Y. Daaka, J. Zhang, S. S. G. Ferguson, S. Hester, S. Miller, M. G. Caron, R. J. Lefkowitz, and L. S. Barak, *J. Biol. Chem.* **273,** 12316 (1998).

[13] C. V. Carman, T. Som, C. M. Kim, and J. L. Benovic, *J. Biol. Chem.* **273,** 20308 (1998).

[14] A. N. Pronin, A. J. Morris, A. Surguchov, and J. L. Benovic, *J. Biol. Chem.* **275,** 26515 (2000).

[15] A. Ruiz-Gomez, J. Humrich, C. Murga, U. Quitterer, M. J. Lohse, and F. Mayor, Jr., *J. Biol. Chem.* **275,** 29724 (2000).

[16] K. Haga and T. Haga, *Biomed. Res.* **10,** 293 (1989).

[17] K. Haga and T. Haga, *J. Biol. Chem.* **267,** 2222 (1992).

inhibitors, 1 liter]. After centrifugation at 30,000 rpm for 20 min, the supernatant is mixed with solid ammonium sulfate to be 30% (weight/volume) followed by stirring for 60 min and centrifugation. The pellet is resuspended in a buffer solution [B: 20 m*M* Tris–HCl (pH 7.5), 2 m*M* EDTA, 2 m*M* EGTA, protease inhibitors, 110 ml], and the suspension is applied to a Ultrogel AcA34 column (4.6 × 71 cm, 1.2 liters), which is prewashed with solution B. The bulk of proteins is eluted in the void volume fraction and separated from mAChR kinase. The fraction containing mAChR kinase is applied to a DEAE-Sephacel column (30 ml). After washing the column with solution B, mAChR kinase is eluted with solution B supplemented with 50 m*M* NaCl. This procedure, using Ultrogel and DEAE-Sephacel columns, is repeated four times, and mAChR kinase fractions are combined. The combined fraction is applied to a CM-Toyopearl column (12 ml), and mAChR kinase is eluted with a linear gradient of 20 and 150 m*M* NaCl in a solution containing 20 m*M* Tris–HCl (pH 7.5), 1 m*M* EDTA, and protease inhibitors (50 × 50 ml). The fraction with high activities (80 ml) is collected, mixed with 20 ml of 50% glycerol and 0.1% CHAPS, and stored at −80° before use. The concentration of mAChR kinase in the final preparation is 27 units/ml, where I unit is defined as the amount of enzyme that transfers 1 pmol of phosphate/min when assayed in the presence of 20 μM ATP, 30 n*M* muscarinic M_2 receptors, 50 n*M* G protein $\beta\gamma$ subunits, and 1 m*M* carbamylcholine. Kinase activity can be measured using rod outer segments containing light-activated rhodopsin as substrates as well as agonist-bound muscarinic M_2 receptors; in both cases, G protein $\beta\gamma$ subunits are added as activators. The activity to phosphorylate muscarinic receptors is eluted in the same fractions as the activity to phosphorylate rhodopsin,[17] and the purified preparation has the ability to phosphorylate adrenergic β_2 receptors.[7] mAChR kinase is thought to be the same as or very similar to GRK2. The kinase activity in each fraction can be measured using tubulin as substrates.

Expression and Purification of GRK2 in Insect Cells

cDNA for GRK2 may be expressed in and purified from Sf9 cells on a larger scale compared with the preparation of mAChR kinase from porcine brain. The following procedure is modified from that described by Kameyama *et al.*[18] Transfected Sf9 cells (1 liter cultured suspension) are homogenized in a buffer solution [C: 20 m*M* HEPES–KOH (pH 7), 2 m*M* $MgCl_2$, 1 m*M* dithiothreitol (DTT), and protease inhibitors, 20 ml]. The homogenate is centrifuged, and the pellet is suspended in solution C supplemented with 0.5 *M* KCl (50 ml), followed by centrifugation. To the supernatant, saturated ammonium sulfate solution is added to a volume of 20%. After centrifugation, saturated ammonium sulfate solution is added to the supernatant to a volume of 30%, the suspension is centrifuged, and

[18] K. Kameyama, K. Haga, T. Haga, O. Moro, and W. Sadee, *Eur. J. Biochem.* **226,** 267 (1994).

the pellet is dissolved in the solution C (15 ml). The suspension is applied to a phenyl Sepharose column (5 ml) equilibrated with solution C supplemented with 1 *M* ammonium sulfate at a flow rate of 1 ml/min. The column is washed with the same solution (10 ml) and then eluted with a linear gradient of solution C containing 1 *M* ammonium sulfate and the solution C (20 × 20 ml). Fractions containing GRK2 are combined, dialyzed against a buffer solution [20 m*M* HEPES-KOH (pH 7.0) and 50 m*M* NaCl], and applied to a heparin column (1 ml) equilibrated with the dialysis buffer. The column is eluted with a linear gradient of 50–500 m*M* NaCl in 20 m*M* HEPES–KOH (pH 7.0) (15 × 15 ml). Fractions containing GRK2 (ca 5 ml) are mixed with an equal volume of glycerol and stored at −80° until use. The fraction shows a single band on SDS–PAGE and contains at least 3 mg and at most 9 mg of purified GRK2. The kinase activity is monitored using tubulin as a substrate in the procedures just described.

Preparation of Substrates

Authentic substrates of GRK2 are GPCRs, particularly adrenergic β_2 receptors, muscarinic receptors, and rhodopsin. The expression levels of GPCRs, except for rhodopsin, are not high enough for their phosphorylation to be detected using membrane preparations of tissues or cultured cells, and hence GPCRs must be purified or at least partially purified in order to be used as substrates of GPCRs. The purification protocol using affinity chromatography has been established for adrenergic β_2 receptors and muscarinic receptors. It is difficult, however, to purify them from tissues because the expression level is low, at most 1 pmol/mg protein and much less in most cases. Muscarinic receptors may be expressed as a fusion protein with a maltose-binding protein in *Escherichia coli* with intact function to activate G proteins,[19] but the expression level is approximately 1 pmol/mg protein, which is 10 times lower than that in Sf9 cells. This section describes purification of muscarinic receptors expressed in Sf9 cells.[18,20]

ROS membranes and tubulin are other substrates of GRK2, which may be prepared more easily as compared with other GPCRs. In addition, fusion proteins with glutathione *S*-transferase (GST) of peptides, including phosphorylation sites, are expressed in *E. coli* and used as substrates of GRK2.

Expression and Purification of Muscarinic M_2 Receptors and Receptor Mutants

Muscarinic M_2 receptors are phosphorylated by mAChR kinase or GRK2 at the central part of the third intracellular loop.[21] We have prepared the human M_2

[19] H. Furukawa and T. Haga, *J. Biochem.* **127,** 151 (2000).

[20] E. M. Parker, K. Kameyama, T. Higashijima, and E. M. Ross, *J. Biol. Chem.* **266,** 519 (1991).

[21] H. Nakata, K. Kameyama, K. Haga, and T. Haga, *Eur. J. Biochem.* **220,** 29 (1994).

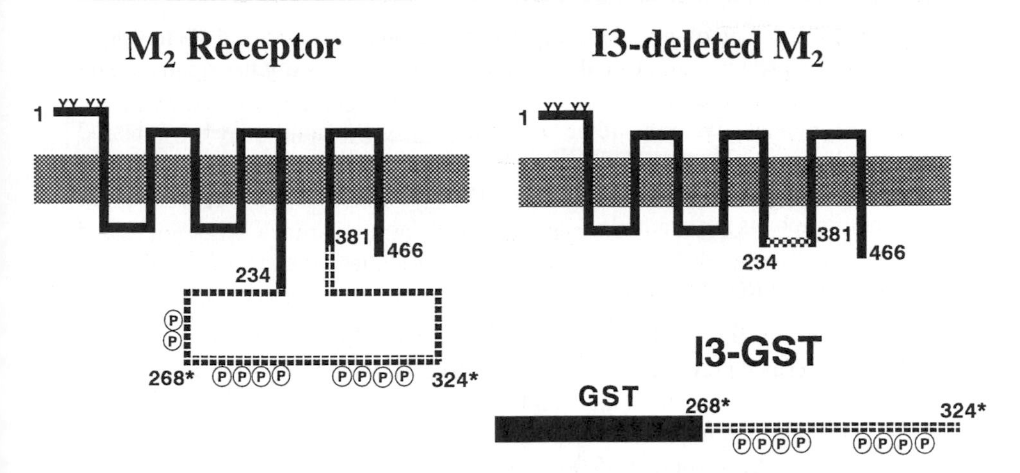

FIG. 1. Schemes of muscarinic M_2, I3-deleted M_2 receptor, and a fusion protein with glutathione *S*-transferase (GST) of the central part of the third intracellular loop of M_2 receptor (I3-GST). Numbers represent those of amino acid residues for the human M_2 receptor. Circled P shows phosphorylation sites by mAChR kinase.

receptor, a human M_2 receptor mutant (I3-deleted M_2), with a deletion in the central part of the third intracellular loop (234–381), including phosphorylation sites, and a fusion protein of the third intracellular loop of M_2 receptors (268–324), which includes phosphorylation sites, with glutathione *S*-transferase (I3-GST)[18] (Fig. 1). M_2 receptors are expressed in and purified from Sf9 cells. The transfected Sf9 cells were collected from 3 to 5 liters of cultured suspension by centrifugation and homogenized in a buffer solution [D: 20 m*M* Tris–HCl (pH 7.5), 1 m*M* EDTA, 1 m*M* EGTA, 2 m*M* $MgCl_2$, protease inhibitors]. The homogenate was centrifuged and the pellet was suspended in the solubilizing buffer [10 m*M* potassium phosphate (pH 7.0), 50 m*M* NaCl, protease inhibitors, 1% digitonin, and 0.3% sodium cholate) (protein concentration 4–8 mg/m)]. After stirring for 1 hr and centrifugation, the supernatant was applied to an affinity column [aminobenztropine (ABT) agarose,[22,23] 400 ml] preequilibrated with a buffer solution (E: 20 m*M* potassium phosphate, 0.1% digitonin). The column was washed with 3–4 column volumes of solution E supplemented with 0.15 *M* NaCl, connected with a small column of hydroxyapatite (1 ml), and eluted with a 3–4 column volume of solution E containing 0.15 *M* NaCl and 100 μM atropine. Muscarinic receptors are discharged from ABT agarose by atropine and bound to the hydroxyapatite column. The hydroxyapatite column is washed with 0.15 *M* potassium phosphate buffer containing 0.1% digitonin, and then muscarinic receptors are eluted from the column by 0.5 *M* potassium phosphate buffer containing 0.1% digitonin. Purified

[22] K. Haga and T. Haga, *J. Biol. Chem.* **258,** 13575 (1983).
[23] K. Haga and T. Haga, *J. Biol. Chem.* **260,** 7927 (1985).

M_2 receptors (1–5 nmol/ml) are frozen at $-80°$ before use. I3-deleted M_2 receptors are also expressed in and purified from Sf9 cells by essentially the same method as used for M_2 receptors.

An M_2 mutant with a 6 histidine tag at the carboxyl terminus may be expressed in Sf9 cells and purified using a Co^{2+}-immobilized gel.[24] The same strategy could be used for expression and purification of GPCRs for which efficient ligand-affinity chromatography is not available.

Preparations of ROS Membranes

Rod out segment (ROS) membranes are prepared from the bovine or porcine retina as described previously.[25] ROS membranes derived from 50 porcine retinas are suspended in 5 *M* urea solution containing 5 m*M* EDTA and 50 m*M* Tris–HCl buffer and sonicated for 4 min, followed by dilution with 50 m*M* Tris–HCl buffer (50 ml) and centrifugation at 35,000 rpm for 30 min. The washing procedure is repeated three times, and the final pellet is suspended in 50 m*M* Tris–HCl (5 ml) and stored at $-80°$. All of these procedures are done in dim light. The concentration of rhodopsin in the suspension will be approximately 6 nmol/ml, which is estimated by measuring the absorption of the digitonin-solubilized solution and using the molar absorbance of 4×10^4 at 500 nm.[17]

Preparations of Tubulin

Tubulin is one of the most abundant proteins in the brain. Partially purified tubulin can be obtained by repetition of polymerization and depolymerization in the presence of glycerol.[26] Porcine brain (4 cerebra, 300 g) is homogenized in a buffer solution [F: 0.1 *M* MES (pH 6.8), 0.5 m*M* $MgCl_2$, 1 m*M* EDTA, and 1 m*M* GTP, 200 ml] followed by centrifugation at 40,000 rpm for 40 min at 4°. To the supernatant (270 ml), one-third volume of glycerol (90 ml) and GTP (0.1 *M*, 3.5 ml) are added. The suspension is incubated at 37° for 40 min and then centrifuged at 30° for 40 min. The pellet is suspended in solution F (20 ml) with a Potter-type homogenizer followed by centrifugation at 4° for 40 min. Polymerization at 37° and depolymerization at 4° are repeated once more. The partially purified tubulin preparation (20 mg/ml, 16 ml) is stored at $-80°$ after addition of one-third volume of glycerol. For further purification, the stocked tubulin solution (12 ml) is thawed, incubated in the presence of 1 m*M* GTP for 40 min at 37°, and centrifuged. The pellet is resuspended and centrifuged at 4°. The supernatant is applied to a phosphocellulose column (50 ml), which has been prewashed with solution F (100 ml). Pure tubulin is recovered in the flow-through fraction, to which

[24] M. K. Hayashi and T. Haga, *J. Biochem.* **120,** 1232 (1996).

[25] D. S. Papermaster and W. J. Dreyer, *Biochemistry* **13,** 2438 (1974).

[26] M. L. Shelanski, F. Gaskin, and C. R. Cantor, *Proc. Natl. Acad. Sci. U.S.A.* **70,** 705 (1973).

one-tenth volume of a solution containing 0.9 *M* MES, 11 m*M* $MgCl_2$, and 1 m*M* EGTA and one-hundredth volume of 50 m*M* GTP are added. The purified tubulin is much more unstable than the partially purified tubulin. The purified tubulin is stored at −80° immediately after preparation.

Partially purified tubulin can be used as the substrate of GRK2, but further purification with an ion-exchange column is recommended to remove contaminated microtubule-associated proteins (MAPs) and endogenous kinases of MAPs. If partially purified tubulin is subjected to phosphorylation in the presence of [^{32}P]ATP, major bands of [^{32}P]phosphorylated proteins are MAPs rather than tubulin, despite the fact that the level of MAPs is much less than that of tubulin.[11] Tubulin is phosphorylated by copurified kinase in a heparin-sensitive manner, whereas phosphorylation of MAPs is not sensitive to heparin, which is a specific inhibitor of GRK2. These results indicate that tubulin is specifically phosphorylated by GRK2, but not by other kinases that phosphorylate MAPs.

Preparation of GST-Fusion Proteins as GRK2 Substrates

A fusion (I3-GST) with GST of the central part of the third intracellular loop of muscarinic M_2 receptors (268–324 in human M_2 receptor) is expressed in *E. coli,* purified using glutathione-Sepharose, and used as a substrate of GRK2.[18] Similarly, fusion proteins with the GST of rat β-tubulin, its mutants with Ala substituted for Ser and Thr residues, which are putative phosphorylation sites, and its part, including phosphorylation sites, are expressed in *E. coli,* purified, and used as substrates of GRK2.

Preparations of Regulators

Purification of G Proteins and G Protein $\beta\gamma$ Subunits from Brain or Lung

G proteins G_o and G_{i1} are solubilized with sodium cholate from porcine brain and purified using a series of column chromatography, including DEAE-Sephacel, Ultrogel AcA 34, heptylamine-Sepharose, and DEAR-Toyopearl.[27,28] Purified preparations are obtained at concentrations of 3–10 nmol/ml in a solution containing 20 m*M* Tris–HCl (pH 8.0), 1 m*M* EDTA, 1m*M* DTT, 0.7% CHAPS, and approximately 100 m*M* NaCl. Lubrol PX was used in the final purification step in the original protocol, but CHAPS is better than Lubrol PX for reconstitution of G proteins with muscarinic receptors because CHAPS is less toxic to receptors than Lubrol PX. Purified G proteins are stored at −80° for more than 8 years, retaining activity to interact with muscarinic receptors and GRK2. G_{i2} may be

[27] P. C. Sternweis and J. D. Robishaw, *J. Biol. Chem.* **259,** 13806 (1984).
[28] K. Haga, T. Haga, and A. Ichiyama, *J. Biol. Chem.* **261,** 10133 (1986).

obtained from brain but tends to be contaminated by G_o. Pure G_{i2} is obtained from porcine lung with essentially the same procedure.[18]

G protein $\beta\gamma$ subunits are prepared as described by Northup *et al.*[29] G protein $\beta\gamma$ subunits are also obtained from G_o by incubating G_o preparations with 50 μM GTPγS and 50 m*M* $MgCl_2$ and then separating Gα and $\beta\gamma$ subunits with DEAE-Toyopearl. Alternatively, G protein $\beta\gamma$ subunits may be prepared more easily from G_s than from Go.[17] G_s is eluted behind G_i and G_o fractions in the heptylamine-Sepharose column. The G_s fraction is diluted with twice the volume of a buffer solution [20 m*M* Tris–HCl (pH 7.4), 1 m*M* EDTA, 1m*M* DTT, 20 m*M* NaCl, 0.9% Lubrol PX] and applied to a DEAE-Toyopearl column (10 ml). The $G_s\alpha$ subunit passes through the column, while the $\beta\gamma$ subunit is bound to the column and is eluted with a linear gradient of 0 and 250 m*M* NaCl in a buffer solution [20 m*M* Tris–HCl (pH 7.4), 1 m*M* EDTA, 1m*M* DTT, and 0.6% Lubrol PX or 0.7% CHAPS]. This procedure to purify $\beta\gamma$ subunits can be applied for G_s preparation but not for G_i or G_o preparations because $G_i\alpha$ and $G_o\alpha$ remain bound to the DEAE-Toyopearl column under the same conditions. The purified $\beta\gamma$ preparation with the concentration of approximately 10 nmol/ml is stored at $-80°$.

Expression and Purification of G Proteins in Sf9 Cells[30]

Sf9 cells are transfected by two recombinant viruses encoding bovine $G\alpha_o$ and $G\beta_1\gamma_2$ subunits together. The two promoters in the vector for β_1 and γ_2 subunits are arranged back to back. Transfected Sf9 cells ($2–3 \times 10^9$ cells) are homogenized in a buffer solution [G: 20 m*M* Tris–HCl (pH 8.0), 1 m*M* EDTA, 1 m*M* DTT, protease inhibitors; 280 ml]. After centrifugation, the pellet is washed with the solution G by centrifugation, resuspended in solution G supplemented with 1% sodium cholate (H : 200 ml), and stirred for 2 hr. After centrifugation, the supernatant is applied to a DEAE-Sephacel column (50 ml) preequilibrated with solution H. The column is washed with 50 ml of solution H and eluted with a gradient of solution H and solution H supplemented with 0.5 *M* NaCl (100×100 ml). The G protein-rich fraction (ca. 30 ml) is diluted five-fold with a buffer solution [I : 20 m*M* HEPES–NaOH (pH 7.0), 1 m*M* EGTA, 1 m*M* DTT, protease inhibitors] and applied to a heptylamine column (20 ml) preequilibrated with solution I supplemented with 0.2% sodium cholate. The column is eluted with a gradient of solution I containing 200 m*M* NaCl and solution I containing 1.75% sodium cholate and 50 m*M* NaCl (70×70 ml). The G protein-rich fraction is applied to a hydroxyapatite column (5 ml) preequilibrated with solution I containing 0.8% sodium cholate and 5 m*M* potassium phosphate. The column is washed with solution I containing 0.8% sodium cholate and 5 m*M* potassium phosphate and eluted with a linear gradient of solution I containing 0.8% sodium cholate and 5 m*M* potassium phosphate and

[29] J. K. Northup, P. C. Sternweis, and A. G. Gilman, *J. Biol. Chem.* **258,** 11361 (1983).

the solution I containing 0.8% sodium cholate and 150 m*M* potassium phosphate. The G protein $\beta\gamma$ subunit is eluted at 5–25 m*M* potassium phosphate, and then the $\alpha\beta\gamma$ trimer is eluted between 50 and 75 m*M* potassium phosphate. Essentially the same procedure can be applied for expression and purification of G_q, G_{11}, G_{14}, and G_{i1}.[30]

Preparation of Calmodulin and S100 Protein

We have searched for proteins with GRK2-inhibiting activity and eventually purified calmodulin and S100 protein.[9] Porcine brain (200 g) is homogenized in a Tris–HCl buffer solution [40 m*M* Tris–HCl (pH 7.5), 2 m*M* EGTA, 1 m*M* DTT, protease inhibitors; a total volume, 1 liter], and the homogenate is centrifuged for 20 min at 30,000 rpm. The supernatant is centrifuged once more after the addition of $CaCl_2$ and NaCl to be 4 m*M* and 0.5 *M*, respectively. The supernatant is applied to a Phenyl-Sepharose CL-4B column (70 ml), which has been equilibrated with a solution containing 40 m*M* Tris–HCl (pH 7.5), 0.5 *M* NaCl, and 1 m*M* $CaCl_2$. The column is washed and then eluted with a solution containing 40 m*M* Tris–HCl (pH 7.5), 0.5 *M* NaCl, and 5 m*M* EGTA. The eluate is dialyzed against a solution containing 20 m*M* Tris–HCl (pH 7.5), 1 m*M* EGTA, and 0.5 m*M* DTT and is then applied to a DEAE-Sephacel column (20 ml). The column is eluted with a gradient of the same solution and the solution supplemented with 0.3 *M* NaCl (100 × 100 ml). S-100 protein is eluted ahead of calmodulin, and they are separated from each other with rechromatography with a DEAE-Sephacel column (10 ml).

Agonist-Dependent Phosphorylation by GRK2 of Muscarinic M2 Receptors

Phosphorylation by GRK2 of Muscarinic M2 Receptors in Vivo[31]

Myc epitope-tagged muscarinic M_2 receptors, GRK2 and a GRK2 mutant with a replacement by tryptophan of lysine in the active site (GRK2-K220W), are transiently expressed in COS 7 cells using a vector pEF-BOS.[32] COS 7 cells are transfected using the calcium phosphate precipitation method with 2 μg of pEF-myc-hm2 together with 8 μg of pEF-BOS, pEF-GRK2, or pEF-GRK2-K220W per 10^7 cells on a 10-cm-diameter dish. Three days after transfection, the cells are incubated with 5 ml of phosphate-free Dulbecco's modified Eagle's medium without serum for 2 hr and then labeled with [^{32}P]orthophosphate (0.5 mCi/dish) for 2 hr. Labeled cells are incubated with or without 10 μM carbamylcholine for

[30] F. Nakamura, M. Kato, K. Kameyama, T. Nukada, T. Haga, H. Kato, T. Takenawa, and U. Kikkawa, *J. Biol. Chem.* **270,** 6246 (1995).

[31] H. Tsuga, K. Kameyama, T. Haga, H. Kurose, and T. Nagao, *J. Biol. Chem.* **269,** 32522 (1994).

[32] S. Mizushima and S. Nagata, *Nucleic Acids Res.* **18,** 5322 (1990).

20 min at 37°, followed by washing five times with 10 ml of ice-cold saline and then homogenization in solution J [20 m*M* HEPES–KOH (pH 7.5), 4 m*M* EDTA, 1 m*M* EGTA, 10 m*M* NaF, protease inhibitors: 1 ml]. After centrifugation and washing, the pellet is solubilized with solution J supplemented with 1% Triton X-100 and 0.05% SDS (1 ml). The extract is incubated with anti-myc-epitope antibody 9E10 (5 μg) at 4° overnight and then with protein A/G PLUS agarose (Santa Cruz Biotechnology; 0.1 ml) for 2 hr at 4°. After the reaction mixture is centrifuged for 2 min using a microfuge, immunoprecipitates are washed five times with solution J supplemented with 1% Triton X-100 and 0.05% SDS and then extracted with 0.1 ml of SDS sample solution containing medium for SDS–PAGE [K: 5% SDS, 20% glycerol, 10% mercaptoethanol, 0.01% bromphenol blue, and 125 m*M* Tris–HCl buffer (pH, 6.8)] with a bath sonicator. The extract is subjected to SDS–PAGE followed by autoradiography and counting of radioactivity. As shown in Fig. 2, M2 receptors are phosphorylated in an agonist-dependent manner, and the agonist-dependent phosphorylation is enhanced or attenuated in the presence of coexpressed GRK2 or GRK2-K220W, respectively. GRK2-K220W acts as a dominant-negative GRK2.

Phosphorylation by GRK2 of Muscarinic M_1 Receptors in Vitro[33]

In original experiments, muscarinic receptors are purified, reconstituted with G proteins in lipid vesicles using a small column of Sephadex G-50, and then subjected to phosphorylation by mAChR kinase or GRK2. This section describes the phosphorylation of muscarinic M_1 receptors. Essentially the same method is used for the phosphorylation of adrenergic β_2 receptors[7] and will be applied for other GPCRs.

Muscarinic M_1 receptors and GRK2 are expressed in and purified from Sf9 cells, and G_o from porcine brain, as described earlier. A lipid suspension (L; 4 mg lipid/ml) is prepared by mixing cholesteryl hemisuccinate (Tris salt, Sigma, 10 mg/ml methanol, 16 μl), egg L-α phosphatidylcholine (type XV-E, Sigma, 20 mg/ml chloroform, 96 μl), and L-α-phosphatidylinositol (ammonium salt, Sigma, 20 mg/ml chloroform, 96 μl), evaporating the solvent under nitrogen, and suspending the residue in 1 ml of solution M [20 m*M* HEPES–KOH buffer (pH 8.0), 1 m*M* EDTA and 160 m*M* NaCl] containing 1% sodium cholate by sonication for 20 min at 4° in a bath-type sonicator. Lipid suspension L (100 μl) is mixed with purified M_1 receptors (60–90 pmol), G protein G_o (300–900 pmol), bovine serum albumin (25 μg), carbamylcholine (1 m*M*), DTT (1–10 m*M*), and $MgCl_2$ (5–10 m*M*) (total volume 200 μl), and the mixture is passed through a small column of Sephadex G-50 (fine)(2 ml) preequilibrated with solution M. The

[33] K. Haga, K. Kameyama, T. Haga, U. Kikkawa, K. Shiozaki, and H. Uchiyama, *J. Biol. Chem.* **271,** 2776 (1996).

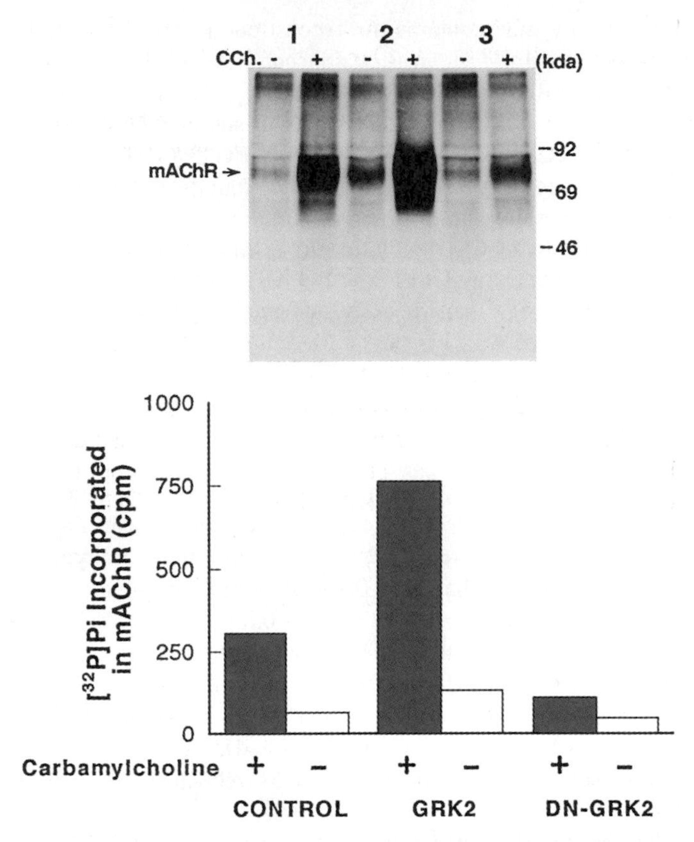

FIG. 2. Agonist-dependent phosphorylation of muscarinic M_2 receptors expressed in COS 7 cells. myc-tagged human muscarinic M_2 receptors are expressed transiently in COS 7 cells without (1) or with (2) GRK2 or the dominant-negative form of GRK2 (DN-GRK2) (3). Transfected cells are preincubated with [^{32}P]orthophosphate, incubated with or without 10 μM carbamylcholine (CCh) for 20 min, and then ^{32}P phosphorylated M_2 receptors (mAChR) are precipitated by anti-myc antibodies and subjected to SDS–PAGE and radioautography or counting of radioactivity. Data from Tsuga *et al., J. Biol. Chem.* **269,** 32522 (1994), with permission.

void volume fraction (400 μl) is mixed with 1.1 ml of solution *M* containing 5 m*M* DTT and 1 m*M* carbamylcholine and then with 0.5 ml of 50% (w/v) polyethylene glycol 6000 (PEG), kept for 10 min at room temperature, and then centrifuged for 30 min at 50,000 rpm. The pellet is resuspended in 400 μl of solution M and used as substrates for phosphorylation by GRK2. The precipitation with PEG is dispensable for muscarinic M_2 receptors, and the void volume fraction from the Sephadex column described earlier can be directly used as substrates of GRK2. The precipitation with PEG may also be skipped for other GPCRs, when they can be prepared as purified and concentrated solutions.

A standard assay tube contains the reconstituted vesicle (1 μl; 1.8–4.0 nM M_1 receptors and 9–40 nM G_o in final concentrations), 1 mM carbamylcholine or 10 μM atropine, 100 μM GTP, 1 or 10 μM [γ-^{32}P]ATP (2×10^5 cpm/tube, 1–5 cpm/fmol), and purified GRK2 (13 nM) in solution M [20 mM Tris–HCl (pH 7.5), 5 mM $MgCl_2$, 2 mM EDTA, 0.5 mM EGTA) (total volume, 40 μl)]. G_o and GTP are added to supply G protein $\beta\gamma$ subunits and may be replaced by $\beta\gamma$ subunits. The phosphorylation reaction is carried out at 30° and is terminated by the addition of 20 μl of SDS solution K, followed by autoradiography. The incorporation of ^{32}P is estimated by cutting the relevant band and counting by use of Cerenkov's effect or by using an image analyzer (Fuji BAS2000, Fuji Film Corp.).

Phosphorylation by GRK2 of Muscarinic M_2 Receptors, Rhodopsin, and Tubulin in Vitro

We may skip not only PEG precipitation but also the Sephadex step for the phosphorylation of concentrated M_2 receptors. Purified muscarinic M_2 receptors are eluted as a solution of more than 2 nmol/ml from the hydroxyapatite column with 0.5 M potassium phosphate and 0.1% digitonin. The eluate is diluted with water by 10-fold, and an aliquot (1–2 μl, 0.2–0.8 pmol) is subjected directly to phosphorylation by GRK2. A standard reaction medium contains M_2 receptors (5–20 nM at a final concentration), G protein $\beta\gamma$ subunits (66 nM), GRK2 (33 nM), 1–10 μM [^{32}P]ATP, 1 mM carbamylcholine or 10 μM atropine, and lipid suspension L (1 μl per tube) (total volume, 30–40 μl).[8] It should be noted that the addition of lipid suspension L is necessary for M_2 receptors to be phosphorylated by GRK2.

Phosphorylation of rhodopsin is observed using urea-washed ROS membranes instead of M_2 receptors.[17] ROS membranes prepared and stored under dark are incubated with GRK2 and [^{32}P]ATP with or without light under the same conditions used for the phosphorylation of M_2 receptors except that carbamylcholine, atropine, and the lipid suspension are omitted. Phosphorylation of tubulin is carried out by the same method as used for the phosphorylation of rhodopsin except that ROS membranes are replaced by tubulin preparations.[11]

Regulation of GRK2

Stimulation of Phosphorylation of Muscarinic Receptors by G Protein $\beta\gamma$ Subunits

The stimulating effect of G protein $\beta\gamma$ subunits on phosphorylation of muscarinic receptors, adrenergic β receptors, and rhodopsin by mAChR kinase[6,17] or GRK2[7] is detected by simply adding G protein $\beta\gamma$ subunits in the reaction mixture. Their phosphorylation in the presence of agonists or light is stimulated 5- to 10-fold by $\beta\gamma$ subunits. Their phosphorylation is dependent on the presence of agonists or light, and virtually no phosphorylation is observed in their absence, irrespective of the presence or absence of $\beta\gamma$ subunits. The stimulatory effect of $\beta\gamma$ subunits is observed by increasing their concentrations from 1 to 100 n*M*.[7,8] Stimulating activity is also observed by the G protein trimer in the presence of GTPγS[6]. The extent of stimulatory effects is similar among G_{i1}, G_{i2}, and G_o[17] but is lower for transducin.[7] The $\beta\gamma$ subunit was suggested to stimulate the phosphorylation of GPCRs by facilitating the translocation of GRK2 into membranes.[34] However, the stimulatory effect of $\beta\gamma$ subunits is also observed for the phosphorylation of I3-GST[8] or tubulin,[11,13] which are soluble substrates of GRK2. This indicates that $\beta\gamma$ subunits not only facilitate the translocation of GRK2, but also stimulate the catalytic activity of GRK2.

The phosphorylation of M_2 receptors by mAChR kinase is inhibited in the presence of excess G protein trimers[16] or excess G protein α subunits over $\beta\gamma$ subunits in the absence of GTPγS.[6] It is likely that G protein $\alpha\beta\gamma$ trimers compete with GRK2 for interaction with agonist-bound M_2 receptors, thereby inhibiting the phosphorylation of M_2 receptors.

Synergistic Activation of GRK2 by Agonist-Bound M_2 Receptors and $\beta\gamma$ Subunits[8,18]

Phosphorylation by GRK2 of muscarinic M_2 receptors is dependent on the presence of muscarinic agonist. It is likely that this is not due to the exposure of phosphorylation sites by agonist binding but to the activation by agonist-bound M_2 receptors of GRK2. Evidence supporting this idea is that the I3-GST containing phosphorylation site in M_2 receptors is not a good substrate of GRK2 and that the phosphorylation of I3-GST is stimulated by agonist-bound M_2 receptors. When I3-GST is phosphorylated by GRK2 in the presence of G protein $\beta\gamma$ subunits and M_2 or I3-deleted M2 receptors, the phosphorylation of I3-GST is much greater in the presence of the muscarinic agonist carbamylcholine than in the presence of the muscarinic antagonist, atropine[18] (Fig. 3). Because M_2 or I3-deleted M_2 receptors, but not I3-GST, bind carbamylcholine, this result indicates that agonist-bound M_2 or I3-deleted M_2 receptors activate GRK2, which phosphorylates I3-GST. This result also indicates that I3-deleted M_2 receptors interact with and activate GRK2 at a site different from phosphorylation sites because they do not contain phosphorylation sites and that M_2 receptors interact with GRK2 at two different sites; phosphorylation sites and activation sites. The activating effect of agonist-bound receptors is clearly observed in the presence of G protein $\beta\gamma$ subunits than in their absence.

[34] J. A. Pitcher, J. Inglese, J. B. Higgins, J. L. Arriza, P. J. Casey, C. Kim, J. L. Benovic, M. M. Kwatra, M. G. Caron, and R. J. Lefkowitz, *Science* **257,** 1264 (1992).

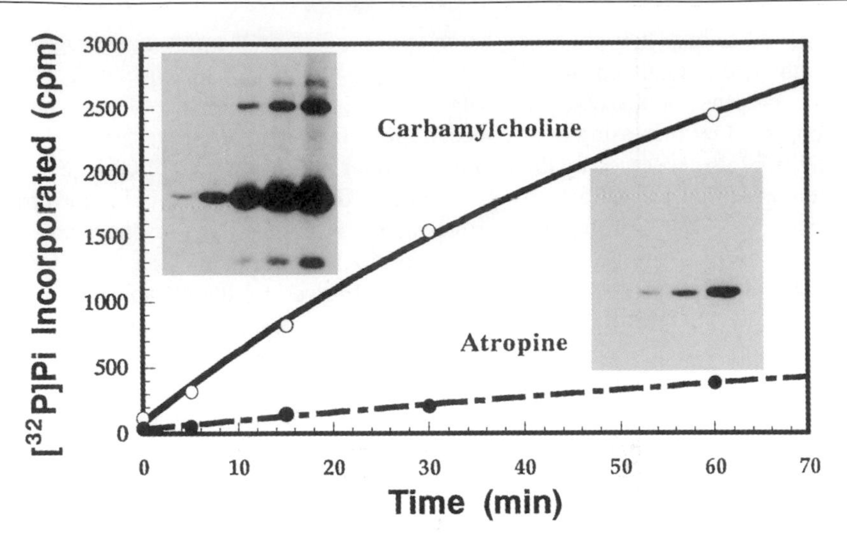

FIG. 3. Agonist-dependent phosphorylation of I3-GST in the presence of I3-deleted M_2 receptors. I3-GST and I3-deleted M_2 receptors are expressed in and purified from *E. coli* and Sf9 cells, respectively. I3-GST (8 μM) is incubated at 30° with 10 μM [γ-^{32}P]ATP and 60 n*M* GRK2 in the presence of 17 n*M* I3-deleted M_2 receptors, 66 n*M* G protein $\beta\gamma$ subunits, and I m*M* carbamylcholine or 10 μM atropine (total volume 30 μl). The reaction is terminated by addition of the SDS solution K, followed by SDS–PAGE and counting of the I3-GST band. (Inset) Autoradiography of phosphorylated bands of I3-GST. Data from Kameyama *et al., Eur. J. Biochem.* **226,** 267 (1994), with permission.

The phosphorylation of I3-GST, as well as of muscarinic M_2 receptors, is also stimulated by mastoparan, a bee venom, which is known to mimic agonist-bound receptors and activate G proteins.[8] The phosphorylation of both I3-GST and M_2 receptors is synergistically stimulated by mastoparan and $\beta\gamma$ subunits (Fig. 4). There is a difference in phosphorylation of the two substrates and the effects of two regulators. Phosphorylation of I3-GST is stimulated to only a limited extent by $\beta\gamma$ subunits alone but is stimulated significantly by mastoparan alone, whereas phosphorylation of M_2 receptors is not stimulated by mastoparan alone but is significantly stimulated by $\beta\gamma$ subunits alone.

Peptides corresponding to sequences adjacent to transmembrane segments of the second and third intracellular loops and the carboxyl terminus of M2 receptors also stimulate the phosphorylation of I3-GST.[8] These regions are assumed to undergo conformational changes by agonist binding and then interact with and activate G proteins. It will be reasonable to assume that GRK2 interacts with and is activated synergistically by both $\beta\gamma$ subunits and those regions that have undergone conformational change. These regions, as well as mastoparan, are very basic peptides. G protein $\beta\gamma$ subunits are known to interact with the carboxyl-terminal

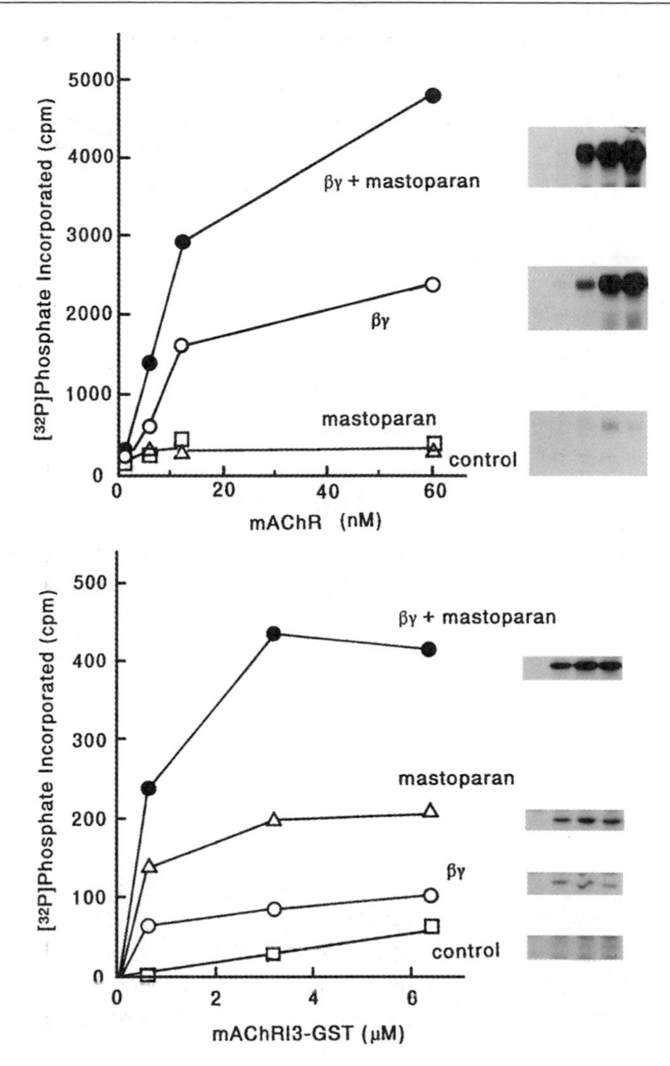

FIG. 4. Synergistic activation by mastoparan and G protein $\beta\gamma$ subunits of phosphorylation of muscarinic M_2 receptors and I3-GST. Different concentrations of muscarinic M_2 receptors (mAChR) or I3-GST (mAChRI3-GST) are incubated for 60 min at 30° with 10 μM [γ-^{32}P]ATP and 60 n*M* GRK2 in the presence of 10 μM mastoparan or 6.6 n*M* G protein $\beta\gamma$ subunits or both. Data from Haga *et al.*, *J. Biol. Chem.* **269,** 12594 (1994), with permission.

TABLE I

Ca^{2+} BUFFER SOLUTIONS WITH BUFFERING ACTION IN A WIDE RANGE OF FREE Ca^{2+} CONCENTRATIONS AND WITH CONSTANT FREE Mg^{2+} CONCENTRATION, pH (7.5), AND IONIC STRENGTH[a]

	Total $MgCl_2$ (m*M*)	Total $CaCl_2$ (m*M*)	NaOH (m*M*)	NaCl (m*M*)	Free Mg^{2+} (m*M*)	Free Ca^{2+}	$-\log[\text{free } Ca^{2+}]$
1	6.5	0	1.0	0.5	5.0	1 n*M*	>9
2	6.45	0.1	1.05	0.45	5.0	4.3 n*M*	8.37
3	6.25	0.5	1.25	0.25	5.0	38.0 n*M*	7.42
4	6.05	0.9	1.45	0.05	5.0	344 n*M*	6.46
5	5.9	1.1	1.50	0	5.0	5.3 μM	5.28
6	5.5	1.5	1.50	0	5.0	47.0 μM	4.33
7	5.1	1.9	1.50	0	5.0	350 μM	3.45
8	5.0	4.0	1.50	0	5.0	2000 μM	2.70

[a] Each solution contains 1 m*M* EDTA and 1 m*M* EGTA. Concentrations of free Ca^{2+} ions are calculated as described in the text.

part of GRK2.[7,35] The carboxyl-terminal of the GRK2 part has been shown to have a positive and negative surface,[36] and $\beta\gamma$ subunits are thought to bind to the positive surface. It is tempting to speculate that the basic peptides interact with the negative surface of the carboxyl terminus of GRK2, thereby activating GRK2 synergistically with $\beta\gamma$ subunits.

Inhibition of GRK2

Ca^{2+} Ion Buffer. The Ca^{2+} ion buffer covering a wide range from 10^{-8} to 10^{-3} *M* of free Ca^{2+} ions is made of a mixture of 1 m*M* EDTA, 1 m*M* EGTA, 0–4 m*M* $CaCl_2$, and 5–6.5 m*M* $MgCl_2$, as shown in Table I. In this system, the concentration of free Mg^{2+} ions is kept constant.

Concentrations of free Ca^{2+} ions ($[Ca^{2+}]$) are calculated according to the following equations:

$$[Ca^{2+}] = [CaCl_2]/[1 + ([EGTA] - [CaCl_2])K_1],$$

when [EGTA] is greater than $[CaCl_2]$

$$[Ca^{2+}] = [CaCl_2]/[1 + ([EGTA] + [EDTA] - [CaCl_2])K_2],$$

when [EGTA] is less than $[CaCl_2]$

[35] W. J. Koch, J. Inglese, W. C. Stone, and R. J. Lefkowitz, *J. Biol. Chem.* **268,** 8256 (1993).

[36] D. Fushman, T. Najmabadi-Haske, S. Cahill, J. Zheng, H. LeVine III, and D. Cowburn, *J. Biol. Chem.* **273,** 2835 (1998).

where [$CaCl_2$], [EGTA], and [EDTA] are total concentrations of $CaCl_2$, EGTA, and EDTA, respectively; K_1 is Kc,g/(1 + [Mg^{2+}]Km,g); K_2 is Kc,d/(1 + [Mg^{2+}]Km,d). Kc,g, K_m,g, Kc,d, and Km,d are apparent equilibrium-binding constants at pH 7.5 for the binding of EGTA and Ca^{2+}, EGTA and Mg^{2+}, EDTA and Ca^{2+}, and EDTA and Mg^{2+} and are $10^{7.72}$, $10^{2.30}$, $10^{7.92}$, and $10^{5.91}$, respectively.

Total concentrations of $MgCl_2$ ([$MgCl_2$]) are adjusted for free Mg^{2+} ion concentration ([Mg^{2+}]) to be constant (5 mM at the present case) according to the following equations:

$$[MgCl_2] = [Mg^{2+}] + [Mg^{2+}][EGTA]Km,g/[1 + [Mg^{2+}]Km,g + [CaCl_2]Kc,g/[1 + ([EGTA] - [CaCl_2])Kc.g/(1 + [Mg^{2+}]Km.g)]], \text{ when [EGTA] is greater than } [CaCl_2],$$

$$[MgCl_2] = [Mg^{2+}] + [Mg^{2+}][EDTA]Km,d/[1 + [Mg^{2+}]Km,d + ([CaCl_2] - [EGTA])Kc,d/[1 + ([EDTA] + [EGTA][CaCl_2])Kc,d/(1 + [Mg^{2+}]Km,d)]], \text{ when [EGTA] is less than } [CaCl_2].$$

Actually the total $MgCl_2$ concentrations are approximated by the following equations:

$$[MgCl_2] = [Mg^{2+}] + [EGTA] + [EDTA] - [CaCl_2], \text{ when [EGTA] + [EGTA]} > [CaCl_2]$$
$$[MgCl_2] = [Mg^{2+}], \text{ when [EGTA] + [EGTA]} < [CaCl_2]$$

NaCl and NaOH are added for ionic strength and pH to be constant. Depending on the concentrations of [$CaCl_2$] and [$MgCl_2$], different concentrations of NaOH are added to adjust the pH. The adjustment of pH may be done when 5 or 10 times concentrated buffer solutions are made. The calculation of NaCl to be added is skipped here. This adjustment will not be necessary unless high concentrations of [EGTA] and [EDTA] are used or the reaction is very sensitive to ionic strength.

The values of K_1 and K_2 are estimated to be 2.62×10^{-7} and 2.05×10^{-4}, respectively, under the present condition. Thus the concentrations of free Ca^{2+} ions around 10^{-7} M are buffered by EGTA, and the concentrations around 10^{-4} M are buffered by EDTA. The Ca^{2+} ion buffer usually uses only EGTA and covers from 10^{-8} to 10^{-6} M. The present system utilizes the Ca^{2+}-buffering action of EDTA at higher concentrations of free Ca^{2+} ions. EDTA has only 10^2-fold higher affinity for Ca^{2+} than for Mg^{2+}, in contrast with 10^6-fold higher affinity of EGTA for Ca^{2+} than for Mg^{2+}. Then the Ca^{2+}-buffering action of EDTA-Ca buffer is affected by Mg^{2+} concentrations, and the buffer compositions affect free Mg^{2+} concentrations. This complication is avoided by using high concentrations of $MgCl_2$ so that EDTA is virtually occupied by Mg^{2+} ions and the concentration of free Mg^{2+} ions is constant. This buffer system will be useful for many purposes requiring a wide range of free Ca^{2+} ion concentrations.

Inhibition by Ca^{2+}-Calmodulin or S-100 Protein.[9] GRK2 is inhibited by calmodulin depending on the free Ca^{2+} ion concentrations. The effect of Ca^{2+}-calmodulin is observed more clearly for the phosphorylation of rhodopsin than for the phosphorylation of M_2 receptors. The effect on phosphorylation of rhodopsin is observed at lower concentrations of calmodulin and free Ca^{2+} ions than on phosphorylation of M_2 receptors. Effects of free Ca^{2+} ion concentrations are shown in Fig. 5. The

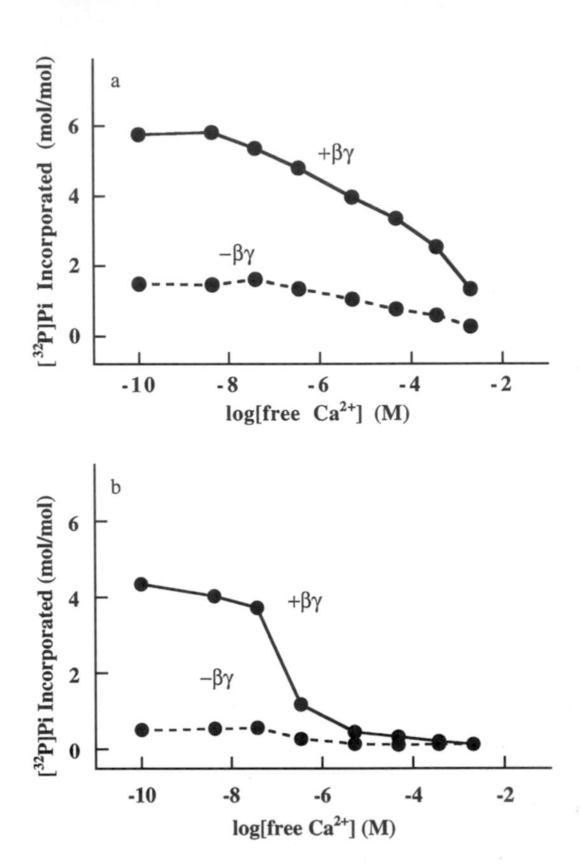

FIG. 5. Effects of free Ca^{2+} ion concentrations on phosphorylation of muscarinic M_2 receptors or rhodopsin in the presence of calmodulin. The phosphorylation reaction of (a) M_2 receptor (7 n*M*) in the presence of 1 m*M* carbamylcholine or (b) rhodopsin (20 n*M*) is carried out with or without 66 n*M* $\beta\gamma$ subunits in the presence of 10 μM [^{32}P]ATP, 33 n*M* GRK2, 25 μM calmodulin, 20 m*M* Tris–HCl buffer (pH 7.5), 1 m*M* EDTA, 1 m*M* EGTA, and different concentrations of $CaCl_2$, $MgCl_2$, NaOH, and NaCl as shown in Table I (total volume 30 μl). Concentrations of free Ca^{2+} ions are calculated as described in the text and are shown in Table I. Virtually no phosphorylation of M_2 receptors and rhodopsin is observed in the absence of carbamylcholine or light, respectively. Data from Haga *et al.*, *Biochemistry* **36,** 1315 (1997), with permission.

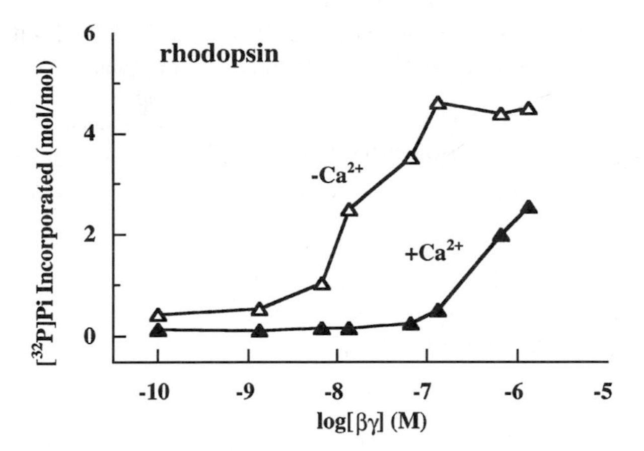

FIG. 6. Effects of various concentrations of $\beta\gamma$ subunits on phosphorylation of rhodopsin in the presence of S-100 protein. Experimental conditions are the same as those described in the legend to Fig. 5, except that 2 m*M* EDTA, 0.5 m*M* EGTA, plus or minus 2 m*M* $CaCl_2$, and 5 m*M* $MaCl_2$ are used instead of Ca^{2+} ion buffer and that different concentrations of $\beta\gamma$ subunits and 25 μM of S100 protein instead of calmodulin are used.

phosphorylation of rhodopsin is sharply inhibited by an increase in free Ca^{2+} concentrations from 10^{-8} to 10^{-6} *M*. In contrast, the phosphorylation of M_2 receptors is gradually inhibited with an increase from 10^{-8} to 10^{-3} *M*. Phosphorylation of both rhodopsin and M_2 receptors is also inhibited by S-100 protein in the presence of Ca^{2+} ions.

The dose–response curves for stimulation by G protein $\beta\gamma$ subunits of phosphorylation of rhodopsin or M_2 receptors are shifted to the right in the presence of Ca-calmodulin or Ca-S100 protein. Figure 6 shows the effect of Ca-S-100 protein on the phosphorylation of rhodopsin. The phosphorylation by GRK2 of I3-GST is not affected by Ca-calmodulin in the absence of activators, but is inhibited greatly in the presence of both G protein $\beta\gamma$ subunits and agonist-bound I3-deleted M2 receptors. These results indicate that Ca-calmodulin interfere with the synergistic activation of GRK2 by $\beta\gamma$ subunits and agonist-bound receptors.

[36] Rhodopsin and Its Kinase

By IZABELA SOKAL, ALEXANDER PULVERMÜLLER, JANINA BUCZYŁKO, KLAUS-PETER HOFMANN, and KRZYSZTOF PALCZEWSKI

Introduction

In photoreceptor cells of the vertebrate retina, the light-sensitive chromophore, 11-*cis*-retinal, is linked covalently to receptor molecules. In rod cells, absorption of a photon causes photoisomerization of 11-*cis*-retinal to all-*trans*-retinal, triggering conformational changes in rhodopsin (Rho). Photolyzed Rho (Rho*) binds and activates G proteins; in turn, activated G proteins initiate subsequent biochemical steps that lead to a decrease in [cGMP] and consequently permeability of cyclic-GMP-gated cation channels in the photoreceptor plasma membrane. Similar reactions are initiated by conformational changes in receptor molecules of cones. In order for the signaling state of photoreceptors to return to the dark condition, Rho* and activated G proteins are turned off and cGMP is resynthesized. Phosphorylation of Rho* and binding of the capping protein arrestin (Arr), result in the termination of physiological responses (Fig. 1). While Rho* must be turned off to prevent continued activation of the entire population of G protein, its phosphorylation and subsequent Arr quenching must be timed properly in order for a given number of G protein molecules to become activated by Rho*. All of these activation and quenching reactions occur within hundreds of milliseconds and are collectively termed phototransduction.[1–3]

Many advances made in the elucidation of molecular details of phototransduction have paved the way for the rapid progress made in understanding of other G protein-coupled receptor (GPCR) systems. Thus, phototransduction is a prototypical model of the signal transduction initiated by other GPCR.

Rho Cycle of Activation, Termination of Activity, and Regeneration of Dark State*

Rho exists in an inactive conformation, which upon absorption of light is converted to an active form, Rho* (Fig. 1). The activation persists for a short time (milliseconds), allowing Rho* to catalyze nucleotide exchange (from GDP to GTP) on a fraction of the G protein pool. The amplification gain for a single Rho* is the activation of several hundred G_t (transducin) protein molecules. Like G_t,

[1] A. Polans, W. Baehr, and K. Palczewski, *TINS* **19,** 547 (1996).
[2] Y. Koutalos and K. W. Yau, *TINS* **19,** 73 (1996).
[3] L. Lagnado and D. Baylor, *Neuron* **8,** 995 (1992).

0076-6879/02 $35.00

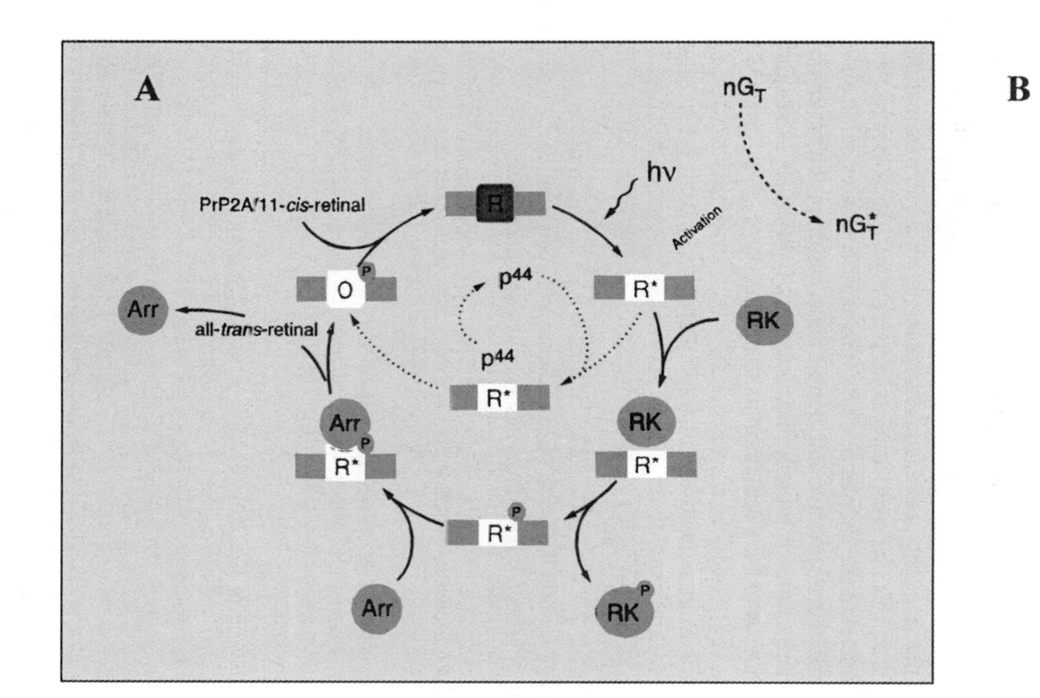

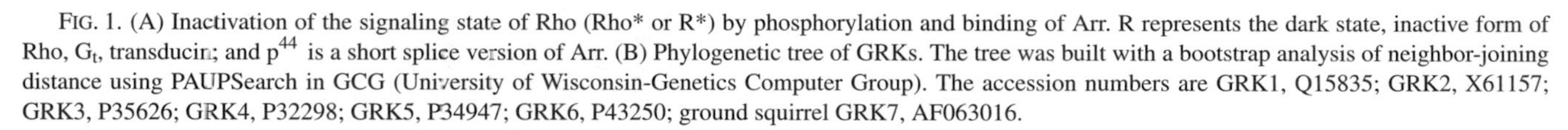

FIG. 1. (A) Inactivation of the signaling state of Rho (Rho* or R*) by phosphorylation and binding of Arr. R represents the dark state, inactive form of Rho, G_t, transducin; and p^{44} is a short splice version of Arr. (B) Phylogenetic tree of GRKs. The tree was built with a bootstrap analysis of neighbor-joining distance using PAUPSearch in GCG (University of Wisconsin-Genetics Computer Group). The accession numbers are GRK1, Q15835; GRK2, X61157; GRK3, P35626; GRK4, P32298; GRK5, P34947; GRK6, P43250; ground squirrel GRK7, AF063016.

rhodopsin kinase (RK, GRK1) also binds to cytoplasmic loops of Rho*, forming a stable complex,[4] and then Rho* is phosphorylated at the C terminus. Phosphorylation may have a small inhibitory effect on G protein activation, but for some receptors, phosphorylation may reduce the activity by as much as 80%.[5] Rho* phosphorylation causes an ~50% decrease in G_t binding.[6] *In vitro,* GPCRs*, including Rho*, were reported to be phosphorylated to a high stoichiometry of seven to nine phosphates per receptor. However, a single phosphate per receptor, as observed *in vivo,*[7] is sufficient for quenching Rho* activation.[8] The affinity of RK for Rho* is weakened by autophosphorylation, allowing RK to dissociate. Next, the regulatory protein Arr, binds to phosphorylated Rho*. In the Rho*–P_i · Arr complex, Rho* is unable to interact with the G protein. Formation of the GPCR*–P_i · Arr complex could be a fast process that results in lowering the effective receptor concentration capable of interacting with G proteins. In vision, this role of Arr is implausible because the Rho concentration is very high, ~3 m*M*. It has been proposed that Rho* could also be quenched by phosphorylation-independent mechanisms that involve p^{44}, a splice form of Arr[9–11] (Fig. 1).

For Rho*, removal of the agonist by reduction of all-*trans*-retinal to all-*trans*-retinol leads to the release of Arr and dephosphorylation of the receptor by a membrane-associated form of protein phosphatase 2A (PP2A).[12,13] Rho is fully regenerated when a new 11-*cis*-retinal binds to opsin.

G Protein Receptor-Kinases (GRKs)

Termination of GPCR signaling by phosphorylation is a well-established phenomenon that involves two types of Ser/Thr protein kinases: G protein-receptor kinases (GRKs) (EC 2.7.1-) and kinases activated by second messengers (e.g., protein kinase A or protein kinase C). Several comprehensive reviews concerning different aspects of receptor phosphorylation have been published.[14–21]

[4] K. Palczewski, *Eur. J. Biochem.* **248,** 261 (1997).

[5] G. S. Kroog, X. Y. Jian, L. Chen, J. K. Northup, and J. F. Battey, *J. Biol. Chem.* **274,** 36700 (1999).

[6] M. Heck, A. Pulvermüller, and K. P. Hofmann, *Methods Enzymol.* **315,** 329 (2000).

[7] H. Ohguro, J. P. Van Hooser, A. H. Milam, and K. Palczewski, *J. Biol. Chem.* **270,** 14259 (1995).

[8] N. Bennett and A. Sitaramayya, *Biochemistry* **27,** 1710 (1988).

[9] K. Palczewski, J. Buczyłko, H. Ohguro, R. S. Annan, S. A. Carr, J. W. Crabb, M. W. Kaplan, R. S. Johnson, and K. A. Walsh, *Protein Sci.* **3,** 314 (1994).

[10] W. C. Smith, A. H. Milam, D. Dugger, A. Arendt, P. A. Hargrave, and K. Palczewski, *J. Biol. Chem.* **269,** 15407 (1994).

[11] K. Palczewski and W. C. Smith, *Exp. Eye Res.* **63,** 599 (1996).

[12] K. Palczewski, P. A. Hargrave, J. H. McDowell, and T. S. Ingebritsen, *Biochemistry* **28,** 415 (1989).

[13] C. Fowles, M. Akhtar, and P. Cohen, *Biochemistry* **28,** 9385 (1989).

[14] J. Inglese, N. J. Freedman, W. J. Koch, and R. J. Lefkowitz, *J. Biol. Chem.* **268,** 23735 (1993).

[15] T. Haga, K. Haga, and K. Kameyama, *J. Neurochem.* **63,** 400 (1994).

Seven members of related mammalian GRKs have been characterized (Fig. 1B). These enzymes are single-subunit (63–80 kDa), water-soluble proteins, with 30–85% amino acid sequence identity between protein pairs (57–96% overall similarity). GRK2, GRK3, and GRK6 are expressed ubiquitously, in contrast to the more specific localization of GRK1, GRK7 (retinal photoreceptor cells and pinealocytes), GRK4 (testis, brain), and GRK5 (heart, lung, and retina). The role of RK in photoreceptors is likely restricted to phototransduction, while its function in the pineal gland is unknown.[22]

Interaction between Rho and Its Kinase*

Binding of ATP to RK[23] increases the affinity for the activated receptor by a factor of ~10 (~10 $\mu M^{-1}\text{sec}^{-1}$). In addition to a weak interaction of phosphorylated region within the catalytic site of RK, interaction between the kinase and the cytoplasmic surface of Rho* provides the stability for the complex. RK recognizes Rho*, or protonated opsin on acidic residues within α-helical or cytoplasmic domains.[24] The Rho*·RK complex involves the N-terminal domain of the kinase and cytoplasmic loops of the receptor.[25] The k_{off} for the Rho*·RK complex is increased when either the kinase or the receptor is phosphorylated.[23] A weak interaction between the C terminus of Rho* and the active site of RK promotes phosphorylation at different sites. Therefore, the Rho*·RK complex may produce several intermediates each with a unique catalytic site positioning of GRK toward the phosphorylation site. Rho* phosphorylation at different sites, including Ser^{334}, Ser^{338}, and Ser^{342}, may play different roles in phototransduction.

The following section describes methods useful in the characterization of GRKs, in particular RK, including assays for enzymatic reactions and interaction with Rho* using spectroscopic methods, isolation, expression, and immunodetection. In addition, it provides short descriptions of methods and references helpful in the identification of phosphorylation sites.

---

[16] R. T. Premont, J. Inglese, and R. J. Lefkowitz, *FASEB J.* **9,** 175 (1995).

[17] M. J. Lohse, K. Bluml, S. Danner, and C. Krasel, *Biochem. Soc. Trans.* **24,** 975 (1996).

[18] S. S. G. Ferguson, L. S. Barak, J. Zhang, and M. G. Caron, *Can. J. Physiol. Pharmacol.* **74,** 1095 (1996).

[19] T. T. Chuang, L. Iacovelli, M. Sallese, and A. DeBlasi, *TIPS* **17,** 416 (1996).

[20] J. A. Pitcher, N. J. Freedman, and R. J. Lefkowitz, *Annu. Rev. Biochem.* **67,** 653 (1998).

[21] K. Palczewski and J. L. Benovic, *TIBS* **16,** 387 (1991).

[22] X. Y. Zhao, F. Haeseleer, R. N. Fariss, J. N. Huang, W. Baehr, A. H. Milam, and K. Palczewski, *Visual Neurosci.* **14,** 225 (1997).

[23] A. Pulvermüller, K. Palczewski, and K. P. Hofmann, *Biochemistry* **32,** 14082 (1993).

[24] J. Buczyłko, J. C. Saari, R. K. Crouch, and K. Palczewski, *J. Biol. Chem.* **271,** 20621 (1996).

[25] K. Palczewski, J. Buczyłko, L. Lebioda, J. W. Crabb, and A. S. Polans, *J. Biol. Chem.* **268,** 6004 (1993).

Experimental Procedures

Preparations of rod outer segments (ROS) are the source of phototransduction enzymes, Rho, and opsin in native membranes.

i. ROS are prepared under dim red light at 5° from fresh bovine retinas obtained locally or from frozen retinas (Schenk Packing Co., Inc.) employing a sucrose gradient procedure[26] and are stored at −80°. Rho, which is a main component of ROS (~85%), as judged by SDS–PAGE, is isolated in 100-mg quantities from bovine retinas (~0.5 mg/eye).

ii. For Rho* phosphorylation, to ensure the completeness of phosphorylation, ROS (~2 mg/ml) are resuspended and homogenized with 0.1 *M* potassium phosphate, pH 7.5. At this concentration, phosphate strongly inhibits PP2A.

iii. For preparation of Rho as RK substrate, soluble and membrane-associated proteins are removed. ROS are suspended at 1 mg of Rho/ml in 5 *M* urea, 10 m*M* BTP (1,3-bis[tris(hydroxymethyl)-methylamino]propane), pH 7.5, at room temperature.[27] Prolonged exposure to urea should be avoided due to potential denaturation of Rho. After 15 min, ROS membranes are centrifuged and washed four times with 67 m*M* potassium phosphate, pH 7.5. Finally, ROS membranes (3.6 mg/ml) are suspended in 10 m*M* BTP, pH 7.5, containing 100 m*M* NaCl and 5 m*M* $MgCl_2$.

iv. For preparation of opsin as RK substrate, urea-washed Rho is prepared. Opsin (0.3 mg/ml) is prepared from urea-washed Rho by thorough bleaching at 0° (typically 15 min) with a 180-W lamp at a distance of 20 cm in 10 m*M* BTP, pH 7.5, containing 50 m*M* NaCl. Opsin is pelleted by centrifugation (37,000*g*, 20 min), suspended in 10 m*M* BTP, pH 7.5, containing 50 m*M* NaCl, and stored frozen at −20° in small aliquots at 3.6 mg/ml. For routine assay of the activity, the removal of all-*trans*-retinal is not necessary. However, if the question is related to the efficiency of opsin activation, removal of all-*trans*-retinal, converted to the oximes, is necessary. In this case, bleaching should be carried out in the presence of 45 m*M* NH_2OH. The oximes of all-*trans*-retinal are extracted with petroleum ether (10 ml/10 ml of the opsin suspension). A brief centrifugation (18,000*g*, 10 min) separates the organic and water layers, and the organic phase is discarded. The extraction is repeated four times. To facilitate opsin sedimentation, NaCl is added to the opsin suspension to a final concentration of 0.5 *M* and centrifuged at 37,000*g* for 35 min. Opsin is suspended with 10 m*M* BTP, pH 7.5, containing 50 m*M* NaCl.

Assays of RK Activity

RK activity is tested using four assays, individually or in combination. Assays measure phosphotransfer reaction from [γ-^{32}P]ATP to (1) Rho*,[27–30] (2) opsin

[26] D. S. Papermaster, *Methods Enzymol.* **81,** 48 (1982).

[27] K. Palczewski, J. H. McDowell, and P. A. Hargrave, *J. Biol. Chem.* **263,** 14067 (1988).

activated by chromophore, all-*trans*-retinal, or analogs,[24] (3) RK in autophosphorylation reaction,[31] and (4) synthetic peptides derived from the C-terminal phosphorylation region of Rho, or other unrelated peptides.[28,32]

i. RK specifically phosphorylates Rho* and has negligible activity toward Rho or opsin. RK activity in fractions of interest is measured using 50 μM [γ-^{32}P]ATP (100–500 cpm/pmol, DuPont NEN) and urea-washed ROS membranes[33] containing 10 μM Rho in 20 m*M* BTP, 1 m*M* $MgCl_2$, pH 7.5, at 30°. The total volume is 200 μl in a 1.5-ml microcentrifuge tube. The reaction is initiated by light and carried out under a 150-W lamp from a distance of 20 cm for 5, 10, and 15 min in triplicates. No specific light requirements are critical. Typically, white light that bleaches 10% Rho/min, as determined spectrophotometrically by the decrease in the absorption of Rho (498 nm),[34] is used. The reaction is stopped by the addition of 10% CCl_3COOH, Rho-containing membranes are pelleted by low-speed centrifugation, and the excess [γ-^{32}P]ATP is removed by repetitive three cycles of washes/centrifugations (Eppendorf centrifuge, Model 5415C, 2 min at maximal speed) with 1.5 ml of 10% CCl_3COOH. Finally, after [γ-^{32}P]ATP is removed, Rho is solubilized with 1 ml of 96–100% HCOOH and mixed with the scintillation cocktail for radioactivity counting. Lower concentrations of HCOOH may not dissolve all membranes or may take a long time (3–4 hr) for solublization. Routinely, a time course (Fig. 2A), RK dose dependence (Fig. 2B), and Rho* dose dependence should determine linear ranges and verify the accuracy of the assay. Typically, the specific activity is 100–700 nmol P_i transfer per minute per 1 mg of purified RK. Due to broad specificities of GRKs and availability of Rho as the receptor, similar assays are employed to monitor activities of all GRKs.

The phosphorylation reaction requires ~1 m*M* $MgCl_2$, whereas a higher concentration inhibits RK.[29] Maximal RK activity is observed between pH 6.0 and pH 8.0, and salts up to 500 m*M* have little effect on the activity.[29] Due to Mg^{2+} coordination, chelators inhibit RK activity. Ca^{2+} and cyclic nucleotides do not affect RK activity.[29] Polycations activate RK (approximately twofold), whereas negatively charged polyanions inhibit RK.[28] Nucleotide specificity has been tested extensively and resulted in the identification of substrates and potent nucleoside inhibitors.[27,35,36] ATP is a preferential donor of the phosphate group, whereas GTP is a poor substrate.

[28] K. Palczewski, A. Arendt, J. H. McDowell, and P. A. Hargrave, *Biochemistry* **28,** 8764 (1989).
[29] K. Palczewski, J. H. McDowell, and P. A. Hargrave, *Biochemistry* **27,** 2306 (1988).
[30] K. Palczewski, *Methods Neurosci.* **15,** 217 (1993).
[31] J. Buczyłko, C. Gutmann, and K. Palczewski, *Proc. Natl. Acad. Sci. U.S.A.* **88,** 2568 (1991).
[32] J. J. Onorato, K. Palczewski, J. W. Regan, M. G. Caron, R. J. Lefkowitz, and J. L. Benovic, *Biochemistry* **30,** 5118 (1991).
[33] H. Shichi and R. L. Somers, *J. Biol. Chem.* **253,** 7040 (1978).
[34] J. H. McDowell, *Methods Neurosci.* **15,** 123 (1993).
[35] K. Palczewski, N. Kahn, and P. A. Hargrave, *Biochemistry* **29,** 6276 (1990).
[36] L. Lebioda, P. A. Hargrave, and K. Palczewski, *FEBS Lett.* **266,** 102 (1990).

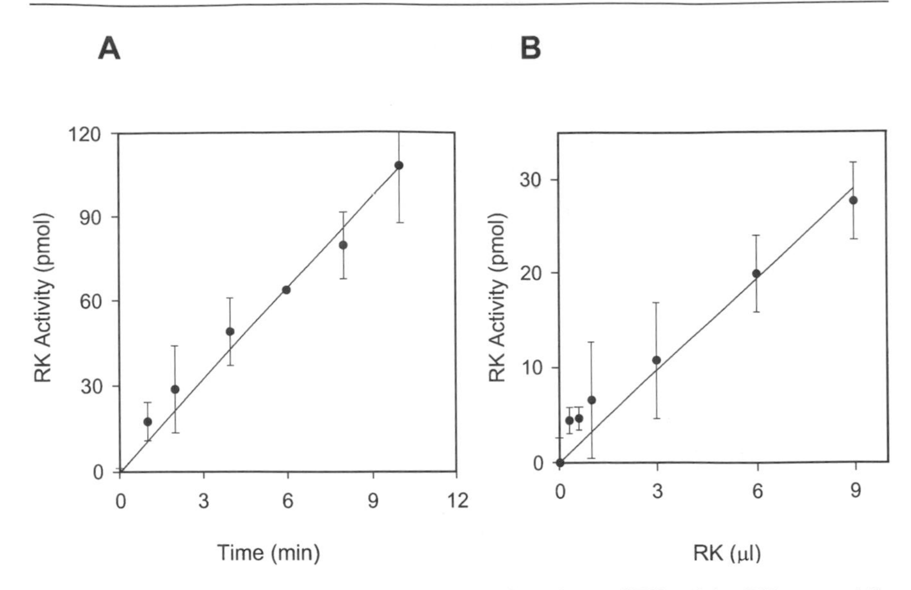

FIG. 2. Assay RK activity. Time (A) and dose (B) dependence of RK activity. RK was partially purified from SF9 cells.

ii. RK activity, and most likely other GRKs, could be tested conveniently using opsin activated by retinoids.[24] Binding of retinoids to a distinct binding pocket causes changes in the opsin structure, converting this receptor from the inactive to the active conformation. This process is similar to the activation of other GPCRs by agonists. There is a specific structural requirement for retinoids for maximal activation. Commercially available all-*trans*-retinal is very potent in the activation. Shortening the polyene chain of retinoid leads to improvement of the activation efficiency, which is followed by a rapid decrease in activation. It appears that the binding pocket can be measured by a "molecular ruler," employing the aldehyde form of retinoids of different lengths, or extended by converting them to oximes (Fig. 3). The reaction does not require light and is steeply pH dependent with maximal activity at lower pH. A practical pH for the assay is 5.5–6.0.[24] The activity is only a fraction (~10%) of the activity obtained using fresh Rho*; however, the opsin/retinoid assay is highly reliable and easy to perform. The assay requires opsin and an ethanolic solution of all-*trans*-retinal (note that this compound is light, oxygen, and temperature sensitive and should be stored at −80° under argon). This assay is not applicable for RK fractions with low enzymatic activities. The specificity of retinoid-dependent activity is demonstrated by employing different lengths of retinoids (Fig. 3). This "molecular ruler" approach suggests that the optimal size of the binding site is for ~C17-long retinoids. Because C17 does not regenerate Rho, these data together suggest that the binding site is different

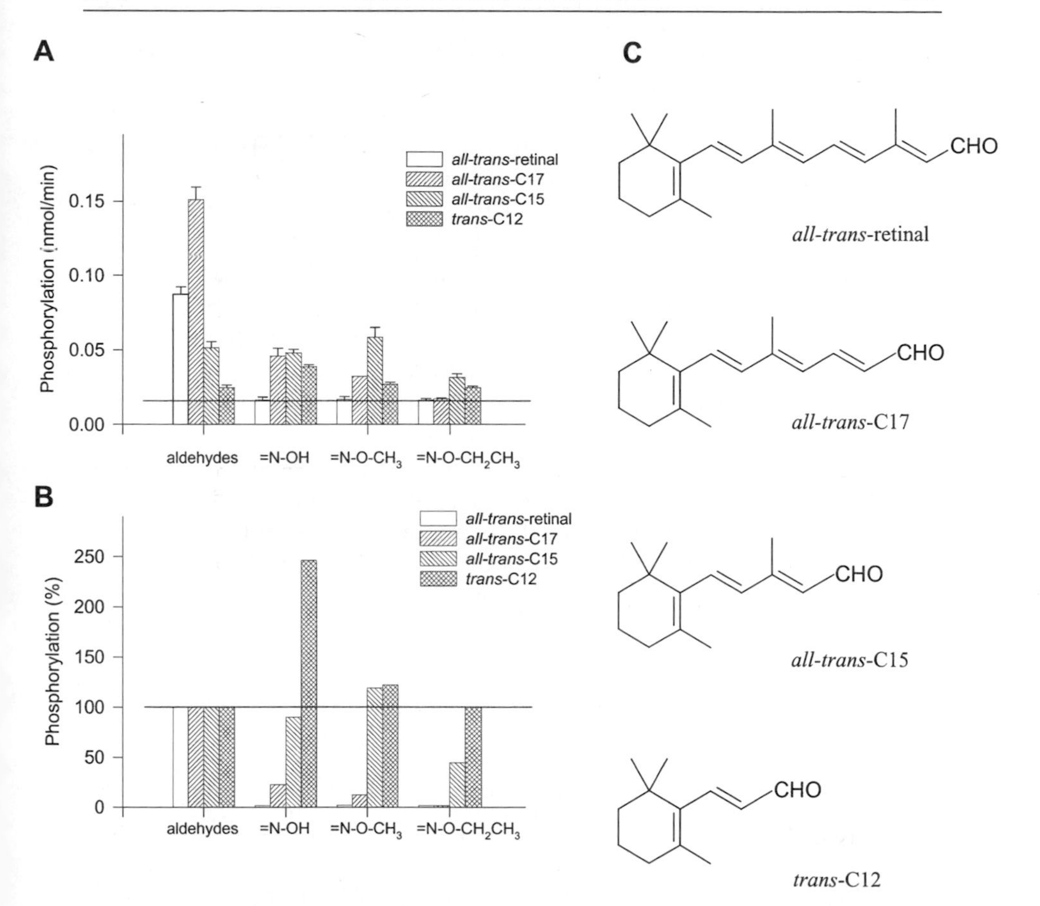

FIG. 3. Molecular rules of opsin activation: phosphorylation of opsin in the presence of all-*trans*-retinal analogs of different lengths and their derivatives with blocked aldehyde groups. (A) Phosphorylation of opsin is carried out in the presence of all-*trans*-retinal, all-*trans*-C17 aldehyde, all-*trans*-C15 aldehyde, *trans*-C12 aldehyde, or their oximes generated with NH_2OH, NH_2OCH_3, or $NH_2OCH_2CH_3$ (8 M excess over 30 μM opsin). The solid horizontal line represents opsin activity. (B) Activity expressed as percentage of maximum phosphorylation for each of the aldehydes (marked as a continuous line). The phosphorylation reaction is carried out in 30 mM BTP, pH 6.5, containing 3 mM $MgCl_2$. Data are reproduced with permission from the American Society for Biochemistry and Molecular Biology, Inc. [J. Buczyłko, J. C. Saari, R. K. Crouch, and K. Palczewski, *J. Biol. Chem.* **271,** 20621 (1996)]. (C) Structures of all-*trans*-retinal, all-*trans*-C17 aldehyde, all-*trans*-C15 aldehyde, *trans*-C12 aldehyde.

from the covalent binding of 11-*cis*-retinal, and likely involves a region around the palmitoylation groups on opsin.[37]

RK activity in fractions of interest is measured using 50 μM [γ-^{32}P]ATP (100–500 cpm/pmol, DuPont NEN) and 10 μM opsin in 20 mM BTP, pH 6.0, at 30°, containing 1 mM $MgCl_2$. The total volume is 200 μl in a 1.5-ml microcentrifuge tube. The reaction is initiated by adding 2 μl of 10 mM all-*trans*-retinal in CH_3CH_2OH and is carried out for 5, 10, and 15 min in triplicate. The reaction is stopped by the addition of 10% CCl_3COOH, and the excess [γ-^{32}P]ATP is removed by three cycles of washes/centrifugations (Eppendorf centrifuge, Model 5415C, 2 min at maximal speed) with 1.2 ml 10% CCl_3COOH. The samples are shaken using an Eppendorf shaker (mixer Model 5432) for 5 min between each wash in room temperature. Finally, Rho is solubilized with 1 ml of 100% HCOOH and mixed with the scintillation cocktail for radioactive counting.

iii. Active RK undergoes intramolecular autophosphorylation.[31,38] The assay is not applicable when RK is prepared using ATP elution from the heparin column. Autophosphorylation is unaffected by the presence of Rho* or P-Rho*.

RK autophosphorylation in fractions of interest is measured using 50 μM [γ-^{32}P]ATP (1000–5000 cpm/pmol, DuPont NEN) in 20 mM BTP, pH 7.5, at 30°, containing 1 mM $MgCl_2$. The total volume is 20 μl in a 1.5-ml microcentrifuge tube. The reaction is initiated by adding [γ-^{32}P]ATP and is carried out for 5, 10, and 15 min in triplicate. The reaction is stopped with 5 μl SDS–PAGE sample buffer. Excess [γ-^{32}P]ATP is separated by electrophoresis using 12% polyacrylamide gels. RK bands, identified by Western blotting, are cut out. The pieces of the gel are solubilized with 300 μl of fresh 30% H_2O_2 at 30° (typically, this step requires 40–60 min incubation) and mixed with the scintillation cocktail for radioactive counting.

iv. RK has low but measurable activity toward peptides corresponding to the phosphorylation region of Rho or unrelated acid peptides containing Ser or Thr residues.[29,32] Phosphorylation of peptides at pH 8.5 is three to four times faster than at pH 6.0. RK preferred acidic residues localize to the carboxyl-terminal side of the Ser residue, whereas other GRKs showed preference for acidic residues at the N terminus of the phosphorylation sites. RK has a much greater K_M for peptide substrates compared to that for intact receptor substrates, suggesting that interactions with Rho* are needed for maximal activity.[4,38–40]

RK activity in fractions of interest is measured using 50 μM [γ-^{32}P]ATP (500–800 cpm/pmol, DuPont NEN) in 20 mM BTP, pH 8.5, at 30°, containing 1 mM

[37] K. Sachs, D. Maretzki, C. K. Meyer, and K. P. Hofmann, *J. Biol. Chem.* **275,** 6189 (2000).

[38] K. Palczewski, J. Buczyłko, J. P. Van Hooser, S. A. Carr, M. J. Huddleston, and J. W. Crabb, *J. Biol. Chem.* **267,** 18991 (1992).

[39] C. Fowles, R. Sharma, and M. Akhtar, *FEBS Lett.* **238,** 56 (1988).

[40] K. Palczewski, J. Buczyłko, M. W. Kaplan, A. S. Polans, and J. W. Crabb, *J. Biol. Chem.* **266,** 12949 (1991).

$MgCl_2$, and 1 mM peptide (e.g., the 327–348 peptide fragment corresponding to bovine Rho). The total volume is 200 μl in a 1.5-ml microcentrifuge tube. The reaction is initiated by the addition of [γ-^{32}P]ATP and is carried out for 30, 60, and 90 min in triplicates. The reaction is stopped with 1% CF_3COOH (100 μl), and the excess [γ-^{32}P]ATP is removed by separation of the peptide on a C18 column (Vydac, 2.2 $\times$ 25 cm) using a 0–40% gradient of CH_3CN containing 1% H_3PO_4. The peptide is collected and mixed with the scintillation cocktail for radioactivity determination. Alternatively, the peptide could be engineered to contain three Arg residues at the N terminus to tightly bind to P-cellulose.[32] After reaction, excess [γ-^{32}P]ATP is washed out from the peptide using P-cellulose paper in 75 mM H_3PO_4.[32]

Phosphorylation of exogenous peptide is stimulated by the presence of catalytic amounts of Rho*, or its truncated forms.[39,40] The half-maximal effect is observed at 5–15 μM Rho*.[40] The efficiency of the peptide phosphorylation, as measured by V_{max}/K_M, increases maximally ~150 times in the presence of Rho*.

Isolation of RK

RK can be purified from native sources or from an expression system using several methods.

i. A combination of classical column chromatographies led to the isolation of highly purified RK.[27] Next, tryptic fragments were obtained and RK cDNA was cloned.[41] This method results in purified RK, but is labor-intense with a low yield.

ii. Efficient methods have been developed using the fact that RK undergoes rapid autophosphorylation (Fig. 4A), which changes the RK affinity for heparin[30,31,42–44] (Fig. 4B). Dephosphorylated RK can be extracted from ROS[30,31] or retina[43] and loaded on to the heparin-Sepharose column. Contaminated proteins are washed off, and ATP/Mg^{2+} is added onto the column to allow RK autophosphorylation. Phosphorylated RK specifically elutes from the column. Critical parameters are precise salt concentrations and the presence of Mg^{2+} in the autophosphorylation reaction on the column. This method and its variations for expressed RK, have been described previously.[44] They are used widely by other laboratories.

[41] W. Lorenz, J. Inglese, K. Palczewski, J. J. Onorato, M. G. Caron, and R. J. Lefkowitz, *Proc. Natl. Acad. Sci. U.S.A.* **88,** 8715 (1991).

[42] K. Palczewski, H. Ohguro, R. T. Premont, and J. Inglese, *J. Biol. Chem.* **270,** 15294 (1995).

[43] X. Y. Zhao, K. Yokoyama, M. E. Whitten, J. Huang, M. H. Gelb, and K. Palczewski, *FEBS Lett.* **454,** 115 (1999).

[44] H. Ohguro, M. Rudnicka-Nawrot, J. Buczyłko, X. Y. Zhao, J. A. Taylor, K. A. Walsh, and K. Palczewski, *J. Biol. Chem.* **271,** 5215 (1996).

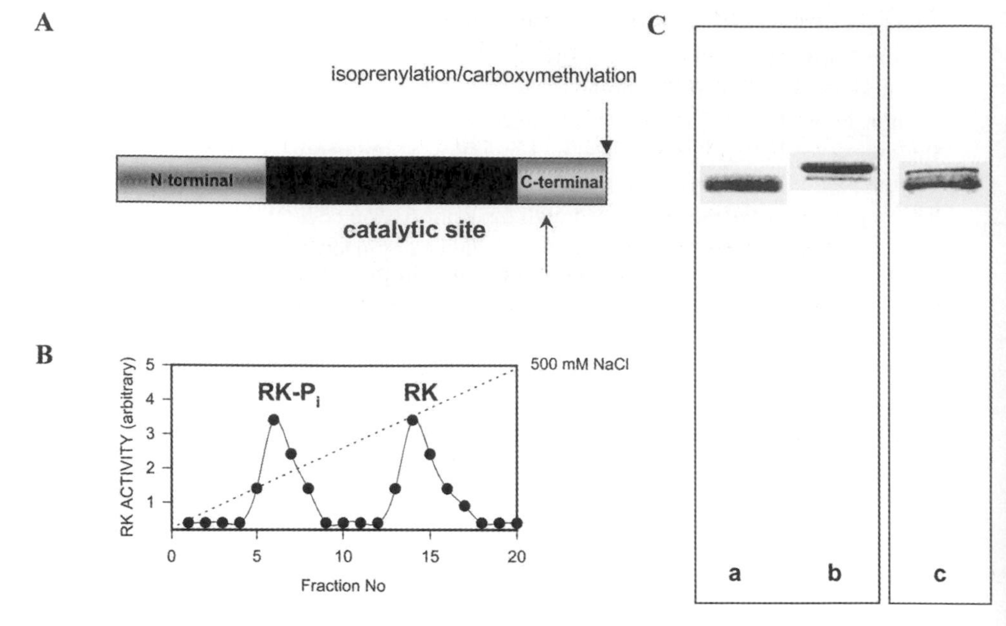

FIG. 4. Domain structure of RK, purification, and immunoblotting. (A) The N-terminal domain (~180 amino acids) may be involved in the interaction with Rho*.[25] The ATP-specific, catalytic domain is found in the middle of the sequence. The C-terminal domain is the site of autophosphorylation on Ser and Thr residues within the sequence FSTVKGV and is also modified by isoprenylation/carboxymethylation. This autophosphorylation changes the affinity for heparin and is used to purify RK. Bound RK is autophosphorylated on the heparin-Sepharose column and eluted at lower salt than is required for the elution of unphosphorylated kinase (see text). (B) Schematic representation of the elution profile for autophosphorylated and dephosphorylated RK from heparin-Sepharose column. At a high salt concentration (~150 m*M* NaCl), unphosphorylated RK is bound to the resin and is eluted specifically with ATP/Mg^{2+} as a result of autophosphorylation. (C) Mobility shift of RK dephosphorylated by PrP2A (lane a) and autophosphorylated in the presence of ATP (lane b). Multiple forms of autophosphorylated RK are visualized by Western blotting of isolated bovine ROS (lane c) probed with monoclonal antibody G8.

Purification of RK from Bovine Retina

RK is purified from bovine retinas according to a modified procedure for RK purification from ROS. All procedures are performed at 4°. To remove soluble proteins that do not interact with Rho, 100 bovine retinas are homogenized in 20 m*M* BTP, pH 7.5, containing 5 m*M* $MgCl_2$ and 1 m*M* benzamidine in a Dounce homogenizer with a motor-driven Teflon pestle (20 ml/100 retinas). The suspension is bleached on ice for 2 min under white light and then centrifuged at 37,000*g* (JA-20 rotor, Beckman) for 10 min. The extraction of soluble contaminating proteins is repeated once more, without bleaching. RK and G_t are bound to Rho*, while the majority of soluble proteins are removed. At this stage of purification, due to a high concentration of other protein kinases and phosphorylations, there is a high phosphorylation background unrelated to RK. Thus, it is imperative that control assays for RK activity are carried out in the dark to substract from light-dependent phosphorylation. Roughly ~15% of RK is lost during this procedure, as determined by immunoblotting.

To extract RK, the pellets are suspended in 25 ml of 20 m*M* BTP, pH 7.5, containing 10 m*M* NH_2OH, 1 m*M* EDTA, 240 m*M* KCl, 1 m*M* benzamidine, and 0.4% Tween 80, and are then further bleached on ice for 10 min under bright white light (150 W from a distance of 20 cm). At this stage, Rho* is inactivated by NH_2OH by a mechanism that involves conversion of photolyzed chromophore all-*trans*-retinal to all-*trans*-retinal oximes. The sample is centrifuged at 37,000*g* for 10 min. Optionally, the membranes can be extracted for the second time under identical conditions. Multiple extractions may cause, however, a progressively higher extraction of contaminating proteins that are difficult to remove during the next stages of purification.

The extract containing RK is dialyzed against 20 m*M* BTP, pH 7.5, containing 0.4% Tween 80, and 1 m*M* benzamidine overnight. During dialysis, endogenous PP2A dephosphorylates RK. The dialyzed extract is loaded onto 20 ml of a DEAE-cellulose column equilibrated with 10 m*M* BTP, pH 7.5, containing 0.4% Tween 80. The column is washed extensively (100 ml) with the same buffer, and RK is eluted with 100 m*M* NaCl in 10 m*M* BTP, pH 7.5, containing 0.4% Tween 80 in a step gradient. Fractions with RK activity are diluted with an equal volume of 10 m*M* BTP and loaded on 5 ml heparin-Sepharose (Pharmacia).

The column is washed with 10 m*M* BTP containing 0.4% Tween 80 until the $A_{280} \leq 0.01$ (~30 ml) at 4°. A linear gradient from 0 to 500 m*M* KCl in 10 m*M* BTP, pH 7.5, containing 0.4% Tween 80 over 40 min is developed at 1.0 ml/min, and 40 × 1-ml fractions are collected. Fractions containing RK activity eluted at 350 m*M* KCl are combined and diluted two-fold with 20 m*M* BTP, pH 7.5, containing 1 m*M* *n*-dodecyl-β-D-maltoside, loaded onto a 5-ml heparin-Sepharose column, and washed with 10 m*M* BTP containing 0.4% Tween 80 and 100 m*M* NaCl. After absorption decreases below <0.01 at 280 nm, the column is equilibrated with 10 m*M* BTP, pH 7.5, containing 1 m*M* *n*-dodecyl-β-D-maltoside, 80 m*M* NaCl, 1 m*M* ATP, and 1 m*M* $MgCl_2$ until ATP begins to elute and then the column is closed for 1 hr to allow autophosphorylation. RK is eluted when the column is open again. RK activity-containing fractions are pooled and diluted three-fold for Q-Sepharose chromatography. After washing the column (Pharmacia HR 5/5) with 5 ml 10 m*M* BTP, pH 7.5, containing 1 m*M* *n*-dodecyl-β-D-maltoside, RK is eluted with a salt gradient from 0 to 500 m*M* NaCl in the same buffer over 20 min at 0.5 ml/min.

iii. RK is also enriched in fraction from the rat pineal gland using affinity chromatography on a peptide from the C-terminal Rho. Due to weak interaction between the phosphorylable region of Rho and RK, the column capacity is low.[45]

[45] K. Palczewski, M. E. Carruth, G. Adamus, J. H. McDowell, and P. A. Hargrave, *Vision Res.* **30,** 1129 (1990).

iv. RK has also been purified using recoverin-Sepharose affinity chromatography.[46]

Immunodetection of RK

RK is detected by a panel of antibodies generated against unique regions of RK and against native purified bovine enzyme. Unfortunately, the polyclonal antibodies are no longer available. They were used successfully to identify the region of RK that is involved in the interaction with Rho* and to immunolocalize the enzyme to rod and cone photoreceptor cells.[25]

High-quality monoclonal anti-RK antibodies are raised against partially purified human RK expressed in *Escherichia coli*.[43,47] Two antibodies are characterized: the G8 antibody displays C-terminal specificity and reacts with bovine (two autophosphorylation forms) (Fig. 4C), human, and chicken RK, and the D11 antibody, which displays N-terminal specificity and reacts strongly only with human RK. The antibodies are useful for immunocytochemistry, immunoprecipitation, and immunoblotting and are available commercially from Santa Cruz Biotechnology, Inc., Santa Cruz, California.

Heterologous Expression of RK

RK has been expressed successfully in heterologous systems, including insect cells and COS cells. RK is inactive when expressed in *E. coli*. The expression of RK in insect cells, in the presence of mevalonolactone, led to heterogeneous incorporations of C-5, C-10, C-15, and C-20 isoprenyl moieties to the C-terminal Cys residue through a thioether linkage.[48] Methods to identify the prenyl group have been described,[49] and detailed analysis of the effect of C-terminal modifications on RK activity has been reported.[50]

i. Insect cell expression. The full-length *Stu*I/*Bam*HI fragment of RK[41] (1847 bp) is subcloned into a baculovirus transfer vector, pVL 1393 (Pharmingen). pVL-RK (3 μg) and a helper virus carrying a lethal deletion (0.5 μg) are cotransfected into insect Sf9 cells (Invitrogen) in a 60-mm tissue culture dish using a BaculoGold (TM) transfection kit (Pharmingen). A single recombinant baculovirus plaque is amplified to produce a virus stock. Three 150-mm plates containing a monolayer of High Five insect cells (Invitrogen) are infected with the recombinant virus. Cells

[46] C. K. Chen, J. Inglese, R. J. Lefkowitz, and J. B. Hurley, *J. Biol. Chem.* **270,** 18060 (1995).
[47] X. Y. Zhao, J. Huang, S. C. Khani, and K. Palczewski, *J. Biol. Chem.* **273,** 5124 (1998).
[48] K. W. Cha, C. Bruel, J. Inglese, and H. G. Khorana, *Proc. Natl. Acad. Sci. U.S.A.* **94,** 10577 (1997).
[49] M. E. Whitten, K. Yokoyama, D. Schieltz, F. Ghomashich, D. Kam, J. R. Yates, K. Palczewski, and M. H. Gelb, *Methods Enzymol.* **316,** 436 (2000).
[50] J. Inglese, J. F. Glickman, W. Lorenz, M. G. Caron, and R. J. Lefkowitz, *J. Biol. Chem.* **267,** 1422 (1992).

are harvested 96 hr after the infection, collected by centrifugation at 5000 rpm, and homogenized with 10 m*M* BTP, pH 7.5, containing 0.4% Tween 80, and the suspension is loaded on a DEAE-cellulose column (1 × 10 cm) equilibrated with the same buffer. The column is washed with 10 m*M* BTP, pH 7.5, containing 0.4% Tween 80, and RK is eluted with 100 m*M* NaCl in the same buffer. Fractions containing RK activity are combined and mixed with 2 μg of PP2A, dialyzed overnight against 10 m*M* BTP, pH 7.5, containing 0.4% Tween 80 and 1 m*M* benzamidine (1 liter), and loaded onto a heparin-Sepharose column (1 × 5 cm) equilibrated with the same buffer. After the column is washed with 10 m*M* BTP, pH 7.5, containing 0.4% Tween 80, 1 m*M* $MgCl_2$, and 125 m*M* NaCl, RK (>98% pure, 300 μg) is eluted with ATP in 10 m*M* BTP, pH 7.5, containing 0.4% Tween 80, 1 m*M* $MgCl_2$, and 100 m*M* NaCl (for details, see Ref. 44). Alternatively, insect cells (SF9, *Sphodoptera frugiperda* ovary cells) are transfected with baculovirus vector (Bacmid) carrying the cloned DNA interest. Insect cells are cultured at 27° in SF-900 II SFM medium (Life Technologies, Gaithersburg, MD) and harvested 96 hr after infection by centrifugation at 1200*g*.

ii. The expression of native RK in COS-7 has been previously described.[51] Several mutants in the autophosphorylation and isoprenylation regions are analyzed using RK in this expression system.[51,42] Frequently, lovastatin and simvastatin are used to improve radiolabeling of the isoprenylation group when ^{14}C-labeled mevalonolactone is used. Application of these hydroxymethyl-glutaryl-CoA reductase inhibitors suppresses the production of endogenous mevalonic acid; however, it may lead to changes in the ratio of isoprenyl groups incorporated into the protein.[49]

Sites of Phosphorylation on Rho

Successes in the analysis of the phosphorylation sites depend on the isolation of phosphopeptides. Several classic chemical methods have been described for the analysis of phosphorylation sites; however, advances in mass spectrometry have revolutionized the field of protein chemistry. As these methods evolve rapidly, we will not focus on actual mass spectrometric techniques, but rather on the isolation of peptides and problems related with their purification. The reader is encouraged to consult with the previous publication, which describes in more detail the problems related to the analysis of the phosphorylation sites on Rho*.[52] The following standard methods are used for the purpose of identifying phosphorylation sites.

i. Phosphorylation and proteolysis. Phosphorylation of Rho is carried out in ROS membrane suspensions containing Rho (3 μ*M*), ATP (2.5 m*M*), [γ-^{33}P]ATP

[51] J. Inglese, W. J. Koch, M. G. Caron, and R. J. Lefkowitz, *Nature* **359,** 147 (1992).

[52] K. Palczewski, J. P. Van Hooser, and H. Ohguro, "Regulation of G Protein-Coupled Receptor Function and Expression" (J. L. Benovic, ed.), p. 69. Wiley-Liss, New York, 2000.

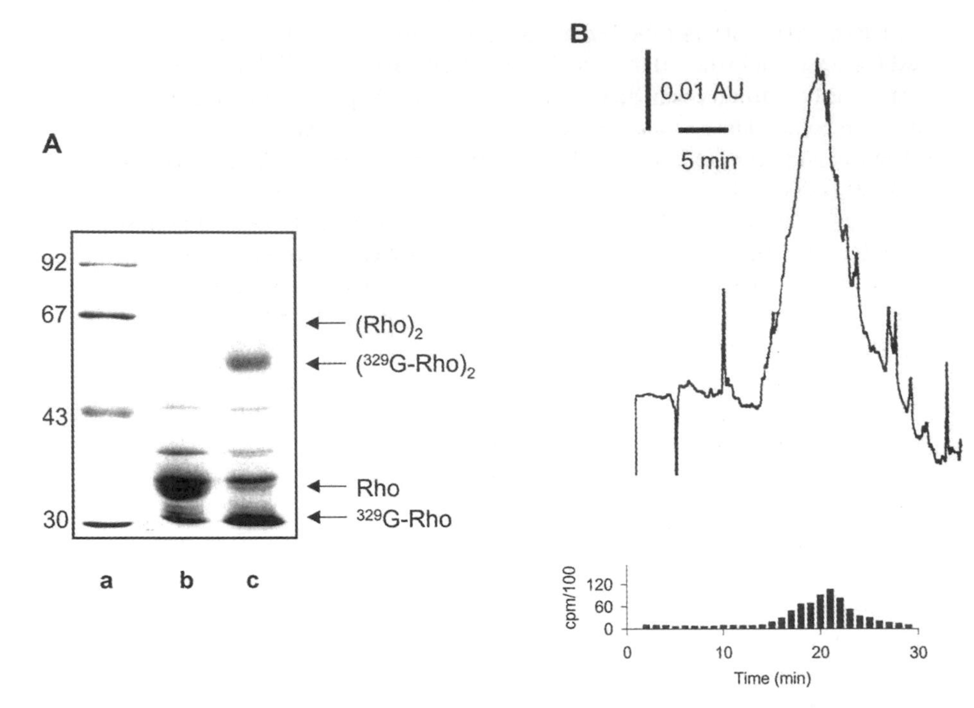

FIG. 5. Proteolysis of Rho and isolation of the C-terminal phosphorylated peptide. (A) SDS–PAGE of standards (lane a in kDa), ^{33}P-labeled Rho (lane b), and Rho digested with endoproteinase Asp-N (lane c). $(Rho)_2$ and $(^{329}G\text{-}Rho)_2$ represent dimers formed during sample preparation. The proteolytic-insoluble nonradioactive ^{329}G-Rho fragment is separated from the soluble C-terminal ^{33}P peptide by pelleting membranes. (B) The C-terminal peptide is initially purified on Ga^{3+}-IMAC and finally on a C_{18} column using HPLC.

(550,000 cpm/nmol, DuPont-New England Nuclear, Boston, MA), $MgCl_2$ (3 m*M*), and 0.1 *M* potassium phosphate, pH 7.5. All manipulations with the samples are done under a dim red light employing a Kodak No. 1 safelight filter (transmittance >560 nm). Radiation from ^{33}P is lower energy and decays slower than ^{32}P. Protection shields for samples containing low radioactivity may not be necessary. The disadvantage is that ^{33}P is more costly than ^{32}P. Phosphorylation is initiated by illumination using a 150-W lamp from a distance of 20 cm at 30°. After 1 hr, the mixture is centrifuged, the supernatant is removed, and the pellet is resuspended in 1 ml of 10 m*M* BTP, pH 7.5, containing 0.1 *M* NaCl for proteolysis.

ii. Proteolysis of Rho with endoproteinase Asp-N. To isolate phosphopeptides, [^{33}P]phosphoopsin (4 mg/ml) is digested with 2 μg endoproteinase Asp-N (Boehringer Mannheim) overnight at room temperature. The progress of digestion is verified by SDS–PAGE (Fig. 5A). The reaction is terminated by adding 300 μl of

CH_3CN and 10 μl of CH_3COOH. Peptides are separated from the insoluble material by ultracetrifugation at 100,000*g* for 10 min. The peptides are purified on HPLC, digested further with trypsin, endoproteinase Glu-C, thermolysin, aminopeptidase, and carboxypeptidase,[52] and the phosphopeptide is separated from unphosphorylated peptide on Ga^{3+}–immobilized metal affinity chromatography (IMAC)[53] (Fig. 5B) as described later.

iii. Ga^{3+}-IMAC[53] of the peptide mixtures is performed to separate phosphorylated/unphosphorylated peptides. Chelating Sepharose gel (1 ml; Pharmacia, #17-0575-01) is packed into a Pasteur pipette plugged with glass wool. The gel is washed with 0.1% CH_3COOH and then charged with 1 ml of 1.1 *M* $GaCl_3$ (Alfa Aesar, #35698). After the column is equilibrated, peptides are loaded on Ga^{3+}-charged resin. Unbound peptides are removed by an exhaustive wash with 0.1% CH_3COOH. ^{33}P-labeled phosphopeptides are eluted with 1 *M* NH_4OH.

iv. HPLC of C-terminal phosphopeptides (Fig. 5B). The Ga^{3+}-IMAC-purified phosphopeptides are injected onto a C_{18} HPLC column (4.6 × 150 mm, Hewlett Packard XDB-C_{18}). Peptides are eluted employing a linear gradient from 0 to 45% CH_3CN with 0.05% TFA for 30 min at a flow rate of 1 ml/min. The absorbance at 220 nm is monitored, and a small aliquot (2 μl) from each fraction is used for counting of ^{33}P activity. Fractions with the highest activity are dried down using a Speed-Vac and are subjected to MS analysis.

Mass Spectrometric Analysis of Sites of Phosphorylation

The analysis of the phosphorylation site is complicated because RK phosphorylates Rho* at different sites.[54] *In vitro*, at least two to four phosphates are incorporated per Rho*, due to the RK mechanism[4] and the high abundance of Ser and Thr residues at the phosphorylation site at the C-terminal region of Rho. Dr. J. Buczylko made the important finding that Rho is cleaved specifically and uniquely at the C terminus by endoproteinase Asp-C[40] (Fig. 5). Soluble peptide could be isolated and analyzed easily by further digestion with other proteolytic enzymes.[44,52,54] Phosphorylation may prevent some proteolytic cleavages, helping in the identification of phosphopeptide.[54,52] We developed complementary methods to isolate phospho-peptides using HPLC and heptafluorobutyric acid as a counter ion.[7,55] In the presence of this acid, we have been able to separate monophosphorylated peptides of Rho into separate pools depending on the position of phosphorylation site.

Analysis of the phosphorylation peptides directly by mass spectrometry, for example, after purification using Ga^{3+}-IMAC as described earlier (Fig. 5), could be difficult to interpret. Each 19 amino acid-long peptide phosphorylated at different

[53] M. C. Posewitz and P. Tempst, *Anal. Chem.* **71,** 2883 (1999).

[54] H. Ohguro, K. Palczewski, L. H. Ericsson, K. A. Walsh, and R. S. Johnson, *Biochemistry* **32,** 5718 (1993).

[55] H. Ohguro and K. Palczewski, *FEBS Lett.* **368,** 452 (1995).

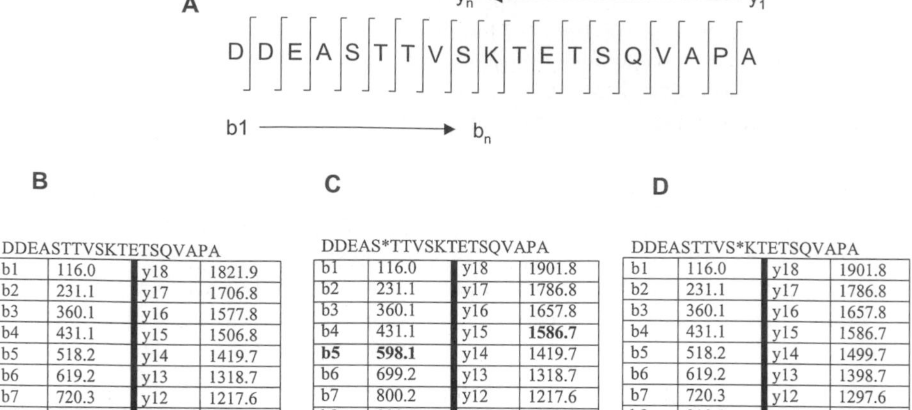

B

DDEASTTVSKTETSQVAPA

b1	116.0	y18	1821.9
b2	231.1	y17	1706.8
b3	360.1	y16	1577.8
b4	431.1	y15	1506.8
b5	518.2	y14	1419.7
b6	619.2	y13	1318.7
b7	720.3	y12	1217.6
b8	819.3	y11	1118.6
b9	906.4	y10	1031.5
b10	1034.5	y9	903.4
b11	1135.5	y8	802.4
b12	1264.6	y7	673.4
b13	1365.6	y6	572.3
b14	1452.6	y5	485.3
b15	1580.7	y4	357.2
b16	1679.8	y3	258.1
b17	1750.8	y2	187.1
b18	1847.9	y1	90.1
b19	1918.9		

C

DDEAS*TTVSKTETSQVAPA

b1	116.0	y18	1901.8
b2	231.1	y17	1786.8
b3	360.1	y16	1657.8
b4	431.1	y15	**1586.7**
b5	**598.1**	y14	1419.7
b6	699.2	y13	1318.7
b7	800.2	y12	1217.6
b8	899.3	y11	1118.6
b9	986.3	y10	1031.5
b10	1114.4	y9	903.4
b11	1215.5	y8	802.4
b12	1344.5	y7	673.4
b13	1445.6	y6	572.3
b14	1532.6	y5	485.3
b15	1660.7	y4	357.2
b16	1759.7	y3	258.1
b17	1830.8	y2	187.1
b18	1927.8	y1	90.1
b19	1998.9		

D

DDEASTTVS*KTETSQVAPA

b1	116.0	y18	1901.8
b2	231.1	y17	1786.8
b3	360.1	y16	1657.8
b4	431.1	y15	1586.7
b5	518.2	y14	1499.7
b6	619.2	y13	1398.7
b7	720.3	y12	1297.6
b8	819.3	**y11**	**1198.5**
b9	**986.3**	y10	1031.5
b10	1114.4	y9	903.4
b11	1215.5	y8	802.4
b12	1344.5	y7	673.4
b13	1445.6	y6	572.3
b14	1532.6	y5	485.3
b15	1660.7	y4	357.2
b16	1759.7	y3	258.1
b17	1830.8	y2	187.1
b18	1927.8	y1	90.1
b19	1998.9		

FIG. 6. Predicated pattern of ions formed from the C-terminal fragment of Rho during mass spectrometrical analysis. The b series and y series of ions formed from (A, B) DDDEASTTVSKTETSQVAPA peptide (C) from unphosphorylated monophosphorylated peptides at Ser-334, or (D) from unphosphorylated monophosphorylated peptides Ser-338.

positions produces a set of b and y ions that differ only in one to three ions. Two examples of ions from monophosphorylated peptide at different positions and nonphosphorylated peptide are shown in Fig. 6. Furthermore, many phosphorylated ions that are formed with low yields, undergo dephosphorylation and elimination reactions, and form complexes with K^+, Na^+, and Ca^{2+}, introducing further heterogeneity. This makes these mass spectrometry techniques, used in published regimes, not suitable for quantitative analysis as discussed previously.[52] To obtain reliable information on the relative distribution of phosphate groups between different sites, radioactive methods are the most favorable *in vitro* and spectroscopic (optical) methods, in conjunction with the separation of phosphorylated peptides into individual species and mass spectrometry, are most suitable for the analysis *in vivo*.[7]

The primary sequences of phosphopeptides are confirmed by electrospray ionization mass spectrometry using a Bruker/Hewlett Packard Esquire~LC ion-trap HPLC/mass spectrometry at the resolution (±1 amu at 14 kDa). The mass spectrometer is configured in direct injection mode, with nubulizer nitrogen at 20 psi, drying nitrogen at 10 liter/min and 300°, capillary and plate voltages at −4000 and −3500, respectively (positive mode), and 3500 and 3000, respectively (negative mode), mass scan range at 80–3000 amu, and the ion charge control (automatic gain control) activated with the target set at 15,000.

Measuring Techniques for Binding Assays of RK to Rho

Several methods were developed or modified to investigate the interaction of RK and Rho*. The most informative are light-scattering measurements and spectroscopic techniques (Fig. 7). These are complementary methods, also used successfully to study Rho* and G_t interaction.

i. Light scattering. Light-scattering changes are measured in a setup similar to the one described previously.[56] The measuring wavelength had to be chosen in the near infrared to avoid interference by both absorption changes and bleaching of Rho. The technique employs a continuous 840-nm incident light beam from a light-emitting diode (Hitachi HLP 60R). The scattered light is collected by two Fresnel lenses and is focused on a solid-state photodetector (Centronics OSD 100 5-T). Using different ring diaphragrams positioned between the Fresnel lenses, the scattering angle θ is selected, in the standard configuration $\theta = 16 \pm 2°$. LS signals are recorded with a 1- to 50-msec dwell time of the A/D converter (Nicolet 400); the cutoff frequency is routinely set to 0.2 of the dwell time. All measurements are performed in 10-mm path cuvettes, in a final volume of 300 μl. LS signals are induced by flash photolysis of Rho with a green flash (500 ± 20 nm), attenuated by appropriate neural density filters.

The LS-binding signal (see Fig. 7A) is interpreted as a gain of protein mass bound to the disk membranes and the LS dissociation signal (see Fig. 7B) as a loss of protein mass from the disk vesicle. A theoretical background for LS methods and the quantitative analysis for the binding and dissociation signal are published elsewhere.[6]

ii. Spectrophotometry. Formation of the photoproduct metarhodopsin II (Meta II, $\lambda_{max} = 380$ nm) is assayed according to the two-wavelength technique.[57] This technique minimizes scattering artifacts by comparing the flash-induced changes in the absorbance at 380 and 417 nm. The absorbance change at 417 nm (Meta I isosbestic to Meta II) serves as a reference for determining the level of Meta II. The measurements are made by using a two-wavelength spectrophotometer (UV 300, Shimadzu Scientific Instruments, Inc., Japan; 2-nm slit) equipped with thermostated cuvettes (2-nm path), temperature regulated (Circulator G/D8, Haake

[56] K. P. Hofmann, A. Schleicher, D. Emeis, and J. Reichert, *Biophysics Struct. Mech.* **8,** 67 (1981).

[57] A. Schleicher, H. Kühn, and K. P. Hofmann, *Biochemistry* **28,** 1770 (1989).

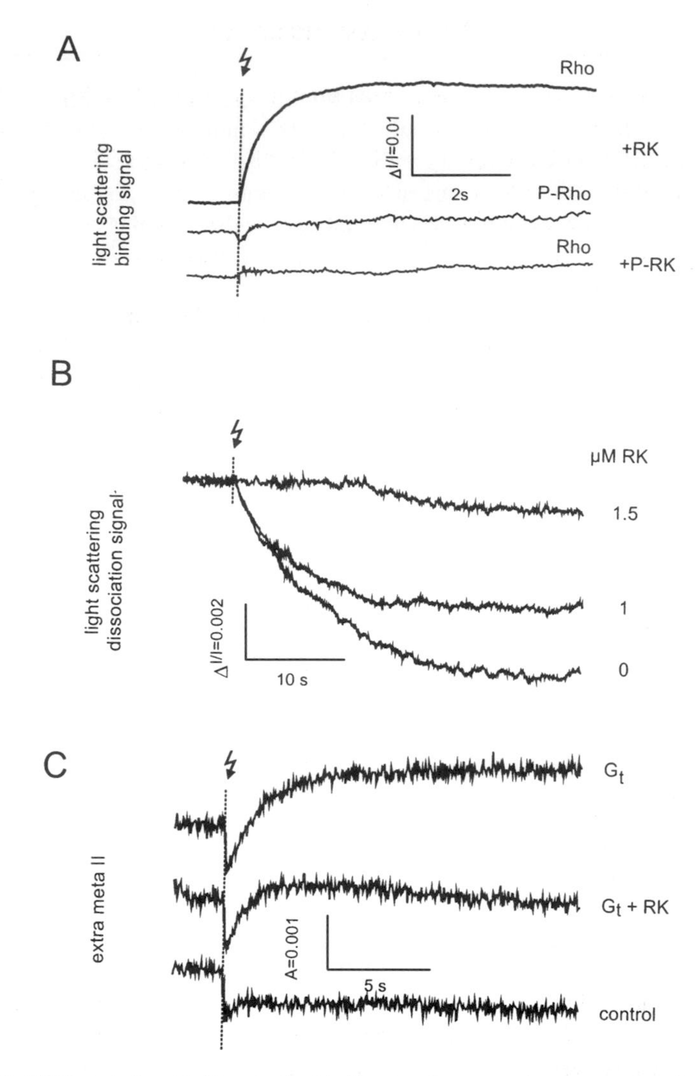

FIG. 7. (A) Flash-induced light-scattering binding signals measured on suspensions of washed disk membranes reconstituted with purified RK. Binding of the proteins is studied using disk membranes containing either unphosphorylated (Rho, upper trace) or (pre)phosphorylated Rho (P-Rho, middle trace), respectively. In the lower trace, autophosphorylated RK (P-RK) is used with Rho. All signals were corrected for control signals measured without added kinase.[23] Measuring conditions: 3 μM Rho and 1 μM RK. The flash illuminated 35% of Rho, the total volume is 300 μl, and the cuvette path length is 10 mm. (B) Inhibition of G_t activation by RK. Dissociation signals induced by flash of intensity $R^*/R = 9 \times 10^{-3}$ in the presence of G_t and GTP. From the lower to the upper trace the amount of RK is increased. Measuring conditions: 3 μM Rho (washed membranes), 0.4 μM purified G_t, and 1 mM GTP. (C) Inhibition of G_t-dependent Meta II stabilization by RK. Upper trace, Meta II formation in the presence of 1 μM G_t; middle trace, suppression of Meta II formed with 1 μM G_t by the addition of 1 μM RK; lower trace, control with Rho alone. Measuring conditions: 2 μM Rho in 100 mM BTP, pH 8.0, containing 90 μM GDP; 4°; flash excitation, 20%; cuvette path length, 2 mm.

GmbH, Karlsruhe, Germany), and green photoflash (filtered to 500 ± 20 nm). The sample is placed next to the photomultiplier, and the temperature is measured with a thermistor (GR 2105, Peltron GmbH, Fürth, Germany). Flash artifacts are avoided using the differential "S-pulse" output of the spectrophotometer for triggering the flash at the beginning of a voltage-free period of photomultiplier, which is additionally protected with a blue-green glass filter (BG 24, 2 mm, Schott, Mainz, Germany).

When Rho*, in its native disk membrane, is cooled to temperatures at which the equilibrium is on the Meta I side (below 5° and pH 8.0),[58] binding of protein to Meta II causes an increase of Meta II (extra-Meta II, Fig. 7C). This so-called extra Meta II provides a kinetic and stoichiometric measure for the complex between Rho* and the interactive proteins.[59] In both methods, the flash intensity is quantified by the amount of Rho* and is expressed in terms of the mole fraction of Rho* (R*/R).

Comments

Surprisingly, competition with G_t for the active receptor [evident from the slowing of G_t activation (Fig. 7B) and from reduced extra MII (Fig. 7C)] occurs with RK in the absence of ATP and phosphorylation. This argues for a direct competition of RK with G_t, which is independent of the second substrate, the enzymatic function of ATP and RK, presumably due to direct steric hindrance by the bound protein (Fig. 8). Rho* shut off by binding of RK (in the absence of phosphorylation) fits to models of photoreceptor saturation,[60] but does not seem to express itself in the single quantum regime.[61] Note that in the parathyroid hormone receptor/inositol phosphate pathway, G protein-coupled receptor kinases can inhibit receptor signaling under nonphosphorylating conditions.[62]

Human Disease and Mutation in RK Gene

The RK locus has been assigned to chromosome 13 band q34[63], and the RK gene is composed of seven exons. Mutation in RK is associated with human conditions that cause desensitization of the visual system after an intense illumination for a longer period of time than wild type. This condition, called Oguchi disease, is a recessively inherited form of stationary night blindness due to the malfunction

[58] J. H. Parkes and P. A. Liebman, *Biochemistry* **23,** 5054 (1984).

[59] K. P. Hofmann, *Biochim. Biophys. Acta* **810,** 278 (1985).

[60] U. Laitko and K. P. Hofmann, *Biophys. J.* **74,** 803 (1998).

[61] C. K. Chen, M. E. Burns, M. Spencer, G. A. Niemi, J. Chen, J. B. Hurley, D. A. Baylor, and M. I. Simon, *Proc. Natl. Acad. Sci. U.S.A.* **96,** 3718 (1999).

[62] F. Dicker, U. Quitterer, R. Winstel, K. Honold, and M. J. Lohse, *Proc. Natl. Acad. Sci. U.S.A.* **96,** 5476 (1999).

[63] S. C. Khani, M. Abitbol, S. Yamamoto, I. Maravic-Magovcevic, and T. P. Dryja, *Genomics* **35,** 517 (1996).

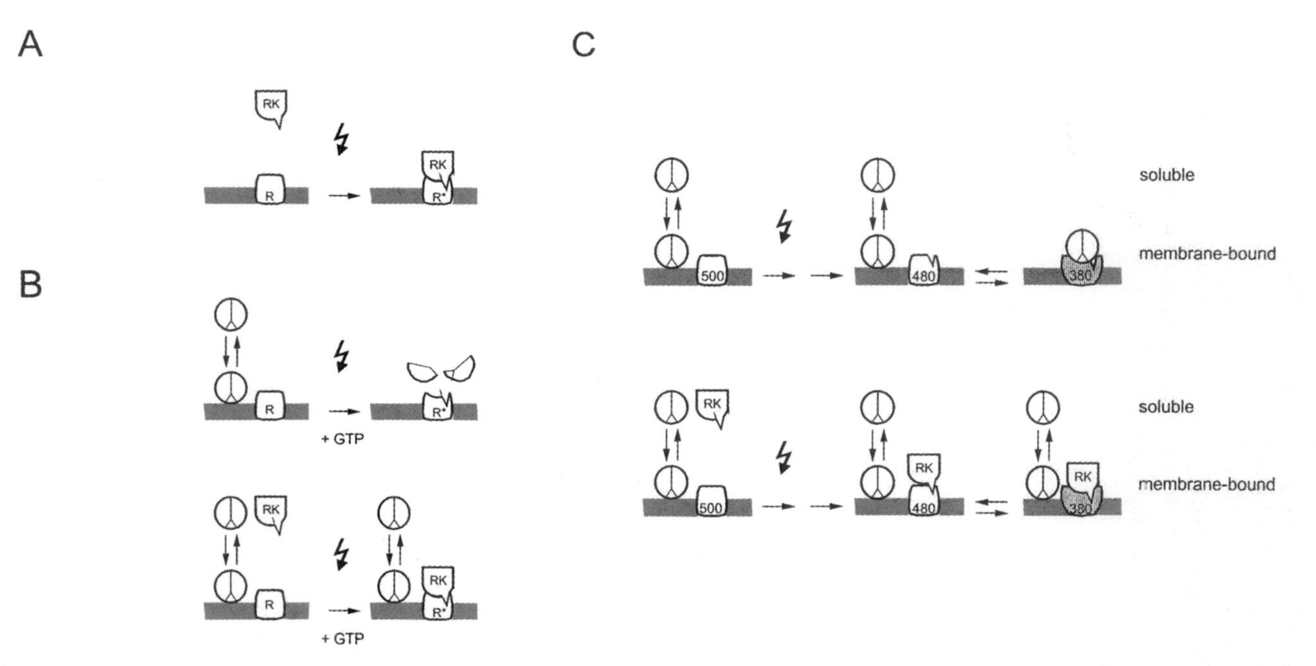

FIG. 8. Model of RK binding to membranes. (A) Binding of RK to Rho*. The model illustrates the direct binding of RK to light R*. For simplification, this model does not distinguish between the different Meta states of Rho*. (B) Activation of G_t and its inhibition by RK. In the presence of GTP, at the preformed complex between R* and G_t, GTP binds to the $G_{t\alpha}$ subunit followed by the dissociation of $G_{t\alpha}$ from both R* and $G_{t\beta\gamma}$. This model shows that activation of G_t by R* is accompanied *in vitro* by a rapid release of the G_t subunits from the disk surface and that this release can be inhibited by the direct binding of RK to R*. It is important to note that only the fraction of G_t present on the membrane at the time of the flash is monitored in the dissociation signal. (C) G_t-dependent MII stabilization and its inhibition by RK. As illustrated in this model, binding of G_t stabilizes only metarhodopsin II (Meta II, $\lambda_{max} = 380$ nm) at the cost of other, noninteractive and spectrophotometrically different forms of Rho*, e.g., metarhodopsin I (Meta I, $\lambda_{max} = 480$ nm). RK can bind to Meta I and Meta II. When G_t and RK are present simultaneously, RK inhibits the complex formation between R* and G_t, resulting in the suppression of the extra-Meta II signal (see Fig. 7C).

of RK in Rho quenching*.[64,65] Many cases are found to be homozygous for mutations of RK leading to a nonfunctional protein. From studies of patients with Oguchi disease, we learned that the role of RK in rods would be to accelerate the inactivation of Rho*, that in concert with regeneration, leads to the normal rate of recovery of sensitivity. Cones may rely mainly on regeneration for the inactivation of photolyzed visual pigments, but RK also contributes to cone recovery. Further analysis of other patients could be carried out using the genetical and physical approaches as published.[64,65]

Other Methods

i. Sodium dodecyl sulfate–polyacrylamide gel electrophoresis. SDS–PAGE is performed using 10% SDS–PAGE gel in a Hoefer (San Francisco, Ca) minigel apparatus. The gel is stained with Coomassie brilliant blue R-250 and destained with 50% (v/v) CH_3OH and 7% (v/v) CH_3COOH.

ii. Immunoblotting. The SDS–PAGE gel is placed in transfer buffer (10 m*M* BTP, pH 8.4) containing 10% CH_3OH for 10 min. Electroblotting onto Immobilon-P (Milipore, Bedford, MA) is performed at 90 V for 1 hr at 4°. After completion of the transfer, the membrane is blocked with 3% gelatin in 20 m*M* Tris–HCl, pH 8.0, containing 150 m*M* NaCl and 0.05% Tween 20, and is incubated with monoclonal antibodies (G8) at dilution 1 : 5000 for 1 hr at room temperature. Secondary antibodies conjugated with alkaline phosphatase (Promega, Madison, WI) are used at 1 : 5000. Antibody binding is detected using 5-bromo-4-chloro-3-indolyl phosphate and nitro blue tetrazolium.

Conclusions

RK is an important enzyme of phototransduction, and its role in human physiology is exquisitely characterized.[65] Many techniques were developed to isolate the enzyme, assay it, and characterize specific interactions with Rho*. Several posttranslational modifications were characterized. With these basic techniques in hand, mechanistical, kinetic, and structural questions can be asked. Toward this goal, structural information on RK, Rho, and Rho* is needed.

Acknowledgments

This research was supported by NIH Grants EY08061 (KP), a National Science Foundation Chemistry Instrumentation Grant (CHE 9807748 to U.W.), and Deutsche Forschungsgemeinschaft Grant Sfb 449 (KPH), and the Fonds der Chemischen industrie (KPH), an unrestricted grant from Research to

[64] S. Yamamoto, K. C. Sippel, E. L. Berson, and T. P. Dryja, *Nature Genet.* **15,** 175 (1997).

[65] A. V. Cideciyan, X. Y. Zhao, L. Nielsen, S. C. Khani, S. G. Jacobson, and K. Palczewski, *Proc. Natl. Acad. Sci. U.S.A.* **95,** 328 (1998).

Prevent Blindness, Inc. (RPB), New York, to the Department of Ophthalmology at the University of Washington, and the E.K. Bishop Foundation. We thank Preston Van Hooser for help during the course of this study, Dr. Rosalie Crouch for synthetic retinoids, and Joshua McBee for critical reading of the manuscript.

[37] Characterization of Arrestin Expression and Function

By STUART J. MUNDELL, MICHAEL J. ORSINI, and JEFFREY L. BENOVIC

Introduction

Arrestins mediate the desensitization and trafficking of numerous G protein-coupled receptors (GPCRs). Agonist-dependent phosphorylation of GPCRs by G protein-coupled receptor kinases (GRKs) promotes high-affinity binding of arrestins that in turn sterically inhibit G protein interaction with the agonist-activated GPCR. There are currently four mammalian arrestins that have been identified. Visual arrestin (termed arrestin-1 in this article) is expressed predominantly in rod cells and acts by preventing light-activated phosphorylated rhodopsin from interacting with the G protein transducin.[1,2] cDNA for bovine arrestin-1 was isolated in 1987 and encodes a protein of 404 amino acids.[3,4] A ubiquitously expressed arrestin-2 (β-arrestin) cDNA was subsequently cloned from bovine brain and found to encode a 418 amino acid protein with 58% identity with arrestin-1.[5] Utilizing various strategies, a third arrestin cDNA (termed arrestin-3 or β-arrestin-2) encoding a 409 amino acid protein with 79% identity with arrestin-2 and 56% identity with arrestin-1 was identified.[6–8] Most recently, a fourth arrestin, termed X-arrestin or arrestin-C (arrestin-4), was cloned and found to be specifically expressed in cone cells, suggesting a role in regulating cone phototransduction.[9,10]

[1] H. Kuhn, S. W. Hall, and U. Wilden, *FEBS Lett.* **176,** 473 (1984).
[2] J. G. Krupnick, V. V. Gurevich, and J. L. Benovic, *J. Biol. Chem.* **272,** 18125 (1997).
[3] K. Yamaki, Y. Takahashi, S. Sakuragi, and K. Matsubara, *Biochem. Biophys. Res. Commun.* **142,** 904 (1987).
[4] T. Shinohara, B. Dietzschold, C. M. Craft, G. Wistow, J. J. Early, L. A. Donoso, J. Horwitz, and R. Tao, *Proc. Natl. Acad. Sci. U.S.A.* **84,** 6975 (1987).
[5] M. J. Lohse, J. L. Benovic, M. G. Caron, and R. J. Lefkowitz, *Science* **248,** 1547 (1990).
[6] B. Rapaport, K. D. Kaufman, and G. D. Chalzenbalk, *Mol. Cell. Endocrinol.* **84,** R39 (1992).
[7] H. Attramadal, J. L. Arriza, T. M. Dawson, J. Codina, M. M. Kwatra, S. H. Snyder, M. G. Caron, and R. J. Lefkowitz, *J. Biol. Chem.* **267,** 17882 (1992).
[8] R. Sterne-Marr, V. V. Gurevich, P. Goldsmith, R. C. Bodine, C. Sanders, L. A. Donoso, and J. L. Benovic, *J. Biol. Chem.* **268,** 15640 (1993).
[9] A. Murakami, T. Yajima, H. Sakuma, M. McLaren, and G. Inana, *FEBS Lett.* **334,** 203 (1993).
[10] C. M. Craft, D. H. Whitmore, and A. F. Wiechmann, *J. Biol. Chem.* **269,** 4613 (1994).

0076-6879/02 $35.00

FIG. 1. Domain architecture of arrestines. Sequences of the four known mammalian arrestins are represented schematically. Solid areas are regions of invariant amino acid sequence. Bifurcation near the C terminus represents divergence in sequence between visual (arrestin-1 and arrestin-4) and nonvisual (arrestin-2 and arrestin-3) arrestins. A, activation-recognition domain; P, phosphorylation-recognition domain; S, secondary hydrophobic interaction domain; C, clathrin-binding domain; AP2, AP2 β subunit-binding domain.

A comparison of the overall amino acid identity and known structural domains of the four mammalian arrestins is depicted in Fig. 1.

Northern analysis demonstrates that both arrestin-1[5] and arrestin-4[9] are expressed most abundantly in the retina and pineal gland. However, reverse transcription-polymerase chain reaction (RT-PCR) has also identified low levels of arrestin-1 in heart, kidney, lung, and brain.[11] Arrestin-2 and arrestin-3 are widely expressed, with the highest levels in the brain, spleen, and prostate.[8,12] In most tissues examined, arrestin-2 appears to be expressed at higher levels than arrestin-3. A notable exception is olfactory epithelium, where arrestin-3 is the major isoform and appears to play a role in the desensitization of odorant receptors.[13]

This article outlines procedures for characterizing and manipulating the expression and function of arrestins in mammalian cell lines.

Methods to Characterize and Manipulate Arrestin Expression

Detection of Arrestins by Immunoblotting

The detection of endogenous arrestins in mammalian cell lines or tissues is best accomplished by immunoblotting using arrestin-specific antibodies. A number of antibodies have been used successfully for such purposes (Table I). Perhaps the best antibody for detecting mammalian arrestins is one that was initially generated against bovine arrestin-1.[14] This antibody (termed F4C1) is a mouse monoclonal that detects the epitope DGVVLVD, a sequence that is present in the four known

[11] W. C. Smith, A. H. Milam, D. Dugger, A. Arendt, P. A. Hargrave, and K. Palczewski, *J. Biol. Chem.* **269,** 15407 (1994).

[12] G. Parruti, F. Peracchia, M. Sallese, G. Ambrosini, M. Masini, D. Rotilio, and A. DeBlasi, *J. Biol. Chem.* **268,** 9753 (1993).

[13] T. M. Dawson, J. L. Arriza, D. E. Jaworsky, F. F. Borisy, H. Attramadal, R. J. Lefkowitz, and G. V. Ronnett, *Science* **259,** 825 (1993).

[14] L. A. Donoso, D. S. Gregerson, L. Smith, S. Robertson, V. Knospe, T. Vrabec, and C. M. Kalsow, *Curr. Eye Res.* **9,** 343 (1990).

TABLE I
ARRESTIN ANTIBODIES

Antibody name	Specificity	Type	Epitope	Reference
mAbF4C1	arr1∼arr2∼arr3	Mouse mAb	DGVVLVD	14
mAbA2G5	arr1 specific	Mouse mAb	PEDPDTAKE	14
mAbC10C10	?	Mouse mAb	RERRGIALD	14
mAbH11A2	arr1>arr3>arr2	Mouse mAb	NLASSTIIKE	14
mAbB11A11	?	Mouse mAb	VFEEFARHNLK	14
S2D2	?	mAb	N-terminal	16
S6D8	?	Mouse mAb	PVDGVVLVDPE	16
S8D8	?	mAb	N-terminal	16
M3A9	?	Rat mAb	EAQEKVPPNSSLTKTLVPL	16
M4H10	?	Rat mAb	RGIALDGKIKH	16
S10H9	?	Mouse mAb	EKIDQEAAMD	16
S1A3	?	Mouse mAb	VFEEFARQNLKD	16
S6H8	arr1∼arr2∼arr3	Mouse mAb	PVDGVVLVDPE	15
S85-63	arr1 specific	Mouse mAb	Bovine arr1 (aa 383–394)	11
β-Arrestin1	arr2	Mouse mAb	Rat arr2 (aa 262–409)	BD Transduction Laboratories
β-Arrestin1	arr2 specific	Rabbit polyclonal	Rat arr2 (aa 331–418)	7
β-Arrestin2	arr3 specific	Rabbit polyclonal	Rat arr3 (aa 333–410)	7
HDH	arr3 specific	Rabbit polyclonal	HDHIALPRPQSA	8
KEE	arr2 specific	Rabbit polyclonal	KEEEEDGTGSPRLNDR	8
C-p44	arr1 (p44 variant)	Rabbit polyconal	CPEDPDTAKESA	11
Arrestin-2	arr2>arr3	Rabbit polyclonal	Bovine arr2 (aa 357–418)	Benovic, unpublished
Arrestin-3	arr3>arr2	Rabbit polyclonal	Bovine arr3 (aa 350–409)	17
β-Arrestin1 (N-19)	arr2>arr3	Goat polyclonal	N-terminal epitope	Santa Cruz Biotechnology
β-Arrestin1 (R-19)	arr2 specific	Goat polyclonal	C-terminal epitope	Santa Cruz Biotechnology
β-Arrestin1 (C-19)	arr2 specific	Goat polyclonal	C-terminal epitope	Santa Cruz Biotechnology
β-Arrestin1 (K-16)	arr2 specific	Goat polyclonal	C-terminal epitope	Santa Cruz Biotechnology
β-Arrestin2 (N-16)	arr3 specific	Goat polyclonal	N-terminal epitope	Santa Cruz Biotechnology
β-Arrestin2 (C-18)	arr3 specific	Goat polyclonal	C-terminal epitope	Santa Cruz Biotechnology

mammalian arrestins. This antibody appears to detect arrestin-1, arrestin-2, and arrestin-3 equally and, thus, it can be used to compare the levels of various arrestins in mammalian cells. Additional antibodies that are specific for arrestin-1,[11,14–16]

[15] J. P. Faure, M. Mirshahi, C. Dorey, B. Thillaye, Y. de Kozak, and C. Boucheix, *Curr. Eye Res.* **3,** 867 (1984).

arrestin-2,[7,8] and arrestin-3[7,17] have also been reported. These antibodies are obviously advantageous for studies that focus on a single arrestin species. This section outlines an immunoblotting protocol used to detect arrestins in numerous cell lines.

Preparation of Cellular Lysate

Although arrestins are primarily cytosolic proteins, we generally use a lysis procedure that extracts both soluble and membrane associated proteins. However, although these extracts are optimal for assessing endogenous (or overexpressed) arrestin levels, these preparations should not be used for measuring arrestin function *in vitro,* as functional effects may be disrupted by the detergent present in the lysis buffer.

1. Prepare 5 ml of lysis buffer containing 20 m*M* Na HEPES, pH 7.5, 1% Triton X-100, 150 m*M* NaCl, 10 m*M* EDTA, 0.5 m*M* phenylmethylsufonyl fluoride (PMSF), 20 μg/ml leupeptin, and 100 μg/ml benzamidine. The volume of lysis buffer can be scaled up or down as needed but should be made fresh and chilled on ice before use.
2. Cells are grown in complete media to 75–90% confluence and then rinsed several times with phosphate-buffered saline (PBS). Ice-cold lysis buffer is then added (~1 ml of buffer per 100-mm dish of cells), and the cells are scraped off using a rubber policeman and transferred to a centrifuge tube.
3. Cells are lysed by freezing in liquid nitrogen or a dry ice/ethanol bath, thawing in a room temperature water bath, and vortexing several times. The cell lysate is centrifuged (40,000*g* for 10 min in an SS34 rotor or equivalent) to remove particulate matter and the supernatant is transferred to a microfuge tube and assayed for protein. The cell supernatant can either be analyzed immediately or aliquoted, frozen in liquid nitrogen, and stored at $-80°$ for later analysis.

Electrophoresis and Western Blot Analysis

Arrestins are found at relatively low concentrations in most nonvisual tissues (typically 0.1–2 pmol/mg protein). However, because several of the arrestin antibodies can detect low nanogram quantities, endogenous arrestins in most mammalian cells are readily detectable by immunoblotting.

1. Incubate 20–40 μg of protein from the cell lysate (prepared as described earlier) with SDS sample buffer for 10 min at room temperature. It is advantageous

[16] A. Razaghi, J. Bonaly, H. Chacun, J. P. Faure, M. Mirshahi, and J. P. Barque, *Biochem. Biophys. Res. Commun.* **233,** 601 (1997).
[17] M. J. Orsini and J. L. Benovic, *J. Biol. Chem.* **273,** 34616 (1998).

to also include samples containing ~1 ng of purified arrestin-2 or arrestin-3[18] as controls as well as a prestained protein standard.

2. Electrophorese the samples on a 10% SDS–polyacrylamide gel at a constant voltage of 120–150 V until the dye front is within a few millimeters of the bottom (~1 hr).

3. Carefully remove the stacking gel and then set up the running gel for transfer to nitrocellulose. This is accomplished by layering a transfer sponge, one piece of Whatman 3 MM paper, the running gel, one piece of nitrocellulose, one piece of Whatman 3 MM paper, and another sponge. Everything should initially be wetted in transfer buffer (25 m*M* Tris base, 192 m*M* glycine, 20% methanol, *do not adjust pH*) before layering. It is also important to make sure that all air bubbles are removed as each layer is added (a 13 × 100-mm plastic test tube can be rolled over the layer at each step to accomplish this). Transfer the protein samples to a nitrocellulose membrane for 1 hr at 100 V (constant voltage) using cold transfer buffer.

4. After transfer, sample loading and transfer efficiency can be checked by staining the nitrocellulose with 0.2% Ponceau S (prepared in deionized water) for ~1 min with the transfer side up. Transferred proteins can then be visualized by rinsing the nitrocellulose with deionized water to remove excess stain.

5. Rinse the nitrocellulose several times in TTBS (20 m*M* Tris–Cl, pH 7.5, 150 m*M* NaCl, 0.05% Tween 20). The membrane is then blocked for 1 hr in 10 ml of TTBS containing 5% (w/v) nonfat dry milk.

6. The nitrocellulose is next incubated with 5 ml of either the mouse monoclonal antibody F4Cl (~1 : 2000 dilution) or an arrestin-specific antibody for ≥1 hr at room temperature. As discussed earlier, F4Cl recognizes an epitope that is common to all mammalian arrestins and thus it can detect arrestin-2 and arrestin-3 as well as arrestin-1 (see Table I).

7. Wash the nitrocellulose three to five times for 10 min each in TTBS.

8. Incubate the nitrocellulose for 1 hr with a 1 : 3000 dilution of affinity-purified goat antimouse IgG conjugated to horseradish peroxidase (Bio- Rad) in TTBS-5% milk.

9. Wash the membrane three to five times for 10 min each in TTBS.

10. Overlay the nitrocellulose with 1–2 ml of enhanced chemiluminescence (ECL) reagent for ~1 min, allow the blot to drip dry, wrap in Saran Wrap, and visualize on X-ray film.

Overexpression of Arrestins in Mammalian Cells

The overexpression of wild-type or mutant arrestins in mammalian cells can provide information regarding the role of arrestins in the desensitization and

[18] O. B. Goodman, Jr., J. G. Krupnick, F. Santini, V. V. Gurevich, R. B. Penn, A. W. Gagnon, J. H. Keen, and J. L. Benovic, *Nature* **383,** 447 (1996).

internalization of a given GPCR. In general, HEK293 and COS cells have been used extensively for such studies because they are widely available, easy to grow, and permit the expression of high levels of GPCRs and arrestins using either transient or stable transfection. The mammalian expression vector pcDNA3, available from Invitrogen, is suitable for such studies. This vector features a convenient multiple cloning site, the strong cytomegalovirus (CMV) promoter, the neomycin resistence gene, and a bovine growth hormone polyadenylation signal. Using 1–5 μg of plasmid DNA, we routinely obtain 25- to 50-fold overexpression of transfected wild-type and mutant arrestins in both cell lines. Both FuGENE (Roche) and LipofectAMINE (Life Technologies, Inc.) cationic lipid-based transfection reagents have yielded comparable results. However, due to its ease of use we routinely use FuGENE, which is the protocol provided.

Cell Culture

1. COS and HEK293 cells are maintained in Dulbecco's Modified Eagles Medium (DMEM) supplemented with 10% fetal calf serum and 100 μg/ml penicillin and streptomycin (complete media). In general, cells are split 1 : 10 twice per week for maintenance. Cells are maintained using ~3 ml of complete medium per 60-mm dish.

2. To split COS cells, aspirate the medium and rinse the monolayer twice with sterile PBS. Incubate with 0.05% trypsin, 0.5 m*M* EDTA (Life Technologies, Inc.) for 1–2 min at room temperature. Aspirate trypsin, tap the plate gently to dislodge cells, and then add ~3 ml of complete medium to quench the remaining trypsin. Occasionally, cells may need a short incubation at 37° for complete trypsinization. HEK293 cells are less adherent than COS and can be split by aspirating the medium, washing gently with PBS lacking Ca^{2+} or Mg^{2+}, and gently tapping the dish or pipetting. Split COS or HEK293 cells a day prior to transfection such that they are 50–80% confluent the following day (generally a 1 : 3 to 1 : 5 split of a previously confluent dish). The degree of confluency required also depends on the toxicity of the transfection reagent used (e.g., in our experience FuGENE works effectively over a broad confluence range, whereas Lipofectamine works optimally when the cells are at higher confluence).

Transfection of Cells

1. Aliquot the plasmid DNA (~3 μg per 60-mm dish) into a 1.5-ml Eppendorf tube. In a separate 15-ml conical tube, add 0.3 ml of warmed DMEM followed by 10 μl of FuGENE added dropwise with swirling. Do not allow the FuGENE to directly contact the sides of the tube. Allow the FuGENE:DMEM mixture to incubate for 5 min at room temperature. We generally maintain a 3 : 1 ratio of FuGENE:DNA; successful transfections in 60-mm dishes have been obtained using a maximum of 5–7 μg of DNA. Additional guidelines for optimizing transfection efficiency depending on the cell type and plasmid are provided in the manufacturers' directions.

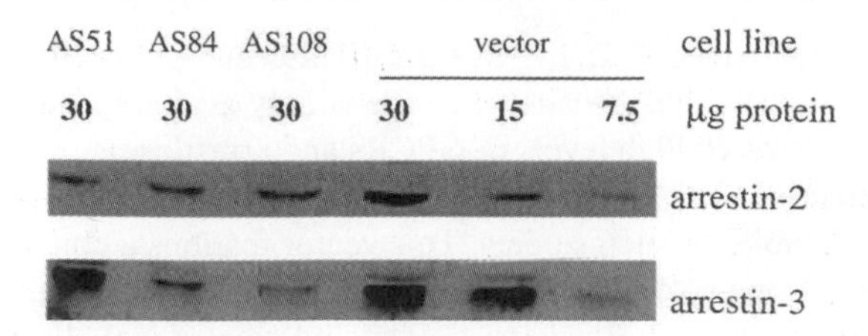

FIG. 2. Analysis of arrestin levels in arrestin antisense mRNA expressing HEK293 cells. Whole cell lysates (7.5–30 μg) were subjected to SDS–polyacrylamide gel electrophoresis followed by electrophoretic transfer and immunoblotting with rabbit polyclonal antibodies selective for arrestin-2 or arrestin-3. Three antisense mRNA expressing cell lines (AS 51, 84, and 108) (30 μg protein) were compared with vector-transfected HEK293-EBNA cells (30, 15, and 7.5 μg protein).

2. Add the DMEM/FuGENE mixture to the DNA dropwise, mix by gentle inversion of the tube, and leave at room temperature for 15 min. The FuGENE:DNA mixture is then added evenly to the cell monolayer dropwise while swirling gently. There is no need to change the existing medium.

3. The transfection mixture can be left on the cells until the time of harvest. The cells are typically harvested 24 to 48 hr posttransfection for Western blot analysis. Cells may also be split 5 hr after transfection, or the following day, for subsequent functional analysis.

Expression of Antisense Arrestin Constructs

The majority of studies to date have employed overexpression of either wild-type[18,19] or dominant-negative[17,19,20] arrestins to elucidate many of the functions of these proteins. While these studies have provided important insight into arrestin function, one of the inherent drawbacks with this approach is the potential for nonspecific effects associated with protein overexpression in a cell. An alternative and perhaps preferable approach is the use of antisense oligodeoxynucleotide strategies to target specific mRNAs and reduce protein expression. We have previously demonstrated that arrestin antisense cDNA constructs can effectively decrease endogenous arrestin expression (Fig. 2).[21]

This section describes the generation of arrestin antisense constructs that effectively reduce endogenous arrestin expression. Although we provide details of the primers used to amplify and subclone these constructs into mammalian expression vectors, the pREP4–arrestin antisense constructs are available from the authors by written request.

[19] S. S. G. Ferguson, W. E. Downey III, A.-M. Colapietro, L. S. Barak, L. Menard, and M. G. Caron, *Science* **271,** 363 (1996).

[20] J. G. Krupnick, O. B. Goodman, Jr., J. H. Keen, and J. L. Benovic, *J. Biol. Chem.* **272,** 15011 (1997).

[21] S. J. Mundell, R. P. Loudon, and J. L. Benovic, *Biochemistry* **38,** 8723 (1999).

While the arrestin antisense constructs can be used for transient transfection, optimal efficiency is obtained when the constructs are stably expressed. The resulting cell lines will maintain an effective antisense concentration level and reduced endogenous arrestin expression. To circumvent potential problems we have encountered when arrestin antisense constructs are stably integrated into the genome, we have used mammalian cell lines that express the pCMV-EBNA vector. This vector helps maintain plasmids such as pREP4 episomally and inhibits integration of the construct into the host cell genome. The reader is expected to have a working knowledge of basic techniques used in molecular biology (i.e., PCR, subcloning) in order to complete this protocol.

Generation of Antisense Expression Constructs

1. Antisense constructs can be amplified by PCR from plasmids containing human arrestin-2 or arrestin-3 DNA using the following pairs of primers. These primers amplify the first 801 bp of the arrestin-2 or arrestin-3 cDNAs, as these were found to be the most effective constructs for reducing arrestin expression.[21]

Oligonucleotides for arrestin-2:

sense: (5′ CAATTCTAGAATGGGCGACAAAGGGACG 3′, contains an *Xba*I site)

antisense: (5′ CAATCTCGAGCGTCGAGCTGGGTGCCAGT 3′, contains an *Xho*I site)

Oligonucleotides for arrestin-3:

sense: (5′CAATGGATCCATGGGGGAGAAACCCGGGACCAGG 3′, contains a *Bam*HI site)

antisense: (5′ CAATAAGCTTGGAGCTGGGAGATACCTGGTC 3′, contains a *Hin*dIII site).

2. PCR amplification is performed using the Expand High Fidelity PCR system (Roche) with 0.3 μM sense and antisense primers, 2 μM dNTPs, and 20 ng of arrestin-2 or arrestin-3 cDNA.

3. To check for the presence of the correct PCR fragment, load a sample from the reaction product given earlier along with appropriate molecular weight markers onto a 1% agarose gel containing ethidium bromide. The required PCR product will be a fragment of ~800 bp. DNA bands on the gel can be visualized under ultraviolet transillumination. By comparing product bands with bands from known molecular weight markers, the antisense arrestin construct should be identified readily. Excise this band and extract and purify the DNA from the gel (we use a QIAquick gel extraction kit).

4. The resulting products should then be digested with *Xba*I/*Xho*I (arrestin-2) or *Bam*HI/*Hin*dIII (arrestin-3) and subcloned into *Xba*I/*Xho*I or *Bam*HI/*Hin*dIII digested pcDNA3 or pREP4.

5. Antisense arrestin pcDNA3 constructs can be used in transient transfections to quickly assess arrestin antisense effectiveness. Seed cells in six-well plates and cotransfect with 2 μg of the antisense construct and 1 μg of either arrestin-2 or arrestin-3. After 48 hr, arrestin protein expression should be assessed by immunoblot as described previously. In contrast, we have found that pREP4 constructs work optimally for stable expression.

Stable Transfection and Expression of Antisense Constructs

1. HEK293-EBNA cells are maintained in complete media in the presence of geneticin (200 μg/ml) to maintain EBNA vector expression. Split cells the day prior to transfection such that they are 50–80% confluent the following day (generally a 1 : 3 to 1 : 5 split of a previously conflunt dish).
2. Transfect the cells with 4 μg of arrestin antisense DNA using FuGENE as described previously.
3. Three days after transfection, split the cells at 10–20% confluence using complete media supplemented with 400 μg/ml hygromycin and 200 μg/ml geneticin.
4. Replace the media every 3 days. Through this period check for the growth of resistant clones.
5. Once resistant clones are apparent, carefully transfer individual colonies into 12-well plates, expand, and check for changes in endogenous arrestin expression by immunoblotting as described previously.
6. Clonal cell lines demonstrating reductions in arrestin expression should be expanded and rechecked for altered arrestin expression.
7. Any clones that still demonstrate reduced arrestin expression should be aliquoted and several aliquots frozen and stored in liquid nitrogen.
8. It is recommended that cells are screened on a regular basis (after every four to five passages) to confirm that reduced arrestin expression is maintained.

Function of Arrestins

Arrestins function by specifically binding to phosphorylated GPCRs, thereby inhibiting GPCR/G protein coupling. An *in vitro* assay for directly measuring arrestin binding to GPCRs has been described previously in depth elsewhere[22] and

[22] V. V. Gurevich, M. J. Orsini, and J. L. Benovic, *in* "Receptor Biochemistry and Methodology: Regulation of G Protein-Coupled Receptor Function and Expression" (J. L. Benovic, ed.), Vol. 4, p. 157. Wiley-Liss, New York, 2000.

therefore will not be reiterated here. In addition, nonvisual arrestins (arrestin-2 and arrestin-3) can augment agonist-promoted internalization of numerous GPCRs. We will therefore concentrate on the measurement of arrestin-mediated GPCR regulation and specifically on the measurement of arrestin-stimulated GPCR internalization using an enzyme-linked immunosorbent assay (ELISA). It should be noted that receptor desensitization can also be measured readily in cells with altered arrestin expression.[21] Protocols to measure receptor desensitization via changes in second messenger production can be found elsewhere.[23]

Effect of Arrestins on Receptor Internalization in Mammalian Cells

For many GPCRs, nonvisual arrestins target the desensitized receptor to clathrin-coated pits for endocytosis.[24] Mechanistic insight into this process has revealed that nonvisual arrestins not only bind to receptors, but also interact with clathrin and AP2, the major proteins found in clathrin-coated pits.[18,20,25] Thus, arrestins can function as adaptor proteins to mediate GPCR uptake into clathrin-coated pits. The receptor ultimately traffics to endosomes where it is dephosphorylated and then either recycled back to the cell surface or trafficked to lysosomes.[26] In general, COS cells, which contain relatively low endogenous arrestin levels, are the best cells in which to assay the *promotion* of internalization by arrestins.[20] Conversely, HEK293 cells, which have higher endogenous arrestin levels, are the best cells in which to observe *inhibition* of internalization by dominant-negative arrestins or antisense approaches.[7,19–21]

This section describes methods to assay receptor internalization by ELISA using the β_2-adrenergic receptor (β_2AR) as an example. Analysis by ELISA is based on the presence of an amino-terminal epitope tag on the receptor, such as the hemagglutinin (HA) or Flag epitope, that is no longer recognized by the cognate antibody once the receptor is internalized.[27]

Internalization Assayed by ELISA

1. Cells are transfected with an epitope-tagged receptor $\pm$ arrestins as described. We generally cotransfect 3 μg of receptor and 1 μg of arrestin per 60-mm dish to observe promotion of internalization in COS and 3 μg of receptor and 1–3 μg of a dominant-negative arrestin or dynamin to observe inhibition of internalization in HEK293 cells. With antisense arrestin-expressing cells, 3 μg of

[23] Methods in Molecular Biology 41, 1995.

[24] J. G. Krupnick and J. L. Benovic, *Annu. Rev. Pharmacol. Toxicol.* **38,** 289 (1998).

[25] S. A. Laporte, R. H. Oakley, J. Zhang, J. A. Holt, S. S. Ferguson, M. G. Caron, and L. S. Barak, *Proc. Natl. Acad. Sci. U.S.A.* **96,** 3712 (1999).

[26] L. Kallal, A. W. Gagnon, R. B. Penn, and J. L. Benovic, *J. Biol. Chem.* **273,** 322 (1998).

[27] D. A. Daunt, C. Hurt, L. Hein, J. Kallio, F. Feng, and B. K. Kobilka, *Mol. Pharmacol.* **51,** 711 (1997).

receptor is transfected while 5 μg of arrestin should be cotransfected to rescue receptor internalization. These amounts can be scaled up or down accordingly depending on the number of cells to be transfected. We have found that internalization assays performed in COS cells are less sensitive to the level of receptor expression, as there is little internalization of the β_2AR in the absence of cotransfected arrestins. Conversely, optimal levels of internalization and effective inhibition of internalization by dominant-negative or antisense arrestins are best observed in HEK293 cells that express 0.5–2 pmol of receptor/mg of protein. Higher expression often lowers the percentage of receptors that are internalized and reduces the efficiency of inhibition by dominant-negative or antisense arrestins, most likely because the ability of the endogenous arrestins to promote internalization of these receptors is exceeded. Prior to splitting the cells, treat wells of a 24-well plate with a solution of 0.1 mg/ml poly-L lysine (PLL) (Sigma). Briefly, dissolve the PLL in sterile water, add ~250 μl to cover the well, wait 1 min, remove the PLL, and then let the well dry (~10 min) (PLL can be reused up to three times). Split the cells into 24-well dishes using 6 wells per transfection (60-mm dish). The cell density should be ~1–3 × 10^5 cells/well. It is convenient to divide the 24-well plate in half vertically, using one side for untreated and the other side for agonist-treated conditions. Each 24-well dish can then accommodate four separate transfections horizontally.

2. Aspirate medium from the wells, wash once with DMEM, and treat triplicate wells for the desired time with DMEM containing 0.1 m*M* ascorbic acid with or without isoproterenol (typically 10–20 μM) at 37°. Isoproterenol is freshly prepared as a 10 m*M* stock solution in 0.1 m*M* ascrbic acid.

3. Aspirate medium and fix the cells with 250 μl/well of 3.7% formaldehyde (Sigma) in TBS (20 m*M* Tris, pH 7.5, 150 m*M* NaCl) for 5 min. The remainder of the assay can be performed at room temperature.

4. Wash the plate three times with TBS (500 μl/well) and then block each well with 500 μl of 1% BSA in TBS for 45 min.

5. Aspirate the blocking solution and add the primary antibody diluted 1 : 1000 in TBS/1% BSA (250 μl/well) and incubate for 1 hr at room temperature. We have successfully used the 101R antihemagglutinin as raw ascites (Babco), as well as the antiflag antibody M1 (Sigma). Because the binding of the M1 antibody is dependent on the presence of calcium, 1 m*M* $CaCl_2$ must be present for this step and all subsequent steps prior to color development.

6. Wash the plate three times with TBS/(Ca^{2+}) (500 μl/well) and reblock with 500 μl of 1% BSA in TBS/(Ca^{2+}) for 15 min.

7. Aspirate the blocking solution and add 250 μl/well of secondary antibody (alkaline phosphatase-conjugated goat antimouse; Bio-Rad) diluted 1 : 1000 in TBS/(Ca^{2+}). Incubate for 1 hr at room temperature.

8. Wash the plate three times with 500 μl/well of TBS/(Ca^{2+}). Develop with the alkaline phosphatase substrate kit (Bio-Rad). Dilute the diethanolamine solution

1 : 5 in water and add one substrate tablet per 5 ml (dissolve the tablet completely by vigorous vortexing). The color of the solution should be colorless to a slight pale yellow. Add 250 μl of developing solution per well and develop until a bright yellow color appears, generally 15–30 min, depending on the efficiency of transfection and receptor expression. If the signal is weak, development is aided by incubation at 37°. It is important not to allow development to proceed beyond the linear range of the ELISA reader. Ideally, OD_{405} readings should fall between 0.5 and 1.2 after stopping the reaction (step 9).

9. Aliquot 100 μl/well of 0.4 *M* NaOH in a 96-well plate. Remove 100 μl from the developed wells and add to the NaOH to stop the development reaction. The color is now stable for several days if the plate is wrapped in Parafilm and stored at 4°.

10. Read plate in ELISA reader at 405 nm. Subtract background from mock-transfected or untransfected control cells and calculate percentage internalization.

[38] Identification of Novel G Protein-Coupled Receptor-Interacting Proteins

By RICHARD T. PREMONT and RANDY A. HALL

The superfamily of G protein-coupled receptors includes proteins that recognize and respond to ligands or agonists as diverse as proteins, peptides, small molecules, ions, and photons. To accommodate this diversity of activators, the G protein-coupled receptor superfamily comprises one of the largest gene families known. These receptor proteins share a conserved seven transmembrane span structure, but can be classified into several distinct groups or families each with unique conserved sequence elements.[1]

Physiological responses to individual agonists are initiated by the binding to specific G protein-coupled receptor proteins on the cell surface. Binding of the agonist to the receptor leads to an activated state that is capable of serving as a guanine nucleotide exchange factor for specific heterotrimeric G proteins.[2] Like all GTP-binding proteins, heterotrimeric G proteins (or more precisely, their α subunits) are active when bound to GTP and inactive when bound to GDP.[3] In

[1] S. Watson and S. Arkinstall, "The G-Protein Linked Receptor Factsbook." Academic Press, San Diego, 1994.

[2] T. Schoneberg, G. Schultz, and T. Gudermann, *Mol. Cell. Endocrinol.* **151,** 181 (1999).

[3] A. M. Spiegel, *Annu. Rev. Physiol.* **58,** 143 (1996).

0076-6879/02 $35.00

the inactive state, each GDP-bound α subunit is associated with a $\beta\gamma$ subunit complex. Activated receptors both promote the release of GDP from the G protein α subunit and catalyze the binding of GTP in its place. Once activated through receptor-catalyzed GDP–GTP exchange, the heterotrimeric G protein undergoes subunit dissociation.

The GTP-bound α subunit and the freed $\beta\gamma$ subunit complex each interact with and modulate the activity of intracellular messenger-generating proteins, or G protein effectors.[4] For example, the G_s α subunit activates adenylyl cyclase to produce cyclic AMP. Specific G protein subtypes and subsequent effector pathways that can be activated by a particular receptor are said to be "coupled" to that receptor. Individual receptor types may couple to a single G protein subtype or to several related or even unrelated G proteins. Individual G protein subtypes activate only a specific set of effectors, which then mediate some subset of the cellular effects initiated by receptor activation. The simultaneous activation of all the G protein effectors coupled to a particular receptor is presumed to give a unique signature that determines the response of the cell to activation by that particular receptor agonist.

G protein-coupled receptor systems are not static, but adapt quickly to the activity state of the cell they reside in. The loss of responsiveness following prolonged or repeated activation is called desensitization.[5] Desensitizing events that affect signaling pathways in addition to that which provoked the desensitization are termed "heterologous" and may include regulation of receptors as well as other components of the signal transduction cascades. Desensitizing events that affect only the signaling ability of the agonist that provoked the desensitization are termed "homologous" and primarily involve regulation of that individual receptor subtype. One major form of receptor activation-dependent or homologous desensitization is the uncoupling of the receptor from the ability to activate G proteins. Numerous studies in several receptor systems have shown that receptor uncoupling is coincident with receptor protein phosphorylation.[5]

A family of G protein-coupled receptor kinases, or GRKs, has been described that strongly prefer the activated or agonist-occupied form of the receptor as a phosphorylation substrate.[6,7] Due to this activation dependence, GRKs serve to initiate the uncoupling of activated G protein-coupled receptors. Two retinal GRKs serve as rhodopsin and cone opsin kinases (GRK1 and GRK7, respectively), while five somatic GRKs (GRKs 2–6) appear to serve to regulate all other G protein-coupled receptors in the body. Receptor phosphorylation in itself does not appear to alter receptor function greatly, but rather targets the activated receptors for uncoupling. The actual uncoupling event requires the action of an additional protein family,

[4] A. J. Morris and C. C. Malbon, *Physiol. Rev.* **79,** 1373 (1999).

[5] N. J. Freedman and R. J. Lefkowitz, *Recent Prog. Horm. Res.* **51,** 319 (1996).

[6] J. A. Pitcher, N. J. Freedman, and R. J. Lefkowitz, *Annu. Rev. Biochem.* **67,** 653 (1998).

[7] M. Bunemann and M. M. Hosey, *J. Physiol.* **517,** 5 (1999).

the arrestins. Arrestin proteins bind to GRK-phosphorylated receptor proteins and prevent activated receptors from further coupling to or activating G proteins.[8]

Thus, there are three classical G protein-coupled receptor-interacting proteins: heterotrimeric G proteins, GRKs, and arrestins. All three appear to recognize a multitude of distinct receptor subtypes and all prefer the agonist-occupied, activated conformation of the receptor. Among these proteins, the general mechanisms of receptor signaling to G proteins and homologous desensitization are accounted for.

The G protein paradigm has been validated innumerable times for a great variety of receptor types. Nonetheless, the fact that individual receptor subtypes exhibit unique signaling properties has reemerged as an important issue after a period of dormancy during which the generality of signaling was the primary focus.[9] That is, many receptor types do not behave quite as expected based on the reductionist assumption that any receptor types coupled to the same set of G proteins should be interchangeable. Because receptors do exhibit such differences, the question has been asked whether there might exist novel proteins that interact with individual receptors or subsets of receptors. Such receptor-interacting proteins might (i) serve direct signaling roles, (ii) localize a receptor to a particular region on the cell surface, (iii) scaffold other defined signaling molecules in association with a receptor, or (iv) assist in targeting the receptor as it cycles from cell surface to intracellular membranes following activation.

The earliest known GPCR-interacting proteins, G proteins, GRKs, and arrestins, were first purified based on their physiological activities, and their direct association with receptors was demonstrated much later. Similarly, receptor activity-modifying proteins (RAMPs) were cloned based on functional studies demonstrating that they are required for proper expression and activity of receptors for adrenomedullin, amylin, and calcitonin gene-related peptide,[10,11] and only following this identification in functional screens were RAMPs shown to associate directly with the receptor proteins they modulate. The converse approach to finding GPCR-interacting proteins, i.e., identification of a protein that associates directly with a given GPCR, followed by functional studies to address the physiological significance of the interaction, has also proven useful in understanding GPCR signaling and regulation. This approach has been particularly useful in identifying cytoplasmic proteins that associate specifically with one or several GPCR subtypes through interaction with subtype-specific structural motifs.

Within the family of G protein-coupled receptors, the highest sequence and function conservation is associated with the ligand-binding and G protein-coupling regions of the membrane spans and juxtamembrane portions of the intracellular

[8] J. G. Krupnick and J. L. Benovic, *Annu. Rev. Pharmacol. Toxicol.* **38,** 289 (1998).

[9] R. A. Hall, R. T. Premont, and R. J. Lefkowitz, *J. Cell Biol.* **145,** 927 (1999).

[10] L. M. McLatchie, N. J. Fraser, M. J. Main, A. Wise, J. Brown, N. Thompson, R. Solari, M. G. Lee, and S. M. Foord, *Nature* **393,** 333 (1998).

[11] S. M. Foord and F. H. Marshall, *Trends Pharmacol. Sci.* **20,** 184 (1999).

loops and carboxyl-terminal tail. G protein-coupled receptors exhibit low sequence conservation in the intracellular loops and carboxyl-terminal tail regions further from the membrane. Integral membrane receptor proteins are notoriously difficult to work with, so using an intact and active receptor protein as a screening tool has been impractical. However, these more unique loop and tail regions can be used as bait to search for novel receptor-interacting proteins using a variety of biochemical techniques.

Receptor Fragment-Fusion Protein Affinity Chromatography

One approach to identifying potential receptor-binding proteins is biochemical, using association with the receptor itself as a form of affinity chromatography. This could be accomplished by immunoprecipitating the full-length receptor protein and searching for coimmunoprecipitated protein bands. Generally, low endogenous receptor expression and lack of adequate antireceptor antibodies often necessitate transfection of epitope-tagged receptors. Further, the interacting proteins of interest may not be present in readily available cell lines, and there can be difficulties in scaling up such an assay. Nonetheless, coimmunoprecipitation has been used in several cases to identify the physical receptor interaction of proteins known to be functionally linked to a given receptor, e.g., the interaction of endothelial nitric oxide synthase with the bradykinin B2 receptor[12] and the interaction of the small GTP-binding proteins ARF and Rho with several GPCRs.[13] The coimmunoprecipitation approach is typically limited, however, to studies aimed at confirming the potential interaction of suspected GPCR-binding partners rather than screens for novel and unsuspected GPCR-interacting proteins.

A method related to coimmunoprecipitation but more useful for large-scale screening efforts is affinity chromatography utilizing immobilized receptor fragments as the bait. This method is technically simple, scalable, and adaptable. Preparation and immobilization of the receptor fragment can be achieved by any of several methods, and tissue or cell lysates from any source and in any amount can be applied. However, this technique may not be amenable to detecting protein interactions that require recognition of overall receptor conformations (e.g., ligand-activated state). Fusion protein affinity chromatography has been used successfully to identify a number of GPCR interactions with cytoplasmic proteins, including the interaction of calmodulin with metabotropic glutamate receptor type 5,[14] interaction of the Na^+/H^+ exchanger regulatory factor with the β_2-adrenergic receptor,[15]

[12] H. Ju, V. J. Venema, M. B. Marrero, and R. C. Venema, *J. Biol. Chem.* **273,** 24025 (1998).

[13] R. Mitchell, D. McCulloch, E. Lutz, M. Johnson, C. MacKenzie, M. Fennell, G. Fink, W. Zhou, and S. C. Sealfon, *Nature* **392,** 411 (1998).

[14] R. Minakami, N. Jinnai, and H. Sugiyama, *J. Biol. Chem.* **272,** 20291 (1997).

[15] R. A. Hall, R. T. Premont, C.-W. Chow, J. T. Blitzer, J. A. Pitcher, A. Claing, R. H. Stoffel III, L. S. Barak, S. Shenolikar, E. J. Weinman, S. Grinstein, and R. J. Lefkowitz, *Nature* **392,** 626 (1998).

interaction of Grb2 and Nck with the dopamine D4 receptor,[16] interaction of SH3p4/endophilin-1 with the β_1-adrenergic receptor,[17] and interaction of 14-3-3 proteins with the α_2-adrenergic receptors.[18] Here we discuss the use of glutathione *S*-transferase (GST) fusion proteins as the affinity bait,[19,20] as utilized in studies aimed at identifying novel β-adrenergic receptor-interacting proteins.

The receptor fragment of interest is prepared by polymerase chain reaction using specific oligonucleotide primers. For human β_1- and β_2-adrenergic receptors, third intracellular loop and carboxyl-terminal tail fusions originally created for use as antigens for antisera production were used. Briefly, receptor fragments are amplified using specific oligonucleotide primers encoding the relevant receptor domains, using the receptor cDNA as template. Sense primers contain *Bam*HI restriction sites and antisense primers *Eco*RI sites for subcloning of the amplified fragments in the pGEX-2TK vector (Amersham Pharmacia). Subcloned inserts are analyzed for the proper DNA sequence. Plasmids bearing the desired fusion inserts are used to transform competent *Escherichia coli* cells for expression. For pGEX series vectors, the BL21 strain of *E. coli* is a suitable host cell. From an overnight culture, 2 ml is used to inoculate 1 liter of LB media containing 100 μg/ml ampicillin, and the cells allowed to grow at 37° until the culture reaches an A_{600} of 0.6 (approximately 3 hr). At this time, expression of the fusion protein is induced by the addition of IPTG to 500 μM final concentration, and the cells are grown for an additional 2 hr at 37°. Induced bacterial cells are collected by centrifugation at 5000*g* in a Sorvall HBB6 rotor for 10 min at 4°. The induced cell pellet is resuspended in 30 ml per liter of original culture volume with phosphate-buffered saline (PBS) containing a cocktail of protease inhibitors (5 μg/ml aprotinin, 150 μg/ml benzamidine, 5 μg/ml leupeptin, 4 μg/ml pepstatin, and 20 μg/ml phenylmethylsulfonyl fluoride (PMSF). After the cells are resuspended, 1 ml of freshly prepared 1 mg/ml lysozyme solution in PBS is added and mixed gently. The resuspended cells are flash-frozen in liquid nitrogen or dry ice–acetone, and stored frozen at −80° until needed.

GST fusion proteins are purified from induced bacterial cell lysates using affinity chromatography on glutathione-agarose[20] (Sigma). Frozen cells are thawed gently and, once thawed, allowed to sit at 4° for 15 min for the lysozyme to digest the cell wall. Lysates are spun at 20,000*g* in a Sorvall SS-34 rotor for 30 min at 4°, and the clarified supernatant is transferred to a 50-ml screw-cap tube. For each liter of original cell culture, 1 ml of a 50% slurry of glutathione-agarose

[16] J. Oldenhof, R. Vickery, M. Anafi, J. Oak, A. Ray, O. Schoots, T. Pawson, M. von Zastrow, and H. H. M. Van Tol, *Biochemistry* **37,** 15726 (1998).

[17] Y. Tang, L. A. Hu, W. E. Miller, N. Ringstad, R. A. Hall, J. A. Pitcher, P. DeCamilli, and R. J. Lefkowitz, *Proc. Natl. Acad. Sci. U.S.A.* **96,** 12559 (1999).

[18] L. Prezeau, J. G. Richman, S. W. Edwards, and L. E. Limbird, *J. Biol. Chem.* **274,** 13462 (1999).

[19] D. B. Smith and K. S. Johnson, *Gene* **67,** 31 (1988).

[20] D. B. Smith, *Methods Enzymol.* **326,** 254 (2000).

beads is added, and the tube is rotated for 2 hr at 4°. The beads are spun at 2000*g* for 5 min at 4° and washed with 40 ml of cold PBS three times. At this point, beads containing bound GST-fusion protein can be used for pull-down assays, and the quality and quantity of the bound fusion protein assessed by SDS–PAGE and Coomassie blue staining. Alternatively, for studies where the GST-fusion is required free of glutathione-agarose beads, the fusion protein is eluted by addition of 10 ml of PBS containing 1 m*M* glutathione and protease inhibitors. The supernatant containing eluted fusion protein is removed from the beads following centrifugation. If needed, the glutathione and/or protease inhibitors can be removed by several repeated cycles of concentration in CentriPrep 30 spin concentrators (Amicon) and dilution with PBS buffer.

Lysates prepared from tissue can contain only soluble proteins (in the absence of detergent) or a mixture of soluble and membrane-associated proteins (extraction with detergent), or may be prepared from just the membrane fraction using detergent extraction. Because separating the soluble and membrane-associated proteins before any affinity chromatography can provide a significant purification in its own right, this may be a favorable approach. From a bovine brain, both types of lysates are easily prepared. First, the brain is minced into roughly 1-cm cubes and crudely homogenized in a blendor using 5 volumes of homogenization buffer: 20 m*M* HEPES (pH 7.4), 250 m*M* NaCl, 1 m*M* EDTA, 1 m*M* dithiothreitol, and protease inhibitors. The homogenate is then more finely disrupted in small aliquots using a large-bore Polytron for 1 min. The homogenate is spun at 50,000*g* for 30 min at 4°, and the resulting supernatant is pooled for use as the "soluble lysate." To prepare membranes, the pellets are homogenized again with the Polytron in the original volume of homogenization buffer and spun at 1000*g* for 10 min at 4° to remove remaining particulates, and the resulting supernatant is then spun at 20,000*g* for 30 min at 4°. After decanting, the membrane pellets are resuspended using a Polytron in 10 volumes of homogenization buffer, which is then supplemented with the desired detergent for extraction, such as 1% NP-40, Triton X-100, or CHAPS. After rotating or stirring for 1 hr at 4° to extract membrane proteins, the lysate is spun at 50,000*g* for 1 hr at 4°. The resulting supernatant is pooled as the "membrane lysate."

To identify receptor fragment-interacting proteins in tissues or cells of interest, a lysate prepared from that tissue is mixed with glutathione-agarose beads still bound to the GST-receptor fragment fusion protein. In the case of β_1- and β_2-adrenergic receptor carboxyl-terminal tails, bovine brain, heart, and kidney membrane lysates prepared using 1% CHAPS are added to the fusion protein beads, or to beads bound to GST bearing no fusion, and rotated for 1 hr at 4°. The beads are washed five times batchwise with homogenization buffer, and bound proteins are eluted in SDS–PAGE sample buffer. SDS–PAGE electrophoresis and Coomassie blue staining reveal a single 50-kDa protein band in samples containing both kidney lysate and GST–β_2-adrenergic receptor tail fusion beads (Fig. 1).

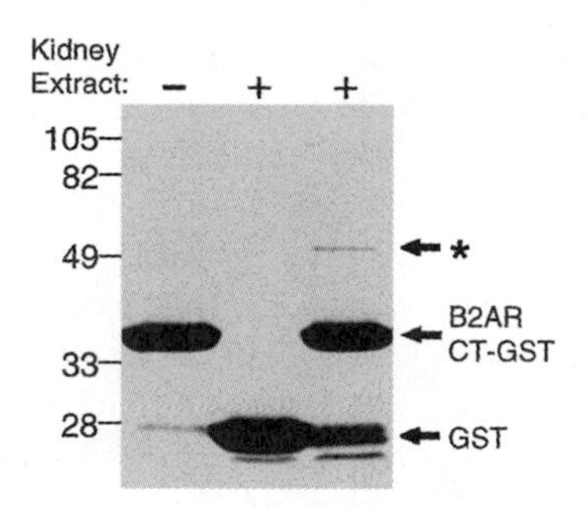

FIG. 1. Purification of a β_2-adrenergic receptor carboxyl-terminal tail (β_2AR CT)-binding protein from bovine kidney. Glutathione beads bound to GST-β_2AR CT fusion protein, or GST protein alone, were incubated in the absence (−) or presence (+) of bovine kidney extract. After extensive washing, bound proteins were eluted with SDS–PAGE sample buffer and separated by SDS–PAGE. Proteins were identified by Coomassie blue staining. The asterisk indicates the ~50-kDa kidney protein (NHERF) that specifically bound to the β_2AR CT. Reprinted by permission from *Nature* **392,** 626. Copyright 1998 Macmillan Magazines Ltd.

This protein, upon microsequencing, was identified as the NHERF protein, which specifically recognizes the carboxyl-terminal DSLL motif of the β_2-adrenergic receptor through a PDZ domain-mediated interaction.[15] In the case of β_1- and β_2-adrenergic receptor third intracellular loops, bovine brain soluble lysate is added to the fusion protein beads or to beads bound to GST bearing no fusion and rotated for 1 hr at 4°. The beads are washed five times batchwise with homogenization buffer, and bound proteins are eluted in SDS–PAGE sample buffer. SDS–PAGE electrophoresis and Coomassie blue staining reveal a single 40-kDa protein band in samples containing both brain lysate and GST–β_1-adrenergic receptor third intracellular loop fusion beads (Fig. 2). This protein, upon microsequencing, was identified as the SH3p4/endophilin 1 protein, which specifically recognizes a polyproline motif in the third intracellular loop of the β_1-adrenergic receptor through an SH3 domain-mediated interaction.[17]

Receptor Fragment-Fusion Protein Overlay

The ability to purify a receptor fragment-interacting protein to sufficient purity and in sufficient quantity to allow direct microsequencing requires a combination of factors. First, the receptor fragment used as the bait must be of sufficient purity that impurities in the preparation itself do not mask any purified proteins. Some fusion proteins may be unstable, and new constructs with slightly altered fusion boundaries may prove more tractable. Second, binding and washing conditions may affect the ability of the fusion to bind to interacting proteins. This is particularly notable for membrane lysates, where the choice of detergent may be critical to efficient extraction and to subsequent binding. Finally, the choice of tissue dictates

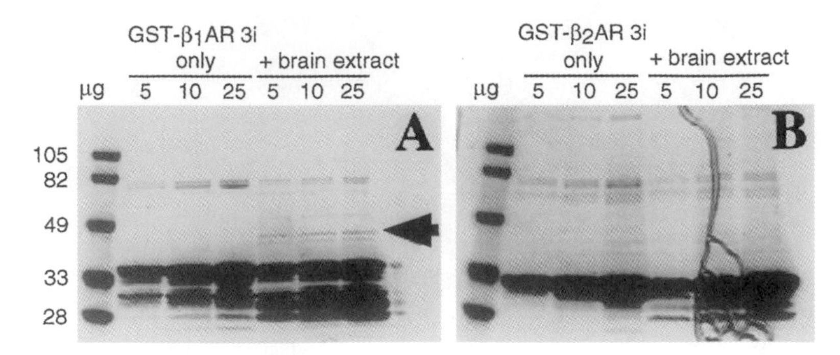

FIG. 2. Purification of a β_1-adrenergic receptor third intracellular loop (β_1AR 3i)-binding protein from bovine brain. Glutathione beads bound to GST-β_1AR 3i fusion protein (A), or to the GST β_2AR 3i fusion protein (B), were incubated in the absence or presence of bovine brain extract, as indicated. After extensive washing, bound proteins were eluted with SDS–PAGE sample buffer, and the indicated total amount of protein was separated by SDS–PAGE. Protein bands were identified by Coomassie blue staining. The arrow indicates the ~40-kDa brain protein (SH3p4, or endophilin 1) that specifically bound to the β_1AR 3i. Reprinted by permission from *Proc. Natl. Sci. U.S.A.* **96,** 12559. Copyright 1999 National Academy of Sciences, U.S.A.

the quantity of interacting protein available to be purified, and if present at too low a level, it may not be detected by Coomassie blue staining. Alternative labeling methods, such as silver staining or trace iodination, may be needed or cells may be labeled with radioactive amino acids prior to preparation of the lysate.

One distinct approach that may help in detect receptor-interacting proteins is to use the affinity of the receptor fragment itself as part of the detection system, such as by "far western" or protein overlay blotting. In this technique, samples to be probed, either tissue lysates or proteins that have bound to immobilized receptor fragment fusions as described earlier, are separated by SDS–PAGE and transferred to a nitrocellulose or nylon membrane. Potential interacting proteins are then detected by first allowing the soluble receptor fragment to bind to proteins on the membrane and then detecting where the soluble fragment has bound to the filter. Detection can be immunological or the probe receptor fragment can be radiolabeled prior to adding it to the membrane. Advantages of the blot overlay approach are that it can be quite sensitive and that it allows for examination of many tissues at one time. The main disadvantage is that the proteins on the blot are denatured and conformationally restrained, which may inhibit many protein–protein interactions. Another disadvantage is that detection of a GPCR-interacting protein via this blot overlay does not automatically lead to identification of the interacting protein. However, blot overlays can be combined effectively with receptor fragment affinity chromatography, as described earlier, to find and then identify GPCR-interacting proteins.

An example of a blot overlay experiment is shown in Fig. 3. The probe in this case is prepared from GST–β_2-adrenergic receptor carboxyl-terminal tail protein

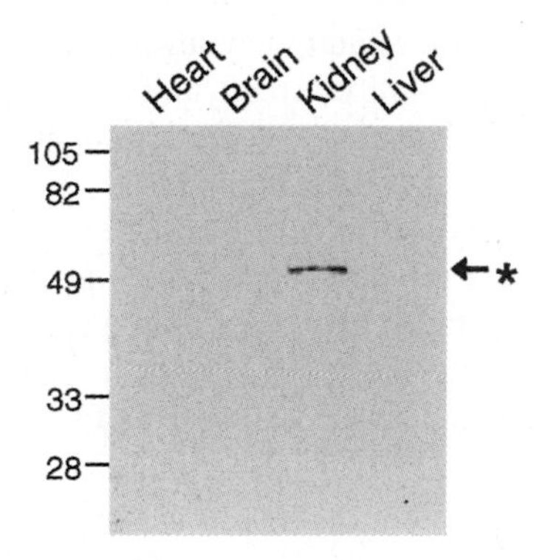

FIG. 3. Identification via blot overlay of a β_2-adrenergic receptor carboxyl-terminal tail (β_2AR CT)-binding protein in bovine kidney. Extracts from the indicated bovine tissues (25 μg) were separated by SDS–PAGE and transferred to nitrocellulose. The filter was blocked with nonfat milk solution and then overlaid with ^{32}P-β_2AR CT protein that had been radiolabeled by incubation with [γ-^{32}P]ATP and protein kinase A and cleaved from the glutathione bead-bound GST using thrombin. The filter was washed extensively and bound β_2AR CT protein visualized by autoradiography. The arrow indicates the ~50-kDa interacting protein band (NHERF) enriched in kidney.

by phosphorylation *in vitro*. The pGEX 2TK vector encodes GST fusions with a protein kinase A site immediately prior to the fusion protein but immediately after a thrombin cleavage site. The glutathione-agarose bead-bound fusion protein (100 μg in 1 ml) is incubated with the protein kinase A catalytic subunit (Promega) and 100 μCi of [γ-^{32}P]ATP for 30 min at 30°. The beads are washed with PBS in the absence of protease inhibitors, and the radiolabeled receptor fragment is removed from the GST and beads by cleavage with thrombin (Novagen Thrombin cleavage kit). The resulting free ^{32}P-labeled β_2-adrenergic receptor carboxyl-terminal tail is then added to the nitrocellulose filter in PBS containing 2% milk and 0.1% Tween-20 and incubated at room temperature with shaking for 1 hr. The blot is washed several times for 5 min with the same buffer lacking radioactive probe and exposed to X-ray film. The radiolabeled receptor fragment specifically binds to a 50-kDa band that is highly enriched in kidney tissue but not detectable in heart, brain, or liver tissue. These overlay studies led to the use of kidney tissue as a starting material for the receptor fragment affinity chromatography studies illustrated in Fig. 1, which led ultimately to the identification of NHERF as a high-affinity binding partner of the β_2-adrenergic receptor.

Receptor Fragment Two-Hybrid Screening

As noted earlier, biochemical approaches to identifying receptor interacting proteins require that the target protein exists in sufficient abundance to be identified after being partially purified. However, individual G protein-coupled receptors themselves are often expressed at quite low levels, on the order of fmol/mg cell protein. A protein that interacts with one particular receptor subtype may be present at quite low levels in tissues compared to G proteins or arrestins, which are present at quantities on scale more with the *sum* of all receptors present. In this case, biochemical approaches may fail to detect important regulatory interactions. A complementary approach is to use genetic screening for protein–protein interactions, such as the two-hybrid or interaction-trap method. Briefly, this technique involves creating a fusion protein between a GPCR fragment and the DNA-binding domain of a transcription factor (e.g., GAL4). This fusion protein is then cotransformed into yeast cells with a library of random cDNAs fused to the activation domain of the transcription factor. Interactions between the GPCR fragment and other proteins are detected when the DNA-binding domain and activation domain are brought together, thereby inducing expression of a reporter gene. The general technical details of two-hybrid screening have been described previously in this series,[21] and required reagents are generally available commercially as kits (e.g., Clontech Matchmaker system) so we will limit the discussion here to particular caveats in screening for receptor-interacting proteins.

There are a few important considerations in applying the yeast two-hybrid system to screens for GPCR-interacting proteins. First of all, because the fusion proteins must be translocated to nuclei of the yeast in order to turn on the reporter genes, a piece of the receptor must be chosen that can exist as a stable, soluble fusion protein. Fusion proteins containing a transmembrane region are unlikely to yield positive results, as the fusion protein probably will not fold properly and may be inserted into the plasma membrane or retained in the endoplasmic reticulum. Second, interactions requiring posttranslational modification of a GPCR (such as phosphorylation) are unlikely to be detected via the yeast two-hybrid system. Third, interactions requiring noncontiguous epitopes or the global conformation of a GPCR will almost certainly not be detected in two-hybrid screens with GPCR fragments. Despite these limitations, yeast two-hybrid screening has proven successful in identifying a number of GPCR interactions that have subsequently been shown to occur in mammalian cells, including the interaction of the eukaryotic initiation factor 2B with adrenergic receptors,[22] the immediate early gene Homer with metabotropic glutamate receptors,[23] the motor protein dynein with rhodopsin,[24] cortactin-binding protein 1 with the somatostatin receptor type 2,[25] SH3p4/endophilin-1 with the β_1-adrenergic receptor,[17] the transmembrane

[21] P. L. Bartel and S. Fields, *Methods Enzymol.* **254,** 241 (1995).

[22] U. Klein, M. T. Ramirez, B. K. Kobilka, and M. von Zastrow, *J. Biol. Chem.* **272,** 19099 (1997).

[23] P. R. Brakeman, A. A. Lanahan, R. O'Brien, K. Roche, C. A. Barnes, R. L. Huganir, and P. F. Worley, *Nature* **386,** 284 (1998).

[24] A. W. Tai, J.-Z. Chuang, C. Bode, U. Wolfrum, and C.-H. Sung, *Cell* **97,** 877 (1999).

[25] H. Zitzer, D. Richter, and H.-J. Kreienkamp, *J. Biol. Chem.* **274,** 18153 (1999).

protein calcyon with the dopamine D1 receptor,[26] and the PDZ protein PICK1 with the metabotropic glutamate receptor type 7.[27]

Summary

Biochemical and genetic methods utilizing G protein-coupled receptor fragments have been used successfully to identify G protein-coupled receptor-interacting proteins. As noted earlier, these methods may be unable to detect interactions that require certain conformations of the native receptor protein, but have nevertheless proven quite useful in expanding our understanding of receptor regulation to include interactions with proteins other than G proteins, G protein-coupled receptor kinases, and arrestins. Undoubtedly, it is likely that all G protein-coupled receptors have their own unique constellations of associated cytoplasmic proteins, and the techniques described here should prove useful in identifying these.

[26] N. Lezcano, L. Mrzljak, S. Eubanks, R. Levenson, P. Goldman-Rakic, and C. Bergson, *Science* **287,** 1660 (2000).

[27] K. K. Dev, Y. Nakajima, J. Kitano, S. P. Braithwaite, J. M. Henley, and S. Nakanishi, *J. Neurosci.* **20,** 7252 (2000).

Author Index

A

B

C

D

E

F

H

K

L

M

N

O

P

Q

R

S

T

U

V

W

Subject Index

A

D

H

I

L

M

N

O

R

T

V

W

X

Y